RADIOCARBON DATING LITERATURE

THE FIRST 21 YEARS, 1947 - 1968

Annotated Bibliography

Compiled by Dilette Polach, A.L.A.A.

RADIOCARBON DATING LITERATURE

THE FIRST 21 YEARS, 1947 - 1968

Annotated Bibliography

Compiled by Dilette Polach, A.L.A.A.

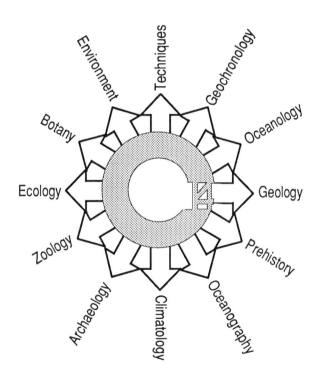

1988

ACADEMIC PRESS

Harcourt Brace Jovanovich, Publishers
London   San Diego   New York   Berkeley   Boston
Sydney   Tokyo   Toronto

ACADEMIC PRESS LIMITED
24/28 Oval Road.
London NW1 7DX

United States Edition published by
ACADEMIC PRESS INC.
San Diego. CA 92101

Copyright © 1988 by Academic Press Limited

All Rights Reserved
No part of this book may be reproduced in any form by photostat, microfilm, or any other means, without written permission from the publishers

ISBN 0–12–559290–6

Printed and Bound in Great Britain by
Thomson Litho Ltd, East Kilbride, Scotland

CONTENTS

Foreword	vii
Preface	ix
Chap. 1 - Bibliographical Works	1
Chap. 2 - Theoretical Works	5
Chap. 3 - Techniques and Instrumentation	43
Chap. 4 - General Geology	82
Chap. 5 - Glacial Geology	139
Chap. 6 - Ocean Studies	176
Chap. 7 - Pleistocene	183
Chap. 8 - Archaeology - Africa	202
Chap. 9 - Archaeology - America	211
Chap.10 - Archaeology - Asia	253
Chap.11 - Archaeology - Europe	262
Chap.12 - Archaeology - Oceania	274
Chap.13 - Conferences	283
Chap.14 - Date Lists	285
Author Index	310
Subject Index	331
Location Index	359

FOREWORD

It is a pleasure to write the foreword to Dilette Polach' annotated bibliography of the first 21 years of Radiocarbon Dating. The vitality of the subject is not measured by its half life but by its doubling time! Between 1946, when Willard Libby recognised that carbon-14 is produced in the upper atmosphere, and 1968, the final year of publications listed in this bibliography, over 2800 articles referring to radiocarbon appeared in the English-language literature, this means that the number of papers doubled every 2.2 years since 1946, which is very fast in any field of science. When papers in other languages are added, the rate is even faster. Consider the position of a student in radiocarbon beginning research 10 years after Libby's first paper: something like 40 papers perhaps when beginning PhD and about 160 by the time \of completion. This number increases to about 700 papers by the time the next student is writing a thesis. What a headache if he or she needed to refer to all of them! This annotated bibliography covers all accessible papers up to 1968, in English, with something like 2800 entries.

The reason for the rapid growth of the fields is that measurements of the natural levels of radiocarbon in materials involved in the carbon cycle provide vital information for archaeologists, geologists, oceanographers, atmospheric physicists and others. This new found facility for measuring the age of mummies, mammoths, water at the bottom of the oceans, as well as the effect of atmospheric testing of nuclear bombs and of the contribution of fossil fuels to urban pollution, has been a remarkable advance. The serious student or user of the method is always at a disadvantage when it becomes as all-pervasive as eating bread, and this is because new discoveries about the joy of bread may go unnoticed amongst the growing number of bread eaters.

This is where the annotated bibliography is important, especially in such a widely applied subject. Advances in techniques, testing, refinement and application of radiocarbon dating became scattered throughout 370 journals, ranging from the international to the local, by 1968. At that time a scholar in any field where the method was important would have been hard pressed to discover all the relevant literature. Nearly 20 years later, it is very much harder to recover material from the first two decades. The tide of more recent literature is terribly distracting, which is a pity because, when distracted, people may not be aware of their debts, or sometimes rediscover the wheel - an unnecessary process! A reading of Chapter 2, Theoretical Work, is most interesting in this respect.

Even a scholarly and devoted mole can miss important things in a literature search, as well as being frustrated by the process itself. An intelligently annotated bibliography is a boon in any field, and in this instance Mrs. Polach has given us a book which is both comprehensive, as to the literature to 1968, and excellently annotated. Many of the annotations are better indication of the contents of the papers than are abstracts provided by the authors themselves. On top of this is a 3-way indexing by author, subject and place, which is uncommonly versatile.

Dilette's husband, Henry Polach, is Head of the Radiocarbon Research Laboratory at the Australian National University, a laboratory of very high international reputation. This book represents but one facet of an enduring commitment to the radiocarbon dating field by a husband and wife team, with a family of four. A careful reading by those who know him will show one of the reasons why Henry Polach has been 'in the front of the game'. He kept abreast of the literature. In fact a large number of the entries in the bibliography are in the Polach's reprint library. The rest of us who use radiocarbon measurements in various ways, stand to gain immediately by having this book on our shelves.

The high quality bibliography of the first 21 years of radiocarbon dating makes regular users of the method, such as myself, wish for something similar covering the last 20 years. This may be quite unrealistic, as this book is the work of one person, and a sequel would have to be by many persons. I don't know whether 'doubling time' calculations are accurate, but if they are, there could be 1 million English-language publications citing radiocarbon measurements at the time of writing.! Even if the figure is one twentieth of this, the task for one person will be impossible, and no conceivable team would take it on (let alone produce something as even and accurate as this book). If I might conclude with a recommendation, it is that someone will shortly take up the project of producing an annotated bibliography of the new technique - accelerator-based radiocarbon dating - with the same care as Dilette Polach has exerted in this volume.

John Chappell
Dept. of Biogeography and Geomorphology
Research School of Pacific Studies
Australian National University

PREFACE

Although the technique of Radiocarbon Dating has been established and used as a research tool in a variety of fields since the late 1940's and is still being developed, refined and expanded, no serious effort has been made to compile a comprehensive annotated bibliography of the literature devoted to radiocarbon dating and its applications. Indeed the few early efforts at bibliographic compilation were selective or, dealt with specific areas of applications (e.g. Levi, 1955 and 1957; Johnson, 1959; Fagan, 1961, 1963, 1965, 1967; Deacon, 1966). Johnson whose bibliography appeared in the first issue of *Radiocarbon* notes the difficulty of compiling such a listing because of the multi-disciplinary aspect of the topic. This crossing of scientific boundaries results in the dispersal of the literature in a very large number of periodicals, not counting the monographs, reports, proceedings etc. Indexing of radiocarbon dating literature is therefore also scattered in a variety of indexing and abstracting services, each dealing with a particular field of science. This makes searching for literature on the general topic of radiocarbon dating very cumbersome to say the least. Dating research is also very dynamic as indicated by the steady growth of the literature. As a consequence, researchers are facing increasing difficulty in selecting from the volume of information available that which is relevant to their work, in particular when they are about to undertake new projects.

Radiometric dating appears as an indexed subject only since 1965 and the computer data bases offer retrospective searches only back to the early 1970's (Inspect-Physics, 1969; Chemical Abstracts, 1970; Geoarchive, 1969; S.C.I. Search, 1974). However a perusal of the current literature shows that early works are still widely referred to, indicating their importance as foundation and framework for new research. Thus the selected range of the bibliography, 1947 (the beginning of usage of naturally produced Radiocarbon as a scientific research tool) to 1968, fills the gap not covered by indexing services or computer information retrieval.

All efforts have been made to be comprehensive. Incorporated are: the literature reporting on the theoretical aspects, the dating techniques, the instrumentation and the research specifically aimed at extending the precision and the age limits, and the interpretation which extends from mathematics of age reporting to validation of radiocarbon dates in various environments (e.g. terrestrial, lacustrine, oceanic), variations of radiocarbon in nature, dendrochronologic calibrations and the interpretation of results within fields of application. Such applications cover, for example: climatology, glaciology, oceanography, pedology, geochemistry, palaeobotany, palaeomagnetism, anthropology, archaeology, pre- and proto-history, etc. The bibliography does not include high countrate radiocarbon tracer studies in their applications to such fields as biology, medicine and industry.

The language is restricted to English. This choice was made on the basis that, over the 14 disciplines examined, some 70% of publications are in the English language, and that original works in other languages often get translated and republished in English, while many original English publications are translated into other languages. Restricting the bibliographical material to English was thus a practical as well as a valid choice. However it must\ be recognised that, by doing so, inevitably some important references would have been missed. This was a self imposed limitation to an otherwise comprehensive compilation of reference material. Some exceptions were made where a very extensive summary was presented.

The method of compilation is based mainly on a search of cited references. The initial source was the extensive private collection of Henry Polach. From this, all relevant references were indexed and checked. The major monographs dealing with the subject were also examined for references and indexed in depth. The main periodicals in which the literature appears were determined and comprehensively indexed. This was augmented by searching various indexing services in the fields of earth sciences, physics, chemistry, nuclear sciences, and also the humanities (for archaeology and prehistory). Each reference was then sighted in order to extract a pertinent annotation which is not a copy of the given abstract, but stresses the use and application of radiocarbon dating. Less than 3% of the references have not been annotated because the source material was not available to the compiler. They have the caption [Not sighted] instead of an annotation. In order to accede to the material for the purpose of writing the annotations, many libraries were visited: in Australia, the National Library of Australia, and the libraries of the Australian National University, the Bureau of Mineral Resources, the Commonwealth

Scientific and Industrial Research Organisation (Canberra Branch), the Forest Research Institute, the Australian Institute of Aboriginal Research, the Universities of Sydney, of New South Wales and of Adelaide, the State Libraries of New South Wales and of South Australia, the Australian Museum (Sydney), the Royal Society of South Australia and the South Australian Museum; in New Zealand, the libraries of the Geological Survey and of The Institute of Nuclear Sciences; in Great Britain, the libraries of the Harwell Radiocarbon Laboratory and the Council of British Archaeology; in Europe, the libraries of the University of Turku (Finland), of the Environmental Radiation Research Institute, GSF-München, FDR and the Bavarian State Library; in the United States, the libraries of the Lamont-Doherty Geological Observatory, Columbia University, the National Bureau of Standards, the University of Miami, Southern Methodist University (Dallas) and the University of California at Los Angeles and at Riverside. Although some use of interlibrary loans was made, it was found in the long run not to be practical because of the sheer volume of material.

The bibliography is divided into 14 chapters. Chapter 1 presents Bibliographical Works. Chapter 2 covers the Theoretical Works and Chapter 3 the Techniques and Instrumentation. Chapter 4 to 7 are devoted to various aspects of the Earth Sciences. The boundaries between those chapters may be slightly blurred, but it was felt that one single large chapter would be too cumbersome for browsing. Chapters 8 to 12 present the Archaeology of the five continents. Finally a short Chapter 13 gathers Conferences, Symposia and reports thereon, and Chapter 14 presents the Date Lists, with short commentaries as to techniques used, period covered, half-life used, materials and projects dated and practices of age reporting. Within the chapters, the entries are in alphabetical order by title. The selection of titles for inclusion within a chapter was done by determining the main thrust of each paper and assigning it to the most appropriate chapter. Each entry is listed only once.

Three indexes offer access to the material by author, subject and geographical names (where relevant). This arrangement was chosen because it offers access at two levels; (i) browsing level, arranged by chapters which enable the user to go directly to the area required, scanning clearly set out titles which convey the desired information, (ii) specific retrieval level through the three indexes.

A Thesaurus of Radiocarbon Dating and Related Terms (which may be published at a later date) was compiled in order to insure consistency in the subject index. The geographical names were checked in the Times Atlas of the World and Gazetteer. The author index is a listing of all authors and co-authors without cross references. The spelling is according to the Oxford Dictionary, the authors' spelling being respected in the titles.

Terms (subjects) and names (authors and geographical places) listed in the indexes are numbered. These numbers refer to the entries in the body of the bibliography. The first digit(s) of these numbers refer to chapters and the next three digits refer to the entries within these chapters. Where there is a series of numbers referring to the same chapter, the first digit(s) have been dropped after the first occurrence, e.g. Arnold, James: **2**.007, 010, 033, 058, 122; **3**.074, 146, 164 etc.

The author wishes to thank Professor Jack Golson, Head of the Department of Prehistory, Research School of Pacific Studies, the Australian National University, for moral support, and Henry Polach, Head of the ANU Radiocarbon Dating Research Laboratory, who gave generously of his time, and whose love, enthusiasm for, and knowledge of the subject encouraged me through the long gestation period. The following members of the Radiocarbon Dating Research Laboratory helped with the preparation of the manuscript for publication: Maureen Powell who did much of the typing and proof reading and Robert Leidl who developed the computer programs and prepared the indexes; their efforts are gratefully acknowledged. Without their generous and competent assistance this work would never have been produced.

As no bibliography however well researched can claim to be absolutely exhaustive, the author would be grateful to be notified of errors or omissions detected by the users of this bibliography. Please send details to:

Dilette Polach   P.O. Box 43 GARRAN A.C.T. 2605 Australia

CHAPTER ONE
14C BIBLIOGRAPHICAL WORKS

1.001   1966
AN ANNOTATED LIST OF RADIOCARBON DATES FOR SUB-SAHARAN AFRICA
Deacon, J.
Annals of the Cape Province Museum, v. 5, 1966: 5-84
The dates published in this list, which covers work executed between 1960 and 1964, have been grouped into three major sections: geological and tree-ring dates, archaeological dates and dates of palaeoenvironmental interest.

1.002   1968
AUSTRALIAN GEOCHRONOLOGY: CHECKLIST 3
Dury, G.H.; Langford-Smith, Trevor
Australian Journal of Science, v. 30, n. 8, Feb. 1968: 304-306
This list consists exclusively of radiocarbon dates. The dates are published with their sources and comments. Some of those dates were not reported in Radiocarbon

1.003   1964
AUSTRALIAN GEOCHRONOLOGY: CHECKLIST I
Dury, G.H.
Australian Journal of Science, v. 27, n. 4, Oct. 1964: 103-109
A list of absolute dates obtained by radiometric and other analysis of Australian samples, giving reference to the laboratory listings and to the primary discussion of significance.

1.004   1966
AUSTRALIAN GEOCHRONOLOGY; CHECKLIST 2
Dury, G.H.
Australian Journal of Science, v. 29, n. 6, Dec. 1966: 158-162
163 dates are listed with comments. Amendments to checklist 1 (1964) are included.

1.005   1962
BIBLIOGRAPHY OF CARBON-14 MEASUREMENTS
Genunche, Ana
Academia R.P.R. Institutul de Fizica Atomica, Bucharest, 1962. pp. 60
Four hundred and eight references to U.S. and foreign journals and books published from 1940 to 1962. Author and subject indexes are included.

1.006   1962
BIBLIOGRAPHY OF CARBON-14 MEASUREMENTS. PART 2 (1961-1963)
Genunche, Ana
Academia R.P.R. Institutul de Fizica Atomica, Bucharest, 1964. pp. 32
One hundred and sixty seven references on chemical, medical, engineering, technological and economic aspects of radiocarbon measurements are given, from reports and journals published from 1961 to 1964. Author index included.

1.007   1957
A BIBLIOGRAPHY OF NEVADA ARCHAEOLOGY
Grosscup, Gordon L.
University of California, Berkley. Department of Anthropology. *Archaeological Survey Reports*, n. 36, February 1957. pp. 55.
A list of radiocarbon dates relevant to Nevada Archaeology is given on p. 51-52.

1.008   1959
BIBLIOGRAPHY OF PAPERS RELATING TO THE CARBON CYCLE IN NATURE AND RADIOCARBON FROM NUCLEAR TESTS
An.
United Nations. Scientific Committee on the Effects of Atomic Radiations, 7 Dec. 1959. pp. 7
A bibliography containing 91 references to papers on the carbon cycle in nature and radiocarbon from nuclear tests.

1.009   1959
BIBLIOGRAPHY OF PRIMARY SOURCES FOR RADIOCARBON DATES
McNutt, Charles H.; Wheeler, Richard P.
American Antiquity, v. 24, n. 3, Jan. 1959: 323-324
List of radiocarbon laboratories and of date lists to date.

1.010   1959
BIBLIOGRAPHY OF RADIOCARBON DATING
Johnson, Frederick
Radiocarbon, v. 1, 1959: 199-214
A selected list of the most significant works from the first ten years.

Ch. 1 Bibliographical Works

1.011   1955
BIBLIOGRAPHY OF RADIOCARBON DATING, COMPILED AT THE COPENHAGEN DATING LABORATORY
Levi, Hilde
Quaternaria, v. 2, 1955: 257-263
A selected list of works for the years 1946 to 1954. Arranged by year.

1.012   1957
BIBLIOGRAPHY OF RADIOCARBON DATING, COMPILED AT THE COPENHAGEN DATING LABORATORY
Levi, Hilde
Quaternaria, v. 4, 1957: 205-210
A selected list of works for the years 1956 and 1957.

1.013   1955
CARBON-14. A LITERATURE SEARCH
Manila, Philippine Atomic Energy Commission, 1965. pp. 140 (PAEC [A] IN-652)
840 references on radiocarbon are listed. The references were abstracted from *Nuclear Science Abstracts* from Jan. 1947 to Apr. 1963. The entries are sorted alphabetically under subject headings, there is an author index. (N.B. Not all entries refer to Radiocarbon dating).

1.014   1966
CARBON-14. A LITERATURE SEARCH
Manila, Philippine Atomic Energy Commission, 1966. pp. 53 (PAEC [A] IN-652)
266 references on radiocarbon are listed. The references were abstracted from *Nuclear Science Abstracts* from Apr. 1963 to date. The entries are sorted alphabetically under subject headings, there is an author index. (N.B. Not all entries refer to Radiocarbon dating).

1.015   1962
AN INDEX OF RADIOCARBON DATES ASSOCIATED WITH CULTURAL MATERIAL
Jelinek, Arthur Jenkins
Current Anthropology, v. 3, n. 5, Dec. 1962: 451-475
After a description of the radiocarbon dating method, the theory on which it is based, and of the publications and lists which are devoted to the presentation of radiocarbon dates, the author presents a compilation of dates published up to 1961 for which cultural associations have been suggested.

1.016   1960
LATE PLEISTOCENE EXTINCTION AND RADIOCARBON DATING
Hester, Jim J.
American Antiquity, v. 26, n. 1, 1960: 58-77
All radiocarbon dates from North America, associated with extinct Late Pleistocene mammals, those from levels stratigraphically later than levels with extinct forms, and dates associated with recent fauna are tabulated alphabetically by site. The dates are analysed and some conclusions presented.

1.017   1967
LIQUID SCINTILLATION COUNTING: SURVEY OF THE LITERATURE
Philips Inc. The Netherlands. pp. 134
This comprehensive survey, covering the field of liquid scintillation counting, has been derived from all relevant material published in *Nuclear Science Abstracts* since 1957. The abstracts are presented in chronological order of appearance and indexed alphabetically by subjects.

1.018   1968
PROGRESS REPORT ON UCLA RADIOCARBON DATES ON THE PACIFIC
Berger, Rainer
Pacific Island Program, Dept. of Anthropology, UCLA. *Bulletin*, n. 13, March 1968. 9 Leaves
Progress report on UCLA radiocarbon dates in the Pacific. Dates up to 1967 presented.

1.019   1957
A PUNCHED CARD SYSTEM FOR RADIOCARBON DATES
Oswalt, Wendell, H.
American Antiquity, v. 23, n. 2, part 1, 1957: 183
A proposed system was presented to the radiocarbon committee of the University of Michigan and the Memorial Phoenix Project provided the funds to carry out the project which is described in this article.

1.020   1954
RADIOACTIVITY IN GEOLOGY AND COSMOLOGY
Kohman, Truman P.; Nobofusa, Saito
Carnegie Institute of Technology. Dept. of Chemistry. *Report to U.S. Atomic Energy Commission*, n. NYO-3627, 1954
The geological and cosmological ramification of radioactivity in nature are surveyed by a review and interpretation of material published from 1951 to 1954. The examination of 592 references provides a basis for discussion on the natural occurence of radionuclides, radioactivity and the measurement of geologic times.

1.021   1954
RADIOACTIVITY IN GEOLOGY AND COSMOLOGY

Kohman, Truman P.; Nobofusa, Saito
Annual Review of Nuclear Science, v. 4, 1954: 401-462
The geological and cosmological ramification of radioactivity in nature are surveyed by a review and interpretation of material published from 1951 to 1954. The examination of 592 references provides a basis for discussion on the natural occurrence of radionuclides, radioactivity and the measurement of geologic times.

1.022   1961
RADIOCARBON DATES FOR SUB-SAHARAN AFRICA (FROM CA 1000 BC), I
Fagan, Brian M.
Journal of African History, v. 2, n. 1, 1961: 137-139
Compiled by the Rhodes - Livingstone Museum. Presents a compilation of radiocarbon dates obtained from various radiocarbon laboratories for the regions of Africa south of the Sahara, and their description.

1.023   1963
RADIOCARBON DATES FOR SUB-SAHARAN AFRICA (FROM CA 1000 BC), II
Fagan, Brian M.
Journal of African History, v. 4, n. 1, 1963: 97-101
Compiled by the Rhodes - Livingstone Museum. Presents a compilation of radiocarbon dates obtained from various radiocarbon laboratories for the regions of Africa south of the Sahara, and their description. All dates that are known to have been released before 1 Aug. 1962 have been included.

1.024   1965
RADIOCARBON DATES FOR SUB-SAHARAN AFRICA (FROM CA 1000 BC), III
Fagan, Brian M.
Journal of African History, v. 6, n. 1, 1965: 107-116
Compiled by the Rhodes - Livingstone Museum. Presents a compilation of radiocarbon dates obtained from various radiocarbon laboratories for the regions of Africa south of the Sahara, and their description. All dates are calculated relative to A.D.1950. The main periods included are: the Rhodesian Iron age, the Kalomo culture in Zambia, Iron age people in Angola, Saharan Neolithic sites occupied by the 4th century BC, the Nok culture in Nigeria.

1.025   1966
RADIOCARBON DATES FOR SUB-SAHARAN AFRICA, IV
Fagan, Brian M.
Journal of African History, v. 7, n. 3, 1966: 495-506
A fourth list of dates including those received before 1 June 1966. Compiled by the Rhodes - Livingstone Museum. Presents a compilation of radiocarbon dates obtained from various radiocarbon laboratories for the regions of Africa south of the Sahara, and their description.

1.026   1967
RADIOCARBON DATES FOR SUB-SAHARAN AFRICA, V
Fagan, Brian M.
Journal of African History, v. 8, n. 3, 1967: 513-527
Contains radiocarbon dates published before 1 May 1967. A discussion of the reliability of radiocarbon dates and of the correlation with tree-ring dating is presented. Compiled by the Rhodes - Livingstone Museum. Presents a compilation of radiocarbon dates obtained from various radiocarbon laboratories for the regions of Africa south of the Sahara, and their description.

1.027   1967
RADIOCARBON DATES OBTAINED THROUGH GEOGRAPHICAL BRANCH FIELD OBSERVATIONS
Andrews, J.T.
Geographical Bulletin, v. 9, n. 2, 1967: 115-162
Radiocarbon dates are given for samples obtained mainly in the Northwest Territories, by Geographical Research parties during the period 1950 to 1966. Not all dates have been made public. A map showing the general location of the sites from which samples were obtained is included.

1.028   1959
RADIOCARBON DATES ON CARDS
Woodburry, Nathalie F.S.
American Antiquity, v. 25, n. 1, July 1959: 145
Interest was manifested, according to a survey, but the cost of the service was such that only a bare minimum of subscriptions were received. A limited program of 5000 cards, to 1961, will be undertaken.

1.029   1967
RADIOCARBON MEASUREMENTS - COMPREHENSIVE INDEX, 1950-1965
Deevey, Edward S.; Flint, Richard Foster; Rouse, Irving
New Haven, Connecticut, 1967. pp. 221
List of dates.

1.030   1964
A REVIEW OF RADIOCARBON MEASUREMENT FROM PERU AND BOLIVIA
Rowe, John Howland
Manuscript, Berkeley, Dittoed, 1964
[Not sighted

1.031 1960
SAMPLING AND MEASURING OF CARBON14 LEVELS IN THE ATMOSPHERE AND BIOSPHERE
United States. Scientific Committee on the Effect of Atomic Radiations. *Atomic Energy Commission Report to the United Nations.*, n. A/AC.82/R (Add.2), 1960
A bibliography is offered on techniques and equipments for the measurement of radiocarbon in the atmosphere and the biosphere. 36 references.

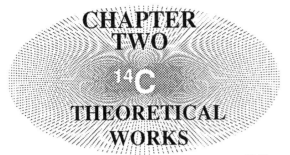

CHAPTER TWO
^{14}C THEORETICAL WORKS

2.001   1966
^{14}C MEASUREMENTS IN THE ATMOSPHERE
Health and Safety Laboratory, New York Operation Office (AEC) N.Y., 1 Jan. 1966. pp. 73
Data on atmospheric ^{14}C measurements are tabulated. Most of the data pertain to aircraft observations, also some from balloon-borne sampling and ground-level collections.

2.002   1955
^{14}C VARIATIONS IN NATURE AND THE EFFECT ON RADIOCARBON DATING
Rafter, T.A.
New Zealand Journal of Science and Technology, section B, v. 37, n. 1, July 1955: 20-38
Variations in the ^{14}C/^{12}C ratios with respect to contemporary wood are shown for air-water, shells, the flesh of shellfish, and the bone carbonate, fats and proteins of animal. Some ^{14}C/^{12}C variations predicted from ^{13}C/^{12}C ratios are shown to hold in nature at least within the confines of the area and specimens examined. The necessity for using the correct contemporary standard to aged archaeological specimens other than wood and charcoals is dicussed.

2.003   1952
THE ABSORPTION RATE OF COSMIC-RAY NEUTRONS PRODUCING C-14 IN THE ATMOSPHERE
Ladenburg, Rudolf
Physical Review, v. 86, n. 1, 1 April 1952: 128
Corroborate the conclusion that no considerable change in the cosmic-ray intensity has occurred in the past several thousand years.

2.004   1956
THE ABUNDANCE OF ATMOSPHERIC CARBON DIOXIDE AND ITS ISOTOPES
Goldberg, Leo
In: Vistas in Astronomy, vol. 2, edited by Arthur Beer. London, Pergamon Press, 1956: 855-863
The average carbon dioxide content of the air above Mount Wilson, California, is 0.040 percent. Any variation in abundance is less than about ± 10 percent. The abundance of the isotopic forms are found to be one half as great as the abundance of the isotopes of carbon and oxygen, but at least part of the discrepancy can be caused by errors in the absorption coefficient

2.005   1963
ACCURACY OF RADIOCARBON DATES
Libby, Willard F.
Science, v. 140, n. 3564, 5 Apr. 1963: 278-280
Apparent discrepancies are analysed for geophysical significance and for a general principle of correction.

2.006   1960
ADVANCES IN GEOCHRONOLOGICAL RESEARCH
Zeuner, Frederick Eberard
In: The Application of Quantitative Methods in Archaeology, edited by Robert F. Heizer and Sherburn F. Cook. Chicago, Viking Fund Publications in Anthropology, n. 28, 1960: 325-343
Describes various methods of dating prehistoric materials, including the radiocarbon dating method which is discussed on the basis of accuracy, validity, the Suess effect, the bomb effect. Also presents a table of dates for Jericho and Jarmo.

2.007   1949
AGE DETERMINATION BY RADIOCARBON CONTENT: RADIOCARBON
Libby, Willard F.; Anderson, Ernest C.; Arnold, James R.
Science, v. 109, n. 2827, 4 Mar. 1949: 227-228
The world-wide uniformity of the radiocarbon assay in the present time results in the logical assumption that this would have been true in ancient times. This leads to dating by the half-life of radiocarbon. A verification test was conducted on wood samples from ancient Egyptian tombs.

2.008   1951
AGE DETERMINATION BY RADIATION METHODS
Fergusson, Gordon J.
New Zealand Science Review, v. 9, n. 1-2, 1951: 1-2
[Not sighted]

2.009   1949
AGE DETERMINATION OF ARCHAEOLOGICAL MATERIAL
Braidwood, Robert J.; Jacobsen, T.; Parker, Richard A.; Weinberg, S.S.
American Journal of Archaeology, v. 54, 1949: 266
A plea from the subcommittee on the Near and Middle East of the American Anthropological Association and the Geological Society of America for samples to enable Drs. Libby and Arnold to test their method of age determination. Type of material, mode of labelling, mode of description and types of projects are described.

Ch. 2 - Theoretical Works

2.010   1949
AGE DETERMINATIONS BY RADIOCARBON CONTENT: CHECK WITH SAMPLES OF KNOWN AGES
Arnold, James R.; Libby, Willard F.
Science, v. 100, n. 2859, 23 Dec. 1949: 678-690
Archaeological and geological samples of known age which had been wood dated by conventional methods were used for testing the radiocarbon method of age determination. The agreement between prediction and observation was seen to be satisfactory (34% of samples were within one standard deviation). Checks on possible contamination were made and nothing significant was found. The results indicate that the two basic assumptions of the radiocarbon age determination, mainly the constancy of the cosmic radiations intensity and the possibility of obtaining unadulterated samples are probably justified for wood up to 4600 years. Dating of sample of Sequoia Gigantea shows that redwood heartwood is truly dead.

2.011   1963
AGE OF THE BAOBAB TREE
Swart, E.R.
Nature, v. 198, n. 4881, 18 May 1963: 708-709
Radiocarbon age measurements from three samples (one from the heart, one from midway between centre and bark, and one directly adjacent to the bark), indicate that the tree grew more slowly over the outer 7.5 feet of diameter. The ages were heart: 1010 ± 100 B.C., midway sample: 740 ± 100 B.C., near bark: A.D. 1890 If it were not for the fact that so many of the really large baobabs are hollow, they might well be ideal samples on which to determine variations in the radiocarbon content of the atmosphere.

2.012   1959
AN ANALYSIS OF THE POSSIBLE CHANGES IN ATMOSPHERIC CARBON DIOXIDE CONCENTRATION
Bray, J. Roger
Tellus, v. 11, n. 2, May 1959: 220-229
Criteria minimising differences in operators, location and time of observation are established for selecting comparative data on atmospheric carbon dioxide concentration during the past 100 years. The resulting selection showed in all cases the period 1907-1956 to have a higher mean than 1857-1907. Several possible explanations for the increase include: (1) an actual atmospheric increase, (2) a coincidence of the influence of microatmosphere, (3) improvement or change in chemical techniques.

2.013   1967
AN ANALYTICAL MODEL OF CARBON-14 DISTRIBUTION IN THE ATMOSPHERE
Schell, William Raymond; Fairhall, A.W.; Harp, G.D.
In: Radioactive Dating and Methods of Low-level Counting. Proceedings of a Symposium organised by the International Atomic Energy Agency (IAEA) in cooperation with the Joint Commission on Applied Radioactivity (ICSU) and held in Monaco, 2-10 March 1967. Vienna, IAEA, 1967. (Proceedings Series): 79-92
An analytic model describing radiocarbon distribution in the atmosphere has been derived. The model assumes that the atmosphere and ocean comprise a closed system with respect to radiocarbon exchange and that the total radiocarbon content of the system is constant during the period under consideration. Under these basic assumptions, the equilibrium of carbon-12, -13 and -14 dioxide in the atmosphere and ocean is evaluated.

2.014   1956
ANOMALOUS CARBON-14 CONTENT OF CARBON DIOXIDE FROM SEWER GAS
Patterson, R.L.; Blifford, I.H.
Science, v. 124, n. 3234, 21 Dec. 1956: 1252
Sewer gas was used as a source of modern carbon in the study of the radiocarbon content of atmospheric carbon. An anomalous high content of radiocarbon in the carbon dioxide fraction of this gas was found.

2.015   1966
ANOMALOUS CARBON-ISOTOPE RATIOS IN NON-VOLATILE ORGANIC MATERIAL
Kaplan, I.R.; Nissenbaum, A.
Science, v. 153, n. 3737, 12 Aug. 1966: 744
Organic materials associated with sulphur deposits in Upper-Pleistocene sand ridges on the coastal plain of Israel yielded ages of $27,750 \pm 500$ and $31,370 \pm 400$ years. $^{13}C/^{12}C$ determinations fell within the $\partial^{13}C$ range of -82.5 to -89.3 per mil relative to the PDB standard, the lowest reported for high molecular weight organic material. The origin of the carbon is probably complex; it must have passed through at least one biological cycle before final deposition.

2.016   1964
THE APPLICABILITY AND RELIABILITY OF THE RADIOCARBON DATING METHOD
Jensen, E.H.
Peking Science Symposium, 1st, 1964. *Proceedings*. 1964: 277-289
[Not sighted]

2.017   1957
APPLICATIONS OF RADIOCARBON TO OCEANOGRAPHY AND CLIMATE CHRONOLOGY
Broecker, Wallace S.
Ph.D. Dissertation. Columbia University, 1957

Two independent studies are included: (1) A study of the deep water flow rates and patterns in the Atlantic Ocean; (2) the provision of an absolute chronology for climatic events over the past 40,000 years and of a correlation between the events in independent climatic-sensitive systems in order to provide boundary conditions which must be met by acceptable theoriees of glaciation.

2.018   1951
ARCHAEOLOGICAL DATING BY RADIOACTIVE CARBON
Zeuner, Frederick Eberard
Scientific Progress, v. 39, 1951: 225-338
Gives a brief synopsis of the development and character of the method and reviews some results, a few of which are controversial.

2.019   1952
ARCHAEOLOGICAL DATING BY RADIOCARBON MEASUREMENTS
Long, J.V.P.
Atomics, v. 3, n. 12, Dec. 1952: 317-319
Dating archaeological material by radiocarbon measurements originated with the work of Professor Libby and his colleagues at the University of Chicago. Since 1946, a considerable amount of progress has been made. A review of the development of the physical and chemical techniques is presented.

2.020   1967
ASTROPHYSICAL PHENOMENA AND RADIOCARBON
Konstantinov, B.P.; Kocharov, G.E.
Translated from Russian for National Aeronautic and Space Administration, Washington DC, 1947. pp. 36 (NASA - TT - F - 11567)
Consideration is given to an expansion of the basis of relationships between astrophysical phenomena and radiocarbon. The radiocarbon content of dendrochronologically dated samples is used to study solar cyclic activity, flares from supernovae, etc. In addition the atmospheric radiocarbon concentration can be used to determine the characteristic time for radiocarbon dispersion into the various reservoirs.

2.021   1961
ATMOSPHERIC ACTIVITIES AND DATING PROCEDURES
Maddock, A.G.; Willis, E.H.
In: Recent Advances in Inorganic and Radio Chemistry, v. 3. New York, Academic Press, 1961
[Not sighted]

2.022   1966
ATMOSPHERIC BOMB RADIOCARBON AS A TRACER IN HUMAN BEINGS
Berger, Rainer; Libby, Willard F.; Alexander, G. W.; Mead, J.F.; Ross, J.F.
In: Advances in Tracer Methodology, v. 3, *A Collection of Papers presented at the 10th Symposium on Tracer Methodology*, edited by Seymour Rothchild. New York, Plenum Press, 1966: 321-329
The results of the measurement of radiocarbon in human tissue reflects the rise in radiocarbon activity in atmospheric carbon in recent year.

2.023   1966
ATMOSPHERIC C-14 CONTENT DURING THE PAST THREE MILLENIA IN RELATION TO TEMPERATURE AND SOLAR ACTIVITY
Bray, J. Roger
Nature, v. 209, n. 5028, 12 March 1966: 1065-1067
Recent summaries of temperature, glacial activity and solar activity provide more exact climatological pattern than previously available, against which to test the change in radiocarbon activity over the past three millenia. These patterns will be used in an attempt to reconcile contradictions in interpretation of radiocarbon and climate relationship.

2.024   1957
ATMOSPHERIC CARBON-14
Patterson, R.L.; Blifford, I.H.
Science, v. 126, n. 3262, 5 Jan. 1957: 26-28
The increase in the radiocarbon content of the atmospheric carbon dioxide from 1952 to 1956 is probably the result of the addition of radiocarbon from thermonuclear sources. The delayed appearance of the radiocarbon increase at ground level may indicate a stratospheric reservoir of this isotope.

2.025   1946
ATMOSPHERIC HELIUM 3 AND RADIOCARBON FROM COSMIC RADIATIONS
Libby, Willard F.
Physical Review, v. 69, n. 11-12, July 1946: 671-672
Nuclear physical data indicate that cosmic ray neutrons produce radiocarbon and tritium from atmospheric nitrogen. Indicate that there must exist a radioactive equilibrium in which the rate of disintegration of radiocarbon is equal to the rate of production.

2.026   1965
ATMOSPHERIC PATHWAYS OF MAN-MADE C14, H3, AND SR90
Lal, D.

Ch. 2 - Theoretical Works

In: International Conference on Radiocarbon and Tritium Dating, Pullman, Washington, Washington State University, June 1-11, 1965. *Proceedings.* United States of America, Atomic Energy Commission, 1965. Conference n. 650652: 541-548

The mean time of exchange between the tropics of the two hemispheres is found to be seven months and essentially independent of seasons. The effective interhemispheric exchange time is found to be fifteen months.

2.027 1960
ATMOSPHERIC RADIOCARBON ACTIVITY IN 1959
Willis, E.H.
Nature, v. 185, n. 4712, 20 Feb. 1960: 552-553

Measurements made in the summer of 1959 indicate a general level of radiocarbon activity greatly in excess of what might have been expected from estimations. The large difference in activity between the summers of 1958 and 1959 must be attributed to the intense nuclear testing programs carried out in late 1958 and its delay in appearing in the atmosphere probably lies in the mechanism of seasonal variations.

2.028 1958
ATMOSPHERIC RADIOCARBON AS A TRACER IN GEOPHYSICAL CIRCULATION PROBLEMS
Rafter, T.A.; Fergusson, Gordon J.
In: International Conference on the Peaceful Use of Atomic Energy, 2nd, Geneva, 1958. *Proceedings.* (A/Conf. 15/P/1123), 1958, v. 2: 526-532

Recent measurements of the radiocarbon specific activity of the atmosphere of the southern hemisphere shows that the radiocarbon specific activity has been increasing at approximately exponential rate since the late 1954. The only explanation is that this is due to the testing of atomic weapons. An exchange time of 3 years for the transfer of carbon dioxide from atmosphere to surface water is indicated by the measurements and also a much greater age for Pacific and Antarctic water than for Atlantic water.

2.029 1960
ATMOSPHERIC RADIOCARBON AS A TRACER IN GEOPHYSICAL CIRCULATION PROBLEMS
Rafter, T.A.; Fergusson, Gordon J.
New Zealand Journal of Science and Technology, Sect. B, v. 18, 1960:

Recent measurements of the radiocarbon specific activity of the atmosphere of the southern hemisphere shows that the radiocarbon specific activity has been increasing at approximately exponential rate since the late 1954. The only explanation is that this is due to the testing of atomic weapons. An exchange time of 3 years for the transfer of carbon dioxide from atmosphere to surface water is indicated by the measurements and also a much greater age for Pacific and Antarctic water than for Atlantic water.

2.030 1957
ATOM BOMB EFFECT. RECENT INCREASE OF CARBON-14 CONTENT OF THE ATMOSPHERE AND THE BIOSPHERE
Rafter, T.A.; Fergusson, Gordon J.
Science, v. 126, n. 3273, 20 Sept. 1957: 557-558

The radiocarbon concentration in the air in New Zealand is measured and shows an increase of 4.9 ± 0.6% indicating rapid exchange between these carbon reservoirs. A comparison of the increase in the atmosphere of the Northern and Southern Hemisphere would yield important meteorologic data on the mixing rate of the atmosphere of the two Hemispheres.

2.031 1957
THE ATOM BOMB EFFECT: RECENT INCREASE IN THE ^{14}C CONTENT OF THE ATMOSPHERE, BIOSPHERE AND SURFACE WATERS OF THE OCEAN
Rafter, T.A.; Fergusson, Gordon J.
New Zealand Journal of Science and Technology, Sect. B, v. 38, n.8, Sept. 1957: 871-883

Recent measurements of the radiocarbon specific activity of the atmosphere of the Southern Hemisphere show that it has been increasing at an approximately exponential rate since late 1954. The radiocarbon specific activity of the contemporary portion of the biosphere has been shown to be increasing at the same rate. The only explanation is the production of radiocarbon from the testing of atomic weapons. An exchange time of 20 months for transfer of carbon dioxide from the atmosphere to the surface water of the oceans has been deduced from measurements of the increase in the radiocarbon specific activity of the surface water of the ocean.

2.032 1958
ATOMIC BOMB EFFECT: VARIATION OF RADIOCARBON IN PLANTS, SHELLS AND SNAILS IN THE PAST FOUR YEARS
de Vries, Hessel
Science, v. 128, n. 3318, 1 Aug. 1958: 250-251

Measurements made between 1953 and 1957 gave evidence to a fast increase of the activity of radiocarbon in the atmosphere due to atom bombs. It also indicates differing radiocarbon concentration in the atmosphere and the oceans.

2.033 1956
BERYLLIUM 10 PRODUCED BY COSMIC RAYS

Arnold, James R.
Science, v. 124, n. 3222, 28 sept. 1956: 584-585
The author investigates the possibility of using beryllium 10, which like radiocarbon, is the product of the bombardment of the nitrogen and oxygen of the atmosphere by cosmic rays, for radioactive age determination.

2.034   1962
BIOLOGICAL HAZARDS OF CARBON-14
Purdom, C.E.
New Scientist, v. 15, n. 298, 2 Aug. 1962: 255-257
Should no further nuclear bomb be exploded, the radiocarbon content of living organisms - as a result of past bomb tests - will be about double the natural level in the next 30 year period; the excess will then diminish at a steady rate as the isotope decays. Although biological hazards from radiocarbon are negligible compared with those from natural radiations or infections, man should not wantonly expose himself to them.

2.035   1959
BOMB CARBON-14 IN HUMAN BEINGS
Broecker, Wallace S.; Schulert, Arthur; Olson, Edwin A.
Science, v. 130, n. 3371, 7 Aug. 1959: 331-332
The concentration of the bomb produced radiocarbon in human beings will lag behind the rising concentration in average atmospheric carbon dioxide

2.036   1951
A BRIEF DISCUSSION OF THE RELATION OF SOME RADIOCARBON DATES TO THE POLLEN CHRONOLOGY
Deevey, Edward S.
American Antiquity, v. 17, part 2 (*Society for American Archaeology, Memoirs*, n. 8), 1951: 56-57
Comments on 31 dates of specific interest to palynology presented in the first part of the memoir. Most radiocarbon dates are in agreement with the palynological dates.

2.037   1968
BRISTLECONE PINE : SCIENCE AND ESTHETIC
Ferguson, C. Wesley
Science, v. 159, n. 3814, 23 Feb. 1968: 839-846
A 7100 year tree ring chronology aids scientists. Radiocarbon analysis of a single, small specimen, that contains a 400 year, high quality ring series, indicate that the specimen is approximately 9000 years old. This holds great promise for the extension of the tree ring chronology farther back in time. Dual dates derived from radiocarbon analysis of tree ring dated bristlecone pine wood show that for material in the period 4000 to 3000 B.C. dates obtained by the conventional radiocarbon method are about 800 years more recent. Discussion of this discrepancy and application of radiocarbon dating to tree ring dating is included.

2.038   1967
BRISTLECONE PINE CALIBRATION OF THE RADIOCARBON TIME SCALE FROM 4100 B.C. TO 1500 B.C.
Suess, Hans E.
In: Radioactive Dating and Methods of Low-level Counting. Proceedings of a Symposium organised by the International Atomic Energy Agency (IAEA) in co-operation with the Joint Commission on Applied Radioactivity (ICSU) and held in Monaco, 2-10 March 1967. Vienna, IAEA, 1967. (Proceedings Series): 143-151.
A bristlecone tree-ring sequence going back some 6700 years was established by C.W. Ferguson of the University of Arizona. For the period 4100 B.C. to 1500 B.C. some 80 bristlecone pine wood samples were analysed for the radiocarbon content. If one assumes that the bristlecone chronology gives the correct age, then the radiocarbon content of wood sample dating from the third millenium B.C. is between 6 and 9 percent higher than that calculated with a half life of radiocarbon of 5730 years. A graphic representation of the deviations and a calibration curve for the conversion of dates to bristlecone pine dates is given. Direct comparison of the radiocarbon content and historically dated samples from ancient Egypt with that of bristlecone pine wood leads to calibrated radiocarbon ages which are compatible with evidence from historical records.

2.039   1967
C-14 CYCLING IN THE ROOT AND SOIL COMPONENT OF A PRAIRIE ECOSYSTEM
Dahlman. R.C.; Kucera, C.L.
O.R.N.L. Report, n. 3082, 1967
[Not sighted]

2.040   1951
C-14 DATING
Collier, Donald
Michigan University. Museum of Anthropology. Anthropological Papers, n. 8, 1951 (Essays on Anthropological Methods. Proceedings of a Conference held under the auspices of the Viking Fund at Michigan University, Ann Arbor): 97- 101
Historical review of radiocarbon dating and summary of the facts and hypothesis which form the basis of radiocarbon dating.

2.041   1956
C-14 DATING
Ralph, Elizabeth K.
Pennsylvania Archaeologist, v. 26, n. 1, June 1956: 27-31
Brief overview of the method.

Ch. 2 - Theoretical Works

2.042   1966
C-14 MEASUREMENTS IN THE ATMOSPHERE
Machta, Lester; Hageman, Frerich T.; Gray, James
Health and Safety Laboratory Report to U.S. Atomic Energy Commission, n. HASL-166, 1966. pp. 73 (Available from Clearing House for Federal Scientific and Technical Information, Springfield, Va.)
Data on atmospheric radiocarbon measurements are tabulated. These include mostly aircraft collections, but also a few balloon born sampling and ground level collections.

2.043   1965
C-14 MEASUREMENTS IN THE ATMOSPHERE - 1953 TO 1964
Machta, Lester; Gray, James; Hageman, Frerich T.
Health and Safety Laboratory Report to U.S. Atomic Energy Commission, n. HASL-159, 1965 (Available from Clearing House for Federal Scientific and Technical Information, Springfield, Va.)
A radiocarbon sampling program has been conducted since 1953. Over 3000 results of both total and bomb produced radiocarbon concentration in carbon dioxide are presented together with auxiliary $\partial^{13}C$ and carbon dioxide concentration determinations. A few radiocarbon determinations are also given and interpreted.

2.044   1953
C-14 MEASUREMENTS ON GEOLOGICAL SAMPLES
Kulp, J. Lawrence
Atomics, v. 4, 1953: 96-98
[Not sighted]

2.045   1968
C-14/C-12 RATIO DURING THE LAST SEVERAL THOUSAND YEARS AND THE RELIABILITY OF C-14 DATES
Olsson, Ingrid U.
In: Means of Correlation of Quaternary Successions, edited by Roger B. Morrisson and Herbert C. Wright. International Association for Quaternary Research (INQUA). Congress, 7th, Boulder, Col., 14 Aug - 19 Sept. 1965. *Proceedings*, v. 8. Salt Lake City, University of Utah Press, 1968: 241-252
The reliability of radiocarbon dates is discussed in terms of two lines of investigation: the $^{14}C/^{12}C$ ratio during the last 2000 years increases the errors to about 200 years. The pretreatment of samples is very important. Samples of carbonates are usually not reliable if they are older than 20,000 years.

2.046   1958
C14/C12 RATIOS IN FRESH WATER SYSTEMS
Broecker, Wallace S.; Walton, Alan
American Geophysical Union. Transactions, v. 39, n. 3, 1958: 509
Abstract of paper presented at the 39th Annual Meeting, Washington D.C., May 5-8, 1958. Radiocarbon age determinations on freshwater carbonate materials are based on the assumption that an accurate estimate can be made of the radiocarbon concentration at the time of deposition. Radiocarbon analyses of materials formed in freshwater allow this assumption to be more adequately defined and also provide information about the carbon dioxide cycle in freshwater systems.

2.047   1952
CALORIMETRIC DETERMINATION OF THE RELATIONSHIP BETWEEN THE HALF-LIFE AND AVERAGE BETA ENERGY OF C-14
Jenks, G.M.; Sweeton, F.H.
Physical Review, ser. 2, v. 86, n. 5, 1 June 1952: 803-804
Half-life of ^{14}C obtained: 6030 years.

2.048   1954
CARBON 13 VARIATIONS IN SEQUOIA RINGS AND THE ATMOSPHERE
Craig, Harmon
Science, v. 119, n. 3083, 19 Jan. 1954: 141-143
The data analysed indicate that the isotopic composition of the atmospheric carbon was constant to at least 1 per mil. during the 2500 years interval from 900 B.C. to A.D. 1600 with a change in mean composition probably less than 0.2 per mil.

2.049   1964
CARBON AND OXYGEN ISOTOPIC COMPOSITION OF MOLLUSK SHELLS FROM MARINE AND FRESH-WATER ENVIRONMENT
Keith, M.L.; Anderson, G.M.; Eichler, R.
Geochimica et Cosmochimica Acta, v. 28, Nov. 1968: 1757-1786
A systematic survey of difference in the isotopic composition of carbon and oxygen of modern mollusc shells from marine and continental environments is presented. It is concluded that the carbon isotope ratio in mollusc shells is considerably influenced by the proportional amount of land-plant derived carbon included in the food of the molluscs or contributed by humus decay to dissolved bicarbonate in the water.

2.050   1951
CARBON CONTENT AND RADIOACTIVITY OF MARINE ROCKS
Burton, Virginia L.; Sullivan, Geraldine R.
American Geophysical Union. Transactions, v. 32, n. 6, Dec. 1951: 881-884
The radioactivity and carbon content of eight geological

formations have been determined. Included were two sandstones, two limestones and four shales. An increasing linear relation was observed between the net beta count and the carbon content, suggesting that there may be a genetic relationship between uranium, thorium and carbon content.

2.051   1957
CARBON DIOXIDE EXCHANGE BETWEEN ATMOSPHERE AND OCEAN AND THE QUESTION OF AN INCREASE OF ATMOSPHERIC CO_2 DURING THE PAST DECADES
Revelle, Roger; Suess, Hans E.
Tellus, v. 9, n. 1, Feb. 1957: 18-27
From a comparison of $^{14}C/^{12}C$ and $^{13}C/^{12}C$ ratios in wood and in marine material and from a slight decrease of the radiocarbon concentration in terrestrial plants over the past 50 years, it can be concluded that the average lifetime of a carbon dioxide molecule in the atmosphere before it is dissolved into the sea is of the order of ten years. This means that most of the carbon dioxide released by artificial fuel combustion since the beginning of the industrial revolution must have been absorbed by the oceans. The increase of atmospheric carbon dioxide from this cause is at present small but may become significant during future decades if industrial fuel combustion continues to rise exponentially.

2.052   1956
THE CARBON DIOXIDE THEORY OF CLIMATIC CHANGE
Plass, Gilbert G.N.
Tellus, v. 8, n. 2, May 1956: 140-154
A study of the variation of carbon dioxide in the atmosphere and its influence on temperature variations. This suggests that changes in carbon dioxide content in the atmosphere can cause appreciable climatic variations. The author presents an overview of the major factors influencing the carbon dioxide balance at the present time.

2.053   1953
CARBON ISOTOPE EFFECTS IN BIOLOGICAL SYSTEMS
Buchanan, Donald L.; Nakea, A.; Edwards, G.
Science, v. 117, n. 3047, 22 May 1953: 541-545
A brief survey of the isotope effects with radiocarbon in biological reactions in which the fractionation factors of ^{14}C and ^{13}C are compared. The need for evaluation of the isotope effect depends upon the type and goal of the research.

2.054   1967
CARBON ISOTOPE FRACTIONATION IN THE SYSTEM CO_2(GAS) - CO_2(AQUEOUS) - HCO_3(AQUEOUS)
Deuser, W.G.; Degens, E.T.
Nature, v. 215, n. 5105, 2 Sept. 1967: 1033-1035
Carbon isotope fractionation between gaseous carbon dioxide and aqueous bicarbonate decreases from 9.2 to 6.8 per thousand over the temperature range 0° - 30°C. This fractionation occurs in the hydration stage, not in the passage of atmospheric carbon dioxide through the air-water interface.

2.055   1954
CARBON-13 IN PLANTS AND THE RELATIONSHIPS BETWEEN CARBON-13 AND CARBON-14 VARIATIONS IN NATURE
Craig, Harmon
Journal of Geology, v. 62, n. 2, Mar. 1954: 115-149
The relationship between ^{13}C and ^{14}C variations in depositional processes of carbonaceous materials are discussed with respect to the relative enrichment of these isotopes in carbonates. An estimated 'age' is derived for the carbon in the ocean.

2.056   1958
CARBON-14 ACTIVITY DURING THE LAST 5000 YEARS
Crowe, C.
Nature, v. 182, n. 4633, 16 Aug. 1958: 470-471
The activity appears to have followed a cycle with a maximum change of about 10 percent in the past 5000 years with a sharp peak approximately 2000 years ago. A rapid decrease in the past 100 years is attributed by Suess to the effect of dilution of the atmospheric reservoir of radiocarbon due to the contribution of fossil fuel.

2.057   1958
CARBON-14 ACTIVITY DURING THE PAST 5000 YEARS
Münnich, K.O.; Östlund, H. Göte ; de Vries, Hessel
Nature, v. 182, n. 4647, 22 Nov. 1958: 1432-1433
The authors argue that the variations in activity as noted by Crowe (*Nature*, 182, 1958: 470) are too great almost by one order of magnitude. The discrepancy seems to be due to the type of standard used by various laboratories.

2.058   1954
CARBON-14 AGE METHOD
Arnold, James R.
In: Nuclear Geology: a Symposium on Nuclear Phenomena in the Earth Sciences, edited by H. Faul. New York, John Wiley, 1954: 349-354
The basis for the theory is presented and the various methods of counting are discussed and compared.

2.059   1961
CARBON-14 AND OTHER SCIENCE METHODS THAT DATE THE PAST

Ch. 2 - Theoretical Works

Poole, Lynn; Poole, E. Grey
New York, Wittlesey House, 1961. pp. 160
[Not sighted]

2.060   1967
CARBON-14 CONCENTRATIONS IN ENVIRONMENTAL MATERIALS AND THEIR TEMPORAL FLUCTUATIONS
Walton, Alan; Baxter, M.S.; Callow, W.J.; Baker, M.J.
In: Radioactive Dating and Methods of Low-level Counting. Proceedings of a Symposium organised by the International Atomic Energy Agency (IAEA) in co-operation with the Joint Commission on Applied Radioactivity (ICSU) and held in Monaco, 2-10 March 1967. Vienna, IAEA, 1967. (Proceedings Series): 41-47
Data show that there is a general correlation between radiocarbon concentration of spirit and atmospheric carbon dioxide samples with the result that the year of manufacture of the spirits is predictable. The 'Suess effect' and the 'bomb effect' are considered in the light of this correlation.

2.061   1960
THE CARBON-14 CONTENT OF URBAN AIRBORNE PARTICULATE MATTER
Lodge, James P.; Bien, George S.; Suess, Hans E.
International Journal of Air Pollution, v. 2, 1960: 309-312
Very large samples of atmospheric particulate matter were collected in the central part of St. Louis, Missouri, and Los Angeles, California. The radiocarbon content of various fractions of the samples was determined and was interpreted in terms of the relative contribution of fossil fuels and contemporaneous sources of carbon to the particulate pollution of the area. The general applicability of the method is discussed.

2.062   1965
CARBON-14 CONTENT OF THE 18TH AND 19TH CENTURY WOOD: VARIATIONS CORRELATED WITH SUNSPOT ACTIVITIES
Stuiver, Minze
Science, v. 149, n. 3683, 30 July 1965: 533-535
A series of radiocarbon measurements of 18th and 19th century wood revealed an excellent correlation between the variations in radiocarbon content and average sunspot activity. It seems probable, therefore, that the predominant cause of the short-term fluctuation in radiocarbon activity is the variability of solar activity.

2.063   1965
CARBON-14 CONTENT OF WOOD FROM DIFFERENT LOCALITIES (Abstract)
Vogel, John C.
In: International Conference on Radiocarbon and Tritium Dating, Pullman, Washington, Washington State University, June 1-11, 1965. *Proceedings*. United States of America, Atomic Energy Commission, 1965. Conference n. 650652: 453
Results reveal difference of up to one percent in the radiocarbon content of wood from various places. This can be explained by the nature of the radiocarbon exchange reservoir.

2.064   1952
CARBON-14 DATES AND ARCHAEOLOGY
Roberts, Frank H.H.
American Geophysical Union. Transactions, v. 33, n. 2, 1952: 170-174
Reviews the development in the field of the application of radiocarbon dating to archaeology and particularly in the establishment of more precise chronologies for the prehistory of Europe and the Near East.

2.065   1957
CARBON-14 DATING
Crane, H.R.
In: The Identification of Non-Artificial Archaeological Materials, edited by Walter. W. Taylor. National Academy of Science. National Research Council. Committee on Archaeological Identification, Publication n. 565, 1957: 54-56
Summary of the present status of radiocarbon dating in the technical sense, with particular emphasis on the method used at Michigan University. Reviews four gas counting methods: proportional counting methods with carbon dioxide, methane and acetylene as the gases to be counted, Geiger counters filled with CO_2/CS_2; the liquid scintillation counting method and the carbon black method.

2.066   1961
CARBON-14 FROM NUCLEAR EXPLOSIONS AS A SHORT TERM DATING SYSTEM: USE TO DETERMINE THE ORIGIN OF HEARTWOOD
Wilson, Alex T.
Nature, v. 191, n. 4789, 12 Aug. 1961: 714
Three outer heartwood rings of a Pinus Radiata were dated. It showed that the activity of the cellulose in a ring corresponds to the activity of the atmosphere when it was the outer ring of the tree and the heartwood extractive corresponds in activity to the cellulose and not to the atmosphere of the time of their conversion from sapwood to heartwood.

2.067   1965
CARBON-14 IN TREE RINGS FROM NORTHERN AND SOUTHERN HEMISPHERES
Jansen, Hans S.
In: Carbon-14 Variations in Nature, New Zealand Institute of Nuclear Science. Contribution n. 10, 1965. pp. 28
[Not sighted]

2.068 — 1967
CARBON-14 MEASUREMENTS IN THE ATMOSPHERE
Hardy, P.H.; Rivera, J.
Health and Safety Laboratory. Quarterly Summary Report, n. HASL - 174, 1967
[Not sighted]

2.069 — 1960
CARBON-14 MEASUREMENTS OF KNOWN AGE
Ralph, Elizabeth K.; Stuckenrath, Robert
Nature, v. 188, n. 4746, 15 Oct. 1960: 185-187
In order to answer a number of fundamental questions concerning the validity of the basic factors of radiocarbon dating, samples of known age, measured in the laboratory at the University of Pennsylvania are examined and checked for deviation against oak standards. Although these few measurements do not cover the whole period of the past 4000 years or answer the specific questions, they do indicate that the maximum error due to meteorological uncertainties could be of the order of approximately 200 years and in some time-intervals is much smaller.

2.070 — 1952
THE CARBON-14 METHOD OF AGE DETERMINATION
Roberts, Frank H.H.
Smithsonian Institution. Annual Report for the Year ended 1951, 1952: 335-350
Describes the main features in the development of the method and discusses some of the results obtained to date.

2.071 — 1952
THE CARBON-14 METHOD OF AGE DETERMINATION
Kulp, J. Lawrence
Scientific Monthly, v. 75, n. 5, 1952: 259-267
Some of the numerous problems of archaeology, geology, ocean circulation, etc, solved by the radiocarbon dating technique, are summarised. The technique is illustrated.

2.072 — 1960
CARBON-14 PRODUCTION FROM NUCLEAR EXPLOSIONS
Latter, Albert L.; Plesset, Milton S.
National Academy of Science. Proceedings, v. 46, n. 2, 15 Feb. 1960: 241-247
Examines the redistribution and decay of excess radiocarbon introduced as a consequence of the atomic explosions in function of the transient responses of the various reservoir systems.

2.073 — 1965
CARBON-14 VARIATIONS IN NATURE, PART 1. A STUDY OF LONG TERM VARIATIONS OF ATMOSPHERIC CARBON-14 CONCENTRATION FROM TREE RINGS
Jansen, Hans S.
Institute of Nuclear Sciences. Department of Scientific and Industrial Research. (Lower Hutt, New Zealand), 1963 (INS.- R 37 Part 10. INS Contribution No. 196). pp. 24
Specimens of kauri, rimu, King William Pine and giant redwood have been dated by the tree ring counting and by radiocarbon measurements. All measurements confirm an increase in atmospheric radiocarbon concentration from about 1000 to 1600 A.D. The explanation for this change in atmospheric radiocarbon concentration over the centuries is either a change in the temperature and/or vertical movement of the ocean water and/or a change in the cosmic ray intenity. Methods for deciding between these alternatives are discussed.

2.074 — 1965
CARBON-14 VARIATIONS IN NATURE, PART 2, INCREASE IN C-14 ACTIVITIES IN THE ATMOSPHERE OF THE SOUTHERN HEMISPHERE FROM THE TESTING OF NUCLEAR WEAPONS
Rafter, T.A.
New Zealand Journal of Science, v. 8, n. 4, Dec. 1965: 472-493. (Institute of Nuclear Sciences Contribution No. 198)
Comparisons are made between Northern and Southern Hemisphere measurements. It is shown that there has been much greater fluctuation in the Northern Hemisphere, particularly since 1958. There appears to be a definite spring and autumn injection of radiocarbon into the troposphere from the stratosphere. Should bomb testing cease, the earth's troposphere would reach a state of uniform radiocarbon distribution within 1-2 years. Rapid changes in radiocarbon activities in the Northern Hemisphere are followed by similar changes at almost the same rate about one year later in the Southern Hemisphere.

2.075 — 1968
CARBON-14 VARIATIONS IN NATURE. PART 3: C-14 MEASUREMENTS IN THE SOUTH PACIFIC AND ANTARCTIC OCEAN
Rafter, T.A.
New Zealand Journal of Science, v. 11, n. 4, Dec. 1968: 551-589
This paper lists 204 measurements of the radiocarbon activity of ocean water collected at the Makara Surface Coastal Water Station, the Radiocarbon Standard Sub-Tropical Surface Water Station, a number of surface ocean water stations and 20 ocean profile studies.

Ch. 2 - Theoretical Works

2.076   1965
CARBON-14 VARIATIONS IN NATURE. PART 5. THE AGE OF THE NEW ZEALAND MOA FROM CARBON-14 MEASUREMENTS
Rafter, T.A.
Institute of Nuclear Sciences. Department of Scientific and Industrial Research. (Lower Hutt, New Zealand), 1965 (INS-R-37 pt.5, INS Contribution No. 196) pp. 16
The present high level of radiocarbon activity in the atmosphere from nuclear weapons tests has enabled the detection of atmospheric carbon dioxide fixation in moa bone specimens so causing apparent decrease in radiocarbon age. This confirms previous evidence on the unreliability of radiocarbon ages for moa bones. Includes a list of all radiocarbon ages reported of archaeological interest to the Moa Hunter - Early Maori period reported from this laboratory.

2.077   1967
CHANGE OF THE EARTH'S MAGNETIC FIELD AND RADIOCARBON DATING
Bucha, V. ; Neustupny, Evzen
Nature, v. 215, n. 5098, 15 July 1967: 261-263
The authors demonstrate graphically the necessity for including the influence of the changes of the Earth's magnetic field in radiocarbon dating.

2.078   1959
CHANGES IN THE CARBON DIOXIDE CONTENT OF THE ATMOSPHERE AND THE SEA DUE TO FOSSIL FUEL COMBUSTION
Bolin, Bert; Eriksson, E.
In: The Atmosphere and the Sea in Motion. Scientific Contribution to the Rossby Memorial Volume. New York Rockefeller Institute Press, 1959: 130-142
The dissociation equilibrium of carbon dioxide in the sea is discussed with particular emphasis on the buffering effect of sea water when changes of partial pressure of carbon dioxide in the gas phase take place (Suess effect).

2.079   1960
THE CHANGING LEVEL OF THE SEA
Fairbridge, Rhodes W.
Scientific American, v. 202, n. 5, May 1960: 70-79
Glacial cycles and the slow sinking of the ocean floor have caused the sea-level to fluctuate from epoch to epoch. Radiocarbon measurements furnish fairly exact dates and by combining radiocarbon dating and palaeobotany it is possible to determine the major changes in climate and sea-level for the past 20,000 years.

2.080   1966
CHARACTERISTICS OF GLOBAL TROPOSPHERIC MIXING BASED ON MAN-MADE C-14, H-3 AND SR-90
Lal, D.; Rama
Journal of Geophysical Research, v. 71, n. 12, 15 June 1966: 2865-2874
Study of global dispersion of radiocarbon, radio-hydrogen and radiostrontium injected into the atmosphere by Soviet nuclear tests supports the existence of the 0 - 30° Hadley meridional circulation cell. The rate of the meridional mixing within the troposphere is evaluated; analysis reveals a strong seasonal dependence for the exchange of air across the 30°N latitude in the troposphere. The annual mass transport of air across the equator is evaluated.

2.081   1968
CLIMATIC CHANGES, SOLAR ACTIVITY, AND THE COSMIC RAY PRODUCTION RATE OF NATURAL RADIOCARBON
Suess, Hans E.
In: Causes of Climatic Changes. American Meterological Society, *Meteorological Monographs*, v. 8, n. 30, Feb. 1968: 146-150
The level of the cosmic ray produced radiocarbon in atmospheric carbon dioxide fluctuates due to changes in the cosmic ray intensity. The radiocarbon measurements so far carried out show that prolonged high solar activity leads to warm winters and periods of a quiet sun lead to cold winters. Future determinations of radiocarbon variations during the past 6000 - 8000 years may supply conclusive evidence regarding the cause for the great ice age.

2.082   1951
COMMENTS ON RADIOCARBON DATING FOR SAMPLES FROM THE BRITISH ISLES
Godwin, Harry
American Journal of Science, v. 249, n. 4, Apr. 1951: 301-307
The results of radiocarbon assay of several samples from the British Isles are in general agreement with expectation. The author discusses some sources of error.

2.083   1953
COMPARATIVE MEASUREMENTS OF STANDARDS FOR CARBON ISOTOPES
Dansgaard, Willi
Geochimica et Cosmochimica Acta, v. 3, n. 5, 1953: 253-256
Measurements on four standards for carbon isotope, two American, one Swedish and one Danish are reported. It is shown that differences in ^{17}O abundance may cause measurable errors.

2.084   1962
COMPARISON BETWEEN RING-DATES AND C14 DATES IN A NEW ZEALAND KAURI TREE

Jansen, Hans S.
New Zealand Journal of Science, v. 5, n. 1, March 1962: 74-84

A 1000 years old Kauri tree from New Zealand has been ring-dated and radiocarbon dated by carbon dioxide proportional counting. From A.D. 1000 to 1550 the radiocarbon enrichment, with respect to 0.95 N.B.S. Oxalic Acid Standard ($D^{14}C$), increased slowly from -35% to -20% then rapidly to +5% by 1650 then slowly to 10% by 1800. From these results tentative correction for ages of New Zealand archaeological samples are derived, ranging up to 340 years. Explanations are suggested for the change in primary cosmic ray flux as a result of variations in the earth's magnetic moment (6.8%) or in solar characteristics.

2.085   1963
A COMPARISON OF CARBON-14 AND TRITIUM AGES OF GROUNDWATER
Münnich, K.O.; Roether, W.
In: *Radioisotopes in Hydrology*, International Atomic Energy Agency (IAEA), Symposium, Tokyo, 1963, *Proceedings*. Vienna, IAEA, 1963: 397-406

Simultaneous measurements of radiocarbon and tritium in groundwater samples from different horizons were undertaken to test whether the decrease in radiocarbon exceeds the radioactive decay and could be attributed to exchange with the aquifer. No increase was shown due to the bomb effect. This could be attributed to a stratification of the decaying organic matter in the soil, so that the very latest material does not contribute appreciably to the solution of lime by the percolating water.

2.086   1965
COMPARISON OF ^{230}TH AND ^{14}C AGES FOR CARBONATE MATERIALS FROM LAKE LAHONTAN AND BONNEVILLE
Kaufman, Aaron; Broecker, Wallace S.
Journal of Geophysical Research, v. 70, n. 16, 15 August 1965: 4039-4054

The reliabilities of the carbon and uranium inequilibrium methods of absolute age determination as applied to lacustrine carbonate have been evaluated. The majority of anomalies for both the ^{14}C and ^{230}Th methods are toward low ages. The results suggest that the radiocarbon method is in error by no more than 25% in the 10,000 to 20,000 year range.

2.087   1950
COMPARISON OF ARCHAEOLOGICAL AND RADIOCARBON DATING
Childe, V. Gordon
Nature, v. 166, n. 4234, 23 Dec. 1950: 1068-1069
On the basis of comparison the author feels that the method will require considerable checking and refinement before it can provide reliable dates for prehistoric events.

2.088   1960
THE CONCENTRATION AND ISOTOPIC ABUNDANCE OF CARBON DIOXIDE IN THE ATMOSPHERE
Keeling, Charles D.
Tellus, v. 12, n. 2, May 1960: 200-203

A systematic variation with season and latitude in the concentration and isotopic abundance of atmospheric carbon dioxide has been found in the Northern Hemisphere. In Antartica, however, a small but persistent increase in concentration has been found. Possible causes for these variations are discussed.

2.089   1966
CONCENTRATION OF C14 IN THE ATMOSPHERE AT THE TIME OF THE TUNGUSKA CATASTROPHE AND ANTIMATTER
Vinogradov, A.P.; Devirts, A.L.; Dobkina, E.I.
Akademiia Nauk U.S.S.R. Doklady. Earth Science Sections, v. 69, n. 1-6, Nov. 1966: 185-188

Analysis of the 1908 and 1909 growth layer of a Siberian larch indicates that the fluctuation in the activity of radiocarbon in the atmosphere following the Tunguska catastrophe was ± 1%, too little to support the speculation that a 7% increase in radiocarbon concentration would be expected if the Tunguska catastrophe was connected with the incursion of antimatter into the atmosphere.

2.090   1962
CONSTANCY OF COSMIC RAYS IN TIME
Heymann, D.; Schaeffer, O.A.
Physica, v. 28, 1962: 1318-1323

Concludes that the ratio of the average cosmic ray intensity in the last 300 years has been 1.02 ± 0.07 times the average cosmic ray intensity during the last 300,000 years. In this sense cosmic ray intensity has been essentially unchanged.

2.091   1953
CONSTANCY OF THE COSMIC RAY FLUX OVER THE PAST 30,000 YEARS
Kulp, J. Lawrence; Volchok, Herbert L.
Physical Review, v. 90, n. 4, May 1953: 713-714

A method was devised to show constancy in the integrated cosmic ray flux from the present back to about the maximum of the last phase of the Wisconsin Glaciation, about 30,000 to 40,000 years ago. This was done by dating layers of mud in a deep sea core by the radiocarbon and the ionium/thorium methods. It seems that the cosmic ray flux has not varied by more than 10-20 percent over the last 35,000 years.

Ch. 2 - Theoretical Works

2.092 1956
THE CONTEMPORARY ASSAY OF MARINE SHELLS
Brannon, H.R.; Daughtry, Perry D.; Whitaker, W.W.; Williams, Milton
Radiocarbon Conference, Andover, Mass., Oct. 1-4 1956. *Papers Presented*, 1956
[Not sighted]

2.093 1956
THE CONTEMPORARY ASSAY OF WOOD
Brannon, H.R.; Daughtry, A.C.; Whitaker, W.W.; Williams, Milton
Radiocarbon Conference, Andover, Mass. October 1-4, 1956. *Papers Presented*, 1956
[Not sighted]

2.094 1960
CONTEMPORARY CARBON-14 IN LEMON GRASS OIL
Hayes, F. Newton; Hansbury, V. N.; Kerr, Vernon N.; Williams, D.L.
Zeitschrift für Physik, v. 158, n. 3, 1960: 374-378
Liquid scintillation counting on p-cymene derived from lemon grass oil has provided information on the world-wide increase in radiocarbon since 1954. By June 1959, the activity in the tropospheric atmosphere and the rapidly equilibrating biosphere is measured to have increased by 26.8 percent in the Northern Hemisphere and 19.7 percent in the Southern Hemisphere. The shape of the relationship between activity and time is essentially linear, with two sections of quite different slopes.

2.095 1960
CONTEMPORARY CARBON-14. THE P-CYMENE METHOD
Hayes, F. Newton; Hansbury, Elizabeth; Kerr, Vernon N.
Analytical Chemistry, v. 32, n. 6, May 1960: 617-620
A world-wide survey program has been set up for investigating radiocarbon in nature. Details of the sampling, the chemical conversions, and the liquid scintillation counting procedures are given. P-cymene of naval stores origin is used as the scintillating solvent.

2.096 1956
THE CONTRIBUTION OF NEUTRONS TO THE BACKGROUND OF COUNTERS USED FOR C-14 AGE MEASUREMENTS
de Vries, Hessel
Nuclear Physics, v. 1, 1956: 477-479
Evidence is given that neutrons produced by cosmic radiation produce an important fraction of the residual background of counters used for radiocarbon age measurements. Neutron monitoring is essential in order to check irregular variations of the neutron flux, as for instance, produced by solar flares. During the solar flare of Feb. 23, 1956, the average background for the whole night was 2.5 times normal.

2.097 1947
COSMIC RADIATION AND NATURAL RADIOACTIVITY OF LIVING MATTER
Grosse, Aristid V.; Libby, Willard F.
Science, v. 106, n. 2743, 25 July 1947: 88
The effect of radiocarbon produced by cosmic radiation on man is compared with the effect due to older sources of activity such as radium and its decay products, or potassium and to the action of cosmic rays.

2.098 1956
COSMIC RADIATIONS DURING THE SOLAR FLARES OF FEBRUARY 23RD AND ITS EFFECTS ON C-14 AGE MEASUREMENTS
de Vries, Hessel
Physica, v. 22, 1956: 357
During the solar flare of February 23rd 1956 an important increase of cosmic radiations occurred. The most pronounced effect was recorded by the BF3 neutron counter. The importance of continuous neutron monitoring is illustrated by the fact that if the effect of the solar flares had not been taken into account, the age calculated for the sample in the counter would have been 6000 years instead of the correct 12,000 years.

2.099 1956
COSMIC RAY INTENSITY AND GEOMAGNETISM
Elsasser, Walter; Ney, E.T.; Winkler, J.R.
Nature, v. 178, n. 4544, 1 Dec. 1956: 1226-1227
Changes of earth magetism in the past have occurred within the period covered by radiocarbon dating. This could mean a need for correction for the radioactive decay equation as given by W. Libby. Discussion based on the remanent magnetism of ancient bricks by E. and O. Thellier.

2.100 1956
COSMIC RAY INTENSITY AND GEOMAGNETISM
Elsasser, Walter
Nature, v. 178, n. 4544, 1 Dec. 1956: 1226-1227
Changes of earth magnetism in the past have occurred within the period covered by radiocarbon dating. This could mean a need for correction for the radioactive decay equation as given by W.L. Libby. Discussion based on surveys of the remanent magnetism of ancient bricks by E. and O. Thellier.

2.101   1962
COSMOGENIC CARBON-14 IN METEORITES AND TERRESTRIAL AGES OF FINDS AND CRATERS
Goel, Parmatma S.; Kohman, Truman P.
Science, v. 136, n. 3519, 8 June 1962: 865
Radiocarbon has been measured in several stone and iron meteorites thus helping the dating of the falls of meteorite 'finds'. Ages up to greater than 21,000 have been recorded and give also valuable information on the relative intensity of cosmic radiations in the past.

2.102   1963
CURRENT DEVELOPMENT IN RADIOCARBON DATING
Olson, Edwin A.
Northwest Science, v. 37, n. 4, Nov. 1963: 168
Abstract of paper presented at the 37th Annual Meeting of the Northwest Scientific Association. Developments in the areas of refinement in theory and in analytical technique and extension of application are presented.

2.103   1958
DATING BY RADIOACTIVE CARBON. (REVIEW OF LIBBY'S BOOK)
Jope, E.M.
Nature, v. 173, n. 4415, 12 June 1954: 1111
The possibility for absolute dating in the range of A.D. 1000 to 20,000 - 30,000 B.C. from organic material would enhance the understanding of the recent past, and the Pleistocene and immediate post-glacial events as well as facilitate intercontinental correlation.

2.104   1957
DATING FOSSIL MAN
Oakley, Kenneth P.
Manchester Literary and Philosophical Society. Memoirs and Proceedings, v. 98, 1957: 75-94
The various types of 'dating' are considered along with the methods used in assessing the antiquity of early human remains. The application of these techniques to the dating of Australopitecus, Java, Peking, Swanscombe, Steinheim, Neanderthal, Cro-Magnon, Piltdown and some other remains is described.

2.105   1951
DATING LATE-PLEISTOCENE EVENTS BY MEANS OF RADIOCARBON
Flint, Richard Foster
Nature, v. 167, n. 4256, 26 May 1951: 833-836
Commenting on radiocarbon dating results obtained for late-Pleistocene events, the author concludes that radiocarbon dating can supply the dates of glacial advances and climatic changes, establish inter-continental correlation, and provide a firm chronology of ancient man. The universal importance is that such dating is absolute and worldwide.

2.106   1950
DATING OF ARCHAEOLOGICAL MATERIAL BY MEANS OF RADIOCARBON
Gates, David M.
South Western Lore, v. 16, n. 1, June 1950: 1-4
[Not sighted]

2.107   1965
DATING OF STONE IMPLEMENTS BY USING HYDRATATION LAYERS OF OBSIDIAN
Katsui, Y.; Kondo, Y.
Japanese Journal of Geology and Geography, v. 36, n. 2-4, Oct. 1965: 45-60
Absolute ages determined by the radiocarbon method were correlated with the thickness of hydration layer of obsidian artifacts. This enabled a calculation of hydration rate.

2.108   1958
DATING THE PAST
Zeuner, Frederick Eberard
London, Methuen, 1958
Radiocarbon dates are used and discussed throughout this classic textbook of archaeology. The radio-carbon method of dating is described and discussed. A table of radiocarbon dates is presented with assessments of each date.

2.109   1950
DATING THE PAST BY RADIOACTIVE CARBON
Zeuner, Frederick Eberard
Nature, v. 166, n. 4227, 4 Nov. 1950: 756-757
The author examines the following assumptions which are implicit to dating by the radiocarbon method: constancy of cosmic radiations, even distribution of radiocarbon in living matter, availability of unadulterated samples (the contamination problem).

2.110   1967
DECAY OF A DISTURBANCE IN THE NATURAL DISTRIBUTION OF CARBON-14
Plesset, Milton S.; Dugas, Doris J.
Journal of Geophysical Research, v. 72, n. 16, 15 Aug. 1967: 4131-4137
The decay of an excess of radiocarbon deposited in the atmosphere as a consequence of a nuclear explosion is calculated.

2.111   1965
DENDROCHRONOLOGY AND RADIOCARBON ANALYSIS (Abstract)

Bannister, Bryant
In: International Conference on Radiocarbon and Tritium Dating, Pullman, Washington, Washington State University, June 1-11, 1965. *Proceedings.* United States of America, Atomic Energy Commission, 1965. Conference n. 650652: 396

The fundamental premises of tree-ring research are presented, the possibilities for world-wide improvement of dendrochronological controls are examined, and direct applications of tree-ring results and techniques of interpretation to radiocarbon analysis are explored.

2.112   1962
DEPLETION OF CARBON-13 IN YOUNG KAURI TREE
Jansen, Hans S.
Nature, v. 196, n. 4849, 19 Oct. 1962: 84-85
It is suggested that the difference in $^{13}C/^{12}C$ ratio of carbon dioxide absorbed by young and old trees lies in the existence of difference in fractionation which arises in the trees themselves.

2.113   1959
DETERMINATION OF ABSOLUTE AGE BY C-14 DATING
Vinogradov, A.P.; Devirts, A.L.; Dobkina, E.I.; Markova, Ñ. G.; Martishchenko, L.V.
Geochemistry International, 1959, n. 7: 815-623
Describes some of the results of determining the age of geological and archaeological material by the radiocarbon method. Determinations were carried out simultaneously on two sets of apparatus by counting the activity of two gases, ethane and carbon dioxide. Includes a table presenting descriptions of samples and suggested ages

2.114   1966
DETERMINATION OF THE AGE OF SWISS LAKE DWELLINGS AS AN EXAMPLE OF DENDROCHRONOGICALLY CALIBRATED RADIOCARBON DATING
Ferguson, C. Wesley; Huber, B.; Suess, Hans E.
Zeitschrift für Naturforschung, v. 21a, 1966: 1173-1177
Comparison of the radiocarbon content of a series of samples of dendrochronologically dated bristlecone pine wood with that from trees for which a so-called floating tree ring chronology has been established makes it possible to determine an empirical age for this floating tree ring series based on the age of the wood used for comparison. For the case of the Swiss Lake Dwellers, the difference between conventional radiocarbon dates and the age values determined in this manner amounts to about 800 years. The age of the floating chronology was determined within a standard error of less than 40 years. The measurements indicate that the dwellings were constructed during the 38th century B.C.

2.115   1968
THE DETERMINATION OF THE HALF-LIFE OF ^{14}C
Bella, F.; Alessio, M.; Fratelli, P.
Nuovo Cimento, Series X, v. 58 B, 11 Nov. 1968: 232-246
Although the most recent measurements of the half-life of radiocarbon have given an average value of 5730 ± 40 years, the decision was taken in 1962 and 1965 to base all the dates published by radiocarbon on the original Libby value of 5570 ± 30 years. New measurements made at the University of Rome give a value of 5660 ± 30 years. This seems to confirm that the actually accepted value is too little.

2.116   1949
A DETERMINATION OF THE HALF-LIFE OF CARBON-14
Jones, W.M.
Physical Review, v. 76, n. 7, Oct. 1949: 885-889
Experiments are described involving the counting of an analysed and diluted sample of carbon dioxide containing radiocarbon in a Geiger counter of known effective volume. The value of the half-life is found to be 5589 ± 75 years.

2.117   1963
DIRECT COMPARISON OF RADIOCARBON AND FALL-OUT MEASUREMENTS IN SEA WATER
Broecker, Wallace S.; Rocco, Gregory G.
In: Nuclear Geophysics. Nuclear Science Series Report. Washington D.C., National Academy of Science. National Research Council. Publication n. 1075, 1963: 150-151
To resolve gross inconsistency between conclusions regarding the rates of vertical mixing in the oceans drawn on the basis of radiocarbon measurements (Broecker et al, 1960 and Bien et al, 1960) and those from ^{90}St and ^{137}Cs, a set of carefully collected water samples from the Caribbean Sea were subjected to both types of analyses. The results, summarised in a table, indicate slow mixing. No ^{137}Cs or ^{90}St were found from 400 to 1300 m.

2.118   1965
DISCRIMINATION BETWEEN NATURAL AND INDUSTRIAL POLLUTION THROUGH CARBON DATING
Rosen, A.A.; Rubin, Meyer
Water Polution Control Federation. Journal, v. 37, n. 9, Sept. 1965: 1302-1307
Organic industrial wastes are distinguishable from the biological materials constituting domestic waste-water, decayed vegetation, and aquatic photosynthetic products by application of the carbon-dating procedure.

2.119 1951
DISCUSSION OF THE GEOLOGIC MATERIAL DATED BY RADIOCARBON
Flint, Richard Foster
American Antiquity, v. 17, part 2 (*Society for American Archaeology, Memoirs*, n. 8), 1951: 54-55
Two general results have emerged from the assay of samples of primarily geologic significance: (1) the dates obtained fall roughly into the same order as the stratigraphic sequence of deposits from which the corresponding samples were collected, from this it is inferred that most of the dates are acceptable as to relative position; (2) the dates obtained are generally more recent than had been believed. Sources of errors are noted.

2.120 1954
DISINTEGRATION RATE OF CARBON-14
Caswell, R.S.; Brabant, J.M.; Schwebel, A.
United States National Bureau of Standards. Journal of Research, v. 53, n. 1, July 1954: 27-28
The energy emission rates of radiocarbon samples have been measured with an extrapolation ionisation chamber. From the energy emission rate, the disintegration rates are determined through knowledge of the average beta-ray energy emitted per disintegration. From earlier data on the isotopic abundance, a value for the half-life of radiocarbon of 5900 ± 250 years is obtained.

2.121 1965
DISTRIBUTION OF CARBON FROM NUCLEAR TESTS
Nydal, Reidar; Lovseth, Knut
Nature, v. 206, n. 4988, 5 June 1965: 1029-1031
The purpose of this communication is to present the latest observations of artificial radiocarbon mainly produced during the thermonuclear tests of 1961 and 1962. The two main purposes were to obtain exact data on terrestrial radiocarbon activity for estimating human hazard and to examine transport processes in the atmosphere.

2.122 1957
THE DISTRIBUTION OF CARBON-14 IN NATURE
Arnold, James R.; Anderson, Ernest C.
Tellus, v. 9, n. 1, Feb. 1957: 28-53
The radiocarbon activity of oceanic materials indicates that the mixing half-time for carbon between the atmosphere and ocean surface waters is one or two decades, while the turnover time for the ocean as a whole is several hundred years. This conclusion is consistent with estimates based on the measured decrease in radiocarbon specific activity of wood grown in recent years compared with 19th century wood.

2.123 1958
DISTRIBUTION OF RADIOCARBON AND TRITIUM, AND THE PRODUCTION OF NATURAL TRITIUM
Craig, Harmon
In: Cosmological and Geological Implications of Isotope Ratio Variation. Proceedings of an Informal Conference, Massachusetts Institute of Technology, June 13-15, 1957. Nuclear Science Series Report n. 23, Washington D.C., Committee on Nuclear Science. Subcommittee on Nuclear Geophysics. National Academy of Science. National Research Council. Publication n. 572, 1958: 135-155
Presents general conclusions resulting from a study on nuclear phenomena in relation to oceanography. Concludes that the atmospheric residence of carbon dioxide, relative to the exchange into the sea, appears to be about seven years.

2.124 1963
EARLY HISTORY OF CARBON-14
Kamen, Martin D.
Science, v. 140, n. 3567, Apr. 1963: 584-590
This is an account of the 'prenatal' history of radiocarbon. The discovery of this supremely important tracer was expected in the physical sense but not in the chemical sense.

2.125 1965
EARLY HISTORY OF CARBON-14
Kamen, Martin D.
In: Advances in Tracer Methodology, v. 2, *A Collection of Papers presented at the 9th and 10th Symposia on Tracer Methodology*, edited by Seymour Rothchild. New York, Plenum Press, 1965: 1-19
Reviews the early studies that led to the discovery of radiocarbon.

2.126 1967
EFFECT OF INDUSTRIAL FUEL COMBUSTION ON THE CARBON-14 LEVEL OF ATMOSPHERIC CO_2
Houtermans, Jan; Suess, Hans E.; Munk, W.
In: Radioactive Dating and Methods of Low-level Counting. Proceedings of a Symposium organised by the International Atomic Energy Agency (IAEA) in co-operation with the Joint Commission on Applied Radioactivity (ICSU) and held in Monaco, 2-10 March 1967. Vienna, IAEA, 1967. (Proceedings Series): 57-68.
The effect of industrial fuel combustion upon the radiocarbon level of the atmosphere can be estimated for the Northern Hemisphere to be in the vicinity of -3%.

2.127 1952
THE EFFECTS OF ISOTOPIC SUBSTITUTION ON THE RATES OF CHEMICAL REACTIONS

Ch. 2 - Theoretical Works

Bigeleisen, Jacob
Journal of Physical Chemistry, v. 56, 1952: 823-828
A review is presented of the general theory of the effect of isotopic substitution on the rates of chemical reaction. A discussion is given of the difference between ^{14}C and ^{13}C substitution for ^{12}C.

2.128   1967
EFFECTS OF THE TUNGUSKA METEOR AND SUNSPOTS ON RADIOCARBON IN TREE RINGS
Lerman, J.C.; Mook, W.G.; Vogel, John C.
Nature, v. 216, n. 5119, 9 Dec. 1967: 990-991
This communication describes new measurements of radiocarbon in tree rings, with higher accuracy. The results show no significant deviations which could be correlated either with the Tunguska meteor or with the sunspot cycles.

2.129   1946
THE ENERGY DISTRIBUTION AND NUMBER OF COSMIC-RAY NEUTRONS IN THE FREE ATMOSPHERE
Korff, S.A.; Hamermesh, B.
Physical Review, v. 69, n. 5-6, March 1946: 155-159
A balloon flight to determine the energy distribution and the number of neutrons in the free atmosphere, produced by cosmic radiations is described. The number and rate of production of the neutron increases rapidly with elevation, in good agreement with previous measurements.

2.130   1963
ESSAY ON RADIOCARBON DATING
Willis, Harry H.
In: Sciences in Archeology, edited by Don Brothwell and Eric Higgs. London, 1963: 35-66
Examines the validity of the assumptions that Libby first made, describes various counting methods and the means by which dates are reported.

2.131   1962
EXTERNAL STANDARD METHOD FOR THE DETERMINATION OF THE EFFICIENCY IN LIQUID SCINTILLATION COUNTING
Higashamura, T.; Yamada. O.; Nohara,N. ; Shidei, T.
International Journal of Applied Radiations and Isotopes, v. 12, 1962: 308-309
A new method, which uses the pulse-height distribution caused by quenching, overcomes the problems of the internal standard method. This present technique uses an external γ source to see if the detection of the γ rays in the scintillation counter could correlate usefully with the efficiency

2.132   1967
FACTORS AFFECTING THE ACCURACY OF THE CARBON-DATING METHOD IN SOIL HUMUS STUDIES
Campbell, C.A.; Paul, E.A.; Rennie, D.A.; McCallum, K.J.
Soil Science, v. 104, n. 4, 1967: 81-85
The precision of the commonly used humus fractionation techniques and of the mass spectrometric analysis of humus carbon does not seem to be a serious limitation to the application of the carbon-dating method for soil humus studies.

2.133   1965
FLUCTUATION OF ATMOSPHERIC C14 DURING THE LAST SIX MILLENIA
Damon, Paul E.; Long, Austin; Grey, Donald C.
In: International Conference on Radiocarbon and Tritium Dating, Pullman, Washington, Washington State University, June 1-11, 1965. *Proceedings*. United States of America, Atomic Energy Commission, 1965. Conference n. 650652: 415-428.
Fluctuations of the initial radiocarbon content of the atmosphere by measurements on accurate, dendrochronologically dated wood specimens were investigated. Various origins for this fluctuation are advanced: (1) storage and release of carbon dioxide from the biosphere with dampened or self-sustained oscillations as a result of the delay between storage and decomposition; (2) alternative periods of slow and faster mixing of the oceans; (3) variations in cosmic rays intensity.

2.134   1968
FURTHER INVESTIGATION ON THE TRANSFER OF RADIOCARBON IN NATURE
Nydal, Reidar
Journal of Geophysical Research, v. 13, n. 2, 15 June 1968: 3617-3635
Using a ten reservoirs box model this study examines the transfer of radiocarbon in nature. Using available radiocarbon data, the author attempts to calculate the exchange time between the atmosphere and the mixed layer of the ocean. There seems to be valid reasons to believe that this exchange time is between 5 and 10 years.

2.135   1958
GASES
Revelle, Roger; Suess, Hans E.
In: The Sea: Ideas and Observations in Progress in the Study of the Seas, vol. 1, Physical Oceanography, edited by H.N. Hill. New York, John Wiley (Interscience), 1962: 313-321
Carbon dioxide is present in ocean water in amounts that exceed that of gaseous carbon dioxide in the atmosphere by a factor of 60. The discovery of radiocarbon is helping to

determine the rate of exchange and uptake of carbon dioxide from the atmosphere into the sea.

2.136   1958
GASES IN GLACIERS
Coachman, L.K.; Hemmingsen, E.; Schlander, P.F.
Science, v. 127, n. 3309, 29 May 1958: 1288-1289
Glacier ice contains gases in sufficient quantity for accurate analysis according to present evidence in well preserved polar ice. A method for using the carbon dioxide content of a Greenland ice sample for radiocarbon dating was devised. It may be thus possible to locate, analyse and date ancient atmosphere in its original state.

2.137   1959
THE GEOCHEMISTRY OF C-14 IN THE FRESH WATER SYSTEM
Broecker, Wallace S.; Walton, Alan
Geochimica et Cosmochimica Acta, v. 16, n. 1/3, 1959: 15-38
The process of controlling the radiocarbon concentration in the dissolved bicarbonate of freshwater systems is investigated in order to allow more precise estimates to be made of the initial $^{14}C/^{12}C$ ratio in material formed in such systems in the past. This helps to reduce the uncertainty attached to radiocarbon ages on random fresh water materials to the level of laboratory errors.

2.138   1953
THE GEOCHEMISTRY OF THE STABLE CARBON ISOTOPES
Craig, Harmon
Geochimica et Cosmochimica Acta, v. 3, n. 2/3, 1953: 53-92
A survey of the variation of the ratio $^{13}C/^{12}C$ in nature is presented. The range of variation in the ratio is 4.5%. Terrestrial organic carbon and carbonate rocks constitute two well defined groups with the carbonates being richer in ^{13}C by some 2%

2.139   1955
GEOLOGICAL CHRONOMETRY BY RADIOACTIVE METHODS
Kulp, J. Lawrence
In: Advances in Geophysics, v. 2, 1955: 179-217
Presents an overview of the radiocarbon method of age determination. It includes the principles, the assumptions, experimental developments and geological applications (p. 198-204).

2.140   1964
GEOPHYSICAL IMPLICATIONS OF RADIOCARBON DATES DISCREPANCIES
Wood, L.; Libby, Willard F.
In: Isotopic and Cosmic Chemistry, edited by Harmond Craig, S.L. Miller and G.J. Wassenburg. Amsterdam, North Holland, 1964: 205-210
The accuracy of radiocarbon dates is directly dependent on the constancy of the geophysical factors controlling the rates of production and mixing. These relations are formulated and explored analytically, and a specific example calculated on a high speed computer.

2.141   1965
GROUND WATER APPLICATIONS OF C-14 DATING (Abstract)
Rubin, Meyer; Hanshaw, Bruce B.; Back, William
In: International Conference on Radiocarbon and Tritium Dating, Pullman, Washington, Washington State University, June 1-11, 1965. *Proceedings*. United States of America, Atomic Energy Commission, 1965. Conference n. 650652: 589
Use of measurement of radiocarbon activity of the bicarbonate in groundwater shows the rate of movement of the groundwater and enables a hydroisochronic map to be drawn, from this a permeability map of the area can be drawn.

2.142   1946
HALF-LIFE DETERMINATION OF CARBON (14) WITH A MASS SPECTROMETER AND LOW ABSORPTION COUNTER
Norris, L.D.; Inghram, Mark G.
Physical Review, ser. 2, v. 70, n. 8, 12 Oct. 1946: 772-773
End window Geiger counters were used with a plateau slope 4%/100v. Half-life results, 5300 ± 80 years.

2.143   1948
HALF-LIFE DETERMINATION OF CARBON (14) WITH A MASS SPECTROMETER AND LOW ABSORPTION COUNTER
Norris, L.D.; Inghram, Mark G.
Physical Review, ser. 2, v. 75, n. 4, 1 Feb. 1948: 350-360
By determining the specific activities of two mass-spectrometrically analysed samples of barium carbonate containing radiocarbon, the half-life of the radiocarbon was found to be 5100 ± 200 years.

2.144   1946
HALF-LIFE OF C-14
Reid, Allan F.; Dunning, John R.; Weinhouse, Sydney; Grosse, Aristid V.
Physical Review, ser. 2, v. 70, n. 5-6, Sept. 1-15 1946: 431
The use of mica end window Geiger counting of solid source produces a half-life of 4700 ± 235 years for radiocarbon

Ch. 2 - Theoretical Works

2.145 1949
THE HALF-LIFE OF C-14
Hawkins, R.C.; Hunter, R.F.; Mann, W.B.; Stevens, W.H.
Canadian Journal of Research, Section. B, v. 27, 1949: 545-554
The efficiency of compensated gas counters filled with carbon dioxide and carbon bisulphide have been investigated for the purpose of determining absolute desintegration rates of $^{14}CO_2$ samples. In addition to comparing compensated counting units of different radii, a special counter has been constructed to determine the efficiency of different radii using a collimated beam of electrons from Tl_2O_4. The $CO_2 + Cs_2$ filler counters have been shown to be at least 97% efficient.

2.146 1948
THE HALF-LIFE OF C-14
Yaffe, L.; Grunlund, Jean H.
Physical Review, v. 74, n. 5, 1 Sept. 1948: 696-697
Using end window Geiger counter, obtained a half-life of 7200 ± 500 years.

2.147 1961
THE HALF-LIFE OF C-14
Mann, W.B.; Marlow, W.F.; Hughes, E.E.
International Journal of Applied Radiations and Isotopes, v. 11, n. 2, 1961: 57-67
A new half-life of 5760 ± 50 years is obtained from very thorough analysis. Difficulties are discussed and the method to overcome them.

2.148 1961
THE HALF-LIFE OF CARBON-14
Watt, D.E.; Ramsden, D.; Wilson, H.W.
International Journal of Applied Radiations and Isotopes, v. 11, n. 2, 1961: 68-74
Specific activity measurements on pure samples of carbon dioxide containing respectively 45% and 2% of radiocarbon by atom, using mass spectrometry and proportional counting have resulted in a new value of 5780 ± 65 years. Description of the method employed and of problems met and overcome is included.

2.149 1964
THE HALF-LIFE OF CARBON-14: COMMENTS ON THE MASS-SPECTROMETRIC METHOD
Hughes, E.E.; Mann, W.B.
International Journal of Applied Radiations and Isotopes, v. 15, n. 3, March 1964: 97-100
An assessment has been made of the mass-spectrometric results in the recent NBS determination of the half-life of radiocarbon. Evidence is presented that the uncertainties due to effusive separation and the dependence of the sensitivity on the isotopic abundance of radiocarbon is small. It is proposed however to reduce the value of the half-life of radiocarbon determined by Mann et al (*International Journal of Applied Radiations and Isotopes*, 11, 1961: 59) to 5745 ± 50 years.

2.150 1950
THE HALF-LIFE OF CARBON FOURTEEN AND A COMPARISON OF GAS PHASE COUNTER METHODS
Miller, Warren W.; Ballentine, Robert; Bernstein, William; Friedman, Lewis; Nier, Alfred O.; Evans, R.D.
Physical Review, v. 77, n. 5, March 1950: 714-715
Two completely separate and independent determinations of the half-life of radiocarbon were carried out. The work does not clear the discrepancy between the two values previously reported but does indicate a basis on which to make a choice of values and possible source of discrepancy.

2.151 1951
[HALF-LIFE OF RADIOCARBON]
Manov, G.G.; Curtiss, L.F.
U.S. National Bureau of Standards. Journal of Research, v. 46, 1951: 328
A pair of Geiger counters filled with $CO_2 + CS_2$ give a half-life of 5370 ± 200 years for radiocarbon.

2.152 1962
HALF-LIFE OF RADIOCARBON
Godwin, Harry
Nature, v. 195, n. 4845, 1962: 984
Text of the recommendation from the 5th Conference on Radiocarbon and Tritium Dating. The following points were made: the half-life to be maintained at the Libby half-life of 5568 years; use of conversion factor of 1.03 to new half-life; confirmation of using 1950 A.D. as zero years age BP.

2.153 1965
HALF-LIFE OF RADIOCARBON
Johnson, Frederick
Science, v. 149, n. 3690, 17 Sept. 1965: 1326
A statement was unanimously approved at the 6th Radiocarbon and Tritium Conference, held at Pullman, Washington, June 11, 1965, to retain the half-life value of 5568 years. The best available mean of other values is 5730 and the conversion factor to apply is 1.03.

2.154 1949
THE HALF-LIFE OF RADIOCARBON (C-14)
Engelkemeir, Antoinette G.; Hamill, W.H.; Inghram, Mark G.; Libby, Willard F.

Physical Review, v. 75, n. 12, June 1949: 1825-1833
The half-life of radiocarbon has been found to be 5720 ± 47 years by means of the use of mass spectrometrically analysed $^{14}CO_2$ as a part of the counter gas in brass wall Geiger counter. Some of the problems of absolute beta counting with internal counters have been investigated, particularly the coincidence and end corrections.

2.155   1967
HISTORY OF RADIOCARBON DATING
Libby, Willard F.
In: Radioactive Dating and Methods of Low-level Counting. Proceedings of a symposium organised by the International Atomic Energy Agency (IAEA) in co-operation with the Joint Commission on Applied Radioactivity (ICSU) and held in Monaco, 2-10 March 1967. Vienna, IAEA, 1967 (Proceedings Series): 3-85.
The development of radiocarbon dating is traced from its birth in curiosity regarding the effects of cosmic radiation on Earth. Aspects discussed are: significance of the initial measurements; the advent of low-level counting; attempts to avoid low-level counting by the use of isotopic enrichment; the gradual appearance of the environmental effects due to the combustion of fossil fuel (Suess effect); proliferation of measurement techniques and the impact of archaeological insight on the validity of radiocarbon dates.

2.156   1959
HOW OLD IS THE EARTH?
Hurley, Patrick M.
Garden City, New York, Anchor, 1959. pp. 160
The radiocarbon dating method is very briefly described and an assessment of its value as a tool for dating is presented.

2.157   1958
AN IMPROVEMENT IN THE AGE DETERMINATION BY THE C-14 METHOD
de Vries, A.E.; Harring, A.
In: International Conference on the Peaceful Use of Atomic Energy, 2nd, Geneva, 1958. Proceedings. 1958, v. 2 (A/Conf. 15/P/1123): 249-250
The age determination of samples containing radiocarbon has been improved by increasing the specific activity in a known way in a thermal diffusion column. The method is described. Samples have been enriched by a factor of 12 in the radiocarbon concentration, which means a gain of about 20,000 years in the age determination.

2.158   1959
INCREASE IN C14 IN THE ATMOSPHERE FROM ARTIFICIAL SOURCES, MEASURED IN A CALIFORNIA TREE
Bien, George S.; Suess, Hans E.

Zeitschrift für Physik, v. 152, 1959: 172-174
A tree from the coastal area of California was investigated in order to measure the increase in radiocarbon due to atomic bomb testing. The results indicate that the increase from 1956 to 1957 was considerably larger than from 1954 to 1955.

2.159   1963
INCREASE IN RADIOCARBON FROM THE MOST RECENT SERIES OF THERMONUCLEAR TESTS
Nydal, Reidar
Nature, v. 200, n. 4903, 19 Oct. 1963: 212-214
It is reasonable to assume that the amount of radiocarbon in the human population will rise to a maximum value, close to that of the final troposphere level and will be 2-3 times the normal activity within a few years. The time lag is due to: (1) the time between photosynthesis of food and consumption by human beings; (2) the length of time the carbon remains in human tissue.

2.160   1967
THE INFLUENCE OF GEOMAGNETIC SHIELDING ON C-14 PRODUCTION
Ramaty, R.
In: Magnetism and the Cosmos. NATO Advanced Study Institute on Planetary and Stellar Magnetism, Newcastle upon Tyne, 1965, edited by W.E. Hindmarsh, F.J. Lowe, P.H. Roberts and S.K. Runcorn. Edinburgh, Oliver and Boyd, 1967: 66-78
Examination of the variations of the rate of production of radiocarbon and of the influence of the changes in the earth magnetic field in these variations.

2.161   1964
INTRODUCTION OF SYSTEMATIC STUDIES OF THE VARIATION OF C14 CONTENTS IN THE ATMOSPHERE FOR THE PAST THOUSAND YEARS
Kigoshi, Kunihiko; Hasagewa, Hiroishi; Yamakoshi, K.; Oda, M.; Shibata, S.; Saito, K.
In: International Conference on Cosmic Rays, vol. 3. Bombay, Tata Institute of Fundamental Research, 1964: 466-469
The long-term variation of the radiocarbon content of the atmosphere was investigated by measuring the radiocarbon content of rings of an old cedar tree taken from Yaku Island, Japan.

2.162   1961
ISOTOPIC COMPOSITION OF ATMOSPHERIC HYDROGEN AND METHANE
Bainbridge, Arnold D.; Suess, Hans E.; Friedman, Irving
Nature, v. 192, n. 4803, 18 Nov. 1961: 648-649

Samples of atmospheric hydrogen dating from 1948 show that artifical tritium does not appear to have influenced the tritium content of the atmosphere prior to 1954. After 1954 the tritium concentration rose with a doubling time of 18 months and its concentration in atmospheric methane is now also very high.

2.163   1962
LATITUDINAL EFFECT IN THE TRANSFER OF RADIOCARBON FROM STRATOSPHERE TO TROPOSPHERE
Tauber, Henrik
In: Radioisotopes in the Physical Sciences and Industry. Vienna, International Atomic Energy Agency, 1962: 67-74
[Not sighted]

2.164   1966
LATITUDINAL EFFECT IN THE TRANSFER OF RADIOCARBON FROM STRATOSPHERE TO TROPOSPHERE
Tauber, Henrik
Science, v. 133, n. 3451, 17 Feb. 1961: 461-462
Latitudinal variations in the descent of bomb-produced radiocarbon from the stratosphere is suggested by difference in tropospheric radiocarbon activity. The magnitude of a similar latitudinal effect may be part of the explanation of the short term oscillations in radiocarbon activity found in tree-rings from the past 1300 years.

2.165   1964
LIMITATION OF RADIOMETRIC DATING
Baadsgaard, H.; Cumming, G.L.; Folinsbee, R.E.; Geoffrey, J.D.
Royal Society of Canada. Special Publication, n. 8, 1964: 20-38
Discusses the validity and limitations of the four principal methods of radiometric dating: uranium - thorium - lead, rubidium - strontium, potasssium - argon and radiocarbon.

2.166   1953
LONG-RANGE DATING IN ARCHAEOLOGY
Heizer, Robert F.
In: Anthropology Today, edited by Kroeber. Chicago, University of Chicago Press, 1953: 2-42
As part of his examination of the methods and results of prehistoric time reckoning as applied to archaeological remains, the author describes the radiocarbon method of age determination (p. 14-16).

2.167   1968
MASS SPECTROMETRIC STUDIES OF CARBON-13 VARIATIONS IN CORN AND OTHER GRASSES
Bender, Margaret M.
Radiocarbon, v. 10, n. 2, 1968: 468-472
Modern and prehistoric corn samples from a variety of locations have been analysed and the $^{14}C/^{12}C$ ratio has been found to be approximately constant; results agree with values reported by other laboratories. Other grains and grasses were analysed and it was found that soybeans, wheat, barley, timothy, and oats all had $\partial^{13}C$ values in the same range as wood. Variations in the $^{13}C/^{12}C$ ratios in grasses depend upon taxonomic relations of the grasses and not upon soil or microclimate. Grasses from Tribes which are members of the *Panicoideae* subgroup may differ as much as +220 years and probably not less than +160 years in radiocarbon age from the average contemporary wood

2.168   1953
MEASUREMENT OF ACTIVITY AND ISOTOPIC ABUNDANCE IN BARIUM CARBONATE
Reynolds, S.A.; Wyatt, E.I.
Oak Ridge National Laboratory. Report, 16 Oct. 1953. pp. 6 (CF - 53 - 10 - 104)
The activity of radiocarbon in carbon dioxide was measured in a calibrated ionisation chamber and electrometer after evolving the carbon dioxide from aqueous carbonate solution. A half-life value of 5800 years was obtained.

2.168   1951
MEASUREMENT OF LOW LEVEL ACTIVITY
Douglas, David L.
Knolls Atomic Power Laboratory. Report to the U.S. Atomic Energy Commission, 1951. pp. 20
Problems involved in measuring extremely low levels of radioactivity are reviewed. Discusses the method used by Professor Willard F. Libby of the Chicago University in determining the age of archaeological and anthropological relics by measuring the extremely small amount of radiocarbon contained in the organic carbon.

2.170   1947
MEASUREMENT OF RADIOACTIVE TRACERS: PARTICULARLY C-14, S-35, T AND OTHER LONG-LIVED, LOW-ENERGY ACTIVITIES
Libby, Willard F.
Annals of Chemistry, v. 19, 1947: 2-6
The importance to the chemist of the detection of soft beta-radiations is pointed out as all the useful long lived beta-radioactive isotopes have soft radiations. Three types of counters are examined: end window counters, screen wall counters, and gas filled counters.

2.171   1963
MEASUREMENTS OF CARBON-14 IN KNOWN AGE SAMPLES AND THEIR GEOPHYSICAL IMPLICATION

Schell, William Raymond
Ph.D. Dissertation. University of Washington, 1963. pp. 115 (Abstract in: *Dissertation Abstracts*. Ann Arbor, Mich., v. 25, n. 2, 1964/65: 821. Order No. 64-6433)
The radiocarbon concentration in the atmosphere has been shown to vary appreciatively since the last glaciation. Two possible explanations for the difference are proposed: (1) incorrect half-life of radiocarbon, (2) change in the $^{14}C/^{12}C$ ratio in the atmosphere due to changing oceanic characteristics. Accurately dated rings of *Sequoia Gigantea* were used in a study to attempt to explain the errors. It was concluded that the effective half-life of atmospheric radiocarbon may, in fact, be the result of a change in oceanic characteristics.

2.172 1965
MEASUREMENTS OF CARBON-14 IN KNOWN AGE SAMPLES AND THEIR GEOPHYSICAL IMPLICATIONS
Schell, William Raymond; Fairhall, A.W.; Harp, G.D.
In: International Conference on Radiocarbon and Tritium Dating, Pullman, Washington, Washington State University, June 1-11, 1965. *Proceedings*. United States of America, Atomic Energy Commission, 1965. Conference n. 650652: 397-414
The data compiled on tree ring deviations and archaeological deviations show that the radiocarbon concentration in the atmosphere was slowly decreasing in a linear manner from some high value. Two possible explanations are advanced: (1) the Libby half-life of 5568 years and the Mann - Marlow - Hughes half-life of 5730 years may both be underestimated; (2) a direct correlation may exist between a change in climate and the $^{14}C/^{12}C$ ratio in atmospheric carbon dioxid

2.173 1967
MODEL EXPERIMENTS FOR ^{14}C WATER AGE DETERMINATIONS
Wendt, Immo; Stahl, W.; Geyh, Mebus A.; Fauth, H.
In: Isotopes in Hydrology. Conference on Isotopes in Hydrology, Vienna, 1-10 March 1967. Proceedings. Vienna, I.A.E.A., 1967: 355-369
The radiocarbon age of water samples is calculated by assuming that fossil carbon is dissolved by biogenic carbon dioxide. Investigations are conducted to check the validity of the theoretical assumption regarding the average radiocarbon activity and the $\partial^{13}C$ value of the total carbon which is dissolved as carbon dioxide and carbonic acid.

2.174 1968
MODERN ASPECTS OF RADIOCARBON DATING
Olsson, Ingrid U.
Earth Science Review, v. 4, n. 3, 1968: 203-218
The radiocarbon method for radioactive-age determination is discussed and recent developments in the fixing of the half-life time of radiocarbon are reviewed. The occurrence and causes of variations in the $^{14}C/^{12}C$ ratio are outlined, as are the possibilities and influences of contamination. The paper aims to give a wide geological public an insight into the background, applicability, and reliability of the method.

2.175 1963
THE NATURAL DISTRIBUTION OF RADIOCARBON: MIXING RATES IN THE SEA AND RESIDENCE TIMES OF CARBON AND WATER
Craig, Harmon
In: Earth Science and Meteoritics, edited by J. Geiss and E.D. Goldberg. Amsterdam, North Holland, 1963: 103-114
The flux of radiocarbon into the deep sea is discussed, and it is shown that at least some of the radiocarbon is brought in by transfer from the mixed layer of the sea. (Harmon, *Tellus*, 1957, 9: 1).

2.176 1951
NATURAL RADIOCARBON 14 MEASUREMENTS AND APPLICATIONS; QUARTERLY PROGRESS REPORTS 4 AND 5
Kulp, J. Lawrence
Lamont Geological Observatory. Columbia University. Quarterly Progress Report, n. 4 and 5, 1951. pp. 18 (ATI - 117710)
Progress is reported in the study of natural radiocarbon measurements and applications. Backgrounds as low as 4.5 cpm were obtained. Modifications in the cosmic ray tubes, anticoincidence circuits and screen wall counters were made. The ages of a number of geologically important samples, including several tree-ring specimens of known age were determined.

2.177 1946
NATURAL RADIOCARBON AND THE RATE OF EXCHANGE OF CARBON DIOXIDE BETWEEN THE ATMOSPHERE AND THE SEA
Suess, Hans E.
In: Nuclear Processes in Geologic Settings, Conference, 1st, Williams Bay, Wisc., Sep. 21-23, 1953. *Proceedings*, edited by William Aldrich. 1953: 52-56
Examines the average life time of a carbon dioxide molecule in the atmosphere and the rate at which a surplus amount of carbon dioxide is absorbed by the sea, in terms of the absorption of carbon dioxide given off by the sea.

2.178 1954
THE NATURAL RADIOCARBON CONTENT OF MATERIALS FROM HARD-WATER LAKES
Deevey, Edward S.; Gross, Marsha S.; Hutchinson, G.E.;

Kraybill, H.L.
National Academy of Science. Proceedings, v. 40, n. 5, 15 May 1954: 285-288
From this study it is apparent that if remains of water plants developed in an ancient hard-water lake were used in radiocarbon dating, spurious ages, up to 2000 years in excess would be obtained. Great care must be exercised when material from high calcareous regions is being examined.

2.179   1947
NATURAL RADIOCARBON FROM COSMIC RADIATIONS
Anderson, Ernest C.; Libby, Willard F.; Weinhouse, Sydney; Reid, Allan F.; Kirschenbaum, A.D.; Grosse, Aristid V.
Physical Review, v. 72, n. 10, Nov. 1947: 931-936
Radiocarbon produced by the nitrogen reaction of cosmic-ray neutrons has been detected and identified. The radiocarbon concentration of biological methane was enriched by factors of up to 60-fold by thermal diffusion to raise the activity to convenient levels for counting. The activity in biological methane was found to be 10.5 disintegration per minute per gram of carbon. This is in reasonable agreement with the value predicted from the estimated values of the cosmic ray neutron flux and the amount of carbon in exchange equilibrium with the atmosphere.

2.180   1966
NEUTRON FLUX AND ITS AFFECT ON RADIOCARBON DATING EQUIPMENT
Polach, Henry A.; Stipp, Jerry J.
Australian National University, Research School of Physical Sciences (available from), pp. 17
Interference with radiocarbon dating operations caused by Tandem Accelerator operations at the Australian National University are investigated and propositions for a shielded counting room which will eliminate this interference are advanced.

2.181   1951
NEW RADIOCARBON METHOD FOR DATING THE PAST
Collier, Donald
Chicago Natural History Museum. Bulletin, v. 22, 1951: 6-

[Not sighted]

2.182   1966
NEW TECHNIQUES IN ARCHAEOLOGY
Rainey, Froelich
American Philosophical Society. Proceedings, v. 110, n. 2: 1966: 145-152
Recent developments of techniques for archaeological dating, exploration and interpretation are reviewed. Applications of radiocarbon and other isotopic dating methods, tree ring dating, thermoluminescence and uranium decay dating of pottery and glass, neutron activation analysis, geophysical and underwater exploration methods are discussed.

2.183   1962
NOTE ON THE EXCHANGE RATE BETWEEN THE NORTHERN AND THE SOUTHERN ATMOSPHERE
Junge, C.
Tellus, v. 14, n. 2, May 1962: 242-246
Difference in carbon dioxide and tritiated methane data for the Northern and Southern Hemisphere are used to estimate the net exchange rate between the Hemispheres. The most likely exchange rate is found to be 0.3 - 0.5 years-1.

2.184   1967
A NOTE ON THE PRESENT STATUS OF RADIOACTIVE CARBON AGE DETERMINATION
Braidwood, Robert J.
Sumer, v. 23, n. 122, 1967: 39-43
Although radioactive carbon age determination is a tool of potentially very great value, it must be used with great care. A fully complete checking of the radiocarbon assay extending back to the beginning of the bristlecone dendrochronology is in process.

2.185   1949
A NOTE ON TWO ASPECTS OF THE GEOCHEMISTRY OF CARBON
Hutchinson, G.E.
American Journal of Science, v. 247, n. 1, Jan. 1949: 27-32
Study of the terrestrial $^{12}C/^{13}C$ ratio and of the methane cycle in the biosphere.

2.186   1966
NUCLEAR CLOCKS
Faul, Henry
United States Atomic Energy Commission. Division of Technical Information, Oak Ridge, Tenn. (Series on Understanding the Atom), 1966. pp. 60
Description of the various methods of dating using radioactive nuclei decay and their applications.

2.187   1958
ON THE AMOUNT OF CO_2 IN THE ATMOSPHERE
Callendar, G.S.
Tellus, v. 10, n. 2, May 1958: 243-248
The average amount of carbon dioxide in the atmosphere obtained by 30 of the most extensive series of observations between 1866 and 1956 is presented and the reliability of the 19th century measurements discussed. Since then the obser-

vations show a rising trend which is similar in amount to the addition from fuel combustion. This result is not in accordance with recent radiocarbon data, but the reasons for the discrepancy are obscure.

2.188   **1963**
ON THE CHEMISTRY OF NATURAL RADIOCARBON
Mackay, Colin; Pandow, Mary; Wolfgang, Richard L.
Journal of Geophysical Research, v. 68, n. 13, 1 July 1963: 3929-3931
Detailed studies with carbon atoms of high kinetic energy confirm that natural radiocarbon initially reacts in the atmosphere to form carbon dioxide. In accord with this, several samples of atmospheric carbon dioxide show a specific activity much higher than would be expected from the predominantly fossil fuel origin of this gas.

2.189   **1960**
ON THE EXCHANGE OF CARBON DIOXIDE BETWEEN THE ATMOSPHERE AND THE SEA
Bolin, Bert
Tellus, v. 12, n. 3, Aug 1960: 274-281
The physical and chemical processes responsible for exchange of carbon dioxide between the atmosphere and the sea are analysed. It is shown that the rate of transfer is considerably decreased due to the finite rate of hydration of carbon dioxide in water. A general agreement is found between the transfer rate deduced in this way and the rate of exchange estimated on the basis of $^{12}C/^{14}C$ ratio in the atmosphere and the sea.

2.190   **1966**
ON THE QUANTITATIVE RELATIONSHIP BETWEEN GEOPHYSICAL PARAMETERS AND THE NATURAL C-14 INVENTORY
Houtermans, Jan
Zeitschrift für Physik, v. 193, 1966: 1-12
The relationship between observed variations of the specific radiocarbon activity in the atmosphere and the possible secular variations in radiocarbon production rate, carbon residence times and carbon reservoir size are calculated. Variations in these geophysical parameters which determine the natural radiocarbon activity in the atmosphere are strongly attenuated. Attenuation coefficients and phase shifts are calculated quantitatively and results are shown graphically.

2.191   **1966**
ON THE RELATIONSHIP BETWEEN RADIOCARBON DATES AND TRUE SAMPLE AGES
Stuiver, Minze; Suess, Hans E.
Radiocarbon, v. 8, 1966: 534-540
A summary of the present knowledge regarding differences between radiocarbon ages and true ages and the present status of the empirical calibration of the radiocarbon time scale.

2.192   **1958**
ON THE USE OF CARBON-14 AS A TOOL IN METEOROLOGY
Machta, Lester
American Geophysical Union. Transactions, v. 39, n. 3, June 1958: 524
Abstract of paper presented at the 39th. Annual Meeting, Washington D.C., May 5-8, 1958. Radiocarbon is produced by the action of neutrons on atmospheric nitrogen and may be produced either through cosmic rays or as a by-product of nuclear tests. Difference in the ratio of radiocarbon to stable carbon in carbon dioxide reflects an increase in stable carbon due to the growth of carbon dioxide from the burning of fossil fuels. Movement of nuclear test produced radiocarbon to and from the stratosphere may yield clues to air circulation between stratosphere and ground level.

2.193   **1963**
ON THE VALIDITY OF RADIOCARBON DATES FROM SNAIL SHELLS
Rubin, Meyer; Likins, Robert C.; Berry, Elmer G.
Journal of Geology, v. 71, n. 1, Jan. 1963: 84-89
Tracer measurements of the incorporation of inorganic carbon in the shells of snails indicate that an uptake of 10-12 percent is possible in an extremely favourable environment. Uncertainties in radiocarbon dates from snail shells can be of the order of one thousand years or somewhat more, and are of the same range as the analytical error in older age samples. Used conservatively, they can be helpful in dating geologic deposits.

2.194   **1967**
ORIGIN AND EXTENT OF ATMOSPHERIC ^{14}C VARIATIONS DURING THE PAST 10,000 YEARS
Stuiver, Minze
In: Radioactive Dating and Methods of Low-level Counting. Proceedings of a Symposium organised by the International Atomic Energy Agency (IAEA) in co-operation with the Joint Commission on Applied Radioactivity (ICSU) and held in Monaco, 2-10 March 1967. Vienna, IAEA, 1967 (Proceedings Series): 27-40.
Study of sedimentation rates in three lakes at widely different latitudes confirms the change in time-scale observed in radiocarbon ages of tree rings and Egyptian known ages around 2500 BP. In addition important synchronous changes are lacking in sedimentation rates between 2500 and 10,000 BP. This implies that there have been neither important variations in sedimentation rates due to world-

wide climatic change nor drastic change in the radiocarbon time-scale during this time span. The present radiocarbon results from lake sedimentation agree with the hypothesis of a 'super-quiet' sun as being the cause for glacial intervals.

2.195   1960
pCO_2 IN SEA WATER AND ITS EFFECT ON THE MOVEMENT OF CO_2 IN NATURE
Kanwisher, John
Tellus, v. 12, n. 2, May 1960: 209-215
A method is described for measuring the percentage of carbon dioxide (pCO_2) in sea water. This has been used to determine the change in pCO_2 produced by variations of temperature and total carbon dioxide. It appears that most of the fossil fuel carbon dioxide released by man has been effective in increasing the percentage of this gas in air.

2.196   1961
PHYSICS AND ARCHAEOLOGY
Aitken, M.J.
New York, Wiley Interscience, 1961. pp. 181
Chapter 6, p. 88-120, introduces and describes the theory, the general principles including the complication caused by the variations in the intensity of cosmic rays and in the size of the exchange reservoir, the Suess effect, the atom bomb effect, and isotopic fractionation; and finally the various methods of measurements.

2.197   1962
PHYSICS APPLIED TO ARCHAEOLOGY. PART I
Aitken, M.J.
Contemporary Physics, v. 3, Feb. 1962: 161-167
Methods of age determination and chemical analysis using physical techniques are discussed. Techniques and limitations of determination from radiocarbon content and thermoluminescence are presented.

2.198   1959
THE POSSIBILITIES OF NATURAL RADIOCARBON AS A GROUND WATER TRACER IN THERMAL AREAS
Fergusson, Gordon J.; Knox, F.B.
New Zealand Journal of Science, v. 2, n. 3, Sept. 1959: 431-441
An evaluation of the factors that affect the radiocarbon concentration of the carbon dioxide coming to the surface with water and steam in the thermal areas, shows that radiocarbon measurements can assist in the study of the underground movement of water. Measurements so far carried out at five locations in the Wairakei thermal area of New Zealand show that the time for appreciable underground circulation of water in this area is less than 40,000 years for travel from the surface to depths of 1500 ft or more.

2.199   1965
POSSIBLE ANTI-MATTER CONTENT OF THE TUNGUSKA METEOR OF 1908
Cowan, Clyde; Atlury, C.R.; Libby, Willard F.
Nature, v. 206, n. 4987, May 1965: 862-865
The effect of the explosion of the Tunguska meteor on atmospheric radiocarbon content is discussed.

2.200   1963
POSSIBLE FLUCTUATION IN ATMOSPHERIC CARBON DIOXIDE DUE TO CHANGES IN THE PROPERTIES OF THE SEA
Eriksson, E.
Journal of Geophysical Research, v. 68, n. 12, July 1963: 3871-3876
Concludes that neither temperature fluctuation nor volume changes of the sea can have been large enough to cause any appreciable change in the atmospheric carbon dioxide content. However as there is a considerable excess of carbon dioxide in the ocean, this could, if released, increase the atmospheric concentration by a factor of 5, and serious disturbances in the past, in the biological circulation of carbon in the ocean, may have caused wide fluctuations in the atmospheric carbon dioxide. Some causes for these disturbances are briefly discussed.

2.201   1960
POST-BOMB RISE IN RADIOCARBON ACTIVITY IN DENMARK
Tauber, Henrik
Science, v. 131, n. 3404, 25 Mar. 1960: 921-922
During the summer of 1958 and 1959 the increase in concentration of bomb-produced radiocarbon in Denmark was several percent higher than the average increase for the Hemisphere. This additional increase is probably a radiocarbon equivalent to the spring peaks in strontium 90 fallout in the North Temperal Zone in the same year and suggests latitudinal variations in radiocarbon contamination.

2.202   1953
PREDICTED ISOTOPIC ENRICHMENT EFFECTS IN SOME ISOTOPIC EXCHANGE INVOLVING C-14
Stranks, D.R.; Harris, G.M.
American Chemical Society. Journal, v. 75, n. 8, 23 April 1953: 2015
The equilibrium constant of the $^{14}C/^{12}C$ exchange is predicted

2.203   1959
PRELIMINARY REPORTS ON RADIOCARBON FROM NUCLEAR TESTS
United Nations Secretariat, 7 Dec. 1959. pp. 17. (A/AC.82/

R.77)
The production of radiocarbon by nuclear tests and its present and future distribution throughout the carbon cycle in nature were studied. The time and latitudinal distribution were calculated. The radiation dose due to naturally occurring radiocarbon and that produced in nuclear tests were also calculated.

2.204   1953
THE PRESENT STATUS OF CHEMICAL METHODS FOR DATING PREHISTORIC BONE
Cook, S.F.; Heizer, Robert F.
American Antiquity, v. 18, n. 4, Apr. 1953: 354-358
A study of the constituents of bone in order to determine the chemical changes occurring during the fossilisation to help the dating of fossil bone. Radiocarbon dating is not at that time deemed to be applicable to bone.

2.205   1966
PROBLEMS AND BASIC ASSUMPTIONS IN C-14 RATING
Britt, Claude
Compass, v. 42, n. 4, 1966: 279-356
[Not sighted]

2.206   1967
PROBLEMS OF THE RADIOCARBON CALENDAR
Ralph, Elizabeth K.; Michael, Henry N.
Archaeometry, v. 10, 1967: 3-11
Investigation of the small discrepancies between radiocarbon years and true ages. There is increasing evidence that these discrepancies are becoming progressively larger in the remote millenia B.C.

2.207   1953
THE PRODUCTION AND DISTRIBUTION OF NATURAL RADIOCARBON
Anderson, Ernest C.
Annual Review of Nuclear Sciences, v. 2, 1953: 63-78
The principal results of the studies of natural radiocarbon are: (1) the absolute specificity is 16.6 ± 0.6 disintegration/min/g; (2) the fractionation factor for radiocarbon between wood and shell carbonate is 1.10 ± 0.02; (3) the extreme latitudinal variation of the cosmic ray neutron flux is not reflected in the distribution of radiocarbon; (4) the cosmic ray neutron flux has been constant for at least the past 10 or 15 thousand years.

2.208   1964
PRODUCTION OF CARBON 14 BY SOLAR PROTONS
Lingenfelter, R.E.; Flamm, E.J.
Journal of Atmospheric Science, v. 21, n. 2, March 1964: 134-140

The rate of production of radiocarbon by interaction of solar - proton - produced neutrons with atmospheric nitrogen is calculated for the period 1956 - 1961. A time average production rate of 0.05 to 0.12 radiocarbon atoms per cm/sec, less than the rate of production by galactic cosmic ray neutrons, is obtained. Solar particle radiation of the intensity observed during the last solar cycle increases the neutron flux at high altitudes by more than an order of magnitude for short periods but does not significantly affect the atmospheric reservoir of radiocarbon.

2.209   1963
PRODUCTION OF CARBON-14 BY COSMIC RAY NEUTRONS
Lingenfelter, R.E.
Review of Geophysics, v. 1, n. 1, 1963: 35-55
The rate of production of radiocarbon by cosmic ray neutrons is calculated by multigroup diffusion as a function of altitude, latitude and time and is normalised to absolute cosmic ray neutron flux measurement. The global average production rate over the last 10 solar cycles is found to be 2.50 ± 0.50 ^{14}C atoms per s/cm/sec.

2.210   1960
THE PRODUCTION OF TRITONS AND C-14 IN THE TERRESTRIAL ATMOSPHERE BY SOLAR PROTONS
Simpson, J.A.
Journal of Geophysical Research, v. 65, n. 5, May 1960: 1615-1616
A third source for the natural production of tritons is identified, namely intense fluxes of energetic solar protons which could produce tritons and radiocarbon in the terrestrial atmosphere at times near the maximum of the solar activity cycle.

2.211   1952
THE PRODUCTION RATE OF COSMIC-RAY NEUTRONS AND C-14
Kouts, H.J.; Yuan, L.C.L.
Physical Review, v. 86, 1952: 128-129
Discusses the production rate of radiocarbon and corrects the calculation made by Anderson and Libby, finding a predicted disintegration rate of carbon due to radiocarbon to be 17.6 disintegration per minute per gram.

2.212   1940
RADIOACTIVE CARBON OF LONG HALF-LIFE
Ruben, Samuel; Kamen, Martin D.
Physical Review, v. 57, n. 5, 1940: 549
Describes early searches for long-lived radiocarbon. Indicates its possible use for quantitative work and that the gas technique for counting is the most sensitive method.

2.213   1964
RADIOACTIVE CONTAMINATION OF THE ENVIRONMENT BY NUCLEAR TESTS
United Nations. Scientific Committee on the Effects of Atomic Radiations. Report, NP-14556, 1964: 11-80
A detailed review is presented of data collected between 1962 and June 1964 on contamination of the environment by fallout from nuclear tests. Topic included: the transport of artificial radionuclides in the earth's atmosphere and mechanisms involved in the deposition on the earth. Inventory and levels are discussed, including those of radiocarbon.

2.214   1968
RADIOACTIVE DATING OF QUATERNARY TEPHRA
Damon, Paul E.
In: Means of Correlation of Quaternary Successions, edited by Roger B. Morrisson and Herbert C. Wright. International Association for Quaternary Research (INQUA). Congress, 7th, Boulder, Col., 14 Aug - 19 Sept. 1965. *Proceedings*, v. 8. Salt Lake City, University of Utah Press, 1968: 195-206
Discusses the accuracy and the limits on the range of radiocarbon method of age determinations in Quaternary tephra and compare the K-Ar and radiocarbon methods. The possibility of reasonable agreement between the 2 methods would be a test for samples between 20,000 and 40,000 years old.

2.215   1966
RADIOACTIVE DECAY CONSTANT AND ENERGIES
Wetherill, George W.
In: Handbook of Physical Constants (rev. ed.) Geological Society of America Memoirs, n. 97, 1966: 513-519 (Sec. 23)
Selected values of the natural radioactive decay constants, measurements of specific gamma activity of potassium using scintillation spectrometers, measurements of ^{87}Rb half life and ^{187}Re half-life, decay constants of ^{14}C and ^{3}H and decay energies of U, Th and K are tabulated. The final table gives several versions of the geologic time scale. Since rocks dated radioactively never fall exactly on the boundary between two geologic periods, some interpolation is necessary. No great accuracy can be claimed for such a process.

2.216   1956
RADIOACTIVE FALLOUT AND RADIOACTIVE STRONTIUM
Libby, Willard F.
Science, v. 123, n. 3190, 20 Apr. 1956: 657-660
Although the main part of the radioactivity from high yield weapons dissipates in the stratosphere, the small, but significant, part that falls out within a few hundred miles of the explosion site constitute a real hazard. However the efforts made to protect against misadventure are entirely adequate and the world-wide health hazards from the present rate of testing are insignificant.

2.217   1963
RADIOACTIVE FALLOUT, ITS DISPERSION, DEPOSITION OVER SOUTH AFRICA AND BIOLOGICAL SIGNIFICANCe
McMurray, W.R.; Stander, L.O.
South African Journal of Science, v. 59, n. 1, Jan. 1963: 19-31
Past nuclear testings have resulted in an increase in a 20% additional production of radiocarbon in the lower atmosphere.

2.218   1963
RADIOACTIVE ISOTOPES AND GEOCHRONOLOGY
Hart, S.R.
American Geophysical Union. Transactions, v. 44, n. 2, June.1963: 523-526
Reviews the developments in the field of the application of radioisotope dating to geochronology. Includes radiocarbon dating.

2.219   1962
RADIOACTIVE METHODS OF AGE DETERMINATION
Eckelman, Walter R.
In: Nuclear Radiations in Geophysics, edited by H. Israel and A. Krebs. New York, Academic Press, 1962: 61-75
The field of absolute age dating has continued to be an active field of research which bears on many geological, geophysical and geochemical problems. The improved decay constants, a greater understanding of discordant age patterns as well as the extended range of radiocarbon and potassium-argon dating continue to make this field of research very important to the understanding of the evolution of the earth.

2.220   1968
THE RADIOACTIVITY OF THE ATMOSPHERE AND HYDROSPHERE
Lal, D.; Suess, Hans E.
Annual Review of Nuclear Sciences, v. 18, 1968: 407-434
Summarises the state of knowledge of three categories of radioactive nuclear species that can be observed in the atmosphere and hydrosphere and discusses their usefulness in various fields of research. Particular mention of radiocarbon p. 414-426.

2.221 1958
THE RADIOACTIVITY OF THE ATMOSPHERE AND THE HYDROSPHERE
Suess, Hans E.
Annual Review of Nuclear Sciences, v. 8, 1958: 243-256
Three classes of radioactive nuclear species that occur on the surface of the Earth have been determined: (1) extremely long lived species that have survived from the time when the elements were formed (primary radionuclides) and their daughter products (secondary radionuclides); (2) radioactive nuclei that are currently formed in nature, primarily by cosmic ray induced processes; (3) man-made radioactive substances, such as those produced through the testing of atomic devices or in nuclear reactors. This paper gives a brief account of the nature of geochemical, meteorological, and oceanographic problems involved in the interpretation of the distribution of radioactivity in the atmosphere and the hydrosphere of the Earth.

2.222 1963
RADIOCARBON ACTIVITY OF SHELLS FROM LIVING CLAMS AND SNAILS
Rubin, Meyer; Taylor, Dwight W.
Science, v. 141, n. 3581, 16 Aug. 1963: 637
Three samples representing growth in lime-poor, fairly soft and lime-rich waters were measured and found to be 10,%, 22% and 32% respectively radiocarbon deficient.

2.223 1962
RADIOCARBON AGE AND OXYGEN-18 CONTENT OF GREENLAND ICEBERGS
Scholander, Per Fredrik; Dansgaard, Willi; Nutt, David C.; de Vries, Hessel ; Coachman, L.K.; Hemmingsen, E.
Meddelelser om Grønland, v. 165, n. 1, 1962. pp. 26
Radiocarbon dating was obtained from carbon dioxide in the gas enclosures in the melting of 6 - 16 tons of ice. Meltwater was also sampled for oxygen-18. The ages of the ice ranged from recent to over 3000 years, the oldest ice having the least ^{18}O, i.e. having been formed at the lowest temperature. Overall rate of movement calculated from age and distance ranged from 110 to 270 m/yr.

2.224 1957
RADIOCARBON AGEING IN NEW ZEALAND
Rafter, T.A.
New Zealand Science Review, v. 15, n. 11-12, 1957: 93-96
Following an outline of principles of the radiocarbon dating method, the enrichment of the atmosphere, of sea-water, and of shell carbonates in radiocarbon due to the 'isotope effect' and the 'industrial effect' of the combustion during the last 100 years of enormous quantities of fossil carbonaceous fuels containing no radiocarbon are discussed. In conclusion, some practical advice on the collection and forwarding of samples for age determination are given.

2.225 1968
RADIOCARBON AND CLIMATE. (A COMMENT ON A PAPER BY H. SUESS)
Damon, Paul E.
In: Causes of Climatic Change. Meteorological Monograph, v. 8, n. 30, Feb. 1968: 151-154. (Papers derived from the INQUA-NCAR Symposium on Causes of Climatic Change, Boulder, Col., 1965)
Atmospheric radiocarbon fluctuation and temperature fluctuation during the Christian era are consistent with a solar activity - solar modulator theory as suggested by Suess. The fivefold greater atmospheric radiocarbon fluctuation in B.C. time are not consistent with this relationship. The direct correlation between $\partial^{14}C$ and atmospheric temperature during B.C. time must be the result of a different mechanism. The assumption of a much higher atmospheric radiocarbon concentration during the last glacial episode is inconsistent with tree-ring data, with the varve chronology, and with the production of rate-decay rate relationships.

2.226 1965
RADIOCARBON AND PALEOMAGNETISM
Libby, Willard F.
In: International Conference on Radiocarbon and Tritium Dating, Pullman, Washington, Washington State University, June 1-11, 1965. *Proceedings*. United States of America, Atomic Energy Commission, 1965. Conference n. 650652: 348-356
Of all possible effects which could cause discrepancies in radiocarbon dates, the most likely is the variation of the earth's magnetic field, i.e. palaeomagnetism. It is quite likely that one of the best ways in which to study palaeomagnetism is to check the accuracy of radiocarbon dates against the historical records and tree rings, and other as yet to be developed scales such as the ionium dates of the deep sea sediment cores.

2.227 1967
RADIOCARBON AND PALEOMAGNETISM
Libby, Willard F.
In: Magnetism and the Cosmos, edited by W.R. Hindmarsh, F.J. Lowes, P.H. Roberts and S.K. Runcorn (NATO Advanced Study, Institute of Planetary and Stellar Magnetism, Newcastle upon Tyne). New York, Elsevier, 1967: 60-67
The possible disturbances in the radiocarbon content of living matter which could cause significant error in radiocarbon dates fall into two classes: disturbance of the rate of production by cosmic rays, and disturbance of the mixing reservoirs of exchangeable carbon with which the radiocarbon produced in the high atmosphere is mixed in the lifetime

bances, palaeomagnetism is the most likely to have been the most important.

2.228   1965
RADIOCARBON AND TRITIUM IN THE UPPER TROPOSPHERE
Fergusson, Gordon J.
In: International Conference on Radiocarbon and Tritium Dating, Pullman, Washington, Washington State University, June 1-11, 1965. *Proceedings.* United States of America, Atomic Energy Commission, 1965. Conference n. 650652: 525-540
Measurements of radiocarbon concentration from samples collected in the upper troposphere show radiocarbon activities up to ten times the pre-bomb ground level activity. This observed activity decreases rapidly with decreasing latitude and altitude.

2.229   1963
RADIOCARBON AND TRITIUM IN THE UPPER TROPOSPHERE
Fergusson, Gordon J.
Johnson Laboratories Inc., Baltimore, Maryland, (1963). pp. 17
A number of measurements have been made of radiocarbon concentration in the upper troposphere. Radiocarbon activity is shown to be 10 times the pre-bomb ground level activity. The observed activity decreases rapidly with decreasing latitude and altitude.

2.230   1955
RADIOCARBON CONCENTRATION IN MODERN WOOD
Suess, Hans E.
Science, v. 122, n. 3166, 2 Sept. 1955: 415-416
Redetermination of the absolute radiocarbon concentration in wood carbon and of its variation since the industrial revolution became widespread in the late 19th century was carried out by means of proportional counting of acetylene. Three interpretations of the results are suggested: (1) pressure of relatively large local variations, (2) world-wide contamination of the earth's atmosphere with artificial carbon dioxide, probably amounting to less than 1%, (3) greater than previously assumed rate at which carbon dioxide exchanges and is aborbed in the ocean.

2.231   1966
RADIOCARBON CONTENT OF MARINE SHELLS FROM THE CALIFORNIA AND MEXICAN WEST COAST
Berger, Rainer; Taylor, R.E.; Libby, Willard F.
Science, v. 153, n. 3738, 19 Aug. 1966: 864
The radiocarbon content of contemporary pre-bomb marine shells from the upwelling environment of the California and the West Mexican coast has been determined. In addition, factors leading to the apparent ages of different magnitude for marine environment are discussed.

2.232   1967
RADIOCARBON CONTENT OF MARINE SHELLS FROM THE PACIFIC COASTS OF CENTRAL AND SOUTH AMERICA
Taylor, R.E.; Berger, Rainer
Science, v. 158, n. 3805, 1 Dec. 1967: 1180-1182
The radiocarbon content of pre-bomb marine shells from the region of upwelling of the Pacific coasts of South America has been determined and found to be somewhat similar to the content of shells from the coast of California and the West coast of Mexico. Various deviations have been observed and are discussed and problems associated with radiocarbon dates based on shells are discussed.

2.233   1951
RADIOCARBON DATABILITY OF PEAT, MARL, CALICHE AND ARCHAEOLOGICAL MATERIALS
Bartlett, H.H.
Science, v. 114, n. 2951, 20 July 1951: 55-56
Except under thoroughly understood conditions, marl and caliche cannot be considered datable. Precautions must be taken to detect possibilities not only of materials having had their carbon radioactivity diminished by entry of dead carbon, but also of having been enriched in radiocarbon content by physical and chemical processes that are constantly taking place.

2.234   1950
RADIOCARBON DATES
Linne, Sigvald
Ethnos, v. 15, n. 3-4, 1950: 206-213
Retraces the history of the radiocarbon method of age determination and its development. A summary of results to date is given.

2.235   1956
RADIOCARBON DATES AND ARCHAEOLOGY
Movius, Hallam L.
Geological Society of America. Bulletin, v. 67, Dec. 1956: 1819
Abstract of paper presented at the meeting of the Geological Society of America in New York, Dec. 26-30, 1956. Examines the radiocarbon method which has so far provided several thousand determinations which form the base for a chronology of the last 40,000 years.

2.236   1963
RADIOCARBON DATING
Libby, Willard F.

Annals of Internal Medicine, v. 59, n. 1, Oct. 1963: 566-578
An overview of the theory and the method of radiocarbon dating and its applications.

2.237   1962
RADIOCARBON DATING
Deevey, Edward S.
Scientific American, v. 186, n. 2, Feb. 1952: 24-28
The author reviews the discovery made by W.F. Libby that radiocarbon in nature could be used in dating the past. He describes the method and its applications as well as its difficulties.

2.238   1963
RADIOCARBON DATING
Willis, E.H.
In: Science in Archaeology, edited by Don Brothwell, Eric Higgins and Graham Clark. London, Thames and Hudson, 1963: 35-46
An expose of the method including the following aspects: the basic assumption, the definition, the techniques, standards and the publication of dates.

2.239   1965
RADIOCARBON DATING
Green, H.H.
Atomic Energy (Australia), v. 8, Jan. 1965: 10-14
Principles of radioactive dating are reviewed. Methods of radiocarbon dating are described. Examples of radiocarbon dating in Australia are considered.

2.240   1968
RADIOCARBON DATING
Gill, Edmund D.
Victorian Naturalist, v. 85, n. 5, 1968: 161-164
A review article on radiocarbon dating including an examination of radiocarbon dates from Victoria.

2.241   1956
RADIOCARBON DATING
Libby, Willard F.
American Scientist, v. 44, n. 1, Jan. 1956: 98-112
The curiosity about possible effects that cosmic rays might have on earth and particularly on the earth's atmosphere led to radiocarbon dating. Several stages of the research on radiocarbon dating are described.

2.242   1959
RADIOCARBON DATING
Barker, Harold
Nature, v. 184, n. 4687, 29 Aug. 1959: 672-674
The author reviews the contribution made by representatives of three British laboratories to a joint session of Physics, Anthropology, and Archaeology of the British Association at York. Main contributions were: picture of the basis of the method and its limitation, trends of future development in instrumentation, capabilities of the method as applied to individual problems in the late Quaternary history of the British Isles.

2.243   1951
RADIOCARBON DATING
Libby, Willard F.
Chicago, University of Chicago Press, 1951. pp. 124 (Also 2nd ed., 1955. pp. 186)
Advances in sampling and measurement techniques for radiocarbon dating. Topics discussed include: world-wide distribution of radiocarbon, half-life of radiocarbon, preparation of samples, measurements of samples and significance of radiocarbon dates. Tables of radiocarbon dates are presented. Special equipment and chemicals for radiocarbon sample preparation, apparatus and materials for screen wall counters are described.

2.244   1961
RADIOCARBON DATING AND THE KEY TO PREHISTORY
Willis, E.H.
The Listener, v. 65, n. 1675, 4 May 1961: 773-774
Discusses the radiocarbon method of age determination and its application, in particular, to the last glaciation in the British Isles.

2.245   1955
RADIOCARBON DATING IN THE LIGHT OF STRATIGRAPHY AND WEATHERING
Hunt, Charles B.
Science Monthly, v. 81, 1955: 240-247
Attempts to review the stratigraphy of late Pleistocene and recent deposits and to examine a few of the conflicts between the known stratigraphy and radiocarbon dates and some aspects of the problem involved in the weathering or preservation of specimens for radiocarbon analysis.

2.246   1951
RADIOCARBON DATING, A SUMMARY
Johnson, Frederick; Rainey, Froelich; Collier, Donald; Flint, Richard Foster
American Antiquity, v. 17, part 2 (*Society for American Archaeology, Memoirs*, n. 8), 1951: 58-65
An evaluation of the method in general. It includes notes on statistical calculations involved and of the various factors to be considered in the evaluation of a radiocarbon date.

2.247   1953
RADIOCARBON DATING, ASSEMBLED BY FREDERICK JOHNSON

Ch. 2 - Theoretical Works

FREDERICK JOHNSON
Krieger, Alex D.
American Journal of Archaeology, v. 57, 1953: 50-52
A review of the book edited by Frederick Johnson. It introduces the method and has chapters by a number of authorities which discuss the dating by region or subject.

2.248   1964
RADIOCARBON DATING: A CASE AGAINST THE PROPOSED LINK BETWEEN RIVER MOLLUSCS AND SOIL HUMUS
Broecker, Wallace S.
Science, v. 143, n. 3606, 7 Feb. 1964: 596-597
The author offers another hypothesis for explaining the observation that modern mollusc shells from rivers can have anomalous radiocarbon age (Keith and Anderson, *Science*, 1963). ^{13}C deficiency should be attributed to the uptake of soil carbon dioxide by ground water and the ^{14}C deficiency to the solution of limestone. Both the apparent correlation between ^{13}C and ^{14}C would be attributed to exchange with atmospheric carbon dioxide. As this exchange is more extensive in lakes than in rivers, lake molluscs might be expected to show small deficiencies of both isotopes

2.249   1963
RADIOCARBON DATING: FICTITIOUS RESULTS WITH MOLLUSC SHELLS
Keith, M.L.; Anderson, G.M.
Science, v. 141, n. 3581, 16 Aug. 1963: 634-636
Evidence is presented to show that modern mollusc shells from rivers can have anomalous radiocarbon ages owing mainly to the incorporation of inactive (radiocarbon deficient) carbon from humus. Various other sources of inactive carbon can as well cause errors as large as several thousand years for river shells.

2.250   1951
RADIOCARBON DATING; A REPORT ON THE PROGRAM TO AID IN THE DEVELOPMENT OF THE METHOD OF DATING
Johnson, Frederick
American Antiquity, v. 17, part 2 (*Society for American Archaeology, Memoirs, n.* 8), 1951. pp. 65
This report was assembled for the Committee on Radioactive Carbon 14 of the American Anthropological Association and the Geological Society of America. It contains an introduction by Frederick Johnson and a corrected list of dates which were published originally by Arnold and Libby in 1951 and ten papers which are indexed separately.

2.251   1957
RADIOCARBON EVIDENCE ON THE DILUTION OF ATMOSPHERIC AND OCEANIC CARBON BY CARBON FROM FOSSIL FUELS
Brannon, H.R.; Daughtry, A.C.; Perry, D.; Whitaker, W.W.; Williams, Milton
American Geophysical Union. Transaction, v. 38, n. 5, Oct. 1957: 643-650
The dilution of atmospheric carbon dioxide by carbon dioxide from fossil fuel is estimated to be about 3 1/2% on the basis of radiocarbon assays of tree rings. Radiocarbon assays of oceanic shells indicate a 1 to 2% dilution of shallow oceanic carbonate by carbon dioxide from fossil fuels. Use of these dates in a simplified mathematical model of atmosphere-ocean yields information on mixing time of the ocean.

2.252   1947
RADIOCARBON FROM COSMIC RADIATIONS
Anderson, Ernest C.; Libby, Willard F.; Weinhouse, Sydney; Reid, Allan F.; Kirschenbaum, A.D.; Grosse, Aristid V.
Science, v. 105, n. 2735, 30 May 1947: 576-577
Neutrons formed by cosmic radiations should form radiocarbon in such amounts that all carbon in living matter should be radioactive to the extent of 1 to 10 disintegrations/minute/gm.

2.253   1966
RADIOCARBON FROM NUCLEAR TESTS
United Nations. Scientific Committee on the Effects of Atomic Radiations. Report, A/AC. 82/R 105, 1966. pp. 56
A review is presented on the distribution of radiocarbon in the carbon cycle and the biological effect of radiocarbon. Topics included are: distribution of carbon on earth, production and distribution of natural radiocarbon, movement of carbon in the carbon cycle, production of radiocarbon by nuclear tests, experimental observation of increased radiocarbon levels, inventory and present distribution of artificial radiocarbon.

2.254   1960
RADIOCARBON FROM NUCLEAR TESTS II
Broecker, Wallace S.; Olson, Edwin A.
Science, v. 132, n. 3429, 16 Sept. 1960: 712-721
The testing of nuclear weapons has injected into the earth's carbon cycle large amounts of radiocarbon. Its concentration at various places in the carbon cycle is rising. This gives the scientist a potential large-scale tracer for the elucidation of the mechanisms and rates of many natural processes which involve the element carbon. An estimate of the total amount of bomb produced radiocarbon, and its present and future distribution is needed to evaluate these tracer possibilities.

2.255   1968
RADIOCARBON FROM NUCLEAR WEAPON TESTS

Journal of Geophysical Research, v. 73, n. 4, 15 Feb 1968: 1185-1200

The distribution in time and space of radiocarbon from the 1961-1962 nuclear weapon tests of the U.S. and USSR is used as a tracer for atmospheric mixing phenomena and exchange of carbon dioxide between the atmosphere and the sea.

2.256   1967
RADIOCARBON IN WOOD FROM REGIONS OF CONTEMPORARY VOLCANICITY
Sulerzhitskiy, L.D.; Forova, V. S.
Akademiia Nauk, USSR. Doklady. Earth Science Sections, v. 171, n. 1-6, July 1967: 232-234
Errors in the radiocarbon dates of plant residues may be caused by dilution of the atmosphere by the volcanic 'dead' carbon dioxide under certain conditions. The appropriate correction may be difficult to make because the dilution effect depends on variations in the intensity of the volcanism in the area, although it appears to be unrelated to the distance from the volcanic source.

2.257   1959
RADIOCARBON MEASUREMENTS ON SAMPLES OF KNOWN AGE
Broecker, Wallace S.; Olson, Edwin A.; Bird, Junius Bonton
Nature, v. 183, n. 4675, 6 June 1959: 1582-1584
Using samples which are historically dated or tree ring dated, the data indicates that, with the present observations, it can only be said that the radiocarbon age of an atmospherically derived sample younger than 2000 years would almost always lie within 250 years of the true age.

2.258   1966
RADIOCARBON VARIATIONS IN THE ATMOSPHERE
Olsson, Ingrid U.; Karlén, Ingvar; Stenberg, Allan
Tellus, v. 18, n. 2-3, 1966: 293-297
The authors discuss the findings of Machta (1966), Hageman et al. (*Science*, 1959), Fergusson (*Journal of Geophysical Research*, 1963) and Münnich and Vogel (*Naturwissenschaften*, 1958) and compare with their own data. They conclude that location for mixing between the stratosphere and the troposphere varies but is usually at a lower latitude than 60°N in the northern hemisphere, thus more to the south than suggested by Münnich and Vogel. The mixing is rather slow, even in the west-east direction

2.259   1960
THE REACTION OF ATOMIC CARBON WITH OXYGEN: SIGNIFICANCE FOR THE NATURAL RADIOCARBON CYCLE
Pandow, Mary; McKay, C.; Wolfgang, Richard L.
Journal of Inorganic Nuclear Chemistry, v. 14, n. 3/4, 1960: 153-158
The nuclear reaction responsible for production of natural radiocarbon in the upper atmosphere can be duplicated in an atomic reactor. Radiocarbon in carbon monoxide is found to account for 90-100 percent of the radiocarbon produced. Since radiocarbon as carbon monoxide cannot be utilised in the photosynthetic process, suggested mechanism for converting carbon monoxide into a chemical form suitable for entrance into the biosphere are reviewed. The relatively long mean-life of atmospheric carbon monoxide leads to the prediction that the specific activity of natural carbon monoxide may quite possibly be greater than that of atmospheric carbon dioxide and therefore the biosphere with which carbon dioxide is in equilibrium.

2.260   1960
RECALIBRATION OF THE NBS CARBON-14 STANDARD BY GEIGER-MULLER AND PROPORTIONAL GAS COUNTERS
Mann, W.B.; Seliger, H.H.; Marlow, W.F.; Medlock, R.W.
Review of Scientific Instruments, v. 31, n. 7, July 1960: 690-696
Compensated internal gas counters have been constructed for the recalibration of the National Bureau of Standards radiocarbon solution standard. Satisfactory agreement has been obtained by counting in both the proportional and the Geiger region.

2.261   1955
RECENT ADVANCES IN DATING THE PAST
Bowen, R.N.C.
Discovery, v. 16, n. 9, 1955: 388-391
Briefly describes the principles and application of radioactive and other methods of determining the absolute geologic age of rocks and fossils.

2.262   1967
RECENT DEVELOPMENTS IN RADIOCARBON DATING: THEIR IMPLICATION FOR GEOCHRONOLOGY AND ARCHAEOLOGY
Dyck, Willy
Current Anthropology, v. 8, n. 4, Oct. 1967: 349-351
Comments upon the following developments in radiocarbon dating techniques: dating of bones and shells, half-life of radiocarbon, and variations in the natural radiocarbon concentration.

2.263   1959
RECENT INCREASE IN THE C14 CONTENT OF THE ATMOSPHERE, THE BIOSPHERE AND THE OCEAN
Münnich, K.O.; Vogel, John C.

Ch. 2 - Theoretical Works

Zweites Physikalisches Institut der Universitat Heidelberg, 14C Laboratory. *Report on Measurements*, August 1959. pp. 10

Measurements of samples of air, groundwater, and biological materials collected in the Federal Republic of Germany during 1958 and 1959 showed an increase in radiocarbon during the period. Data on the radiocarbon content of samples of ocean water collected at three locations in the North Atlantic is also included.

2.264   1959
RECENT STUDIES OF GASES IN GLACIER ICE, A SUMMARY
Nutt, David C.
Polar Notes (An Occasional Publication of the Stefansson Collection), v. 1, Nov. 1959: 57-65

Reviews results and techniques. A technique for radiocarbon dating is contributed by H. de Vries.

2.265   1964
RECENT VARIATIONS OF ATMOSPHERIC CONTENTS OF C-14 IN TOKYO AND TRANSFER PROBLEM OF THE ATMOSPHERIC CARBON DIOXIDE
Kigoshi, Kunihiko
Journal of Radiation Research (Japan), v. 5, June 1964: 120-130

Data are presented on the variation of atmospheric radiocarbon concentration in Tokyo since the end of 1959. The observed results show big random fluctuation compared with those obtained by Fergusson (1963) in California. Based on the variations of the troposphere content of radiocarbon since 1955, the transport of carbon dioxide from the atmosphere into the sea water is discussed using a parallelism between the supply of radiocarbon and other radionuclides produced by nuclear tests from the stratosphere into the troposphere.

2.266   1964
REDEPOSITED POLLEN IN LATE-WISCONSIN POLLEN SPECTRA FROM EAST-CENTRAL MINNESOTA
Cushing, Edward J.
American Journal of Science, v. 262, n. 9, Nov. 1964: 1075-1088

Radiocarbon dates from Andree Bog support the relatively young age of the sediments there and confirm the assumption of rapid deposition of the silty sediments at the base of the organic sediment.

2.267   1958
REDUCTION OF THE ATMOSPHERE RADIO CARBON CONCENTRATION BY FOSSIL FUEL CARBON DIOXIDE AND THE MEAN LIFE OF CARBON DIOXIDE IN THE ATMOSPHERE
Fergusson, Gordon J.
Royal Society, London. Proceedings, section A, n. 243, 1958: 561-574

Measurements of the radiocarbon concentration in sets of wood samples from the northern and the southern hemispheres show that the radiocarbon specific activity of atmospheric carbon dioxide has decreased by $2.3 \pm 0.15\%$ over the period 1950 - 1954 and that the present day difference between the decrease in the northern and the southern hemispheres is less than 0.50%. Two deductions are made: (1) the mean life of carbon dioxide in the atmosphere is less than 7 years, probably of the order of 2 years; (2) the exchange time for mixing the atmospheres of the two hemispheres is less than 2 years.

2.268   1966
RELATION BETWEEN THE ^{14}C PRODUCTION RATE AND THE GEOMAGNETIC MOMENT
Wada, Masami; Inoue, Aoi
Journal of Geomagnetism and Geoelectricity, v. 18, 1966: 485-488

The secular variation of the geomagnetic moment affects the intensity of primary cosmic rays and the intensity of secondary neutrons whose capture rate in the atmosphere is equivalent to the production of radiocarbon. Presents a relation between radiocarbon production rate and the geomagnetic moment deduced from the neutron capture rates.

2.269   1958
THE RELIABILITY OF C14 CHRONOLOGY
Immamura, Gakuro
Zeitschrift für Geomorphologie, N. F. v.2, n. 3, 1958: 230-240

Computations of variations in previously published radiocarbon data show a general increase in variation with age and a sudden increase at 12,000 years, which is therefore considered the limit of reliability.

2.270   1965
REMARKS ON C-14 DATING OF SHELL MATERIAL IN SEA SEDIMENT
Olsson, Ingrid U.; Eriksson, K. Gösta
In: Progress in Oceanography, v. 3, edited by Mary Searles. Oxford, Pergamon Press, 1965: 253-266

The investigation has shown that in the Mediterranean Sea different grain sizes in samples of sediments may produce considerable variations in a given radiocarbon dating. It has also been shown that large errors may be introduced during the pre-treatment of the material unless the greatest care is taken to avoid contamination by atmospheric carbon dioxide.

2.271 1964
REPLACEMENT RATES FOR HUMAN TISSUES FROM ATMOSPHERIC RADIOCARBON
Libby, Willard F.; Berger, Rainer; Mead, J.F.; Ross, Alexander; Ross, J.F.
Science, v. 146, n. 3648, 27 Nov. 1964: 1170-1172
Carbonate derived from the testing of nuclear weapons in the atmosphere of the Northern Hemisphere during 1961-62 has been found in human tissues, including the brain in amounts which reflect the atmospheric concentration of radiocarbon as of several months earlier. In collagen of cartilage the rate of uptake of radiocarbon is much slower than in other tissues. As individuals from the Southern Hemisphere show little increase in the radiocarbon of their tissues, detailed tests with individuals travelling from the Southern to th.e Northern Hemisphere will allow closer scrutiny of the tissue replacement rate

2.272 1965
RESULTS OF WATER SAMPLE DATING BY MEANS OF THE MODEL OF MUNNICH AND VOGEL
Geyh, Mebus A.; Wendt, Immo
In: International Conference on Radiocarbon and Tritium Dating, Pullman, Washington, Washington State University, June 1-11, 1965. *Proceedings*. United States of America, Atomic Energy Commission, 1965. Conference n. 650652: 597-603
Confirms Vogel's hypothesis that all aquifer waters in Central Europe are younger than the last glaciation about 10,000 years ago. This indicates that apparently no exchange between dissolved bicarbonate and the rock-carbonate occurs.

2.273 1961
THE ROLE OF TRACE ELEMENTS. GEOCHEMISTRY IN ARCHAEOLOGY
Turekian, Karl K.
Yale Scientific Magazine, v. 35, n. 4, 1961
[Not sighted]

2.274 1958
SAMPLE CONTAMINATION AND RELIABILITY OF RADIOCARBON DATES
Olson, Edwin A.; Broecker, Wallace S.
New York Academy of Sciences. Transactions, Series 2, v. 20, n. 7, May 1958: 593-604
The reasons for anomalous results in radiocarbon determinations are discussed, particularly those that can be explained by sample contamination, in which case the basic assumptions of the radiocarbon age method do not apply. Three main theoretical mechanisms by which a sample's isotopic composition can be altered are examined.

2.275 1961
SECULAR CHANGES IN THE CONCENTRATION OF ATMOSPHERIC RADIOCARBON
Suess, Hans E.
In: Problems Related to Interplanetary Matters. Proceedings of an Informal Conference, Highland Park, Ill., June 20-22, 1960. Committee on Nuclear Science. Subcommittee on Nuclear Geophysics. Nuclear Sciences Series Report, n. 33. Washington D.C., National Academy of Science. National Research Council. Publication n. 845, 1961: 90-95
The method of dating is based on the following premises: (1) constancy of the cosmic ray flux; (2) constancy of the various carbon exchange reservoirs. However the existence of deviations has shown that one or both these assumptions are not precisely correct. The author examines the reasons for the fluctuation of radiocarbon with time.

2.276 1965
SECULAR CHANGES OF NATURAL RADIOCARBON AND THEIR INTERPRETATION
(Abstract)
Suess, Hans E.
In: International Conference on Radiocarbon and Tritium Dating, Pullman, Washington, Washington State University, June 1-11, 1965. *Proceedings*. United States of America, Atomic Energy Commission, 1965. Conference n. 650652: 439
Changes of the radiocarbon activity of wood on a time scale of a hundred years are due to changes in the production rate of radiocarbon caused by changes in the cosmic ray intensity. These have coincided with periods of cold climate and involve only the atmospheric carbon reservoir. Changes with a time period of more than a thousand years involve the total carbon inventory on the earth's surface and could be attributed to changes in the magnetic field of the earth.

2.277 1965
SECULAR VARIATION OF ATMOSPHERIC RADIOCARBON CONCENTRATION AND ITS DEPENDANCE ON GEOMAGNETISM
Kigoshi, Kunihiko
In: International Conference on Radiocarbon and Tritium Dating, Pullman, Washington, Washington State University, June 1-11, 1965. *Proceedings*. United States of America, Atomic Energy Commission, 1965. Conference n. 650652: 429-439
The radiocarbon concentration of atmospheric carbon dioxide depends on the production rate of radiocarbon by cosmic rays in the stratosphere and also on the rates of interchange between the various components of the earth's carbon cycle. These variations were studied by means of measurements on atmospheric radiocarbon concentration.

2.278   1966
SECULAR VARIATIONS IN THE 14C CONCENTRATIONS OF DOUGLAS PINE TREE RINGS
Dyck, Willy
Canadian Journal of Earth Sciences, v. 3, n. 1, Feb. 1966: 1-7
The experimental data collected show definitely that variations in the radiocarbon concentration in the biosphere exist. These variations can be correlated with temperature fluctuations. Although changes in the production rate of radiocarbon could explain these variations, material balance data indicate that the influence of biospheric activity and temperature on the carbon dioxide content of the atmosphere and hence radiocarbon concentration, can be appreciable.

2.279   1965
SECULAR VARIATIONS IN THE C-14 CONCENTRATION OF DOUGLAS FIR TREE RINGS
Dyck, Willy
In: International Conference on Radiocarbon and Tritium Dating, Pullman, Washington, Washington State University, June 1-11, 1965. *Proceedings*. United States of America, Atomic Energy Commission, 1965. Conference n. 650652: 449-451
The experimental data collected show definitely that variations in the radiocarbon concentration in the biosphere exist and that they can be correlated with temperature fluctuation. A mechanism to explain these variations is proposed.

2.280   1966
SECULAR VARIATIONS OF ATMOSPHERIC RADIOCARBON CONCENTRATION AND ITS DEPENDANCE ON GEOMAGNETISM
Kigoshi, Kunihiko; Hasagewa, Hiroishi
Journal of Geophysical Research, v. 71, n. 4, Feb. 1966: 1065-1071
The variation in atmospheric radiocarbon concentration during the last 1800 years was measured in a giant tree grown at Yaku Island in the southern part of Japan. The observed gradual decrease of radiocarbon concentration in the past agrees with the computed variation based on the palaeomagnetic data, which shows that variation of the geomagnetic field strength has a dominant effect on the variation of atmospheric radiocarbon concentration.

2.281   1965
SECULAR VARIATIONS OF THE COSMIC-RAY-PRODUCED CARBON-14 IN THE ATMOSPHERE AND THEIR INTERPRETATION
Suess, Hans E.
Journal of Geophysical Research, v. 70, n. 24, 15 Dec. 1965: 5937-5952
The radiocarbon content of about 150 wood samples, covering the past 2000 years, dated by dendrochronology, was measured. The measurements show that the radiocarbon activity of the atmospheric carbon dioxide has not been entirely constant in the past but has varied by several percent. Two types of variations can be recognised: (1) those that occurred on a time scale of the order of 100 years, occurring only in the atmospheric radiocarbon reservoir; (2) those with a time constant of more than 1000 years and involving the whole of the radiocarbon inventory, whose cause cannot yet be established conclusively.

2.282   1956
SEPARATION OF $^{14}C^{16}O$ AND $^{13}C^{16}O$ BY THERMAL DIFFUSION
de Vries, A.E.; Harring, A.; Slats, W.
Physica, v. 22, 1956: 247-248
Uses the thermal diffusion column of Clusius and Dickel to investigate the properties of the two molecules.

2.283   1956
A SKEPTIC VIEW OF RADIOCARBON DATES
Hunt, Charles B.
University of Utah. Department of Anthropology. Anthropological Papers, n. 26, Dec. 1956: 35-46
While there is good reason to believe that carbon samples give valid ages when the samples have been preserved under conditions simulating those of a sealed test tube, there is equally good reason for expecting erroneous ages where the carbon is subject to loss or addition. As it stands at present, discrepancies in the dates are numerous and the causes of the discrepancies are not fully understood. This means that dates must be viewed as suspect.

2.284   1960
SOME PRELIMINARY MEASUREMENTS OF THE TRITIUM AND CARBON-14 CONTENT OF THE STRATOSPHERE OVER ENGLAND
Goldsmith, P.; Jelley, J.V.; Barklay, F.R.; Elliott, M.J.; Osborne, A.R.
United Kingdom. Atomic Energy Authority. Research Group. Atomic Energy Research Establishment (A.E.R.E.), Harwell, England, and *United Kingdom. Atomic Energy Authority. Weapons Group. Report*, Aldermaston, England. Atomic Weapons Research Establishment, 1960, pp. 11 (A.E.R.E. - R - 3271)
A technique is described whereby samples of water and carbon dioxide from heights of about 90,000 ft. were obtained. Measurements of the tritium content, together with the radiocarbon content of one sample, are presented and discussed.

2.285   1967
THE STATE OF THE ART IN ¹⁴C DATING
Libby, Willard F.
American Nuclear Society. Transactions, v. 10, June 1967: 20
Presented at the 13th annual meeting of the American Nuclear Society, June 11- 15, 1967. An overview of the latest developments in radiocarbon dating.

2.286   1959
STRATOSPHERIC CARBON-14, CARBON DIOXIDE AND TRITIUM
Hageman, Frerich T.; Grey, James; Machta, Lester; Turkevich, Anthony
Science, v. 130, n. 3375, 4 Sept. 1959: 542-552
Results of measurements made between September 1953 and September 1958 of the radiocarbon, carbon dioxide and tritium in the stratosphere to obtain information on the stratospheric concentration of radiocarbon and tritium produced by the explosion of nuclear devices.

2.287   1962
A STUDY OF SOME PROBLEMS CONNECTED WITH C-14 DATING: (INAUGURAL DSSERTATION)
Olsson, Ingrid U.
Acta Universitatis Uppsaliensis. Abstracts of Uppsala Dissertations in Science, v. 14, 1962: 1-6
Deals with the problems encountered in the determination of present activity and the half-life of radiocarbon.

2.288   [1967 ?]
SUNSPOT AND CARBON DATING IN THE MIDDLE AGES
Grey, Donald C.; Damon, Paul E.
In: International Conference on Application of Science to the Medieval Archaeology, University of California Los Angeles, Oct. 1967. [1967 ?]
[Not sighted]

2.289   1963
A SURVEY OF NATURAL ISOTOPES OF WATER IN SOUTH AFRICA (WITH DISCUSSION)
Vogel, John C.; Ehhalt, D.; Roether, W.
In: Radioisotopes in Hydrology, International Atomic Energy Agency (IAEA), Symposium, Tokyo, 1963, Proceedings. Vienna, IAEA, 1963: 407-415
The semi-arid climate conditions cause the isotopic composition of rain, river and groundwater to differ considerably from that in Europe and the northern United States. Results show that evaporation markedly changes the isotope ratio of the precipitation. Discontinuous recharge of the groundwater supplies is probably responsible for the observed deviations in this reservoir.

2.290   1962
A SURVEY OF RADIOACTIVE FALLOUT FROM NUCLEAR TESTS
Machta, Lester; List, R.J.; Telegadas, K.
Journal of Geophysical Research, v. 67, n. 4, April 1962: 1389-13400
A study to track the passage of the radioactive debris, results of fallout from nuclear tests by the Soviet Union, through the atmosphere, to delineate the geographical areas likely to receive fallout and the levels of fallout which may be deposited next spring.

2.291   1967
THOSE LATE CORN DATES: ISOTOPIC FRACTIONATION AS A SOURCE OF ERROR IN CARBON- 14 DATES
Hall, Robert L.
Michigan Archaeologist, v. 13, n. 4, Dec. 1967: 171-180
Archaeologists have often intentionally sought out carbonised corn and grass thatch for radiocarbon dating to avoid the kind of error which may arise when inner tree rings are disproportionately represented in a carbonised wood sample. Corn and starch are shown to be highly susceptible to isotope fractionation, favouring the enrichment of heavy carbon isotopes, an effect which results in radiocarbon ages centuries later than the true age if the calculated ages are not normalised for fractionation. A table of trial normalisations of seven Canadian Iroquoian dates is presented.

2.292   1960
TRANSIENT EFFECTS IN THE DISTRIBUTION OF CARBON-14 IN NATURE
Plesset, Milton S.; Latter, Albert L.
National Academy of Science. Proceedings, v. 46, n. 2, 15 Feb. 1960: 232-241
Develops a transient solution based on a six reservoir model, dividing the atmosphere into two reservoirs, the stratosphere and the troposphere. It is hoped that observation of the radiocarbon transients may give further information on the effective exchange rates between troposphere and stratosphere.

2.293   1966
TRANSPORT AND FALLOUT OF STRATOSPHERIC RADIOCARBON
Feely, Herbert W.; Seitz, H.; Lagomaximo, R.J.; Biscaye, P.E.
Tellus, v. 18, n. 2-3, 1966: 316-328
During the two years following the explosion of high yield nuclear weapons in the atmosphere, tests showed that the stratospheric residence half-time for the radiocarbon pro-

stratospheric residence half-time for the radiocarbon produced artificially by those explosions increased with time, but averaged after about 18 months.

2.294   1961
TRITIUM AND CARBON-14 IN THE TREE RINGS
Kigoshi, Kunihiko; Tomikura, Yoshio
Chemical Society of Japan. Bulletin, v. 34, Nov. 1961: 1739-1740
Determines the annual mean concentration of the atmospheric radiocarbon and of the tritium in rain water by measuring the concentration of radiocarbon and tritium in tree rings.

2.295   1951
TRITIUM IN NATURE
Grosse, Aristid V.; Johnston, William M.; Wolfgang, Richard L.; Libby, Willard F.
Science, v. 113, n. 2923, 5 Jan. 1951: 1-2
The influence of natural radiocarbon in the atmosphere leads to research in the possibility of cosmic radiation produced tritium in surface waters.

2.296   1963
USE OF LIGHT NUCLEIDES IN LIMNOLOGY
Deevey, Edward S.; Stuiver, Minze; Nakai, Naboyuki
In: Radioecology. Conference on Radioecology, 1st., Fort Collins, Colorado, 1961, Proceedings, New York, Reinhold, 1963: 471-475
Results are reported from measurements of the distribution of ^{13}C, ^{14}C and ^{34}S in lakes. The contribution of ^{14}C from fallout, groundwater and aquatic plants in lakes is considered.

2.297   1965
VALIDITY OF RADIOCARBON DATES ON BONES
Tamers, Murray A.; Pearson, Frederick Joseph
Nature, v. 208, n. 5015, 11 Dec. 1965: 1053-1055
The validity of radiocarbon dates on bones is investigated by determining the natural radiocarbon content of samples of known ages. According to the results of the experiment, the majority of radiocarbon dates on bones are in error. The errors are in the same direction, giving falsely young ages. It is suggested that a more realistic use of bone dates would be to take them only as lower limits and to precede them with a sign indicating 'greater than or equal to'.

2.298   1967
VARIATION IN ATMOSPHERIC CARBON-14 ACTIVITY RELATIVE TO A SUNSPOT - AURORAL SOLAR INDEX
Bray, J. Roger
Science, v. 156, n. 3775, 5 May 1967: 640-642
Radiocarbon activity was negatively correlated with a sunspot - auroral index during 25 designated periods from 129 B.C. to A.D. 1964. Change in radiocarbon activity during these periods was inverse to change in solar activity in 22 of the 24 instances. Contamination from radiocarbon formed during previous solar cycles may lessen the value of radiocarbon as a climatic or solar index

2.299   1958
VARIATION IN CONCENTRATION OF RADIOCARBON WITH TIME AND LOCATION ON EARTH
de Vries, Hessel
Koninklijke Nederlandse Akademie van Wetenschappen. Proceedings, section B, v. 61, n. 2, 1958: 1-9
Radiocarbon measurements on samples of known ages (tree-rings) have demonstrated that the concentration of radiocarbon in the atmosphere varies considerably in the course of a few centuries and also with location on earth. The variations are of the order of one percent. Evidence is presented that the fluctuations are due to the variations in vertical mixing in the ocean which brings old (less active water) in contact with the atmosphere. It is not yet possible to understand the origin of the variations with location on earth. Implications on radiocarbon dating are discussed.

2.300   1966
VARIATIONS IN ATMOSPHERIC C-14
Grey, Donald C.; Damon, Paul E.; Long, Austin
American Geophysical Union. Transactions, v. 47, n. 3, Sept. 1966: 495
Abstract of Paper presented at the Sixth Western National Meeting of the American Geophysical Union, Sept. 1-9, 1966, U.C.L.A., section of Volcanology, Geochemistry and Petrology. Continuing radiocarbon measurements of dendrochronologically dated wood have added more detailed knowledge of the fluctuation of natural radiocarbon during the last 6 millenia.

2.301   1966
VARIATIONS IN C-14 CONCENTRATION IN THE ATMOSPHERE DURING THE LAST SEVERAL YEARS
Nydal, Reidar
Tellus, v. 18, n. 2-3, 1966: 271-279
There seems to be a nearly uniform concentration of radiocarbon in the stratosphere from early 1963 in the region 34°N to 71°N at 20 km altitude. Simultaneously occurring radiocarbon peaks in the troposphere in summer at various northern latitudes seem to be partly due to a simultaneous removal of radiocarbon from the stratosphere to the troposphere at the respective latitudes. It turns out that the radiocarbon from polar nuclear explosions is transferred to the

lower and southern latitudes both through the troposphere and through the stratosphere. Radiocarbon transfer through the last one seems to be important.

2.302   1961
VARIATIONS IN RADIOCARBON CONCENTRATION AND SUNSPOT ACTIVITY
Stuiver, Minze
Journal of Geophysical Research, v. 66, n. 1, Jan. 1961: 273-276
Variations in cosmic ray intensities will produce variations in radiocarbon production in the atmosphere. A comparison is made between variations in sunspot activity and fluctuation in radiocarbon concentration during the past 13 centuries. Although a definite conclusion is not reached, the evidence given suggests some correspondence between sunspot activity and radiocarbon concentration in the atmosphere.

2.303   1965
VARIATIONS IN RADIOCARBON CONCENTRATION IN TREE RINGS IN NORTH AMERICA DURING THE 18TH AND 19TH CENTURIES
(Abstract)
Stuiver, Minze
In: *International Conference on Radiocarbon and Tritium Dating*, Pullman, Washington, Washington State University, June 1-11, 1965. Proceedings. United States of America, Atomic Energy Commission, 1965. Conference n. 650652: 452
Detailed tree ring analyses of a Douglas fir from Arizona confirm basically the de Vries series of minima and maxima in the atmospheric radiocarbon concentration.

2.304   1965
VARIATIONS IN THE ATMOSPHERIC RADIOCARBON CONCENTRATION OVER THE PAST 1300 YEARS
Willis, E.H.; Tauber, Henrik; Münnich, K.O.
Radiocarbon, v. 2, 1960: 1-4
The experiment described in this article demonstrates that, over the past 1200 years, the fundamental assumptions of the radiocarbon dating method are empirically correct to about 1.5%. Whereas the implications of an error of this magnitude might be disturbing for very recent samples, with older samples the effect might be expected to be of very little significance. The experiment was conducted on a section of Sequoia tree which had been tree ring counted and was fully documented.

2.305   1959
VARIATIONS IN THE C-14 CONTENT DURING THE LAST YEARS
Münnich, K.O.; Vogel, John C.
Radiocarbon Dating Conference, Groningen, Sept. 1959. *Papers Presented*, 1959
[Not sighted]

2.306   n.d.
VARIATIONS IN THE PRODUCTION OF RADIONUCLIDES
Aegerter, S.K.; Loosli, H.H.; Oeschger, Hans
Physikalisches Institut der Universitat Bern. *Preprint* (SM-87/32), n. d.
A critical review of two theories concerning the variations of the concentration of radiocarbon, namely that radiocarbon variations are caused by climatic changes, and that they are caused by variations of the production rate. It concludes that short term variations are caused by varying solar activity.

2.307   1952
VARIATIONS IN THE RELATIVE ABUNDANCE OF CARBON ISOTOPES IN PLANTS
Wickman, Frans E.
Geochimica et Cosmichimica Acta, v. 2, n. 4, 1952: 243-254
The $^{13}C/^{12}C$ ratio of 105 plants representing all major systematic groups has been determined. There are in principle no systematic difference between the groups. Characteristic differences occur between plants grown in different biotopes and these differences are related to the varying intensity of the local carbon cycle.

2.308   1961
VARIATIONS IN THE RELATIVE ABUNDANCE OF THE CARBON ISOTOPES
Nier, Alfred O.; Gulbransen, E.A.
American Chemical Society. Journal, v. 83, 1961: 697-698
The equilibrium constant of the $^{14}C/^{12}C$ exchange is predicted using a procedure as given by H.C. Urey (1947).

2.309   1952
VARIATIONS IN THE RELATIVE ABUNDANCE OF THE CARBON ISOTOPES IN PLANTS
Wickman, Frans E.
Nature, v. 169, n. 4312, 21 June 1952: 1051
The rate of assimilation of the light and heavy carbon dioxide molecules is influenced by the 'local carbon dioxide cycle' which works as an isotope enrichment process. In places of intense cycle the isotope effect is larger than in places where the cycle is almost absent. This is of some interest in connection with the radiocarbon dating method for age determination.

2.310   1966
VARIATIONS OF "C CONTENT IN THE ATMOSPHERE DURING THE PAST 2000 YEARS

International Conference on Cosmic Rays, 9th., London, 1966. Proceedings, v. 1, edited by A.C. Stickland. London, The Institute of Physics and the Physical Society, 1966: 597-600

In order to study the secular variations of primary cosmic rays, the radiocarbon concentration in tree rings of an old cedar tree in southern Japan was measured systematically. The variations are consistent with that expected from the change of magnetic dipole moment of the earth known from palaeomagnetic studies.

2.311   1960
VARIATIONS OF RADIOCARBON CONCENTRATIONS IN MODERN WOOD
Kigoshi, Kunihiko; Tomikura, Yoshio
Chemical Society of Japan. Bulletin, v. 33, 1960: 1576-1580
Presents the results of the measurements on the change of the atmospheric radiocarbon content using the carbon in tree rings from Honshu Island. The results show a constant concentration during 1782 and 1950, the depletion between those dates seem to be less than 2%, the variation of radiocarbon concentration in the atmosphere of the southern hemisphere seems to follow that of the northern hemisphere with a time lag of several months.

2.312   1961
VARIATIONS OF THE ATMOSPHERIC CARBON-14 IN RECENT YEARS IN TOKYO
Kigoshi, Kunihiko; Endo, Kunihiko
Chemical Society of Japan. Bulletin, v. 34, Nov. 1961: 1739-1740
Results of measurements made of the atmospheric radiocarbon over Tokyo.

2.313   1968
VERIFICATION OF RADIOCARBON DATING OF GROUND WATER BY MEANS OF TRITIUM ANALYSIS (Abstract)
Handshaw, Bruce B.; Rubin, Meyer; Stewart, Gordon; Friedman, Irving
American Geophysical Union. Transactions, v. 49, n. 1, 1968: 166
A study to establish the validity of a correction equation that utilises $^{13}C/^{12}C$ ratio both to adjust radiocarbon ages by accounting for solution of non- radiogenic limestone and to test for isotopic exchange between the limestone and dissolved carbonate species.

2.314   1967
VERY HIGH ^{14}C ACTIVITY IN ABISKO, SWEDEN, DURING SUMMER 1965
Olsson, Ingrid U.; Stenberg, Allan
In: Radioactive Dating and Methods of Low-level Counting.
Proceedings of a Symposium organised by the International Atomic Energy Agency (IAEA) in co-operation with the Joint Commission on Applied Radioactivity (ICSU) and held in Monaco, 2-10 March 1967. Vienna, IAEA, 1967. (Proceedings Series): 69-78
Report on very high radiocarbon activity which reached peak probably due to special radiocarbon experiments in Abisko, Sweden.

2.315   1966
THE WORLD OF RADIOISOTOPES
Gregory, J.N.
Sydney, Angus and Robertson, A.A.E.C., 1966. pp. 204
In chapter 8, p.160-169, the radioisotope clock: radiocarbon dating, a brief description and assessment of the method and its applications is presented.

2.316   1951
WORLD-WIDE DISTRIBUTION OF RADIOCARBON
Anderson, Ernest C.; Libby, Willard F.
Physical Review, v. 81, n. 1, Jan. 1951: 64-69
The natural radiocarbon content of biological materials collected from locations widely scattered over the earth's surface have shown no evidence of the latitudinal variation which would be expected were turnover time for the biosphere comparable to the mean life of radiocarbon.

CHAPTER THREE
¹⁴C TECHNIQUES & INSTRUMENTATION

3.001
ABSOLUTE DATING AND THE HISTORY OF MAN
Pecora, William T.; Rubin, Meyer
In: Time and Stratigraphy in the Evolution of Man. National Academy of Science. National Research Council. Publications, n.1469: 43-56
Evaluates a numbers of methods applicable to the dating of geologic events of the past few million years, particularly with reference to man or to hominid remains. This includes radiocarbon dating.

3.002   1964
ABSOLUTE DETERMINATION OF THE ACTIVITY OF TWO C-14 DATING STANDARDS
Karlén, Ingvar; Olsson, Ingrid U.; Kallberg, Per; Kilicci, Serap
Arkiv för Geofysik, v. 2, n. 22, 1964: 465-471
An absolute age determination of the activity of the two most common radiocarbon standards (oxalic acid from NBS and 'Wilhelm' from Heidelberg) has been performed with the same proportional counter and electronic equipment as was used in the determination of the half-life of radiocarbon. A series of measurements giving the ratio between the activities of the two standards has been made with another proportional counter.

3.003   1962
ACCURACY IN C-14 DATING
Ambrose, W.R.
New Zealand Archaeological Association Newsletter, v. 5, n. 1, March 1962: 19-20
Examines the various factors which can influence the validity of radiocarbon dates.

3.004   1963
THE ACCURACY OF RADIOCARBON DATES
Libby, Willard F.
Antiquity, v. 37, n. 147, Sept. 1963: 213-219
The author examines the apparent discrepancies in radiocarbon dating for their geophysical significance and for a general principle of correction. (Reproduced from *Science,* v. 140, n. 3564, 19 Apr. 1963: 278-280).

3.005   1953
ACTIVITY MEASUREMENT OF SAMPLES FOR RADIOCARBON DATING
Fergusson, Gordon J.
New Zealand Journal of Science and Technology, Section B, v. 35, n. 1, July 1953: 90-108
The equipment and procedure that have been developed for the measurement of the low radioactivity samples involved in radiocarbon dating are described. An internal-sample Geiger-Muller counter, 22 in long and 3.5 in in diameter fitted with an automatic sample changer, is used with automatic voltage control. The counting rates are continuously recorded by a method that allows a statistical analysis and continuous check of the operation.

3.006   1950
AGE DETERMINATION BY RADIOCARBON CONTENT
Movius, Hallam L.
Antiquity, v. 24, n. 94, June 1950: 99-100
The author reports on, examines and evaluates, the radiocarbon method developed by Professor W.F. Libby. He indicates the great potential for the unravelling of imprecise chronologies in the field of pre- and proto-history, Late-Glacial and Early Post-Glacial geology.

3.007   1957
THE AMATEUR SCIENTIST
Scientific American, v. 196, n. 2, Feb. 1957: 159-174
How a young amateur mastered the radiocarbon method of dating ancient organic material: a description of the scintillation counter, and the method of preparation used by Fred Schatzman, Highland Park, N. J. Some dates are compared with dates obtained by official radiocarbon dating laboratories.

3.008   1964
THE ANTHROPOGENIC FACTOR IN VEGETATIONAL HISTORY, I. TREGARON AND WHIXALL MOSSES
Turner, Judith
New Phytologist, v. 63, 1964: 73-90
Vertical series of pollen samples from raised bogs at Tregaron, Cardiganshire and Whixall Moss, Shropshire, have been analysed and several of the crucial horizons dated by radiocarbon measurements. This shows some of the vegetational changes which have occurred in the two regions and their approximate duration. These changes are probably the result of human intervention. The different ways and the extent to which man modified or determined the vegetation are discussed.

3.009 — 1955
APPARATUS FOR CARBON-14 DATING
Ballario, C.; Beneventano, M.; de Marco, A.; Cortesi, Cesare
Science, v. 121, n. 3143, 25 Mar. 1955: 409-412
The method is fundamentally the same as Libby's: measurements on elementary carbon in a screen-wall counter. There is a modification in the technique of preparing radiochemically pure carbon. A four elements counter (Geiger) has been developed.

3.010 — 1961
APPARATUS FOR METHANE SYNTHESIS FOR RADIOCARBON DATING
Fairhall, A.W.; Schell, William Raymond; Takashima, Y.
Review of Scientific Instruments, v. 32, n. 3, March 1961: 323-325
A simple apparatus is described whereby any quantity of carbon dioxide up to several moles can be converted to methane in one step by catalytic hydrogenation using a ruthenium catalyst. The conversion is very rapid (3 hours), the overall yield greater than 98% and the methane is of high purity.

3.011 — 1962
APPARATUS USED IN THE LABORATORY OF ABSOLUTE GEOCHRONOLOGY FOR DETERMINING THE AGE OF SAMPLES BY THE ^{14}C PROCEDURE
Moscicki, Wldzimierz; Zastawny, Andrezej
Nukleonica, v. 7, 1962: 801-817
A description of radiocarbon dating laboratory equipment is given, comprising a proportional counter for carbon dioxide measurements.

3.012 — 1965
THE APPLICABILITY AND RELIABILITY OF THE RADIOCARBON DATING METHOD
Jansen, Hans S.
Peking Symposium of Natural Science, 1st. Proceedings, 1965: 277-289
The principles of radiocarbon dating are described. The significance of the method as a tool to help study the past, soil geology and ocean and atmosphere research is demonstrated.

3.013 — 1954
APPLICATION OF THE CAPILLARY TUBE METHOD TO THE DETERMINATION OF RADIOCARBON
Reinharz, M.; Vanderhaeghe, G.
Nuovo Cimento, v. 12, n. 2, 1 Aug. 1954: 243-249
An application of the capillary tube method to the determination of radiocarbon (C-14) in liquids or dissolved substances is described. The absolute sensitivity of the method has been found to be of the order of 10-15 curie per cm tube length. The high sensitivity is useful when dealing with very small amounts of a feebly active substance.

3.014 — 1965
APPLIED GEOCHRONOLOGY
Hamilton, E.I.
London, Academic Press, 1965: 33-46
A chapter on the radiocarbon method which describes the theory, the method for dating, the various types of counters used, the distribution of radiocarbon in nature and the applications of this method of dating.

3.015 — 1963
THE APPLIED SCIENCE CENTER FOR ARCHAEOLOGY
Rainey, Froelich
American Journal of Archaelogy, v. 67, 1963, 294-295
The purpose of ASCA is described. The successful application of radiocarbon dating has made it clear that other scientific techniques for aid in archaeological research should be investigated and developed or adopted.

3.016 — 1953
ARCHAEOLOGICAL DATING BY CHEMICAL ANALYSIS OF BONE
Cook, S.F.; Heizer, Robert F.
Southwestern Journal of Anthropology, v. 9, n. 2, Summer 1953: 231-238
To investigate the validity of using a chemical system of dating on bone, the authors have presented analytical results on a large number of bones, comparing the ages so calculated to radiocarbon dates and to tree ring dates. The results are not very conclusive.

3.017 — 1955
ARCHAEOLOGICAL DATING METHODS
Porter, James Warren
Wisconsin Archaeologist, v. 36, n. 3, 1955: 69-10
Describes various methods used by archaeologists to date their material. Includes radiocarbon dating on p. 82-86.

3.018 — 1966
ARCHAEOLOGY AND ITS NEW TECHNOLOGY
Rainey, Froelich; Ralph, Elizabeth K.
Science, v. 153, n. 3743, 23 Sept. 1966: 1481-1491
Archaeology comes of age with an interdisciplinary approach in expanding its research horizons. This article emphasises the new techniques, one of which is the radiocarbon method for absolute dating of ancient material.

3.019   1962
AUSTRALIA'S FIRST CARBON-14 LABORATORY
Focken, Charles M.
Nature, v. 193, n. 4814, 3 Feb. 1962: 420-421
The laboratory, opened within the Institute of Applied Science of Victoria, on May 9th 1961, and is the 40th laboratory recognised on the international list. It is run by Miss A. Bermingham and uses a carbon dioxide gas proportional counter. It was tested on samples of known ages. The first archaeological results were for the Clarence Valley, northern New South Wales. A date of 3880 ± 120 years BP is the earliest evidence of Aboriginal occupation in this part of Australia.

3.020   1952
AN AUTOMATIC COUNTER FOR AGE DETERMINATION BY THE C14 METHOD
Crane, H.R.; McDaniel, E.W.
Science, v. 116, n. 3014, 30 Oct. 1952: 342-347
A discussion of the physics of low background and a complete description of the apparatus and technique used at the Harrison M. Randall Laboratory of Physics, University of Michigan.

3.021   1953
THE BACKGROUND AND ^{14}C DETECTION EFFICIENCY OF A LIQUID SCINTILLATION COUNTER
Audric, B.N.; Long, J.V.P.
Journal of Scientific Instruments, v. 30, n. 12, Dec. 1953: 467-469
Experiments with scintillation counters using p-terphenyl in benzene are described. Radiocarbon detection efficiency is approximately 40%.

3.022   1957
THE BACKGROUND OF COUNTERS AND RADIOCARBON DATING
Moljk, A.; Drever, R.W.P.
Royal Society, London. Proceedings, section A, v. 259, n. 1219, 9 April 1957: 433-445

In standard low activity equipment with a Geiger or proportional counter, the residual background is usually due to the detection of secondary particles produced in the wall of the counter by uncharged radiations. A new type of proportional counter, in which the wall effect is reduced, is described.

3.023   1966
THE BALANCED QUENCHING METHOD FOR COUNTING CARBON-14
Wright, E.; Castle, L.
International Journal of Applied Radiations and Isotopes, v. 17, n. 3, March 1966: 193-195
A discussion of the approach by H.H. Ross (*International Journal of Applied Radiations and Isotopes*, 15, 1964: 273) indicates that, at the balance point, the counting rate is not independent of the concentration of quenching agent until a critical concentration has been exceeded. This has been tested for many different quenchers.

3.024   1967
THE BALANCED QUENCHING METHOD FOR COUNTING CARBON-14
Ross, H.H.
International Journal of Applied Radiations and Isotopes, v. 18, n. 5, May 1967: 335-336
Response to Wright and Castle criticism (*International Journal of Applied Radiations and Isotopes*, 17, 1966: 193). Reviews briefly the stages required to select the optimum balanced quenching point.

3.025   1961
BENZENE SYNTHESIS AIDS C-14 DATING
Hood, Donald W.; Isbell, A.T.; Noakes, John E.; Stipp, J.J.
Chemical Engineering News, v. 39, n. 41, 9 oct. 1961: 48-49
A total synthesis which uses the carbon present in sea water is described. The method includes a routine performed on board ship to trap carbon dioxide in a potassium hydroxide solution, followed by a laboratory routine. The synthesis yield is about 50%

3.026   1961
BENZENE SYNTHESIS AT LOW TEMPERATURES FOR RADIOCARBON DATING
Noakes, John Edward; Hood, Donald W.; Isbell, W.S.
Oak Ridge, Tennessee, Oak Ridge Institute of Nuclear Studies, 1961
[Not sighted]

3.027   1963
BENZENE SYNTHESIS BY LOW TEMPERATURE CATALYSIS FOR RADIOCARBON DATING
Noakes, John Edward; Isbell, A.F.; Stipp, Jerry J.; Hood, Donald W.
Geochimica et Cosmochimica Acta, v. 27, n. 7, July 1963: 797-804
A method is described for the ambient temperature synthesis of pure benzene from acetylene in 50-60 percent yields that is suitable for low-level liquid scintillation counting. This extends the sensitivity of the carbon dating method while requiring only standard commercially available counting systems.

3.028 1967
BONE ORGANIC MATTER AND RADIOCARBON DATING
Haynes, C. Vance
In: Radioactive Dating and Methods of Low-level Counting. Proceedings of a Symposium organised by the International Atomic Energy Agency (IAEA) in co-operation with the Joint Commission on Applied Radioactivity (ICSU) and held in Monaco, 2-10 March 1967. Vienna, IAEA, 1967. (Proceedings Series): 163-168
Unsatisfactory results with the radiocarbon dating of bone organic matter have caused bone radiocarbon dates to be considered unreliable. Even bone collagen dates are erroneous for samples more than a few thousand years old. The research reported here is directed toward understanding why bone yields erroneous results, what is the nature of the contaminant, and what can be done to obtain accurate dates from bone organic matter. Results suggest that collagen free of humic acid is the only reliable fraction for dating fossil bones.

3.029 [1961]
BUBBLE CHAMBERS FOR ULTRA LOW-LEVEL COUNTING
Fergusson, Gordon J.; Johnston, William H.
Johnston Laboratories, Inc., Report. Baltimore, Maryland, [1961]
The characteristics of a bubble chamber are investigated with a view to use it as a detector of low-level activity. The results are encouraging. The advantages are that the sample is in liquid form and there is no theoretical limitations on size as the detection mechanism is continuous throughout the liquid.

3.030 1963
C-14 AND H-3 SPECIFIC ACTIVITIES BY BOMB COMBUSTION AND LIQUID SCINTILLATION COUNTING
McFarlane, A.S.; Murray, K.
Analytical Biochemistry, v. 6, n. 3, Sept. 1963: 284-287
Description of a steel oxygen bomb used to give quantitative combustion of small amounts of non-volatile solids.

3.031 1957
A C-14 DATING STATION USING THE CO_2 PROPORTIONAL COUNTING METHOD
Olsson, Ingrid U.
Arkiv för Fysic, v. 13, n. 3, 1957: 37-60
Reviews briefly the principles of formation and the occurrence of radiocarbon in nature and describes and compares the various methods of counting. The Uppsala station which uses a proportional counter, counting carbon dioxide is described.

3.032 1955
C-14 DATING WITH A METHANE PROPORTIONAL COUNTER
Burke, W.H.; Meinschein, W.G.
Review of Scientific Instruments, v. 26, n. 12, Dec. 1955: 1137-1140
Apparatus and procedures for conversion of sample carbon to carbon dioxide, quantitative hydrogenation of carbon dioxide to methane, and radiocarbon assay of the methane are described.

3.033 1968
CALIBRATION OF THE RADIOCARBON TIME SCALE
Walton, Alan; Baxter, M.S.
Nature, v. 220, n. 5166, 2 Nov. 1968: 475-476
The authors warn about the hasty application of 'correction factors' to radiocarbon ages, particularly in view of certain discrepancies which have been observed in existing data. The need to state clearly all standards used in calculation and wherever possible to report the $\partial^{14}C$ is emphasised.

3.034 1955
CARBON DIOXIDE AS A SUBSTITUTE FOR SOLID CARBON IN ^{14}C AGE MEASUREMENTS
Rafter, T.A.
New Zealand Journal of Science and Technology, section B, v. 36, n. 4, Jan. 1955: 363-370
Techniques developed for the preparation of carbon dioxide gas for use in radiocarbon age measurements are described. The gas can be produced free from electro-negative contaminants and all but traces of radon. Such gas is suitable as a filling for an eight litres proportional counter filled with carbon dioxide.

3.035 1947
CARBON DIOXIDE FILLED GEIGER-MULLER COUNTERS
Brown, Sarborn C.; Miller, Warren W.
Review of Scientific Instruments, v. 18, n. 7, July 1947: 496-499
Efficient detection of the long period carbon isotope, radiocarbon, can be achieved when this isotope is oxidised to carbon dioxide gas which may then be used in a Geiger-Muller counter filling. Problems are discussed.

3.036 1963
CARBON DIOXIDE PROPORTIONAL COUNTERS: EFFECTS OF GASEOUS IMPURITIES AND GAS PURIFICATION METHODS
Srdoc, D.; Sliepcevic, A.
International Journal of Applied Radiations and Isotopes,

v. 14, n. 10, Nov. 1963: 481-488
Systematic investigations have been made of the influence of the addition of small quantities of gaseous impurities in the operation of carbon dioxide filled proportional counters. A series of diagrams show the effect of contamination on counting properties of carbon dioxide filled proportional counters. A method of purification, using a vacuum technique, is presented.

3.037   1957
CARBON DIOXIDE PROPORTIONAL COUNTING FOR NATURAL RADIOCARBON MEASURE-MENTS
Östlund, Göte H.
Arkiv för Kemi, v. 12, n. 6, 1957: 69-78
A modified method for the preparation and purification of carbon dioxide for use in a proportional counting tube, and the apparatus for radiocarbon age determination are described. The stability and reproducibility of the apparatus is recorded, and statistical aspects on the statement of errors and the limits are given. $^{13}C/^{12}C$ ratios for groups of common materials for dating have been measured, and an objective test for the reliability of the whole radiocarbon dating method by measurements of sequoia tree rings is reported.

3.038   1968
CARBON ISOTOPE EFFECTS IN METHANE PRODUCTION BY THERMAL CRACKING
Sackett, W.M.; Nakaparksin, S.; Dalrymple. D
Advances in Organic Geochemistry, v. 3, 1968: 37-53
[Not sighted]

3.039   1954
CARBON-14 AGE RESEARCH
Broecker, Wallace S.; Kulp, J. Lawrence
Geological Society of America. Bulletin, v. 64, n. 12, Dec. 1954: 1934
Abstract of paper presented at the November Meeting of the Geological Society of America, in Los Angeles, Dec. 1954. The radiocarbon age method has been extended to 45,000 years at the Lamont Observatory by the introduction of large volumes of acetylene proportional counters with consequent increase in precision. This is applied to sea-level studies in the Mississippi delta region. Various other researches in China and Canada are also reported.

3.040   1965
CARBON-14 AND THE UNWARY ARCHAEOLOGIST
Stuckenrath, Robert
In: International Conference on Radiocarbon and Tritium Dating, Pullman, Washington, Washington State University, June 1-11, 1965. *Proceedings*. United States of America, Atomic Energy Commission, 1965. Conference n. 650652: 304-317
A review of the various misunderstandings by the archaeologists with regard to the use of radiocarbon dates and the problems caused by faulty collection and scanty description. Suggestions are made to overcome these problems.

3.041   1965
CARBON-14 DATING OF IRON, A NEW ARCHAEOLOGICAL TOOL
Van der Merwe, Nicholas J.
Current Anthropology, v. 6, n. 6, Oct. 1965: 475
Describes a method of radiocarbon dating of iron and assesses it for validity and accuracy.

3.04   1960
CARBON-14 DATING WITH LIQUID SCINTILLATION COUNTER: TOTAL SYNTHESIS OF THE BENZENE SOLVENT
Tamers, Murray A.
Science, v. 132, n. 3428, 2 Sept.1960: 668-669
Samples are analysed for natural radiocarbon content by total synthesis of benzene from their organic constituents. The instrument used permits 15 grams of carbon to be counted with an efficiency of 40% and a background of 13 cpm.

3.043   1953
CARBON-14 SAMPLE PREPARATION AND COUNTING TECHNIQUES, I: GAS COUNTING METHODS
Neville, O.K.
Atomics, v. 3, n. 12, Dec. 1953: 309-316
As a result of its low-energy beta rays, radiocarbon presents a number of problems for radioassay work. One of the two main methods that have been adopted for radiocarbon measurements, gas counting on the one hand and solid or liquid counting on the other hand, is reviewed.

3.044   1965
CARBON-14 VARIATIONS IN NATURE. PART I - TECHNIQUES OF ^{14}C PREPARATION, COUNTING, AND REPORTING RESULTS
Rafter, T.A.
New Zealand Journal of Science, v. 8, n. 4, Dec. 1965: 451-471 (INS Contribution No 196)
Describes the chemical procedures, the counting equipment, and the methods of reporting radiocarbon ages and radiocarbon enrichment as operating at the Carbon-14 Laboratory of the Institute of Nuclear Sciences, New Zealand. The system consists of an eight litres proportional counter, and carbon dioxide is used as the counting gas.

Ch. 3 - Techniques and Instrumentation

3.045   1958
CENTRAL AGENCY FOR CARBON-14 DETERMINATIONS
Crane, H.R.
Radiocarbon, v. 3, 1961: 46
Abstract of a mimeographed paper describing the CO_2 - CS_2 Geiger counter system in operation at the University of Michigan Radiocarbon Dating Laboratory. (Available from the author).

3.052   1959
CO_2 COUNTER TECHNIQUE FOR C14 MEASUREMENTS
Brownell, Gordon L.; Lockhart, Helen S.
Massachusetts Institute of Technology Laboratory. Nuclear Science and Engineering Technique. Report, n. 30, 1959
[Not sighted]

3.053   n.d.
CO_2 IONIZATION CHAMBER TECHNIQUES FOR C-14 MEASUREMENTS
Brownell, Gordon L.; Lockhart, Helen S.
Massachusetts Institute of Technology Laboratory. Nuclear and Engineering Technique Reports, n. 56. (NP - 3510)
The utilisation of carbon dioxide filled ionisation chambers as sensitive detection of radiocarbon is described. The primary advantage of these chambers is in the larger permissible size of the sample. The theory of the ionisation chamber is reviewed with respect to the ^{14}C - CO_2 combination.

3.054   1960
A CO_2 PROPORTIONAL COUNTER OF SMALL VOLUME AND HIGH EFFICIENCY FOR LOW LEVEL BETA COUNTING
Alessio, Marisa; Allegri, Lucia
Ricerca Scientifica, v. 30, n. 12, Dec. 1960: 1-3
Description of a proportional counter designed mainly for radiocarbon dating and used at the Radiocarbon Dating Laboratory of the University of Rome. Its fundamental characteristic is its small volume, and hence the possibility of dating when the material is scanty.

3.055   1965
COBALT MOLYBDATE CATALYST FOR AMBIENT TEMPERATURE ANALYSIS OF BENZENE FOR LIQUID SCINTILLATION COUNTING
Noakes, John Edward; Kim, Stephen M.; Akers, L.K.
United States Atomic Energy Commission. Report, ORINS - 50, 1965
A method for synthesising acetylene for liquid scintillation counting of radiocarbon was developed. Catalysis is accomplished using a cobalt molybdate alumina catalyst which requires no pretreatment other than dehydration. Isotope fractionation does not occur in this method.

3.056   1966
COLLECTION OF SPECIMENS FOR RADIOCARBON DATING AND INTERPRETATION OF RESULTS
Polach, Henry A.; Golson, J.
Institute of Aboriginal Studies. Manual, n. 2, 1966. pp. 42
A handbook for archaeologists. Covers the following areas: basis of the method; validity and sources of error; interpretation of radiocarbon ages; submission of samples and publication of results.

3.057   1966
COMBUSTION OF SAMPLES FOR LIQUID SCINTILLATION
Ragland, James B.
Nucleus, n. 20, Jan.1966: 1-11
The conventional combustion methods, wet, sealed-tube, oxygen-train, oxygen bomb, and oxygen flask are all discussed from the standpoint of use for combustion of biological samples containing ^{14}C, ^3H, ^{35}Sr, ^{32}P for determination by scintillation counting.

3.058   1966
COMPUTER CALCULATIONS OF C-14 DETERMINATIONS
Olsson, Ingrid U.
Uppsala University. Institute of Physics. Report, UUIP-477, May 1966
Lecture presented at the Radiocarbon and Tritium Dating Conference, June 7- 11, 1965, Pullman, Washington. Description of the computer system, including Fortran program and flow chart.

3.059   1965
COMPUTER CALCULATIONS OF C-14 DETERMINATIONS
Olsson, Ingrid U.
In: International Conference on Radiocarbon and Tritium Dating, Pullman, Washington, Washington State University, June 1-11, 1965. *Proceedings*. United States of America, Atomic Energy Commission, 1965. Conference n. 650652: 383-392
A description of the computer process for the calculation of radiocarbon dates. The flow chart and the Fortran program for the various stages of the calculations are presented.

3.060   1961
CONVERSION OF CARBON DIOXIDE TO BENZENE FOR CARBON DATING BY LIQUID SCINTILLATION COUNTING (Abstract)

Noakes, John Edward; Isbell, A.F.; Stipp, Jerry J.; Hood, Donald W.
Journal of Geophysical Research, v. 66, part 3, n. 7, July 1961: 2550
The conversion of carbon dioxide to benzene in over-all yields of better than 30% has been accomplished through the use of acetylene as an intermediate. The synthesis of benzene from acetylene is accomplished at atmospheric temperature by use of diborane impregnated aluminium oxide catalyst. Yields in excess of 50% are obtained. The benzene is used for liquid scintillation spectrometry.

3.061   1955
CORRECTION FOR THE EFFECT OF COSMIC RADIATIONS ON FIELD MEASUREMENTS OF THE RADIOACTIVITY OF SOILS
McCallum, G. John
New Zealand Journal of Science and Technology, section B, v. 37, n. 2, Sept. 1955: 172-178
The background countrate in the apparatus used for soil radioactivity measurements in the field is analysed. A formula for calculating the intensity of the cosmic ray background under varying depths of soil, soil densities, and barometric pressures, is developed. An estimate is made of the countrate to be expected from given concentrations of potassium, uranium and thorium.

3.062   1957
THE COSMIC RAY FLARE OF 23 FEBRUARY 1956 AND ITS EFFECTS ON THE NEW ZEALAND RADIOCARBON DATING EQUIPMENT
Fergusson, Gordon J.; McCallum, G. John
New Zealand Journal of Science and Technology, section B, v. 38, n. 6, May 1957: 577-587
The cosmic ray flare of 23 February 1956 was detected by the New Zealand radiocarbon dating equipment as an increase in the background counting rate. It was shown that ± 2.6% of the normal background rates is caused by the nucleonic component of the cosmic radiation. A neutron monitor has been installed to monitor the intensity of the nucleonic component.

3.063   1963
COUNTING C14O2 WITH A LIQUID SCINTILLATION COUNTER
Jeffay, Henry
In: *Advances in Tracer Methodology*, v. 1. Proceedings of the Fifth Annual Symposium in Tracer Methodology, held on Oct. 20, 1961, and selected papers of the first four Annual Symposia and from published issues of Atomlight, edited by Seymour Rothchild. New York, Plenum Press, 1963: 113-120
Presents a method in which radiocarbon is oxydised to $^{14}CO_2$ and the gas trapped in an etholamine solution of glycol monomethyl ether. An aliquot of the carbonate solution is added to toluene containing a scintillator, cooled to 0°C and counted in a liquid scintillation spectrometer.

3.064   1954
THE COUNTING OF LOW LEVEL ACTIVITY
Crathorn, A.R.
Atomics, v. 5, 1954: 99-104
Methods of reducing the background for the measurement of low specific activities are discussed. The particular case of radiocarbon when counted as gas in a proportional counter is examined in detail.

3.065   1963
THE COUNTING OF NATURALLY OCCURRING RADIOCARBON IN THE FORM OF BENZENE IN A LIQUID SCINTILLATION COUNTER
Leger, Concèle; Tamers, Murray A.
International Journal of Applied Radiations and Isotopes, v. 14, Feb. 1963: 65-70
A counting system is described for radiocarbon dating with a benzene scintillator solution. The single photomultiplier used is relatively simple and reliable. Different sizes and forms for the containers of the counting solution have been investigated. The calibration and operation of the counter are discussed. Where the laboratory personnel is more skilled in chemical manipulations than in electronic technique, it is suggested that standard gas counting be used.

3.066   1950
DATING BY THE C-14 METHOD
Cornwall, I.W.
The Archaeological News Letter, n. 2, n. 11, April 1950: 177-178
A brief summary of the method and its scientific background. Presents some results indicating that, where stratigraphic evidence was available, the date agreed with the relative chronology.

3.067   1968
DATING OF IRON
Van der Merwe, Nicholas J.; Stuiver, Minze
Current Anthropology, v. 9, n. 1, Feb. 1968: 48-53
Report on a project to develop a laboratory method for the radiocarbon dating of iron. It involves the extraction of carbon from iron alloys by direct combustion, followed by radiocarbon determinations. The applicability is determined by the type of alloy, the size of the iron specimen and the type of fuel used for smelting. The method was tested on samples of known age. A discussion of sample requirement and archaeological applications follows.

Ch. 3 - Techniques and Instrumentation

3.068   1967
[DATING OF IRON]
Van der Merwe, Nicholas J.
Current Anthropology, v. 4, n. 4, Oct. 1967: 375-376
The writer asks for suitable samples of known age to test the technique of dating iron by the radiocarbon method.

3.069   1965
THE DATING OF MORTARS BY THE CARBON-14 METHOD
Delibrias, Georgette; Labeyrie, Jean
In: International Conference on Radiocarbon and Tritium Dating, Pullman, Washington, Washington State University, June 1-11, 1965. *Proceedings*. United States of America, Atomic Energy Commission, 1965. Conference n. 650652: 344-347
The radiocarbon method applied to the dating of mortars, if used with caution, will be of some use to archaeologists.

3.070   1964
THE DATING OF OLD MORTARS BY THE CARBON-14 METHOD
Labeyrie, Jean; Delibrias, Georgette
Nature, v. 201, n. 4920, 15 Feb. 1964: 742
The radiocarbon dating method may be applied to the dating of old mortars in the same way as if these mortars were remains of organic substances. Because calcareous sands are sometimes used, a microscopic examination is needed.

3.071   1957
DATING OF RELICS BY RADIOCARBON ANALYSIS
Crane, H.R.
Nucleonics, v. 9, n. 6, 1957: 16-23
The radiocarbon method of dating relics is discussed and a specially designed Geiger - Müller tube is described, in which the carbon sample is coated within the cathode sleeve. Background counts are reduced to ca 24/day/cm^3 of sample. Techniques of purifying, spreading and coating are briefly described.

3.072   1963
DATING OF SKELETAL MATERIAL
Oakley, Kenneth P.
Science, v. 140, n. 3566, 3 May 1963: 488
A nitrogen test has become ancilliary to the radiocarbon dating of bones.

3.073   1965
(DATING THE EVENTS)
Hole, Frank; Heizer, Robert F.
In: An Introduction to Prehistoric Archaeology. New York, Holt, Rinehart and Winston, 1965: 145-174
Chap. 10 covers dating by physical chemical methods and includes radiocarbon dating and other radioactive methods. Other methods of dating are discussed in chapters 11 and 12.

3.074   1953
DATING WITH CARBON-14
Kulp, J. Lawrence
Journal of Chemical Education, v. 30, Sept. 1953: 432-435
A description of the principles, techniques, results, problems and the future possibilities of dating by the radiocarbon determination method.

3.075   1954
DATING WITH NATURAL RADIOACTIVE CARBON
Carr, D.R.; Kulp, J. Lawrence
New York Academy of Science. Transactions, series 2, v. 2, n. 4, Feb. 1954: 175-181
The principles and the technique of radiocarbon dating are presented. The types of samples and problems are examined. The application to deep sea sediments and to ocean waters are discussed, and finally the results obtained for shell dating are assessed.

3.076   1955
DATING: A SUMMARY OF METHODS AND SOME RECENT DEVELOPMENTS
Harding, J.R.
Polynesian Society. Journal, v. 64, n. 1, March 1955: 102-112
Various methods of dating applicable to archaeology are examined. The radiocarbon method of dating is described and assessed.

3.077   1965
DESIGNATION OF RADIOCARBON DATES
Johnson, Frederick
American Antiquity, v. 31, n. 2, part 1, Oct. 1965: 311
A discussion of the use of the year 1958 A.D. as a point of reference for reporting radiocarbon dates and the importance of providing the laboratory number of the sample dated and the range of error in order to avoid errors of interpretation.

3.078   1960
DETERMINATION OF LIQUID SCINTILLATION EFFICIENCY BY PULSE HEIGHT SHIFT
Baillie, L.A.
International Journal of Applied Radiations and Isotopes, v. 8, n. 1, May 1960: 1-7
A method has been developed by which liquid scintillation counting efficiency can be determined simultaneously with

the count. The two scalers of a liquid scintillation spectrometer are set to count simultaneously different parts of the spectrum. The ratio between the counts on the two scalers can be used as an accurate index of the counting efficiency of one of them. This saves time compared with the internal standard method and appears to be of equal efficiency.

3.079   1963
DETERMINATION OF LIQUID SCINTILLATION EFFICIENCY BY PULSE HEIGHT SHIFT
Baillie, L.A.
In: *Advances in Tracer Methodology*, v. 1. Proceedings of the Fifth Annual Symposium in Tracer Methodology, held on Oct. 20, 1961, and selected papers of the first four Annual Symposia and from published issues of Atomlight, edited by Seymour Rothchild. New York, Plenum Press, 1963: 86-92
[See entry n. 3.078]

3.080   1948
DETERMINATION OF RADIOACTIVE CARBON IN SOLID SAMPLES
Armstrong, W.D.; Schubert, Jack
Analytical Chemistry, v. 20, n. 3, March 1948: 270-271
The method of collecting a precipitate of barium carbonate is described. It yields reproducible results in the determination of the activity of both thick and thin precipitates.

3.081   1964
DETERMINATION OF RADIOCARBON AND TRITIUM IN BLOOD AND OTHER BIOLOGICAL MATERIALS
Tamers, Murray A.; Diez, M.
International Journal of Applied Radiations and Isotopes, v. 15, 1964: 697-702
A chemical conversion for the treatment of organic samples that are not amenable to direct measurement in the liquid scintillation counter is presented.

3.082   1955
DETERMINATION OF SOURCES OF PARTICULATE ATMOSPHERIC CARBON
Clayton, G.D.; Arnold, James R.; Patty, F.A.
Science, v. 122, n. 3173, 21 Oct.1955: 751-753
In this study of the origin of the carbon constituent of atmospheric particulate matter, radiocarbon is used to distinguish fossil carbon from carbon which comes from biological sources. Carbon is found to be a minor constituent of the total particulate load and contemporary carbon constitutes less than 20% of the total carbon. Application of the method for use as tracer experiment is discussed. The sampling instrument and measurement procedures are also discussed.

3.083   1962
DETERMINATION OF THE HALF-LIFE OF C-14 WITH A PROPORTIONAL COUNTER
Olsson, Ingrid U.; Karlén, Ingvar; Turnbull, S.H.; Prosser, N.J.D.
Arkiv för Fysic, v. 22, n. 4, 1962: 237-255
The method to determine the half-life of radiocarbon with high accuracy is described. The determination resulted in the value of 5680 ± 40 years, where the indicated uncertainty denotes an estimated probable error of the results.

3.084   1951
DETERMINATION OF TOTAL CARBON AND ITS RADIOACTIVITY
Van Slyke, Donald D.; Steele, Robert; Plazin, John
Journal of Biological Chemistry, v. 129, n. , Oct. 1951: 769-805
Methods for the determination of total carbon and carbon radioactivity in a sample are described.

3.085   1961
DETERMINATION OF TRITIUM AND CARBON-14 IN BIOLOGICAL SAMPLES BY RAPID COMBUSTION TECHNIQUES
Buyske, D.A.; Kelley, R.; Florini, J.; Gordu, S.
Atomlight, n. 20, Dec. 1961: 1-6
Methods for the preparation of any biological sample so that the tritium or radiocarbon content can be measured by liquid scintillation technique are described. Samples up to 300 mg are combusted in a Schoniger oxygen flask. Heavier samples or bone samples are combusted in a simple furnace of special design. Illustrations included.

3.086   1958
DIFFICULTIES IN THE APPLICATION OF C-14 RESULTS IN ARCHAEOLOGY
Tauber, Henrik
Archaeologia Austriaca, v. 29, 1958: 59-69
Some problems in the interpretation of radiocarbon dates applied to archaeology are discussed.

3.087   1964
THE DIRECT CONVERSION OF WOOD CHARCOAL TO LITHIUM CARBIDE IN THE PRODUCTION OF ACETYLENE FOR RADIOCARBON DATING
Swart, E.R.
Experientia, v. 120, 1964: 47-48
Description of a new method to convert charcoal, in a steel furnace at a temperature of 800°C, directly with lithium carbide to produce acetylene for radiocarbon determination. The method offers a notable economy of time in the preparation of the samples, and the conversion is found to be better than 95% thus eliminating any possibility of fractionation.

Ch. 3 - Techniques and Instrumentation

3.088   1968
DIRECT MEASUREMENT OF $^{14}CO_2$ IN A LIQUID SCINTILLATION COUNTER
Horrocks, D.L.
International Journal of Applied Radiations and Isotopes, v. 19, n. 12, Dec. 1968: 859-864
Samples of carbon dioxide were dissolved in a liquid scintillation solution for the measurement of radiocarbon activity with counting efficiency greater than 90% above an appropriate low-energy bias. As much as 50 mL (STP) of carbon dioxide per mL of scintillator solution produced no measurable quenching.

3.089   1958
A DIRECT-PLATING METHOD FOR THE PRECISE ASSAY OF CARBON-14 IN SMALL LIQUID SAMPLES
McCready, C.C.
Nature, v. 181, n. 4620, 17 May 1958: 1406
Incorporation of liquid samples of compounds containing radiocarbon in a layer of gel produces counting samples which will dry to an even film, thus ensuing reproducible self-absorption and counter geometry. The procedure is described in detail.

3.090   1960
EDITORIAL
Daniel, Glyn
Antiquity, v. 34, n. 133, March 1960: 4-5
Some comments on radiocarbon dates and their acceptability.

3.091   1960
EDITORIAL
Daniel, Glyn
Antiquity, v. 34, n. 135, Sept. 1960: 161-162
Comments on the delay in the reporting of radiocarbon dates and on the validity of the dates.

3.092   1949
EDITORIAL NOTES
Johnson, Frederick
American Antiquity, v. 53, 1949: 286
Note of the report by the Committee on Radioactive Carbon-14 on the testing of the method on Egyptian specimens.

3.093   1950
EDITORIAL NOTES
Johnson, Frederick
American Antiquity, v. 54, 19450: 236
Report on the continuing development of the method of dating remains by radiocarbon. One of the problems is the question of validity, and suitability of the samples submitted.

3.094   1949
EDITORIAL NOTES: DATING ARCHAEOLOGICAL SPECIMENS BY MEANS OF THEIR RADIOCARBON CONTENT
Crawford, O.G.S.
Antiquity, v. 23, n. 91, Sept. 1949: 113-114
The note records the appearance of the new method of dating dead pieces of formerly living substances by means of their radiocarbon content.

3.095   1964
EDITORIAL STATEMENT
Deevey, Edward S.; Flint, Richard Foster; Rouse, Irving
Radiocarbon, v. 6, 1964
The following decisions were agreed on at the Fifth Radiocarbon Dating Conference, Cambridge 1962: (1) all radiocarbon dates published in U.S. are based on the Libby value of 5570 ± 30 years for the half-life of radiocarbon, (2) the mean of three new determinations of the half-life, 5730 ± 40 years is regarded as the best value now obtainable, (3) conversion of published dates to this basis is accomplished by multiplying by 1.03, (4) standard year of reference is fixed at 1950.

3.096   1968
ELECTRONIC DATA PROCESSING FOR RADIOCARBON DATING
Taylor, R.E.; Berger, Rainer; Dinsdale, B.
American Antiquity, v. 33, n. 2, April 1968: 180-184
The need for a more flexible index for radiocarbon dates with shorter retrieval times has led to a suggestion for the development of an electronic or punch-card data retrieval system. A suggested format for coding radiocarbon data on punch cards and two techniques for the rapid retrieval of desired information are discussed.

3.097   1962
ELECTRONIC INSTRUMENTATION FOR RADIOCARBON DATING
Bell, J.; Neuhaus, John William George; Green, J.H.
Institution of Radio and Electronic Engineers (Australia). Proceedings, 1962: 718-721
The sciences of geology, archaeology, ethnology, soils and climatology require accurate datum points on the time-scale. Of the methods available for this purpose, radiocarbon is the most suitable for the period ranging up to 40,000 years ago. In systems using proportional counters and carbon dioxide gas, it is necessary to use high-gain non-overloading linear amplifiers, electronic pulse-stretching, delaying, sorting, blocking, discriminating channel recording. The instrumental features of the unit at the University of New South Wales are described.

3.098 1950
END AND WALL CORRECTIONS FOR ABSOLUTE COUNTING IN GAS COUNTERS
Engelkemeir, Antoinette G.; Libby, Willard F.
Review of Scientific Instruments, v. 21, n. 6, June 1950: 550-557
For flat end brass wall Geiger counters containing active gas samples the end loss presumed due to inhomogeneity of the fields has been measured for ^{37}Ar and ^{85}Kr. The loss was found to be approximately the same for them as for radiocarbon. It shows the loss to be inversely proportional to diameter.

3.099 n.d.
EQUIPMENT FOR THE MEASUREMENT OF RADIOCARBON (C-14) AND TRITIUM (H-3) IN THE GASEOUS PHASE
Philips Industrial Equipment Division, n. d.
Description of the instrumentation.

3.100 1950
ERROR OF COMBUSTION OF COMPOUNDS FOR C14 ANALYSIS
Armstrong, W.D.; Singer, Leon; Zbarski, S.H.; Dunshee, Bryant
Science, v. 112, n. 2914, 3 Nov. 1950: 531-533
The precipitation of xanthydrol ureide affords a simple method for the isolation of urea from urine and has been used for this purpose in tracer studies with radiocarbon. Because of the probable different rate of oxidation of carbons in an organic compound, care should be taken to achieve complete oxidation and mixing of the resulting carbon dioxide before removal of samples for carbon analysis.

3.101 1964
ESTIMATION OF $^{14}CO_2$ BY LIQUID SCINTILLATION
Pande, G.S.
Indian Journal of Chemistry, v. 2, July 1964: 287-289
A simple and easy to manipulate glass vacuum system for collecting and estimating the radioactivity of small amounts of $^{14}CO_2$ by a liquid scintillation method was developed. Results obtained with this system are quite reproducible. However the efficiency of counting decreases with the increasing amounts of hyamine used.

3.102 1965
ESTIMATION OF RADIOCARBON BY GAS-PHASE COUNTING
Soman, S.D.; Iyengar, T.S.; Sadarangani, S.H.; Vaze, P.K.
Indian Journal of Pure and Applied Physics, v. 3, May 1965: 170-172
The characteristics of a gas-phase counting system for the estimation of radiocarbon are described. It uses carbon dioxide and a commercial gas fuel as the counting and quenching gases. The sample to be assayed is converted to carbonate before its radiocarbon content is estimated by the system.

3.103 1965
ETHNOLOGICAL DATING WITH RADIOACTIVE CARBON
Neuhaus, John William George
M.Sc. Dissertation. Sydney, University of New South School of Chemistry, 1965. (Manuscript available from the library, New South Wales University)
Substituting methane for carbon dioxide and altering the anticoincidence shield is the technique briefly reviewed. The following aspects are discussed: (1) the design and development of dating apparatus at the University of New South Wales; (2) statistical accuracy in view of the uncertainty of prehistorical samples; (3) preparation of oxalic acid dating standard; (4) the de Vries effect.

3.104 1955
EVALUATION OF THE ACCURACY OF NEW ZEALAND RADIOCARBON DATING RESULTS
McCallum, John
New Zealand Journal of Science and Technology, section B, v. 37, n. 3, Nov. 1955: 370-381
Operation of the New Zealand radiocarbon dating equipment is described. The method of evaluating results and sources of possible error in reporting dates is given. The inherent difficulties of the measurements are discussed and future improvements to the apparatus are described.

3.105 1947
EXCHANGE OF CARBON DIOXIDE BETWEEN BARIUM CARBONATE AND THE ATMOSPHERE
Armstrong, W.D.; Shubert, Jack
Science, v. 106, n. 2756, 24 Oct. 1947: 403-404
It was observed that precipitate of barium carbonate used as standard for radiocarbon measurements appeared to lose a significant amount of radioactivity when stored for several weeks in contact with air.

3.106 1967
EXPERIENCE GATHERED IN THE CONSTRUCTION OF LOW-LEVEL COUNTERS
Geyh, Mebus A.
In: Radioactive Dating and Methods of Low-level Counting. Proceedings of a Symposium organised by the International Atomic Energy Agency (IAEA) in co-operation with the Joint Commission on Applied Radioactivity (ICSU) and held in Monaco, 2-10 March 1967. Vienna, IAEA, 1967. (Proceeding Series): 575-591

Ch. 3 - Techniques and Instrumentation

Three different types of counters have proved successful in the low level technique, namely the quartz tube, the Oeschger type counter, and the plastic scintillation counter. Difficulties arise if high factors of merit and excellent plateau properties are desired. Detailed informations on this subject are given and both the advantages and disadvantages on the individual constructions are compared.

3.107 1952
EXTENSION OF THE CARBON-14 AGE METHOD
Kulp, J. Lawrence; Tryon, Lansing E.
Review of Scientific Instruments, v. 23, 1962
The use of a 1 inch mercury shield around the screen wall counter inside the anticoincidence ring of cosmic-ray counters makes it possible to reduce the background from 5 cpm to 2 cpm for an active counting volume 8 inch in length and 2 inch in diameter. This makes it possible to extend routine radiocarbon age measurements from about 25,000 to 30,000 years.

3.108 1953
FAST COINCIDENCE CIRCUITS FOR H3 AND C14 MEASUREMENTS
Hiebert, R.D.; Watts, R.J.

Nucleonics, v. 11, n. 12, Dec. 1953: 38-41
Description of an equipment which proved capable of counting tritium with an efficiency of 10-15% and radiocarbon with an efficiency of 60-70%.

3.109 1958
FINAL REPORT ON RADIOCARBON DATING TO THE OFFICE OF NAVAL RESEARCH
Deevey, Edward S.
Yale University. Geochronometric Laboratory, 1958. pp. 17
[Not sighted]

3.110 1965
A FORTRAN III COMPUTER PROGRAM FOR THE PROCESSING OF RADIOCARBON DATA
Aldous, K.J.
Institute of Nuclear Sciences. Department of Scientific and Industrial Research. (Lower Hutt, New Zealand) *Lab. Notes*, n. 25, 1965 (I.N.S. Contribution No. 191) pp.34
A computer program, written in Fortran III for the IBM650 computer at D.S.I.R. which facilitates the processing and reporting of results. It will handle all of the various cases which arise in the cause of processing radiocarbon data.

3.111 1957
FURTHER ANALYSIS OF THE NEUTRON COMPONENT OF THE BACKGROUND OF COUNTERS USED FOR C-14 AGE MEASUREMENTS
de Vries, Hessel

Nuclear Physics, v. 3, 1957: 65-68
The background of counters, used for radiocarbon age measurements, has been reduced appreciably by a special arrangement of blocks of paraffin wax (mixed with boracic acid) inside the iron shield. The neutron background has been reduced by a factor of about 7. The relative increase of the neutron component in the radiocarbon counter during the solar flare of February 23rd (1956) has been the same as in the neutron monitor.

3.112 1968
FURTHER INVESTIGATIONS OF STORING AND TREATMENT OF FORAMINIFERA AND MOLLUSCS FOR C-14 DATING
Olsson, Ingrid U.; Göksu, Yeter; Stenberg, Allan
Geologiska Foreningens i Stockholm. Forhandlingar, v. 90, 1968: 417-426
Laboratory experiments concerned with the contamination during storing of crushed and uncrushed samples for radiocarbon dating as well as decontamination experiments with deliberately contaminated shell samples suggest that the contamination is dependent on the grain size and storing conditions. Recommendations are made for reducing or detecting contamination.

3.113 1968
FURTHER REFINEMENTS OF THE RADIOCARBON METHOD
Zavel'skiy, F.S.
Akademiia Nauk SSSR. Doklady. Earth Science Section, v. 180, 1968: 82-85
A discussion of the radiocarbon dating method and of the various factors affecting the precision and reliability of the dates.

3.114 1954
GAS COUNTING OF NATURAL RADIOCARBON
Crathorn, A.R.; Loosemore, W.R.
In: Physical Science and Industrial Applications, edited by J.E. Johnston. Radioisotope Conference, 2nd, Oxford, 1954. *Proceedings*, vol.II. London, Butterworth, 1954: 123-133
Description of the gas counting technique using acetylene as the counting gas. Gas preparation and counter design are included.

3.115 1960
GAS-PROPORTIONAL COUNTING OF CARBON-14 AND TRITIUM AND THE DRY COMBUSTION OF ORGANIC COMPOUNDS
Christman, David R.; Paul, Catherine M.
Analytical Chemistry, v. 32, Jan. 1960: 131-132
An anticoincidence counter, operating with a ring of propor-

tional counters and with two proportional sample counting channels was used for radiocarbon and tritium gas proportional counting.

3.116   1955
GEIGER COUNTING ON CARBON DIOXIDE
Broda, E.
Journal of Inorganic and Nuclear Chemistry, v. 1, Dec. 1955: 411-412
A Geiger counter, employing an external quenching circuit (Neher - Pickering), has been adapted for radiocarbon determinations. The sensivity of the method is comparable with other routine assay methods.

3.117   1948
A GEIGER-MULLER COUNTING UNIT AND EXTERNAL QUENCHING EQUIPMENT FOR THE ESTIMATION OF C-14 IN CARBON DIOXIDE
Mann, W.B.; Parkinson, G.B.
Review of Scientific Instruments, v. 20, n. 1, Jan. 1948: 41-47
A counter unit and external quenching circuit for use with carbon dioxide and carbon disulphide are described. Using a compensated Geiger-Muller counter and a multi-vibrator quenching circuit have made it possible to obtain highly reproducible measurements. Both are described.

3.118   1958
GEOLOGIC DATING IN PREHISTORY
Wright, Herbert E.
Archaeology, v. 11, n. 1, March 1958: 19-25
Although radiocarbon dating has enabled a reasonable accuracy for the last 40,000 years with possibility to extend to 75,000 years, it has not solved all problems of dating, specially for Palaeolithic sites which yield very little material for dating. Proposes and discusses a geologic method of dating.

3.119   1968
THE GEOLOGICAL SURVEY OF CANADA RADIOCARBON DATING LABORATORY
Dyck, Willy
Ottawa. Geological Survey of Canada (Department of Energy, Mines and Resources). Papers, n. 66-45, 1968. pp. 45
The apparatus, sample preparation system, sample pretreatment and counting, and results from the Laboratory of the Geological Survey of Canada are described and evaluated.

3.120   1966
GROUND WATER AGES AND FLOW RATES BY THE CARBON 14 METHOD
Pearson, Frederick Joseph
Ph.D. Dissertation. University of Texas, 1966. pp. 105 (Abstract in: *Dissertation Abstracts*, Ann Arbor, Mich., Sec. B, v. 27, n. 8, 1967: 2749B. Order No. 66-114,424)
Description of a method for direct determination of ground water ages and flow rates based on the radiocarbon content of carbonate dissolved in the water. The various origins of the carbonate and their influence on radiocarbon content are discussed.

3.121   1962
GULBENKIAN RADIOCARBON DATING LABORATORY: NOTES FOR THE GUIDANCE OF PERSONS WISHING TO SUBMIT SAMPLES FOR RADIOCARBON AGE MEASUREMENTS
Robins, P.A.; Swart, E.R.
University College of Rhodesia and Nyassaland, 1962
Laboratory manual.

3.122   1963
THE HALF-LIFE OF C14 AND THE PROBLEMS WHICH ARE ENCOUNTERED IN ABSOLUTE MEASUREMENTS IN BETA -DECAYING GASES
Olsson, Ingrid U.; Karlén, Ingvar
In: Radioactive Dating. Proceedings of a Symposium on Radioactive Dating held in Athens, 19-23 November 1962. Vienna, International Atomic Energy Agency, 1963. (Proceedings Series): 3-11
New determinations of the half-life of radiocarbon have been performed at the National Bureau of Standards, Washington D.C., U.S.A., at the Atomic Weapons Research Establishment, Aldermanston, England and at the Institute of Physics, University of Uppsala, Sweden. The results are given as 5760 ± 50, 5780 ± 65 and 5685 ± 35 years respectively. The method used at the Uppsala Laboratory is described.

3.123   1964
THE HALF-LIFE OF CARBON-14; COMMENTS ON THE MASS-SPECTROMETRIC METHOD
Hughes, E.E.; Mann, W.B.
International Journal of Applied Radiations and Isotopes, v. 15, 1964: 97-100
An assessment has been made of the mass spectrometer results in the recent NBS determination of the half-life of radiocarbon. Evidence is presented to show that the uncertainties due to effusive separation and the dependence of the sensitivity constant on the isotopic abundance of radiocarbon are small. It is proposed to reduce the value of the half-life of radiocarbon to 5745 ± 50 years.

3.124   1947
HIGH EFFICIENCY COUNTING OF LONG LIVED CARBON AS CO_2
Miller, Warren W.

Ch. 3 - Techniques and Instrumentation

Science, v. 105, n. 2718, 31 Jan. 1947: 123-124
Adoption of the Geiger-Muller counter to the measurement of long-lived radiocarbon.

3.125   1964
HIGH SENSITIVITY COUNTING TECHNIQUES
Watt, D.E.; Ramsden, D.
Oxford, Pergamon Press, 1964
A summary of developments in the method of radiocarbon determination is given. A number of important and interessant dates are presented in the following areas: geology, palaeontology, prehistory, late Quaternary terraces, radiocarbon chronology of the Wisconsin glacial, the Würm (Weichsel) glacial and a combined radiocarbon and ^{18}O chronology of the Pleistocene.

3.126   1961
HIGH SENSITIVITY DETECTION OF NATURALLY OCCURRING RADIOCARBON. I - CHEMISTRY OF THE COUNTING SAMPLE
Tamers, Murray A.; Stipp, Jerry J.; Collier, J.
Geochimica et Coschimica Acta, v. 24, 1961: 266-276
Presents a means by which the entire counting solution can be synthesised from the sample to be dated. The solvent used is benzene which is one of the most suitable material for liquid scintillation counting since it contains 92% carbon and shows no scintillation quenching properties. The chemistry involved in the complete synthesis of benzene is described in detail.

3.127   1963
HIGH-VOLUME SAMPLER FOR ATMOSPHERIC CARBON DIOXIDE
Fergusson, Gordon J.
Review of Scientific Instruments, v. 34, n. 4, April 1963: 403-406
A new sampler successfully used in jet aircrafts for collection of atmospheric carbon dioxide at altitudes up to 50,000 feet is described. Ten litres of carbon dioxide which is sufficient for accurate radiocarbon measurements can be obtained in one hour.

3.128   1952
HOW TO GET SAMPLES DATED
Nucleonics, v. 10, n. 8, August 1952: 21
Determination of the age of samples by the radiocarbon method has proved to be quite useful. The requirements as to sample preparation of four laboratories equipped to perform dating services are listed.

3.129   1965
THE IMPACT OF RADIOCARBON DATING
Johnson, Frederick
In: International Conference on Radiocarbon and Tritium Dating, Pullman, Washington, Washington State University, June 1-11, 1965. *Proceedings*. United States of America, Atomic Energy Commission, 1965. Conference n. 650652: 762-784
Traces the history of the foundation and emergence of the radiocarbon chronology which has had such profound effect upon archaeology.

3.130   1955
IMPORTANCE OF SOLVENT IN LIQUID SCINTILLATION COUNTING
Hayes, F. Newton; Rogers, Betty S.; Sanders, Phyllis
Nucleonics, v. 12, n. 1, January 1955: 46-48
The significance of the solvent was shown by measuring the following: photon near-free-path in 44 solvents, relative pulse height of 2,5 - diphenyloxazol solutions, effect of solute concentration on pulse height peak, solvent concentration curves in toluene.

3.131   1966
IMPROVED SYNTHESIS TECHNIQUES FOR METHANE AND BENZENE RADIOCARBON DATING
Polach, Henry A.; Stipp, Jerry J.
International Journal of Applied Radiations and Isotopes, v. 18, 1967: 359-364
A chemical preparation system based on a dual purpose reaction vessel is shown in detail.

3.132   1964
IMPROVED TECHNIQUE FOR DETERMINATION OF C14 AND H3 BY FLASK COMBUSTION
Baden, Howard P.
Analytical Chemistry, v. 34, April 1964: 960
Description is given of a new inexpensive combustion and collection system that simplifies the counting of evolved $^{14}CO_2$ and $^{3}H_2O$.

3.133   1967
AN IN SITU GAS EXTRACTION SYSTEM FOR RADIOCARBON DATING GLACIER ICE
Oeschger, Hans; Langway, Chester C.; Alder, B.
U.S. Army Material Command. Cold Regions Research and Engineering Laboratory. Research Report, v. 236, 1967. pp.
In March 1966 at the Tuto ice tunnel, Greenland, a team from USA CRREL and the University of Bern tested a new down-borehole device which would allow gas to be extracted from within shallow or deep boreholes. The tunnel ice was unfractured and its temperature was constant at 10°C. A location where, in 1964, radiocarbon dates had

been obtained was used as a check point for the down borehole tests. Comparative samples show good agreement and indicate a main value of 5120 years BP for the age of ice at this location. The simplicity of the down-borehole gas extraction system enables application of the carbon dating method to any natural, undisturbed glacier ice mass which can be sampled by boring. The gas extraction apparatus and field experiments are described.

3.134   1967
AN 'IN SITU' GAS-EXTRACTION SYSTEM TO RADIOCARBON DATE GLACIER ICE
Oeschger, Hans; Alder, B.; Lanfway, C.C.
Journal of Glaciology, v. 6, n. 48, Oct. 1967: 939-942
A new bore-hole instrument to extract atmospheric gases entrapped in glacier-ice was designed, developed and tested in a Greenland ice tunnel. Radiocarbon measurements made on the carbon dioxide thus extracted agree with results obtained from carbon dioxide collected from the same tunnel using a vacuum vessel melting technique.

3.135   1958
INCREASED ACTIVITY OF SILICA -ALUMINA CATALYST
Weiss, H.G.; Shapiro, I.
American Chemical Society. Journal, v. 80, n. 3, 8 July 1958: 3195-3198
Modification of the active sites on silica alumina with diborane results in a marked increase in the activity of the catalyst toward cyclisation of acetylene to benzene. Isotope studies show the reaction to be rapid, with no intermediates remaining absorbed on the surface of the catalyst. The results of several reactions on silica gel alumina and silica alumina tend to support the concept of alumina as the Lewis acid site in silica alumina catalyst.

3.136   1958
THE INFLUENCE OF GEOLOGICAL HISTORY AND SAMPLE PREPARATION ON RADIOCARBON DATING, WITH COMMENTS ON THE METHOD ITSELF
Vaugham, David Evan
Ph.D. Dissertation. Institute of Archaeology, London, Sept. 1958. pp. 142 (Available from Institute of Archaeology Library, University of London)
Carefully selected samples have been dated and the reliability of the dates tested in terms of environmental information. It was found that the geological history of the samples, particularly the soils in which they were buried interferred appreciably with the relativity of the dates. Acidity, alkalinity, temperature of the soils and the nature of the elutriation are the causes of interference. Sample preparation with alkali and acids produce comparable results. A survey of various methods of radiocarbon dating indicates that proportional counting on acetylene provides a very good radiocarbon dating method.

3.137   1965
INTERDISCIPLINARY APPRAISAL OF RADIOCARBON DATES IN ARCHAEOLOGY
Davis, E. Mott
In: International Conference on Radiocarbon and Tritium Dating, Pullman, Washington, Washington State University, June 1-11, 1965. *Proceedings*. United States of America, Atomic Energy Commission, 1965. Conference n. 650652: 294-303
In applying the radiocarbon technique to a chronological problem, the archaeologist and the radiochemist must treat the work as a joint piece of research. Dates are derived through the appraisal, jointly by those two researchers, of a series of assays on selected samples.

3.138   1968
AN INTENAL GAS PROPORTIONAL COUNTER FOR MEASURING COSMIC-RAY -PRODUCED RADIONUCLIDES (Abstract)
Schell, William R.
American Chemical Society Meeting, 155th., San Francisco, 1968. *Proceedings*. Division of Nuclear Chemistry, entry n. 60, 1968
Describes a membrane proportional counter which permits both sides of the sealed membrane to be filled simultaneously. The limit of counting sensibility for radiocarbon dating is equivalent to a sample age greater than 60,000 years.

3139   1959
ISOLATION OF GELATIN FROM ANCIENT BONES
Sinex, F. Marrott; Faris, Barbara
Science, v. 129, n. 3354, 10 Apr. 1959: 969
The use of geatin from ancient bones for radiocarbon dating may improve the accuracy of the dating procedure because gelatin is not likely to be contaminated by extraneous carbon. The isolation and characteristics of gelatin from a 12,000 years old antler is described.

3.140   1955
ISOLATION OF ORGANIC CARBON FROM BONES
May, Irving
Science, v. 121, n. 3144, 1 April 1955: 508-509
A method for the isolation of organic carbon from bone samples in which secondary carbonates were present as white incrustations and impregnations of exposed porous material is described.

3.141   1965
ISOTOPE EFFECTS IN THE BENZENE SYNTHESIS FOR RADIOCARBON DATING
Tamers, Murray A.; Pearson, Frederick Joseph
Nature, v. 205, n. 4977, 20 March 1965: 1205-1207
The authors conclude that in the benzene liquid scintillation method, variations in the measured activities due to causes other than the random nature of the disintegration process amount to less than 1.0% and probably less than 1.5%. This is insignificant for all purposes of radiocarbon dating.

3.142   1949
ISOTOPIC CARBON: TECHNIQUES IN ITS MEASUREMENT AND CHEMICAL MANIPULATION
Calvin, Melvin; Heidelberger, Charles; Reid, James C.; Tolbert, Bert M.; Yankwich, Peter E.
New York, John Wiley, 1949. pp. xiii, 376
This work, designed as a sort of laboratory manual for use in any group engaged in work involving carbon isotopes, gives detailed descriptions of every syntheses with isotopic carbon reported up to April 1948.

3.143   1966
LABORATORY PRODUCTION OF DIBORANE AND ACTIVATION OF SILICA-ALUMINA CATALYST FOR CONVERSION OF $^{14}C_2H_2$ TO $^{14}C_6H_6$: RADIOCARBON DATING BY LIQUID SCINTILLATION SPECTROMETRY
McDowell, L.L.; Ryan, M.E.
International Journal of Applied Radiation and Isotopes, v. 17, 1966: 175-183
A cntinuous process method is described for the routine laboratory production of diborane and the activation of the silica alumina catalyst employed in the synthesis are described. The diborane activated silica alumina is used in the conversion of acetylene to benzene for use in counting natural radiocarbon by liquid scintillation method.

3.144   1953
LARGE THIN-WALL GEIGER COUNTERS
Sugihara, Thomas T.; Wolfgang, Richard L.; Libby, W. F.
Review of Scientific Instruments, v. 24, n. 7, July 1953: 511-512
A new thin-walled Geiger counter of large area sensitive to soft beta radiation has been developed. The sensitive areas can be as large as desired and the wall thin enough for radiocarbon. The cylindrical geometry renders the counters specially suitable for accurate assays of large solid samples of low specific activity and low energy.

3.145   1955
LIQUID SCINTILLATION - II. RELATIVE PULSE HEIGHT COMPARISON OF SECONDARY SOLUTES
Hayes, F. Newton; Ott, Donald G.; Kerr, Vernon N.
Nucleonics, v. 14, n. 1, January 1956: 42-45
Double solute liquid scintillators can give greater pulse height and less absorption than single solute systems. The best secondary solutes are POPOP and BBO. Confirming data for 24 compounds and large detector results for some are presented. This study can also serve as a useful guide for the choice of efficient scintillators.

3.146   1968
LIQUID SCINTILLATION ANALYSES - COMPUTER PROCESSING
Osburn, J.O.
International Journal of Applied Radiations and Isotopes, v. 19, n. 11, Nov. 1968: 821-822
A Fortran IV program tailored for two channel use in an instrument using automatic external standardisation, capable of reporting raw dates as counts per minute, minus background, is presented.

3.147   1961
A LIQUID SCINTILLATION COINCIDENCE COUNTER FOR RADIOCARBON
Nygaard, K.J.
Applied Science Research, v. 6, 1961: 89-92
Measurements are made of the radiocarbon counting performance of a liquid scintillation coincidence counter used in conjunction with an anticoincidence shield counter. When operated at room temperature, the system has a radiocarbon dating efficiency of 59% at a background of 16 cpm. The activity of ethanol made from contemporary wood is determined to 13.1 ± 0.6 disintegration per minute per gram of carbon.

3.148   1954
A LIQUID SCINTILLATION COUNTER FOR CARBON-14
Dietrich, Jacob E.; Kennedy, William R.
Atomic Energy Project. University of California, Los Angeles. Report, 9 Nov. 1954. pp. 18 (UCLA - 315)
A prototype liquid scintillation counter designed for counting radiocarbon and possibly tritium is described. The instrument uses coincidence circuitry and incorporates a circuit which rejects pulses originating from external radiations. A method of counting radiocarbon in sodium carbonate and barium carbonate is described.

3.149   1956
LIQUID SCINTILLATION COUNTER FOR CARBON-14 EMPLOYING AUTOMATIC SAMPLE ALTERNATION
Weinberger, Arthur J.; Davidson, Jackson B.; Ropp, Gus A.
Analytical Chemistry, v. 28, Jan. 1956: 110-112

A scintillation counter for radiocarbon, which uses solution phosphore is described. It employs alternative automatic counting to reduce possible effects of instrument drift and changing background.

3.150   1958
LIQUID SCINTILLATION COUNTING FOR TRITIUM AND CARBON-14
Haigh, C.P.
Nuclear Power, v. 3, Dec. 1958: 585-587
Problems involved in measuring low energy beta emitters by liquid scintillation counters and their future application in science and medicine are outlined.

3.151   1964
LIQUID SCINTILLATION COUNTING OF C14 USING A BALANCED QUENCHING TECHNIQUE
Ross, H.H.
International Journal of Applied Radiations and Isotopes, v. 15, May 1964: 273-277
A new method was developed for liquid scintillation counting of radiocarbon at a constant known efficiency in quenched samples. Ideally suited for routine counting of similar types of samples in any activity range.

3.152   1961
LIQUID SCINTILLATION COUNTING OF CARBON-14. USE OF ETHANOLAMINE - ETHYLENE GLYCOL MONOMETHYL ETHER - TOLUENE
Feffay, Henry; Alvarez, Julian
Analytical Chemistry, v. 33, April 1961: 612-615
A scintillation method for measuring radiocarbon is described. It consists of the oxydation of an organic compound to carbon dioxide and the trapping of gas as the ethanolamine carbonate in the ethylene glycol monomethyl ether. A portion of the ethanolamine salt solution is transferred to a vial containing toluene and the scintillator and then counted in a liquid scintillation counter.

3.153   1953
LIQUID SCINTILLATION COUNTING OF NATURAL C14
Hayes, F. Newton; Williams, D.L.; Rogers, Betty S.
Physical Review, v. 92, n. 2, 15 Oct. 1953: 512-513
Describes investigation of the use of liquid scintillation counting for radiocarbon dating.

3.154   1956
LIQUID SCINTILLATION COUNTING OF NATURAL RADIOCARBON
Hayes, F. Newton; Anderson, Ernest C.; Arnold, James R.
In: International Conference on the Peaceful Use of Atomic Energy, 1st, Geneva, 8-20 Aug. 1955. *Proceedings*, v. 14. New York, United Nations, 1956: 188-192
A comparison of counting methods is presented. The liquid scintillation counting method is described in some details. The following aspects are examined: instrumentation, chemistry of sample preparation and an assessment of the results.

3.155   1964
LIQUID SCINTILLATION COUNTING, 1957 - 1963: A REVIEW
Rapkin, E.
International Journal of Applied Radiations and Isotopes, v. 15, n. 2, Feb. 1964: 67-88
Briefly reviews the literature highlights in the field of liquid scintillation counting in the years before 1963. The use for radiocarbon dating, although not yet widely prevalent, is examined, indicating that the method could be ideal for that purpose.

3.156   1968
LIQUID SCINTILLATION COUNTING: AUTOMATED MATHEMATICAL FITTING AND USE OF CHANNEL RATIO METHOD BY COMPUTER PROGRAM
Blanchard, F.R.; Wagner, Marie R.
In: Advances in Tracer Methodology, v. 4. Proceedings of the 11th Annual Symposium in Tracer Methodology, edited by Seymour Rothchild. New York, Plenum Press, 1968: 133-143
Presents a computer program for handling the calculation needs of liquid scintillation counting, taking the data directly from the counter. It includes the computation of channel ratios

3.157   1955
LIQUID SCINTILLATION TECHNIQUES FOR RADIOCARBON DATING
Pringle, R.W.; Turchinetz, W.
Review of Scientific Instruments, v. 26, n. 9, Sept. 1955: 859-865
A study of the merits of liquid scintillation methods for radiocarbon dating.

3.158   1955
LIQUID SCINTILLATORS - I. PULSE HEIGHT COMPARISON
Hayes, F. Newton; Ott, Donald G.; Kerr, Vernon N.; Rogers, Betty S.
Nucleonics, v. 13, n. 12, Dec. 1955: 38-41
Presents a scintillating ability ranking of 102 selected compounds, in toluene. The implied relationships between structure and scintillation are explored.

3.159   1956
LIQUID SCINTILLATORS - II. ATTRIBUTES AND APPLICATIONS
Hayes, F. Newton
International Journal of Applied Radiations and Isotopes, v. 1, Jul. 1956: 46-56
Liquid scintillators are evaluated and characterised. Liquid scintillation detectors are described, specifically small-volume detectors used for radiocarbon and tritium.

3.160   1957
LIQUID SCINTILLATORS - III. THE QUENCHING OF LIQUID SCINTILLATOR SOLUTION BY ORGANIC COMPOUND
Kerr, Vernon N.; Hayes, F. Newton; Ott, Donald G.
International Journal of Applied Radiations and Isotopes, v. 4, Jan. 1957: 284-288

A sudy has been made of the quenching characteristics of a large number of selected organic compounds in a few commonly encountered liquid scintillation solutions. From this it should be possible to anticipate the quenching difficulties associated with homogeneous liquid scintillation counting.

3.161   1960
LIQUID SCINTILLATORS FOR RADIO- CARBON DATING IN ARCHAEOLOGY
Starik, I.E.; Rudenko, S.I.; Artemiev, V.V.; Butomo, S.V.; Drozhzhin, V.M.; Romanova, E.N.
In: Conference on the Use of Radioisotopes in Physical Sciences and Industry, Copenhagen, 6-10 September 1960. *Proceedings.* 1960. pp.10
Liquid scintillation counters were used for measuring the activity of natural carbon in archaeological specimens. The carbon for investigation is added in the form of a solvent or diluent.

3.162   1961
LIQUID SCINTILLATORS FOR RADIO-CARBON DATING IN ARCHAEOLOGY
Starik, I.E.; Rudenko, S.I.; Artemiev, V.V.; Butomo, S.V.; Drozhzhin, V.M.; Romanova, E.N.
International Journal of Applied Radiations and Isotopes, v. 9, 1961: 193-194
Liquid scintillation counters were used for measuring the activity of natural carbon in archaeological specimens. The carbon for investigation is added in the form of a solvent or diluent.

3.163   1952
LOW ENERGY COUNTING WITH A NEW LIQUID SCINTILLATION SOLUTE
Hayes, F. Newton; Hiebert, R.D.; Schuch, R.L.
Science, v. 116, n. 306, 8 Aug. 1962: 140
Radiocarbon and tritium scintillation counting makes use of both coincidence circuitry and refrigeration to decrease dark current noise. A solution of 30 mL of 0.3% 5-diphenyloxazol as a primary scintillant in toluene is used.

3.164   1958
LOW LEVEL COUNTING
Suess, Hans E.
Scripps Institute of Oceanography, La Jolla, California. Interim Progress Report, 15 August 1958. pp. 23 (AECU - 4164)
Progress is reported on radiocarbon increase from artificial radiocarbon measured in a Californian tree. A list of 27 radiocarbon dates is appended.

3.165   1963
LOW-LEVEL COUNTING METHODS
Oeschger, Hans
In: Radioactive Dating. Proceedings of a Symposium organised by the International Atomic Energy Agency (IAEA), and held in Vienna, 1962. Vienna, IAEA, 1963: 13-34
A review of naturally occurring radioisotopes, which are of interest for radioactive dating giving the physical properties of their decay. The problems of measuring these isotopes and the several special types of low-level counting devices are discussed.

3.166   1952
LOW-LEVEL COUNTING TECHNIQUES
Freedman, Arthur J.; Anderson, Ernest C.
Nucleonics, v. 10, n. 8, August 1952: 57-59
With the same statistical error, counting time was reduced by a factor of three, using pulse-height discrimination with a $^{14}CO_2$ filled proportional counter. Equations relating sample-to-background ratios to statistical error are derived.

3.167   1967
LOW-LEVEL GAS COUNTER WITH A PLASTIC SCINTILLATOR AS COINCIDENCE SHIELD
Huber, B.; Heinrich, F.; Buttlar, H., von
Nuclear Instruments and Methods, v. 52, n. 1, June 1967: 104-108
The construction of a low-level counter is described, the wall of which is made of plastic material serving as the anti-coincidence shield. Its background and performance are examined. The memory effect due to the solubility of ethane in the plastic material is small.

3.168 1954
MAN, TIME AND FOSSILS
Moore, Ruth
London, Cape, 1954. pp. 383
The third section of the book deals with ways of dating the organic evolution and fossil record of man. It gives an up to date account of the development of radiocarbon for dating the last 25,000 years.

3.169 n.d.
MANUAL FOR THE OPERATION OF THE CARBON-14 ELECTRONICS
Jansen, Hans S.
Institute of Nuclear Sciences. Department of Scientific and Industrial Research. (Lower Hutt, New Zealand) *Manual*, n. INS M4 , n. d. 27 leaves
Describes manipulation that occur in connection with the electronic operation of the radiocarbon dating equipment. Includes diagrams.

3.170 1962
MASS SPECTROMETER MEASUREMENTS IN THE THERMAL AREAS OF NEW ZEALAND, PART 2 - CARBON ISOTOPIC RATIOS
Hulston, John R.; McCabe, William J.
Geochimica et Cosmochimica Acta, v. 26, 1962: 399-410
The variation of the $^{13}C/^{12}C$ ratio of gas and water samples from pools, bores and fumaroles of the North Island of New Zealand have been studied. In general the isotopic composition is in the range $\partial^{13}C = 0$ to -7‰ with respect to the PDB Belemnite Standard.

3.171 1961
MASS-SPECTROMETER ANALYSES OF RADIOCARBON STANDARDS
Craig, Harmon
Radiocarbon, v. 3, 1961: 1-3
Analyses for radiocarbon content of carbon dioxide from combustion of the NBS oxalic acid radiocarbon standards from 13 laboratories were made in order to provide comparative data to normalise the counting results.

3.172 n.d.
MATHEMATICS OF C-14 AGE DETERMINATIONS
Jansen, Hans S.
Institute of Nuclear Sciences. Department of Scientific and Industrial Research. (Lower Hutt, New Zealand) *Lab. Notes*, n. 13, n. d. pp. 9 (I.N.S. Contribution No. 146)
Development of the mathematical formula for the calculation of radiocarbon ages.

3.173 1959
MEASUREMENT AND USE OF NATURAL RADIOCARBON
de Vries, Hessel
In: *Researches in Geochemistry*, edited by P.H. Abelson. New York, Wiley, 1959: 169-189
Discusses some of the more recent improvements in methods, reviews highlights of some older chronologies in North America and Europe and discusses some new measurements of the variation of radiocarbon with time and location on earth.

3.174 1959
THE MEASUREMENT OF CARBON 14 ACTIVITIES WITH CARBON DIOXIDE FILLED GEIGER COUNTERS
Melhuish, W.H.
Institute of Nuclear Sciences. Department of Scientific and Industrial Research. (Lower Hutt, New Zealand) *Lab. Notes*, n. 3, 1963. pp. 14
The counter described is used in the Geiger, not the proportional region since the purity of the sample and regulation of the high voltage are then unimportant.

3.175 1960
THE MEASUREMENT OF CARBON-14 AND TRITIUM ACTIVITIES IN GAS-FILLED GEIGER COUNTERS
Melhuish, W.H.
New Zealand Journal of Science, v. 3, n. 3, Sept. 1960: 549-558
The construction of Geiger counters and a vacuum system for filling them with ^{14}C-carbon dioxide or ^{3}H-hydrogen is described. The counters must be quenched electronically. The effect of the pressure of the filling gas and ambient temperature on the counter is investigated. Likely errors are discussed and are found to be consistent with the experimental reproducibility.

3.176 1951
MEASUREMENT OF LOW LEVEL RADIO-CARBON
Anderson, Ernest C.; Arnold, James R.; Libby, Willard F.
Review of Scientific Instruments, v. 22, n. 4, Apr. 1951: 225-230
Techniques are described for chemical purification and measurement of radiocarbon at natural levels.

3.177 1954
THE MEASUREMENT OF LOW SPECIFIC ACTIVITY C-14 BY LIQUID SCINTILLATION COUNTER
Audric, B.N.; Long, J.V.P.

Ch. 3 - Techniques and Instrumentation

In: Physical Science and Industrial Applications, edited by J.E. Johnston. Radioisotope Conference, 2nd, Oxford, 1954. *Proceedings*, vol.II. London, Butterworth, 1954: 134-139

Investigates the possibility of using acetylene as the solute in a liquid scintillation counter.

3.178   1954
THE MEASUREMENT OF LOW SPECIFIC ACTIVITY C14 BY LIQUID SCINTILLATION COUNTING
Arrhenius, Gustaf; Kjellberg, G.; Libby, Willard F.
Oxford Radioisotope Conference. Proceedings, v. 2, 1954: 134-
[Not sighted]

3.179   1950
THE MEASUREMENT OF RADIOACTIVE CARBON BY GAS COUNTING OF CARBON DIOXIDE
Audric, B.N.; Long, J.V.P.
Chemical Research Laboratory, Teddington. Report, 1950. pp. 7 (Report CRL/AE 51)
Copper cathod tubes have given satisfactory counting characteristics when filled with CO_2 - CS_2 mixture and used in conjunction with a quench probe unit.

3.180   1953
THE MEASUREMENT OF RADIOACTIVE CARBON IN GAS COUNTERS
Hemon, A.F.
British Journal of Applied Physics, v. 4, July 1953: 217-219
An AC coupled quench circuit is used to quench counters containing non-self-quenching carbon dioxide - carbon disulphide mixtures for the measurement of radioactive carbon as carbon dioxide.

3.181   1965
THE MEASUREMENT OF RADIOCARBON ACTIVITY AND SOME DETERMINATIONS OF AGES OF ARCHAEOLOGICAL SAMPLES
Agrawal, D.P.; Kusumgar, Sheela; Lal, D.
Current Science, v. 34, n. 13, 5 July 1965: 394-397
A new simple, rapid and quantitative method of synthesising methane gas from carbon atoms is described. Results of 15 dates of archaeological samples are presented.

3.182   1950
MEASUREMENT OF RADIOCARBON AS CARBON DIOXIDE INSIDE GEIGER -MULLER COUNTERS
Eidinoff, Maxwell Leigh
Analytical Chemistry, v. 22, n. 4, Apr. 1950: 529-534
The measurement of radiocarbon activity as carbon dioxide admixed with carbon disulphide vapor is studied quantitatively over the pressure range 2 to 175 cm of mercury. The construction of Geiger-Muller counter tubes and auxiliary equipment is described. The counting rate was found to be directly proportional to the quantity of active gas sample.

3.183   1954
MEASUREMENT OF THE AGE BY THE C14 TECHNIQUE
de Vries, Hessel; Barendsen, G.W.
Nature, v. 174, n. 4421, 18 Dec. 1954: 1138-1141
A method for removing radon is discussed: a layer of mercury between the counter and the anticoincidence ring proves very effective in reducing the background to 2.64 counts/min. Dating of the lowering of the Dutch coast and bronze age man in Europe on marine shell, charred grain and wood, and charcoal samples by the Groningen laboratory are used.

3.184   1961
MEASUREMENTS OF THE WATER VAPOUR TRITIUM AND CARBON-14 CONTENT OF THE MIDDLE STRATOSPHERE OVER SOUTHERN ENGLAND
Brown, F.; Goldsmith, P.; Green, H.F.; Parham, A.G.
Tellus, v. 13, 1961: 407-416
Measurements of the water vapour, tritium and radiocarbon content of the stratosphere at heights between 8000 and 10,000 feet, made over England during the years 1956-60 are described. The concentrations are higher than expected from natural production due to cosmic radiations. These concentrations are a result of the bomb effect.

3.185   1952
MEASURING LOW-LEVEL RADIOACTIVITY
Douglas, David L.
General Electric Review, v. 55, n. 5, Sept. 1952: 16-20
Description of an ionisation chamber used in radiocarbon dating. Two samples were dated. The first one from a ruin in the Southwest gave an age in agreement with the age determined by the tree ring dating method, the second one dated wood from La Brea Tar Pit in Los Angeles which was presumed contemporaneous with the sabre-toothed tiger and dire wolves. The ages of 16,500 and 16,400 ± 2000 is the first estimate with any claim to reliability of the age of the late Pleistocene beasts.

3.186   1967
THE METALLURGICAL HISTORY AND CARBON 14 DATING OF IRON
Van der Merwe, Nicholas J.
Ph.D. Dissertation. Yale University, 1967. pp. 172 (Ab-

stract in: *Dissertation Abstracts*. Ann Arbor, Mich. v.28, v. 1 ser. B: 36. Order n. 67-8429)
Describes the successful development of a laboratory method for the radiocarbon dating of iron artifacts and presents a demonstration of the application of radiocarbon measurements to the dating of iron slag. Examines the application of these methods to provide solutions to the specific problem in Iron Age chronology.

3.187   1954
METHANE PROPORTIONAL COUNTER METHODS FOR C-14 AGE DETERMINATION
Burke, W.H.; Meinschein, W.G.
Physical Review, v. 93, n. 4, 15 Feb. 1954: 915
A method has been developed for making radiocarbon age determinations on samples containing 0.8 gram of carbon by convesion to methane and counting the radiocarbon beta radiations in a proportional counter filled with the sample methane.

3.188   1956
A METHANE PROPORTIONAL COUNTER SYSTEM FOR NATURAL RADIOCARBON MEASUREMENTS
Diethorn, Ward
Ph.D. Dissertation. Carnegie Institute of Technology. United States Atomic Energy Commission, report, NYO-6628, 1956. pp. 146
A methane gas sample system for dating by radiocarbon is developed to explore the potentialities of methane for this purpose. The study shows that it is practical to measure natural radiocarbon with a methane filled counter at high pressure. This would prove to be the most practical, highly sensitive method of natural radiocarbon measurements. Contemporary radiocarbon counting rates are comparable to that realised by the best scintillation counter.

3.189   1951
A METHOD FOR SIMULTANEOUS ANALYSIS OF CARBON 14 AND TOTAL CARBON
Buchanan, Donald L.; Nakao, Akira
Argonne National Laboratory, April 1951. pp. 31 (AECU - 1385; UAC - 387)
An analytical method is described whereby a single sample of either organic or inorganic carbon-containing material may be assayed for total carbon content as well as radiocarbon content. The carbon analysis is carried out by wet oxydation with subsequent purification of carbon dioxide in a vacuum system. Carbon dioxide is measured manometrically and then tranferred to a gas proportional counter for radiocarbon determination.

3.190   1954
MODERN METHODS OF DATING

Ch. 3 - Techniques and Instrumentation

Cole, Sonia Mary
South African Archaeological Bulletin, v. 9, n. 33, 1954: 18-24
Examine the various methods of relative and absolute dating of archaeological material. Indicates that the radiocarbon method of age determination is probably the most valuable means of absolute dating for the prehistorians.

3.191   1964
NATURAL CARBON-14 ACTIVITY OF ORGANIC SUBSTANCES IN STREAMS
Rosen, A.A.; Rubin, Meyer
Science, v. 143, n. 3611, 13 Mar. 1964: 1163
Radiocarbon measurements made on organic contaminants extracted from streams show percentages of industrial waste and domestic sewage. The method, used previously for studies of the atmosphere, can be used in studies of pollution sources.

3.192   1965
NATURAL RADIOCARBON AND TRITIUM IN RETROSPECT AND PROSPECT
Libby, Willard F.
In: International Conference on Radiocarbon and Tritium Dating, Pullman, Washington, Washington State University, June 1-11, 1965. *Proceedings*. United States of America, Atomic Energy Commission, 1965. Conference n. 650652: 745-751
A talk on the early research projects which led to the establishment of the method of radiocarbon and tritium age determination and the various applications to which they have been put and the prospects for future use.

3.193   1954
NATURAL RADIOCARBON MEASUREMENTS BY ACETYLENE COUNTING
Suess, Hans E.
Science, v. 120, n. 3105, 2 July 1954: 5-7
This is a description of the technique involving acetylene counting. It includes the chemical steps in the preparation of acetylene gas.

3.194   1963
NATURAL RADIOCARBON MEASUREMENTS BY LIQUID SCINTILLATION COUNTING
Noakes, John Edward
Ph.D. Dissertation. College Station, Texas Agricultural and Mechanical College, 1963. pp. 144
A synthesis of high purity benzene in 50 to 60% yields that is suitable for low-level liquid scintillation counting is described. The method involves a low temperature catalytic conversion of acetylene to benzene which was at first reported by I. Shapiro and H.G. Weiss in 1957. Application

of this method to radiocarbon dating essentially eliminates previous problems encountered in the use of liquid scintillation counting for this purpose. The method extends the sensitivity of the radiocarbon dating method and yet requires only standard commercially available counting systems. Fourteen archaeological samples, fifteen geological samples and two marine water samples were used to test this method.

3.195   **1959**
NATURAL RADIOCARBON MEASUREMENTS OF SURFACE WATER FROM THE NORTH ATLANTIC AND THE ARCTIC SEA
Fonselius, Stig; Östlund, Göte H.
Tellus, v. 11, n. 1, 1959: 77-82
An apparatus for the extraction of carbon dioxide from 100 litres of sea water, using carbonate free sodium hydroxide solution as absorbant liquid, is described and its performance is examined. A test series of six sea water samples is presented, on which natural radiocarbon determinations are made, and the results obtained are briefly discussed in relation to other age measurements on similar waters. The correlation between salinity and radiocarbon activity is shown.

3.196   **1964**
THE NATURE OF CHARCOAL EXCAVATED AT ARCHAEOLOGICAL SITES
Cook, S.F.
American Antiquity, v. 29, v. 4, Apr. 1964: 514-517
Samples of charcoal recovered from excavated sites were treated with strong acid, followed by strong base. This process removes most of the inorganic carbon. The fixed carbon content of the specimens from four sites varied widely and indicated different sources of organic matter. Each charcoal, therefore, has a characteristic chemical composition.

3.197   **1966**
NEW APPLICATION OF RADIOCARBON DATING TO COLLAGEN RESIDUE IN BONES
Sellstedt, H.; Engstrand, Lars G.; Gejvall, V. G.
Nature, v. 212, n. 5062, 5 Nov. 1966: 572-574
The authors describe a technique for preparing bone samples for radiocarbon analysis which reduces it to collagen. A number of experiments result in close agreement between archaeological and radiocarbon age. This presents the archaeologist with an elegant and versatile tool which is likely to find wide application as bones are found at most excavation sites.

3.198   **1968**
MODERN ASPECTS OF RADIOCARBON DATING
Olsson, Ingrid U.
Earth Science Review, v. 4, n. 3, Sept. 1968: 203-218
The radiocarbon method for radioactive age determination is discussed and recent developments in the fixing of the half-life time of radiocarbon are reviewed. The occurrence and cause of $^{12}C/^{14}C$ ratio variations are explored and the possibilities and influences of contamination. The background, applicability and reliability of the methods are discussed.

3.199   **1968**
NEW DEVELOPMENTS IN THE DATING OF CERAMIC ARTIFACTS
Taylor, R.E.; Swain, J.L.; Berger, Rainer
In: International Congress of Americanists, 38th, Stuttgart-Munnich, 12-18 Aug. 1968. *Proceedings*, v. 1. 1968: 55-60
Radiocarbon dating in ceramics is examined as well as dating by the thermoluminescence technique.

3.200   **1965**
NEW EXPERIENCE ON C-14 DATING OF TESTS OF FORAMINIFERA
Olsson, Ingrid U.
In: International Conference on Radiocarbon and Tritium Dating, Pullman, Washington, Washington State University, June 1-11, 1965. *Proceedings*. United States of America, Atomic Energy Commission, 1965. Conference n. 650652: 319-331
Three possible sources for errors in radiocarbon dating of Foraminifera are being examined at the radiocarbon laboratory at Uppsala: (1) contamination by foreign material in sediments which show no traces of re-worked material; (2) contamination during the preparation of the samples; (3) contamination during the storage of the samples.

3.201   **1953**
A NEW QUENCHING EFFECT IN LIQUID SCINTILLATORS
Pringle, R.W.; Black, L.D.; Funt, B.L.; Sobering, S.
Physical Review, v. 92, n. 6, 15 Dec. 1953: 1582
Liquid scintillator studies initiated in connection with a radiocarbon liquid scintillator dating project, have revealed the existence of a considerable quenching effect attributable to dissolved oxygen. This is discussed. Suggestion is made that the removal of oxygen is necessary in all liquid scintillator experiments before complete reproducibility is obtained.

3.202   **1952**
A NEW TECHNIQUE FOR THE MEASUREMENT OF AGE BY RADIOCARBON
de Vries, Hessel ; Barendsen, G.W.
Physica, v. 18, n. 8-9, Aug.-Sept. 1952: 652

Describes briefly a new technique for routine radiocarbon age measurement which uses carbon dioxide as the filling gas of a proportional counter.

3.203   1965
NEW TECHNIQUES OF WATER SAMPLING FOR CARBON-14 ANALYSIS
Crosby, James W.; Chatters, Roy M.
Journal of Geophysical Research, v. 70, v. 12, 15 June 1965: 2839-2844
Equipment has been developed, using anion exchangers. Analytical results, by means of which resin techniques and standard methods are compared have proved to be within the normal errors expected in the radiocarbon method and may actually decrease error by greatly reducing the possibility of atmospheric contamination.

3.204   1951
NOTES AND NEWS: ADDITIONAL RADIO-CARBON DATES
American Anthropology, v. 17, n. 2, Sept. 1951: 174-175
A report on recent dates and on new radiocarbon laboratories organised at Yale, Columbia, Michigan and California Universities.

3.205   1963
NOTES FOR THE OPERATION OF C-14 DUPLICATION AND VACUUM FILLING SYSTEM
Aldous, K.J.
Institute of Nuclear Sciences. Department of Scientific and Industrial Research. (Lower Hutt, New Zealand) *Report*, n. INS-M7, 1963. 7 leaves (I.N.S. Contribution n.147)
Description of system and procedure. Includes diagrams.

3.206   1955
NUCLEAR RADIATION DETECTORS
Sharpe, Jack
Methuen Monograph on Physical Subjects. London, Methuen, 1955. pp. 237
General descriptions of various types of detectors and their characteristics.

3.207   196
NUCLEAR STUDIES IN CHEMISTRY AND BIOLOGY
Rafter, T.A.
Journal of the New Zealand Institute of Chemistry, v. 29, n. 2, 1965: 64-73
In this address to the Institute, the author examines how the knowledge that certain elements have isotopes can be used to help solving various problems.

3.208   1968
THE OLD STONE AGE
Bordes, Francois
London, World University Library, 1968: 18-19
The author presents the radiocarbon method of age determination and discusses the validity of the dates so obtained.

3.209   1958
ON A NEW METHOD FOR INTRODUCING C-14 IN A LIQUID SCINTILLATION COUNTER
Leger, Concèle; Delibrias, Georgette; Pichat, L.; Baret, C.
In: Liquid Scintillation Conference, North Western University. New York, Pergamon Press, 1958: 261-267
Use of paraldehyde as a compound extracted from the sample to be dated by the radiocarbon method in a liquid scintillation counter.

3.210   1965
ON THE CARE AND FEEDING OF RADIO-CARBON DATES
Stuckenrath, Robert
Archaeology, v. 18, n. 4, Winter 1965: 281-285
The author reviews some of the basic hypothesis of radiocarbon dating and the problems associated with the method, such as: non-representative samples, apparent discrepancies in the dates of different materials, the pre-treatment of samples, the standard deviation.

3.211   1949
ON THE EFFICIENCY OF GAS COUNTERS FILLED WITH CARBON DIOXIDE AND CARBON DISULPHIDE
Hawkins, R.C.; Hunter, R.F.; Mann, W.B.
Canadian Journal of Research, Section B, v. 27, 1949: 555-564
[Not sighted]

3.212   1953
ON THE USE OF CO_2 + CS_2 FILLED G. M. COUNTERS FOR AGE DETERMINATION
Moscicki, Wldzimierz
Acta Physica Polonica, v. 12, n. 3-4, 1953: 238-240
Experiments were performed to inquire into the possibility of using Geiger - Müller counters filled with carbon dioxide + carbon disulfide for geochronometrical purposes.

3.213   1958
ON THE USE OF G. M. COUNTERS FILLED WITH A MIXTURE OF CO_2 + CS_2 FOR THE ACTIVITY OF NATURAL CARBON, PART 1 AND 2
Moscicki, Wldzimierz
Acta Physica Polonica, v. 17, n. 3-4, 1958: 311-343
In Part 1, comparisons are made of techniques of measuring the radioactivity of natural radiocarbon, using (a) solid

carbon and screen wall counter, (b) gas samples and proportional counters. The gas technique is shown to be more advantageous in spite of the complicated electronics required and lack of control of the background level. Part 2 presents reports of a number of experiments conducted with the Geiger-Müller counter filled with carbon dioxide and carbon disulfide.

3.214   1962
OPERATION AND MAINTENANCE OF THE ^{14}C COUNTER FILLING SYSTEM FOR AGE MEASUREMENTS
Muir, Janis A.
Institute of Nuclear Sciences. Department of Scientific and Industrial Research. (Lower Hutt, New Zealand) *Manual*, n. INS M6, 1962. 7 leaves (I.N.S. Contribution n.133)
Description of a flexible vacuum system used for filling and evacuating a 7.7 litre proportional counter from 21 litre copper cylinders. Includes diagrams.

3.215   1962
PARR BOMB COMBUSTION OF TISSUES FOR CARBON-14 AND TRITIUM ANALYSIS
Sheppard, Herbert; Rodegker, Waldtraut
Atomlight, n 22, February 1962: 1-3
The Parr metal oxygen bomb was used for the combustion of biologcal samples for the determination of low levels of radiocarbon and tritium. The system is outlined.

3.216   1951
PEAT SAMPLES FOR RADIOCARBON ANALYSIS: PROBLEMS IN POLLEN STATISTICS
Deevey, Edward S.; Potzger, John E.
American Journal of Science, v. 249, n. 7, July 1951: 473-511
Various problems connected with pollen analysis are discussed. The need to obtain large samples of peat for radiocarbon determination is discussed and an apparatus for this purpose is described.

3.217   1962
PHYSICAL METHODS OF ARCHAEOLOGICAL RESEARCH
Aitken, M.J.
Research, v. 15, April 1962: 145-151
Six important applications of physics to archaeological research are described, one of which is radiocarbon dating. Some results are briefly described.

3.218   1963
POLYETHYLENE CONTAINERS FOR LIQUID SCINTILLATION SPECTROMETRY
Rapkin, E.; Gibbs, J.A.
International Journal of Applied Radiation and Isotopes, v. 14, 1963: 71-74
Medium density polyethylene vials have been compared to glass vials for use in liquid scintillation counting. Lower backgrounds and increased efficiencies were observed with plastic vials. Glass is superior for storage and heating of samples.

3.219   1965
POTENTIALITIES OF RADIOCARBON DATING IN THE CONNECTED FIELD OF SOIL SCIENCE AND ARCHAEOLOGY
Dimbleby, G.W.
In: International Conference on Radiocarbon and Tritium Dating, Pullman, Washington, Washington State University, June 1-11, 1965. *Proceedings*. United States of America, Atomic Energy Commission, 1965. Conference n. 650652: 287-293
Surveys the potential applications of radiocarbon dating to soil organic matter.

3.220   1957
PRACTICAL ASPECTS OF INTERNAL - SAMPLE LIQUID SCINTILLATION COUNTING
Davidson, Jack D.; Feigelson, Philip
International Journal of Applied Radiations and Isotopes, v. 2, n. 1, April 1957: 1-18
The advantages and disadvantages of the method are discussed in detail. The applicability of liquid scintillation counting to any counting problem should be evaluated specifically. Liquid scintillation counting seems to have become the method of choice for radiocarbon and tritium samples of low activity.

3.221   1959
PRECISION OF THE DATING METHOD. STANDARDIZATION OF THE CALCULATION OF THE ERRORS AND THE MAXIMUM AGE IN THE ^{14}C METHOD
Crevecoeur, E.H.; Vander Stricht, A.; Capron, P.C.
Academie Royale de Belgique, Bulletin, Classe des Sciences 5 series, v. 5, n. 45, 1959: 876-890
Because of the large variety in recording the precision of results, it seems necessary to standardise the method of calculation. The relative importance of several factors, such as background radiation, counting time and activity has been analysed concerning the value of maximum age, as well as precision of the measurements.

3.222   1966
PREPARATION OF C14 STANDARDS FOR LIQUID SCINTILLATION COUNTING
Williams, D.L.; Hayes, F. Newton; Kandel, R.J.; Rogers,

W.H.
Nucleonics, v. 14, n. 1, January 1966: 62-64
The procedures for preparing various radiocarbon standards for liquid scintillation counting are described.

3.223   1953
PREPARATION OF CARBON FOR C-14 AGE MEASUREMENTS
Rafter, T.A.
New Zealand Journal of Science and Technology, section B, v. 35, n. 1, July 1953: 64-89
Describes the techniques used for the production of carbon from carbonaceous material. The samples produced are free from detectable radioactivities other than radiocarbon, contain 1% or less of ash, and the carbon percentage varies within limits that cannot affect the counting rate. A full description is given of technical difficulties associated with carbon preparation and its deposition as a uniform layer on the inside of copper cylinders.

3.224   1968
PREPARATION OF LIQUID SCINTILLATION MIXTURE FOR THE MEASUREMENT OF ^{14}C AND ^{3}H SAMPLES
Roberts, W.A.
Laboratory Practice, v. 17, June 1968: 703-706
Selection of suitable materials for liquid scintillation is described. Mixtures for a wide range of radiocarbon and tritium labelled samples are given. The most efficient preparation technique is combustion and oxydation, leading to the collection of tritium as water and radiocarbon as carbon dioxide.

3.225   1963
PREPARATION OF WATER SAMPLES FOR CARBON-14 DATING
Feltz, H.R.; Hanshaw, Bruce B.
United States Geological Survey. Circular, n. 483, 1963. pp. 3
For most natural water, a large sample is required to provide the 3 g of carbon needed for a radiocarbon determination. A field procedure for isolating total dissolved carbonate species is described. Carbon dioxide gas is evolved by adding sulphuric acid to the water sample. The gas is then cooled in a sodium hydroxide trap by recycling in a closed system. The trap is then transported to the dating laboratory.

3.226   1965
THE PRESERVATION AND DATING OF COLLAGEN IN ANCIENT BONES
Krueger, Harold W.
In: International Conference on Radiocarbon and Tritium Dating, Pullman, Washington, Washington State University, June 1-11, 1965. *Proceedings*. United States of America, Atomic Energy Commission, 1965. Conference n. 650652: 332-337.
A discussion of the problems involved in dating bone remains. The following conclusions are made: (1) the dating of bone samples does not involve difficult or tedious laboratory procedures; (2) dates on the collagen content of bones are almost always reliable; (3) dates on the carbonate portion of bones which have been significantly altered will be too young and provide only a minimum age; (4) dates on the carbonate portion of bones that show no evidence of alteration appear to be generally reliable; (5) collagen is recoverable from bones as old at 40,000 years when the bones have been preserved under circumstances which do not favour oxidation.

3.227   1962
PRETREATMENT OF WOOD AND CHAR SAMPLES
Schultz, Hyman
Ph.D. Dissertation. University of Pennsylvania, 1962. pp. 85 (Microfilm copy from *University Microfilms Inc.*, Ann Arbor, Mich, v. 23, n. 11, 1963: 4084. Order n. 63-4019)
Description of a treatment to extract cellulose from 'cellulosic' samples by treating the sample with Schweizer's reagent to obtain crude fibre and extracting the cellulose from the fibre.

3.228   1963
PRETREATMENT OF WOOD AND CHAR SAMPLES
Schultz, Hyman; Currie, L.A.; Matson, F.R.; Miller, Warren W.
Radiocarbon, v. 5, 1963: 342 (Abstract)
A treatment to extract cellulose from 'cellulosic' samples by treating the samples to obtain crude fibre and extracting the cellulose from the fibre with Schweizer's reagent.

3.229   1958
PROBLEM OF HUMIC ACID CONTAMINATION IN RADIOCARBON DATING
Olson, Edwin A.
Geological Society of America Bulletin, v. 69, n. 12, Dec. 1958: 1625
Abstract of paper presented at the November 1958 meeting of the Geological Society of America in St. Louis. A radiocarbon age is a true age only if the assumption of the radiocarbon dating method are met. For buried samples exposed to soil solutions, the most critical assumption is that foreign carbon is absent. Humic acid is considered the most likely contaminant for buried samples. The authors give examples of difference in datation results.

3.230 1963
THE PROBLEM OF SAMPLE CONTAMINATION IN RADIOCARBON DATING
Olson, Edwin A.
Ph.D. Dissertation. Columbia University, 1963. pp. 339 (Abstract in: *Dissertation Abstracts*, Ann Arbor, Mich., v. 28, Section, B, n. 4, 1967: 158. Order n. 65-7466)
Contamination of radiocarbon samples can occur at any time from the moment of original emplacement to final measurements in counting chambers. Five broad mechanisms of natural process are examined and a number of conclusions made.

3.231 1965
PROBLEMS IN THE ESTABLISHMENT OF A CARBON-14 AND TRITIUM LABORATORY
Rafter, T.A.
In: International Conference on Radiocarbon and Tritium Dating, Pullman, Washington, Washington State University, June 1-11, 1965. *Proceedings.* United States of America, Atomic Energy Commission, 1965. Conference n. 650652: 752-761
A description of the early problems of the establishment of a radiocarbon and tritium dating laboratory in New Zealand.

3.232 1957
PROBLEMS OF RADIOCARBON DATING
Deevey, Edward S.
Yale Scientific Magazine, v. 31, n. 7, Apr. 1957: 42-52
What is radiocarbon and why use it? Its application, problems confronting the men using this method, and its accuracy in regard to other methods of calibrating chronologies, are explained and discussed.

3.233 1963
PROCESSING OF C-14 DATA
Jansen, Hans S.
Institute of Nuclear Sciences. Department of Scientific and Industrial Research. (Lower Hutt, New Zealand) *Lab. Notes*, n. 11, 1963. pp.17 (I.N.S. Contribution n. 137)
A guide to process radiocarbon data in order to find radiocarbon ages, radiocarbon enrichments. Includes examples of calculations.

3.234 1948
A PROGRESS REPORT ON THE DATING OF ARCHAEOLOGICAL SITES BY MEANS OF RADIOACTIVE ELEMENTS
Merrill, R.S.
American Antiquity, v. 13, n. 4, Apr. 1948: 281-286
A discussion of the various methods of dating archaeological sites including the radiocarbon method which appears to be very promising and capable of development into a usable method.

3.235 1965
PROPORTIONAL COUNTER EQUIPMENT FOR SAMPLE DATING WITH AGES EXCEEDING 60,000 YEARS BP WITHOUT ENRICHMENT
Geyh, Mebus A.
In: International Conference on Radiocarbon and Tritium Dating, Pullman, Washington, Washington State University, June 1-11, 1965. *Proceedings.* United States of America, Atomic Energy Commission, 1965. Conference n. 650652: 29-35.
Description of method.

3.236 1959
A PROPORTIONAL COUNTER FOR LOW-LEVEL COUNTING WITH HIGH EFFICIENCY
de Vries, Hessel; Stuiver, Minze; Olsson, Ingrid U.
Nuclear Instruments and Methods, v. 5, 1959: 111-114
Two small nearly identical counters have been built in the radiocarbon dating laboratories in Groningen and Uppsala, from essentially the same material. The counting space is lined with quartz covered with a conducting layer; it is enclosed in a copper cylinder at ground potential. The backgrounds are 1.33 and 0.9 cpm. The total volumes are 0.55 L with efficient volumes of about 0.46 L.

3.237 1967
PROPORTIONAL COUNTERS FOR DETERMINATION OF LOW-LEVEL ^{3}H AND ^{14}C IN SMALL SAMPLES
Tykva, R.; Kokta, L.
Nuclear Instruments and Methods, v. 55, n. 1, 1967: 381-382
The construction of a low-background internal gas proportional counter with anticoincidence shielding of a plastic scintillator is described. If natural gas (98% vol CH_4) is used as the flow gas, the background of the counter (effective vol. 52.4 mL) is 0.20 ± 0.03 cpm under a 10 cm lead shield.

3.238 1955
PROPORTIONAL COUNTING OF CARBON DIOXIDE FOR RADIOCARBON DATING
Brannon, H.R.; Taggart, M.S.; Williams, Milton
Review of Scientific Instruments, v. 26, n. 3, March 1955: 269-273
Work on a procedure similar to that used by de Vries and Barendsen (*Nature*, 1954) for radiocarbon assay is described. It differs from the latter method by using carbon dioxide at a higher pressure and purifying carbon dioxide by the use of hot calcium oxide. The procedures are described.

3.239 1965
PROPORTIONAL COUNTING TECHNIQUES FOR LOW LEVEL ^{14}C ACTIVITIES

Moreno y Moreno, Augusto
Revista Mexicana Fisica, v. 14, 1965: 1-16
Preliminary work performed for dating archaeological samples and determining the content of radiocarbon at low level using a proportional counting technique is reported. Includes descriptions of the acetylene method, electronic equipment, counter, and breakdown tests.

3.240   1962
PROPORTIONAL COUNTING TECHNIQUES FOR RADIOCARBON MEASUREMENTS
Nydal, Reidar
Review of Scientific Instruments, v. 33, n. 12, Dec. 1962: 1313-1320
A proportional counting system for radiocarbon dating is described. The main improvements were made in the following fields: reducing dead volume and improving end effect in the carbon dioxide counter; construction of a multi-anode proportional counter for anti-coincidence shielding; lowering the counter background and barometric effect.

3.241   1955
PURIFICATION OF CO_2 FOR USE IN A PROPORTIONAL COUNTER FOR ^{14}C AGE MEASUREMENTS
de Vries, Hessel
Applied Science Research, Section B, v. 5, 1955: 387-400
Description of a physical purification to remove impurities in carbon dioxide which is to be used in a proportional counter. Special attention is paid to the inert gases of which Radon is important because of its radioactivity.

3.242   1962
QUENCHING AND ADSORPTION IN LIQUID SCINTILLATION COUNTING
Dobbs, Horace E.
United Kingdom. Atomic Energy Authority. Research Group (AERE - M - 1075), 1962
The possibility of increasing the counting efficiency for radiocarbon in a coincidence liquid scintillation system by optimising the voltages on the photomultipliers for all concentrations of quenching agents is demontrated.

3.243   1962
QUENCHING CIRCUIT FOR SELF-QUENCHING GEIGER-MULLER ARRAY
Caini, Vasco; Olsson, Ingrid U.
Arkiv för Fysik, v. 22, n. 13, 1962: 225-235
For low-level measurement it is necessary to surround the main counter with an array of long Geiger-Müller tubes to reduce the background due to the penetrating radiations. Various problems inherent to these tubes are examined, mainly of short-life, difficulty of replacement and changes of characteristics. A suitable method of extending the life of the long organic quenched counters is the use of an auxiliary circuit.

3.244   1964
THE QUENCHING OF CARBON-14 AND TRITIUM BY ORGANIC SOLVENTS IN TWO COMMON LIQUID SCINTILLATION SOLUTIONS
Shapiro, I.L.; Kritchevsky, D.
International Journal of Applied Radiations and Isotopes, v. 15, 1964: 325-330
The quenching effects exerted on two liquid scintillation solutions, toluene - PPO - dimethyl POPOP, and Bray's solution, by various organic solvents was studied.

3.245   1957
RADIOCARBON AGE ESTIMATES OBTAINED BY AN IMPROVED LIQUID SCINTILLATION TECHNIQUE
Pringle, R.W.; Turchinetz, W.; Funt, B.L.; Danyluk, S.S.
Science, v. 125, n. 3237, 11 Jan. 1957: 69-70
Trimethyl borate prepared from methanol is used as the labelled component for routine radiocarbon dating.

3.246   1962
RADIOCARBON ANALYSIS
Buist, A.G.
New Zealand Archaeological Association Newsletter, v. 5, n. 4, Dec. 1962: 249-51
Review of a Report by L. Lockerbie on the state of radiocarbon analysis in New Zealand.

3.247   1959
RADIOCARBON ANALYSIS OF OCEANIC CO_2
Broecker, Wallace S.; Tucek, C.S.; Olson, Edwin A.
International Journal of Applied Radiation and Isotopes, v. 7, 1959: 1-18
Variations in the radiocarbon concentration of oceanic bicarbonate offer clues to large-scale ocean circulation patterns as well as to operation of the terrestrial carbon dioxide cycle. This article presents a description and evaluation of the method used. There are four steps to the procedure: (1) shipboard release and collection of carbon dioxide from a sea water sample; (2) laboratory purification of the carbon dioxide; (3) assay of the radiocarbon concentration in the carbon dioxide; (4) determination of the $^{13}C/^{12}C$ ratio in the carbon dioxide for use in monitoring isotopic fractionation.

3.248   1952
RADIOCARBON CONTAMINATION
Bliss, Wesley L.
American Antiquity, v. 17, n. 3, Jan. 1952: 250-251
The author examines sources of errors in radiocarbon dating

and concludes that after considering the many factors and possibilities for contamination he is convinced that many of the published dates cannot be considered as valid and that no attempt should be made to re-evaluate indiscriminately the archaeological chronology using the radiocarbon dates until the possible factors for contamination have been eliminated or minimised.

3.249 1951
RADIOCARBON DATABILITY OF PEAT, MARL, CALICHE AND ARCHAEOLOGICAL MATERIAL
Bartlett, H.H.
Science, v. 114, n. 2949, July 1951: 55-56
The problems associated with the radiocarbon dating of peat, marl and caliche are examined in view of the uncertainty about the probable composition of the material, and subsequent contamination.

3.250 1952
RADIOCARBON DATES
Jennings, Jesse D.
American Antiquity, v. 18, n. 1, July 1952: 89-90
The author comments on radiocarbon dates, contamination and other sources of error and the price of radiocarbon dates.

3.251 1953
RADIOCARBON DATES - A SUGGESTION
Lee, Abel
American Antiquity, v. 19, n. 2, Oct. 1953: 158
The author criticised the practice of describing radiocarbon dates as years BP, P meaning 1950, because it will become cumbersome in the future. He makes a plea for using the established calendar.

3.252 1964
RADIOCARBON DATES OF ARCHAEOLOGICAL SAMPLES
Agrawal, D.P.; Kusumgar, Sheela; Sarna, R.P.
Current Science, v. 33, 1964: 40-42
The measured radiocarbon dates of several archaeological samples are reported. The method used is the acetylene synthesis (Suess, 1964). The dates are discussed.

3.253 1961
RADIOCARBON DATING
Libby, Willard F.
Science, v. 133, n. 3453, 3 March 1961: 621-629
A review of the first decade of radiocarbon dating, of its use in the dating of archaeological and geological samples, and an evaluation of the accuracy of the results.

3.254 1968
RADIOCARBON DATING
Polach, Henry A.
Etruscan, v. 17, n. 1. March 1968: 3-7
A description for the layman of the radiocarbon dating method and its applications, particularly in Australia.

3.255 1960
RADIOCARBON DATING
Libby, Willard F.
Chemical Society. Proceedings, May 1960: 164-171
A description of the method, the measurement techniques, the natural distribution of radiocarbon and some applications of the method.

3.256 1954
RADIOCARBON DATING
Libby, Willard F.
Endeavour, v. 13, n. 49, 1954: 9-17
The basic principles of the radiocarbon dating technique are presented and discussed. In addition, evidence bearing on the validity of the absolute date obtained is discussed. The techniques of measurement is described and the types of materials acceptable for measurements are given. A partial list of the radiocarbon dates so far obtained is included.

3.257 1965
RADIOCARBON DATING (EDITORIAL NOTE)
Deacon, H.J.
South African Archaeological Bulletin, v. 20, n. 77, part 1, March 1965: 43-44
Notes for users of radiocarbon dating services offered by Geochron Laboratories. Discusses some of the factors involved which must be considered when submitting samples for dating.

3.258 1958
RADIOCARBON DATING AND ARCHAEOLOGY
Antiquity, v. 32, n. 127, Sept. 1958: 193-194
The validity of radiocarbon dates is discussed. The punch card system developed by the Committee for Distribution of Radiocarbon Dates is announced. New possible sources of error are noted.

3.259 1965
RADIOCARBON DATING AND FAR NORTHERN ARCHAEOLOGY
Campbell, John M.
In: International Conference on Radiocarbon and Tritium Dating, Pullman, Washington, Washington State University, June 1-11, 1965. *Proceedings.* United States of America, Atomic Energy Commission, 1965. Conference n. 650652: 179-186
Characteristics of the far north are discussed. Some but not all of those are more or less unique to the higher latitudes

and, either in theory or in practice, affect the validity of radiocarbon dates as they relate to the ages of archaeological remains from those regions.

3.260   1960
RADIOCARBON DATING AND QUATERNARY HISTORY IN BRITAIN. THE CROONIAN LECTURE
Godwin, Harry
Royal Society. Proceedings, series B, v. 153, n. 952, 3 Jan. 1961: 287-320
The lecture is an attempt to show the way in which research upon the Quaternary Period in Britain is affected by the application to it of radiocarbon dating.

3.261   1953
RADIOCARBON DATING BY A PROPORTIONAL COUNTER FILLED WITH CARBON DIOXIDE
de Vries, Hessel; Barendsen, G.W.
Physica, v. 19, 1953: 987-1003
A technique is described for the measurement of the natural activity of radiocarbon in a carbon dioxide filled proportional counter.

3.262   1964
RADIOCARBON DATING BY LIQUID SCINTILLATION COUNTING: SYNTHESIS OF BENZENE (C_6H_6)
McDowell, L.L.; Ryan, M.E.
United States. Department of Agriculture. Agricultural Research Service. Publications, n. ARS 41-88, Dec. 1964. pp. 19
Describes in detail the chemical synthesis of benzene from the original carbon sample. Benzene of scintillation quality is produced in this synthesis with an overall carbon yield of 50 to 60% and can be synthesised and counted in one week.

3.263   1968
RADIOCARBON DATING BY THE SCINTILLATION METHOD
Arslanov, Kh. A.; Gromova, L.I.; Polevaya, V. I.
Geochemistry International, v. 5, n. 1, 1968: 148-157
A method for the chemical preparation of samples submitted for radiocarbon dating is described. Age determinations of 32 samples (ranging from 300 to >62,000 years) are reported. Most of the age determinations are compared with palynological results.

3.264   1968
RADIOCARBON DATING FOR MEDIEVAL TIMBER BUILDINGS
Fletcher, John
Antiquity, v. 42, n. 167, Sept.1968: 230-231
This note is to indicate the procedure capable of giving relatively accurate building dates for medieval timber buildings: (1) use heartwood; (2) measure radiocarbon age and, if possible, $^{14}C/^{13}C$ ratio; (3) correct the radiocarbon age to the true age by application of the Stuiver - Suess relationship; (4) allow for the growth allowance (i.e. the number of tree rings between the sample and the bark).

3.265   1957
RADIOCARBON DATING IN TRONDHEIM
Nydal, Reidar; Sigmond, R.S.
Applied Scientific Research, Section B, v. 6, 1957: 393-400
Naturally occurring radiocarbon is registered in a proportional counter filled with pure carbon dioxide. The carbon dioxide gas is purified in almost the same manner as developed by de Vries and Barendsen (*Physica*, 19, 1953). Checking of gas purity is described. A grid proportional counter with an effective volume of 2.5 L with gas pressure 1 to 4 atm is used. At 4 atm net count for modern wood is 58 cpm above a bkg of 15 cpm. Dating may be extended to 35,000 years by 24 hours counting at 4 atm.

3.266   1960
RADIOCARBON DATING LABORATORY, MUSEUM OF APPLIED SCIENCES OF VICTORIA
Focken, Charles M.
Australian Journal of Science, v. 23, n. 4, 21 Oct. 1960: 127-138
Report on the initial work of the new Radiocarbon Laboratory established at the Museum of Applied Sciences of Victoria. The reliability and accuracy of the equipment have been established. The limit of age within which measurements can be made without increasing the time of counting abnormally is about 37,000 years.

3.267   1956
RADIOCARBON DATING LISTS AND THEIR USE
Johnson, Frederick
American Antiquity, v. 21, n. 3, Jan. 1956: 312-313
The degree of success of the method of dating depends upon the continuing research in the laboratories and meticulous attention to details by collectors and very clear and precise description of samples.

3.268   1954
THE RADIOCARBON DATING METHOD
Focken, Charles M.
Australian Journal of Science, v. 17, n. 1, 1954: 10-11
A brief summary of the method and of present investigations of the age of fossil and sub-fossil carbonaceous material.

3.269   1965
RADIOCARBON DATING OF ANCIENT MORTAR AND PLASTER

Stuiver, Minze; Smith, C.S.
In: International Conference on Radiocarbon and Tritium Dating, Pullman, Washington, Washington State University, June 1-11, 1965. *Proceedings*. United States of America, Atomic Energy Commission, 1965. Conference n. 650652: 338-343
Inadequate burning of lime is suggested as an additional source of error in the radiocarbon dating of ancient mortar and plaster

3.270   1967
RADIOCARBON DATING OF BIOGENIC OPAL
Wilding, L.P.
Science, v. 156, n. 3771, 7 Apr. 1967: 66-67
Approximately 75 grams of biogenic opal were isolated from 45 kilograms of soil by employing a gross particle size and sink-float specific gravity fractionation procedure. After pretreatment of the sample to remove extraneous organic and inorganic carbon contaminants, the carbon occluded within opal phytoliths was dated at $13,300 \pm 450$ years BP, therefore biogenic opal is stable for relatively long periods.

3.271   1964
RADIOCARBON DATING OF BONE AND SHELL FROM THEIR ORGANIC COMPONENT
Berger, Rainer; Horney, Amos G.; Libby, Willard F.
Science, v. 144, n. 3621, 22 May 1964: 999-1001
A method of dating bone and shell by radiocarbon content has been developed. The mineral is removed by mild acid treatment and the residual carbon is dated in the usual manner.

3.272   1967
RADIOCARBON DATING OF BONES
Barker, Harold
Nature, v. 213, n. 5074, 28 Jan. 1967: 415
A discussion and refutation of the article by Sellstedt and others (*Nature*, 212: 572), indicating that according to Münnich (*Science*, 126: 194), Olson and Broecker (*Radiocarbon*, 3: 141), Berger and Libby (*Radiocarbon*, 8: 467) and other researchers, no general conclusion can safely be drawn about the suitability of bones as dating material.

3.273   1965
RADIOCARBON DATING OF FOSSIL HOMINIDS
Oakley, Kenneth P.
In: International Conference on Radiocarbon and Tritium Dating, Pullman, Washington, Washington State University, June 1-11, 1965. *Proceedings*. United States of America, Atomic Energy Commission, 1965. Conference n. 650652: 277-286

The absolute or chronometric dating of early human remains provides the ultimate framework for hominid evolution. Four orders of absolute dating are discussed.

3.274   1966
RADIOCARBON DATING OF ICE
Oeschger, Hans; Alder, B.; Loosli, H.H.; Langway, Chester C.; Renaud, André
Earth and Planetary Science Letters, v. 1, n. 2, 1966: 49-54
A counter is described for use in radiocarbon dating of ice samples containing only 20 to 100 mg of carbon. It was developed for the dating of ice samples from the Tuto tunnel, Thule, Greenland. The counter is described, illustrated and its characteristics are tabulated. Trapping of carbon dioxide by molecular sieves was successful in addition to the commonly used technique of precipitation from melting ice by sodium hydroxide. Both methods of preparation yielded similar age results. Age data are tabulated.

3.275   1965
RADIOCARBON DATING OF ICE
Oeschger, Hans; Alder, Bernhard; Loosli, H.H.; Langway, Chester C.; Renaud, André
In: International Conference on Radiocarbon and Tritium Dating, Pullman, Washington, Washington State University, June 1-11, 1965. *Proceedings*. United States of America, Atomic Energy Commission, 1965. Conference n. 650652: 93-102
A special counter which allows measurements of a sample of only 20 to 50 mg of carbon was developed in order to be able to date polar ice by the radiocarbon technique.

3.276   1964
RADIOCARBON DATING OF MUSSEL SHELLS: A REPLY
Keith, M.L.; Anderson, G.M.
Science, v. 144, n. 3620, 15 May 1964: 890
In this reply, the author indicates that the relative contribution of the several carbon reservoirs should be more exactly defined by a detailed study of ^{13}C and ^{14}C content of bicarbonate, food web and mollusc flesh in biologic communities of rivers (See Broecker, *Science*, 143: 596 and Keith and Anderson, *Science*, 141: 634).

3.277   1968
RADIOCARBON DATING OF THE ORGANIC PORTION AND WATTLE-AND-DAUB HOUSE CONSTRUCTION MATERIALS OF LOW CARBON CONTENT
Taylor, R.E.; Berger, Rainer
American Antiquity, v. 33, n. 3, July 1968: 363-366
Radiocarbon determinations on a series of low carbon content ceramic and wattle-and-daub samples were made to

determine the validity of radiocarbon dates based on these types of sample materials. Good agreement between radiocarbon dates obtained from the ceramic samples and from the charcoal samples stratigraphically associated with the ceramics suggest that radiocarbon dates obtained from low carbon content ceramic materials are reliable if appropriate precautions are observed. The confidence which can be placed on radiocarbon dates obtained on wattle-and-daub sample material is, at present, somewhat less secure. Problems in the use of these sample materials are discussed.

3.278 — 1955
RADIOCARBON DATING SYSTEM
Fergusson, Gordon J.
Nucleonics, v. 13, n. 1, Jan. 1955: 18-23
Dating range has been extended to 45,000 years and accuracy of 'living' carbon assay increased to 0.3% by carefully instrumented system centered on a 7.7 L. proportional counter filled up to 3 atmospheres of pure carbon dioxide.

3.279 — 1959
A RADIOCARBON DATING SYSTEM USING SCINTILLATION TECHNIQUES
Delaney, C.F.G.; McAuley, I.R.
Royal Dublin Society of Science. Proceedings. Series A, v. 1, n. 1, 1959: 1-20
The development of a radiocarbon dating system using scintillation techniques is described. The sample to be dated is converted to methyl alcohol and mixed with a liquid scintillator. A discussion of the chemical synthesis is given and some results are presented.

3.280 — 1959
A RADIOCARBON DATING SYSTEM: MEASUREMENT OF C-14 ACTIVITY OF SEQUOIA RINGS
Dorn, Thomas Felder
Ph.D. Dissertation. University of Washington, 1965 (Microfilm copy from: *University Microfilms Inc.*, Ann Arbor, Mich., 1959 - Mic 59-1219)
A methane proportional counter for measurement of natural radiocarbon activity was used to measure the radiocarbon activity of several samples of sequoia rings. The counting system produced meaningful results to 45,000 years.

3.281 — 1963
RADIOCARBON DATING TECHNIQUES
Kusumgar, Sheela; Lal, D.; Sharma, v. K.
Indian Academy of Science. Proceedings, v. 58, Section A, n. 3, Sept. 1963: 125-140
The radiocarbon method and the specific techniques which are employed at the Radiocarbon Laboratory at the Tata Institute of Fundamental Research are described.

Ch. 3 - Techniques and Instrumentation

3.282 — 1958
RADIOCARBON DATING UP TO 70,000 YEARS BY ISOTOPIC ENRICHMENT
Haring, A.; de Vries, A.E.; de Vries, Hessel
Science, v. 128, n. 3322, 29 Aug. 1958: 472-473
A method of isotopic enrichment of radiocarbon for dating samples which are too old, and therefore have too small an activity, is tested and discussed. A date of 64,000 ± 1100 years of a test sample is considered as fitting fairly well with the established chronology.

3.283 — 1957
RADIOCARBON DATING WITH LIQUID CO_2 AS DILUENT IN A SCINTILLATION SOLUTION
Barendsen, G.W.
Review of Scientific Instruments, v. 28, June 1957: 430-432
The use of the scintillation technique for radiocarbon dating has been retarded by the difficulties encountered in the chemical procedure. This paper shows that it is feasible to use the carbon dioxide obtained by combustion of the sample, without any further chemical conversion, as a diluent in a scintillation solution. Up to 80% by weight and possibly more liquid carbon dioxide can be dissolved in toluene plus 5 g/L diphenyl-oxazol (PPO).

3.284 — 1958
RADIOCARBON DATING: ITS SCOPES AND LIMITATIONS
Barker, Harold
Antiquity, v. 32, n. 128, Dec. 1958: 253-263
In a reply to criticism published by various archaeologists, the principles underlying the technique of radiocarbon dating are restated to help bridge the gap between the archaeologist and the physicist.

3.285 — 1953
RADIOCARBON DATING: LARGE SCALE PREPARATION OF ACETYLENE FROM ORGANIC MATERIAL
Barker, Harold
Nature, v. 172, n. 4379, 3 Oct. 1953: 631-632
The British Museum Laboratory studies a new method of dating by the conversion of carbon to acetylene and the counting of radiocarbon activity in a proportional counter. Description of the acetylene production using lithium. (Arrol - Glascock method, *Nature*, 159, 1947: 810).

3.286 — 1956
THE RADIOCARBON METHOD OF AGE DETERMINATION
Broecker, Wallace S.; Kulp, J. Lawrence
American Antiquity, v. 22, n. 1, July 1956: 1-11
The principle of the method is described and the possible

sources of error in radiocarbon determinations and the limitations they place upon the method are discussed. Whenever anomaly is present, further field and laboratory work must be done. As a radiocarbon determination gives an estimate of the time at which a sample was removed from the carbon dioxide cycle, only careful determination in the field will ascertain whether a date represents the time of an event in which the archaeologist is interested.

3.287 1966
RADIOCARBON SAMPLES: CHEMICAL REMOVAL OF PLANT CONTAMINANTS
Haynes, C. Vance
Science, v. 151, n. 3716, 18 March 1966: 1391-1392
Roots and similar plant debris can be efficiently removed from charcoal samples by nitration and acetone leaching after preliminary removal of humic acid and lignine by standard procedures.

3.288 1968
RADIOCARBON: ANALYSIS OF INORGANIC CARBON OF FOSSIL BONE AND ENAMEL
Haynes, C. Vance
Science, v. 161, n. 3842, 16 Aug. 1968: 687-688
Carbon dioxide from calcium carbonate in fossil bone can be selectively separated from carbon dioxide in bone apatite by hydrolising the sample first in acetic acid and then in hydrochloric acid. Radiocarbon analysis of the inorganic carbon dioxide in three samples of known age clearly show calcium carbonate in fossil to be secondary and the carbonate of bone apatite to be indigenous and suitable for dating in some cases. Agreement between dates on collagen - bone apatite pairs increase the level of confidence.

3.289 1959
RADIOISOTOPES IN THE DATING OF GEOLOGICAL AND ARCHAEOLOGICAL EVENTS
Giletti, Bruno J.; Lambert, R. St. J.
Research (London), v. 12, n. 10-11, 1959: 369-373
Reviews the methodology of radioisotope dating. Describes the radiocarbon method for dating the past and discusses the other radioisotopic methods for dating the distant past.

3.290 1965
A RADON EFFECT WITH A COPPER COUNTING TUBE
Freundlich, J.
Prepared for: *International Conference on Radiocarbon and Tritium Dating*, Pullman, Washington, Washington State University, June 1-11, 1965. 1965
A low, but quantitatively measurable, alpha count rate is observed with a proportional counter made of high purity copper tubing.

3.291 1968
A RAPID METHOD OF ^{14}C COUNTING IN ATMOSPHERIC CARBON DIOXIDE
Povinec, P.; Saro, S.; Chudy, M.; Seliga, M.
International Journal of Applied Radiations and Isotopes, v. 19, Dec. 1968: 877-881
A simple method of collecting samples and an apparatus for the preparation and activity measurement of samples of atmospheric carbon dioxide is described. Using this technique, the time for purification of carbon dioxide for radiocarbon measurements was abour 6 hours. The radiocarbon activity observed in the atmosphere around the nuclear power station at Trnava, Czechoslovakia, during Autumn 1967 (before the station started operating) are reported.

3.292 1948
RAPID WET COMBUSTION METHOD FOR CARBON DETERMINATION WITH PARTICULAR REFERENCE TO ISOTOPIC CARBON
Lindenbaum, Arthur; Schubert, Jack; Armstrong, W.D.
Analytical Chemistry, v. 20, n. 11, Nov. 1948: 1120-1121
Description of a convenient wet combustion procedure for the determination of the total and radioactive content of the same sample of animal tissues, excreta and other materials.

3.293 1961
RECENT DEVELOPMENT IN C-14 DATING
Tauber, Henrik
In: International Association of Quaternary Research (INQUA), Congress, 6th, Warsaw, 1961. *Report*, 1961: 729-741
Developments due to an intensive program of investigation carried out in the Radiocarbon Laboratories and to checks with independent dating methods, are described. New and more reliable dating methods for the sediment in ocean cores have provided a most needed check on the validity of long range radiocarbon dates.

3.294 1961
RECENT DEVELOPMENT IN THE INTERPRETATION AND REPORTING OF CARBON-14 ACTIVITY MEASUREMENTS FROM NEW ZEALAND
Rafter, T.A.
In: Radioactive Tracers in Oceanography, I.U.G.G. Monograph No 20, 1961
Paper presented at the 10th Pacific Science Congress, Aug. 20 - Sept. 6, 1961. The recent work at the Carbon-14 Laboratory of the New Zealand Institute of Nuclear Sciences is reported and the problems met are discussed, in particular the difficulties in reporting, assessing and comparing results for oceanic samples.

3.295   1967
RECENT IMPROVEMENT IN BENZENE CHEMISTRY FOR RADIOCARBON DATING
Noakes, John Edward; Kim, Stephen M.; Akers, Lawrence
Geochimica et Cosmochimica Acta, v. 31, n. 6, 1967: 1094-1096
Benzene radiocarbon dating is discussed in view of recent improved chemistry. A high yield benzene catalyst is described which eliminates concern for carbon isotope fractionation, benzene purity, and extensive sample preparation time.

3.296   1950
A REDETERMINATION OF THE RELATIVE ABUNDANCE OF THE ISOTOPES OF CARBON, NITROGEN, OXYGEN, ARGON AND POTASSIUM
Nier, Alfred O.
Physical Review, v. 77, v. 6, 15 March 1950: 789-793
Carbon, nitrogen, oxygen, argon and potassium were investigated and new values given of the relative abundance of the isotopes.

3.297   1948
REDUCTION OF CO_2 TO METHANOL BY LITHIUM ALUMINUM HYDRIDE
Nystrom, R.F.; Yanko, W.H.; Brown, W.C.
American Chemical Society. Journal, v. 70, n. 1, 3 Feb. 1948: 441
Carbon dioxide is rapidly absorbed by lithium aluminium in ether solution.

3.298   1955
REFERENCE SAMPLES OF ISOTOPIC ABUNDANCE
Mohler, F.L.
Science, v. 122, n. 3164, 19 Aug. 1955: 334
List of reference samples of isotopic abundance for mass-spectrometric analysis available from the United States National Bureau of Standards.

3.299   1963
REGIONAL REPORTS, 12, NEW ZEALAND
Wilkes, Owen
Asian Perspectives, v. 7, v. 1-2, Summer-Winter 1963: 65-73
Notes the use of radiocarbon dating to establish chronological data and of tree ring dating and the discrepancies between the two methods of dating, indicating that the corrections, as established by Jansen, are not used by the N.Z. Institute of Nuclear Sciences until the discrepancies are confirmed by further work.

3.300   1956
THE REMOVAL OF RADON FROM CO_2 FOR USE IN ^{14}C AGE MEASUREMENTS
de Vries, Hessel
Applied Science Research, section B, v. 6, 1956: 461-470
It is shown that traces of radon can be extracted from carbon dioxide by pumping off from solid carbon dioxide. The method can be used only in those cases where the volatility of the contaminant is not too much lower than that of carbon dioxide. This purification is much faster than the chemical purification described previously (de Vries and Barendsen, *Nature*, 174: 1138).

3.301   1956
RESPONSIBILITIES OF THE ARCHAEOLOGIST IN USING THE RADIO CARBON METHOD
Meighan, Clement W.
Utah University. Anthropological Papers, v. 26, 1956: 48-53
Various problems concerning collaboration between archaeologists and radiocarbon dating specialists are examined. In particular the need to relate the date to a culture, avoid reliance on a single date, and adequate definition of sample.

3.302   1959
REVIEW ARTICLE ON MEASUREMENT AND USE OF NATURAL RADIOCARBON
de Vries, Hessel
In: Research in Geochemistry, edited by P.H. Abelson. New York, Wiley Press, 1959
[Not sighted]

3.303   1965
ROUTINE ^{14}C DATING USING LIQUID SCINTILLATION TECHNIQUES
Tamers, Murray A.
Acta Scientifica Venezolana, v. 16, n. , 1965: 156-162
Development of radiocarbon counting techniques for radiocarbon dating are reviewed and some problems are discussed. The advantages of liquid scintillation counting for routine radiocarbon dating are enumerated.

3.304   1965
ROUTINE CARBON-14 DATING USING LIQUID SCINTILLATION TECHNIQUES
Tamers, Murray A.
In: International Conference on Radiocarbon and Tritium Dating, Pullman, Washington, Washington State University, June 1-11, 1965. *Proceedings*. United States of America, Atomic Energy Commission, 1965. Conference n. 650652: 53-67

Description of the liquid scintillation counting method using benzene as the scintillation solvent. The following laboratory errors are discussed: isotope effect, radon, quenching, counter reproducibility. The performance of various counters and the advantages of the method are listed.

3.305   1959
THE ROUTINE COUNTING OF CARBON-14
Peisach, M.
South African Chemical Institute Journal, v. 12, 1959: 57-61
An improved method of counting radiocarbon labelled barium carbonate is described. Barium carbonate is centrifuged on to a metal plate to produce a thin deposit over a large area. The method permits rapid and easy preparation of samples for analysis. The coefficient of variation of the results is better than 2.5%.

3.306   1965
ROUTINE METHOD FOR DETERMINATION OF ^{14}C BY LIQUID SCINTILLATION
Edwards, B.; Kitchener, J.A.
International Journal of Applied Radiation and Isotopes, v. 16, n. 7, July 1965: 445-446
Organic materials labelled with radiocarbon were oxydised and the $^{14}CO_2$ collected in a scintillation counter for determination of the radiocarbon activity.

3.307   1965
A SAMPLER FOR RADIOCARBON IN SURFACE AIR
Drobinski, J.C.; Goldin, A.S.; Shleien, Bernard
In: *International Conference on Radiocarbon and Tritium Dating*, Pullman, Washington, Washington State University, June 1-11, 1965. *Proceedings*. United States of America, Atomic Energy Commission, 1965. Conference n. 650652: 107-116
A sampling equipment has been constructed to selectively obtain the carbon dioxide from surface air which, after purification, is counted as carbon dioxide in the proportional region. The apparatus and the counting method are described.

3.308   1965
A SAMPLER FOR RADIOCARBON IN SURFACE AIR
Drobinski, J.C.; Goldin, A.S.; Shleien, Bernard
Journal of Geophysical Research, v. 70, n. 24, 15 Dec. 1965: 6043-6046
Compact, automatic equipment for the collection of carbon dioxide from surface air at ambient temperature is described. The sample is collected over a period of 24 to 28 hours, producing approximately 10 litres of carbon dioxide, equivalent to 5 grams of carbon. The radiocarbon content of bi-weekly samples is measured. The experiment shows that the specific activity of the carbon increased over the period January 1964 to December 1966, indicating that the radiocarbon from a stratospheric reservoir was continually entering the troposphere.

3309   1965
SAMPLING POLAR ICE FOR RADIOCARBON DATING
Langway, Chester C.; Oeschger, Hans; Renaud, André; Alder, Bernhard
Nature, v. 206, n. 4983, 1 May 1965: 500-501
The main object of the study was to develop the methods and techniques of sampling, handling and processing glacier ice and to extract the carbon dioxide component of the occluded atmospheric gases. Procedures and investigations are described.

3.310   1955
SCINTILLATING TECHNIQUES FOR THE DETECTION OF NATURAL RADIOCARBON
Funt, B.L.; Sobering, S.; Pringle, R.W.; Turchinetz, W.
Nature, v. 175, n. 4467, 11 June 1955: 1042-1043
Description of a synthesis and use of toluene as solvent in a liquid scintillator.

3.311   1956
A SCINTILLATION COUNTER FOR THE ASSAY OF RADIOACTIVE GASES
Stranks, D.R.
Journal of Scientific Instrumentation, v. 33, Jan. 1956: 1-4
A glass chamber with a plastic scintillator serving as one wall may be used to assay gaseous samples of weak $\int$ emitting nuclides, (notably ^{14}C). The scintillation counter assembly can be readily adapted to the assay of both liquid and solid samples.

3.312   1960
SCINTILLATION COUNTING OF CARBON-14
Jenkinson, David S.
Nature, v. 186, v. 4725, 21 May 1960: 613-614
A precise method is described for the radioassay of radiocarbon in aqueous solutions. The method was developed for use in work on the decomposition in soil of plant material. The radioactivity of the weak beta emitter is determined by counting the pulses of light emitted by a plastic phosphor in contact with a solution of the isotope.

3.313   1954
SCINTILLATION COUNTING OF NATURAL RADIOCARBON: THE COUNTING METHOD
Arnold, James R.

Science, v. 119, n. 3083, 29 Jan. 1954: 155-157
The scintillation counter with moderate efficiency allows the effective use of samples and has greater sensitivity than both screen wall counter (Libby) and gas proportional counters. The author describes the counting technique developed in the Laboratory of the Institute of Nuclear Studies, University of Chicago.

3.314   1961
SCINTILLATION TECHNOLOGY FOR COUNTING NATURAL RADIOCARBON AND ITS USE FOR DETERMINATION OF ABSOLUTE AGE
Starik, I.E.; Arslanov, Kh. A.; Zharkov, A.P.
Soviet Radiochemistry (Radiokhimiya), v. 2, n. 182, 1961: 67-68
A method developed for liquid scintillation counting of natural radiocarbon, designed to determine absolute age, is described. A coincidence scintillation counter was developed. Use of benzene and ethylbenzene as liquid scintillator solvents in conjunction with a highly effective counter made it possible to determine absolute age up to 37,000 years for ethylbenzene and to 42,000 years for benzene.

3.315   1966
SENSITIVITY ENHANCEMENT FOR LOW LEVEL ACTIVITIES BY COMPLETE SYNTHESIS OF LIQUID SCINTILLATION SOLVENT
Tamers, Murray A.
In: Symposium on Organic Scientillators. Argonne, Ill., 1966, 1966. pp. 27
A procedure for synthesising a large part of the solvent for liquid scintillation measurements for low-level activity from the material in the sample to be counted has been applied for three natural radioactivity dating methods, in particular ^{14}C. Benzene was found to be the most suitable compound.

3.316   1958
THE SIGNIFICANCE OF DIFFERENCES BETWEEN RADIOCARBON DATES
Spaulding, A.C.
American Antiquity, v. 23, n. 3, Jan. 1958: 309-311
As radiocarbon dates are derived from estimates of the true rate of emission of electrons from radioactive carbon statistically expressed, the author suggests methods of using the statistical information to make decisions about the meaning of the difference between two or more radiocarbon dates.

3.317   1957
SIMPLE LABORATORY METHOD FOR PRODUCING ENRICHED CARBON 13
Bernstein, Richard B.
Science, v. 126, n. 3264, 19 July 1957: 119-120

Ch. 3 - Techniques and Instrumentation

The reaction of dehydration, of formic acid with sulphuric acid was found to be a suitable one for ^{13}C enrichment by the high-cut method.

3.318   1965
SIMULTANEOUS GAS-PROPORTIONAL COUNTING OF ^3H AND ^{14}C
Jordan, Pierre
Nucleonics, v. 23, Nov. 1965: 46-49
Attention is given to the question of the strong dependence of proportional counter amplification on the gas filling so that the counters become useful as proportional devices only after calibration. Topics discussed include counting precision, simultaneous measurements, sample preparation and the performing of the measurements.

3.319   1958
SINGLE PHOTOTUBE FOR INTRODUCING C-14 IN A LIQUID SCINTILLATION COUNTER
Bernstein, William; Bjerkenes, Clara; Steele, Robert
In: Liquid Scintillation Conference, North Western University, 1958. *Proceedings*. New York, Pergamon Press, 1958: 74-75
By means of a good optical coupling between sample vial and phototube, and the use of quartz instead of glass sample vials to lower the background, radiocarbon can be counted, using a single phototude, by the liquid scintillation method at about 80% efficiency with background counting rate of 80-90 cpm.

3.320   1965
SINGLE-STAGE HIGH-FIELD HYDROGENATION OF CO_2 TO METHANE USING THE HYDROGEN OF WATER
Lal, D.
In: International Conference on Radiocarbon and Tritium Dating, Pullman, Washington, Washington State University, June 1-11, 1965. *Proceedings*. United States of America, Atomic Energy Commission, 1965. Conference n. 650652: 487-490
The technique used for sensitive measurement of radiocarbon and tritium activities is described.

3.321   1956
SOME ASPECTS OF THE QUANTITATIVE APPROACH IN ARCHAEOLOGY
Heizer, Robert F.; Cook, Sherburne F.
Southwestern Journal of Anthropology, v. 12, n. 3, Autumn 1956: 229-248
Reviews the progress which has been made in quantitative analysis and the procedures and techniques available. Indicates that for pure dating and chronology the radiocarbon method is the most valid.

9.322   1961
SOME HIGHLIGHTS FROM THE NATURAL RADIOCARBON DATING OF THE LA JOLLA LABORATORY
Hubbs, Carl L.
National Academy of Science. Annual meeting, Proceedings, 29 Oct. - 1st Nov.1961:-
[Not sighted]

3.323   1949
SOME OBSERVATIONS ON EXCHANGE OF CO_2 BETWEEN $BaCO_3$ AND CO_2 GAS
Samos, George
Science, v. 110, n. 2868, 16 Dec., 1949: 663-665
Previous studies on the exchange of carbon dioxide between barium carbonate and atmospheric carbon dioxide have shown that exchange takes place in the presence of moisture and that the amount of exchange can be reduced by heat. An attempt is made to confirm this by using ^{14}C. (Armstrong and Schubert, *Science,* 106: 403)

3.324   1963
SOME PROBLEMS IN CONNECTION WITH C-14 DATING OF TESTS OF FORAMINIFERA
Eriksson, K. Gösta; Olsson, Ingrid U.
Geological Institutions of the University of Uppsala. Bulletin, v. 42, 1963: 1-13
A radiocarbon dating of foraminiferal tests has been made on deep-sea core material collected from the Western Mediterranean Sea. During the course of the experiments some difficulties arose concerning the preparation of samples. A systematic investigation of the reagents employed demonstrated the importance of using water completely free of carbon dioxide for the preparation of the samples.

3.325   1952
SOME PROBLEMS IN RADIOCARBON DATING
Anderson, Ernest C.; Levi, Hilde
Det Kongelige Danske Videnskabernes. Selskab Matematisk - Fysisk. Meddelelser, v. 27, n. 6, 1952: 1-22
A detailed comparison of the screen wall counter with a gas sample counter with respect to the problem of radiocarbon is given. Curves are presented, showing the accuracy and range of the method as functions of the background rate and the counting time. The errors due to the intrusion of extraneous carbon are presented and discussed, and certain improvements in the method are described.

3.326   1958
SOME USES OF PHYSICS IN ARCHAEOLOGY
Hall, V. T.
Physical Society. Year Book, 1958: 22-34
One of the methods mentioned is direct dating by radiocarbon dating. A new approach to radiocarbon dating is described.

3.327   1958
SPECIAL LOW-LEVEL COUNTERS
Geiss, J.; Gefeller, G.; Hontermans, F.G.; Oeschger, Hans
In: *International Conference on the Peaceful Use of Atomic Energy,* 2nd, Geneva, 1958. *Proceedings.* (A/Conf. 15/P/1123), 1958, v. 21: 147
The anti-coincidence volume and the counting volume are separated by a foil of adapted thickness or very thin wire. Three different counters were constructed based on this principle, in particular one for gas samples of low specific activity, e.g. radiocarbon. Full description of the performance of the counters is given.

3.328   1957
SPECIFICATION SURVEY FOR RADIOCARBON DATING
Krueger, Paul
American Antiquity, v. 23, n.. 1, Oct. 1957: 182
The result of a survey made to ascertain the interest in having a commercial laboratory for radiocarbon dating available and the specifications required are presented.

3.329   1961
STABILIZATION OF THE OVER-ALL COUNTING SENSITIVITY OF GAS-FILLED PROPORTIONAL COUNTING SYSTEM
Carnan, R.D.
Australian Journal of Science, v. 23, n. 10, 21 April 1961: 340-343
A stabilising system developed from G.J. McCallum suggestions is described. It uses the constancy of the meson pulse amplitude distribution as a reference. (McCallum, *N.Z. Journal of Science and Technology,* 37: 370)

3.330   1968
STRATIGRAPHY AND GEOLOGIC TIME
Harbaugh, John W.
Dubuque, Iowa, Wm. C. Brown, 1968
A text in historical geology. The sixth chapter deals with radiometric age dating - radioactive decay and its application to quantitative measurement of geologic time.

3.331   1960
STRATOSPHERIC MIXING FROM RADIOACTIVE FALLOUT
Libby, Willard F.; Palmer, C.E.
Journal of Geophysical Research, v. 65, n. 10, Oct. 1960: 3307-3317

Some general conclusions about the residence time of stratospheric debris are suggested: latitude, altitude, season and even year of injection are all variables.

3.332   1957
A SUITABLE METHOD OF MEASURING RADIOCARBON
Apelgot, S.
Journal of Physical Radiations, v. 16, n. 7, 1957: 78-89
[Not sighted]

3.333   1965
SYSTEM DESIGN IN LOW BACKGROUND INTERNAL GAS SAMPLE COUNTING OF CARBON-14 AND TRITIUM
Sharp, Rodman A.; Ellis, John G.
In: International Conference on Radiocarbon and Tritium Dating, Pullman, Washington, Washington State University, June 1-11, 1965. Proceedings. United States of America, Atomic Energy Commission, 1965. Conference n. 650652: 17-28
Good results in low background internal gas sample counting of radiocarbon and tritium depend on careful optimisation of an integrated system including design of detector, cosmic ray guard counter, shielding and choice of counting gas. Recent advances leading to a simplified and economical high performance internal gas sample counting system are described. The relative merits of internal gas sample counting and liquid scintillation counting for low level radiocarbon and tritium measurements are compared.

3.334   1965
A SYSTEM FOR METHANE SYNTHESIS
Olson, Edwin A.; Nickoloff, Nick
In: International Conference on Radiocarbon and Tritium Dating, Pullman, Washington, Washington State University, June 1-11, 1965. Proceedings. United States of America, Atomic Energy Commission, 1965. Conference n. 650652: 41-52
Description of a system for easy, efficient methane synthesis building upon the work of Fairhall, Schell and Takashima, (*Review of Scientific Instruments*, 32: 323). Use is made of ruthenium catalyst and of a batch reactor. The yield is 95% and the time for processing is 6 hours.

3.335   1952
TECHNIQUES OF NATURAL CARBON-14 DETERMINATION
Kulp, J. Lawrence; Tryon, Lansing E.; Feely, Herbert W.
American Geophysical Union. Transactions, v. 33, n. 2, April 1952: 183-192
Description of the techniques in use at the Lamont Geological Observatory for the preparation and counting of the radiocarbon in natural carbon-bearing samples.

3.336   1965
TECHNIQUES OF METHANE PRODUCTION FOR C-14 DATING
Long, Austin
In: International Conference on Radiocarbon and Tritium Dating, Pullman, Washington, Washington State University, June 1-11, 1965. Proceedings. United States of America, Atomic Energy Commission, 1965. Conference n. 650652: 37-40
The Smithsonian Radiocarbon Dating Laboratory uses a modified commercial methane system with two flow type ruthenium catalyst reactors. Problems of tritium contamination can be avoided by proper selection of hydrogen and pretreatment of catalyst. Radon contamination can be removed on activated charcoal at -40°C. Error due to fractionation could become significant only if reaction yield is less than 90%.

3.337   1958
TELLTALE RADIOACTIVITY IN EVERY LIVING THING IS CRACKING THE RIDDLE OF AGE
Briggs, Lyman J.; Weaver, Kenneth, F.
National Geographic Magazine, August 1958: 235-255
Describes how radiocarbon dating has helped reveal the date of past events and the many applications in which the method has been used.

3.338   1965
TEN YEARS TRIAL AND ERROR WITH THE CO, PROPORTIONAL COUNTING TECHNIQUE IN TRONDHEIM
Nydal, Reidar
In: International Conference on Radiocarbon and Tritium Dating, Pullman, Washington, Washington State University, June 1-11, 1965. Proceedings. United States of America, Atomic Energy Commission, 1965. Conference n. 650652: 1-16
Paper presenting the development and improvements in the following fields: construction and shielding of proportional counters, construction of anticoincidence ring counters, arrangement of counting and electronic equipment, combustion and purification systems.

3.339   1963
TERRESTRIAL AGES OF METEORITES FROM COSMOGENIC C14
Kohman, Truman P.; Goel, Parmatma S.
In: Radioactive Dating. Proceedings of a Symposium on Radioactive Dating held in Athens, 19-23 November 1962. Vienna, International Atomic Energy Agency, 1963. (Pro-

ceedings Series): 395-411

Techniques have been developed for the isolation and measurement of cosmogenic radiocarbon in meteorites. For stones, up to 10 to 40 g samples are decomposed with an oxidising flux in a closed system. For irons, 150 to 400 g samples are dissolved in nitric acid in a stream of oxygen, sweeping outbound and cosmogenic carbon but leaving graphite. The radiocarbon is counted as highly purified carbon dioxide in small low level proportional counters.

3.340   1967
TRITIUM AND CARBON-14 BY OXYGEN FLASKS
Davidson, Jackson D.; Oliviero, Vincent T.
In: Advances in Tracer Methodology, v. 3. Proceedings of the 10th Annual Symposium in Tracer Methodology, edited by Seymour Rothchild. New York, Plenum Press, 1968: 67-69

Provides a cookbook version of the entire technique of oxygen flask combustion for the handling of biological samples.

3.341   1962
UNPURIFIED CARBON DIOXIDE AS A FILLING GAS FOR COUNTERS
Kalab, B.; Broda, E.
International Journal of Applied Radiation and Isotopes, v. 13, 1962: 191-203

Using suitable electronics it is possible to obtain good plateaux, with counters filled with unpurified carbon dioxide without addition of foreign gases. The simplicity of the procedure described and the quality of the plateau make possible the rapid, precise and sensitive routine assay of radiocarbon by this method.

3.342   1963
UPPER TROPOSPHERIC CARBON-14 LEVELS DURING SPRING 1962
Fergusson, Gordon J.
Journal of Geophysical Research, v. 68, n.. 13, 1st July 1963: 3933-3941

Measurements of the radiocarbon activity of atmospheric carbon dioxide at an altitude of 12 km and at latitude 30°N have shown wide variations during the spring of 1962 owing to the transfer of $^{14}CO_2$ from the stratosphere to the upper troposphere. The observed increased activity during the period January to July 1962 was 14%. The technique for sample collection is described. The aim is to investigate the possibility for the use of $^{14}CO_2$ from nuclear weapon tests as a gaseous tracer of atmospheric movements.

3.343   1953
USE OF AN ACETYLENE FILLED COUNTER FOR NATURAL RADIOCARBON
Crathorn, A.R.
Nature, v. 172, n. 4379, 3 Oct. 1953: 632-633

A mild steel counter of 3 litres volume surrounded by 11 Geiger counters used in anticoincidence enclosed in a 4 m thick lead shield is used. Three milligrams of carbon are introduced, filled to 1 atmosphere pressure with acetylene.

3.344   1965
USE OF C13/C12 RATIOS TO CORRECT RADIOCARBON AGES OF MATERIALS INITIALLY DILUTED BY LIMESTONE
Pearson, Frederick Joseph
In: International Conference on Radiocarbon and Tritium Dating, Pullman, Washington, Washington State University, June 1-11, 1965. *Proceedings*. United States of America, Atomic Energy Commission, 1965. Conference n. 650652: 357-366

Where the active datable carbon is plant-derived and the diluent is limestone, as in terrestrial snail shells, the carbonate dissolved in groundwater, caliche, and mortar and water, the $^{13}C/^{12}C$ ratio of the sample material should reflect the proportion of plant to limestone carbon, and provide a factor for correcting the measured radiocarbon activity to determine the true sample age.

3.345   1955
THE USE OF CARBON DIOXIDE FILLED PROPORTIONAL COUNTERS FOR RADIOCARBON DATING
Fergusson, Gordon J.
Nucleonics, v. 13, n. 1, Jan. 1955: 18-23

Dating range has been extended to 45,000 years and accuracy of 'living carbon' assay increased to 0.3% by a carefully instrumented system centred on a 7.7 L proportional counter filled with up to 3 atmospheres of pure carbon dioxide. Description of the instrumentation includes the proportional counter, the anti-coincidence Geiger counter ring, the filling system, the shield and the associated electronics.

3.346   1954
THE USE OF CARBON DIOXIDE FILLED PROPORTIONAL COUNTERS FOR RADIOCARBON DATING
Fergusson, Gordon J.
New Zealand. Department of Scientific and Industrial Research. Dominion Physical Laboratory. Research Report, n. 225, 1954: 18-23
[Not sighted]

3.347   1954
USE OF DISSOLVED ACETYLENE IN LIQUID SCINTILLATION COUNTERS FOR THE MEASUREMENT OF CARBON 14 OF LOW SPECIFIC ACTIVITY
Audric, B.N.; Long, J.V.P.
Nature, v. 173, n. 4412, 22 May 1954: 992-993
A 2 : 5 diphenyl oxazol solution at a concentration of 3 mg/L in toluene containing 2% ethyl alcohol is used as a phosphor solution for dissolving acetylene (10 lb of acetylene equivalent to 11 mg of carbon in a 100 mL counter). Efficiency should be 50%.

3.348   1957
VALIDITY OF RADIOCARBON DATES IN ORGANIC SAMPLES WITH AGES GREATER THAN 25,000 YEARS (Abstract)
Olson, Edwin A.; Broecker, Wallace S.
Geological Society of America. Bulletin, v. 68., n. 12, Dec. 1957: 1775-1776
Two types of contamination of organic materials are discussed: permeation by humic acid, intrusion of macroscopic animal or plant materials. The handling of such contamination is discussed. Samples in the age range 25,000 to 50,000 years are dated and discussed.

3.349   1962
VERY LOW BACKGROUND PROPORTIONAL COUNTER ASSEMBLIES
Ramsden, D.
United Kingdom Atomic Energy Authority. Weapons Group. Atomic Weapons Research Establishment, Aldermaston, England, August 1962. pp. 29 (AWRE - NR/P - 3/62)
Design features are described of very low background proportional counter assemblies for use in radiocarbon dating. For an optimum background, a large screen wall counter was used in the low level tank in conjunction with boron wax shielding. For higher activity, with 24 hours counting, a plastic phosphor counter gave sufficient accuracy.

3.350   1951
WESTERN PREHISTORY IN THE LIGHT OF CARBON-14 DATING
Cressman, L.S.
Southwestern Journal of Anthropology, v. 7, n. 3, Autumn 1951: 289-313
Compares the dates suggested by archaeologists, based on stratigraphical and archaeological observations, to the dates determined by radiocarbon analysis. Raises the problem of the validity of radiocarbon dates on wet material and suggests the type of checking to be done. Indicates that radiocarbon dates show unexpected antiquity and long persistence for certain technologies.

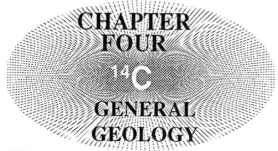

CHAPTER FOUR
14C GENERAL GEOLOGY

4.001   1967
^{14}C DATES OF CALCAREOUS SAMPLES, HERON ISLAND, GREAT BARRIER REEF
Wolf, Karl H.; Östlund, Göte H.
Sedimentology, v. 8, n. 3, 1967: 249-251
Two samples of plant rich limy sand collected from the 9 and 11 ft level of the Heron Island cay were submitted for radiocarbon dating. Uncertainty related to secondary changes and contamination and the unknown transportation history of the woody material prior to accumulation in combination with the relatively young age of the carbon did not result in radiocarbon value useful for correlation and absolute dating purposes.

4.002   1967
A 3,300 YEAR OLD THYLACINE (*MARSUPIALIA: THYLACINIDAE*) FROM THE NULLARBOR PLAIN, WESTERN AUSTRALIA
Partridge, Jeannette
Royal Society of Western Australia. Journal, v. 50, part 2, Dec. 1967: 57-59
An almost complete skeleton of the Thylacine *Thylacinus cynocephalus (Harris)* was recovered from the Murra-el-elevyn cave on the Nullarbor Plain, Western Australia. Analysis of dried tissue attached to the bones gave a radiocarbon date of 3280 ± 70 years BP, making this specimen the youngest Thylacine yet recovered from the mainland of Australia.

4.003   1962
A.N.Z.A.A.S. COMMITTEE FOR THE INVESTIGATION OF THE QUATERNARY STRANDLINE CHANGES (SECTIONS C AND P)
Gill, Edmund D.
Australian Journal of Science, v. 25, n. 5, 1962: 202-205
Recent progress in the correlation of marine and terrestrial Quaternary deposits throughout the world have been aided by advances in research on deep sea sediments, radiocarbon dating and oxygen isotope measurements. Data on the results of shoreline investigation in New South Wales, Victoria, Tasmania, South and Western Australia and New Zealand are reported.

4.004   1962
ABSENCE OF CARBON-14 ACTIVITY IN DOLOMITE FROM FLORIDA BAY
Deffeyes, Kenneth S.; Martin, E.L.
Science, v. 136, n. 3518, 1 June 1962: 782
A sample of dolomite crystals concentrated from recent carbonate sediments in Florida Bay gave a radiocarbon age greater than 35,000 years. Since recent sedimentation began less than 4000 years ago, the dolomite must be derived from older rock and Taft's hypothesis that dolomite is forming today is incorrect.

4.005   1965
ABSOLUTE AGE OF HOLOCENE DEPOSITS OF THE USSR
Neustadt, M.I.
In: International Association for Quaternary Research (INQUA). Congress, 7th, Boulder, Col., 14 Aug - 19 Sept. 1965. *Abstracts of General Sessions*. Boulder, Colorado, 1965: 357
Absolute chronological determinations of peat and lacustrine deposits of Holocene age were made be means of radiocarbon; this allowed an estimate on the chronological position of the Pleistocene-Holocene boundary.

4.006   1967
ABSOLUTE CHRONOLOGY OF THE CENTRAL URAL HOLOCENE
Khotinskiy, n. A.; Devirts, A.L.; Markova, N. G.
Akademiia Nauk SSSR. Doklady. Earth Science Sections, v. 171, n. 1-6, June 1967: 131-123
Palinological and radiocarbon dates are interpreted in terms of the ecological succession of the areal vegetation. The newly determined age of the lower part of the section (about 10,000 years) is about one half of the earlier interpretation.

4.007   1963
ABSOLUTE CHRONOLOGY OF THE LATE QUATERNARY BALTIC
Fromm, E.
Baltica, v. 1, 1963: 46-59
[Not sighted]

4.008   1966
ACCUMULATION OF DIATOMACEOUS SILICA IN THE SEDIMENTS OF THE GULF OF CALIFORNIA
Calvert, S.E.
Geological Society of America. Bulletin, v. 77, n. 6, June 1966: 569-596
The contribution of biogenous silica to the sediments of the Gulf of California and the factors controlling the formation of the deep-water sediment facies have been determined by

4.009   1967
THE ACTIVITY OF COASTAL LANDSLIDES RELATED TO SEA-LEVEL
Emery, Kenneth Orriz
Revue de Geographie Physique et Geologie Dynamique, ser. 2, v. 9, n. 3, 1967: 177-180
Coastal landslides require a sea cliff that has been steepened through undercutting by wave erosion. General considerations supported by some radiocarbon dates indicate that coastal landslides were much intensified about 5000 years ago when sea-level rose to near its present position.

4.010   1954
ADDITIONAL POLLEN PROFILES FROM SOUTHEASTERN ALASKA
Heusser, Calvin John
American Journal of Science, v. 252, n. 2, Feb. 1954: 106-119
In a further investigation of the post-glacial development of the Coast Forest Formation in Alaska, pollen and peat stratigraphies of three additional muskeg sections taken from the mainland of southeastern Alaska, north of Alexander Archipelago, were examined. An estimated date for the Wisconsin recession, obtained by Heusser (1952), through radiocarbon measurements, was used.

4.011   1968
ADDITIONAL RADIOCARBON DATES, TYRRELL SEA AREA
Wagner, Frances J.E.
Maritime Sediments, v. 3, n. 4, 1967 (1968): 100-104
A table of radiocarbon dates for raised beaches around Hudson Bay is presented. From the dates plotted relative to position above present sea-level, three curves follow the pattern of ice retreat and subsequent uplift in the northern central, and southern areas. The latter was freed from ice earlier: its more scattered samples of corresponding ages are higher than in the northern parts.

4.012   1963
AGE AND ACCUMULATION RATE OF DOLOMITE BEARING CARBONATE SEDIMENTS IN SOUTH AUSTRALIA
Skinner, H.C.W.; Skinner, B.J.; Rubin, Meyer
Science, v. 139, n. 3552, 25 Jan. 1963: 335-336
Radiocarbon measurements indicate that dolomite is being formed today by direct precipitation from saline waters in the Coorong, a shallow arm of the sea in South Australia, at an approximate rate of 0.2 mm per annum.

4.013   1968
AGE AND COMPARATIVE DEVELOPMENT OF DESERT SOILS AT THE GARDNER SPRING RADIOCARBON SITE, NEW MEXICO
Gile, L.H.; Hawley, J.W.
Soil Science Society of America. Proceedings, v. 32, n. 5, 1968: 709-716
Radiocarbon ages were obtained from seven buried charcoal horizons in a desert area (along the San Andres Mts.). Describes the progression of soil development with increasing age. Notes that soils of Pleistocene age can have distinct argillic horizons and strong horizons of carbonate accumulation.

4.014   1951
AGE DETERMINATION OF PACIFIC CHALK OOZE BY RADIOCARBON AND TITANIUM CONTENT
Arrhenius, Gustaf; Kjellberg, G.; Libby, Willard F.
Tellus, v. 3, n. 4, Nov. 1951: 222-229
Measurements have been carried out on the average radiocarbon activity of a sediment core of cupelagic chalk ooze and the age of the lower boundary was calculated by integration of the radiocarbon decay functions, thus if constancy of the rate of sedimentation can be proved the titanium content, calibrated in terms of absolute age by means of radiocarbon content, can be used for dating.

4.015   1964
AGE OF A WIDESPREAD LAYER OF VOLCANIC ASH IN THE SOUTHWESTERN YUKON TERRITORY
Stuiver, Minze; Borns, Harold W.; Denton, George H.
Arctic, v. 17, n. 4, Dec. 1964: 259-261
Radiocarbon dates pertaining to a widespread layer of volcanic ash in the southwestern Yukon Territory are reported. This should provide workers with a limiting age for Little Ice Age deposits in this region.

4.016   1966
AGE OF ALLUVIAL CLAYS IN THE WESTERN GEZIRA, REPUBLIC OF SUDAN
Williams, M.A.J.
Nature, v. 211, n. 5046, 16 July 1966: 270-271
From radiocarbon dated shell beds in silt terraces it would seem that aggradation of the main Nile began between 30,000 and 15,000 years ago and reached its peak about 10,000 years BP.

4.017   1965
THE AGE OF CLEAR LAKE, OREGON

Benson, G.T.
Ore Bin, v. 27, n. 2, 1965: 37-40

Clear Lake, at the head of the McKenzie River in Linn County, Oregon, was lava-dammed by a flow from the Sand Mountain line of craters. In order to date the various lava flows of the area, an unsuccessful attempt was made to find charcoal formed by burning by the lava. Instead, two samples of wood from a tree in a drowned forest in the lake were dated and gave ages of 3200 ± 220 years and 2705 ± 200 years, thus providing an age of about 3000 years for the lake and indirect ages for the lava flows of 3000 years or less.

4.018 1957
AGE OF GLACIER-PEAK ERUPTION AND CHRONOLOGY OF POST-GLACIAL PEAT DEPOSITS IN WASHINGTON AND SURROUNDING AREAS
Rigg, G.B.; Gould, H.R.
American Journal of Science, v. 255, n. 5, May 1957: 341-363

The age of the ash layer, as determined by radiocarbon dating of peat immediately underlying the ash at two localities, is 6700 years. This age is an absolute chronology of post-glacial events over a wide area in the Pacific North-West.

4.019 1966
AGE OF PUMICE DEPOSITS IN GUATEMALA
Bonis, Samuel; Bohnanberger, Otto; Stoiber, R.F.; Decker, R.W.
Geological Society of America. Bulletin, v. 77, n. 2, 1966: 211-212

Charcoal samples from deposits of the glowing avalanche type in two pumice filled basins in Guatemala have radiocarbon ages of 31,000 ± 3000 and 35,000 ± 3000 years. The avalanche deposits overlie airborne pumice deposits with little or no weathering along the contact suggesting that both originated from the same explosive eruption. Explosive volcanism in Central America may have been the source for other Equatorial Pacific Ash of about the same age.

4.020 1968
AGE OF ROTOITI BRESCIA
Thompson, B.N.
New Zealand Journal of Geology and Geophysics, v. 11, n. 5, Dec. 1968: 1189-1191

A carbonised wood sample collected from a section of the basal siltstone was dated at greater than 41,000 years BP. This indicates a minimum age for the commencement of rhyolitic volcanism in Okataina rhyolite centre.

4.021 1962
THE AGE OF SALT MARSH PEAT AND ITS RELATION TO RECENT CHANGES IN SEA LEVEL AT BARNSTABLE, MASSACHUSETTS
Redfield, Alfred C.; Rubin, Meyer
National Academy of Sciences. Proceedings, v. 48, n. 10, Oct. 1962: 1728-1735

The measurements reported of the radiocarbon age of peat recovered from various depths in salt marshes on Cape Cod were made to determine the rate of vertical accretion of the high marsh peat and thus provide a chronology of the development of these marshes. The results are interpreted to indicate the change of the relative elevation of the sea and land which has occurred in the Cape Cod region during the past 3700 years.

4.022 1965
AGE OF THE BURIED SOIL IN THE SYDNEY, OHIO, AREA
Forsyth, Jane L.
American Journal of Science, v. 263, n. 7, Summer 1965: 571-597

The interval represented by the Sydney soil is fairly long, as shown by the nature of the buried soil, by the series of plants recorded in the pollen profile and radiocarbon dates. The dates show also that the soil-forming interval ended about 22,000 years ago.

4.033 1965
THE AGE OF THE MONDNEDO FORMATION AND THE MASTODON FAUNA OF MOSQUERA (SABENA DE BOGOTA)
Hammen, Th. van der
Geologie en Mijnbouw, v. 4, n. 11, Nov. 1965: 384-350

The upper Mondñedo formation, Columbia, is principally late-glacial, the middle Mondñedo is at least older than the Allerød and probably not older than early Würm; the lower is probably older than early Würm. Two radiocarbon dates are presented and discussed.

4.024 1955
AGE OF THE OBSIDIAN FLOW AT GLASS MOUNTAIN, SISKIYOU COUNTY, CALIFORNIA
Chesterman, C.W.
American Journal of Science, v. 253, n. 7, July 1955: 418-424

Samples of wood from the tree trunks standing in a bed of pumice at Glass Mountain, Eastern Siskiyou County, California, give a maximum age of 1660 ± 300 years, a minimum of 1107 ± 380 years and an average of 1360 ± 240 years when dated by the radiocarbon method. As the trees were killed by the pumice fall, the age of the pumice is no more than 1660±300 years and no less than 1107±300 years. The age of the ryolite obsidian and composite flows that overlie the pumice cannot exceed 1660±300 years and more probably are 700 to 1000 years old.

4.025   1960
AGE OF THE RIO GRANDE IN SOUTHERN NEW MEXICO
Ruhe, Robert V.
Geological Society of America. Bulletin, v. 71, n. 12, Dec. 1960: 1962-1963.
Abstract of paper presented at the November 1960 meeting of the Geological Society of America. The stratigraphy and fossil content of gravel bordering the Rio Grande Valley indicate a Kansan age. The youngest of the stepped sequence cut below the Jordana - La Mesa sequence is dated at 2620 ± 200 years old. This would indicate cut and extension northward in early mid-Pleistocene.

4.026   1962
AGE OF THE SPIT AT BARROW, ALASKA
Péwé, T.L.; Church, R.E.
Geological Society of America. Bulletin, v. 73, 1962: 1287-1291
The ages of three driftwood specimens recovered from three different stratigraphic levels in the spit at Barrow are discussed. The validity of using these ages to date the spit is questioned as the driftwood may have been re-worked from some older sediments.

4.027   1967
AGE OF TRAPPED SEA WATER AT BOTTOM OF LAKE TUBORG, ELLESMERE ISLAND, NORTH WEST TERRITORIES (Abstract)
Long, Austin
In: Geological Association of Canada - Mineral Association of Canada, Int. Meeting, 1967. Abstracts of Papers, 1967: 51-52
[See entry No. 4.028]

4.028   1967
AGE OF TRAPPED SEA WATER AT THE BOTTOM OF LAKE TUBORG, ELLESMERE ISLAND, NORTH WEST TERRITORIES (Abstract)
Long, Austin
American Geophysical Union. Transactions, v. 48, n. 1, 1967: 136
$^{14}C/^{13}C$ and $^{13}C/^{12}C$ analyses on dissolved bicarbonate from the lake and from a nearby fjord indicate that the salt water in the bottom of the lake has been trapped there for about 3000 years.

4.029   1957
THE AGES OF SOME QUATERNARY SEDIMENTS FROM WANGANUI DISTRICT
Fleming, C.A.
New Zealand Journal of Science and Technology, series B, v. 38, n. 7, 1957: 726-731
The sequence and correlation of late Pleistocene-Holocene formations in the Wanganui district are discussed in the light of four radiocarbon dates. The coastal plain formations of the Hawera series in its type area are older than 45,000 years and thus older than the later substages of the Last Glaciation.

4.030   1968
ALLUVIAL CHRONOLOGY OF THE TESUQUE VALLEY, NEW MEXICO
Miller, John P.; Wendorf, Fred
Journal of Geology, v. 66, n. 2, Mar. 1968: 177-194
Charcoal from a depth of 13 feet in the high terrace alluvium yielded a radiocarbon date of 2230 ± 250 years. This information suggests that rates of sediment production during the period of high-terrace deposition were similar in magnitude to modern rates in this area.

4.031   1963
ALLUVIAL CHRONOLOGY OF THE THOMPSON CREEK WATERSHED, HARRISON COUNTY, IOWA
Daniels, R.B.; Rubin, Meyer; Simonson, G.H.
American Journal of Science, v. 261, n. 5, May 1963: 473-487
Radiocarbon dates indicate that from about 14,000 years to sometime before 2000 years ago the Soetmelk and Watkins members were deposited with only an apparent temporary pause about 11,000 years ago.

4.032   1968
AN ANCIENT LANDSLIDE ALONG THE SAGUENAY RIVER, QUEBEC
La Salle, Pierre; Chagnon, Jean-Yves
Canadian Journal of Earth Science, v. 5, n. 3, 1968: 548-549
The largest landslide (8 square miles) yet reported in eastern Canada is described and two radiocarbon dates (560 ± 90 years BP, 420 ± 90 years BP) believed to be related to the time of its occurrence are presented.

4.033   1965
ANCIENT OYSTER SHELLS ON THE ATLANTIC CONTINENTAL SHELF
Merrill, S. Arthur; Emery, Kenneth Orriz; Rubin, Meyer
Science, v. 147, n. 3656, 22 Jan. 1965: 398-400
Shells of long-dead *Carssostrea virginica* are reported at 71 stations in depth of 42 to 82 meters. The depths exceed those

of the estuaries where the species flourished. Radiocarbon measurements indicate that the oysters were alive 8000 to 11,000 years ago. It is concluded that the oysters lived in lagoons or estuaries which became submerged when the sea level rose at the end of the latest glacial epoch.

4.034   1968
THE ANCYLUS TRANSGRESSION IN THE SKEDE MOSE AREA
Königsson, Lars König
Geologiska Föreningen i Stockholm. Förhandlingar, v. 90, part 1, n. 532, 1968: 5-36
The study concerns the establishment of a time table for the Ancylus transgression of central Öland based on detailed pollen analyses and showing how it is possible to record the beginning of a transgression in a beach covered peat. Radiocarbon datings are used to establish a tentative chronology. It is however noted that sole radiocarbon datings are of limited value if not accompanied by modern pollen investigations.

4.035   1967
APPLICABILITY OF THE CARBON-DATING METHOD OF ANALYSIS TO SOIL HUMUS STUDIES
Campbell, C.A.; Paul, E.A.; Rennie, D.A.; McCallum, K.J.
Soil Science, v. 104, n. 3, 1967: 217-224
The natural radiocarbon activity of the organic matter fractions suggests that the humus was much more stable than the mobile humic acids and the latter more resistant to decomposition than the fulvic acids. Humic fractions of the podzolic gray wooded soil were less stable than their chernozemic counterparts. It is concluded that the radiocarbon dating method supported by other chemical methods of analysis is a useful research procedure in soil organic matter investigations.

4.036   1963
ARCTIC SOILS
Tedrow, John Charles Fremont
In: *Permafrost International Conference*, 11-15 November 1963, LaFayette, Indiana. *Proceedings*. 1963: 50-55
Uses radiocarbon dating to establish a genesis chronology for tundra soils.

4.037   1964
ASPECTS OF THE EARLY POST-GLACIAL FOREST SUCCESSION IN THE GREAT LAKES REGION
Wright, Herbert E.
Ecology, v. 45, n. 3, Summer 1964: 439-448
The rate of deposition of an early post-glacial sequence is estimated from radiocarbon dates.

4.038   1965
AUSTRALASIAN RESEARCH IN QUATERNARY SHORELINES
Gill, Edmund D.
Australian Journal of Science, v. 28, n. 11, 1965: 407-410
Reports on progress with research in New Zealand, Papua - New Guinea, New South Wales, Queensland, Victoria, South Australia and Tasmania. A table of relevant radiocarbon dates for Victoria is presented; the work in Tasmania is also based on radiocarbon dates.

4.039   1954
AUSTRALASIAN RESEARCH ON EUSTATIC CHANGES OF SEA-LEVEL
Gill, Edmund D.
Australian Journal of Science, v. 16, n. 6, June 1954: 227-229
Notes that changes of sea-level can now be dated by radiocarbon wich provides absolute dating, enabling intercontinental correlation to be made.

4.040   1956
AUSTRALIAN AND NEW ZEALAND RESEARCH IN EUSTASY - PART I
Gill, Edmund D.; Gage, Maxwell; Jennings, J.N.
Australian Journal of Science, v. 19, n. 1, Aug. 1956: 17-23
The role of radiocarbon dating in eustatic research is described. Three main points are made: (1) radiocarbon dates lead to the conclusion that eustatic changes of sea-level and accompanying climatic events took place much faster than was formerly supposed, (2) eustatic movements are worldwide and synchronous, and (3) radiocarbon dating will make possible the elucidation of the much debated post-glacial changes.

4.041   1956
AUSTRALIAN AND NEW ZEALAND RESEARCH IN EUSTASY - PART 2
Whitehouse, F.P.; Gill, Edmund D.; Banks, Maxwell R.; Tindale, Norman B.
Australian Journal of Science, v. 19, n. 2, Oct. 1956: 54-58
Work in Queensland, Victoria, Tasmania, South Australia and Western Australia is described. Some dates are based on radiocarbon measurements.

4.042   1955
THE AUSTRALIAN 'ARID PERIOD'
Gill, Edmund D.
Australian Journal of Science, v. 17, n. 6, June 1955: 204-206
With the help of radiocarbon dating it is possible to state that the last glacial period was wetter than the present, the arid period was drier and the period coinciding with the 'Little

Ice Age' was wetter than the present in Victoria.

4.043   1965
AVIAN SPECIATION IN THE QUATERNARY
Selander, Robert K.
In: The Quaternary of the United States: a review volume for the 7th Congress of the International Association of Quaternary Research, edited by H.E. Wright and David G. Frey. Princeton. N.J., Princeton University Press, 1965: 527-542
Estimates of average longevity of avian species are based on radiocarbon averages and geological evidence.

4.044   1967
BRIDGE RIVER ASH AND SOME RECENT ASH BEDS IN BRITISH COLUMBIA
Nasmith, H.; Mathews, W.H.; Rouse, G.E.
Canadian Journal of Earth Science, v. 4, n. 1, Feb. 1967: 163-169
The character, distribution and age of the Bridge River ash from post-glacial sites in British Columbia and Alberta are discussed. The radiocarbon dates enable to calculate an accumulation rate of 0.07cm of peat per year.

4.045   1962
A BURIED BIOTIC ASSEMBLAGE FROM AN OLD SALINE RIVER TERRACE AT MILAN, MICHIGAN
Kapp, Ronald O.; Kneller, William A.
Michigan Academy of Science, Arts, and Letters. Papers, v. 47, 1962: 135-145
An organic deposit bearing several types of plant and animal remains was recovered from beneath a terrace of the Saline River, in southeastern Michigan. Careful study of the geologic setting, ecological interpretation of the biotic assemblage, and radiocarbon dates of buried wood, suggest that the burial may correlate with certain post-glacial events in this region, including changes in the level of Lake Erie. The date of 4080 ± 200 years BP confirms a Nipissing time.

4.046   1967
BURIED SOIL PROFILE UNDER APRON ON AN EARTH FLOW
Eden W.J.
Geological Society of America. Bulletin, v. 78, n. 9, 1967: 1183-1184
An earth flow in sensitive clays of the Saint Lawrence Valley, near Ottawa, covered and preserved an old soil profile. The preserved soil profile includes a layer of peaty material dated about 1140 years BP. An abundance of organic debris, a preferred orientation of the debris pointing away from the scar, the intimate mixing of flow and organic debris, the very flat slopes involved, and slope stability studies indicate very wet conditions at the time of the earth flow and that the former wooded area probably had been struck by an onrushing wave of liquid clay from the flow.

4.047   1967
C14 DATES OF A FOSSIL ELEPHANT TUSK FROM SIRSA VALLEY, HIMACHAL PRADESH
Mohapatra, G.C.
Current Science, v. 36, n. 11, June 1967: 296-297
Report on the date of fossil elephant tusk found nearly 30 ft below the surface of a terrace in the Sirsa Valley. The date is 4070 ± 95 years BP.

4.048   1959
C14 DATING OF CAVE FORMATIONS
Broecker, Wallace S.; Olson, Edwin A.
National Speleologic Society. Bulletin, v. 21, 1959: 43
[Not sighted]

4.049   1963
C14 DATING OF THE UPPER ANTHROPOGENE IN THE U.S.S.R. (with discussions)
Vinogradov, A.P.
In: Radioactive Dating. Proceedings of a Symposium organised by the International Atomic Energy Agency (IAEA), and held in Vienna, 1962. Vienna, IAEA, 1963: 245-268
Reports the results of radiocarbon dating of Holocene peats, Black Sea silts, and other materials from various localities in the USSR. Some of the data are supported with results of pollen analyses.

4.050   1962
C14 DATINGS REFERRING TO SHORELINES, TRANSGRESSIONS, AND GLACIAL SUBSTAGES IN NORTHERN NORWAY
Marthinussen, M.
Norway, Geologiske Undersøkelse. Årbok, 1961. Published as Skrifter, n. 215, 1962: 37-66
Reports preliminary datings of wood, peat and sea shells from a score of locations. Discusses their significance in interpreting the oscillations in the shore levels during the past 13,000 years. Evidence of changed shorelines is described and the nature of the radiocarbon materials and their dates considered for each locality.

4.051   1965
C14 DETERMINATION OF SEA-LEVEL CHANGES IN STABLE AREAS
Shepard, Francis P.; Curray, Joseph R.
In: International Association for Quaternary Research (INQUA). Congress, 7th, Boulder, Col., 14 Aug - 19 Sept. 1965. *Abstracts of General Sessions.* Boulder, Colorado, 1965: 424

A series of radiocarbon dates on salt marsh peat found along low but stable coasts such as Florida and Holland appear to give an almost complete history of the rising sea-level over the past 6000 years, casting doubts on the long-held opinion that the sea stood several meters above the present at one or more periods since the retreat of the ice sheets.

4.052   1968
A C-14 DATE FOR THE MUSCOTAH MARSH
McGregor, Ronald L.
Kansas Academy of Sciences. Transactions, v. 71, n. 1, 1968: 85-86
A raised artesian marsh at the edge of the Delaware River flood plain is investigated. From the bottom of a 36 ft core, a large sample of Abies needles gave a radiocarbon date of 15,500 ± 1500 years. Although its pollen occurs in all but the upper 10 feet, macro remains are not found above the 20 ft level, which supports evidence for transition from boreal forest to prairie between 11,000 and 7000 BP.

4.053   1964
C-14 DATINGS OF QUATERNARY DEPOSITS IN ICELAND. (Icelandic, English summary)
Kjartansson, Gudmundur; Thorarinsson, Sigurdur; Einarsson, T.
Reykjavik. Museum of Natural History. Miscellaneous Papers, n. 39, 1964
Lists radiocarbon ages obtained for peat and other Quaternary materials collected in Iceland during 1950-64, with data on the localities of the significance of the dates.

4.054   1960
C-14 DETERMINATION OF DEEP GROUNDWATERS
Münnich, K.O.; Vogel, John C.
International Association of Scientific Hydrology. Publications, n. 52, 1960: 537-541
The radiocarbon content of bicarbonate in recent groundwater is relatively well defined. This examines the variations in content in samples from various parts of Germany and their ages. A table of radiocarbon content and ages by location is presented together with a figure showing the carbon isotope ratio of the bicarbonate in ground water under varying conditions.

4.055   1960
C-14 MEASUREMENTS AND ANNUAL RINGS IN CAVE FORMATIONS
Broecker, Wallace S.; Olson, Edwin A.; Orr, Phil C.
Nature, v. 185, n. 4706, 9 Jan. 1960: 93-94
The radiocarbon method has been used to test the hypothesis that the prominent rings found in certain calcite deposits at Moaning Cave, California, represent annual growth variations.

4.056   1956
CAINOZOIC HISTORY OF MOWBRAY SWAMP AND OTHER AREAS OF NORTH-WESTERN TASMANIA
Gill, Edmund D.; Banks, Maxwell R.
Queen Victoria Museum, Launceston, Tasmania. Reports, n. 6, 15 July 1956: 9-42
The Mowbray Swamp is described. Peat and sandy peat accumulation between sand ridges accumulated, according to radiocarbon dates, more than 37,000 years ago. Peat and marl formed in a swamp at Pulbeena gave an age of 13,500 years.

4.057   1967
CARBON 14 AGES AND FLOW RATES OF WATER IN CARRIZO SAND, ATESCOSA COUNTY, TEXAS
Pearson, Frederick Joseph; White, D.E.
Water Resources Research, v. 3, n. 1, 1967: 251-261
The radiocarbon in carbonate dissolved in groundwater is derived from plant-produced carbon dioxide in the soil of the recharge area and is usually diluted by radiocarbon-free carbonate dissolved from minerals in the soil and in the aquifer. Techniques based on ratios of stable isotopes and overall carbonate chemistry of the water can be used to correct for this dilution and allow calculation of true water ages. In Texas, ages of water samples were dated by this method from wells of the Eocene Carrizo Sand. Flow rates calculated from available hydrologic data are in agreement with these results.

4.058   1964
CARBON-14 CONTENT OF AQUATIC PLANTS, MONTEZUMA WELL, ARIZONA
Haynes, C. Vance; Damon, Paul E.; Cole, G.A.
American Geophysical Union, Transactions, v. 45, n. 1, 1964: 117
Abstract of paper presented at the 46th Annual Meeting, Washington D.C., April 21-24, 1964. Montezuma Well, Arizona, a lake in the Montezuma National Monument, Yavapai County, Arizona, represents an extreme, but probably not unique, case of photosynthesis using 'fossil' carbon dioxide from artesian water, and serves to emphasise the importance of a careful evaluation of the initial radiocarbon content in radiocarbon dating. Radiocarbon analyses were made of organic material in lake bottom core, modern aquatic and non-aquatic plants and the carbon dioxide evolved from water.

4.059   1956
CARBON-14 DATE FOR A MARINE TERRACE AT SANTA CRUZ, CALIFORNIA
Bradley, C.W.

Geological Society of America. Bulletin, v. 67, n. 5, May 1956: 675-677
On the basis of radiocarbon analysis it is concluded that the emergence which ultimately led to the formation of the 100 ft marine terrace began at some time prior to 39,000 years ago.

4.060 1968
CARBON-14 DATES FROM THE BROAD RIVER VALLEY, MT. FIELD NATIONAL PARK, TASMANIA
Caine, N.
Australian Journal of Science, v. 31, n. 3, Sep. 1968: 119-120
The surficial sands and gravels of the valley bottom have been dated as post-glacial by radiocarbon dating of vegetation remains from within those deposits.

4.061 1965
CARBON-14 DATING OF GROUND WATER IN AN ARID BASIN
Bennett, Richard
In: *International Conference on Radiocarbon and Tritium Dating*, Pullman, Washington, Washington State University, June 1-11, 1965. *Proceedings*. United States of America, Atomic Energy Commission, 1965. Conference n. 650652: 590-596
Work to date indicates that the radiocarbon dating of groundwater in the Tucson Basin is not only compatible with the known hydrologic characteristics, but will also be the first quantitative data from which groundwater velocities and relative recharge contributions of streams can be determined.

4.062 1963
CARBON-14 DATING OF MEDIEVAL ALLUVIUM IN LYBIA
Vita-Finzi, Claudio
Nature, v. 198, n. 4883, 1963: 880
A radiocarbon date of 610 ± 100 years BP was obtained for charcoal from the middle part of a 10-foot high terrace in the lower Ganima wadi in Tripolitania, indicating that the deposition of the alluvium continued during the 14th century. The Ganima and most of the streams in this area are now eroding the medieval alluvium. Alluvial deposits of similar age are known to occur elsewhere in North Africa.

4.063 1958
CARBON-14 DATING OF SOME ARCTIC SOILS
Tedrow, John Charles Fremont; Douglas, L.A.
New Brunswick, N J., Rutgers University. Department of Soils, 1958. pp. 6 (Mimeographed)
Gives results of radiocarbon dating of two samples of upland tundra soils and one of arctic brown from North Alaska. Age determination may throw a light on how it was deposited. An arctic brown sample from Point Barrow was found of great age (2900 ± 13 years BP). It is therefore in equilibrium and not likely to develop into a podzol.

4.064 1963
CARBON-14 DATING OF THE FOSSIL DUNE SEQUENCE, LORD HOWE ISLAND
Squires, Donald F.
Australian Journal of Science, v. 25, n. 9, 1963: 412-413
Radiocarbon dating of calcium carbonate shell material from sediments underlying fossil dune deposits of Lord Howe Island (New South Wales) has permitted identification of two distinct episodes in the Pleistocene section of the island.

4.065 1963
CARBON-14, CARBON DIOXIDE AND CLIMATE
Damon, Paul E.
In: *International Union of Geodesy and Geophysics. General Assembly*, 13th, Berkeley, California, 1963. *Proceedings*. 1963
[Not sighted]

4.066 1965
CARBONATE EQUILIBRIA AND RADIOCARBON DISTRIBUTION RELATED TO GROUNDWATER FLOW IN THE FLORIDAN LIMESTONE AQUIFER, U.S.A.
Hanshaw, Bruce B.; Back, William; Rubin, Meyer
In: *International Association of Scientific Hydrology, Symposium*, Dubrovnik, 1965. *Proceedings*. 1965: 601-614
Study of radiocarbon concentration helps determine the principal areas of recharge in the Floridan aquifer as well as allow calculation of apparent groundwater velocities.

4.067 1966
CHANGES IN LEVEL OF LAKE NAIVASHA, KENYA, DURING POST-GLACIAL TIMES
Richardson, J.L.
Nature, v. 209, n. 5020, 15 Jan. 1966: 290-291
Core analyses of diatoms, pollen, chemistry and mineralogy indicate rate of sedimentation and radiocarbon give absolute date and time span for the various phases of water level. Comparison with climatic patterns in North Eastern Angola and Chad suggests that widespread climatic phenomena, particularly the warmer temperature of the climatic optimum, were at last partially responsible for the changes in size of Lake Naivasha.

4.068   1961
CHANGES IN THE LEVEL OF THE SEA RELATIVE TO THE LAND IN AUSTRALIA DURING THE QUATERNARY ERA
Gill, Edmund D.
Zeitschrift für Geomorphologie, Supplement n. 3, 1961: 73-79

In Australia, there is evidence from radiocarbon analysis of samples from both the East and the West coasts of the mainland that favours a post-glacial higher sea-level. The dates are presented and analysed.

4.069   1965
THE CHANGING RATE OF CALCIUM CARBONATE SEDIMENTATION ON THE EQUATORIAL ATLANTIC FLOOR AND ITS RELATION TO CONTINENTAL LATE QUATERNARY STRATIGRAPHY
Wiseman, John D.A.
Swedish Deep Sea Expedition. Reports, v. 7, n. 7 *Sediment Cores from the North Atlantic Ocean*: 289-349

Tentative correlations have been made between the sedimentary ages of the secondary oscillations in carbonate sedimentary rates, and the radiocarbon ages of climatic changes derived from palynological and glaciological studies as well as from glacio-eustatic investigations. There is good agreement between sedimentary ages for the secondary oscillation and the radiocarbon dated continental climatic changes.

4.070   1968
A CHECK ON GEOLOGIC ESTIMATES OF THE AGE OF OLD COASTAL FORMATIONS OF THE BALTIC SEA
Devirts, A.L.; Markova, N G.; Serebryannyy, I.R.
Akademiia Nauk SSSR Doklady Earth Science Section, v. 182, n. , 1968: 108-111

The Holocene history of the Baltic sea rests on a group of age measurements - archaeologic, varve measurements, palaeontologic and others - on the basis of which a chronology has been developed. Application of the radiocarbon method of chronology makes it possible to test the accuracy of the existing scale and to bring the necessary revisions. A table of radiocarbon dates for organic matter buried under Littorina sediments in the Baltic region is presented.

4.071   1967
CHRONOLOGY OF POST-GLACIAL POLLEN PROFILES IN THE PACIFIC NORTHWEST (U.S.A.)
Hansen, Henry P.
Review of Palaeobotany and Palynology, v. 4, n. 1-4, (*Quaternary Palynology and Actuopalynology*), 1967: 103-105

Composite pollen profiles from any peat sections in the Pacific Northwest showing post-glacial forest sequences are correlated in a table with radiocarbon dated pumice and ash. The 'thermal interval' is shown from 8000 to 4000 years ago. A bibliography of papers by the author, which are relevant to this subject, is added.

4.072   1961
CLIMATIC CHANGES SINCE THE LAST INTERGLACIAL
Flint, Richard Foster; Brandtner, Friedrich
American Journal of Science, v. 259, n. 5, May 1961: 321

Six curves from European and American localities, showing inferred fluctuation of climate during various segments of the time since the last interglacial, are compared. The curves are time calibrated by radiocarbon ages. It is concluded that a broad pattern of fluctuation of climate within the time range considered is becoming evident.

4.073   1961
A CLIMATIC SEQUENCE FOR TWO NEVADA CAVES
Sears, Paul B.; Roosma, Aino
American Journal of Science, v. 259, n. 9, Nov. 1961: 669-678

The caves were cut in granite by pluvial Lake Lahontan. By combining radiocarbon dates on a study of the sediment as well as statistical analysis of the included pollen, a profile consonant with known climatic history since about 15,000 years BP has been obtained. There is indication that the caves became habitable at about the time of the Two Creek interval and were abandoned only about 2,000 years ago.

4.074   1967
THE CLIMATIC SIGNIFICANCE OF THE HOSTERMAN'S PIT LOCAL FAUNA, CENTRE COUNTY, PENNSYLVANIA
Guilday, John E.
American Antiquity, v. 32, n. 2, 1967: 231-232

Charcoal associated with mammal bones found beneath sealed sinkholes in Hosterman's Pit, a limestone cave, was radiocarbon dated at 7290 ± 1000 B.C. The Hosterman's Pit local fauna, here listed, represents the earliest known date for a 'Recent' temperature fauna in northwestern North America: the New Paris No. 4 local fauna, from a cave 65 miles to the southwest, dated at 9300 ± 1000 B.C. represents a boreal fauna. Thus, within a probable time span of 2000 years (or a maximum of 4000 years), the mammalian fauna of central Pennsylvania changed from an assemblage of cold weather species characteristic of central Canada today to the temperate fauna that still lives in the area.

4.075 1960
THE COASTAL PLAIN AT NOME, ALASKA: A LATE CENOZOIC TYPE SECTION FOR THE BERING STRAIT REGION
Hopkins, David Moody; MacNeil, F.S.; Leopold, Estella B.
In: Chronology and Climatology of the Quaternary. International Geological Congress, 21st, Copenhagen, 1960. *Report,* part 4, 1960: 46-57
The molluscan fauna, pollen and foraminifera in marine and non-marine sediments were examined and radiocarbon analyses were made. The nearly complete record of Quaternary sea-level fluctuation and climatic events obtained is evaluated.

4.076 1966
COASTAL SAND DUNES OF GUERRERO NEGRO, BAJA CALIFORNIA, MEXICO
Inman, D.L.; Ewing, G.C.; Corliss, J.B.
Geological Society of America. Bulletin, v. 77, n. 8, 1966: 787-802
Barchan ridges in a dune field of about 40 km have moved inland from a barrier beach to form a sand partition across large lagoons. Observations over 3.5 years show that the dunes moved with a mean velocity of 18 m per year and that the sand discharge rate is compatible with the volume of sand required to build the dune field across the lagoons during the 1800 year existence of the barrier beach, as dated by radiocarbon.

4.077 1960
COASTS AND FJORD AREAS OF FINNMARK
Marthinussen, M.
Norges Geologiske Undersøkelse, n. 208, *The Geology of Norway,* edited by Olaf Holtendhal, 1960: 416-429
A table of radiocarbon dated samples of peat and driftwood from Finnmark is presented.

4.078 1967
COLLAPSE STRUCTURE NEAR SASKATOON, SASKATCHEWAN
Christiansen, E.A.
Canadian Journal of Earth Science, v. 4, n. 5, 1967: 757-767
A structural depression called the Saskatoon Low, part of the Upper Colorado Group, is examined. It is shown that it was well developed at the time of the advent of the Pleistocene. The final collapse is radiocarbon dated at about 12,000 years ago.

4.079 1968
COMPARISON OF VARVE CHRONOLOGY, POLLEN ANALYSIS AND RADIOCARBON DATING: INCLUDING AN INVESTIGATION OF THE A- AS A SYNCHRONOUS LEVEL IN SWEDEN
Wenner, Carl Gösta
Stockholm Contributions in Geology, v. 18, n. 3, 1968: 75-97
In Quaternary geology there are three common dating methods: varve chronology, pollen analysis and radiocarbon dating. These are compared by radiocarbon dating the beginning of the Alnus curve in a pollen diagram and comparing the age with the age of the same curve as counted by Liden in the inorganic delta varves to establish a connection between the present time and de Geer Swedish time scale. A discrepancy of 400 years was found between the radiocarbon age and the varve age. Corrections for sources of error have to be made.

4.080 1951
CONCLUSIONS FROM C-14 AND DE GEER' CHRONOLOGY, DANI - GOTIGLACIAL, WITH DATING
de Geer, Ebba Hult
Geologiska Föreningen i Stockholm. Förhandlingar, v. 73, n. 4 (467), 1951: 557-570
The contemporaneity of the last Pleistocene glaciations in north Europe and in North and South America is discussed. Radiocarbon dates verify G. de Geer's chronology down to the very oldest dates. A chronologic correlation table of the main glacial stages in North America and Europe is presented.

4.081 1968
A CONSIDERATION OF POLLEN, DIATOMS AND OTHER REMAINS IN POST-GLACIAL SEDIMENTS
Dodd, John; Webster, Ruth M.; Collins, Gary; Wehr, Larry
Iowa Academy of Science. Proceedings, v. 75, 1968: 197-209
The significance of pollen and diatoms in post-glacial sediments from a shallow bay of Lake Okoboji, Iowa, is considered. Problems encountered and discussed include recognition of Chrysophyte statospores, redeposition, interpretation of data from lower levels and reliability of radiocarbon dates applied to organic matter from aquatic vegetation.

4.082 1958
CONTRIBUTION TO CANADIAN PALYNOLOGY, N. 1, PART 2: NON-GLACIAL DEPOSITS IN THE ST LAWRENCE LOWLANDS, QUEBEC; PART 3: NON-GLACIAL DEPOSITS ALONG MISSINAIBI RIVER, ONTARIO
Terasmae, Jaan
Canada. Geological Survey Bulletin, n. 36, 1958: 13-34
Uses radiocarbon dating to determine the age of the various sequences.

4.083 1960
CONTRIBUTION TO CANADIAN PALYNOLOGY, N. 2, PART 1: A PALYNOLOGICAL STUDY OF POST-GLACIAL DEPOSITS IN THE ST LAWRENCE LOWLANDS; PART 2: A PALYNOLOGICAL STUDY OF PLEISTOCENE INTERGLACIAL BEDS AT TORONTO, ONTARIO
Terasmae, Jaan
Canada. Geological Survey Bulletin, n. 56, 1960: 1-40
Radiocarbon dating is used to date the sequences. The pollen zones, correlated with a radiocarbon based chronology, are presented in table form.

4.084 1967
CONTRIBUTION TO THE POST-GLACIAL VEGETATIONAL HISTORY OF NORTHERN TURKEY
Beng, Hans-Jürgen
In: *Quaternary Palaeoecology*, edited by M.J. Cushing and H.E. Wright. International Association for Quaternary Research (INQUA). Congress, 7th, Boulder, Col., 14 Aug - 19 Sept. 1965. *Proceedings*, v. 7. New Haven, Yale University Press, 1967: 349-356
Pollen analytical investigations on lakes Yenikaya and Abant in the northern Anatolian mountains are presented. Radiocarbon dates indicate that deposition must have been quite rapid.

4.085 1958
CONTRIBUTIONS TO THE QUATERNARY HISTORY OF THE NEW ZEALAND FLORA. 2. PLANT REMAINS FROM A BURIED PEAT LAYER AT BAVENDALE, CHRISTCHURCH
Moar, n. T.
New Zealand Journal of Science, v. 1, n. 3, 1958: 480-486
Plant fragments from a peat layer three feet thick and buried by five feet of silt are listed and are briefly discussed. Wood from the base of the profile has been assigned an age of 940 ± 70 years by the radiocarbon dating method.

4.086 1965
COPROLITES IN WESTERN CAVES
Berger, Rainer; Ting, William S.; Libby, Willard F.
In: *International Conference on Radiocarbon and Tritium Dating*, Pullman, Washington, Washington State University, June 1-11, 1965. *Proceedings*. United States of America, Atomic Energy Commission, 1965. Conference n. 650652: 731-744.
Radiocarbon dating and pollen analysis of the Amberat coprolites, excrements of the pack rat of Pintwater cave indicate that some 16,000 years ago the micro-climate conditions around Pintwater cave were of an arid environment moving in the direction of even more arid conditions. These climatic changes to a more severe environment caused the pack rats to abandon their nests to which they did not return even if the climate improved for shorter periods.

4.087 1964
CORING OF FROZEN GROUND, BARROW, ALASKA, SPRING 1964
Sellman, P.V.; Brown, J.
U.S. Army. Cold Regions Research and Engineering Laboratory. Special Report, n. 81, 1964. pp. 8
Describes objectives, field methods, sample processing, cost and preliminary observations of this program. Accurate correlation of Pleistocene sedimentary units in the area appears possible over distances of 4 km and organic materials were radiocarbon dated in excess of 25,000 years BP.

4.088 1966
CORRELATION AND STRATIGRAPHY OF DACITIC ASH-FALL LAYERS IN NORTH-EASTERN PAPUA
Ruxton, B.P.
Geological Society of Australia. Journal, v. 13, part 1, 1966: 41-66
Radiocarbon dates and the buried soil profiles suggest a discontinuous pattern of activity with eruptive sequence and periods of quiescence for Mount Lamington. This pattern is estimated.

4.089 1962
COSMIC RAY INDUCED RADIOACTIVIES IN METEORITES - IV: CARBON-14 IN A STONE AND IN AN IRON
Goel, Parmatma S.; Kohman, Truman P.
In: *Research on Meteorites*, edited by C.B. Moon. New York, Wiley, 1952: 100-106
The detection and measurement of cosmic rays produced ^{14}C in the chondrite meteorite Richardson and the iron meteorite Henbury are discussed. The Henbury carbon specific activity is about the same as that of contemporary terrestrial carbon, therefore the possibility cannot be excluded that part or all of the radiocarbon is a result of contamination. The Richardson carbon specific activity is too high to admit of the same explanation. Results are given in table form.

4.090 1961
CYCLES AND GEOCHRONOLOGY
Hansen, Henry P.
California Academy of Sciences. Occasional Papers, n. 31, 1961. pp. 24
Discusses several dating methods, including radiocarbon dating, and presents a general chronology of the Pleistocene and correlations of late-glacial and post-glacial stratigraphy in European and North American areas as determined by

4.091   1962
A DATE FOR THE CHERANGANI HILLS OF WESTERN KENYA
Van Zinderen Bakker, Edward M.A.
Current Anthropology, v. 3, n. 2, Apr. 1962: 218
This date indicates that temperature fluctuations during the Quaternary were identical and contemporaneous in Europe and in the Equatorial Andean mountains with those observed from the Cherangani Hills of Western Kenya.

4.092   1961
DATING DESERT GROUNDWATER
Thatcher, Leyland; Rubin, Meyer; Brown, G.F.
Science, v. 134, n. 3472, 14 July 1961: 105-106
Tritium in Arabian rainfall has followed the trend observed in North America with peaks in 1958 and the spring of 1959. Water from wadi gravels average 10 years old. Radiocarbon measurements of deep water indicate ages of several thousand years.

4.093   1967
DATING OF GROUNDWATER WITH TRITIUM AND ^{14}C
Münnich, K.O.; Roether, W.; Thilo, L.
In: Isotopes in Hydrology. Conference on Isotopes in Hydrology, Vienna, 1-10 March 1967. *Proceedings*. Vienna, I.A.E.A., 1967: 305-320
Shallow groundwater can be dated with some accuracy on the basis of its bomb tritium. Radiocarbon increase in groundwater is delyed compared to tritium, the reason being delay in the biological system and exchange with carbonate in the soil.

4.094   1964
DATING OF HUMUS PODZOLS BY RESIDUAL RADIOCARBON ACTIVITY
Perrin, R.; Willis, E.H.; Hodge, C.A.H.
Nature, v. 202, n. 4928, 11 Apr. 1964: 165
The authors discuss the use of the radiocarbon method of age determination as applied to the dating of humus podzols and the unreliability of dates so obtained.

4.095   1967
DATING OF SHORE FEATURES OF LAKE GEORGE, NEW SOUTH WALES
Galloway, R.W.
Australian Journal of Science, v. 29, n. 12, June 1967: 477
A date on charcoal collected from a coarse sand lens, 4-5 feet below the top of the exposed surface in a gravel pit, 5 miles S.S.W. of Collector, N.S.W., has been dated at 15,100 ± 300 years BP. It indicates that the lake was 400 feet deep at or shortly after the maximum of the last major phase of the Wisconsin glaciation.

4.096   1963
DATING OF TROPICAL COASTAL REGRESSION
Boughey, A.S.
Nature, v. 200, n. 4906, 9 Nov. 1963: 600
On the basis of radiocarbon dates, the author establishes that on the coast of tropical Africa, the whole geomorphological cycle from the creation of one salt lagoon to the next must be at least 560 ± 100 years. This would give an average minimum regression of 3 yards a year. This would mean a landward retreat of a minimum of 45 miles during the present Quaternary climatic cycle of 30,000 years. The location for the study is Estoril Beach, four miles east of Beira in Mozambique.

4.097   1964
DATING ON THE BANKS OF THE POTOMAC
Rubin, Meyer
Washington Academy of Sciences. Journal, v. 54, 1964: 356-357
Generally materials from terraces of the Potomac are not suitable for radiocarbon dating, being either too young or too old.

4.098   1958
DATING THE LATEST MOVEMENT OF THE QUATERNARY SEA-LEVEL
Fairbridge, Rhodes W.
New York Academy of Science. Transactions, Ser. 2, v. 20, n. 6, 1958: 471-482
The data so far available suggests that coastal terraces are controlled chiefly by eustatic oscillations. Radiocarbon dating offers a solution to the problem of dating sea-level changes over the period of the last 45,000 years. A correlation chart and graph of sea-level oscillation of the last 12,000 years, based on radiocarbon dates, is presented.

4.099   1951
DE GEER'S CHRONOLOGY CONFIRMED BY RADIOACTIVE CARBON, C-14
de Geer, Ebba Hult
Geologiska Föreningen i Stockholm. Förhandlingar, v. 73, n. 3 (466), 1951: 517-518
The contemporaneity of the glaciations in Europe and North America is discussed. The Hult - de Geer - Linden chronology is absolute, and de Geer's teleconnections are valid.

4.100   1968
DECEMBER 1964, A 400 YEAR FLOOD IN NORTHERN CALIFORNIA
Helley, Edward J.; La Marche, Valmore C.

United States Geological Survey. Professional Paper, 600-D, 1968 (Geological Survey Research, Chap. D): D34-D37.
Twice in the past 13 years, record breaking floods have occurred over large areas of northern California. The true long-term recurrence intervals of these destructive floods is difficult to estimate by conventional flood-frequency. Radiocarbon analysis, supplemented by tree-ring counts, established a date about 400 years ago of a flood event that had approximately the same order of magnitude as the devastating floods of December 1964.

4.101   1966
THE DEEP AND THE PAST
Ericson, David B.
London, Jonathan Cape, 1966. pp. 292
Based on deep-sea cores the author seeks to establish a chronology of the Quaternary. Radiocarbon dating is used throughout the study for dating within the time range of 35,000 years.

4.102   1966
DEEP LAYER OF SEDIMENTS IN ALPINE LAKE IN THE TROPICAL MID-PACIFIC
Woodcock, Alfred H.
Science, v. 154, n. 3749, 4 Nov. 1966: 647-648
Sediments from a unique high-altitude lake on Hawaii indicate ash falls and other airborne and waterborne materials for a period estimated to extend into the Pleistocene. This conclusion was made after two radiocarbon dates were made on samples from a core at 1 meter and 2 meter depth, giving ages of 2270 ± 500 years and 7160 ± 500 years respectively. The result of the age depth curve, so determined, extrapolated suggesting 30,000 years as the maximum age of the deepest sediments.

4.103   1961
DEEP-SEA CORES AS AN AID TO ABSOLUTE DATING IN THE QUATERNARY PERIOD
Wiseman, John D.A.
In: *International Association of Quaternary Research* (INQUA), *Congress*, 6th, Warsaw, 1961. *Report*. 1961: 743-764
The difficulties inherent to the use of deep sea cores to unravel the record of Quaternary climatic change is examined. The radiocabon age determinations of the top part of cores are discussed.

4.104   1965
DEEP-SEA SEDIMENTS IN THE WESTERN MEDITERRANEAN SEA
Eriksson, K. Gösta
In: *International Association for Quaternary Research* (INQUA). *Congress*, 7th, Boulder, Col., 14 Aug - 19 Sept. 1965. *Abstracts of General Sessions*. Boulder, Colorado, 1965: 135
Detailed investigations of sediment cores collected in the southern part of the Western Mediterranean Sea by a Swedish expedition in 1948 show that some of the cores contain sediments as old as the Riss / Würm interglacial and Middle Würm. The stratigraphic succession is correlated with the continental Quaternary chronology by means of radiocarbon dates and other methods.

4.105   1953
THE DETERMINATION OF GEOLOGICAL AGE BY MEANS OF RADIOACTIVITY
Curran, S.C.
Chemical Society (London). Quarterly Review, v. 7, n. 7, 1953: 1-18
Outlines methods of measuring geologic time by determining the lead-uranium, helium, rubidium-strontium and potassium-argon content of minerals and by measurement of radiocarbon content of substances.

4.106   1965
DETERMINATION OF THE ABSOLUTE AGE OF THE FOSSIL REMAINS OF MAMMOTHS AND WOOLY RHINOCEROS FROM THE PERMAFROST IN SIBERIA
Heintz, Anatol; Garutt, V. E.
Norsk Geologisk Tidsskrift, v. 45, n. 1, 1965: 73-79
Six samples of soft tissues of the mammoth (*Mamuthus primigenius*) and one of the wooly rhinoceros (*Coelodonta antiquitatis*) discovered in permafrost in Siberia were dated by the radiocarbon method. The ages range from about 44,000 years to about 33,500 years, with the age of a mammoth sample at 11,450 years. The ages are discussed.

4.107   1965
DIFFERENTIAL POLLEN DISPERSION AND THE INTERPRETATION OF POLLEN DIAGRAMS
Tauber, Henrik
Danmarks Geologiske Undersøgelse, Series 2, n. 189, 1965. pp. 69
The author presents a model of the likely course and mechanisms of pollen transfer and applies this model to parts of a diagram from a former lake at at Weier. The various transitions are radiocarbon dated.

4.108   1965
THE DISPLACEMENT OF DEPOSITS FORMED AT SEA-LEVEL, 6500 YEARS AGO IN SOUTH BRITAIN
Churchill, D.M.
Quaternaria, v. 7, 1965: 239-249
The stratigraphic succession reflecting a change upwards

from inorganic marine, estuarine or brackish water and the radiocarbon dates indicate a gradual change in level. It is however difficult to distinguish the separate effects of uplift and downwarping.

4.109   1964
DISTRIBUTION OF NATURAL ISOTOPES OF CARBON IN LINSLEY POND AND OTHER NEW ENGLAND LAKES
Deevey, Edward S.; Stuiver, Minze
Limnology and Oceanography, v. 9, n. 1, Jan. 1964: 1-11
Lacustrine carbon is isotopically fractionated with respect to the carbon in air, in water, and in rock, and ^{13}C and ^{14}C, obeying partly different rules, are fractionated with respect to each other. Both isotopes show differential distribution between epilimnion and hypolimnion to a degree that vary with the seasons. The results of the study show that the carbon cycle of a small lake is dominated by the hydrologic cycle, within which lacustrine is an episode.

4.110   1966
DOLOMITE SOFT SEDIMENTS FROM PLUVIAL LAKES, MOUND, LYNN AND TERRY COUNTIES, TEXAS (Abstract)
Parry, W.T.; Reeves, C.C.
Geological Society of America. Special Papers, n. 87, 1966: 123
The dolomite is dated at greater than 37,000 years BP for the top layer, other evidence indicates also recent growth of the top layer dolomite.

4.111   1968
DRAMATIC CHANGES IN THE RECENT SEDIMENTARY ENVIRONMENT OF CHOCTAWHATCHEE BAY, FLORIDA (Abstract)
Goldsmith, Victor
Geological Society of America. Special Papers, n. 101, 1968: 359-360
A change in the environmental conditions to the present highly reducive condition is dated at 3302 ± 65 years BP. The measurement was made on a very large gastropode shell (*Busycon contrarium*) recovered from the sub-bottom layer.

4.112   1954
EARLY PRE-CAMBRIAN CARBON OF BIOGENIC ORIGIN FROM THE CANADIAN SHIELD
Rankama, Kalervo
Science, v. 119, n. 3094, 16 Apr. 1954: 506-507
Isotopic data supported by geologic evidence indicate that the carbon present in at least some of the pre-Cambrian carbonaceous slates from the Canadian shield is of biogenic origin.

4.113   1963
ECOLOGICAL AND RADIOCARBON CORRELATIONS IN SOME COLORADO MOUNTAIN LAKES AND BOG DEPOSITS
Pennack, Robert W.
Ecology, v. 44, n. 1, Winter 1963: 1-15
Complete organic deposit cores were taken from two small Colorado mountain lakes and two dry bogs. Studies were made of the microfossils, phosphate and organic contents and radiocarbon of various strata. Conclusion is drawn that during the 3000 years immediately past, the area between elevation 2300 to 3500 metres have been covered by stable and characteristic climax vegetation.

4.114   1968
AN ECOLOGICAL HISTORY OF THE LAKE VICTORIA BASIN
Kendall, R.L.
Ph.D. Dissertation. Duke University, Durham, North Carolina, 1968. pp. 194 (Abstract in: *Dissertation Abstracts*, Ann Arbor, Michigan, v. 29, n. 12B, 1969: 4704. Order No. 69-09067)
Two dated sedimnt cores from Pilkington Bay, northern Lake Victoria, have provided an absolute chronology covering the last 15,000 years. Twenty eight radiocarbon dates from the longer core have allowed the calculation of fossil sedimentation rates and the precise dating of environmental changes.

4.115   1964
EFFECTS OF COSMIC RAYS ON METEORITES
Honda, Masatake; Arnold, James R.
Science, v. 143, n. 3603, 17 Jan. 1964: 203-212
Because change occurring in the nuclides produced by cosmic ray bombardment is not well understood, it is difficult to decipher the fossil record of cosmic radiations. Radiocarbon is an exception and produces very good evidence for the approximate constancy of the intensity of the cosmic ray striking the atmosphere over the last few half-lives of this nuclide. However, meteorites are more suitable than terrestrial targets for this study.

4.116   1961
ENGINEERING GEOLOGY PROBLEMS IN THE YUKON-KOYUKUK LOWLAND, ALASKA
Weber, Florence Robinson; Péwé, Troy L.
In: Short Papers in the Geologic and Hydrologic Sciences. United States Geological Survey. Professional Papers, n. 424-D, 1961: 371-373
Study of the flood plain in the Yukon-Koyukuk lowlands. A radiocarbon date of 8140 ± 300 years was obtained from organic material taken from the scalloped phase of the flood plain.

4.117   1968
EOLIAN PERIGLACIAL SAND IN NORTHERN NOVA SCOTIA (Abstract)
Swift, Donald J.P.; Byers, Douglas S.; Kingsley, David H.
Geological Society of America. Special Papers, n. 101, 1968: 279
Excavation at a palaeoindian site at Debert, Nova Scotia, has revealed an aeolian, periglacial sand body. Thirteen radiocarbon dates from cultural material in the soil zone average 8635 ± 47 years B.C.

4.118   1965
EOLIAN SEDIMENTATION IN THE PACIFIC OCEAN OFF NORTHERN MEXICO
Bonatti, Enrico; Arrhenius, Gustaf
Marine Geology, v. 3, 1965: 337-348
The sediment cores were examined and discussed for source and transport mode. Radiocarbon dating of the immediately overlying sediment indicates the age to be more than 40,000 years.

4.119   1964
ESTIMATION OF AGES AND RATE OF MOTION OF GROUNDWATER BY THE C14 METHOD
Ingerson, Earl; Pearson, Frederick Joseph
In: Recent Research in the Field of Hydrosphere, Atmosphere and Nuclear Geochemistry. Festival, volume in honor of Prof. Ken Sugiwara, edited by Tadsimo Koyamara. Tokyo, Maruzen, 1964: 263-283
A method of estimating rates of motion and absolute ages of groundwater in aquifers, using the radiocarbon method of age determination is described.

4.120   1961
EUSTASY AND THE YARRA DELTA, VICTORIA, AUSTRALIA
Gill, Edmund D.
Royal Society of Victoria. Proceedings, v. 74, part 2, 1961:125-133
Application of local radiocarbon dates to a temperature graph for the last 10,000 years, based on oxygen isotope measurements, supports the conclusion that in the Yarra delta region two warmer periods of high sea-level were separated by a long colder period of lower sea-level, during which considerable erosion of the delta took place. There is no evidence that the emergence was due to tectonic activity.

4.121   1961
EUSTATIC SHORELINES ON PACIFIC ISLANDS
Stearns, Harold T.
Zeitschrift für Geomorphologie, Supplement 3, 1961: 3-16
Using radiocarbon based chronology for the recognised major glaciations and three sub-glaciations within the Wisconsin the author examines shoreline terraces on Pacific Islands. He concludes that the world-wide concordance of terraces at approximately 600 feet are highly suggestive that this level is a eustatic shore line.

4.122   1958
EVIDENCE FOR A LOW SEA-LEVEL 9000 YEARS AGO
Te Punga, Martin T.
New Zealand Journal of Geology and Geophysics, v. 1, n. 1, Feb. 1958: 92-94
It is tentatively submitted that radiocarbon dating of Podocarpus wood from a well at Foxton (New Zealand) suggests that sea-level may have risen about 150 feet during the last 9900 years. But there is no decisive evidence to indicate whether the land at the well site has been stable or tectonically depressed in recent times.

4.123   1960
EVIDENCE FOR AN ABRUPT CHANGE IN CLIMATE CLOSE TO 11,000 YEARS AGO
Broecker, Wallace S.; Ewing, Maurice
American Journal of Science, v. 259, n. 6, 1960: 429-448
The radiocarbon age determinations suggest that the changes, characterised by decrease in sedimentation rate, shrinking of lakes and warm period in north western Europe, occurred in less than 1000 years, close to 11,000 years ago.

4.124   1966
EVIDENCE FROM ALASKAN GLACIERS OF MAJOR CLIMATIC CHANGES
Goldthwait, Richard P.
In: World Climate from 8000 to 0 B.C. International Symposium, London, 1966. *Proceedings*. London, Royal Meteorological Society, 1966: 40-53
Climatic changes are dated by radiocarbon measurements from samples from Glacier Bay in Southern Alaska and the Icefield Ranges in the Yukon. Neoglaciation was climaxed in a series of pulses mostly between A.D. 1500 and 1850.

4.125   1968
EVIDENCE OF CLIMATIC CHANGES IN RED SEA CORES
Herman, Y.
In: Means of Correlation of Quaternary Successions, edited by Roger B. Morrisson and Herbert C. Wright. International Association for Quaternary Research (INQUA). Congress, 7th, Boulder, Col., 14 Aug - 19 Sept. 1965. *Proceedings*, v. 8. Salt Lake City, University of Utah Press, 1968: 325-348
Micropalaeontologic and lithologic studies of seven piston cores and six plankton cores coupled with radiocarbon age determinations have been used to reconstruct climatic events in the Red Sea region. A table of 11 radiocarbon age

determinations is presented.

4.126 **1966**
THE EVOLUTION AND DEVELOPMENT OF PART OF THE NORTHERN FLORIDA COAST
Schnable, J.N.
Ph.D. Dissertation. Florida State University, 1966. pp. 244 (Abstract in: *Dissertation Abstracts,* Ann Arbor, Mich, v. 27, n. 4B, 1966. Order No. 66-9085)
Radiocarbon dates, stratigraphic relationships and environmental interpretation, suggest that there was a relatively high stand of the sea, near present level, sometime between 24,000 and 40,000 years BP. Sea-level was approximately 10 to 15 feet below present level between 4000 and 5000 years ago, followed by a rapid rise.

4.127 **1968**
THE EVOLUTION OF THE COASTAL AREA OF SURINAME (SOUTH AMERICA)
Zonneveld, J.I.S.
In: Means of Correlation of Quaternary Successions, edited by Roger B. Morrisson and Herbert C. Wright. International Association for Quaternary Research (INQUA). Congress, 7th, Boulder, Col., 14 Aug - 19 Sept. 1965. *Proceedings,* v. 8. Salt Lake City, University of Utah Press, 1968: 577-589
Pedologic investigations and radiocarbon dates have yielded more details on the palaeogeographic evolution of the coastal area of Suriname. Present data indicate both rising and lowering of sea-level in the Holocene with a high stand during the mid-Holocene.

4.128 **1967**
EVOLUTION OF THE VEGETATION IN HIGH BELGIUM AND ITS RADIOCARBON CHRONOLOGY
Mullenders, W.; Gilot, E.; Ancion, N.; Capron, P.C.
In: Quaternary Palaeoecology, edited by M.J. Cushing and H.E. Wright. International Association for Quaternary Research (INQUA). Congress, 7th, Boulder, Col., 14 Aug - 19 Sept. 1965. *Proceedings,* v. 7. New Haven, Yale University Press, 1967: 333-339
Twenty eight radiocarbon dates allow to set up a chronology and to synchronise the main palynological phases with those of neighbouring countries. The radiocarbon dates are presented in table form.

4.129 **1968**
EXTINCTION OF MASTODONS IN EASTERN NORTH AMERICA: TESTING A NEW CLIMATIC - ENVIRONMENT HYPOTHESIS
Dreimanis, Aleksis
Ohio Journal of Science, v. 68, n. 6, Nov. 1968: 257-269
More than three quarters of the radiocarbon dates of mastodon sites are between 9000 and 12,000 years BP with a rapid decline noticeable by about 10,000 BP. An abrupt change in pollen diagrams is recorded between 10,000 to 11,000 years ago from spruce-dominated to high-pine maxima, with the onset of drier climate. It is inferred that mastodons preferred spruce areas, that their decrease in abundance was preceded by a reduction and disappearance of spruce forest and that their migration northward was hampered by the development of a wide belt of pine and hardwood forest. Hence their decline and disappearance.

4.130 **1962**
FAULTED PLEISTOCENE STRATA NEAR JACKSON, NORTH WESTERN WYOMING
Love, J.D.; Taylor, Dwight W.
United States Geological Survey. Professional Papers, n. 450-D, article n. 160, 1962: D136-D139
Two sedimentary units which contain molluscs in some abundance are discussed. They are important in as much as they represent widely divergent environments of deposition, lacustrine, and aeolian, they are cut by normal faults, the younger unit contains molluscs within the radiocarbon age range which also provide correlation with similar formations elsewhere. According to the radiocarbon dates, the lacustrine deposit may be about 27,000 years old, the loess is between 15,300 and 14,000 years old, a recent episode of normal faulting took place between the last 13,000 - 15,000 years.

4.131 **1958**
FINAL REPORT ON AIR SAMPLING AND CARBON-14 ANALYSIS OF LOS ANGELES ATMOSPHERE
Truesdail, R.W.
API and S & F project n. 13. American Petroleum Institute, New York, 1958
[Not sighted]

4.132 **1965**
FISSURE ERUPTION NEAR BEND, OREGON
Nichols, Robert L.; Stearns, Charles E.
In: State of Oregon Lunar Geology Field Conference, 1965. *Guidebook.* Oregon. Department of Geology and Mineral Industries. *Bulletin,* v. 57, 1965: 8-10
Eight separate basaltic flows down the slope from Newberry Volcano are assumed to be younger than the Mount Mazama eruption. The oldest of these has been radiocarbon dated from a tree mould at 6150 ± 210 years BP.

4.133 **1959-60**
FIVE RADIOCARBON DATING OF POST-GLACIAL SHORELINES IN CENTRAL SPITSBERGEN
Feyling-Hanssen, Rolf W.; Olsson, Ingrid U.

Norsk Geografisk Tidsskrift, v. 17, n. 1-4, 1959: 122-131
The dates, inserted in a height-time diagram, provide an excellent illustration of the main trend in the negative shift of shoreline during post-glacial time. The dates are discussed.

4.134   1966
FOREST HISTORY OF OHIO I. RADIOCARBON DATES AND POLLEN STRATIGRAPHY OF SILVER LAKE, LOGAN COUNTY, OHIO
Ogden, J. Gordon
Ohio Journal of Science, v. 66, n. 4, July 1966: 387-400
Radiocarbon and pollen analysis of a sediment core from Silver Lake in central Ohio indicate that deposition began in late Woodfordian time (11,000 - 14,000 radiocarbon years ago) and has been continuous to the present. A total of 19 radiocarbon dates document the climatic and environmental changes inferred from the sediment record.

4.135   1960
THE FOREST SEQUENCE AT THE HARTSTOWN BOG AREA IN WESTERN PENNSYLVANIA
Walker, Philip C.; Hartman, Richard T.
Ecology, v. 41, n. 3, July 1960: 461-474
Radiocarbon datings of various pollen studies relating to the forest sequence studied are discussed.

4.136   1967
FRESHWATER PEAT ON THE CONTINENTAL SHELF
Emery, Kenneth Orriz; Wigley, R.L.; Bartlett, Alexandra S.; Rubin, Meyer; Barghoorn, E.S.
Science, v. 158, n. 3806, 8 Dec. 1967: 1301-1307
Freshwater peats from the continental shelf off northeastern United States contain the same general pollen sequence as peats from ponds that are above sea-level and that are of comparable radiocarbon ages. These peats indicate that during glacial times of low sea-level terrestrial vegetation covered the region that is now the continental shelf in an unbroken extension from the adjacent land areas to the north and west.

4.137   1965
THE FUNDAMENTAL STUDY OF THE APPLIED SEDIMENTS IN KUWANA DISTRICT, MIE PREFECTURE [Japanese, English summary]
Watanabe, Kazue
Journal of Geology (Tokyo), v. 64, n. 3 (697), n. 4 (698), 1965: 71-86, 135-146
Correlation of the tertiary and quaternary formations and identification of the Pliocene - Pleistocene boundary in the Kuwana District, Japan, have been made on the basis of radiocarbon age measurements. Application of isogeochronologic data obtained by radiocarbon dating is an aid to locating porcelain and lower grade clays in the region.

4.138   1960
GENERAL REMARKS ON CHRONOLOGY, ICE RECESSION AND SHORELINE DISPLACEMENT
Holdetahl, Olaf
Norges Geologiske Undersøkelse, n. 208, *The Geology of Norway*, edited by Olaf Holdetahl. 1960: 369-373
Absolute chronology based on radiocarbon dating indicates that the Allerød period covers the time about 12,000 - 10,800 years BP and the Younger Dryas 10,800 - 10,000 years BP in excellent agreement with de Geer's dating.

4.139   1968
GEOCHRONOLOGY OF LATE QUATERNARY ALLUVIUM
Haynes, C. Vance
In: Means of Correlation of Quaternary Successions, edited by Roger B. Morrisson and Herbert C. Wright. International Association for Quaternary Research (INQUA). Congress, 7th, Boulder, Col., 14 Aug - 19 Sept. 1965. *Proceedings*, v. 8. Salt Lake City, University of Utah Press, 1968: 591-631
New alluvial chronology studies in the southwestern United States require that the older chronology be altered. Five different units, some of which contain human artifacts are discused. At least 100 radiocarbon dates show that these are time stratigraphic units.

4.140   1967
GEOLOGIC AGE OF SEDIMENTS IN THE OLD MOLOGA - SHIKSHA LAKE
Arslanov, Kh. A.; Gromova, L.I.; Zarrina, Ye.P.; Krasnov, I.I.; Novskiy, V. A.; Rudnev, Yu.P.; Spiridonova, Ye.A.
Akademiia Nauk SSSR. Doklady. Earth Science Sections, v. 172, n. 1-6, August 1967: 12-16
An ancient interglacial age is indicated palynologically, and by the new radiocarbon date, with an inference that the limit of the Valayan glaciation was west of the Rybinsk ore. All other reported radiocarbon dates of sediments are shown to be unreliable because of contamination with younger material.

4.141   1953
GEOLOGICAL BACKGROUND OF THE IYATAYET ARCHAEOLOGICAL SITE, CAPE DENBIGH, ALASKA
Hopkins, David Moody; Giddings, J.L.
Smithsonian Institution. Miscellaneous Collection, v. 121, n. 11, 11 June 1953. pp. 33 (Publication n. 4110)
The suggested correlation between features of the Iyatayet valley and late Quaternary deposits elsewhere in Alaska is

presented. A number of ages were determined by radiocarbon dating.

4.142   **1962**
GEOLOGICAL RADIOCARBON DATINGS FROM THE STOCKHOLM STATION
Lundqvist, Gösta
Sweden. Geologiska Undersøknig. Avhandlingar och Uppsatser, Series C, n. 589, 1962. pp. 23
Sumarises the geological dating carried out from 1955 to 1962, at the Stockholm Natural ^{14}C Station, including an outline of the author's unpublished results.

4.143   **1955**
GEOLOGICAL SETTING OF NEW ORLEANS
Fisk, H.H.
Geological Society of America. Bulletin, v. 66, n. 12, Dec. 1955: 1559.
Abstract of paper presented at the 1955 meeting of the Geological Society of America in New Orleans. The sequence of marine and deltaic deposits beneath the levee and lowlands, on which New Orleans stands, are interpreted as to environment of deposition and their age determined from radiocarbon analyses. Marine sands and clays were deposited less than 10,000 years ago. The existing conditions began less than 1000 years ago.

4.144   **1957**
GEOLOGICAL TESTS OF THE VARVE AND RADIOCARBON CHRONOLOGY
Antevs, Ernst
Journal of Geology, v. 65, n. 2, Mar. 1957: 129-148
In appraising radiocarbon dates, it is essential to keep in mind that radiocarbon content in an organic layer, such as the Two Creek forest bed, can be expected to have undergone about equal extra changes so that several analyses check only one another but not the actual age of the sample. Each geological radiocarbon date should be checked for general reasonableness and weighted against the geological knowledge. They are not absolute to be accepted blindly. Time estimates here made generally support the varve chronology but suggest that for radiocarbon dates the most older dates become progressively too young. The radiocarbon chronology and its correlation with Europe have unreasonable geologic and climatic implications, whereas the varve chronology synchronises the major temperature ages in North America and Europe.

4.145   **1968**
GEOLOGY AND FOUNDATION TREATMENT OF SENSITIVE SEDIMENTS - WORLD'S FAIRWAY COMPLEX, NEW YORK
Baskerville, Charles A.
Geological Society of America. Engineering Geology Case Histories, n. 6, 1968: 55-68
As part of an engineering geology survey, peat layers were dated by radiocarbon measurements at 700 to 900 years BP.

4.146   **1967**
GEOLOGY AND SLOPE FAILURE IN THE MAYBESO VALLEY, PRINCE OF WALES ISLAND, ALASKA
Swanston, Douglas.
Ph.D. Dissertation. Michigan State University, 1967. pp. 233 (Dissertation in: *Dissertation Abstracts,* Ann Arbor, Mich., Series B, v. 28, n. 12, 1968: 5085B-5086B. Order No. 68-4221)
A late Wisconsinan advance is represented in the Maybeso valley by extensive deposits of a compacted blue-grey till cemented with $CaCO_3$. A radiocarbon date from molluscs in a raised marine beach directly overlying the blue-grey till indicates deposition of this till prior to 9510 ± 289 years BP.

4.147   **1966**
GEOLOGY OF THE ADAM WEISS PEAK QUADRANGLE, HOT SPRINGS AND PARK COUNTIES, WYOMING
Rohrer, Willis L.
United States Geological Survey. Bulletin, n. 1241A, 1966: A1-A39
Exposed formations of the Adam Weiss Peak quadrangle range in age from Late Cretaceous to Recent: unexposed formations penetrated by drilling include rocks as old as Mississippian in age. Thin alluvial deposits of late Pleistocene to Recent age are present: Pleistocene alluvium was dated by radiocarbon methods on wood at $34,700 \pm 2200$ years BP.

4.148   **1967**
GEOLOGY OF THE FLOOR OF BERING AND CHUKCHI SEAS. AMERICAN STUDIES
Creager, Joe S.; McManus, Dean A.
In: The Bering Land Bridge, edited by David M. Hopkins. Stanford, Calif., Stanford University Press, 1967: 7-31
Rate of sedimentation is established on the basis of radiocarbon dating which also places the absolute age of the sediment as Post-Wisconsin. A table of radiocarbon dates is presented.

4.149   **1958**
GEOLOGY OF THE HARPER AND AVOCA MIDDLE RIVER VALLEYS, MID CANTERBURY, NEW ZEALAND
Suggate, R.P.
New Zealand Journal of Geology and Geophysics, v. 1, n. 1, Feb. 1958: 31-46

Wood embedded in a marine-like deposit dated the deposit as post-glacial and suggests that the recent fault scarp of the Bruce Fault is post-glacial.

4.150   1968
GEOLOGY OF THE TERRACES AT COCHRANE, ALBERTA
Stalker, S. MacS.
Canadian Journal of Earth Science, v. 5, n. 5, 1968: 1455-1466
An examination of the prominent terraces north of Bow River at Cochrane is presented. They are dated 19,000 to 15,000 years ago during the maximum phase of classical Wisconsin. A valley fill called Bighill Creek Formation and rich in vertebrate fossils gives a radiocarbon age of about 11,000 years.

4.151   1967
GEOLOGY OF THE USA CRREL PERMAFROST TUNNEL, FAIRBANKS, ALASKA
Sellman, Paul V.
U.S. Army Material Command Cold Regions Research and Engineering Laboratory. Technical Report, n. 199, 1967. pp. 22
A perennially frozen Quaternary silt section in the tunnel near Fairbanks, Alaska, proved to be late Wisconsin, with a maximum radiocarbon date of 33,700 (+2500, -1000). There are two unconformities, one at a depth of 30 feet indicated by: (1) a jump in radiocarbon dates from 14,000 to 30,000, (2) a 20 fold increase in chemical concentration with depth, (3) a sudden occurrence of large wedge structures showing second cycle growth. The upper break at 10 feet suggested by radiocarbon dates and a small truncated ice wedge, may indicate a warming period.

4.152   1968
THE GEOMAGNETIC FIELD AND RADIOCARBON DATES
Green, R.
Australian Journal of Science, v. 31, n. 1, July 1968: 42-43
Examines anomalous radiocarbon dating results and discusses the reason for these anomalies, in particular the effect of the geomagnetic field.

4.153   1966
GEOMORPHIC AND VEGETATIONAL STUDIES IN MESTERS VIG DISTRICT, NORTHEAST GREENLAND - GENERAL INTRODUCTION
Washburn, A.L.
Meddelelser om Gronland, v. 166, n. 1, 1965. pp. 60.
Included in this introduction is a summary of the general nature of the Mesters Vig district. The bedrock geology is reviewed and glacial geology and changes of level discussed. The experimental sites are described and radiocarbon dated shells and driftwood listed.

4.154   1965
GEOMORPHOLOGY AND SOILS OF THE LOWER IOWAN KANSAN BORDER AREA, TAMA COUNTY, IOWA
Hall, George Frederick
Ph.D. Dissertation. Iowa State University, 1965. pp. 300 (Dissertation in: *Dissertation Abstracts*, Ann Arbor, Mich. v.26, n. 10, 1966: 5624. Order No. 66-3876)
Study to determine the time and space relationships of the geologic materials and the relation of the soils to the landscape across a segment of the Iowan-Kansan border in Tama County, Iowa. Radiocarbon dates of organic material from the base of the calcareous loess are 29,000 ± 3500 and 18,300 ± 500 years. The upper Tazewell loess was deposited between 1800 and 4000 years ago.

4.155   1965
GLACIATION OF THE NORTHWESTERN CANADIAN ARCTIC ISLANDS
Fyles, J.G.; Blake, Weston
In: International Association for Quaternary Research (INQUA). *Congress*, 7th, Boulder, Col., 14 Aug - 19 Sept. 1965. *Abstracts of General Sessions*. Boulder, Colorado, 1965: 156
Radiocarbon dated marine shells indicate that the northwestern part of Victoria Island became ice-free ca 12,000 years ago. Several dates show that much of coastal Bathurst Island was ice-free by 9700 years BP, as was Byam Martin Island.

4.156   1964
GROSS-ATMOSPHERIC CIRCULATION AS DEDUCED FROM RADIOACTIVE TRACERS
Bolin, Bert
In: Research in Geophysics, vol. 2, *Solid Earth Phenomena*, edited by Hugh Odishaw. Cambridge, Mass., M.I.T. Press, 1964: 479-508
Examines the way in which tracers may aid in the attempt to improve knowledge about atmospheric motion. An examination of the radiocarbon content of atmospheric carbon dioxide is included.

4.157   1966
GROUND WATER RECHARGE AS REVEALED BY NATURALLY OCCURRING RADIOCARBON: AQUIFERS OF CORO AND PARAGUANA, VENEZUELA

Tamers, Murray A.
Nature, v. 212, n. 5061, 29 Oct. 1966: 487-492
This investigation illustrates the usefulness of radiocarbon measurements of groundwaters for the evaluation of the extent of aquifer and water movements.

4.158   1968
HIGH FREQUENCY ACOUSTIC PROFILES ON THE CHUKCHI SEA CONTINENTAL SHELF
(Abstract)
Holmes, Mark L.; Creager, Joe S.; McManus, Dean A.
American Geophysical Union. Transactions, v. 49, n. 1, 1968: 207
Description of the profile of the Chukchi Sea. Radiocarbon dates from a core in the southeast Chukchi Sea have shown an average depositional rate of 70 mg /cm² /year, indicating a more rapid rate of accumulation in the early Holocene sea.

4.159   1968
HISTORY OF HOLOCENE TRANSGRESSION IN THE GULF OF PANAMA
Golik, Abraham
Journal of Geology, v. 76, n. 5, Sep. 1968: 497-507
Radiocarbon dates on samples from mud and shells from sediment cores give uniform dates between 11,000 and 12,800 years BP, thus placing the transgression in this area in Holocene time. The dates are discussed in terms of the material dated

4.160   1967
HOLOCENE CHANGES IN SEA-LEVEL: EVIDENCE IN MICRONESIA
Shepard, Francis P.; Curray, Joseph R.; Newman, W.A.
Science, v. 157, n. 3788, 1967: 542-544
Radiocarbon dating of 11 samples so far suggests a former, slightly higher sea-level, but this is debatable. Other evidence indicate no change in sea-level.

4.161   1968
HOLOCENE POLLEN ASSEMBLAGES FROM THE TIGER HILLS, MANITOBA
Ritchie, J.C.; Lichti-Federovitch, Sigrid
Canadian Journal of Earth Sciences, v. 5, n. 4, part 1, Aug. 1968: 873-880
Coring of the three kettle lakes in the moraine area known as the Tiger Hills, Manitoba, yielded sections of sediments which span the Holocene for the region. Five main pollen assemblages are suggested. Zone V, the oldest, is interpreted as a spruce dominated vegetational association. Radiocarbon age determinations suggest that the area was occupied by Zone V assemblage from about 12,800 years BP. A table of radiocarbon dates is presented.

4.162   1961
HOLOCENE SEA-LEVEL CHANGES IN THE NETHERLANDS (Ph.D. Thesis, Leiden)
Jelgersma, Saskia
Mededelesen Geologisk Stitching, Ser. C, 1961: 6-7
[Not sighted]

4.163   1968
HOLOCENE SEDIMENTATION IN TANNER BASIN, CALIFORNIA CONTINENTAL BORDERLAND
Gorsline, Donn S.; Drake, David E.; Barnes, Peter W.
Geological Society of America. Bulletin, v. 79, n. 6, 1968: 659-674
Piston cores from Tanner Basin, on the California Continental Borderland, are examined. Radiocarbon and palaeontologic dates provide markers for computation of sedimentation rates which show that contributions of biogenic and terrigenous fines have been constant for the past 7500 years, reflecting a greatly reduced rate of sea-level rise. Carbonate contributions, have been nearly constant over the past 12,000 years. This reflects little change in the oceanographic circulation.

4.164   1965
HOLOCENE SUBMERGENCE OF THE EASTERN SHORE OF VIRGINIA
Newman, Walter S.; Rusnak, Gene A.
Science, v. 148, n. 3676, 11 June 1965: 1464-1465
Radiocarbon ages of basal peats 4500 years old or younger and the thickness of salt-marsh peat in the lagoon east of Wachapreague, Virginia, are nearly the same as those of equivalent samples from New Jersey and Cape Cod. This suggests that these coasts have had similar submergence histories. Data obtained from the coasts of Connecticut and North Eastern Massachusetts indicate that the Atlantic coast of the United States has been differentially warped during the late Holocene.

4.165   1956
HYDROCARBON IN SEDIMENTS OF GULF OF MEXICO
Stevens, Nelson P.; Bray, Ellis E.; Evans, Ernest D.
American Association of Petroleum Geologists. Bulletin, v. 40, n. 5, May 1956: 975-983
Radiocarbon determination shows that the organic matter in the recent muds of the Gulf of Mexico ranged in age from 3120 ± 220 to 9360 ± 600 years.

4.166   1958
HYDROCARBONS IN SEDIMENTS OF THE GULF OF MEXICO
Stevens, Nelson P.; Bray, Ellis E.; Evans, Ernest D.

In: Habitat of Oil, a Symposium conducted by the American Association of Petroleum Geologists, edited by Lewis G. Weeks. Tulsa Oklahoma, American Association of Petroleum Geologists, 1958: 779-789

Radiocarbon age determinations show that the organic matter in the recent muds of the gulf used in this study ranged in age from 3120 ± 220 to 9360 ± 600 years.

4.167   1968
IDENTIFICATION OF PALEOSOLS IN LOESS DEPOSITS IN THE UNITED STATES
Ruhe, Robert V.
In: Loess and Related Earlier Deposits of the World, edited by C. Bertrand Schultz and John C. Frye. International Association for Quaternary Research (INQUA). Congress, 7th, Boulder, Col., 14 Aug - 19 Sept 1965. *Proceedings,* v. 12. Lincoln, Neb., University of Nebraska Press, 1968: 49-65.

Radiocarbon dates associated with the Sangamon, Farmdale and Brady paleosols are presented and discussed.

4168   1967
INCOMPLETE CONTINENTAL GLACIAL RECORD OF ALBERTA, CANADA
Bayrock, L.A.
In: The Bering Land Bridge, edited by David M. Hopkins. Stanford, Calif., Stanford University Press, 1967: 99-103
adiocarbon dates indicate that the Saskatchewan gravels do not belong to the 'classical' Wisconsin stage which mean that the areas where the gravels occur were not glaciated during the Illinoian and Kansian glacial ages.

4.169   1967
INFLUENCE OF ISLAND MIGRATION ON BARRIER ISLAND SEDIMENTATION
Hoyt, John H.; Henry, Vernon J.
Geological Society of America. Bulletin, v. 78, n. 1, 1967: 77-86

Migrating barrier islands along coasts with a dominant longshore current have distinctive deposition characteristics. Changes in sea level and sedimentation rates result in a variety of sediment-body shapes from shoestring to blanket deposits. Eight radiocarbon dates of Recent sediments from the margin of Sapelo Island, Ga., are included.

4.170   1965
INTRODUCTION
Flint, Richard Foster
In: The Quaternary, v. I, edited by Kalervo Rankama. New York, Interscience, 1965: xi-xxiii

The Quaternary system, a major time-stratigraphic unit, encompasses the youngest strata in the whole rock sequence and includes sediments being deposited now. Radiometric dating, particularly radiocarbon dating, is of prime importance in fixing, in time, the position of points of distinctive deposits within the glacial and non-glacial physical units.

4.171   1959
INVESTIGATION OF ALASKAN VOLCANOES: GEOLOGY OF UNMAK AND BOGOLOF ISLANDS, ALEUTIAN ISLANDS, ALASKA
Byers, F.M.
United States Geological Survey. Bulletin, 1028-L, 1959: 267-369

A radiocarbon determination on charcoal at the Nikolski site which is on top of one of the old gravel beach ridges gives 3018 ± 230 years. By extrapolation, the oldest deposit at this site is estimated to be 5000 years or more, meaning that the beach ridge may have been formed during the thermal maximum.

4.172   1967
INVESTIGATION OF GROUNDWATER FLOW WITH RADIOCARBON
Vogel, John C.
In: Isotopes in Hydrology. Conference on Isotopes in Hydrology, Vienna, 1-10 March 1967. *Proceedings.* Vienna, I.A.E.A., 1967: 355-369

An interesting potentiality of the radiocarbon dating of groundwater is the deduction of recharge rates of aquifers. Natural radiocarbon has been used to study flow pattern and flow rates of subterranean water in the Netherlands.

4.173   1964
INVESTIGATION OF SOIL HUMUS UTILISING CARBON DATING TECHNIQUES
Paul, E.A.; Campbell, D.A.; Rennie, R.J.; McCallum, K.J.
In: International Congress of Soil Science, 8th, Bucharest, 1964. *Proceedings.* v. 3, 1964: 201-208
[Not sighted]

4.174   1963
INVESTIGATIONS OF MERIDIONAL TRANSPORT IN THE TROPOSPHERE BY MEANS OF CARBON-14
Münnich, K.O.; Vogel, John C.
In: Radioactive Dating. Proceedings of a Symposium on Radioactive Dating held in Athens, 19-23 November 1962. Vienna, International Atomic Energy Agency, 1963. (Proceedings Series): 189-197

The seasonal varying injection of bomb radiocarbon from the stratosphere gives rise to a pile-up in higher latitudes in the northern hemisphere. This brings about seasonal variations with the maximum of the tropospheric radiocarbon level in summer. Investigation of the decay of radiocarbon concentration during autumn gives information about the meridional transport of air masses.

4.175   1966
THE INYO CRATER LAKES - A BLAST IN THE PAST
Rinehart, C. Dean; Hubee, N. King
California. Division of Mines and Geology. Mineral Information Service, v. 18, n. 4, May 1966: 164-172
The craters mark a recent, relatively small, event in recurrent volcanic eruptions that built the imposing Mono Craters and Mammoth Mountain. Charred fragments in the pumice date that eruption at about A.D. 365.

4.176   1966
IOWAN DRIFT PROBLEM, NORTH EASTERN IOWA (Abstract)
Ruhe, Robert V.; Dietz, W.T.; Fenton, T.E.; Hall, George Frederick
Geological Society of America. Special Papers, n. 87, 1966: 144
The erosion surface was formed during a period between 29,000 and 14,000 years ago, according to radiocarbon dates.

4.177   1962
ISHANGO
Heinzelin, J. de
Scientific American, v. 206, n. 6, July 1962: 105-118
A volcanic eruption which upset the carbon isotopic ratio in Lake Ishango made the aquatic shells of this area unfit for the application of radiocarbon dating.

4.178   1962
ISOTOPIC COMPOSTION OF SULFUR COMPOUNDS IN THE BLACK SEA
Vinogradov, A.P.; Grivenko, V. A.; Ustinov, V. I.
Geochemistry, n. 10, 1962: 973-997
The isotopic constitution of sulphur, determined in several series of water samples and sediments from the bottom of the Black Sea, USSR, is presented, and the radiocarbon ages of sediments from four localities are given and used to estimate the time of beginning of hydrogen sulphide production in sea waters. The results are discussed with regard to the hydrogeochemical evolution of the Black Sea.

4.179   1966
LAKE ARKONA - WHITTLESEY AND POST-WARREN RADIOCARBON DATES FROM 'RIDGETOWN ISLAND' IN SOUTHWESTERN ONTARIO
Dreimanis, Aleksis
Ohio Journal of Science, v. 66, n. 6, Nov. 1966: 582-586
Three radiocarbon dates from raised beaches along the 'Ridgetown Island' support the age assignment of 13,000 years BP for the beginning of Lake Whittlesey in Ohio, and the termination of Lake Warren before 12,000 years BP, as concluded from post-Warren data in Ohio.

4.180   1963
LANDFORM STUDIES IN THE MIDDLE HAMILTON RIVER AREA, LABRADOR
Morrison, A.
Arctic, v. 16. n.4, Dec. 1963: 273-275
Three samples from bogs around Grand Falls were dated. Allowing time for the deposition of silt below the dated peats, the best estimate of deglaciation of the area around Grand Falls appears to be 5750 ± 350 BP. The radiocarbon dates were: 5255 ± 200, 5450 ± 220, 5575 ± 250 BP.

4.181   1963
THE LAST TEN THOUSAND YEARS, A FOSSIL POLLEN RECORD OF THE AMERICAN SOUTH WEST
Martin, Paul S.
Tucson, Arizona, University of Arizona Press, 1963. pp. 87
Most of the pollen profiles are radiocarbon dated. Some of the conclusions of the examination of pollen profiles is that the altithermal was not hot and dry, but relatively wet, at least in summer, and that drought cannot be advanced as an explanation for the extinction of larger mammals, 8000 to 10,000 years ago.

4.182   1968
LATE HOLOCENE SEA-LEVEL FLUCTUATIONS IN MICRONESIA (Abstracts)
Curray, Joseph R.; Bloom, Arthur L.; Newman, W.A.; Newell, Norman D.; Shepard, Francis P.; Tracey, J.I.; Veeh, H.H.
Geological Society of America. Special Papers, n. 115, 1968: 40-41
Radiocarbon dating of coral from various islands indicate that a late Holocene sea-level higher than present probably did not occur in these islands.

4.183   1965
LATE QUATERNARY AND MODERN POLLEN RAIN AT SEARLES LAKE, CALIFORNIA
Leopold, Estella B.
In: International Association for Quaternary Research (INQUA). Congress, 7th, Boulder, Col., 14 Aug - 19 Sept. 1965. *Abstracts of General Sessions*. Boulder, Colorado, 1965: 289
Three subsurface cores from Searles Lake (Mohave Desert) were used to prepare pollen diagrams. The cores penetrate as far as 130 feet, and, according to radiocarbon evidence, range from recent to 40,000 years BP. A comparison with

modern pollen rain samples indicate that the Searles Lake pollen records do not seem to reflect the climatic events of the late Quaternary as recorded locally by other lines of evidence. The sporadic fluctuation of fossil pollen may record changes in distance of the shoreline from the fossil localities or other factors affecting pollen distribution.

4.184   1955
LATE QUATERNARY DELTAIC DEPOSITS OF THE MISSISSIPPI RIVER
Fisk, H.H.; McFarland, E.
Geological Society of America. Special Paper, n. 62, part 2, 1955: 279-302
The late Quaternary river mouth deposits of the Mississippi were laid down during the cycle of the sea-level change that has occurred since the beginning of the Late Wisconsin glacial epoch. All data from the coastal Louisiana marshlands and the adjacent continental shelf, together with cores and samples from the Gulf floor permit generalisations as to the nature, distribution, depositional history and volume of these deltaic deposits. Radiocarbon analyses of wood and shells provide dates for younger deposits of the cycle.

4.185   1958
LATE QUATERNARY DEPOSITS OF THE CHRISTCHURCH METROPOLITAN AREA
Suggate, R.P.
New Zealand Journal of Geology and Geophysics, v. 1, n. 1, Feb. 1958
The youngest of the Christchurch formation (marine, estuarine and fluvatile deposits) was formed during the postglacial rise of sea-level and radiocarbon dates record the rise of sea-level as follows: 73 feet - 9400 years ago, to 12 feet - 6100 years ago.

4.186   1959
LATE QUATERNARY EUSTATIC CHANGES IN THE SWAN RIVER DISTRICT
Churchill, D.M.
Royal Society of Western Australia. Journal, v. 42, part 2, June 1959: 53-55
Pollen analysis has been made on a submerged fresh water peat from 68 feet below sea-level. The radiocarbon age is 9850 ± 130 years BP. The implication of the change in sea-level are discussed. Late Quaternary shore lines to the west of Fremantle show that Rottnest and Garden Islands have been isolated from the mainland since about 5000 B.C.

4.187   1968
LATE QUATERNARY GEOLOGIC AND CLIMATIC HISTORY OF SEARLES LAKE, SOUTHERN CALIFORNIA
Smith, G.I.
In: Means of Correlation of Quaternary Successions, edited by Roger B. Morrisson and Herbert C. Wright. International Association for Quaternary Research (INQUA). Congress, 7th, Boulder, Col., 14 Aug - 19 Sept. 1965. *Proceedings*, v. 8. Salt Lake City, University of Utah Press, 1968: 293-310
The subsurface deposits down to 875 ft indicate a succession of long, permanent lakes and small, saline or dry lakes. Absolute ages of less than 45,000 years are based on 81 published radiocarbon dates. Absolute ages of more than 45,000 years are estimated from the rates of mud sedimentation determined within the radiocarbon dated section.

4.188   1965
LATE QUATERNARY GLACIAL AND VEGETATIONAL SEQUENCE, SIERRA NEVEDA DEL COCUY, COLUMBIA (Abstract)
Gonzales, E.; Hammen, Th. van der ; Flint, Richard Foster
In: International Conference on Radiocarbon and Tritium Dating, Pullman, Washington, Washington State University, June 1-11, 1965. *Proceedings*. United States of America, Atomic Energy Commission, 1965. Conference n. 650652: 730
Pollen samples were calibrated by nine radiocarbon dates from organic materials from samples of lake sediments stratigraphically related to four bodies of glacial drifts in a high Andean valley. The results support the view that major climatic events in high-altitude, tropical South America during at least the last 12,000 years were synchronous with those in mid- and high-latitudes of North America and Europe.

4.189   1965
LATE QUATERNARY HISTORY, CONTINENTAL SHELVES OF THE UNITED STATES
Curray, Joseph R.
In: The Quaternary of the United States: a Review Volume for the 7th Congress of the International Association of Quaternary Research, edited by H.E. Wright and David G. Frey. Princeton. N.J., Princeton University Press, 1965: 723-735
A table presenting late Quaternary fluctuations of sea-level from compilation of published and unpublished radiocarbon dates and other geologic evidence suggests an interstadial high stand of sea-level sometimes between about 22,000 and 35,000 BP.

4.190   1967
LATE QUATERNARY MARINE FOSSILS FROM FROBISHER BAY (BAFFIN ISLAND, N. W.T., CANADA)
Matthews, Barry
Palaeogeography, Palaeoclimatology, Palaeoecology, v. 3, n. 2, 1967: 243-263

Eighten species of fossil marine pelecypods and eighteen species of foraminifera from raised beaches (27, 48 and 77 feet above sea-level) yield evidence of post-glacial changes in the marine environment. Radiocarbon dates of approximately 6100 to 6500 years BP on shells at the 48- and 11-ft levels are determined. Frobisher Bay was deglaciated at least 6500 years ago and the other dates suggest rebound of about 1 ft per century for 6000 years.

4.191   1961
LATE QUATERNARY SEA LEVEL: A DISCUSSION
Curray, Joseph R.
Geological Society of America. Bulletin, v. 71, n. 7, July 1961: 1113-1120
Critical review of recent literature and new evidence on the position of sea-level during the late Quaternary. New radiocarbon dates have come to hand to support various hypotheses.

4.192   1964
LATE QUATERNARY SEA-LEVEL CHANGES AND CRUSTAL RISE AT BOSTON, MASSACHUSETTS, WITH NOTES ON THE AUTOCOMPACTION OF PEAT
Kaye, C.A.; Barghoorn, E.S.
Geological Society of America. Bulletin, v. 75, n. 1, Jan. 1964: 63-80
The compression of peat beneath its own weight is discussed and it is shown that because of this process radiocarbon-dated samples of salt-marsh peat or peaty sediment other than very thin samples cut from the base of the deposit cannot be correlated with sea-level without construction of a sea-level change curve from other type of data. Sixteen radiocarbon dates from the Boston area are used to construct a sea-level curve going back to 14,000 years BP.

4.193   1968
LATE QUATERNARY SEDIMENT CORES FROM BJØRNØYA; APPENDIX, RADIOCARBON ANALYSES OF LAKE SEDIMENT SAMPLES FROM BJØRNØYA
Olsson, Ingrid U.
Geografiska Annaler, Ser. A, v. 50, n. 4, 1968: 246-247
The samples are discussed and their treatment described. The results are given in table form and range from $11,200 \pm 500$ years BP at ca 90 cm, to $6000 \pm 3100/2200$ years BP at ca 20 cm.

4.194   1966
LATE QUATERNARY VEGETATION IN EASTERN BLEKINGE, SOUTH-EASTERN SWEDEN: A POLLEN ANALYTICAL STUDY, I, LATE-GLACIAL TIME.
Berglund, B.E.
Opera Botanica, v. 12, n. 1, 1966: 1-180
Ch. VII, p. 113-119 of this study presents an analysis of the reliable radiocarbon samples dating the Zone boundaries in southern Scandinavia, included in tables for each boundary of dates known from other parts of the area. A table of radiocarbon dated samples for the Blekinge area is also presented. (Dates: Uppsala IV and Stockholm VII).

4.195   1966
LATE QUATERNARY VEGETATION IN EASTERN BLEKINGE, SOUTH-EASTERN SWEDEN: A POLLEN ANALYTICAL STUDY, I, LATE-GLACIAL TIME.
Berglund, B.E.
Lund University. Institute of Mineral Paleontology and Quaternary Geology. Publications, n. 134, 1966. pp. 171
[See entry n. 4.194]

4.196   1966
LATE QUATERNARY VEGETATION IN EASTERN BLEKINGE, SOUTH-EASTERN SWEDEN: A POLLEN ANALYTICAL STUDY, II, POST-GLACIAL TIME
Berglund, B.E.
Opera Botanica, v. 12, n. 2, 1966: 1-190
Radiocarbon analyses enable to establish an absolute chronology of post-glacial times in the Blekinge area. Discussions of the radiocarbon dates are on p. 155-157 of the study. (Dates: Uppsala V and Stoxkholm VII).

4.197   1966
LATE QUATERNARY VEGETATION IN EASTERN BLEKINGE, SOUTH-EASTERN SWEDEN: A POLLEN ANALYTICAL STUDY, II, POST-GLACIAL TIME
Berglund, B.E.
Lund University. Institute of Mineral Paleontology and Quaternary Geology. Publications, n. 135, 1966. pp. 182
[See entry n. 4.196]

4.198   1967
LATE QUATERNARY VEGETATION IN THE MOHAVE DESERT (U.S.A.)
Mehringer, Peter J.
Review of Palaeobotany and Palynolgy, v. 2, n. 1, 1967: 319-320
Pollen and radiocarbon dates in Nevada and California from alluvium, playa cores, and ancient spring deposits show that the present vegetation zones were reduced by 1000 m during the maximum Wisconsin glaciation. Southward vegetational shifts of several hundred kilometers are shown. Other

Ch. 4 - General Geology

vegetational changes of less magnitude also are discussed.

4.199   1964
A LATE WISCONSIN BURIED SOIL, NEAR AITKIN, MINNESOTA, AND ITS PALEOBOTANICAL SETTING
Farnham, R.S.; Mc Andrews, J.H.; Wright, H. E.
American Journal of Science, v. 262, n. 3, March 1964: 393-412
A palaeosol buried by 3 in of calcareous sand, overlain by 3 feet of lacustrine marl and clay, under 3.5 feet of herb peat is dated by radiocarbon at 11,635 years, these formed during the Two Creeks interval when the lake drained.

4.200   1968
LATE-GLACIAL AND POST-GLACIAL SHORELINE DISPLACEMENT IN DENMARK
Krog, Harald
In: Means of Correlation of Quaternary Successions, edited by Roger B. Morrisson and Herbert C. Wright. International Association for Quaternary Research (INQUA). Congress, 7th, Boulder, Col., 14 Aug - 19 Sept. 1965. *Proceedings*, v. 8. Salt Lake City, University of Utah Press, 1968: 421-435
Molluscs from late-glacial deposits from northern Jylland (Denmark) indicate an interval of milder climate which is correlated with the Bølling oscillation, based on radiocarbon datings.

4.201   1968
LATE-GLACIAL AND POST-GLACIAL SHORELINES IN IRELAND AND SOUTH WEST SCOTLAND
Stephens, N.
In: Means of Correlation of Quaternary Successions, edited by Roger B. Morrisson and Herbert C. Wright. International Association for Quaternary Research (INQUA). Congress, 7th, Boulder, Col., 14 Aug - 19 Sept. 1965. *Proceedings*, v. 8. Salt Lake City, University of Utah Press, 1968: 437-456
The earliest post-glacial transgressional shorelines of Ireland and southwest Scotland form a metachronous series from north to south, from southwest Scotland along the east coast of Ireland. Archaeological and radiocarbon datings for several important sites demonstrate that the highest post-glacial shoreline cannot be everywhere synchronous. A table of data, including radiocarbon dates, is presented.

4.202   1967
LATE-RECENT ALLUVIUM IN WESTERN NORTH DAKOTA
Hamilton, Thomas M.
North Dakota Geological Survey. Miscellaneous Series, n. 30 *(Glacial geology of the Missouri Coteau*. Midwest Friends of the Pleistocene, Field Conference 1967, *Guidebook)* 1967: 151-158
The late-recent alluvium of western North Dakota can be divided into five individual units, two of which are palaeosols. By radiocarbon dating and tree age estimates it was established that the upper 15-20 feet of the sediment was probably deposited between 1775 and 1936. The age of the alluvium makes possible a tentative correlation with similar alluvial units studied in Wyoming and Nebraska.

4.203   1964
LATE-RECENT RISE OF SEA-LEVEL
Coleman, James M.; Smith, William G.
Geological Society of America. Bulletin, v. 75, n. 3, Sept. 1964: 833-840
Radiocarbon dating of marsh peat in south-central Louisiana permits interpretations of relationships between former positions of land and sea.

4.204   1962
THE LATEST ERUPTIONS FROM MOUNT RAINIER VOLCANO
Hopson, Clifford A.; Waters, Aaron, C.; Bender, V. R.; Rubin, Meyer
Journal of Geology, v. 70, n. 6, Nov. 1962: 635-647
The youngest blanket of pumice and ash from Mount Rainier consists of two or more distinct ash falls. Both tree ring and radiocarbon methods indicate that it was deposited about 550-600 years ago, confirming geologic evidence that there is nothing to substantiate recent activity.

4.205   1967
THE LOWER HOLOCENE BOUNDARY
Neustadt, M.I.
In: Quaternary Palaeoecology, edited by M.J. Cushing and H.E. Wright. International Association for Quaternary Research (INQUA). Congress, 7th, Boulder, Col., 14 Aug - 19 Sept. 1965. *Proceedings*, v. 7. New Haven, Yale University Press, 1967: 415-425
The stratigraphy of some peat bogs and lakes has been studied in the European part of the USSR. Radiocarbon dating shows that the oldest of the Holocene deposits is 11,975 years old, leading to the conclusion that the Holocene begins with the Allerød.

4.206   1968
MAMMOTHS FROM THE MIDDLE WISCONSIN OF WOODBRIDGE, ONTARIO
Churcher, C.S.
Canadian Journal of Zoology, v. 46, n. 2, March 1968: 219-221
Parts of a tusk and lower molar of the woolly mammoth, *Elephas (Mammuthus) primigenius*, have been recovered from above the early Wisconsin Sunnybrook till (near

Woodbridge, York County, Ontario). A mid-Wisconsin (Port Talbot Interstade) age of 40,000 to 50,000 years is suggested for these specimens. The estimates are based on a radiocarbon date on peat and wood below the till of greater than 49,700 years BP. A preliminary date on wood from within the Sunnybrook till, about 60 cm below the surface, suggests an age range within the Port Talbot Interstade.

4.207 1966
MAN AND GEOMORPHIC PROCESS IN THE CHEMUNG RIVER VALLEY, NEW YORK AND PENNSYLVANIA
Nelson, J.G.
Association of American Geographers. Annals, v. 56, n. 1, March 1966: 24-34
Observation of flood effects, and stratigraphic and other evidence, including a radiocarbon date, indicate that overbank deposition on the Chemung floodplain has accelerated in very recent geologic times, principally in the last one thousand years.

4.208 1968
MAN THE DESTROYER: LATE QUATERNARY CHANGES IN THE AUSTRALIAN MARSUPIAL FAUNA
Merrilees, D.
Royal Society of Western Australia. Journal, v. 51, 1968: 1-24
The late Quaternary extinction of Australian marsupials may have been part of large scale world wide processes such as the climatic change linked with the waning of the large high latitudes ice sheets. Existing Australian evidence presents as an alternative hypothesis the mid-Recent aridity and man-made fires as a cause. But more testing needs to be made to prove this hypothesis. A table of late Quaternary times, noting existing radiocarbon dates is presented.

4.209 1958
THE MANGROVE AND SALT-MARSH FLATS OF THE AUCKLAND ISTHMUS
Chapman, V. J.; Donaldson, J.W.
New Zealand. Department of Scientific and Industrial Research. Bulletin, n. 125, 1958. pp. 75
Shallow bores from the swamps and salt-marshes provide evidence of buried peat and of blue mud deposits, indicating past fluctuations of sea-level. Radiocarbon analysis show the peat to be more than 20,000 years old, and may well be older than the main Monasterian transgression.

4.210 1960
MARINE DEPOSITS OF THE OSLOFJORD - ROMERIKE DISTRICT
Holdetahl, Olaf

In: Geology of Norway, edited by Olaf Holdetahl. *Norges Geologiske Undersøkelse*, n. 208, 1960: 374-389
Thick layers of Quaternary deposits in the Oslofjord area are studied. Various radiocarbon dates were made on a number of deposits.

4.211 1961
MARINE TRANSGRESSION IN THE ENGLISH FENLANDS
Willis, E.H.
New York Academy of Science. Annals, v. 95, 1961: 368-376
A marine transgression is dated at 3000 year B.C., followed by a regression about 2200 B.C.

4.212 1959
A MARINE TRANSGRESSION OF BOREAL AGE
Gabrielsen, Gunnar
Nature, v. 183, n. 4675, 6 June 1959: 1616
Dating made at Trondheim (Norwegian Institute of Technology) place the apparition of *Alnus* in the Gothenburg area at 6620 ± 125 B.C. Boreal material resting directly under the main deposit indicates a transgression at or about that date.

4.213 1962
THE MASTODONS AND MAMMOTHS OF MICHIGAN
Skeels, Margaret Anne
Michigan Academy of Science, Arts, and Letters. Papers, v. 47, 1962: 101-126
According to radiocarbon dates, presented in table form, the American mastodon became extinct sometimes after 6000 years ago (approximately the beginning of recorded history).

4.214 1967
MASTODONS, THEIR GEOLOGIC AGE AND EXTINCTION IN ONTARIO, CANADA
Dreimanis, Aleksis
Canadian Journal of Earth Science, v. 4, n. 4, August 1967: 663-675
Most Canadian occurrences of mastodons are from southern Ontario. About 4/5 of them have been found below Lake Warren shore, this being younger than 12,400 years BP. The youngest radiocarbon date is 8910 ± 150 years BP. Change of climate and therefore of food availability is one of the reasons advanced for mastodon extinction in the area.

4.215 1967
MAZAMA ASH FROM THE CONTINENTAL SLOPE OFF WASHINGTON
Royse, Chester F.
Northwest Science, v. 41, n. 3, 1967: 103-109

Sedimentologic studies of a core from a bench in Willapa submarine canyon off southern Washington showed an ash layer at a depth of 240 cm. The core represents a sequence of continuous deposition. The age of the ash, calculated from radiocarbon dates elsewhere in the core, is about 7360 years and considered to be Mazama on the basis of age and refractive index, and extends the distribution of this ash about 100 miles westward.

4.216   1968
MAZAMA ASH IN THE NORTHEASTERN PACIFIC
Nelson, C. Hans; Kulm, L.D.; Carlson, Paul R.; Duncan, J. R.
Science, v. 161, n. 3836, 5 July 1968: 47-49
Volcanic glass in marine sediments off Oregon and Washington correlates with continental deposits of Mount Mazama ash by stratigraphic position, refractive index, and radiocarbon dating. Ash deposited in the abyssal regions by turbidity currents is used for tracing of the dispersal routes of glacial sediments and for evaluation of marine sedimentary processes.

4.217   1968
MEANS OF TIME-STRATIGRAPHIC DIVISION AND LONG DISTANCE CORRELATION OF QUATERNARY SUCCESSIONS
Morrison, Robert B.
In: Means of Correlation of Quaternary Successions, edited by Roger B. Morrisson and Herbert C. Wright. International Association for Quaternary Research (INQUA). Congress, 7th, Boulder, Col., 14 Aug - 19 Sept. 1965. *Proceedings*, v. 8. Salt Lake City, University of Utah Press, 1968: 1-113
Examines the application of radiocarbon dating for absolute dating as a means of correlation (p.90-91).

4.218   1965
THE MESA VERDED LOESS
Arhenius, Gustaf; Bonatti, Henrico
American Antiquity, v. 31, n. 2, part 2, Oct. 1965: 92-100 (*Contributions of the Wetherill Mesa Archaeological Project. Society for American Archaeology, Memoirs*, n. 19, 1965)
The age relationship of the loess is established by radiocarbon dating. The earliest preserved loess is probably pre-Wisconsin, possibly Sangamon age and post-pluvial time. Two stages of positive loess accumulation are distinguished, one beginning more than 4000 years ago and the other more than 3000 years ago.

4.219   1968
MID-WISCONSIN INTERSTADIAL PEATS NEAR LONG BEACH, NORTH CAROLINA (Abstract)
Whitehead, Donald R.; Doyle, Michael V.
Geological Society of America. Special Papers, n. 115, 1968: 505-506
Two thin peat horizons overlain and separated by dune sand are dated at 36,000 years BP and 35,800 years BP. This, and other evidence, suggest that this barrier complex should be correlated with the mid-Wisconsin Port Talbot Interstadial.

4.220   1968
MINERAL ZONATION OF WOODFORDIAN LOESSES OF ILLINOIS
Frye, John C.; Glass, H.D.; William, H.R.
Illinois Geological Survey. Circular, n. 427, 1968. pp. 44
Woodfordian loesses of central and western Illinois are zoned by their clay-mineral composition. Data suggest that surface soils in west-central Illinois loess formed during $12,000 \pm 1000$ radiocarbon years; those in the central Illinois Valley formed during $14,000 \pm 1000$ years.

4.221   1964
MODERN DOLOMITE FROM SOUTH AUSTRALIA
Borch, C.C. von der; Rubin, Meyer; Skinner, Brian J.
American Journal of Science, v. 262, n. 9, Nov. 1964: 1116-1118
Radiocarbon measurement of a sediment composed of partially ordered and relatively pure dolomite, forming in a small lake near the Coorong in southeastern South Australia, suggests that an ordered form of dolomite is forming there today.

4.222   1956
MODERN HYDROCARBONS IN TWO WISCONSIN LAKES
Judson, Sheldon; Murray, Raymond C.
American Association of Petroleum Geologists. Bulletin, v. 40, n. 4, Apr. 1956: 447-476
Dating of the hydrocarbons in modern sediment indicates generation of hydrocarbons in the modern environment of the lakes, not contamination by dead hydrocarbons.

4.223   1959
MUMMIFIED SEAL CARCASSE IN THE McMURDO SOUND REGION
Péwé, T.L.; Rivard, N.; Llano, G.A.
Science, v. 130, n. 3377, 18 Sept. 1959: 716
Mummified carcasses of seal preserved in the dry arid environment of the Antarctic have been dated at 1600 to 2600 years old.

4.224   1968
A NEW CHRONOLOGY FOR BRAIDED STREAM SURFACE FORMATION IN THE LOWER MISSISSIPPI VALLEY
Saucier, Roger T.
Southeastern Geology, v. 9, n. 2, 1968: 65-76
Existing late Quaternary chronology for the Lower Mississippi Valley indicates that widespread braided-stream surfaces were formed during waning stages of Late Wisconsin glaciation and mark the end of an alluviation period that preceded development of modern meanders. Recent stratigraphic investigations suggest that some surfaces were affected by a period of waxing glaciation following their formation. This evidence and two radiocarbon dates indicating a 29,000 to 31,000-year age for one valley surface, necessitates a new chronology for valley events.

4.225   1965
A NEW MAMMOTH FIND FROM NORWAY AND A DETERMINATION OF THE AGE OF THE TUSK FROM TOTEN BY MEANS OF C14
Heintz, Anatol
Norsk Geologisk Tidsskrift, v. 45, n. 2, 1965: 227-230
A large fragment of a mammoth tusk, discovered in a gravel pit near Kvam in the Gudbrandsdalen valley, Norway, is described. A radiocarbon age of about 19,000 years obtained for a tusk from Toten is considered a minimum age, thus refuting the suggestion that mammoths were present in Norway in post-glacial times.

4.226   1962
NEW RADIOCARBON DATES OF NILE SEDIMENTS
Fairbridge, Rhodes W.
Nature, v. 196, n. 4850, 13 Oct. 1962: 108-110
Radiocarbon dates on fossil wood, charcoal from human camps and shells of fossil molluscs bring the whole sedimentary regime of the Nile into mass-time perspective. The aim of the research was to examine the records of the complex problem of climatic change in Africa which will help shed light on the world problem of climatic change.

4.227   1967
A NEW SPECIES OF FOSSIL CHLEMYS FROM WRIGHT VALLEY, McMURDO SOUND, ANTARCTICA
Turner, Ruth D.
New Zealand Journal of Geophysics, v. 10, n. 2, 1967: 446-455
Radiocarbon dating of specimens from the Pecten deposit indicates hat these shells are more than 35,000 years old. Uranium-Thorium measurements suggest an age over 800,000 years. Early to middle Pleistocene is suggested.

4.228   1965
NEWBERRY VOLCANO AREA FIELD TRIP - GEOLOGIC SUMMARY
Peterson, N. V.
In: State of Oregon Lunar Geology Field Conference, 1965. *Guide book*. Oregon, Department of Geology and Mineral Industries. *Bulletin*, v. 57, 1965: 11-18
Dating of charred logs in the topmost layer of pumice, part of the core separating two lakes in the caldera of Newberry prehistoric volcano show that the latest eruptions were no more than 2054 ± 230 years ago.

4.229   1967
NON SORTED STEPS IN THE MT. KOSCIUSKO AREA, AUSTRALIA
Costin, Alec B.; Thom, Bruce G.; Wimbush, D.J.; Stuiver, Minze
Geological Society of America. Bulletin, v. 78, n. 8, Aug. 1967: 979-992
Radiocarbon dates from buried organic lenses near the margin of each step indicate that steps 50 to 400 feet below the mountain crest are 2000 to 3000 years old and that further upslope episodic movement occurred during the last 300 years.

4.230   1961
A NOTE ON HIGH-LEVEL MARINE SHELLS ON FOSHEIM PENINSULA, ELLESMERE ISLAND, N.W.T.
Sim, Victor Wallace
Geographical Bulletin, n. 16, Nov. 1961: 120-123
Fragments and unbroken shells of marine molluscs found at elevations considerably higher than reported for similar shell occurences were radiocarbon dated at $19,500 \pm 1100$ years. Several modes of deposition are possible.

4.231   1966
NOTES ON THE ALLUVIAL HISTORY OF THE LAMPASAS RIVER, TEXAS
Cheatum, E.P.; Slaughter, Bob H.
Graduate Research Center Journal, v. 35, n. 1, 1966: 48-54
A floodplain and two terraces in the Lampasas River valley are described. The radiocarbon age of molluscan shells from clay in one of the terraces is 5000 years BP. Moluscan fauna recovered from near the base of this terrace is comparable to the Ben Franklin fauna in Delta County, dated at 9550 ± 375 years BP. The higher terrace seems to be related to others of known Wisconsin age.

4.232   1955
NOTE ON THE DATING OF TERRACES IN THE LAKE MELVILLE DISTRICT, LABRADOR

Blake, Weston
Science, v. 121, n. 3134, 21 Jan. 1955: 112
Fossil trees in a bank of the Crooked River, 25 to 30 ft above present sea-level with sand above and part sand and clay below, with marine molluscs present at this lower level, were dated by radiocarbon at an average age of 1915 ± 127 years. This indicates a slow rise of the land in this area of 25 ft in 2000 years as against 3 ft per century in Hudson Bay.

4.233   1966
OCCURRENCE AND ORIGIN OF QUATERNARY DOLOMITE OF SALT FLAT, WEST TEXAS
Friedman, Gerald M.
Journal of Sedimentary Petrology, v. 36, n. 1, 1966: 263-267
Salt Flat graben is an intermontane basin in Hudspeth and Culbertson Counties, western Texas. The basin is filled with Quaternary sediments that include dolomite distributed among evaporite minerals. Radiocarbon dating indicates an age of $20,300 \pm 825$ years for one dolomite. Dolomite was formed during periods of intense evaporation.

4.234   1952
THE OCCURRENCE OF HYDROCARBON IN RECENT SEDIMENTS FROM THE GULF OF MEXICO
Smith, Paul V.
Science, v. 116, n. 3017, 24 Oct. 1952: 437-439
A radiocarbon age determination helps demonstrate that hydrocarbons found in recent sediment samples in the Gulf of Mexico were either deposited with, or generated in, the sediments themselves (radiocarbon dates: Kulp-Lamont).

4.235   1956
ON THE GEOCHEMICAL STUDY OF CARBON-14. THE OZAGHARA PEATS
Shima, Makoto
Chemical Society of Japan. Bulletin, v. 29, n. 4, 1956: 443-447
Determines the age of the peat at Ozaghara. A layer 2 m beneath the surface is 1000 years old, another layer 3 to 4 m beneath is 5000 years old. The rate of sedimentation and the compression rate of the peat are also determined. The laboratory procedure is described.

4.236   1966
OOLITES ON THE GEORGIA CONTINENTAL SHELF EDGE
Pikley, Orrin H.; Schnitker, Detmar; Pevear, D.R.
Journal of Sedimentary Petrology, v. 36, n. 2, 1966: 462-467
Aragonitic oolites (ooids) are present as a minor constituent of Georgia outer continental shelf and upper slope sediments. They are found at depths between about 35 and 150 m and are most abundant in a north-south band at the continental shell break. Oolites make up at a maximum, about 14 percent of the total sediment. Associated shallow water foraminifera and a radiocarbon date of 26,000 years BP indicate the origin of these oolites at a time of lowered Pleistocene sea-level.

4.237   1968
ORGANIC SEDIMENTS AND RADIOCARBON DATES FROM CRATER LAKES IN THE AZORES
Fries, Magnus
Geologiska Föreningen i Stockholm. Förhandlingar, v. 90, part 3, n. 534, 1968: 360-368
Great irregularities in the chronology of the sedimentation are shown by radiocarbon dating of the lowest organic layer possible to reach with the corer. Radiocarbon dates of two buried juniper stems shed more light on the history of volcanism of the western São Miguel.

4.238   1966
ORIGIN OF DIATOM-RICH, VARVED SEDIMENTS FROM THE GULF OF CALIFORNIA
Calvert, S.E.
Journal of Geology, v. 74, n. 5, Sept. 1966: 546-565
The laminae cannot be distinguished on the basis of the diatom assemlage, only on the total content of biogenous and terrigenous materials. Rates of deposition, determined by radiocarbon dating of organic and carbonate carbon, in selected core sections, demonstrate that one light-coloured lamina and one dark-coloured lamina are deposited in a year, and a couple of laminae constitute a varve.

4.239   1966
ORIGIN OF THE BEERI (ISRAEL) SULFUR DEPOSIT
Nissenbaum, A.; Kaplan, I.R.
Chemical Geology, v. 1, 1966: 295-316
A sulphur deposit from Upper Pleistocene sandstone in Beeri, Israel, is described. Organic material in the deposit and associate organic mats have a radiocarbon age of 30,000 years BP. The suggested depositional environment is a lagoon in which the sulphate was completely reduced to sulphide.

4.240   1967
ORIGIN OF THE SEDIMENTS AND SUBMARINE GEOMORPHOLOGY OF THE INNER CONTINENTAL SHELF OFF CHOCTAWHATCHEE BAY, FLORIDA
Hyne, Norman J.; Goodell, H. Grant
Marine Geology, v. 5, n. 4, 1967: 299-313
Part of the inner continental shelf in the northern Gulf of

Mexico exhibits sinusoidal, submarine ridges and troughs oriented roughly 70° to the strand line. It is proposed that the sand bodies originated as barrier islands and/or spits during the Late Wisconsin regression. Fluvial action cut the ridge and trough topography into the terrace, and the Wisconsin sea level rise modified the topography to its present form. Radiocarbon dates the youngest aspects of the 70 ft deep sand body as at least 5000 years BP.

4.241   1965
OTHER INVERTEBRATES: AN ESSAY IN BIOGEOGRAPHY
Frey, David G.
In: The Quaternary of the United States: a review volume for the 7th Congress of the International Association for Quaternary Research, edited by H.E. Wright and David G. Frey. Princeton. N.J., Princeton University Press, 1965: 613-631
Shells of shallow water species at present water depth of 110-115 m radiocarbon dated at 17,000 - 19,000 years BP indicate a lowering of this level by this amount during classical Wisconsin.

4.242   1961
PALAEOCLIMATOLOGY AND ARCHAEOLOGY IN THE NEAR EAST
Solecki, Ralph S.; Leroi-Gourhan, A.
New York Academy of Science. Annals, v. 95, art. 1, 1961: 729-739
Results of palynological, trace element, and radiocarbon analyses of soil samples from middle and upper Palaeolithic and later occupation layers at the Shanidar cave and nearby Zawi Chemi Shanidar village sites in the Zagros mountains, Iraq, are reported which permit the establishment of a relatively complete record of climatic changes in the region for the past 50,000 years.

4.243   1966
PALEOECOLOGY OF AN ARCTIC ESTUARY
Faas, Richard W.
Arctic, v. 19, n. 4, 1966: 343-348
An alternating sequence of black silt and clay units, and coarse sand and gravel units, has been found to depths of 30 feet below sea-level in the sediments underlying an arctic estuary (Nerravak [Esatkuat] Lagoon, Alaska). The black silt and clay units were deposited in an anaerobic environment while the estuary was isolated from the ocean by a gravel bar across its mouth. The sand and gravel units resulted from the destruction and spreading of the gravel bar inland under marine conditions. A radiocarbon date at the base of the sequence indicates that these conditions first occurred 6450 years ago.

4.244   1964
PALEOTEMPERATURE ANALYSIS OF FOSSIL SHELLS OF MARINE MOLLUSCS (FOOD REFUSE) FROM THE ARENE CANDIDE CAVE, ITALY, AND THE HAUA FTEAH CAVE, CYRENAICA
Emiliani, Cesare L.; Cardini, L.; Mageda, T.; McBurney, C.B.M.; Tongiorgi, E.
In: Isotopic and Cosmic Chemistry, edited by Harmon Craig and others. Amsterdam, North Holland, 1964 : 133-156
Radiocarbon dates ranging from greater than 59,000 to 2500 ± 250 years BP covering cultural succession from Pre-Aurignacian to late Greek are correlated with palaeotemperatures obtained through $^{18}O/^{16}O$ analysis on marine shells collected from the archaeological layers. The results indicate that the temperature maximum correlates with the main portion of the Atlantic zone and the subsequent minor minima with the beginning of the sub-boreal zone.

4.245   1959
PALYNOLOGICAL STUDIES OF THE BARNSTABLE MARSH, CAPE COD, MASSACHUSETTS
Butler, Patrick
Ecology, v. 40, n. 4, Oct. 1959: 735-737
A table of radiocarbon ags and physical characteristics of the Barnstable Marsh peat is presented. From this, a mean rate of rise of sea-level of slightly less than 6 in per century is deduced. This is consistent with values from radiocarbon measurements on peat and shell material from the coast of Holland.

4.246   1968
THE PARINGA FORMATION, WESTLAND, NEW ZEALAND
Suggate, R.P.
New Zealand Journal of Geology and Geophysics, v. 11, n. 2, June 1968: 345-355
Radiocarbon dated shells from the Paringa formation show that 13,400 years ago the ice had retreated almost halfway from the last Otira glaciation maximum towards the head of the valley. Uplift since that date, predominently tectonic, amounts to about 470 ft half a mile from the alpine fault.

4.247   1966
PATTERN OF COASTAL UPLIFT AND DEGLACIERIZATION, WEST BAFFIN ISLAND, N. W.T.
Andrews, John T.
Geographical Bulletin, v. 8, n. 2, 1966: 174-193
An isostatic curve based on seven radiocarbon dates and adjusted for tilt and eustatic sea-level changes is presented.

Ch. 4 - General Geology

4.248 1966
PEAT RESOURCES OF MINNESOTA - REPORT OF INVENTORY N.3, RED LAKE BOG, BELTRAMI COUNTY, MINNESOTA
Farnham, R.S.; Grubich, Donald N.
St. Paul, Minnesota. Office Iron Range Resources and Rehabilitation, 1966. pp. 24
The Red Lakes phagnum peat bog, located in one of the largest continuous peat land regions of the world, is a raised area of more than 3500 acres with steep slopes that create good natural drainage; thickness ranges from five to ten feet. A radiocarbon analysis taken from the bottom of the strata at the eight to nine foot level was dated 2250±120 years BP. This represents a growth rate of one foot in 250 years, a fast rate of accumulation. Analyses of peat from 80 sites are tabulated.

4.249 1965
PERMAFROST IN THE RECENT EPOCH
Nichols, D.R.
In: International Symposium on Permafrost. Proceedings. National Academy of Science. Research Advisory Board, 1965
Ages of the permafrost in the Copper River Basin was determined by the radiocarbon method to be of recent age at two places, that is within the limits of the post-glacial thermal maximum.

4.250 1963
PLANT REMAINS ASSOCIATED WITH MASTODON AND MAMMOTH REMAINS IN CENTRAL MICHIGAN
Oltz, Donald F.; Kapp, Ronald O.
American Midland Naturalist, v. 70, n. 2, Oct. 1963: 339-346
A radiocarbon date of 10,700±400 years BP from one of the mastodons, plus a pollen diagram from the burial site, provide a partial picture of the late glacial flora of central Michigan.

4.251 1966
PLUVIAL LAKE BASINS OF WEST TEXAS
Reeves, C.C.
Journal of Geophysical Research, v. 74, n. 3, May 1966: 269-291
Various pluvial lake basins in Texas are examined and their formation studied. Radiocarbon dates, sedimentation rates and palaeoclimatic studies indicate that the basins have received little fill since the last pluvial period.

4.252 1964
POLLEN ACCUMULATION RATES: ESTIMATES FROM LATE GLACIAL SEDIMENTS OF ROGER LAKE
Davis, Margaret B.; Deevey, Edward S.
Science, v. 145, n. 3638, 18 Sept. 1964: 293-295
Absolute pollen deposition in a Connecticut lake over a 4000 year interval has been estimated from pollen frequencies in a core of late-glacial sediment dated by radiocarbon techniques.

4.253 1967
POLLEN ANALYSIS AND VEGETATIONAL HISTORY OF THE ITASKA REGION, MINNESOTA
McAndrews, John F.
In: Quaternary Palaeoecology, edited by M.J. Cushing and H.E. Wright. International Association for Quaternary Research (INQUA). Congress, 7th, Boulder, Col., 14 Aug - 19 Sept. 1965. *Proceedings*, v. 7. New Haven, Conn., Yale University Press, 1967: 219-236
A reconstruction of the vegetational history of the various landforms and associated vegetational formations in the Itasca region of Minnesota. The chronology is based on radiocarbon ages.

4.254 1965
POLLEN ANALYSIS FROM FOUR LAKES IN THE SOUTHERN MAYA AREA OF GUATEMALA AND EL SALVADOR
Tsukada, Matsuo; Deevey, Edward S.
In: Quaternary Palaeoecology, edited by M.J. Cushing and H.E. Wright. International Association for Quaternary Research (INQUA). Congress, 7th, Boulder, Col., 14 Aug - 19 Sept. 1965. *Proceedings*, v. 7. New Haven, Conn., Yale University Press, 1967: 303-331
The assignment of zones in the pollen diagram to classic, post-classic and post-colonial are supported by two radiocarbon dates.

4.255 1964
POLLEN ANALYSIS FROM THE DEPOSITS OF SIX UPLAND TARNS IN THE LAKE DISTRICTS
Pennington, W.
Royal Society (London). Transactions, Series B, v. 248, 1964: 205-244
The Ulmus horizon has been dated by radiocarbon analysis at ca 3000 B.C. in many parts of North and West Europe.

4.256 1957
POLLEN ANALYSIS OF A VALLEY FILL NEAR UMIAT, ALASKA
Livingstone, D.A.
American Journal of Science, v. 255, 1957: 254-260
Extends a three zones pollen stratigraphy already established for the central Brookes Range. The boundary between the lower two zones has a radiocarbon age of 7500 to

8000 years, that between the birch and alder zones, of approximately 6000 years. This indicates that the postglacial thermal maximum occurred at about the same time in arctic Alaska as in the rest of the world.

4.257   1963
POLLEN ANALYSIS OF FOUR SAMPLES FROM THE RIVER ANNAN, DUMFRIESSHIRE
Moar, N. T.
Dumfriesshire and Galloway Natural History Antiquarian Society. Transactions, Ser. 3, v. 40, 1963: 133-135
[Not sighted]

4.258   1963
POLLEN ANALYSIS OF SURFACE MATERIAL AND LAKE SEDIMENT FROM THE CHUSKA MOUNTAINS, NEW MEXICO
Bent, A.M.; Wright, Herbert E.
Geological Society of America. Bulletin, v. 74, n. 4, Apr. 1963: 491-500
In this comparison between modern and past vegetation and climate through pollen analysis, the inferred Pleistocene age of sediments studied in the Chuska Mountain is confirmed by three radiocarbon dates.

4.259   1965
POLLEN ANALYTIC CORRELATION IN THE COASTAL BARRIER DEPOSITS NEAR THE HAGUE (THE NETHERLANDS)
Zagwijn, W.H.
Netherlands Geologishe Stichting. Meddelingen, N. S., n. 17, 1965: 83-88
Pollen analysis and radiocarbon dating of tidal flats and superjacent coastal barrier deposits near The Hague have shown the tidal flats to be of Atlantic and the coastal barrier deposits to be of Sub-boreal ages.

4.260   1957
A POLLEN ANALYTICAL INVESTIGATION OF THE FLORISBAD DEPOSITS (SOUTH AFRICA)
Van Zinderen Bakker, Edward M.A.
In: Panafrican Congress on Prehistory and Quaternary Study, Congress, third, Livingston. *Proceedings*, 1955. London, Chatto and Windus, 1957: 56-57
Deals with the prehistoric vegetational background of the site of the Florisbad skull of *Homo helmei* and other fossil material. It adds to the understanding of climatic changes which took place there. Pollen analytical results are presented from samples throughout the profile, from basal layer to the topmost layer IV. Radiocarbon dating (Chicago V, Libby, 1954) indicates the following ages: layer I, older than 41,000 years BP; layer II, 9104 ± 420 years BP; layer III, 6700 ± 500 years BP. The pollen spectra reveal the following climatic conditions: arid climate when layer I was buried, oscillatory climatic conditions towards the middle part of the sedimentary sequence, indicative of longer cycle variations other than seasonal; semi-arid towards the top, leading to recent conditions. Libby states that methane in spring water could have affected the dates which then may be too old.

4.261   1967
POLLEN ANALYTICAL STUDIES IN EAST AND SOUTHERN AFRICA
Coetzee, J.A.
In: Palaoecology of Africa and the Surrounding Islands and Antarctica, vol.3. Cape Town, A.A. Balkema, 1967. pp. 146
The pollen analysis is supported by a number of radiocarbon dates which are presented on p. 131.

4.262   1968
POLLEN ANALYSIS IN THE THULE SPRINGS SITE, NEVADA
Mehringer, Peter J.
Ph.D. Dissertation. University of Arizona, 1968. pp. 176
(Abstract in: *Dissertation Abstracts*, Ann Arbor, Mich., v. 29, n. 4, 1968: 1406B-1407B. Order n. 68-13,544)
Radiocarbon dates and pollen analysis of the Thule Springs site are used to establish the geochronology of the deposits and association of artifacts, and to relate these to the late Quaternary vegetation history of the Las Vegas Valley.

4.263   1962
A POLLEN DIAGRAM FROM EQUATORIAL AFRICA, CHERANGANI, KENYA
Van Zinderen Bakker, Edward M.A.
Geologie en Mijnbouw, v. 43, n. 3, 1962: 123-128
The examination of a pollen diagram indicates that climate correlation with Europe is convincing and that the temperature fluctuations during the Quaternary were cosmic in origin. A radiocarbon date helps calculate the age of the section and indicates that the zones described correspond remarkably closely in age with the zones of the climatic scale in Europe.

4.264   1961
A POLLEN DIAGRAM FROM SOUTHEASTERN CONNECTICUT
Beetham, N. ; Niering, W.A.
American Journal of Science, v. 259, 1961: 69-75
A pollen diagram has been constructed from cores taken from Red Maple Swamps, in the Connecticut Arboreum Natural Area at the Connecticut College, New London. Radiocarbon dates help interpret several pollen zones.

Ch. 4 - General Geology

4.265   1963
POLLEN DIAGRAMS FROM OGOTURUH CREEK, CAPE THOMPSON, ALASKA
Heusser, Calvin John
Grana Palynologica, v. 4, n. 1, 1963: 149-159
Discusses the significance of the pollen diagrams in the post-glacial vegetational history at Cape Thompson and in relation to pollen diagrams from other places in arctic Alaska. Various zones are dated from correlation with radiocarbon dated sites at Nome and Umiat.

4.266   1963
POLLEN DIAGRAMS IN THE MACKENZIE DELTA AREA, N. W.T.
Mackay, J. Rodd; Terasmae, Jaan
Arctic, v. 16, n. 4, Dec. 1963: 229-238
The geomorphological interpretation of the exposure is based on radiocarbon dates. The two pollen diagrams are also controlled by radiocarbon dates.

4.267   1961
A POLLEN PROFILE FROM THE GRASSLAND PROVINCE
Sears, Paul B.
Science, v. 134, n. 3495, 22 Dec. 1961: 2038-2040
Radiocarbon dating of lake sediment from Hackberry Lake together with pollen profiles suggest that the climate was cooler and more humid 5,000 years BP. Dating: Yale/Stuiver.

4.268   1967
POLLEN STRATIGRAPHY AND AGE OF AN EARLY POST-GLACIAL BEAVER SITE NEAR COLUMBUS, OHIO
Garrison, Gail C.
Ohio Journal of Science, v. 67, n. 2, March, 1967: 96-105
Pollen analyses indicate that the beaver occupied the site more than 12,000 years ago and that the site was abandoned prior to the increase in oak and other hardwood pollens which marks the beginning of the Hypsithermal Interval. This was substantiated by radiocarbon dates of peat and wood from the site.

4.269   1968
POLLEN, SEED AND MOLLUSC ANALYSIS OF A SEDIMENT CORE FROM PICKEREL LAKE, NORTH-EASTERN SOUTH DAKOTA
Watts, W.A.; Bright, R.C.
Geological Society of America. Bulletin, v. 79, n. 7, July 1968: 855-876
A radiocarbon date suggests that Pickerel Lake was formed on the Coteau des Prairies, Northeastern South Dakota, prior to 10,670 ± 140 years ago.

4.270   1969
A POSSIBLE LATE-QUATERNARY CHANGE IN CLIMATE IN SOUTH AUSTRALIA
Twidale, C.R.
In: Quaternary Geology and Climate. International Association for Quaternary Research (INQUA). Congress, 7th, Boulder, Col., 14 Aug - 19 Sept. 1965. *Proceedings*, v.16. Boulder, Col. 1968: 43-48
Various evidence suggest climatic change in the Adelaide region, the southern Mt. Lofty ranges and the Lake Eyre depression. Fossils which suggest a more humid climate in the late Pleistocene were dated at 14,000 ± 225 years.

4.271   1956
POST-BOREAL POLLEN DIAGRAMS FROM IRISH RAISED BOGS (STUDIES IN IRISH QUATERNARY DEPOSITS NO. 11)
Mitchell, G.F.
Royal Irish Academy. Proceedings, v. 57, sect.B, 1954-1956: 185-251
New Irish pollen-diagrams are presented in order to correlate the dating evidence they present. A number of those diagrams have been dated by radiocarbon.

4.272   1964
POST-GLACIAL CHANGE OF SEA-LEVEL IN NEW HAVEN HARBOR, CONNECTICUT
Upson, Joseph E.; Leopold, Estella B.; Rubin, Meyer
American Journal of Science, v. 262, n. 1, Jan. 1964: 121-132
A radiocarbon and pollen analyses were made of samples from a depth of 30 to 40 feet below sea-level. The age is 5900 ± 200 BP. A relative rise of sea-level at an average rate of slightly more than half a foot per century is suggested.

4.273   1966
POST-GLACIAL DELEVELLING IN SKELDAL, NORTHEAST GREENLAND
Lasca, N. P.
Arctic, v. 19, n. 4, 1966: 349-353
Thirteen radiocarbon dates of shell material were used to establish a rate of emergence in Skeldal, Mesters Vig district, Northeast Greenland, as delevelling occurred. The dates indicate that Skeldal was partially open to the sea by ca 8500 BP. Early emergence (8000-7000 BP) was approximately 3 m per century. Emergence is related almost entirely to isostatic adjustment due to glacial unloading.

4.274   1967
THE POST-GLACIAL HISTORY OF VEGETATION AND CLIMATE AT ENNADAI LAKE, KEEWATIN AND LYNN LAKE, MANITOBA (CANADA)

Nichols, Harvey
Eiszeitalter und Gegenwart, v. 18, 1967: 176-197
The vegetational history of those two lakes (Ennendai and Lynn) appear to be mutually consistent and is explicable in climatic terms. A radiocarbon dated diagram is presented.

4.275   1966
THE POST-GLACIAL MARINE TRANSGRESSION IN N. IRELAND; CONCLUSIONS FROM ESTUARINE AND RAISED BEACH DEPOSITS; A CONTRAST
Singh, Gurdip; Smith, A.G.
Paleobotanist (Lucknow), v. 15, n 1-2, Sep. 1966: 230-234
Stratigraphical, palynological and radiocarbon investigations of estuarine deposits in County Down indicate that the maximum of the Post-Glacial relative sea-level rise was shortly before 3000 BP, when high water level reached to 12 ft above the present mean sea-level.

4.276   1960
POST-GLACIAL RISE OF SEA-LEVEL IN THE NETHERLANDS
Jelgersma, Saskia; Pannekock, A.J.
Geologie en Mijnbouw, v. 39, n. 6, June 1960: 201-207
About 40 radiocarbon dates have been made on peat samples directly overlying the Pleistocene surface at various depths in the coastal region of the Netherlands. These datings serve to establish a curve of the relative rise of the sea-level. It shows a well known strong rise before 6000 BP and a slower rise afterward. This slow rise may be partly a consequence of tectonic subsidence. The remaining fluctuations are only small, which is not in accordance with Fairbridge's curve. Additional radiocarbon dates give the times of later marine transgressions and regressions, which were partly determined by the offshore bars.

4.277   1968
POST-GLACIAL SEA-LEVEL RISE IN THE CHRISTCHURCH METROPOLITAN AREA, NEW ZEALAND
Suggate, R.P.
Geologie en Mijnbouw, v. 47, n. 4, April 1968: 291-297
Radiocarbon dated samples are discussed in relation to the local stratigraphic sequence which records post-glacial transgression of the sea followed by regression during progradation of the shoreline. A table of Christchurch Metropolitan area dates is presented.

4.278   1962
POST-GLACIAL UPLIFT IN NORTH AMERICA
Farrand, William R.
American Journal of Science, v. 260, n. 3, 1962: 181-199
Tilted beaches around the Great Lakes and raised marine features throughout Arctic Canada which have been dated by radiocarbon analyses furnish sufficient data for the construction of curves of post-glacial uplift vs time for eleven areas. The curves are presented and discussed.

4.279   1955
POST-GLACIAL VOLCANIC ASH IN THE ROCKY MOUNTAIN PIEDMONT, MONTANA AND ALBERTA
Horberg, Leland; Robie, Richard A.
Geological Society of America. Bulletin, v. 66, n. 8, Aug. 1955: 949-955
Radiocarbon dates provide additional evidence to help propose a post-glacial chronology in the area.

4.280   1967
POSTGLACIAL CHANGE IN SEA LEVEL IN THE WESTERN NORTH ATLANTIC OCEAN
Redfield, Alfred C.
Science, v. 157, n. 3789: 1967: 687
Radiocarbon determinations on peat indicate that in Bermuda, southern Florida, North Carolina and Louisiana relative sea-level has risen at a rate of about 2.5×10 ft (0.76×10 m) per year during the past 4000 years; this is tentatively taken to be the rate of eustatic change in sea-level. The sea-level rise has been much faster along the northeast coast of the United States, indicating local subsidence.

4.281   1967
POSTGLACIAL CHRONOLOGY AND FOREST HISTORY IN THE NORTHERN LAKE HURON AND LAKE SUPERIOR REGION
Terasmae, Jaan
In: *Quaternary Palaeoecology*, edited by M.J. Cushing and H.E. Wright. International Association for Quaternary Research (INQUA). Congress, 7th, Boulder, Col., 14 Aug - 19 Sept. 1965. *Proceedings*, v. 7. New Haven, Yale University Press, 1967: 45-58
Radiocarbon dates and palynology are used to establish a chronology. A map of sites studied showing the radiocarbon dates is presented.

4.282   1967
POSTGLACIAL PLANT SUCCESSION AND CLIMATIC CHANGES IN A WEST GREENLAND BOG
Fredskild, Bent
Review of Palaeobotany and Palynology, v. 4, n. 1-4, 1967 (*Quaternary Palynology and Actuopalynology*): 113-127
A series of radiocarbon datings has shown that the humid conditions responsible for the change into the first Sphagnum bog was contemporaneous with RY III in northwestern

Europe (about 600 B.C.), while the second growth of Sphagnum starts about A.D. 400.

4.283   1968
POSTGLACIAL STRATIGRAPHY AND MORPHOLOGY OF CENTRAL CONNECTICUT. TRIP A-1 IN GUIDEBOOK FOR FIELD TRIPS IN CONNECTICUT
Bloom, Arthur L.
In: Connecticut Geological and Natural History Survey Guidebook, 2, *Intercollegiate Geology Conference, Annual. Meeting*, 60th. New Haven, Conn. 1968. pp. 7 (Originally published in 1965)
Three significant paludal environments - estuarine freshwater marsh, former deep bay or lagoon, and shallow coastal marsh can be distinguished along the Connecticut coast. Positions and radiocarbon dates of peat from the deep bay or lagoon environment are included: ages range from 910 to 11,000 years BP.

4.284   1968
PRELIMINARY REPORT ON THE LATE QUATERNARY GEOLOGY OF THE SAN PEDRO VALLEY, ARIZONA
In: Southern Arizona Guidebook 3. Geological Society of America, Cordilleran Section, Annual Meeting, 64th, Tucson, 1968. Tucson, Arizona, Geological Society, 1968: 79-96
Cycles of deposition and erosion starting with an oldest unit of lacustrine deposits of Wisconsinan age followed by various episodes of erosion and of deposition are dated by radiocarbon at from 30,000 years BP to 10,000 years BP.

4.285   1960
PRELIMINARY REPORTS ON THE PALYNOLOGICAL FIELDWORK 1960, INCLUDING TWO RADIOCARBON DATINGS
Jacobsen - McGill Arctic Research Expedition to Axel Heiberg Island. Preliminary Reports of 1959-1960, edited by Barbara S. Muller. Montreal, McGill University, 1961: 201-208
A peat profile south of Expedition River near the base camp on Axel Heiberg Island was dated by radiocarbon on two samples: (1) the lowest level containing a considerable amount of organic matter (depth 250 - 260 cm) at 4210 ± 100 years, (2) the upper layer (depth 130 - 140 cm) at 2900 ± 120 years.

4.286   1958
A PRELIMINARY STUDY OF $^{13}C/^{12}C$ RATIOS IN THE THERMAL AREAS
Hulston, John R.; Melhuish, W.H.; McCabe, William J.
Institute of Nuclear Sciences. Department of Scientific and Industrial Research. (Lower Hutt, New Zealand) *Lab. Notes*, n. 2, 1958. pp.17
A study undertaken to trace sources of carbon dioxide and carbonates in streams and bores in thermal areas.

4.287   1968
PREQUAKE RECENT VERTICAL DISPLACEMENT IN THE ZONE AFFECTED BY TECTONIC DEFORMATION DURING THE ALASKA EARTHQUAKE. (Abstract)
Pflaker, George
Geological Society of America. Special Papers, n. 101, 1968: 327-328
Radiocarbon dates from 3 of 5 pre-quake terraces on Middleton Island in the uplifted zone indicate that 40 m of the relative uplift occurred in at least five major upward pulses during the last 4470 ± 250 years and that the last pulse occurred 700 ± 250 years BP. Radiocarbon dated drowned forests along the coast of the uplifted zone indicate pre-quake subsidence of as much as 4.3 m in the last 390 ± 160 years.

4.288   1965
PROBLEMS OF DATING ICE-CORED MORAINES
Ostrem, Gunnar
Geografiska Annaler, v. 47A, 1965: 1-38
Several methods are used and checked against each other: radiocarbon, tritium, pollen analysis. A description of the method for dating ice cores by radiocarbon measurements is included.

4.289   1961-1962
PROBLEMS OF RADIOCARBON DATING OF RAISED BEACHES BASED ON EXPERIENCE IN SPITZBERGEN
Olsson, Ingrid U.; Blake, Weston
Norsk Geografisk Tidsskrift, v. 18, n. 1-2, 1961-1962: 47-64
Twenty two samples of driftwood, whole bones, and shells collected from raised beaches in Spitzbergen were dated by the Uppsala radiocarbon laboratory. Problems associated with the laboratory procedures and the geological background are discussed. The aim of the project was to establish an absolute chronology of events leading to the partial deglaciation of the island of Nordaustlandet.

4.290   1961
PROBLEMS OF RADIOCARBON DATING OF RAISED BEACHES, BASED ON EXPERIENCE IN SPITZBERGEN
Olsson, Ingrid U.; Blake, Weston
Ohio University. Institute of Polar Studies. Contributions, n. 7, 1961
Problems connected with laboratory procedures as well as

4.291 1966
PUMICE AT SUMMER LAKE, OREGON
Allison, Ira S.
Geological Society of America. Bulletin, v. 77, n. 3, 1966: 329
A lake-land layer of sandy pumice was incorrectly correlated with Mt. Mazama pumice more than 30,700 radiocarbon years old. True Mazama air-land pumice 6600 years old lies on the lake flat.

4.292 1966
PUMICE DEPOSITS OF THE TOWADA CALDERA IN THE VICINITY OF KOSAKA TOWN, AKITA PREFECTURE, JAPAN
Satoh, Hiroguki
Geological Society of Japan. Journal, v. 72, n. 9, 1966: 405-411
Radiocarbon dating on peat and charcoal, and pollen analysis on peat enable to date the geological sequence in Pleistocene and Holocene times, from 32,000 years BP to 2000 years BP. A table of dates is presented and a columnar section showing the relationship of the samples.

4.293 1962
PYROCLASTIC DEPOSITS OF RECENT AGE AT MOUNT RAINIER, WASHINGTON
Crandell, Dwight R.; Mullineaux, Donald H.; Miller, Robert; Rubin, Meyer
United States Geological Survey. Professional Papers, n. 450-D, article n.138, 1962: D64-D68
The ages of the various pyroclastic deposits are dated by radiocarbon age determination of organic matter associated with the deposits. It is presented in table form.

4.294 1967
QUATERNARY CHRONOLOGY OF GOULBURN VALLEY SEDIMENTS AND THEIR CORRELATION IN SOUTHEASTERN AUSTRALIA.
Bowler, J.M.
Geological Society of Australia. Journal, v. 14, part 2, 1967: 287-292
Radiocarbon dates from the Goulburn Valley, Victoria, throw new lights on the age of Quaternary tectonics, ancestral streams, fluviatile sediments and associated soils. A list of radiocarbon dates from sediments on the Riverine Plain, Victoria is given in table form and a radiocarbon chronology in relation to other stratigraphic divisions in the K-cycle-ground surface system.

4.295 1965
QUATERNARY CLIMATIC RECORDS IN RED SEA CORES
Herman, Y.
In: International Association for Quaternary Research (INQUA). Congress, 7th, Boulder, Col., 14 Aug - 19 Sept. 1965. Abstracts of General Sessions. Boulder, Colorado, 1965: 208
Good correlation between cores based on faunal evidence and two radiocarbon dates suggest the following succession of climates: (1) Post-glacial, (2) Last Glacial, (approximate end of this is 16,000 BP, approximate duration 100,000 years), (3) Last Interglacial, (4) Penultimate Glaciation.

4.296 1968
QUATERNARY DEPOSITS AROUND HOLSTEINBORG
Weidick, Auker
Grönlands Geologie Undersögelse Rap., No.15 (Report of activities, 1967), 1968: 23-24
The high outer coastal terrain is cut by an east-west depression around Ikertoq Fjord, which continues offshore as a deep submarine canyon and was a drainage channel for the Inland Ice during the Pleistocene. The highest Quaternary deposits are 110-140 m above sea level. Radiocarbon dates on shells in deposits below 70 m indicate a trend of uplift resembling that described for the Mestersvig area in East Greenland. Ice margin deposits were laid down before the fjord stages but later than Wisconsin - Würm moraines of offshore banks; these moraines represent a stage designated the Taserqat stage.

4.297 1968
QUATERNARY DEPOSITS OF THE EAST FORK OF THE TRINITY RIVER, NORTH-CENTRAL TEXAS. (Abstract)
Thurmond, John T.
Geological Society of America. Special Papers, n. 101, 1968: 340
The extensive terrace system of the Trinity River in Texas gives promise of providing a means of direct physical and faunal correlation between the marine Quaternary sequence of the Gulf Coast and temporally equivalent continental deposits of the Great Plains. Three terraces were mapped. The lower yielded a radiocarbon date of 10,000 years BP, the middle terrace is dated at 37,000 years BP and the upper terrace whose age is outside the radiocarbon dating range is tentatively assigned to the Sangamon Interglacial.

4.298 1965
THE QUATERNARY ERA IN THE NORTHWEST SAHARA
Alimen, Marie-Henriette
In: International Studies in the Quaternary. Papers presented on the Occasion of the 7th Congress of the International Association for Quaternary Research (INQUA),

Ch. 4 - General Geology

Boulder, Colorado, 1965, edited by H.E. Wright and David G. Frey. New York, Geological Society of America, 1965: 273-291 (*Geological Society of America Special Paper* No. 84, 1965)

Stratigraphically the following phases are detected: Villafranchian, middle Quaternary and late Quaternary. A radiocarbon analysis dates a Saourien (late Quaternary) level at 20,000 ± 1000 BP and the maximum of Guirien sedimentation at 6160 ± 320 or Neolithic age.

4.299   1965
QUATERNARY GEOLOGY OF THE TULE SPRINGS AREA, CLARK COUNTY, NEVADA
Haynes, C. Vance
Ph.D. Dissertation. University of Arizona, 1965. (Dissertation in: *Dissertation Abstract*, Ann Arbor, Michigan, 26/03/1592, n. 65-9854)

Analysis of 80 radiocarbon samples provides geochronological control, permits geochemical evaluation of tufa formation, supports correlation of the Tule Springs beds with some other areas of the Southwest, and indicates that aquifers of the Las Vegas basin contain 'fossil' water.

4.300   1965
QUATERNARY GEOLOGY, RADIOCARBON DATING, AND THE AGE OF AUSTRALIA
Gill, Edmund D.
In: International Studies in the Quaternary. Papers presented on the Occasion of the 7th Congress of the International Association for Quaternary Research (INQUA), Boulder, Colorado, 1965, edited by H.E. Wright and David G. Frey. New York, Geological Society of America, 1965: 415-452 (*Geological Society of America Special Paper* No. 84, 1965)

An attempt is made to date 14 Australites found at Port Campbell on the south coast of Western Victoria, Australia by their stratigraphic occurrence. Radiocarbon determinations were made on seven samples stratigraphically associated with the Australite. The dates which indicate that the Australites do not seem, on geological ground, to be older than late Pleistocene, are discussed.

4.301   1956
QUATERNARY HISTORY AND THE BRITISH FLORA
Godwin, Harry
Advancement of Science, v. 13, n. 50, Sept. 1956: 118-124
Radiocarbon enables to determine absolute dating for temperature and climate variations for the period extending back to 45,000 years.

4.302   1965
THE QUATERNARY OF DENMARK
Hansen, Sigurd
In: The Quaternary, v. I, edited by Kalervo Rankama. New York, Wiley Interscience, 1965: 1-90
Presents a short review of the radiocarbon dates applied to the Quaternary of Denmark and compares them to varve counts. There seems to be reasonable agreement, taking into consideration the statistical error inherent to radiocarbon dates.

4.303   1967
THE QUATERNARY OF FRANCE
Alimen, Marie-Henriette
In: The Quaternary, v. II, edited by Kalervo Rankama. New York, Interscience, 1967: 89-239
A review article which draws on many preceding studies. A radiocarbon chronology of the Quaternary of France is presented in table form.

4.304   1967
THE QUATERNARY OF GERMANY
Woldstedt, Paul
In: The Quaternary, v. II, edited by Kalervo Rankama. New York, Interscience, 1967: 239-300
A review article on the Quaternary in Germany. A concentrated and revised text of the book Das Eiszeitalter, II, Stuttgart, 1958

4.305   1965
THE QUATERNARY OF SWEDEN
Lundqvist, Jan
In: The Quaternary, v. I, edited by Kalervo Rankama. New York, Interscience, 1965: 139-198
A review of the Quaternary of Sweden which uses radiocarbon dating as one of the methods of establishing a Quaternary evolution. A table of the subdivisions of the Quaternary in Sweden where the ages are corrected on the basis of radiocarbon dates is presented.

4.306   1967
THE QUATERNARY OF THE NETHERLANDS
de Jong, Jan D.
In: The Quaternary, v. II, edited by Kalervo Rankama. New York, Interscience, 1967: 1-88
In this review article of the Quaternary of the Netherlands, radiocarbon dates are cited throughout and various tables of chronologies and stratigraphies based on radiocarbon dates are presented.

4.307   1965
QUATERNARY PALEOPEDOLOGY
Ruhe, Robert V.
In: The Quaternary of the United States: a Review Volume for the 7th Congress of the International Association for

Quaternary Research, edited by H.E. Wright and David G. Frey. Princeton. N.J., 1065
Methods of determining which features of a soil are related to the past and which are related to the present by dating organic carbon, inorganic carbon or calcium carbonate of various layers is presented.

4.308   1968
QUATERNARY PHENOMENA IN THE WESTERN CONGO
Ploey, J. de
In: Means of Correlation of Quaternary Successions, edited by Roger B. Morrisson and Herbert C. Wright. International Association for Quaternary Research (INQUA). Congress, 7th, Boulder, Col., 14 Aug - 19 Sept. 1965. *Proceedings*, v. 8. Salt Lake City, University of Utah Press, 1968: 501-517
Discusses the origin and chronology of surficial deposits in the western Congo Republic. The results of the investigations are summarised in a stratigraphic chart. Radiocarbon dates provide a good evidence of the contemporaneity of Lupemban and Tshitolian cultures in west central Africa.

4.309   1964
QUATERNARY SHORELINE IN AUSTRALIA
Gill, Edmund D.
Australian Journal of Science, v. 26, n. 12, June 1964: 388-391
Discusses radiocarbon dates in terms of other factors available.

4.310   1968
QUATERNARY SHORELINE RESEARCH IN AUSTRALIA
Gill, Edmund D.
Australian Journal of Science, v. 31, n. 1, 1968: 106-111
A review article on Quaternary shorelines, reporting research in all states of Australia. Includes tables of radiocarbon dates obtained from sediments between Bowen and Innisfail in north Queensland and from the Yarra delta in Victoria.

4.311   1966
QUATERNARY TECTONICS AND THE EVOLUTION OF THE RIVERINE PLAIN NEAR EUCHUCA, VICTORIA
Bowler, J.M.; Harford, L.B.
Geological Society of Australia. Journal, v. 13, part 2, 1966: 339-354
Important late Quaternary tectonic movements near Euchuca initiated drainage diversions and modifications from which a detailed geomorphic sequence is established. Carbon dates from charcoal fragments were 6800 ± 150 and 4200 ± 130 providing a minimum age for the movement on the fault which occurred little more than 7000 years ago.

4.312   1967
QUATERNARY VEGETATIONAL HISTORY OF ARCTIC ALASKA
Colinvaux, Paul A.
In: The Bering Land Bridge, edited by David M. Hopkins. Stanford, Calif., Stanford University Press, 1967: 207-231
The lowlands or arctic Alaska and of the islands in the Bering sea have never been completely glaciated and thus have supported vegetation throughout the Pleistocene. Peat deposits have survived, leaving pollen records. This enables to reconstruct the history of the vegetation of northern regions by means of pollen analysis. Radiocarbon dating of the pollen spectra allows arrangement by chronological order. A table and dated diagrams ar presented. The pollen record suggests that the Bering Strait region has never supported forest.

4.313   1964
RADIO-CARBON DATING OF AUSTRALITE OCCURRENCES, MICROLITHS, FOSSIL GRASSTREE AND HUMUS PODZOL STRUCTURES
Gill, Edmund D.
Australian Journal of Science, v. 27, n. 10, 1964: 300-301
Radiocarbon dating and other evidence indicate that australites recovered at Port Campbell, Victoria, fell either in late Pleistocene or late Holocene.

4.314   1963
RADIOCARBON AGE OF THE TWO-CREEK FOREST BED, WISCONSIN
Broecker, Wallace S.; Farrand, William R.
Geological Society of America. Bulletin, v. 74, n. 6, June 1963: 795-802
In an attempt to clarify the correlation between events marking the closing phases of the last glacial period in Europe and North America, radiocarbon ages of lignin and cellulose, separated from six wood samples from the Two Creeks forest bed were carefully determined. The results were within 140 years of 11,850 BP suggesting that the forest is 450 years older than assumed previously. This would suggest an inter-continental correlation of the Two Creeks with the Bølling warm interval in Europe (not the Allerød). It also denies correlation of the Two Creeks forest and the Champlain sea.

4.315   1968
RADIOCARBON AGES FROM THE BOTTOM DEPOSITS OF LAKE SARKKILANJARVI, SOUTH-WEST FINLAND
Alhonen, Pentti

Geological Society of Finland. Bulletin, n. 40, 1968: 65-70
A contribution to the post-Pleistocene (Holocene) chronology of south-west Finland, based on pollen stratigraphy and radiocarbon dating, is presented.

4.316   n.d.
RADIOCARBON AGES OF GROUND WATER IN AN ARID ZONE UNCONFINED AQUIFER
Tamers, Murray A.
Instituto Venezolano de Investigaciones Cientificas. Dept. di Chimica. Caracas, Venezuela, n. d.
The dissolved carbon dioxide and bicarbonate of water are extracted and used as material for radiocarbon dating. A typical arid zone unconfined aquifer has been studied (Macaraibo). It was found that the system is no longer recharged and the water dates from the pluvial conditions prevailing at the end of the last glacial period. The effect of the dilution of the original carbon by dissolved limestone is discussed with regard to the extent of the error involved in corrections for this factor.

4.317   1966
RADIOCARBON AGES OF GROUNDWATER FLOWING INTO A DESICCATING LAKE
Tamers, Murray A.; Thielen, C.
Acta Cientifica Venezolano, v. 17, 1966: 150-157
Studies of groundwater surrounding the lake of Valencia in Venezuela were carried out using naturally occurring radiocarbon. It was found that water flows into the lake rapidly from the north, but very slowly from the east. The apparent age of the water is 5000 to 6000 year at its entrance to the lake. The original cause for desiccation is the climate change which took place 6000 years ago but this shows up in the lake only 500 years ago due to the slowness of water movement from the east. A table of radiocarbon ages is presented.

4.318   1965
RADIOCARBON AGES OF OOLITIC SANDS ON GREAT BAHAMA BANK
Martin, E.L.; Ginsburg, R.N.
In: International Conference on Radiocarbon and Tritium Dating, Pullman, Washington, Washington State University, June 1-11, 1965. *Proceedings*. United States of America, Atomic Energy Commission, 1965. Conference n. 650652: 705-719
The results of age measurements on nine samples from widely separated localities prove that at least the outermost 10 percent by volume of the ooliths in the most abundant size range has been formed in less than 160 years. Age determinations of the innermost parts of ooliths and the dating of entire samples of oolitic sands provide data on the ages of the sand and their rates of formation.

4.319   1962
RADIOCARBON AGES OF POSTGLACIAL LAKE CLAYS NEAR MICHIGAN CITY, INDIANA
Winkler, Ebhard M.
Science, v. 137, n. 3529, 17 Aug. 1962: 528-529
Two radiocarbon date were obtained for post-glacial clays. One date, about 11,000 years or older, indicates a Glenwood age for the lower pebbly clay; the other date of 5475 years is for the upper blue clays which are of shallow swamp origin and were deposited during the late low water state of Lake Shippewa.

4.320   1967
RADIOCARBON AND POLLEN EVIDENCE FOR A SUDDEN CHANGE IN CLIMATE IN THE GREAT LAKES REGION APPROXIMATELY 10,000 YEARS AGO
Ogden, J. Gordon
In: Quaternary Palaeoecology, edited by M.J. Cushing and H.E. Wright. International Association for Quaternary Research (INQUA). Congress, 7th, Boulder, Col., 14 Aug - 19 Sept. 1965. *Proceedings*, v. 7. New Haven, Yale University Press, 1967: 117-127
Two main lines of evidence must be considered in postulation of a dramatic rather than gradual change in climate in late-Wisconsin time. The first line of evidence is based upon the correlation of radiocarbon dates with pollen stratigraphy. The major conclusion on the evidence is that, according to the data, the distance covered within the change and the time available is too short to support the theory of a gradual inversion of deciduous forest with established boreal forest community. The only mechanism sufficient to produce the rapid change described would be a dramatic and rapid change in temperature and/or precipitation approximately 10,000 years ago. A table of radiocarbon dates of the transition from spruce to hardwood pollen zone in the northeast North America is presented.

4.321   1965
RADIOCARBON AND SOILS, EVIDENCE OF FORMER FOREST IN THE SOUTHERN CANADA TUNDRA
Bryson, Reid A.; Irving, William N.; Larsen, James A.
Science, v. 147, n. 3653, 1 Jan. 1965: 46-48
Radiocarbon dating of charcoal on podzols along a transect reaching 280 km north of the present tree line from Ennadai Lake in the Hudson Bay area indicates that former forests were burnt about 3500 years ago and again about 900 years ago. These forests were probably associated with periods of relatively mild climates.

4.322 1965
RADIOCARBON CHRONOLOGIES OF LAKES LAHONTAN AND BONNEVILLE: A STRATIGRAPHIC EVALUATION
Morrison, Robert B.
In: International Association for Quaternary Research (INQUA). *Congress*, 7th, Boulder, Col., 14 Aug - 19 Sept. 1965. *Abstracts of General Sessions*. Boulder, Colorado, 1965: 347
In this evaluation, the author contests the stratigraphy as suggested by Broecker and Kauffman (1964, 1965) by showing that the radiocarbon dates do not fit with the observed stratigraphy.

4.323 1958
RADIOCARBON CHRONOLOGY OF LAKE LAHONTAN AND LAKE BONNEVILLE
Broecker, Wallace S.; Orr, Phil C.
Geological Society of America. Bulletin, v. 69, n. 8, Aug. 1958: 1009-1032
Radiocarbon measurements of fresh-water carbonates have been used to determine the absolute chronology of the two largest fossil lakes in the Great Basin. The possibility of systematic error due to exchange and to low initial radiocarbon concentration have been considered with the conclusion that most of the measurements reported have not been affected by more than 10%. Conclusion is made that as lake level response to climatic change is sufficiently rapid, lake-level chronology can be used as climate chronology for the Great Basin.

4.324 1965
RADIOCARBON CHRONOLOGY OF LAKE LAHONTAN AND LAKE BONNEVILLE II, GREAT BASIN
Broecker, Wallace S.; Kaufman, Aaron
Geological Society of America. Bulletin, v. 76, n. 5, May 1965: 557-566
The climate chronology proposed by Broecker and Orr (*Geological Society of America, Bulletin*, 1958) for the two major Great Basin pluvial lakes has been confirmed by compiling 80 new radiocarbon measurements with uranium-series isotope measurements. The lake-level chronology has been extended to the postpluvial period (<8,000 years ago) by radiocarbon measurements and to the pre-radiocarbon period (>30,000 years ago) by the ^{230}Th-^{234}U measurement. The results suggest that both lakes were relatively high four times during the late-Wisconsin.

4.325 1965
THE RADIOCARBON CHRONOLOGY OF THE LAST GLACIATION AND OF THE POST-GLACIAL INTERVAL IN SIBERIA
Kind, N.V.
In: International Association for Quaternary Research (INQUA). *Congress*, 7th, Boulder, Col., 14 Aug - 19 Sept. 1965. *Abstracts of General Sessions*. Boulder, Colorado, 1965: 269
An absolute chronology is suggested for the principal events and climatic fluctuations during Late Quaternary time in Siberia, based on more than 30 radiocarbon dates, the majority of which refer to the well studied area of Northern Yenissei.

4.326 1962
RADIOCARBON CONTENT AND TERRESTRIAL AGE OF TWELVE STONY METEORITES AND ONE IRON-ORE METEORITE
Suess, Hans E.; Wanke, H.
Geochimica et Cosmochimica Acta, v. 26, 1962: 475-480
The radiocarbon content of twelve stony meteorites, six with observed time-of-fall and the others finds, and that of an iron, have been determined in the La Jolla Radiocarbon Laboratory. The average contemporaneous radiocarbon content of the falls was found to correspond to 48.2 disintegration/min. A much lower value of 5.4 disintegration/min was found to be Pleistocene. The average radiocarbon content of the finds investigated indicates that terrestrial ages of the order of a thousand years are quite common.

4.327 1956
A RADIOCARBON DATE FROM SMOKY LAKE, ALBERTA
Gravenor, C.P.; Elwood, B.
University of Alberta. Research Council of Alberta. Preliminary Report, 1956: 56-59
[Not sighted]

4.328 1965
A RADIOCARBON DATED POLLEN SEQUENCE FROM SILVER LAKE, LOGAN COUNTY, OHIO
Gooding, Ansel M.; Ogden, J. Gordon
Ohio Journal of Science, v. 65, n. 1, Jan. 1965: 1-11

A radiocarbon date of 12,000 ± 450 has been obtained from wood beneath a mastodon in North Central Indiana. The sediment sequence in which it is situated is described. Correlation of this sequence with other radiocarbon dated pollen sequences indicates that the mastodon died prior to the replacement of spruce pollen by oak pollen approximately 10,500 years ago.

4.329 1965
A RADIOCARBON DATED POLLEN SEQUENCE FROM THE WELLS MASTODON SITE NEAR ROCHESTER, INDIANA

Gooding, Ansel M.; Ogden, J. Gordon
Ohio Journal of Science, v. 65, n. 1, 1965: 1-11
Wood found below the bones of a *Mastodon americanus* at the Wells Mastodon Site gave a radiocarbon date of 12,000 ± 450 years. The stratigraphic sequence of the site is described. The wood is older than the mastodon bones which are estimated to be between 12,000 and 11,000 years old, probably nearer the latter.

4.330   1963
RADIOCARBON DATES AND NOTES ON THE CLIMATIC AND MORPHOLOGICAL HISTORY
Müller, Fritz
In: Axel Heiberg Research Reports, McGill University, Montreal. Preliminary Reports 1961 - 1962. Montreal, McGill University,1963: 169-172
A sample from organic material collected from the interface between an ancient moraine and a more recent one, eroded by the Between River, give a date of 240 ± 100 years BP, indicating that the last advance of the White Glacier beyond its present position probably occurred during the first half of the 18th century. A sample from the older morainic outwash fan give a date of 2219 ± 205 years BP. Peat from the Expedition River give a date of 4210 ± 100 years BP. Various other radiocarbon dates on shells from raised marine beaches are listed and discussed.

4.331   1966
RADIOCARBON DATES FROM A STONE-BANKED TERRACE IN THE COLORADO ROCKY MOUNTAINS, U.S.A.
Benedict, James B.
Geografiska Annaler, v. 48, n. 1, 1966: 24-31
Niwot Ridge, west of Boulder, Colorado, has near its center a group of stone-banked terraces on a snow-accumulation slope. Beneath the stones is a layer of platy loam, and below that the original A-C soil profile. Radiocarbon dates of the entire thickness of the A horizon suggest that the terrace originated 3000 to 2500 years ago, during a waning of the Temple Lake glacial epoch.

4.332   1955
RADIOCARBON DATES IN CENTRAL IOWA
Ruhe, Robert V.; Scholtes, W.H.
Journal of Geology, v. 63, n. 1, Jan. 1955: 82-92
This paper aims to clarify the stratigraphy of the deposits of the Central Iowa radiocarbon samples and to establish the late Pleistocene chronology of the horizons dated by radiocarbon.

4:333   1968
RADIOCARBON DATES OF AN OOLITIC SAND COLLECTED FROM THE SHELF OFF THE EAST COAST OF INDIA
Naidu, A.S.
National Institute of Science of India. Bulletin, n. 38 (*Symposium on Indian Ocean*. Part I), 1968: 467-471
A composite core sample consisting of calcareous oolites and littoral shells off the east coast of India was dated by radiocarbon at 10,800 ± 155 years BP. This suggests an abrupt change of climate around that date, in conformity with Broecker et al conclusions.

4.334   1961
RADIOCARBON DATES OF LATE QUATERNARY DEPOSITS, SOUTH LOUISIANA: A DISCUSSION
Broecker, Wallace S.
Geological Society of America. Bulletin, v. 72, n. 1, Jan. 1961: 159-162
In this discussion of McFarland article (*Geological Society of America. Bulletin*, 1961) the author contests two of the conclusions: (1) a still-stand in level between 35,000 and 18,000 years ago (2) a rejection of the suggestion made by Ewing (*Science*, 1956), Ericson / Broecker (*Science*, 1956), Broecker / Orr (*Science*, 1958) that an abrupt change in climate occurred close to 11,000 years ago. The author bases his rejection on the possibility of sample contamination, uncertainties in the subsidence corrections, and the scarcity of data.

4.335   1968
RADIOCARBON DATES ON, AND DEPOSITIONAL ENVIRONMENT OF, THE WASAGA BEACH (ONTARIO) MARL DEPOSIT
Farrand, William R.; Miller, B.B.
Ohio Journal of Science, v. 68, n. 4, July 1968: 235-239
Five radiocarbon dates from the marl vary widely and show the effect of contamination by the addition of dead carbon from adjacent Palaeozoic carbonate rocks. Other evidence indicate a Nipissing Great Lakes stage.

4.336   1966
RADIOCARBON DATES RELATING TO SOIL DEVELOPMENT, COAST-LINE CHANGES AND VOLCANIC ASH DEPOSITION IN SOUTH-EAST SOUTH AUSTRALIA
Blackburn, G.
Australian Journal of Science, v. 29, n. 2, Aug. 1966: 50-52
Six dates are presented and discussed.

4.337   1960
RADIOCARBON DATES RELATING TO THE GUBIK FORMATION, NORTHERN ALASKA
Coulter, H.W.; Hussey, K.M.; O'Sullivan, J.B.
United States Geological Survey. Professional Papers, n..400-B, Article 160, 1960: B350-B351
Radiocarbon dates indicate that deposition of the upper

member of the Gubik formation near Barrow was initiated prior to 38,000 years BP and was terminated prior to 9100 years BP.

4.338   1954
RADIOCARBON DATING AND CHANGE OF SEA LEVEL AT VELZEN (NETHERLANDS)
Van Straaten, L.M.J.U.
Geologie en Mijnbouw, v. 16, n. 6, June 1954: 247-253
In the reconstruction of the relative movements of sea-level two sets of data are required: age determination and determination of the corresponding sea-levels. These are discussed and the various interpretations which are possible are presented and discussed.

4.339   1956
RADIOCARBON DATING FOR GLACIAL VARVES IN AUSTRALIA
Gill, Edmund D.
Australian Journal of Science, v. 19, 1956: 80
Based on one radiocarbon date of 26,480 ± 800, a site at Gormanston, near Queenstown, presenting a thick series of varves, could be ascribed to the commencement of the Würm glaciation or early Wisconsin age. Physiography seems to agree with this.

4.340   1965
RADIOCARBON DATING FROM A POST-GLACIAL BEACH IN WESTERN WELLINGTON
Fleming, C.A.
New Zealand Journal of Geology and Geophysics, v. 8, n. 6, Dec. 1965: 1222-1223
A bed consisting of plant debris encased in sediments and exposed by wave action at Paekakeriki, N.Z. was dated at 5140 ± 90 years, which indicates that the date of the maximum stand of the sea is later than 5140 years ago.

4.341   1962
RADIOCARBON DATING OF CALIFORNIA BASIN SEDIMENTS
Emery, Kenneth Orriz; Bray, Ellis E.
American Association of Petroleum Geologists. Bulletin, v. 46, Oct. 1962: 1839-1856
Radiocarbon age determinations on cores from the basin off Southern California yield rates of deposition of sediments ranging from 5 to 160 cm/1000 years with greater rates inshore and indicate that the present rates of deposition are similar to those which existed during the Pliocene. This indicates a continuation of the same kinds of environment.

4.342   1966
RADIOCARBON DATING OF COASTAL PEAT, BARROW, ALASKA
Brown, Jerry; Sellman, Paul V.
Science, v. 153, n. 3733, July 1966: 299-300
A buried, frozen section of peat from sea-level yielded radiocarbon dates between 700 and 2600 B.C., it suggests burial by a transgressing sea.

4.343   1959
RADIOCARBON DATING OF FOSSIL SOILS AT OBER FELLABRUN
de Vries, Hessel
Koninklijke Nederlandske Akademie van Wetenschappen. Proceedings, series B, v. 62, 1959: 84-91
The investigation made on the fossil soils at Ober Fellabrun (Austria) indicates that the dating of loess profiles, with or without fossil soils is possible. The ages obtained are compatible with radiocarbon dates from other areas.

4.344   1968
RADIOCARBON DATING OF LACUSTRINE STRANDS IN ARCTIC ALASKA
Carson, Charles E.
Arctic, v. 21, n. 1, Mar. 1968: 12-26
Present exposure of lacustrine shelter 10 to 12 miles inland from the arctic coast of Alaska were dated at less than 3500 years old. It suggests that the lacustrine expansion reached a maximum near the end of the hypsithermal, around 5000 to 4000 years ago and that the onset of the post-hypsithermal cooling phase corresponds in time with the initial period of draining. Dating was done mainly on peat.

4.345   1962
RADIOCARBON DATES RELATING TO A WIDESPREAD VOLCANIC ASH DEPOSIT, EASTERN ALASKA
Fernald, Arthur T.
United States Geological Survey. Professional Papers, n. 450-B, 1962: B29-B30
Radiocarbon dates from layers bracketing an ash deposit confirm the estimated age of 1400 years (between 1750 and 1520 BP). The age of the ash provides an important reference point in the interpretation of the surficial geology of the Upper Nenana River Valley, particularly the flood-plain deposits.

4.346   1961
RADIOCARBON DATING OF LATE QUATER-NARY DEPOSITS, SOUTH LOUISIANA
McFarland, E.
Geological Society of America. Bulletin, v. 72, n. 1, Jan. 1961: 129-158
The late Quaternary deposits of southern and offshore Louisiana record a complete cycle of sea-level fluctuation which is associated with major changes in the volume of ice

on the continent since the beginning of the last glacial stage. A eustatic curve, based on the radiocarbon age determination of 122 samples, supports previous estimates from geological data that the sea during the early part of the cycle fell to a position at least 450 feet below its present level. The eustatic curve implies that the ice sheets of the major glacial stage had begun to retreat before 35,000 years ago after reaching their maximum extension.

4.347   1965
RADIOCARBON DATING OF PAST SEA LEVELS IN S.E. AUSTRALIA
Gill, Edmund D.
In: International Association for Quaternary Research (INQUA). Congress, 7th, Boulder, Col., 14 Aug - 19 Sept. 1965. *Abstracts of General Sessions.* Boulder, Colorado, 1965: 167-168
In South East Autralia, twelve emerged shell beds from across three sunklands and three horsts traversed by the Victorian coastline were surveyed and dated on marine shells which gave ages from 3980 to 6500 years BP. The dates correspond approximately to the post-glacial control theory, higher sea levels would be expected.

4.348   1956
RADIOCARBON DATING OF LATE QUATERNARY SHORELINES IN AUSTRALIA
Gill, Edmund D.
Quaternaria, v. 3, 1956: 133-138
Radiocarbon dating allows the sifting out of eustatic from tectonic change and make possible intercontinental correlation.

4.349   1961
RADIOCARBON DATING OF RAISED BEACHES IN NORDAUSLANDET, SPITZBERGEN
Blake, Weston
In: Geology of the Arctic, edited by G.O. Raasch. International Symposium on Arctic Geology, 1st., Calgary, Alberta, Jan. 11-13, 1960. *Proceedings.* Toronto, University of Toronto Press, 1961: 133-145
Radiocarbon determinations on imbedded driftwood, shells and whale bones, as well as a tentative correlation of pumice fragments with dated pumice in Denmark and Norway, have provided a means of dating certain beach levels. Three tables of radiocarbon dates on driftwood, shells and whale bones are presented.

4.350   1959
RADIOCARBON DATING OF RAISED BEACHES IN NORDAUSLANDET, SPITZBERGEN
Blake, Weston
Canadian Oil and Gas Industry, v. 12 (International Symposium on Arctic Geology, 1st., Calgary, Alberta, January 1960. *Abstracts of Papers Presented).* 1959: 51
[See entry n. 4.349]

4.351   1953
RADIOCARBON DATING OF RANGITIKEI RIVER TERRACE
Te Punga, Martin T.
New Zealand Journal of Science and Technology, v. 35, n. 1, sect. B, July 1953: 45-48
Radiocarbon dating of fossil wood from the Ohakea terrace in the Rangitikei Valley indicates that this terrace was formed 3050 ± 200 years ago. The geological significance of this dating is discussed.

4.352   1966
RADIOCARBON DATING OF RECENT SEDIMENTS IN SAN FRANCISCO BAY
Story, James A.; Wessels, Vincent E.; Wolfe, John A.
California. Division of Mines and Geology. Mineral Information Service, v. 19, n. 3, 1966: 47-50
In 1962, 107 tests holes were drilled through the San Francisco Bay mud, and radiocarbon dates were determined on oyster shells and peat from five representative holes. Ages range from 2300 to 7360 years according to depth: the shell is found above 32 feet, but peat extends down to 50 feet. An idealised cross section shows the inferred relations. It is concluded that the bay mud formed during and after melting of the Wisconsin glaciers, sea level rose slowly from 6150 to 4685 years ago, and the main biostrome formed 2500 years ago.

4.353   1960
RADIOCARBON DATING OF SOIL HUMUS
Tamm, C.O.; Östlund, H. Göte
Nature, v. 185, n. 4714, 5 Mar.1960: 706-707
The radiocarbon dating method is used to determine the approximate time of residence in the soil of the organic matter in samples of a podzol profile. The relatively low age of the organic matter in the B horizon suggests that the organic matter in this horizon is continually broken down and re-synthesized from substances coming from above.

4.354   1967
RADIOCARBON DATING OF TERRACES OF THE ULA RIVER, LITHUNIAN S.S.R.
Shuliya, K.S.; Luyanas, V. Yu.; Kibilda, Z.A.; Genutene, I.K.
Akademiia Nauk SSSR Doklady. Earth Science Sections, v. 175, n. 1-6, Feb. 1967: 33-36
Radiocarbon and spore-pollen data on peat from fluviatile, glacial and lake sediments justify a comparison of Ula interstage (Holocene) with the Brandenburg - Pomeranian.

4.355   1955
RADIOCARBON DATING OF THE LATE QUATERNARY IN SOUTHERN LOUISIANA
McFarland, E.
Geological Society of America. Bulletin, v. 66, n. 12, Dec. 1955: 1594.
Abstract of paper presented at the November 1953 meeting of the Geological Society of America. Dating of both wood and shells from beneath the Mississippi deltaic plain and adjacent portions of the continental shelf indicate that the sea had risen to within 100 feet of its present level approximately 10,000 years ago.

4.356   1953
RADIOCARBON DATING OF THE THERMAL MAXIMUM IN SOUTHEASTERN ALASKA
Heusser, Calvin John
Ecology, v. 34, n. 3, July 1953: 637-640
A more precise dating of the post-glacial thermal maximum in southeastern Alaska, based on radiocarbon assays on wood buried in muskeg, is presented. The Little Ice Age which terminated this xerothermic interval is dated at 1790 ± 285 years.

4.357   1956
RADIOCARBON DATING OF THE WISCONSIN - RECENT CLIMATIC CHANGE FROM OCEAN FLOOR SEDIMENTS
Broecker, Wallace S.
American Geophysical Union. Transactions, v. 37, n. 3, June 1956: 338
Abstract of paper presented at the 37th. Annual Meeting, Washington D.C., April 26 - May 4, 1956. Radiocarbon measurements on cores from the Atlantic and the Caribbean indicate that the Wisconsin to Recent climate change occurred rather sharply about 11,000 years ago. Also shows that the sedimentation rate during the Wisconsin was rather uniform, increased just prior to 11,000 years ago and decreased sharply towards the present.

4.358   1957.
RADIOCARBON DATING OF WAVE-FORMED TUFAS FROM THE BONNEVILLE BASIN
Feth, J.H.; Rubin, Meyer
Geological Society of America. Bulletin, v. 68. n.12, Dec. 1957: 1827
Abstract of paper presented at the April 1957 meeting of the Geological Society of America in Los Angeles. Radiocarbon dates have been determined for 10 wave-formed tufas from shore-lines of pluvial Lake Bonneville and other specimens from sediments of the same lake. The data suggest that: (1) the Stansbury shoreline is older than some higher stands and (2) the history of rising and falling lake stages is complex.

4.359   1965
RADIOCARBON DATING, BARROW, ALASKA
Brown, Jerry
Arctic, v. 18, n. 1, March 1965: 36-48
Various groups of samples, acquired in support of several scientific disciplines, were collected and processed for radiocarbon age determination, from the Barrow area, Alaska. These dates are presented in an attempt to clarify several aspects of the complex pedologic and geomorphic history of this arctic region. A table of radiocarbon dates is presented.

4.360   1967
RADIOCARBON DETERMINATIONS APPLIED TO GROUNDWATER HYDROLOGY
Hanshaw, Bruce B.; Rubin, Meyer; Back, William; Friedman, Irving
In: Isotope Techniques in the Hydrologic Cycle, edited by Glenn E. Stout. Washington, D.C., American Geophysical Union, 1967 (Geophysical Monograph n.11): 117-118
Studies have indicated that the use of radiocarbon in hydrology and geochemistry can be an important technique for the hydrologist and geochemist. Some of the problems are discussed.

4.361   1968
RADIOCARBON DETERMINATIONS FOR ESTIMATING GROUND WATER FLOW VELOCITIES IN CENTRAL FLORIDA
Hanshaw, Bruce B.; Back, William; Rubin, Meyer
Science, v. 148, n. 3669, 23 Apr. 1968: 494-495
Radiocarbon activity was determined from HCO_3 in samples of ground water obtained from the principal aquifers in Florida. From these data, the age of water obtained from a series of wells, each progressively farther down the gradient on the pizometric surface, was established. Relative radiocarbon ages indicated a velocity of ground water movement of 23 feet (7 m) per year for about 85 miles (137 km) of travel.

4.362   1962 (1963)
RADIOCARBON DETERMINATIONS FROM THE BOG PROFILE OF LAPANEVA, KIHNIÖ, WESTERN FINLAND
Salmi, Martti
Society Geologic of Finland. C. R., n. 34, 1962 (1963): 195-205
The radiocarbon ages of three peat samples from the Lapaneva bog, as well as the radiocarbon age of a pine trunk

from the Aitonava bog, are presented and discussed with respect to results of pollen and diatom analysis from the sampling sites and to the post-glacial chronology of Finland.

4.363   **1967**
RADIOCARBON DETERMINATIONS OF SEDIMENTATION RATES FROM HARD AND SOFT WATER LAKES IN NORTHERN NORTH AMERICA
Ogden, J. Gordon
In: Quaternary Palaeoecology, edited by M.J. Cushing and H.E. Wright. International Association for Quaternary Research (INQUA). Congress, 7th, Boulder, Col., 14 Aug - 19 Sept. 1965. *Proceedings*, v. 7. New Haven, Conn., Yale University Press, 1967: 175-183
A series of radiocarbon dates and pollen-stratigraphic analyses of core samples from post-glacial kettle lakes in Ohio and Massachusetts has demonstrated difference in rate of sediment accumulation since the close of the Valderan substage. The radiocarbon dates from hard water lakes were corrected for contamination by carbonate from adjacent palaeozoic limestone.

4.364   **1959**
RADIOCARBON FROM NUCLEAR TESTS
Broecker, Wallace S.; Walton, Alan
Science, v. 130, n. 3371, 7 Aug. 1959: 309-314
During the past 4 years man has been producing radiocarbon about 15 times faster than nature. Some conclusions: radiocarbon concentration of tropospheric carbon dioxide in the Northern Hemisphere up 5% p.a. beetween 1955 and 1958. Horizontal mixing, less than 2 years. Only 10% of radiocarbon entered the oceans.

4.365   **1960**
RADIOCARBON MEASUREMENTS AND ANNUAL RINGS IN CAVE FORMATIONS
Broecker, Wallace S.; Olson, Edwin A.
Nature, v. 185, n. 4706, 9 Jan. 1960: 93-94
The results suggest that the rings observed in the travertine are annual, probably reflecting seasonal changes in the mode of deposition. Two samples from a presumably active section of travertine covering a human femur had approximately 1,400 rings. The accumulation period for the entire 8.8 cm of travertine is estimated at 1400 ± 50 years based on the assumption of constant initial $^{14}C/^{12}C$ ratios and of a linear growth rate.

4.366   **1963**
RADIOCARBON STUDIES OF RECENT DOLOMITE FROM DEEP SPRING LAKE, CALIFORNIA
Peterson, M.N.A.; Bien, George S.; Berner, R.A.
Journal of Geophysical Research, v. 68, n. 24, 15 Dec. 1963: 6493-6505
Age determination by the radiocarbon method have shown that dolomite in surficial sediments of Deep Spring Lake, California, is virtually recent. Conditions in the lake favourable for the formation of dolomite were initiated near the close of the pluvial period.

4.367   **1958**
RADIOCARBON STUDIES OF THE BAHAMA BANKS
Thurber, David; Purdy, Edward; Broecker, Wallace S.
Geological Society of America. Bulletin, v. 69, n. 12, Dec. 1958: 1652.
Abstract of paper presented at the November 1958 meeting of the Geological Society of America in St. Louis. Radiocarbon analyses of Bahamian sediments revealed that they range in age from 700 to 2000 years. An underlying peat deposit was deposited at 4370 ± 100 years. The data also suggests that the active zone of oolite formation is the margin of the bank.

4.368   **1968**
A RADIOCARBON STUDY OF THE AGE AND ORIGIN OF CALICHE (Abstract)
Rightmire, Craig T.
Geological Society of America. Special Papers, n. 115, 1968: 184
A method of determining the age and origin of freshwater carbonate deposits, more specifically caliche, has been developed, based on their radiocarbon content and stable isotope composition.

4.369   **1960**
RADIOCARBON-DATED ORGANIC SEDIMENT NEAR HERBERT, SASKATCHEWAN
Kupsh, W.O.
American Journal of Science, v. 258, n. 4, Apr. 1960: 282-292
Fauna and flora in the sediment indicate a change from forest to grassland at the time the sediment was deposited and is dated by radiocarbon to 10,050 ± 300 years BP.

4.370   **1958**
RADIOCARBON-DATED POLLEN SEQUENCES IN EASTERN NORTH AMERICA
Deevey, Edward S.
Geobotanischer Institut Rubel, Zurich. Veroff., v. 34, 1958: 30-37
Reports a number of sequences in eastern North America and their probable correlations, made by the classic method of pollen stratigraphy. Its absolute time-scale is only approximate and is arrived at by considering (but not blindly relying on) radiocarbon dates and their means. The dates are

presented in table form.

4.371   1965
A RADIOCABON-DATED PEAT DEPOSIT NEAR HORSNUND, VESTSPITZBERBEN, AND ITS BEARING ON THE PROBLEM OF LAND UPLIFT
Blake, Weston; Olsson, Ingrid U.; Srodon, Andrzej
Norsk Polarinstitut. Årboc, 1963, 1965: 173-180
A radiocarbon determination shows that the basal peat in a bog 12 m above sea-level near Hornsund is 1390 ± 70 years old. The peat at 55-60 cm depth, did not start to accumulate until after the site had emerged from the sea. Uplift less than 1 m/century (not 2.3 m/century for the last 350 years as previously suggested) would suffice to raise the bog to its present elevation.

4.372   1962
RADIOCARBON-DATED POST-GLACIAL DELEVELLING IN NORTH-EAST GREENLAND AND ITS IMPLICATIONS
Washburn, A.L.; Stuiver, Minze
Arctic, v. 15, n. 1, 1962: 66-73
Radiocarbon dates on shells and driftwood collected from emerged marine deposits at Mesters Vig, Northeast Greenland, indicate that emergence at that locality is probably due to isostatic adjustment with a high rate of emergence of the order of 9 m/1000 years, and decreased approximately exponentially to about 0.6 m/100 years for the interval 9000 BP to 6000 BP.

4.373   1960
RATE OF CLAY FORMATION AND MINERAL ALTERATION IN A 4000-YEAR OLD VOLCANIC ASH SOIL ON ST. VINCENT B.W.I.
Hay, Richard L.
American Journal of Science, v. 258, n. 5, May 1960: 254-268
The soil studied was dated by radiocarbon measurement on unburned wood from the mud-flow portion of the upper deposit of the fan at 3890 ± 300 years BP.

4.374   1958
RATE OF FORMATION OF A DIAGENETIC CALCAREOUS CONCRETION
Pantin, H.M.
Journal of Sedimentology, v. 28, n. 3, 1958: 366-371
Age determinations by the radiocarbon dating method were made on the matrix and enclosed fauna of a Quaternary calcareous concretion from the continental shelf southeast of Cape Campbell, New Zealand. It was formed within the last 19,500 years and its formation probably occupied 7500 years or less.

4.375   1966
RATES OF DENUDATION IN THE PLATEAUS OF SOUTHWESTERN UTAH
Eardley, A.J.
Geological Society of America. Bulletin, v. 72, n. 7, 1966: 777-779
From two radiocarbon dates on wood from lake sediments that accumulated behind landslide dams in deep canyons of Zion National Park, Utah, the rate of denudation of the related drainage basin is found to be about 33 inches per 1000 years. This is compared to denudation rates in other areas.

4.376   1960
RATES OF SUBMERGENCE OF COASTAL NEW ENGLAND AND ACADIA
Lyon, C.J.; Harrison, W.
Science, v. 132, n. 3422, 29 July 1960: 295-296
Altitudinal and radiocarbon age determination of in-place *Pinus Strobus* stumps of drowned forests in Odiorne Point, N. H. and Grand Pré and Fort Lawrence, Nova Scotia, yield apparent average ages of subsidence of 3.1, 14.5 and 20.3 feet per 1000 radiocarbon years respectively. Rate difference are assessed in terms of eustatic rise of sea-level, crustal movements and tidal effects.

4.377   1968
RATES OF WEATHERING OF QUATERNARY VOLCANIC ASH IN NORTH-EAST PAPUA
Ruxton, B.P.
In: International Congress of Soil Science, 9th., 1968. *Transactions*, v. 4. Sydney, Angus and Robertson, 1968: 367-376
Radiocarbon dating of charred wood fragments in weathered dacitic ash-fall layers derived from Mount Lamington volcano in north-east Papua provide a means of assessing rates of weathering. A table of stratigraphy and chronology based on radiocarbon dates made by the ANU Radiocarbon Dating Research Laboratory is presented.

4.378   1963
RECENT C14 DATA FROM UPPER TROPOSPHERE
Fergusson, Gordon J.
In: Nuclear Geophysics. Washington DC, National Academy of Science. National Research Council, 1963: 240-244
The radiocarbon activity of the upper troposphere was investigated for information on atmospheric circulation and estimation of the amount of radiocarbon introduced to the troposphere in a given time period. An estimate of the carbon dioxide movement out of the troposphere and into the ocean and biosphere was also made.

4.379 1967
RECENT DELTAIC DEPOSITS OF THE MISSISSIPPI RIVER - THEIR DEVELOPMENT AND CHRONOLOGY
Frazier, David E.
Gulf Coast Association Geological Association. Transactions, v. 17, *Symposium on the Geological History of the Gulf of Mexico, Antillean - Caribbean Region,1967*: 287-315
More than 100 radiocarbon age determinations on discrete delta - plain peats have been used to establish the chronology of the delta lobes. These data together with the facies relationship indicate that the development of each delta complex was not simple.

4.380 1966
RECENT EMERGED BEACH IN EASTERN MEXICO
Behrens, E. William
Science, v. 152, n. 3722, 29 Apr. 1966: 642-643
A bluff on the eastern coast of Mexico reveals a cross section through an ancient beach deposit now lying 4 m above sea-level. Radiocarbon dates on the shells within the deposit reveal an age of 1840 years. The deposit appears to be valid evidence for submergence greater than that of the present, but whether that submergence was due to a higher eustatic stand of the sea or whether there has been an uplift of the land since that time cannot be determined.

4.381 1964
RECENT HIGH SEA-LEVEL STAND NEAR RECIFE, BRAZIL
Van Andel, Tj.H.; Laborel, Jacques
Science, v. 145, n. 3632, 7 Aug. 1964: 580-581
Radiocarbon dates for Vermatidae limestone from the edge of the Brazilian shield at Cape San Agostinho, Brazil, indicate sea-level stands of up to 2.60 m above the present position, 3660, 2790 and 1190 years ago.

4.382 1965
RECENT HISTORY OF THE UPPER TANANA RIVER LOWLAND, ALASKA
Fernald, Arthur T.
United States Geological Survey. Professional Papers, n. 525-C, 1965: 124-127
Twenty radiocarbon analyses provide dates for events which post-dated the late Quaternary glaciation and aeolian activity: (1) stabilisation of dunes completed more than 5000 years ago; (2) alluvial-colluvial filling along lowland borders began 10,500 to 6000 years ago; (3) deposition of fluvial-lacustrine sediments in the lowlands widespread within the last 3000 years.

4.383 1967
RECENT SUBMERGENCE OF SOUTHERN FLORIDA: A COMPARISON WITH ADJACENT COASTS AND OTHER EUSTATIC DATA
Scholl, David W.; Stuiver, Minze
Geological Society of America. Bulletin, v. 78, n. 4, Apr. 1967: 437-454
Data indicate that approximately 4400 years ago sea-level was about 4 metres lower than present level. Sea-level rose at high rate until 3500 BP then slowed down considerably. The Florida submergence curve shows that sea-level has risen steadily during the last 4400 years. Radiocarbon dates on which this conclusion is based are discussed. A table presents the data (p. 442-443).

4.384 1967
RECENT SUBMERGENCE OF SOUTHERN FLORIDA: A DISCUSSION
Smith, W.G.; Coleman, James M.
Geological Society of America. Bulletin, v. 78, n. 9, Sept. 1967: 1191-1194
Many of the dated samples reported by Scholl and Stuiver (*Geological Society of America. Bulletin,* Apr. 1967) probably cannot be as closely related to past sea-level as their small margin of error would suggest. More precise determinations seem necessary before accepting the area as a standard of comparison for other coasts.

4.385 1967
RECENT SUBMERGENCE OF SOUTHERN FLORIDA: A REPLY
Scholl, David W.; Stuiver, Minze
Geological Society of America. Bulletin, v. 78, n. 9, Sept. 1967: 1195-1197
An answer to the discussion by Smith and Coleman (*Geological Society of America. Bulletin,* Sept. 1967) in which the authors discuss the validity of the radiocarbon dates used to establish their sea-level curve in Florida and present a new recently completed radiocarbon age determination which substantiates the Florida curve presented by Scholl and Stuiver (*Geological Society of America, Bulletin,* Apr. 1967).

4.386 1965
RECORD OF SEA-LEVEL CHANGES IN SOUTHERN FLORIDA OCCURRING OVER THE LAST 4400 YEARS
Scholl, David W.; Stuiver, Minze
In: International Association for Quaternary Research (INQUA). *Congress,* 7th, Boulder, Col., 14 Aug - 19 Sept. 1965. *Abstracts of General Sessions.* Boulder, Colorado, 1965: 412
The character and rate of marine submergence has been

determinated by radiocarbon dates on marine shells, brackish-water mangrove peat and fresh-water fine grained calcareous and organic matter. The following notes of sea-level raise have been determined: 1.0 feet/100 years between 4300 and 3500 years ago with a level 5.3 feet below present at 3500 years BP; from 3500 to 1700, the rate became 0.2 feet/100 years and 0.09 feet/100 years during the last 1700 years.

4.387   1965
RELATION OF CARBON-14 CONCENTRATIONS TO SALINE WATER CONTAMINATIONS OF COASTAL AQUIFER
Hanshaw, Bruce B.; Back, William; Rubin, Meyer
Water Resource Research, v. 1, n. 1, 1965: 109-114
Radiocarbon was used to determine the origin of saline water in the Ocala limestone aquifer near Brunswick.

4.388   1960
REMARKS ON SOME WESTERN AND NORTHERN PARTS OF NORWAY
Holdetahl, Olaf
In: The Geology of Norway, edited by Olaf Holdetahl. *Norges Geologiske Undersøkelse*, n. 208, 1960; 409-416
An overview of research made on a very strongly dissected part of Norway, presenting some radiocarbon dates.

4.389   1957
REPORT OF THE A.N.Z.A.A.S. COMMITTEE FOR THE INVESTIGATION OF QUATERNARY STRANDLINE CHANGES
Gill, Edmund D.
Australian Journal of Science, v. 20, n. 1, July 1957: 5-9
The report covers New Zealand and Australia state by state. A number of radiocarbon dates are cited, along with other evidence and references.

4.390   1964
RETROGRESSIVE VEGETATIONAL SUCCESSION IN THE POST-GLACIAL
Iversen, Johannes
Journal of Ecology, v. 52, Supplement (British Ecological Society Jubilee Symposium, London, 28-30 March 1963, edited by A. MacFadyen and P.J. Newbould): 59-70
A Mor-profile in Draved Forest, in South Jutland, included radiocarbon dates, indicating a sharp change from oak-beech forest to Calluna heath by 740 ± 100 A.D.

4.391   1962
RIVERINA PLAINS CHRONOLOGY
Langford-Smith, Trevor
Australian Journal of Science, v. 25, 1962: 96-97
Eight radiocarbon dates on driftwood indicate vigorous stream discharge in the late Pleistocene, beginning before 25,000 years BP in the Riverina plains of central New South Wales (Australia) and renewed activity during the Holocene up to at least ca 1500 BP which could lead to the assumption that major Quaternary climate oscillations were synchronous in both hemispheres as supported by Flint and Brantner (*American Journal of Science*, 1961).

4.392   1963
SEA LEVEL AND CRUSTAL (?) MOVEMENTS AT CHESAPEAKE BAY ENTRANCE: 10,000-15,000 BP
Harrison, W.; Rusnak, Gene A.
Geological Society of America. Special Paper, n. 73, Abstracts for 1962, 1963: 297-298.
Abstract of paper presented at the 1962 meeting of the Geological Society of America in Houston, Texas. The dating of peat 9 feet above the Miocene-Pleistocene contact gives the following dates: $14,870 \pm 200$ years BP at -89 feet, $11,180 \pm 150$ years BP at -85 feet, 9930 ± 130 years BP at -82 feet. Assuming crustal stability and allowing for a maximum of 12 feet of peat consolidation after loading by overlying sediments, it would seem that the sea-level had risen to -70 feet by 9930 years BP.

4.393   1961
SEA LEVEL AND THE HOLOCENE BOUNDARY IN THE EASTERN UNITED STATES
Fairbridge, Rhodes W.; Newman, Walter S.
In: International Association of Quaternary Research (INQUA), Congress, 6th, Warsaw, 1961. *Report*. 1961: 397-418
On the evidence presented, which includes a 'standard' eustatic curve for the past 20,000 years, based on radiocarbon dates, and various other radiocarbon dates, it is quite apparent why the southern New England and New Jersey coastlines do not exhibit evidence of emerged mid-Holocene or inter-glacial shore lines. This section of the east coast has been subsiding at a discernible rate through much of Holocene.

4.394   1960
SEA LEVEL FLUCTUATION DURING THE LAST 40,000 YEARS AS RECORDED BY A CHENIER PLAIN, FIRTH OF THAMES, NEW ZEALAND
Schofield, J.C.
New Zealand Journal of Geology and Geophysics, v. 3, n. 3, Aug. 1960: 467-485
Radiocarbon ages for seven samples of shell permit the drawing of a reasonably accurate time sea-level curve which correlates favourably with transgressional periods recorded along the European coast and possibly with changes in European climate.

Ch. 4 - General Geology

4.395   1961
SEA LEVEL RISE DURING THE PAST 20,000 YEARS
Shepard, Francis P.
Zeischrift für Geomorphologie, Supplement n. 3, 1961: 30-35
The rise of sea-level is indicated by radiocarbon dates of organisms and plants that lived close to sea-level in relatively stable areas.

4.396   1968
SEA LEVELS DURING THE PAST 35,000 YEARS
Milliman, John D.; Emery, Kenneth Orriz
Science, v. 162, n. 3858, 6 Dec. 1968: 1121-1123
A sea-level curve of the past 35,000 years for the Atlantic continental shelf of the United States is based on more than 80 radiocarbon dates, 15 of which are older than 15,000 years. Materials include shallow water molluscs, oolites, coralline algae, beach rock and salt-marsh peat. Sea-level 30,000 to 35,000 years ago was near the present one. Subsequent glacier growth lowered sea-level to about -130 meters 16,000 years ago. Holocene transgressions probably began about 14,000 years ago, and continued rapidly to about 7000 years ago. Dates from most shelves of the world agree with this curve, suggesting that it is approximately the eustatic curve for the period.

4.397   1967
SEA LEVELS, 7,000 TO 20,000 YEARS AGO
Emery, Kenneth Orriz; Garrison, Louis E.
Science, v. 157, n. 3789, 11 Aug. 1967: 684-687
Relative sea-levels for early Pleistocene times are best known from radiocarbon dates of sediments from the continental shelves off Texas and off the northeastern United States. Differences in indicated rates of the rise of relative sea-level and in depth of shelf-breaks reveal differential vertical movement of the shelves during this time with the result that the Atlantic shelf has sunk with respect to the Texas shelf. Radiocarbon age measurements enable to calculate the rate of movement.

4.398   1960
THE SEA OFF SOUTHERN CALIFORNIA: A MODERN HABITAT OF PETROLEUM
Emery, Kenneth Orriz
New York, John Wiley, 1960. pp. 366
Radiocarbon measurements of the rates of deposition and data on the organic content of sediment show that the loss of organic matter to the bottom is approximately 0.3 million tonnes per year (dry weight). Absolute rates of deposition is determined by radiocarbon dating (p.179) A discussion of the method and assessment of the results is presented (p.249-253).

4.399   1965
THE SEDIMENT CORE N. 210 FROM THE WESTERN MEDITERRANEAN SEA
Eriksson, K. Gösta
Swedish Deep Sea Expedition. Reports, v. 8, *Sediment core from the Mediterranean Sea*, n. 7, 1965: 397-594
The chronology represented by the sediments of core 210 is mainly based on radiocarbon dates of Foraminifera tests. A description of the material and a discussion of the sources of error are included.

4.400   1963
SEDIMENTATION IN THE MODERN DELTA OF THE RIVER NIGER, WEST AFRICA
Allen, J.R.L.
International Sedimentation Congress, 6th, the Netherlands and Belgium. 1963. pp. 7 (Preprint)
Radiocarbon measurements date the older sand deposits in off shore coastal areas of the delta at $12,250 \pm 240$ to $10,750 \pm 150$ years BP, that is pre late-glacial and the shallower sands at 3380 ± 150 to 1190 ± 150 years BP, that is late-glacial to Holocene.

4.401   1958
SEDIMENTS AND TOPOGRAPHY OF THE GULF OF MEXICO
Ewing, Maurice; Ericson, David B.; Heezen, Bruce C.
In: Habitat of Oil, a Symposium conducted by the American Association of Petroleum Geologists, edited by Lewis G. Weeks. Tulsa, Oklahoma, American Association of Petroleum Geologists, 1958: 995-1053
Micro-palaeontological correlation and radiocarbon dating have established the abrupt transition at the base of the ooze as the Pleistocene-Recent boundary (11,000 years BP).

4.402   1966
SHELL BEDS AND CHANGE IN SEA-LEVEL ALONG THE NORTH-WEST COAST OF WESTERN AUSTRALIA
Kriewaldt, M.
Western Australia Geological Survey. Annual Report, 1965, 1966: 57-58
Beds which contain the taxodont pelecypod *Anandara* extend intermittently at the same elevation along the northwestern coast of Western Australia. A sample at 15 ft above Indian Spring low water mark give a radiocarbon date of 2080 ± 80 years BP. The sea-level at the time of deposition was possibly at least 2 ft above the present level, assuming stable land mass.

4.403   1965
SHORELINE DISPLACEMENT IN CENTRAL SPITZBERGEN

Feyling-Hanssen, Rolf W.
Norsk Polarinstitut. Meddelingen, v. 93, 1965: 1-5
Results of radiocarbon dating are correlated with the previously established sequence of late glacial and younger raised shorelines in central Spitzbergen.

4.404   1967
SIGNIFICANCE OF AITAPE (NEW GUINEA) RADIOCARBON DATES OF EUSTASY AND TECTONICS
Gill, Edmund D.
Australian Journal of Science, v. 30, n. 4, Oct. 1967: 142
Dating of shell and wood samples in a lenticle of Holocene intertidal deposit that contained Aitape cranial fragments indicate that the remains are much younger than previously thought and that the north coast of New Guinea has been one of the most mobile in the world; in the order of 1/2 inch per annum.

4.405   1960
THE SIGNIFICANCE OF COILING RATIOS IN THE FORAMINIFERA *GLOBIGERINA PACHYDERMA (EHRENGERG)*
Bandy, Orvill C.
Journal of Palaeontology, v. 34, 1960: 671
Dextral forms of these foraminifera have persisted for about 11,000 years and were preceded by sinistral populations in late Pleistocene as determined by radiocarbon dating. Pleistocene sediments were deposited at rates which were 25 to 500 times faster than those in recent times.

4.406   1967
SIGNIFICANCE OF RADIOCARBON DATES FROM BOTANY BAY ISLAND, SOUTH CAROLINA
Hoyt, John H.; Hails, John R.
South Carolina Division of Geology. Geology Notes, v. 10, n. 4, 1967: 61-65
The presence of marine shells above normal high tide level is explained. Radiocarbon analyses of shells taken from the layers date the individual shells rather than the period of accumulation of the layers. Shells of different ages and several environments are strewn over these beaches and may be concentrated in layers during temporary high water.

4.407   1968
SOILS OF PRINCE PATRICK ISLAND
Tedrow, John Charles Fremont; Bruggemann; Walton, G.F.
Arctic Institute of North America. Research Paper, n. 44, 1968. pp. 82
The age of the two tundra soils were determined at 5600 and 8460 years BP by radiocarbon measurements of organic matter buried in the soils.

4.408   1965
SOLIFLUCTION FEATURES IN THE KOSCIUSKO AREA
Costin, Alec B.
In: International Association for Quaternary Research (INQUA). Congress, 7th, Boulder, Col., 14 Aug - 19 Sept. 1965. *Abstracts of General Sessions*. Boulder, Colorado, 1965: 78
Solifluction lobes and terraces in the Kosciusko area of New South Wales (Australia) are studied to obtain information on their mode of origin, age, and palaeo-climatic significance. They are believed to have originated under periglacial conditions. Samples from one terrace gave dates ranging from 2980 ± 180 to 1540 ± 160, indicating that solifluction was active through a time known elsewhere as the Little Ice Age. A younger group dated at 170 ± 100 and 120 ± 130 years suggests continuous activity at least into the 18th century A.D.

4.409   1954
SOME ASH SHOWERS OF THE CENTRAL NORTH ISLAND
Baumgart, I.L.
New Zealand Journal of Science and Technology, v. 35, n. 5, Sect. B, Mar. 1954: 456-467
The sequence of recent ash showers in the Taupo district is discussed, and four members are described in detail. Radiocarbon age determinations have been used to establish a time-scale over the sequence. The ash beds are considered as evidence of the history of volcanicity.

4.410   1967
SOME ASPECTS OF THE GEOLOGICAL DEPOSITS OF THE SOUTH AND OF LAKE AGASSIZ BASIN
Brophy, John A.
In: Life, Land, and Water - Conference on Environmental Studies of the Glacial Lake Agassiz Region, 1966. *Manitoba University, Department of Anthropology. Occasional Papers*, n. 1, 1967: 91-105
In the area of the Sheyenne Delta, near Fargo, North Dakota, four units which can be recognised from road cuts and boreholes are discussed. Carbonaceous material 28 feet under the surface and overlying over 70 feet of clay deposited over the glacial till is 9900 years old. The details of the relationship of this sequence to the glacial lake, glacial advance, and post-glacial erosion are discussed.

4.411   1964
SOME QUATERNARY EVENTS OF NORTHERN ALASKA
Tedrow, John Charles Fremont; Walton, G.F.
Arctic, v. 17, n. 4, 1964: 268-271

Reports that radiocarbon dating of an organic deposit in a glacial terrace on the upper Killik River containing well preserved alder and willow necessitates an acceptance of a climate warmer or as warm as the present at a date of 5650 ± 230 years BP.

4.412   1966
SOME RADIOCARBON DATES AND THEIR GEOMORPHOLOGICAL SIGNIFICANCE, EMERGED REEF COMPLEX OF THE SUDAN
Berry, L.; Whiteman, A.J.; Bell, S.V.
Zeitschrift für Geomorphologie, v. 10, n. 2, 1966: 119-143
Sevn samples of reef-dwelling *Tridactna gigas* from emerged reefs on the Red Sea coast of the Sudan were radiocarbon dated. The samples are described and the dates discussed. A general conclusion was drawn: (1) most of the emerged reef material at all levels on the Sudan coast in the area studied was formed in pre-Würm times; (2) there appears to have been one main period of reef formation shoreward of the present coastline; (3) at later times benches have been cut into this emerged reef complex.

4.413   1956
SOME RADIOCARBON DATES IN THE POST-GLACIAL VEGETATION HISTORY OF THE NORTHERN NETHERLANDS
Van Zeist, Willem
Geobotanisher Institut Rubel, Zurich. Veroff., v. 34, 1958: 160-165
Radiocarbon dates for materials from different levels in a raised bog of the southern Drenthe region, northwest Netherlands, fix the boundary between Pre-Boreal and Boreal at about 6700 B.C. and the Sub-Boreal - Sub-Atlantic boundary at about 800 B.C., and brings other dates in the vegetational history in agreement with those based on archaeologic data. Notes on the regional post-glacial climate are included.

4.414   1963
THE SQUIBNOKET CLIFF PEAT: RADIOCARBON DATES AND POLLEN STRATIGRAPHY
Ogden, J. Gordon
American Journal of Science, v. 261, n. 4, Apr. 1963: 343-353
Pollen stratigraphy and radiocarbon dates are combined to date a peat deposit on the island of Martha's Vineyard.

4.415   1968
THE STRATIGRAPHIC DISTRIBUTION OF VERTEBRATE FOSSILS IN QUATERNARY EOLIAN DEPOSIT IN THE MID-CONTINENT REGION OF NORTH AMERICA
Schultz, Bertrand C.
In: Loess and Related Aeolian Deposits, edited by C. Bertrand Schultz and John C. Frye. International Association in Quaternary Research (INQUA). Congress, 7th, Denver, Col., 14 Aug - 19 Sept 1965. *Proceedings*, v. 12. University of Nebraska Press, 1968: 115-138
The central great plains and adjacent areas have provided an ideal place in North America for the establishment of a faunal sequence for the Quaternary. Radiocarbon dates of depositional layers are used to establish a chronology.

4.416   1955
STRATIGRAPHICAL POSITION OF LIBBY'S RADIOCARBON SAMPLE 358
Mitchell, G.F.
American Journal of Science, v. 253, n. 5, May 1955: 306-307
For a sample of peat from a peat bog at Clonsast, Ireland, Libby obtained a radiocarbon date of 5824 ± 300 years. Pollen counts from the section indicate that the sample is of early Atlantic rather than late Boreal age as previously reported.

4.417   1964
STRATIGRAPHY AND CHRONOLOGY OF LATE QUATERNARY VOLCANIC ASH IN TAUPO, ROTORUA AND GISBORN DISTRICTS
Healy, James; Vucetich, C.G.; Pullar, W.A.
New Zealand. Geological Survey Bulletin, n. 73, 1964. pp. 88
A two-part study of the Taupo ash sequence (Holocene) of the Taupo -Rotorua - Gisborne region, here redefined as the Taupo subgroup. The first part examines radiocarbon dating results for the younger volcanic eruptions of the Taupo region (by Healy). The second part examines the stratigraphy of the Holocene ash of the Rotorua and Gisborne districts (by Vucetich and Pullar). The type section at the Terraces pumice pit is described in detail, as well as sections exposed in highway cuts. A report on the mineralogy of the Taupo pumice (by A. Steiner) is appended.

4.418   1967
STRATIGRAPHY AND CHRONOLOGY OF QUATERNARY DEPOSITS OF ASSINIBOINE RIVER VALLEY AND ITS TRIBUTARIES. IN: REPORT OF ACTIVITY, PART B, NO.V 1966 TO APR. 1967
Klassen, R.W.
Canada. Geological Survey Paper, n. 67-1, part B, 1967: 55-60
Information that refines and confirms present concepts concerning local and regional Quaternary events was obtained by stratigraphic drilling, study of exposures and radiocarbon dating of deposits within and adjacent to Assin-

iboine, Qu'Appelle and Shell River Valley in southern Manitoba and south eastern Saskatchewan. From the dates obtained it appears that the last glaciation of this area was the main or late Wisconsin advance and was preceded by at least two and probably three major intervals of glaciation.

4.419 1967
STRATIGRAPHY AND PALYNOLOGY OF BURIED ORGANIC DEPOSITS FROM CAPE BRETON ISLAND, NOVA SCOTIA
Mott, R.J.; Prest, V. K.
Canadian Journal of Earth Science, v. 4, n. 4, Aug. 1967: 709-724
Pollen analysis of four non-glacial sub-till deposits yielded pollen asemblages indicative of climates cooler than the present in the area. New radiocarbon dates from Hillsborough (>51,000 years), Bay St. Lawrence (>38,300 years), and Whycocomagh (>44,000 years) place these deposits in pre-classical Wisconsin time.

4.420 1958
STRATIGRAPHY AND RADIOCARBON DATES AT SEARLES LAKE, CALIFORNIA
Flint, Richard Foster; Gale, W.A.
American Journal of Science, v. 256, n. 10, Dec. 1958: 689-714
Radiocarbon dates for samples from the upper units of the strata beneath Searles dry lake show that the later of two pluvial climates lasted from before 23,000 years BP to approximately 10,000 years BP. It was therefore contemporary with the classical Wisconsin glaciation of the central United States.

4.421 1968
STRATIGRAPHY OF PLUVIAL LAKE DOLOMITES, WEST TEXAS (Abstract)
Reeves, C.C.
Geological Society of America. Special Papers, n. 115, 1968: 376
Correlation with radiocarbon dating, palynological studies and associated mineralogy indicate that deposition, at the rate of about 1 in per 133 years, resulted from dessication of the lakes.

4.422 1967
STRATIGRAPHY, CLIMATE SUCCESSION AND RADIOCARBON DATING OF THE LAST GLACIAL IN THE NETHERLANDS
Hammen, Th. van der ; Maarleveld, Gerardus Cornelis; Vogel, John C.; Zagwijn, W.H.
Geologie en Mijnbouw, v. 46, n. 3, March 1967: 79-95
Field data, pollen analyses and some 30 radiocarbon dates have been the basis for the construction of a climatic curve for the Last Glacial in the Netherlands. There is a close resemblance of the climatic curve with paleotemperature curves obtained from deep-sea sediments. A list and a table of radiocarbon dates for Pleni-glacial beds in the Netherlands are presented.

4.423 1965
STRATIGRAPHY, NON-MARINE MOLLUSCA AND RADIOMETRIC DATES FROM QUATERNARY DEPOSITS IN KOTZEBUE SOUND AREA, WESTERN ALASKA
McCulloch, David S.; Taylor, Dwight W.; Rubin, Meyer
Journal of Geology, v. 73, n. 3, May 1965: 442-453
Radiocarbon dating and Uranium-Thorium dating were made of a sequence of interbedded marine and non-marine Quaternary sediments in the Kotzebue Sound area. This records marine transgression of Yarmouth and Sangamon age, glaciation of Illinoian age, and two periods (Sangamon and late Wisconsin to early Recent) when the climate was warmer than at present.

4.424 1961
STUDIES OF DEEP SEA CORES
Olausson, Eric
Swedish Deep Sea Expedition. Reports, v. 8, *Sediment core from the Mediterranean Sea*, n. 6, 1961: 337-391
The correlation between cores from the eastern Mediterranean is based on radiocarbon dates which are presented in table form (p.349), and confirm that the age recorded by Emiliani *(In: Isotopic and Cosmic Chemistry, 1964)*, of the beginning of the last rise of temperature, may be ca 13,000 years.

4.425 1963
STUDIES ON THE VEGETATIONAL HISTORY OF THE KUUSAMO DISTRICT (NORTH-EAST FINLAND) DURING THE LATE QUATERNARY PERIOD, II: RADIOCARBON DATINGS, PRELIMINARY REPORT
Vasari, Y.
Archivium Societatis Zoologicae Botanicae Fennicae, 'Vanamo', v. 18, n. 2, 1963: 121-127
Presents the radiocarbon dates on peat cores from several sites in the area studied in order to support a tentative correlation presented in a study published previously. The results are not entirely satisfactory. This is discussed, and a table of ages and age deviations is presented. Dating by Tauber (Copenhagen) and Isotopes Inc. (New Jersey).

4.426 1966
A STUDY IN LATE QUATERNARY PLANT -BEARING BEDS IN THE ISORTOQ VALLEY, BAFFIN ISLAND, N. W.T.

Terasmae, Jaan; Webber, P.J.; Andrews, J.T.
Arctic, v. 19, 1966: 296-318
Buried plant-bearing beds along the Isortoq River at the northern end of Barnes Ice Cap on Baffin Island have been radiocarbon dated at more than 38,830 and 40,000 years BP. A palynological and palaeobotanical study has indicated the presence of species which now occur several hundred kilometres south of this locality. Because of the inferred climatic conditions, more favourable than the present, an interglacial age (Sangamon) is assigned to the Isortoq plant bearing beds.

4.427   1965
A STUDY OF THE CHARLESWORTH LANDSLIDES NEAR GLOSSOP NORTH DERBYSHIRE
Johnson, R.H.
Institute of British Geographers. Transactions, v. 33, 1965: 111-126
The landslides in the Derbyshire Pennines are Post-glacial, according to radiocarbon dates.

4.428   1968
SUBAERIAL LAMINATED CRUSTS OF THE FLORIDA KEYS
Multer, H.G.; Hoffmeister, J.E.
Geological Society of America. Bulletin, v. 79, n. 2, 1968: 183-192
Exposed Pleistocene marine limestones of the Florida Keys are often coated by laminated 1-6 cm thick calcitic crusts. Fourteen datings of five different crust samples reveal a time of formation (within the last 4395 ± 90 years) during which the lad surface was above sea-level. Field relationship and laboratory evidence also indicate subaerial origin.

4.429   1965
SUBMARINE PEAT IN THE SHETLAND ISLANDS
Hoppe, Gunnard; Fries, Magnus; Quennerstedt, N.
Geografiska Annaler, v. 47, ser. A, n. 4, 1965: 195-203
The radiocarbon dates agree fairly well and there is a discussion of the value of organic pollen analysis.

4.430   1965
A SUBMERGED LATE QUATERNARY DEPOSIT AT RODDANS PORT ON THE NORTH-EAST COAST OF IRELAND
Morrisson, M.E.S.; Stephens, N.
Royal Society (London). Philosophical Transactions, Series B, v. 249, n. 758, 1965: 221-255
The sequences studied are identified on the basis of pollen stratigraphy and other evidence as deposits of the Older and Younger Dryas and the Allerød interstadial of the northwest European late-glacial chronology. An appendix deals with radiocarbon dates.

4.431   1965
A SUBMERGED PEAT DEPOSIT OFF THE ATLANTIC COAST OF THE UNITED STATES
Emery, Kenneth Orriz; Wigley, L.; Rubin, Meyer
Limnology and Oceanography (Alfred C. Redfield 75th Anniversary Volume) sup., 1965: R97-R102
A sample of salt-marsh peat from a depth of 59 m at the northwestern margin of George Bank has a radiocarbon age of 11,000 ± 350 years. No greater depth or age of salt-marsh peat lying exposed on the ocean floor elsewhere in the world are known to the authors. The age corresponds well with the date of ice retreat from the Gulf of Maine and with the ages of oyster and other shells that mark the advance of the ocean across the bank and the adjacent continental shelf.

4.432   1968
SUBMERGENCE ALONG THE ATLANTIC COAST OF GEORGIA
Wait, Robert L.
United States Geological Survey. Professional Papers, n. 600-D, 1968: D38-D41
Cypress stumps recovered from river terrace material near Brunswick, Ga., may indicate submergence of the Atlantic coastal area during Holocene geologic time. The older material, found at a depth of from 9 to 17 feet below mean sea-level was dated by radiocarbon at 3,670 ± 300 years (BP 1950), the younger, found 1 foot above mean sea level and buried in 3 feet of marsh silt and clay, was dated at 2780 ± 250 years BP. Presence of the cypress stumps may indicat that fresh water once discharged from the Turtle River, now a drowned estuary.

4.433   1963
SUBMERGENCE OF THE CONNECTICUT COAST
Bloom, Arthur L.; Stuiver, Minze
Science, v. 139, n. 3552, 25 Jan. 1963: 332-334
Radiocarbon dated samples show that the Connecticut coast has submerged about 9 feet in the last 3000 years and about 33 feet in the last 700 years. The rate of submergence is similar to rates reported from other coasts. The finding strengthens the hypothesis that a world-wide post glacial rise of sea-level is the cause.

4.434   1963
SUBMERGENCE OF THE NEW JERSEY COAST
Stuiver, Minze; Daddario, Joseph J.
Science, v. 142, n. 3594, 15 Nov. 1963: 951
A series of five radiocarbon dates obtained from samples taken along the base of the lagoon between Brigantine City Barrier (New Jersey) and the mainland indicates a rate of submergence of 3 metres per millenium between 6000 and 2600 years BP. During the last 2600 years the average submergence has slowed down to only 1.2 to 1.4 metres per

millenium. The general picture of a rapid rise and the subsequent slackening is in agreement with results published for the New England area.

4.435   **1962**
SUBSURFACE STRATIGRAPHY OF LATE QUATERNARY DEPOSITS, SEARLES LAKES, CALIFORNIA: A SUMMARY
Smith, George I.
United States Geological Survey. Papers, n. 450-C, article n.82, 1962: C65-C69
A diagrammatic section of Searles Lake evaporites showing position and name of stratigraphic units and radiocarbon dates is presented. These dates indicate that the approximate ages of three contacts are 10,000, 23,000, and 34,000 BP.

4.436   **1968**
SUPRAMARINE LITORINA LAYERS AT SEGEVANGEN, MALMÖ
Welinder, Stig
Geologiska Föreningens i Stockholm. Förhandligar, v. 90, 1968: 126-128
[Not sighted]

4.437   **1963**
SURFICIAL GEOLOGY AND GEOMORPHOLOGY OF THE LAKE TAPPS QUADRANGLE, WASHINGTON
Crandell, Dwight R.
United States Geological Survey. Professional Papers, n. 388-A, 1963. pp. 84
Stratigraphic chronology is supplied by radiocarbon dates. A summary of the stratigraphy and Quaternary history in the Puget Sound lowland south and south-east of Seattle is presented. Ages are based on radiocarbon dates.

4.438   **1966**
SURFICIAL GEOLOGY OF DUNCAN AND SHAWNIGAN MAP AREAS, BRITISH COLUMBIA
Halstead, E.C.
Canada. Geological Survey Paper, n. 65, 1966. pp. 3
The last major ice sheet moved south to southeast across the map areas, and during deglaciation split into a network of glaciers flowing down the valleys to the sea. Surficial deposits are commonly below an elevation of 500 feet and include laminated silts and clays, gravels, and stony till. Organic matter from near Skutz Falls yielded a radiocarbon date of 21,000 ± 270 years and material from Cowichan Bay 24,560 ± 800 years.

4.439   **1960**
SURFICIAL GEOLOGY OF NORTH-CENTRAL DISTRICT OF McKENZIE, NORTHWEST TERRITORIES
Craig, Bruce Gordon
Canada. Geological Survey Paper, n. 60-18, 1960. pp. 8
Dating of shells at four localities, indicates a sequence of events that agrees with the general pattern of deglaciation. The dates were discussed with regard to their special relationship and stratigraphy.

4.440   **1956**
SURFICIAL GEOLOGY OF THE HORNE LAKE AND PARKSVILLE MAP AREAS, VANCOUVER ISLAND, BRITISH COLUMBIA
Fyles, J.G.
Ph.D. Dissertation. Ohio State University,1956. pp. 119
(Abstract in: *Dissertation Abstracts*, Ann Arbor, Michigan, v. 16, n. 11, 1956: 2136. L.C. number: mic.56-3394)
Peat from basal beds of the Quandra group of marine-shore and fluvial origin are dated at more than 30,000 years. This is a non-glacial interval, probably equivalent to the Sangamon age.

4.441   **1964**
THE TAUPO QUATERNARY PUMICE SEQUENCE AND ITS BEARING ON THE PREDICTION OF FUTURE ERUPTIONS
Ewart, Anthony
New Zealand Journal of Geology and Geophysics, v. 7, n. 1, Feb. 1964: 101-105
Study of the thick series of pumice deposits of the Taupo area indicates that there have been at least ten major eruptions during the last 10,000 years, most of which have been dated by radiocarbon.

4.442   **1957**
TEMPERATURE AND AGE ANALYSIS OF DEEP-SEA CORES
Emiliani, Cesare L.
Science, v. 1215, n. 3244, 1 Mar. 1957: 383-387
Deep-sea cores have been analysed by different methods: oxygen-isotope analysis (Emiliani), radiocarbon analysis (Rubin and Suess; Broecker, Kulp and Tucek). The results have been interpreted differently. The purpose of this article is to analyse these differences and draw the possible conclusions from the published evidence.

4.443   **1962**
TERRACE CHRONOLOGY AND SOIL FORMATION ON THE SOUTH COAST OF N.S.W.
Walker, Philip H.
Journal of Soil Science, v. 13, n. 2, 1962: 178-186
Radiocarbon dates for four terrace sites show that the K3 (K cycle system of Butler, 1959) commenced 29,000 years ago, the K2 commenced 3740 years ago and the K1 cycle, 390 years ago. The present K0 cycle of erosion dates from

Ch. 4 - General Geology
0 to 120 years ago.

4.444   1968
THE THIRTY FOOT RAISED BEACH AT
RAPAHOE, NORTH WESTLAND
Suggate, R.P.
New Zealand Journal of Geology and Geophysics, v. 11, n. 3, Aug. 1968: 648-650
The altitude of a post-glacial coastal terrace is revised to 29 ± 1 feet above sea-level. A log dated at 4720 ± 70 years BP was found in estuarine gravel 7 feet below the terrace surface, 1/2 mile inland. Tectonic uplift is certain, at a rate of about 6 ft/1000 years.

4.445   1958
THREE POLLEN DIAGRAMS FROM CENTRAL
MASSACHUSETTS
Davis, Margaret B.
American Journal of Science, v. 256, 1958: 540-570
The three pollen diagrams are believed to record the vegetational sequence from the time of retreat of the pre-Valders ice to a time just before European colonisation. A number of radiocarbon dates help to interpret several pollen zones.

4.446   1960
TIME OF THE LAST DISPLACEMENT ON THE
MIDDLE PART OF THE GARLOCK FAULT,
CALIFORNIA
Smith, George I.
United States Geological Survey. Professional Papers, n. 400-B, Article 128, 1960: B280
Radiocarbon dates indicate that the sediment deposited during the last period of overflow of lake Searle is at least 50,000 years old and that the tufa which is just below is of about the same age, as it was clearly found, later than the scarp. This is evidence that the Garlock fault is older than 50,000 years, not recent as hitherto believed.

4.447   1958-61
TIME-STRATIGRAPHIC SUBDIVISION OF THE
QUATERNARY AS VIEWED FROM NEW
ZEALAND
Gill, Edmund D.
Quaternaria, v. 5, 1958-61: 5-17
The Quaternary deposits of New Zealand, formed in various widely different tectonic environments and as a result of numerous different geological processes, can nevertheless be broadly correlated if climatic fluctuation is accepted, as a basis for Quaternary subdivision. Radiocarbon dating is used to place the various deposits in their time sequence.

4.448   1957
TOPOGRAPHIC STUDY OF LACUSTRINE
TERRACES AND CRUSTAL MOVEMENT
AROUND LAKE KUTCHARA, HOKKAIDO,
JAPAN
Horie, Shoji
Japanese Journal of Geography and Geolgoy, v. 28, n. 1-3, 1957: 1-10
Lake Kutcharo, regarded as a caldera lake, has abruptly undergone depression three times in recent times, resulting in the formation of three terraces. The middle terrace is dated by a fragment of earthware collected from the surface. This fragment is regarded as belonging to the age of 'Kasoria E' which is as old as 2500 years B.C. according to the radiocarbon dating method.

4.449   1964
TREES AND CLIMATIC HISTORY IN SCOT-
LAND: A RADIOCARBON TEST AND OTHER
EVIDENCES
Lamb, H.H.
Royal Meteorological Society. Quarterly Journal, v. 90, 1964: 382-394
The author explains and describes the radiocarbon dating method and applies it to the Badentarbet wood samples. It indicates that the trees and roots found in that bog grew between 2770 ± 100 years B.C. and therefore belong to the latter part of the main post-glacial warm epoch. This is confirmed by circumstantial evidence.

4.450   1962
TREES BURIED IN HUERFJALL AND HEKLA
TEPHRA (Icelandic, English Summary)
Thorarinsson, Sigurdur
Reykjavik. Museum of Natural History. Miscellaneous Papers, n. 35, 1962
Hollows left by trees have been found in a layer of tephra from the explosion crater Hverfjall in the Myvatn region, north Iceland, which was deposited about 2500 years ago, and in the most extensive tephra layer of Hekla volcano. Radiocarbon dating of charcoal in a buried-tree hollow in the Hekla tephra layer gave an age of 2820 ± 70 years, of underlying peat an age of 2700 ± 130 years, and of charcoal from an overlying layer an age of 2660 ± 80 years.

4.451   1966
TROPICAL LAKES, COPROPEL, AND OIL
SHALE
Bradley, W.H.
Geological Society of America. Bulletin, v. 77, n. 12, 1966: 1333-1337
Four shallow lakes have been found (two in Florida, two in

Africa) that are producing the kind of organic ooze that is judged to be a modern analog of the precursors of rich oil shale such as the Green River Formation. The ooze in all four is predominantly algal, entirely in the form of minute fecal pellets, and does not decay in warm wet oxidizing environments. The algal ooze at a depth of three feet below the mud-water interface has a radiocarbon age of 2280±200 years.

4.452 1965
TROPOSPHERIC C-14 VALUES IN THE PACIFIC NORTHWEST AND IN THE ARCTIC BASIN DURING 1964
Young, I.J.; Erickson, N.; Fairhall, A.W.
In: Atomic Energy Commission Conference on Radioactive Fallout, 2nd., Germantown, Mc., Nov. 1964. Proceedings. 1965: 422-427
Report based on carbon dioxide samples collected in the state of Washington and the Arctic Basin, with results expressed in % above normal. Tropospheric radiocarbon levels reached 115% in the summer of 1963, fell sharply to 88% in the fall, a level maintained till May 1964, to rise again in the summer. Factors affecting these changes are discussed.

4.453 1961
TUNDRA SOILS OF ARCTIC ALASKA
Douglas, L.A.; Tedrow, John Charles Fremont
International Congress of Soil Sciences, 7th, 1961. Transactions, v. 4: 291-304
Radiocarbon dates on alkali insoluble organic matter well within the permafrost, i.e. not affected by present day pedologic processes, gave ages that fell between 8150 and 10,600 years, a period which, according to evidence presented, had a climate somewhat warmer than the present day climate.

4.454 1962
TRANSATLANTIC CLIMATIC AGREEMENT VERSUS C-14 DATES
Antevs, Ernst
Journal of Geology, v. 70, n. 2, 1962: 194-205
Radiocarbon dates indicate reversed climatic trends in North America and Europe during post-glacial time and are considered too unreliable to refute the conclusion that a general climatic conformity obtained during episodes of deglaciation.

4.455 1965
UPPER MISSISSIPPI VALLEY
Black, Robert F.; Reed, E.C.
International Association for Quaternary Research (INQUA), Congress, 7th., Boulder, Col., 15 Aug. - 19 Sept. 1965. *Guide Book for Field Conference C.* Boulder, Col., INQUA, 1965. pp. 126
Throughout the description for the various stops made during this field, mention is made of the radiocarbon dates which support the stratigraphy in dating the various Quaternary deposits.

4.456 1965
USE OF NATURALLY OCCURING -C TO MEASURE THE PERSISTENCE OF ORGANIC COMPONENTS IN SOIL
Campbell, C.A.
Ph.D. Dissertation. University of Saskatchewan, Saskatoon, Sas.,1965. pp. 163
Applicability of the radiocarbon dating method of analysis to study in soil science is investigated. Factors affecting the accuracy of the method, e.g. precision, isotopic fractionation, bomb produced radiocarbon, 'Suess effect', 'de Vries effect', were found to cause either insignificant errors, or errors for which correction could easily be made.

4.457 1963
THE USE OF THE CARBON ISOTOPE IN GROUNDWATER STUDIES
Vogel, John C.; Ehhalt, D.
In: Radioisotopes in Hydrology. Symposium, Tokyo, 1963. International Atomic Energy Agency (IAEA), 1963: 383-395 (Discussions: 404- 406)
The bicarbonate in groundwater is dissolved in the upper layers of the soil by the action of biogenic carbon dioxide on limestone. As a result of this solution mechanism the bicarbonate in shallow groundwater appears to have a quite characteristic ^{13}C and ^{14}C content so that these isotopes can be used to study isotopic exchange with, and storage time in, the aquifer. Measurements on water from deeper levels indicate that low radiocarbon concentrations are mainly a result of radioactive decay and thus give true residence times. Results of investigations of waters from oases and artesian wells in Egypt and polders areas in the Netherlands are cited as examples.

4.458 1964
VALIDITY OF C14 DATES IN VICTORIA LAND, ANTARCTICA (Abstract)
Berg, Thomas E.; Black, Robert F.
Geological Society of America. Special Papers, n. 76, 1964: 12-13
[Not sighted].

4.459 1963
VARVED MARINE SEDIMENTS IN A STAGNANT MARINE BASIN
Gross, M.G.; Gugluer, S.M.; Creager, Joe S.; Dawson, W.A.

Science, v. 141, n. 3584, 6 Sept. 1963: 918-919
Varved sediments from Saanich Inlet, Vancouver Island, British Columbia, are examined and dated according to the thickness of each pair of laminae. Radiocarbon dating of a piece of bark estimated to be 600 years on the basis of varves is found to be 500 ± 150, thus establishing the varved nature of this sediment.

4.460   1967
VERTICAL TECTONIC DISPLACEMENTS IN SOUTH-CENTRAL ALASKA DURING AND PRIOR TO THE GREAT 1964 EARTHQUAKE
Plafker, George; Rubin, Meyer
Journal of Geoscience, v. 10, art. 1-7, March 1967: 53-66
Radiocarbon dates from three of five terraces on Middleton Island in the uplifted zone indicate that 40 meters of relative emergence occurred in at least five major upward pulses during the last 4470 ± 250 years. Relative submergence also occurred in the same zone. Radiocarbon dated drowned forests indicate pre-quake submergence of as much as 4.3 meters over a period of 390 ± 160 years. This is interpreted as direct evidence of a significant downward-directed component of regional strain preceding the earthquake.

4.461   1964
VOLCANIC ASH FROM MOUNT MAZAMA (CRATER LAKE) AND FROM GLACIER PEAK
Powers, H.A.; Wilcox, Ray E.
Science, v. 144, n. 3624, 12 June 1964: 1334-1336
A radiocarbon date of charred wood in the ashflow of the Mazama eruption places it at 6600 years BP.

4.462   1967
VOLCANIC ASH LAYERS OF RECENT AGE AT BANFF NATIONAL PARK, ALBERTA, CANADA
Westgate, J.A.; Dreimanis, Aleksis
Canadian Journal of Earth Science, v. 4, n. 1, Feb. 1967: 155-161
The chemical and mineralogical composition of three distinctive volcanic ash layers, in post-glacial sediments at Banff National Park, and radiocarbon age of associated charcoal, indicate the presence of Bridge River ash, derived from the heart of the southern Coast Mountains of British Columbia, and Mazama ash, from Mount Mazama at Crater Lake, Oregon. The age for Bridge River ash is 2395 ± 275 years BP. Mazama ash, established as about 6600 years has a slightly younger date of about 6020 years in Banff Park. Source and age of the third ash is not definitely known, but is thought be St. Helens Y ash, derived from Mount St. Helens, Washington.

4.463   1959
THE VOLCANOES OF IHUMATAO AND MANGERE, AUCKLAND, NEW ZEALAND
Searle, E.J.
New Zealand Journal of Geology and Geophysics, v. 2, 1959: 870-888
The volcanic features of the Mangere - Ihumatao district near the city of Auckland are described and their relative ages and eruptive histories discussed. In all cases eruptions in the area occurred during the Flandrian transgression, before sea-level had attained its present height and have been dated by radiocarbon at $29,000 \pm 1500$ years ago.

4.464   1965
WATER DATING TECHNIQUES AS APPLIED TO THE PULLMAN - MOSCOW GROUND WATER BASIN
Crosby, James W.; Chatters, Roy M.
Washington State University, Pullman, Washington. Technical Extension Services. College of Engineering Bulletin, n. 296, 1965. pp. 21
Radiocarbon dating of the Pullman - Moscow groundwater basin indicates that the groundwaters are distinctly stratified and display a well defined relationship between water age and elevation. The bulk of the water appears to have been placed by the closing phase of the Pleistocene and there has been no measurable recharge in recent times.

4.465   1967
WINTER HARBOUR MORAINE, MELVILLE ISLAND, NORTHWEST TERRITORY. IN: REPORT OF ACTIVITIES: FIELD, 1966. COMPILED BY S.E. JENNESS PART A, MAY TO OCT. 1966.
Fyles, J.G.
Canada. Geological Survey Paper, n. 67, part A, 1967: 8-9
Marine shells from silt on the inland side of the moraine have yielded radiocarbon dates of $10,340 \pm 150$ and $10,900 \pm 160$ years, whereas shells close to the marine limit on the seaward side have radiocarbon age of 9550 ± 160 years. On the basis of this, the ice-sheet margin is inferred to have stood in the vicinity of the moraine 10,000 to 11,000 years ago.

4.466   1962
WORLD SEA LEVEL AND CLIMATIC CHANGES
Fairbridge, Rhodes W.
Quaternaria, v. 6, 1962: 111-134
The hypothesis that the 3 m terrace of coral sand was contemporaneous with the Climatic Optimum about 4000 to 6000 years BP is confirmed by a radiocarbon date of 5190 ± 130 years BP at West Australian Point Peron raised beach.

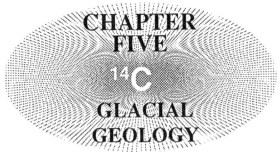

CHAPTER FIVE
14C GLACIAL GEOLOGY

5.001   1955
ABSOLUTE CHRONOLOGY OF THE LAST GLACIATION
Suess, Hans E.
Science, v. 123, n. 3192, 2 Mar. 1955: 355-357
The author makes an evaluaton of the role of radiocarbon dating in establishing the chronology of the Wisconsin stage of glaciation in North America. There is evidence of the two main types of climatic fluctuations on two different timescales. Apparent contradictions in radiocarbon dates are at least in part the result of miscorrelation between events on these two time scales.

5.002   1965
ABSOLUTE CHRONOLOGY OF THE MAIN-STAGES OF LAST GLACIATION AND POST-GLACIAL PERIODS IN SIBERIA
Kind, N. V.
In: International Association for Quaternary Research (INQUA). *Congress,* 7th, Boulder, Col., 14 Aug - 19 Sept. 1965. *Abstracts of General Sessions.* Boulder, Colorado, 1965: 267-269
An absolute chronology scheme of the principal geological events and climatic fluctuations during Upper Quaternary times for the territory of Siberia is presented. It is based on more than 30 radiocarbon datings.

5.003   1965
ABSOLUTE DATE AND THE ASTRONOMICAL THEORY OF GLACIATION
Broecker, Wallace S.
Science, v. 151, n. 3708, 21 Jan. 1966: 299-304
Changes in climate occur in response to periodic variations in the earth's tilt and precession. Radiocarbon dating provides a detailed chronology of the changes in climate which have taken place during the last 25,000 years. This, plus the oceanic record for this period, shows that 11,000 years ago marks the mid-point of a rapid transition from glacial to inter-glacial which also coincides with the last insolation maximum.

5.004   1962
AGE OF DEGLACIATION IN THE RUWENZORI RANGE, UGANDA
Livingstone, D.A.
Nature, v. 194, n. 4831, 2 June 1962: 859-860
The first abolute date of a major glacial event in East Africa indicates that in Equatorial Africa glaciers were retreating 15,000 years ago at the same time as the Würm glacier of Europe and the Wisconsin glacier of Northern America. Pleistocene climatic changes in the tropics have been of the same sign as those in the rest of the world and synchronous with them.

5.005   1959
AGE OF MARGINAL WISCONSIN DRIFT AT CORRY, NORTHWESTERN PENNSYLVANIA
Droste, John B.; Rubin, Meyer; White, George W.
Science, v. 130, n. 3391, 25 Dec. 1959: 1760
Marl began to accumulate about 14,000 years ago, as determined by radiocarbon dating, in a pond in a kettle hole in Kent drift at Corry, Pa., 9 miles inside the Wisconsin drift margin. This radiocarbon age represents the minimum time since the disappearance of the ice from Corry and confirms an assignment of Cary age to the drift.

5.006   1962
AGE OF MORAINES IN VICTORIA LAND, ANTARTICA
Péwé, Troy L.
Journal of Glaciology, v. 4, n. 31, March 1968: 93-100
Radiocarbon dating of mummified seal carcasses lying near glacier fronts indicate that the glaciers have not been more extensive for at least 1000 years. The youngest moraines in the area are ice-cored moraines to the existing glaciers. Radiocarbon dates of algae from extinct ephemeral ponds in the ice-cored moraines indicate the moraine to be at least 6000 years old.

5.007   1967
AGE OF THE IRISH SEA GLACIATION OF THE MIDLANDS
Shotton, Frederick William
Nature, v. 215, n. 5108, 23 Sept. 1967: 1366
Through work in stratigraphy, palaeontology and radiocarbon dating, it can be demonstrated that the advance of the Irish Sea Glacier to the Wolverhampton line occurred in Late-Würm times, that is probably centred around a date of about 25,000 years ago.

5.008   1968
AGE OF THE UPPER BOULDER CLAY GLACIATION IN THE MIDLANDS

Ch. 5 - Glacial Geology

Poole, E. Grey
Nature, v. 217, n. 5234, 23 March 1968: 1137-1138
A date of 30,000 to 36,000 years for peaty samples obtained from gravels occurring beneath the Upper Boulder Clay of the Four Ashes area helps standardise the chronology of the Shropshire - Cheshire basin.

5.009   1965
AGE OF WOOD FROM WISCONSIN TERMINAL MORAINE NEAR ADELPHI, ROSS COUNTY, OHIO
Forsyth, Jane L.
Ohio Journal of Science, v. 65, n. 3, May 1965: 159-160
Wood from a silt lens in till was dated at $17,292 \pm 436$ years, suggesting that the late Wisconsin ice reached its maximum position at roughly the same time along the entire Wisconsin boundary from north of Adelphi to west of Chillicott.

5.010   1967
ALTITHERMAL TIMBERLINE ADVANCE IN WESTERN UNITED STATES
La Marche, Valmore C.; Mooney, H.A.
Nature, v. 213, n. 5080, 11 Mar. 1967: 980
A combination of radiocarbon dating and counting of annual rings of the remains of trees has been used to establish the time of retreat of forests from sub-alpine mountains in California and Nevada. The information obtained can serve as palaeoclimatic indicator.

5.011   1965
ALTONIAN AND WOODFORDIAN (WISCONSINAN) DEPOSITS OF NORTHERN ILLINOIS
Kempton, John P.; Hackett, J.E.
In: *International Association for Quaternary Research (INQUA). Congress*, 7th, Boulder, Col., 14 Aug - 19 Sept. 1965. *Abstracts of General Sessions*. Boulder, Colorado, 1965: 262-263
Subsurface tills in Northern Illinois are assigned to the Wisconsinan (Altonian and Woodfordian Substages) because of their stratigraphic relations to associated deposits and radiocarbon dates. One till that overlies a dated organic zone indicates that glaciers advanced in the area after $35,000 \pm 2500$ years BP. A radiocarbon date from a peat layer above the same till indicated that the ice had retreated from the area before $25,600 \pm 800$ years BP.

5.012   1962
ANALYSIS OF SOME STRATIGRAPHIC OBSERVATIONS AND RADIOCARBON DATES FROM TWO PINGOS IN THE MACKENZIE DELTA AREA, N. W.T.
Müller, Fritz
Arctic, v. 15, n. 4, Dec. 1962: 279-288
After describing the pingos and the stratigraphic situation of the organic material which was dated, the author gives an estimate of the time of glacierisation of the Mackenzie Delta as lasting from 25,000 BP to 15,000 BP, therefore late Wisconsin.

5.013   1962
AN ARCTIC FOREST IN THE TUNDRA OF NORTHERN UNGAVA, QUEBEC
Maycock, P.F.; Matthews, Barry
Arctic, v. 19, n. 2, June 1962: 114-144
Radiocarbon dates on shells collected 8 miles inland from Deception Bay have given an age of 6740 ± 140 years BP indicating that ice had vacated the Deception Bay area by that time at the latest to allow the sea to invade.

5.014   1968
ATLANTIC / EARLY SUB-BOREAL GLACIATION IN NORWAY
Page, Neil R.
Nature, v. 219, n. 5155, 17 Aug. 1968: 694-697
There is evidence which indicates that Norwegian glaciers attained their maximum post-glacial extent in the middle of the thermal or climatic optimum. Comparison with radiocarbon dates from other glaciated areas helps establish the correlation.

5.015   1966
BAFFIN ISLAND REFUGIA OLDER THAN 54,000 YEARS
Löken, Olav H.
Science, v. 153, n. 3742, 16 Sept. 1966: 1378
Two radiocarbon determinations on marine shells from the east coast of Baffin Island give ages exceeding 50,000 years. These findings indicate the existence of unglaciated areas (refugia) between fiords occupied by outlet glaciers flowing toward Baffin Bay, from the central part of the Wisconsin ice sheet, over the Foxe Basin, Hudson Bay area.

5.016   1958
BEGINNING OF THE NIPISSING PHASE OF LAKE HURON
Dreimanis, Aleksis
Journal of Geology, v. 66, n. 5, Sept. 1958: 591-594
The Nipissing beach bar east of Sarnia, Ontario, covers material re-worked by waves, including several logs and other remains of plants and molluscs, and, at one place, soil below it. Radiocarbon dates of two logs are 4650 ± 200 and 4610 ± 210. A theoretical time/uplift curve suggests the beginning of the Nipissing phase of Lake Huron at 4300 ± 270 years ago.

5.017   1967
BERING LAND BRIDGE: EVIDENCE OF SPRUCE IN LATE WISCONSIN TIMES
Colinvaux, Paul A.
Science, v. 156, n. 3773, 21 Apr. 1967: 380-383
A 14 metre core from a crater lake on Saint Paul island in the Pribilofs has been examined by pollen analysis. Radiocarbon dating indicates that the core spans more than 10,000 years and probably more than 18,000 years. A spruce pollen maximum about 10,000 years ago suggests that spruce advanced to the flanks of the southern coast of the Bering land bridge toward the close of the land bridge period.

5.018   1965
BIOTA OF A LATE GLACIAL ROCKY MOUNTAIN POND
Kauffman, Erle G.; McCulloch, David S.
Geological Society of America. Bulletin, v. 76, n. 11, 1965: 1203-1232
The sediments of a sag pond, in the Huerfano Park area, south central Colorado, have yielded wood with a radiocarbon age of 9600 ± 200 years and a varied biota consisting of vertebrate remains, terrestrial and freshwater molluscs, sponges, pollen and diatoms, on which an evaluation of the climatic conditions can be made.

5.019   1968
BOTANICAL DATING OF RECENT GLACIAL ACTIVITY IN WESTERN NORTH AMERICA
Viereck, Leslie A.
In: Arctic and Alpine Environments, edited by H. E. Wright and W. M. Osburn. International Association in Quaternary Research (INQUA). Congress, 7th, Boulder, Col., 14 Aug - 19 Sept 1965. *Proceedings*, v. 10. Indiana University Press, 1968: 189-204
The radiocarbon dating method can be conveniently used for determining dates of glacial activity during late-glacial or early post-glacial and the Hypsithermal interval, but are of little use for dating moraines of the past several hundred years.

5.020   1961
BOTANICAL EVIDENCE FOR QUATERNARY CLIMATES IN AFRICA
Van Zinderen Bakker, Edward M.A.
Annales of the Cape Provincial Museums, v. 2, Aug. 1961: 16-31
Radiocarbon dates support the late part of the chronology of the four major glaciations in Europe. A table of the Würm chronology and radiocarbon dates is presented.

5.021   1961
C-14 DATING ON SOIL OF THE WEICHSEL GLACIATION IN SCHLESWIG-HOLSTEIN
Ducker, Alfred
In: International Association of Quaternary Research (INQUA), Congress, 6th, Warsaw, 1961. *Report*. 1961: 205
Radiocarbon dating on soil of the Weichsel glaciation shows that this glaciation needed a long start, called the 'Ana-Phase', characterised by a continual fight between periglacial phenomena and a constantly advancing vegetation.

5.022   1967
A CARBON DATED POLLEN DIAGRAM FROM THE CAPE BRETON PLATEAU, NOVA SCOTIA
Livingstone, D.A.; Estas, A.H.
Canadian Journal of Botany, v. 45, n. 3, 1967: 339-359
Radiocarbon dating of organic sediments in a lake near the edge of Cape Breton Plateau indicates that sedimentation began about 9000 years ago and that deglaciation began much more recently there than at any other Nova Scotia localities.

5.023   1964
CARBON ISOTOPIC DISTRIBUTION AND CORRELATED CHRONOLOGY OF SEARLES LAKE SEDIMENTS
Stuiver, Minze
American Journal of Science, n. 262, Mar. 1964
A purpose of this investigation is to study in detail the sequence that led to the striking succession of evaporites and mud, especially the climatic phenomena and the possible changes in carbon isotopic distribution in the sediments (Flint and Gale, *American Journal of Science*, 1958).

5.024   1952
CARBON-14: NEW APPROACH TO THE GLACIAL AGE
Thwaites, Frederick T.
Wisconsin Magazine of History, v. 35, n. 4, Summer 1952: 277-279
Description of the method of age determination by radiocarbon dating. An example of its application for determining the age of the Two Creek Forest beds on the banks of lake Michigan is presented.

5.025   1959
CENOZOIC HISTORY OF THE BERING LAND BRIDGE
Hopkins, David Moody
Science, v. 129, n. 3362, 5 June 1959: 1519-1528
Many facts of palaeontology and biogeography indicated that the Old and New Worlds have sometimes been connected by a continuous land route that extended from Alaska across shallow floors of the Bering and Chukchi seas to Siberia. The application of radiocarbon dating to the stratigraphic and oceanographic study of Pleistocene marine

sediments has helped defining detailed positions of sea-level during the last 10,000 years and up to 30,000 years.

5.026   1955
CENTRAL TEXAS COAST SEDIMENTATION: CHARACTERISTICS OF SEDIMENTARY ENVIRONMENTS, RECENT HISTORY AND DIAGENESIS
Shepard, Francis P.; Moore, David G.
American Association of Petroleum Geologists. Bulletin, v. 39, n. 8, Aug. 1955: 1463-1593
Radiocarbon determinations by Magnolia Petroleum show that the deeper deposits under the bays are contemporaneous in part with the rise of sea-level following the glacial period.

5.027   1960
CHANGES IN WISCONSIN GLACIAL STAGE CHRONOLOGY BY C-14 DATING
Rubin, Meyer
American Geophysical Union. Transactions, v. 41, n. 2, June 1960: 288-289
Discusses the revisions to the previously accepted models of glacial method of age determination made necessary by the dates obtained by the radiocarbon method of age determination

5.028   1967
THE CHRONOLOGY OF DEGLACIATION AROUND EQE BAY AND LAKE GILLIAN, BAFFIN ISLAND, N. W.T.
King, Cuchlaine A.M.; Buckley, Jane T.
Geographical Bulletin, v. 9, n. 1, 1967: 20-32
The stages of deglaciation in the area between Ikpik Bay on the Foxe Basin coast of Baffin island and the inner part of Lake Gillian are traced and related to sea-level. Absolute dating is provided by radiocarbon dates on shells.

5.029   1967
CHRONOLOGY OF NEOGLACIATION IN THE NORTH AMERICAN CORDILLERA
Porter, Stephen C.; Denton, George H.
American Journal of Science, v. 265, n. 3, Mar. 1967: 205-210
Geological, botanical and historical evidence, together with critical radiocarbon dates, permit reconstruction of a reasonably comprehensive picture of glacier fluctuations in western North-America during the last three millenia.

5.030   1964
THE CLASSIFICATION OF THE WISCONSIN GLACIAL STAGE
Wright, Herbert E.
Journal of Geology, v. 72, n. 5, Sept. 1964: 629-637
Comments on the classification of the Wisconsinan stage by Frye and Willman (*Geological Stociety of America. Bulletin:*, 1963, 74: 501) and maintains that although the intervals Tazewell, Cary, and Port Huron of the Lake Michigan lobe cannot be defined or traced satisfactorily on a stratigraphic basis, they should be termed phases and serve as convenient radiocarbon dated subdivisions of a 9000 years fluctuating retreat of a central lobe in the Great Lakes region to which the fluctuations in Minnesota and elsewhere can be compared.

5.031   1960
CLASSIFICATION OF THE WISCONSIN STAGE IN THE LAKE MICHIGAN GLACIAL LOBE
Frye, John C.; Willman, H.B.
Illinois State Geological Survey. Circular, n. 285, 1960. pp. 16
Extrapolation from available radiocarbon dates suggests that Wisconsin time started 50,000 to 70,000 radiocarbon years ago.

5.032   1960
THE CLASSIFICATION OF THE WISCONSIN GLACIAL STAGE OF NORTH CENTRAL UNITED STATES
Leighton, Morris M.
Journal of Geology, v. 68, n. 5, Sept. 1960: 529
Names are proposed for the interglacial sub-stages within the Wisconsin glacial stage and a re-interpretation and absolute dating of the glacial and interglacial sub-stages based on radiocarbon dates are presented.

5.033   1965
CLIMATE AND POST-GLACIAL FOREST HISTORY IN SOUTH-WESTERN KEEWATIN, N. W.T.: 1. SOILS AND RADIOCARBON EVIDENCE.
Bryson, Reid A.; Larsen, James A.; Irving, William N.
In: *International Association for Quaternary Research (INQUA). Congress*, 7th, Boulder, Col., 14 Aug - 19 Sept. 1965. *Abstracts of General Sessions*. Boulder, Colorado, 1965: 53
Buried charcoal - podzol combinations in the southern tundra give evidence of two episodes when the forest border was farther north than the present, these were dated as terminating at about 1500 B.C. for one and 1000 A.D. for the other.

5.034   1960
CLIMATIC CHANGE AND RADIOCARBON DATING IN THE WEICHSELIAN GLACIAL OF DENMARK AND THE NETHERLANDS
Anderson, Sv.Th.; de Vries, Hessel; Zagwijn, W.H.
Geologie en Mijnbouw, n. s., v. 22, n. 1, Jan. 1960: 38-42

A table of radiocarbon dates and chronological comparison between Denmark and the Netherlands is presented.

5.035   1961
CONTINENTAL GLACIATION IN RELATION TO MCFARLAND SEA-LEVEL CURVES FOR LOUISIANA
Frye, John C.; Willman, H.B.
Geological Society of America. Bulletin, v. 72, n. 6, June 1961: 991-99
McFarland (*Geological Society of America. Bulletin*, 1961) sea-level curve for Louisiana is compared with an interpretative glacial advance curve for the Wisconsinan of the lake Michigan glacial lobe. A simple ice withdrawal during Farmdalian time is not recorded in the sea-level curve. World-wide glacial fluctuation should be reflected in sea-level variations when sufficient data become available from both environments. Both curves are based on well documented radiocarbon dates.

5.036   1961
CONVERGENCE OF EVIDENCE ON CLIMATIC CHANGE AND ICE AGES
Fairbridge, Rhodes W.
New York Academy of Science. Annals, v. 95, part 6, 1961: 542-579
Presents a theory of the ice ages. Time parameters are provided by radiocarbon dates. A eustatic curve based on radiocarbon dates is presented.

5.037   1960
THE COOK INLET, ALASKA, GLACIAL RECORD AND QUATERNARY CLASSIFICATION
Karlstrom, Thor N.V.
United States Geological Survey. Professional Papers, n. 400-B, Article 153, 1960: B330-B332
Quaternary deposits in Cook Inlet, Alaska, record five major Pleistocene glacial oscillations and related depositional changes are now closely dated by more than 50 radiocarbon dates.

5.038   1965
THE CORDILLERAN ICE SHEET OF THE NORTHERN ROCKY MOUNTAINS AND RELATED QUATERNARY HISTORY OF THE COLUMBIA PLATEAU
Richmond, Gerald M.; Fryxell, Roald; Naff, George E.; Weis, Paul E.
In: The Quaternary of the United States: a Review Volume for the 7th Congress of the International Association for Quaternary Research, edited by H.E. Wright and David G. Frey. Princeton. N.J., Princeton University Press, 1965: 231-254

The last catastrophic flood from glacial lake Missoula is described, and dated by radiocarbon at 38,700 ± 9000 years although this could be too old as the date was obtained from transported woods in the deposit of the flood at Vantage which could be derived from older deposits.

5.039   1959
THE CORRELATION BETWEEN UPPER PLEISTOCENE PLUVIAL AND GLACIAL STAGES
Maarleveld, Gerardus Cornelis; Hammen, Th. van der
Geologie en Mijnbouw, n. s., v. 21, n. 3, February 1959: 40-45
Botanical, archaeological, geological and pedological data seem to point to a positive correlation between the more humid climatics of lower latitudes (pluvials) and glacial stages of the temperate climatic belt of the northern hemisphere. A correlation table of cultures, strandlines, pluvials and glaciations is presented. The chronology is based on radiocarbon dates.

5.040   1968
CORRELATION OF MIDWESTERN LOESSES WITH THE GLACIAL SUCCESSION
Frye, John C.; Willman, H.B.; Glass, H.D.
In: Loess and Related Aeolian Deposits, edited by C. Bertrand Schultz and John C. Frye. International Association in Quaternary Research (INQUA). Congress, 7th, Denver, Col., 14 Aug - 19 Sept 1965. *Proceedings*, v. 12. University of Nebraska Press, 1968: 3-21
The correlation of the loess stratigraphy to the glacial succession is helped by a number of radiocarbon dates.

5.041   1966
CORRELATION OF ROCKY MOUNTAINS AND LAWRENTIDE GLACIAL CHRONOLOGY IN SOUTHERN ALBERTA, CANADA
Wagner, William Philip
Ph.D. Dissertation. University of Michigan, 1966 (Abstract in: *Dissertation Abstracts*, Ann Arbor, Mich., v. 27, n. 10, April 1967: 3572-B. Order n. 67-1821)
Previous interpretation at Kipp, west of Lethbridge, based on radiocarbon dates is invalidated by the present study.

5.042   1960
CORRELATION OF WISCONSIN DRIFTS IN ILLINOIS, INDIANA, MICHIGAN AND OHIO
Zumbergé, James H.
Geological Society of America. Bulletin, v. 71, n. 8, Aug. 1960: 1177-1188
Radiocarbon dates are considered to be a means of establishing an absolute chronology of events that is based on independent field evidence and not as a primary tool for

5.043   1968
THE COWICHAN ICE TONGUE, VANCOUVER ISLAND
Halstead, Carl
Canadian Journal of Earth Sciences, v. 5, n. 6, Dec. 1968: 1409-1415
The nature and distribution of unconsolidated surficial deposits in southeastern Vancouver Island suggest that glacier ice occupied the Cowichan valley during the late Pleistocene. Radiocarbon dates indicate that the Cowichan ice tongue moved into the valley less than 19,000 years BP, reached its maximum about 18,000 years BP, and that the area was free of ice by 12,800 years BP.

5.044   1967
CRUSTAL MOVEMENT AND DATING OF GLACIAL STAGES IN THE SPITZBERGEN REGION
Grosswald, M.G.; Devirts, A.L.; Dobkina, E.I.
Geochemistry International, v. 4, n. 1, 1967: 30-35
Four radiocarbon datings of driftwood from raised shores in south-eastern Spitzbergen were obtained. They range from 5070 ± 200 to 2400 ± 120 years BP. The curve of crustal uplift for this part of the archipelago has been drawn and is concave, indicating an uplift rate similar to that of Canada. Two stages of young glacier advances have been established and dated: Hornsund stage, about 10,000 years BP and Treskelen stage, less than 800 years BP. A table listing new radiocarbon dates for Spitzbergen is given.

5.045   1952
THE DATING OF A DEPOSIT CONTAINING AN ELK SKELETON FOUND AT NEASHAM NEAR DARLINGTON, COUNTY DURHAM
Blackburn, Kathleen B.
New Phytologist, v. 51, 1952: 364-377
Datings on a series of mud and peat samples believed to be of Allerød age collected from Europe are presented. The radiocarbon dating results give an average antiquity for the Allerød period of 10,800 years which correlates very well with varve dating.

5.046   1952
DATING OF PLUVIAL LAKES IN THE GREAT BASIN
Allison, Ira S.
American Journal of Science, v. 250, n. 12, 1952: 907-909
The uppermost pluvial lake shorelines in south-central Oregon are considered pre-Wisconsin in age. On the pollen and pumice record, a low-level beach in Chewacan - Abert basin, Oregon, is assigned to Mankato time and correlated with the Stansbury of lake Bonneville phase which is dated by the radiocarbon method.

5.047   1961
DATING OHIO'S GLACIERS
Forsyth, Jane L.
Ohio. Division of Geological Survey. Information Circular, n. 30. Columbus, Ohio, 1961.
Describes the method of dating as applied to the Pleistocene glaciers in Ohio. Presents a map showing the locations and dates of Pleistocene wood samples from Ohio and a table of radiocarbon dates, followed by a map showing the succession of estimated positions of the ice front as the last Wisconsin glacier advanced south across Ohio, based on these dates.

5.048   1963
DATING THE LITTLE ICE AGE IN GLACIER BAY, ALASKA
Goldthwait, Richard P.
In: Contributions to Discussions. International Geological Congress, 21st Session, Norden, 1960 *Reports*, part 27. Copenhagen, Berlingske Bogtrykkeri, 1963: 37-46
Thirty nine radiocarbon dates relevant to the area are presented in table and graph form and discussed in terms of the contribution they make to the glacial history of the area.

5.049   1968
DEFINITION OF WISCONSINAN STAGE
Faye, John C.; Willman, H.B.; Rubin, Meyer; Black, R. F.
U.S. Geological Survey. Bulletin, 1274-E, 1968. pp. 22 (Contributions to Stratigraphy)
The Wisconsinan stage and its Altonian, Farmdalian, Woodfordian, Twocreekan, and Valderan substages are described as time-stratigraphic units of the Pleistocene of Illinois and Wisconsin.

5.050   1968
DEGLACIATION AND DIFFERENTIAL POST-GLACIAL REBOUND IN THE APPALACHIAN REGION OF SOUTH-EASTERN QUEBEC
McDonald, Barrie Clifton
Journal of Geology, v. 76, n. 6, Nov. 1968: 664-677
Correlation of glacial sequences in the Appalachian region with those of the American Midwest have been hindered by different processes of deglaciation in these two broad regions. This is examined. Ten radiocarbon dates are available. The Champlain Sea episode dated on shells at $12,000 \pm 230$ and $11,500 \pm 160$ years BP may be partly correlated with glacial lake Iroquois and with the Two Creek interstadial.

5.051 1964
DEGLACIATION DATES IN CENTRAL LABRADOR - UNGAVA: A REAPPRAISAL
Morrison, A.
American Association for the Advancement of Science. Annual Meeting, Montreal 1964. Papers presented. 1964. pp. 14
Proposes three hypotheses, from which dated deglaciation of this area of central Labrador may be fixed with some certainty, formulated on the basis of three radiocarbon dates from samples of pollen collected in 1961 from bogs near Grand Falls.

5.052 1969
DEGRADATION OF THE LAST EUROPEAN ICE SHEET
Gerasimov, I.P.
In: Quaternary Geology and Climate. International Association for Quaternary Research (INQUA). Congress, 7th, Boulder, Col., 14 Aug - 19 Sept. 1965. *Proceedings*, v. 16. Boulder, Col., 1968: 72-78
The development and extinction of the last continental glaciation is examined. A table presenting correlation and radiocarbon chronology of marginal zones of the last glaciation in Northern Europe is included.

5.053 1967
THE DENEKAMP AND PAUDORF INTERSTADIALS
Vogel, John C.; Hammen, Th. van der
Geologie en Mijnbouw, v. 46, n. 5, May 1967: 188-194
The stratigraphical position, pollen diagrams and radiocarbon dates for the Denekamp Interstadial are discussed. The radiocarbon dates for the Paudorf Interstadial are compared with those of the Denekamp Interstadial. The conclusion is drawn that the end of both interstadials is contemporaneous.

5.054 1963
DEVELOPMENT OF WISCONSINAN CLASSIFICATION IN ILLINOIS RELATED TO RADIOCARBON CHRONOLOGY
Frye, John C.; Willman, H.B.
Geological Society of America. Bulletin, v. 74, n. 4, Apr. 1963: 501-505
The evolution of the classification of the Wisconsinan stage in Illinois is shown in graphic form. More than 75 radiocarbon dates from the midwest region are used to place the various named units in a uniform chronology.

5.055 1962
DISCOVERY OF A TREE STUMP IN FRONT OF ENGABREEN, SVARTISEN
Liestøl, Olav
Norsk Polarinstitutt. Årbok 1960, 1962: 64-65
Tree stumps standing at their place of growth in a moraine gave a radiocarbon date of 350 ± 100 years, indicating a glacier advance about 1600 A.D. which is in good agreement with previously formed pictures of glacier variations.

5.056 1966
DISTRIBUTION AND AGES OF PINGOS OF INTERIOR CANADA
Holmes, G.W.; Foster, H.L.; Hopkins, David Moody
In: International Conference on Permafrost, 1963. Proceedings. 1966: 88-93
The distribution of pingos is mapped. The age of the pingos was determined by examination of soil profiles, vegetation communities and, in one case, by radiocarbon dating. They range from a few decades to 7000 years BP.

5.057 1962
EARLY POST-GLACIAL BEAVERS IN SOUTHEASTERN NEW ENGLAND
Kaye, C.A.
Science, v. 138, n. 3543, 23 Nov. 1962: 906-907
Radiocarbon dates indicate that beavers entered the region 12,000 years ago and occupied most low lying places. Beaver dams have probably upset the pollen stratigraphy and created ponds and later peat deposits, causing the early post-glacial stratigraphy of Eastern North America to be less regular than the European. The presence of charred wood dated at $12,700 \pm 300$ years BP would indicate drier climate in post-glacial times.

5.058 1965
AN EARLY POST-GLACIAL POLLEN PROFILE FROM FLÅMSDALEN, A TRIBUTARY VALLEY TO THE SOGNEFJORD, WESTERN NORWAY
Klovning, Ivar; Hafsten, Ulf
Norsk Geologisk Tidsskif, v. 45, n. 3, 1965: 333-338
Pollen analysis and radiocarbon measurements of nekton-mud in the lower part of Flåmsdalen valley shows that this part of the valley was free of ice before 7000 B.C.

5.059 1967
ELEPHANT TEETH FROM THE ATLANTIC CONTINENTAL SHELF
Whitmore, Frank C.; Emery, Kenneth Orriz; Cooke, H.B.S.; Swift, Donald J.P.
Science, v. 156, n. 3781, 16 June 1967: 1477-1481
Teeth of mastodons and mammoths have been recovered from at least 40 sites on the continental shelf, as deep as 120 metres. Radiocarbon dates and the presence of submerged shoreline features indicate that elephants and other large mammals ranged this region during the glacial stage of low sea-level of the last 25,000 years.

Ch. 5 - Glacial Geology

5.060   1966
END MORAINE AND DEGLACIATION CHRONOLOGY IN NORTHERN CANADA, WITH SPECIAL REFERENCE TO SOUTHERN BAFFIN ISLAND, CANADA
Blake, Weston
Canada. Geological Survey Paper, v. 66-26, 1966. pp.31
Numerous radiocarbon dates help establish a chronology of ice movement and deglaciation for this area, showing in particular that all or nearly all of Anderson Bay was filled by ice 9000 years ago but was ice free by 8000 years ago.

5.061   1966
EVIDENCE FOR AN EARLY RECENT WARM INTERVAL IN NORTHWESTERN ALASKA
McCulloch, David; Hopkins, David Moody
Geological Society of America. Bulletin, v. 77, n. 10, 1966: 1089-110
A warm interval beginning at least 10,000 years BP when climates in other parts of the world became non-glacial, and lasting until at least 8300 years BP, is recorded in coastal tundra covered northwestern Alaska by eight radiocarbon dates associated with fossil wood of tree size or tree species, fossil beaver gnawed wood beyond modern beaver range, evidence of ice-wedge melting, buried soils, and soils below modern permafrost table.

5.062   1964
EVIDENCE OF CLIMATIC FLUCTUATIONS ON AXEL HEIBERG ISLAND, CANADIAN ARCTIC ARCHIPELAGO
Müller, Fritz
Paper presented at International Geographical Congress, 1964. Montreal, McGill University, 1964. pp. 11
Various radiocarbon determinations enable to date the movement of ice, the climatic optimum (6000 years BP), an extensive glaciation (20,000 to 10,000 years BP) and a Wisconsin ice maximum. Indicates that climatic fluctuation appear to have a lesser amplitude than in temperate latitudes.

5.063   1957
EXTENSIVE PROGLACIAL LAKE OF WISCONSIN AGE IN THE COPPER RIVER BASIN, ALASKA
Ferrians, Oscar J.; Schmoll, H.R.
Geological Society of America. Bulletin, v. 68, n. 12, Dec. 1957: 1726
Abstract of Paper presented at the November 1957 meeting of the Geological Society of America in Atlantic City. Strand-lines and glacio-lacustrine deposits in the northeastern Copper River Basin demonstrate the existence of a proglacial lake with a maximum extent of more than 4000 square miles. Stratigraphic evidence, supported by radiocarbon dates, indicates Wisconsin age.

5.064   1966
FARMDALE GLACIATION IN NORTHERN ILLINOIS AND SOUTHERN WISCONSIN
Leighton, Morris M.; Brophy, John A.
Journal of Geology, v. 74, n. 4, July 1966: 478-499
The authors cite various evidences for assigning the Pecatorica drift to the Farmdale substage, including the age determined by radiocarbon analysis at 29,000 to 31,000 years BP.

5.065   1960
FINITE RADIOCARBON DATES OF THE PORT TALBOT INTERSTADIAL DEPOSITS IN SOUTHERN ONTARIO
de Vries, Hessel; Dreimanis, Aleksis
Science, v. 131, n. 3415, 10 June 1960: 1738-1739
Three new finite radiocarbon dates suggest that the thermal maximum of the Port Talbot Interstadial occurred prior to 47,000 years BP and that the interstadial deposits were overridden by a glacial advance approximately 44,000 years BP.

5.066   1957
A FIRST C-14 DATE FOR THE WURM I CHRONOLOGY ON THE ITALIAN COAST
Blanc, Alberto Carlo; de Vries, Hessel; Folliari, M.
Quaternaria, v. 4, 1957: 83-93
The age of wood samples for the lower Würm (Pleistocene) peat exposed in a channel through the Pontine Marshes, Italy, is shown to be about 55,000 years, on the basis of radiocarbon tests

5.067   1967
FURTHER EVIDENCE FOR A MIDDLE WURM INTERSTADIAL AND A MAIN WURM GLACIATION OF SOUTH-WEST WALES
John, Brian S.
Geological Magazine, v. 104, n. 6, 1967: 630-633
A radiocarbon age determination on wood fragments from a glacial outwash sand at Cil-maenllwyd (Cardiganshire) gives an age of 33,750±2500/1900 years BP, indicating that the wood fragments are probably middle Würm age and that the last glaciation of St. Georges Channel occured during the main Würm.

5.068   1953
GEOCHRONOLOGY OF THE DEGLACIAL AND NEOTHERMAL AGES
Antevs, Ernst
Journal of Geology, v. 61, n. 3, May 1953: 195-230

This is a review of the chronologies based on varved sediments and a revision of the estimates of the gap in the varve data. The author contests the validity of the radiocarbon dates for the Two Creek Forest Beds and the Valders - Mankato maximum and discusses the reasons for the erroneous radiocarbon dates. Change in the $^{12}C/^{14}C$ ratio by physical contamination with younger organic matter and old carbonates and by decomposition is the main source of error.

5.069   1954
GEOCHRONOLOGY OF THE DEGLACIAL AND NEOTHERMAL AGES: A DISCUSSION: TELE-CORRELATION OF VARVES, RADIOCARBON CHRONOLOGY AND GEOLOGY
Antevs, Ernst
Journal of Geology, v. 62, n. 5, Sept. 1954: 516-521
The author, in replying to Mrs. de Geer recent re-statement of her husband tele-correlation (*Journal. of Geology*, 1954, 62: 514), investigates and refutes the assertion that two tele-correlation dates and the radiocarbon dates for the Two Creek Forest Beds are in agreement and discusses the geologic radiocarbon chronology.

5.070   1954
GEOCHRONOLOGY OF THE DEGLACIAL AND NEOTHERMAL AGES: A DISCUSSION: DE GEER'S CHRONOLOGY OR A STRETCHED ONE WITH INTERRUPTIONS
de Geer, Ebba Hult
Journal of Geology, v. 62, n. 5, Sept. 1954: 514-516
Examines and criticises the chronology of the deglacial and neothermal ages as presented by E. Antevs (*Jl. of Geology*, 1953, 61: 195) and confirms the reliability of the radiocarbon dates which form one of the bases of the de Geer's chronology.

5.071   1955
GEOLOGIC - CLIMATIC DATING IN THE WEST
Antevs, Ernst
American Antiquity, v. 20, n. 4, April 1955: 317-335
Includes a discussion of the radiocarbon dates which correlate the various ages of retreat and advance of the pre-Cochrane ice retreat in North America.

5.072   1967
GEOLOGIC MAP OF NORTH TRURO QUADRANGLE, BARNSTABLE COUNTY, MASSACHUSETTS
Koteff, Carl; Oldale, Robert N.; Hartshorn, J.H.
United States Geological Survey. Quadrangle Maps, GQ 599, 1967
On Cape Cod, the source of the Wellfleet, Truor, and Highland plains deposits was the South Channel ice lobe of the last advance that lay east of Cape Cod. Radiocarbon ages of marine shell fragments in the Wellfleet and Highland plain deposits are greater than 25,000 BP. Valleys were probably cut during late-glacial time.

5.073   1966
GEOLOGICAL AND PALYNOLOGICAL STUDIES OF EARLY LAKE ERIE DEPOSITS
Lewis, C.F.M.; Anderson, T.W.; Berti, A.A.
Michigan University. Great Lakea Research Division, Publications, n. 15, Great Lakes Research, Conference, 9th, Chicago, 1966. *Proceedings*. 1966: 176-191
Coring and echo sounding of Lake Erie bottom sediments indicate a thin lag concentrate of sand, with fossils in places, underlying recent silty clay muds, and overlying clay till or late glacial lacustrine clays. Radiocarbon dates of 10,200 and 11,300 years BP, on organic matter, suggest that Early Lake Erie came into existence about 12,400 years ago.

5.074   1966
THE GEOLOGY AND PALEONTOLOGY OF A LATE PLEISTOCENE BASIN IN SOUTHWEST KANSAS
Schultz, Gerald Edward
Ph.D. Dissertation, University of Michigan, 1966. pp. 123
(Abstract in: *Dissertation Abstracts*, Ann Arbor, Mich., Sect. B, v. 27, n. 7, 1967: 2417B-2418B. Order No. 66-14,590)
Reconstruction of the geologic history of a sinkhole collapse basin that formed during late Pleistocene times in Meade County, Kansas: the Butler Spring Basin. A radiocarbon date of 11,000 ± 390 years BP was obtained by dating shells of *Succinea ovalis Say*. Most of the mammals also present are characteristic of marshy conditions. This indicate that during late Wisconsin glacial, the climate had cooler summers and more effective moisture in the area than today.

5.075   1967
GEOLOGY AND PALEONTOLOGY OF PLEISTOCENE DEPOSITS IN SOUTHWESTERN MANITOBA
Klassen, R.W.; Delorme, L.D.; Mott, R.J.
Canadian Journal of Earth Science, v. 4, n. 3, 1967: 433-447
Fossils and pollen in Pleistocene deposits under tills in the Duck Mountain and Riding Mountain of southwestern Manitoba indicate a cool-warm-cool climatic sequence, which began more than 37,760 years BP. Bones were recovered from an intertill silt on Riding Mountain. Grass associated with the bones was dated as more than 31,300 radiocarbon years BP. The sediments may be correlative with Port Talbot interstadial in the Lake Erie region, though it is inferred they are older and correlative either with an

early Wisconsin interstadial or with the Sangamon interglacial in the mid-western United States.

5.076 **1958**
THE GEOLOGY AND VEGETATION OF LATE-GLACIAL RETREAT STAGES IN SCOTLAND
Donner, J.J.
Royal Society of Edinburgh. Transactions, v. 63, part II, 1956-58: 221-264
The radiocarbon dating of critical organic beds in North America has permitted to correlate the last major retreat stages with those in Europe.

5.077 **1967**
GEOLOGY OF GLACIAL LAKE AGASSIZ
Elson, John A.
In: Life, Land and Water. Conference on Environmental Studies of the Glacial Lake Agassiz Region, 1966. *Proceedings.* Manitoba University. Department of Anthropology. Occasional Papers, n. 1, 1967 : 36-96
The topographic setting and deposits in the lake basin are outlined, and stratigraphic, radiocarbon, and geomorphic interpretations are presented. A hypothetical history of the lake, with four high-water stages and three low-water stages, is outlined. This four episode history is a substantial advance over the previously held two phase lake.

5.078 **1958**
GEOLOGY OF THE GREAT LAKES
Hough, Jack L.
Urbana, University of Illinois Press, 1958
Chapter 16 gives a radiocarbon chronology of great lakes history. An absolute time-scale is based on radiocarbon dates which are listed and evaluated. The scale is in radiocarbon years BP. As few events in the lakes history have been dated and the position of many of the boundaries shown in the table are located by interpolation, the time-scale is only an approximation.

5.079 **1968**
GEOMORPHOLOGY OF INGLEFIELD LAND, NORTH WEST GREENLAND (Abstract)
Nichols, Robert L.
Geological Society of America. Special Papers, n. 115, 1968: 281-282
Radiocarbon measurements indicate that the deglaciation of the coastal areas near Rensselaer Bay occurred before 8200 ± 300 years, and near Dallas Bay, before 6180 ± 200 years BP.

5.080 **1965**
GLACIAL AND NON-GLACIAL STRATIGRAPHY IN SEA CLIFFS OF WHIDBEY ISLAND, FOSSILIFEROUS GLACIAL MARINE DRIFT
Easterbrook, Don J.
In: Pacific Northwest. International Association for Quaternary Research (INQUA), Conference, 7th, Boulder, Col., 15 Aug. - 19 Sept. 1965. *Guide Book for Field Conference J.* Boulder, Col., INQUA, 1965: 68-75
A table of stratigraphic sequence on Whidbey island supported by radiocarbon dates is presented.

5.081 **1957**
GLACIAL AND PLEISTOCENE GEOLOGY
Flint, Richard Foster
New York, John Wiley, 1957: 298-300
Evaluates the radiocarbon dating method. Indicates that the method has altered the existing concept of date and duration of the last major glaciation by implying that this event is embraced within the last 30,000 years. A fairly detailed and consistent calendar of glacial events can also be established and it makes possible true time correlation of strata that are disconnected and separated by long distances.

5.082 **1962**
GLACIAL AND POST-GLACIAL GEOMORPHOLOGY OF THE SUGLUK - WOSTENHOLME AREA, NORTHERN UNGAVA
Matthews, Barry
McGill Sub-Arctic Laboratory. Research Papers, n. 12, May 1962: 17-46
By referring to radiocarbon dates from other areas, the author suggests a date older than the climatic optimum for the raised beaches at 281 feet on the west side of Deception Bay.

5.083 **1960**
GLACIAL CHRONOLOGY OF THE LAGUNA SAN RAPHAEL AREA, SOUTHERN CHILE (Abstract)
Müller, Ernest H.
Geological Society of America. Bulletin, v. 71, n. 12, Dec. 1960: 2106
Radiocarbon dates on organic material collected in the Laguna San Raphael area, Southern Chile, afford a chronology for comparison with other mid-latitude, west coast marine locations.

5.084 **1965**
GLACIAL CHRONOLOGY OF WESTERN TROMS, NORTH NORWAY
Andersen, Björn G.
In: International Studies in the Quaternary. Papers presented on the Occasion of the 7th Congress of the International Association for Quaternary Research (INQUA), Boulder, Colorado, 1965, edited by H.E. Wright and David G. Frey. New York, Geological Society of America, 1965: 35-54 (Geological Society of America Special Paper No. 84,

1965)
An analysis of the glacial features on the narrow submarine continental shelf in Troms suggest that the Würm ice sheet covered the shelf. A total of 13 radiocarbon dates have been made on shells from marine deposits corresponding to the Tromsö - Lyngen moraine suggest a division of the Tromsö - Lyngen substage into two glacial phases which most likely represent Older Dryas and Younger Dryas periods. The radiocarbon dates are presented in table form.

5.085   1958/59
GLACIAL DRAINAGE CHANNELS AS INDICATORS OF LATE-GLACIAL CONDITIONS IN LABRADOR - UNGAVA: A DISCUSSION
Ives, John David
Cahiers de Geography de Quebec, v. 3, n. 5, Oct. 1958 - March 1959: 57-72
Initial report on a broad regional study of glacial drainage features to ascertain conditions prevailing during deglaciation. Results of radiocarbon and palynological datings are included in the discussion.

5.086   1967
GLACIAL FEATURES OF THE NORTH-CENTRAL LAKE SUPERIOR REGION, ONTARIO
Zoltai, S.C.
Canadian Journal of Earth Sciences, v. 4, n. 3, June 1967: 515-528
Catalogues and interprets the glacial features of the region. Stratigraphic correlations with radiocarbon dates suggest that the Nakima moraine was built some 9400 years ago and that the ice disappeared before 6390 years ago.

5.087   1965
GLACIAL FEATURES OF THE QUETICO-NIPIGON AREA, ONTARIO
Zoltai, S.C.
Canadian Journal of Earth Sciences, v. 2, June 1965: 247-269
A reconstruction of the sequence of events based on morphological features allows a tentative correlation of glacial Lake Agassiz in the west with glacial lake Minong stages in the Superior basin. A radiocarbon date of 9380 ± 150 years BP was obtained from wood buried in a post-Minong beach.

5.088   1966
GLACIAL GEOLOGY OF MUIR INLET, SOUTHEAST ALASKA
Haselton, George M.
Ohio State University. Institute of Polar Studies. Report, n. 18, 1966. pp. 34
[Not sighted]

5.089   1956
GLACIAL GEOLOGY OF NORTH-CENTRAL KEEWATIN, NORTH WEST TERRITORIES, CANADA
Taylor, R.S.
Geological Society of America. Bulletin, v. 67, n. 12, Aug.1956: 943-956
In this examination of glacial movements in the Keewatin region, a radiocarbon age determination of 4140 ± 150 years was made on peat buried by the last ice advance.

5.090   1958
GLACIAL GEOLOGY OF THE LAKE GENEVA AREA; SOUTH-EAST WISCONSIN
Black, Robert F.
Geological Society of America. Bulletin, v. 69, n. 12, Dec. 1958: 1536
Abstract of paper presented at the November meeting of the Geological Society of America in St. Louis. A radiocarbon dated log from a deposit underlying Cary ice deposits indicate that the deposit in which it was found was left by the original advance of ice and that the Marengo moraine would be a recessional moraine rather than the product of a distinct readvance.

5.091   1963
GLACIAL GEOMORPHOLOGICAL INVESTIGATIONS IN NORTHERN UNGAVA, QUEBEC
Matthews, Barry
Ice, v. 12, July 1963: 9-10
Preliminary results of radiocarbon datings on shell deposits suggest that probably part of the southern coastal area of the Hudson Strait was free of glacier ice by about 10,500 years BP and that a general marine invasion occurred approximately 7000 to 8000 years ago at the onset of the classical Hypsithermal period. The rate of glacio-isostatic uplift was also calculated on the basis of radiocarbon dates

5.092   1951
GLACIAL HISTORY AND RADIOCARBON
Flint, Richard Foster
In: Glaciological Conference, 2nd, New York City, Jan. 16-17, 1951. *Proceedings*. 1951: 22-
[Not sighted]

5.093   1961
THE GLACIAL HISTORY OF ALASKA : ITS BEARING ON PALAEOCLIMATIC THEORY
Karlstrom, Thor N.V.
New York Academy of Science. Annals, v. 95, Part IV, 1961: 290-340
Two tables of radiocarbon dates are presented: Table 1 - Radiocarbon samples dating the advance and retreat of the

Lake Huron glacial lobe during Wisconsin times; Table 2 - Bonneville and Lahontan basin radiocarbon dates.

5.094   **1966**
GLACIAL HISTORY OF NORTHERN ONTARIO I. THE COCHRANE - HEARTS AREA
Boissonneau, A.N.
Canadian Journal of Earth Sciences, v. 3, n. 5, Oct. 1966: 559-578
Surficial deposits, ice movement, and glacial lakes are described for an area of 42,000 square miles in northeastern Ontario. Glacial chronology is based on radiocarbon dates.

5.095   **1968**
GLACIAL HISTORY OF NORTHERN ONTARIO II. THE TIMISKAMING - ALGOMA AREA
Boissonneau, A.N.
Canadian Journal of Earth Sciences, v. 5, n. 1, Feb. 1968: 97-109
Surficial deposits, ice movement, and glacial lakes are described for an area of 34,500 square miles in northeastern Ontario. Glacial chronology of the area is supported by radiocarbon dates.

5.096   **1965**
THE GLACIAL HISTORY OF WESTERN WASHINGTON AND OREGON
Crandell, D.P.
In: The Quaternary of the United States: a Review Volume for the 7th Congress of the International Association for Quaternary Research, edited by H.E. Wright and David G. Frey. Princeton, N.J., Princeton University Press, 1965: 342-352
Peat and wood interbedded with non-glacial deposits give an Olympia interglaciation age of 15,000 to 35,000 years.

5.097   **1966**
GLACIAL REBOUND AND THE DEFORMATION OF THE SHORELINES OF PROGLACIAL LAKES
Broecker, Wallace S.
Journal of Geophysical Research, v. 71, n. 20, 15 Oct. 1966: 4777-4883
A simple isostatic model explaining the pattern of deformation of the shorelines of proglacial lakes has been developed, whereby the rate of glacial retreat can be derived from the curvature of its uplifted portion. The rate so calculated for the retreat preceding the formation of Lake Algonquin is not in conflict with the radiocarbon chronology for the interval.

5.098   **1960**
GLACIAL RETREAT IN THE NORTH BAY AREA, ONTARIO
Terasmae, Jaan; Hughes, O.L.
Science, v. 131, n. 3411, 13 May 1960: 1444-1446
Geological and palynological studies in Ontario and Quebec, supported by radiocarbon dates, suggest that the opening of the North Bay outlet and the initiation of the Stanley - Chippewa stages in the Huron and Michigan basins took place 10,000 to 11,000 years ago.

5.099   **1967**
GLACIAL STAGES AND RADIOCARBON DATES IN SCOTLAND
Sissons, J.B.
Scottish Journal of Geology, v. 3, part 3, 1967: 375-381
New radiocarbon dates and previously published ones related to Scottish glacial events are discussed. It is suggested that the Perth readvance maximum occurred between 13,500 and 13,000 years ago and that the British Isles were largely or entirely free of ice during the interstadial.

5.100   **1968**
GLACIATION AND SOLAR ACTIVITY SINCE THE FIFTH CENTURY B.C. AND THE SOLAR CYCLE
Bray, J. Roger
Nature, v. 220, n. 5168, 16 Nov. 1968: 672-674
A combination of geophysical, biological and glaciological information supports the idea of a 2600 year solar cycle. There is a remarkable similarity between the ice advance patterns and variations in activity of radiocarbon, the basis of which may be the relationship of both glaciation and radiocarbon variations to solar activity.

5.101   **1965**
GLACIATION IN THE NABESNA RIVER AREA, UPPER TANANA RIVER AREA, ALASKA
Fernald, Arthur T.
In: Geological Survey Research 1965. United States. Geological Survey Professional Papers, 525-C, 1965: C120-C123
Morainal deposits define two major glaciations: the Black Hill Glaciation of Illinoian age and the Jatahmund Late Glaciation of Wisconsin age. Non-glacial terrace deposits, consisting of an upper unit of Wisconsin age, and a lower unit thought to be of Sangamon age are correlated with the glacial sequence.

5.102   **1965**
GLACIATION OF MINNESOTA AND IOWA
Wright, Herbert E.; Ruhe, Robert V.
In: The Quaternary of the United States: a Review Volume for the 7th Congress of the International Association for Quaternary Research, edited by H.E. Wright and David G. Frey. Princeton, N.J., Princeton University Press, 1965: 29-

41
The Wisconsin glacial history in Minnesota is described in phases of ice advance, correlated by radiocarbon dates with the Cary (possibly Tazewell) to possibly Port Huron advances of the Lake Michigan lobe.

5.103 1968
GLACIATION OF POSSIBLE SCOTTISH READVANCE IN NORTH-WEST WALES
Saunders, G.E.
Nature, v. 218, n. 5136, 6 Apr. 1968: 76-78
An investigation of the glacial deposits of the Llcyn Peninsula has revealed that the stratigraphical succession is more complicated than previously thought. Evidence and radiocarbon dating support this chronology and correlation.

5.104 1967
GLACIATION OF THE CHAGVAN BAY AREA, SOUTHWESTERN ALASKA
Porter, Stephen C.
Arctic, v. 20, n. 4, 1967: 227-246
Glaciers, originating from the Ahklun Mountains, covered the area as lobes from piedmont glaciers at least four times. Radiocarbon dates for the Chagvan glaciation are >45,000 years and for the Unaluk drift are 8910 ± 110 years. A late Tertiary submergence of the Bering Shelf is indicated, followed by eustatic changes resulting from Pleistocene glaciation.

5.105 1959
THE GLACIATION OF THE KING VALLEY, WESTERN TASMANIA
Ahmad, N. ; Bartlett, H.A.; Green, D.H.
Royal Society of Tasmania. Papers and Proceedings, v. 93, 1959: 15-16
A date by Gill (*Australian Journal of Science*, 1956, 119: 80) of wood from varves in the Gormanston Moraine, if actually from this locality, would date the initial advancing phase of King Glacier at $26,480 \pm 800$ years, corresponding with the beginning of the Wisconsin Glaciation in North America.

5.106 1965
GLACIATION OF THE ROCKY MOUNTAINS
Richmond, Gerald M.
In: The Quaternary of the United States: a Review Volume for the 7th Congress of the International Association for Quaternary Research, edited by H.E. Wright and David G. Frey. Princeton, N.J., Princeton University Press, 1965: 217-230
Radiocarbon dates are of assistance in relating events younger than 45,000 years.

Ch. 5 - Glacial Geology

5.107 1958
GLACIATION ON THE ARCTIC SLOPE OF THE BROOK RANGE, NORTHERN ALASKA
Detterman, R.L; Bowsker, A.L.; Dutro, J.T.
Arctic, v. 11, 1958: 43-61
A tentative chronology based on three dated samples of peat is given. It is somewhat different from those given by Péwé (*U.S. Geological Survey*, 1953, 65: 149) and Karlstrom (*Science*, 1956, 125: 73).

5.108 1954
GLACIO-MARINE CHRONOLOGY IN THE THULE AREA, GREENLAND
Krinsley, Daniel Bernard
American Geophysical Union. Transactions, v. 35, n. 2, 1954: 383
Abstract of paper presented at the 35th Annual Meeting of the Union, Washington D.C., May 3-5, 1954. Radiocarbon analysis of samples collected in the Thule area of north-west Greenland make it possible to date a glaciation in that area (ca 32,000 years), and place a maximum date on the development of raised marine terraces and on the disappearance of the glacier that occupied Wostenholme Fjord, north of Thule (younger than ca 8500 years)

5.109 1963
GLACIOLACUSTRINE DIAMICTON DEPOSITS IN THE COOPER RIVER BASIN, ALASKA
Ferrians, Oscar J.
United States Geological Survey. Professional Papers, n. 475-C, Article 91, 1962: C121-C125
At Gakona in the north-eastern part of the Cooper River Basin, the age of sediments that were deposited in this proglacial lake is bracketed between a maximum of greater than 38,000 years BP and a minimum date of 9400 ± 300 years BP, indicating that the last major glaciation in the Cooper River Basin is comparable in age to the last major glaciation (Wisconsin) of central North America.

5.110 1964
GUBIK FORMATION OF QUATERNARY AGE IN NORTHERN ALASKA
Black, Robert F.
United States Geological Survey. Professional Papers, n. 302-C, 1964: 59-91
Evidence for age of most of the Gubik formation is inconclusive or conflicting. The various radiocarbon dates, ranging from older than 38,000 years to 3550 ± 300 years BP do not agree with the geomorphic interpretation.

5.111 1963
ILLINOIAN AND WISCONSIN GLACIATIONS, S.E. INDIANA AND ADJACENT AREAS

Ch. 5 - Glacial Geology

Gooding, Ansel M.
Journal of Geology, v. 71, n. 6, Nov. 1963: 665-682
Uses radiocarbon dates to correlate the order of succession and classification of Illinoian, Sangamon and Wisconsin units. A table of dates is presented.

5.112   1965
ILLINOIAN DRIFT IN SOUTHEASTERN SOUTH DAKOTA
Steece, F.V.
South Dakota Academy of Science. Proceedings, v. 44, 1965: 62-71
[Not sighted]

5.113   1958
IMPORTANT ELEMENTS IN THE CLASSIFICATION OF THE WISCONSIN GLACIAL STAGE
Leighton, Morris M.
Journal of Geology, v. 66, n. 3, May 1958: 288-309
Several interpretations of radiocarbon dates are discussed, in so far as the evaluation must be accompanied by the practice of critically examining all pertinent evidence, be it stratigraphical, geomorphological, geochemical, mineral, palaeontological or other. It is impossible to adopt any one as invariably decisive.

5.114   1958
IMPORTANT ELEMENTS IN THE CLASSIFICATION OF THE WISCONSIN GLACIAL STAGE: A DISCUSSION
Ruhe, Robert V.; Scholtes, W.H.
Journal of Geology, v. 67, n. 5, Sept. 1958: 585-593
A rediscussion of the interpretation of radiocarbon dates relative to the Wisconsin glacial stage in Iowa in relationship with the stratigraphy of the area.

5.115   1959
IMPORTANT ELEMENTS IN THE CLASSIFICATION OF THE WISCONSIN GLACIAL STAGE: A REPLY
Leighton, Morris M.
Journal of Geology, v. 67, n. 5, Sept. 1959: 594-598
Radiocarbon dating has not only created problems but stimulated inquiries into aspects of glacial history. Discussion of various radiocarbon dates.

5.116   1966
INSECT FAUNAS OF THE LAST GLACIATION FROM THE THAMES VALLEY
Cooper, G.R.; Sando, C.H.S.
Royal Society of London. Proceedings, series B, v. 165, n. 1000, 1966: 389-412
An examination of the remains of insect faunas in peaty lenses occurring at the base of terrace gravels resting upon Keuper marl in the Thames Valley. The radiocarbon age is $32,160 \pm 1780/1450$ BP. A local climatic environment for this fauna is reconstructed.

5.117   1960
THE INTERGLACIAL OOZE AT PORSI IN LAPLAND
Lundqvist, Gösta
Sweden Geologiska Undersokning Avhandlingar ock Uppsatser, ser. C, n. 575, 1960. pp. 26
The complete core of an organic layer between two large moraines at Porsi, Lapland, was examined. The deposit is radiocarbon dated at 40,000 years BP.

5.118   1962
AN INTERGLACIAL OR INTERSTADIAL DEPOSIT AT GALLEJAURE, NORTHERN SWEDEN
Magnusson, E.
Geologiska Föreningens i Stockholm, Förhanlingar, v. 84, n. 4, Nov./Dec. 1962: 363-371
A deposit of muddy silt partly intercalated with varved fine-grained sediments under about 20 m of till and other glacial deposits in Lake Gallejaure is examined. It may be a fragmentary interglacial deposit but most likely can be correlated with some Early Weichselian interstadial. The radiocarbon dating of the deposit gave an age of more than 35,000 years.

5.119   1962
INTERGLACIAL PEAT IN THE ROVANIEMI AREA
Korpela, K.
Geologi, v. 14, n. 2, 1962: 30
Peat found between two beds of morainic material on a construction site at 66.29°N 26.08°E has a radiocarbon dating of greater than 35,000 years.

5.120   1965
AN INTERGLACIAL SOIL AT TEINLAND, MORAYSHIRE
FitzPatrick, E.A.
Nature, v. 207, n. 4997, 7 Aug. 1965: 621-622
The buried soil is dated $28,140 \pm 480/450$ years BP and marks the onset of a glacial readvance.

5.121   1961
AN INTERSTADIAL (RADIOCARBON DATED) AND THE SUBSTAGES OF THE LAST GLACIATION IN SWEDEN
Brotzen, F.
Geologiska Föreningens i Stockholm, Förhanlingar, v. 83, n. 2, Mar./Apr. 1961: 144-150

Research on the stratigraphy of the marine Quaternary in the valley of the river Gota in Sweden have established the existence of an hitherto unknown marine interstadial. Radiocarbon dates from core samples gave two groups of dates; an older one related to the interstadial, averaging between 26 - 30,000 years and a younger one between 10 - 16,000 years BP. The difference between these two groups is always more than 10,000 years and this demonstrates that below 56 m there exists a large break in sedimentation. The radiocarbon dates are presented in table form.

5.122   1965
THE IOWA QUATERNARY
Ruhe, Robert V.
In: Guidebook for Field Conference C, Upper Mississippi Valley. International Association of Quaternary Research (INQUA), Congress, 7th., Boulder, Col., 1965. Lincoln, Nebraska, Nebraska Academy of Science, 1955: 110-126
Iowa is a classic region of continental glaciation. Pre-Wisconsin drifts and loess are confined generally to the southern half of the state, and Wisconsin drifts to the northern half. In this field conference the stratigraphy is studied, and radiocarbon ages of various tills and loesses are given.

5.123   1962
ISOBASES ON THE WISCONSIN MARINE LIMIT IN CANADA
Farrand, William R.; Gajda, R.T.
Geography Bulletin. Canada, Department of Mines and Technical Surveys, 1962
Radiocarbon dates are used to help interpret the isobases.

5.124   1957
LAKE AGASSIZ AND THE MANKATO - VALDERS PROBLEM
Elson, John A.
Science, v. 126, n. 3281, 15 Nov. 1957: 999-1002
This article is a brief review of data on Lake Agassiz in the light of radiocarbon dates. It shows that the Valders ice border probably lay well inside the margin of the Canadian shield in western Ontario and northern Manitoba and that the Cary - Mankato retreat and readvance were minor compared with the Mankato - Valders marginal fluctuations.

5.125   1964
LAKE LAHONTAN; GEOLOGY OF SOUTHERN CARSON DESERT, NEVADA
Morrison, Robert B.
United States Geological Survey. Professional Papers, n. 401, 1964. pp.156
A stratigraphic study of the Cenozoic geology of part of the basin of Lake Lahontan, one of the great late Pleistocene lakes of the Western United States. Published radiocarbon dates are listed in table form with their relations to deposits and archaeologic cultures. On the basis of these dates, the archaeologic chronology determined by Heizer, the estimates by Antevs and the writer's correlation, the Fallon time is estimated to have lasted from about 2000 B.C. to the present, the Toyeh interval from 3000 to about 2000 B.C. and the Turupah time from 5500 to 300 B.C.

5.126   1966
LAKE LUNDY TIME
Sears, Paul B.
Science, v. 152, n. 3720, 16 April 1966: 386
A date of 8513 ± 500 years for the ice-dammed lake Lundy which was assigned an elevation of 620 feet (190 m) has been rejected.

5.127   1964
LAKE WARREN AND TWO CREEKS INTERVAL
Dreimanis, Aleksis
Journal of Geology, v. 72, n. 2, March 1964: 247-250
The recently published radiocarbon dates of Lake Iroquois, a new date of early Lake St Clair and absence of unconformities between the early and the late Lake Warren deposits in Southern Ontario suggest that, not only the early Lake Warren, but also the late Lake Warren, Lake Grassmere, Lake Lundy, early Lake Algonquin, and the initial phase of early Lake St Clair are older than the Two Creek interval.

5.128   1963
THE LAST 10,000 YEARS, A FOSSIL POLLEN RECORD OF THE AMERICAN SOUTHWEST
Marin, Paul S.
Tucson, University of Arizona Press, 1963. pp. 87
Chapter 7, p. 56-59 examines the radiocarbon dates associated with the pollen profiles and their validity and presents two correlation tables of radiocarbon dates directly and indirectly associated with the pollen profiles. This enables to form a post-pluvial chronology and to determine the major features of post-pluvial climatic history.

5.129   1968
THE LATE ALTONIAN (WISCONSINAN) GLACIAL SEQUENCE IN NORTHERN ILLINOIS
Kempton, John P.; Hackett, J.E.
In: Means of Correlation of Quaternary Successions, edited by Roger B. Morrisson and Herbert C. Wright. International Association for Quaternary Research (INQUA). Congress, 7th, Boulder, Col., 14 Aug - 19 Sept. 1965. *Proceedings,* v. 8. Salt Lake City, University of Utah Press, 1968: 535-546
Radiocarbon dates from peat and organic silts show that the late Altonian till occurs just below Farmdale organic deposits and just above an older peat zone. The oldest Farmdalian

date is 26,900 ± 1600/1300 (L 1625) years BP. The maximum date of peat beds below the Altonian till is 41,000 ± 1500 (GrN 4468) years BP and the minimum is 32,600 ± 520 (GrN 4408) years BP A till between these two dated zones demonstrates a late Altonian glacial advance into Illinois between about 32,600 and 26,900 years ago after a significant interval of ice withdrawal.

5.130   1960
LATE GLACIAL AND POST GLACIAL HUDSON BAY SEA EPISODE
Lee, H.A.
Science, v. 131, n. 3413, 27 May 1960: 1609-1611
Geological investigations, archaeological studies and radiocarbon dates indicate a similarity of events around Hudson Bay, commencing at the time Hudson Bay was freed of glacier ice. The sea that then spread around Hudson Bay 7000 to 8000 years ago is named 'Tyrrell Sea'. The subsequent rate of land emergence decreased exponentially.

5.131   1963
LATE GLACIAL DEPOSITS NEAR LOCKERBIE, DUMFRIESSHIRE
Bishop, W.W.
Dumfriesshire and Galloway Natural History and Antiquarian Journal, v. 40, 1963: 117-132
Dates from a deposit at Lockerbie contribute to the evaluation of a minimal age of the Late-glacial Interstadial. (Dates are: 19,940 ± 250 years BP)

5.132   1966
LATE GLACIAL MOLLUSCAN FAUNA NORTH OF LAKE SUPERIOR, ONTARIO
Zoltai, S.C.; Herrington, H.B.
Journal of Paleontology, v. 40, n. 2, 1966: 439-446
The fossil molluscan fauna found at 38 different localities is described. Its distribution establishes the extent of late glacial lakes north of Lake Superior. Ages range from 9000 to about 5000 years BP.

5.133   1963
A LATE GLACIAL SITE AT LOCH DROMA, ROSS AND CROMARTY
Kirk, W.; Godwin, Harry
Royal Society of Edinburgh. Transactions, v. 65, 1963: 225-249
A late-glacial deposit at Loch Droma, dated at 12,815 ± 155 years BP provides minimal age for the onset of the late-glacial interstadial.

5.134   1963
THE LATE POST-GLACIAL BOUNDARY ACCORDING TO RADIOCARBON DATES AND THE DATA OF POLLEN ANALYSIS
Vinogradov, A.P.
Geokhimija (Geochemistry), v. 11, 1963: 1009-1019
Late glacial and post-glacial bog and lake sediments of the Somino Lake area, Yaroslav province, USSR, were investigated for their pollen assemblages and were dated by the radiocarbon method. The results are used to discuss the evolution of the flora and climatic changes and to date the late glacial - post-glacial boundary.

5.135   1967
LATE QUATERNARY LAND EMERGENCE IN NORTHERN UNGAVA, QUEBEC
Matthews, Barry
Arctic, v. 20, n. 3; 1967: 176-202
Twenty-one radiocarbon dates of [mainly mollusc] material from Late Quaternary marine terraces are used to construct an isostatic uplift curve. The phase of rapid uplift, 7000 to 6000 years BP, averaged about 26 feet per 100 years. For the past 5200 years uplift was just under 1 foot per 100 years. Radiocarbon dates indicate general deglaciation of northern Ungava about 7000 to 8000 years ago. Fauna of raised beaches (40 ft and 55 ft strandlines) suggests optimal marine conditions from about 5230 to 3900 radiocarbon years ago during a possible marine transgression.

5.136   1965
LATE WEICHSELIAN GLACIATION IN THE CHESHIRE - SHROPSHIRE BASIN
Boulton, G.S.; Worsley, P.
Nature, v. 207, n. 4998, 14 Aug. 1965: 704-706
A radiocarbon date on two shells obtained from coarse sand at the base of a till/sand complex, named the Upper Boulder Clay, near Northwich, part of the Cheshire drift sequence, place the Bar Hill - Whitchurch - Wrexham advance at the late Weichselian glaciation, younger than 28,000 and older than 10,000 years BP.

5.137   1966
LATE WEICHSELIAN GLACIATION IN THE CHESHIRE - SHROPSHIRE BASIN
Poole, E. Grey
Nature, v. 211, n. 5048, 30 July 1966: 507
This is a reappraisal of the date of a composite morainic suite in the Wrexham - Bar area. According to a new radiocarbon date and the stratigraphic evidence, it predates the late Weichselian Glaciation and is considerably older than 40,000 years.

5.138   1961
LATE WISCONSIN AGE OF TERRACE ALLUVIUM ALONG THE NORTH LOUP RIVER, CENTRAL NEBRASKA: A REVISION
Miller, Robert D.; Scott, Glenn R.

Geological Society of America. Bulletin, v. 72, n. 8, Aug. 1961: 1283-1284

A radiocarbon age of 10,500 ± 250 years BP from shell material near the base of a terrace along the North Loup River, Howard County, Nebraska, dates the alluvial sequence of silts and soils as late Wisconsin. The alluvium containing the shell material was previously reported as Kansan and Yarmouth in age.

5.139　　　　　　　　　　　　　　　　1961
LATE WISCONSIN AND RECENT HISTORY OF THE MATANUSKA GLACIER, ALASKA
Williams, John R.; Ferrians, Oscar J.
Arctic, v. 12, n. 2, June 1961: 83-90

According to radiocarbon dates the glacier has not advanced during the last 8000 years and the period of retreat preceding readvance to a position within a mile of the glacier is logically correlated with the altithermal.

5.140　　　　　　　　　　　　　　　　1961
A LATE WISCONSIN BURIED PEAT AT NORTH BRANCH, MINNESOTA
Fries, Magnus; Wright, Herbert E.; Rubin, Meyer
American Journal of Science, v. 259, n. 9, Nov. 1961: 279-293

A radiocarbon date on the peat and the wood contained within, which began to accumulate after the Grantsburg lobe (Mankato substage) started to retreat, indicate that the Mankato substage preceded the Two Creek interstadial rather than followed it.

5.141　　　　　　　　　　　　　　　　1956
LATE WISCONSIN CHRONOLOGY OF THE LAKE MICHIGAN BASIN CORRELATED WITH POLLEN STUDIES
Zumbergé, James H.; Potzger, John E.
Geological Society of America. Bulletin, v. 67, n. 3, March 1956: 271-288

Discussion of previous chronologies established on the basis of radiocarbon dates noting that, without the stratigraphic information necessary for reconstructing geologic history, radiocarbon dates are useless and that geologists must not depend too much on radiocarbon dates which are obviously anomalous. A table of radiocarbon dates supporting the proposed chronology is presented.

5.142　　　　　　　　　　　　　　　　1965
LATE WISCONSIN END MORAINES IN NORTHERN CANADA
Falconer, George; Andrews, John T.; Ives, Jack D.
Science, v. 147, n. 3658, 5 Feb. 1965: 608-610

A system of end moraines nearly 2240 km long has been investigated by field explorations and aerial photography. It extends through north eastern Keewatin, Melville Peninsula and Baffin Island and marks the border of a late Wisconsin sheet centred over Foxe Basin and Hudson Bay. Radiocarbon measurements suggest a minimum age of 7930 ± 140 years BP for marine molluscs associated with terminal moraines.

5.143　　　　　　　　　　　　　　　　1962
A LATE-GLACIAL AND POST-GLACIAL CORRELATION BETWEEN EAST AFRICA AND EUROPE
Van Zinderen Bakker, Edward M.A.
Nature, v. 194, n. 4824, 14 Apr. 1962: 201-203

It appears that the changes of temperature which occurred in the Cherangani Mountains in northwest Kenya are in general contemporaneous with the climatic chronology of the late-glacial and post-glacial of Europe. A radiocarbon dated horizon (Groningen date) falls in the Oldest Dryas time of Europe and a cold period of Cherangani Mountains.

5.144　　　　　　　　　　　　　　　　1958
LATE-GLACIAL AND POST-GLACIAL VEGETATION FROM GILLIS LAKE IN RICHMOND COUNTY, CAPE BRETON ISLAND, NOVA SCOTIA
Livingstone, D.A.; Livingstone, B.G.R.
American Journal of Science, v. 256, n. 5, May 1958: 341-359

Nine zones of vegetation have been determined. The upper limit of the late-glacial zones has been dated by radiocarbon at 10,340 ± 220 years, which suggests that this part of Nova Scotia was deglaciated shortly before Two Creek times and has been ice-free since.

5.145　　　　　　　　　　　　　　　　1963
THE LATE-GLACIAL AND THE POST-GLACIAL BOUNDARY ACCORDING TO RADIOCARBON DATES AND THE DATA OF POLLEN ANALYSIS
Vinogradov, A.P.; Devirts, A.L.; Markova, N. G.; Khotinskiy, N. A.
Geokhimiya (Geochemistry), 1963, v, 11: 1009-1019

The boundary between late-glacial and post-glacial sediments of Lake Somino (Pereyaslav District, Yaroslav Province) was determined by radiocarbon dating as 10,260 ± 330 years BP, in excellent agreement with dates from Denmark and England for the same boundary.

5.146　　　　　　　　　　　　　　　　1957
THE LATE-GLACIAL CHRONOLOGY OF EUROPE: A DISCUSSION
Wright, Herbert E.
American Journal of Science, v. 255, n. 7, Summer 1957: 447-460

The correlation between radiocarbon dates and pollen

5.147   1967
LATE-GLACIAL CLIMATE IN NORTHERN UNITED STATES: A COMPARISON OF NEW ENGLAND AND THE GREAT LAKES REGION
Davis, Margaret B.
In: Quaternary Palaeoecology, edited by M.J. Cushing and H.E. Wright. International Association for Quaternary Research (INQUA). Congress, 7th, Boulder, Col., 14 Aug - 19 Sept. 1965. *Proceedings, v.* 7. New Haven, Conn., Yale University Press, 1967: 11-58
Pollen analysis documents changes of climate in late-glacial times. Radiocarbon dating enables to establish a chronology of these changes.

5.148   1963
LATE-GLACIAL DEPOSITS NEAR LOCKERBIE, DUMPHRIESSHIRE
Bishop, W.W.
Dumfriesshire and Galloway Natural History and Antiquarian Society. Transactions, v. 40, 1963: 117-132
Dates from a deposit at Lockerbie contribute to the evaluation of a minimal age for the onset of the Late-glacial Interstadial. Dates are: $12,949 \pm 250$ years BP.

5.149   1963
LATE-GLACIAL DEPOSITS ON THE CHALK OF SOUTH-EAST ENGLAND
Kerney, M.P.
Royal Society, London. Philosophical Transactions, Series B, v. 246, n. 730, Apr. 1963: 203-254
Subaerial deposits of the glacial period 12,000 to 8300 B.C. of the last glaciation are described at a number of sites in Kent, Surrey an Sussex. The climatic improvement of Zone II or Allerød Oscillation, 10,00 to 8000, is reflected stratigraphically and the age confirmed by radiocarbon.

5.150   1967
LATE-GLACIAL PLANT MACROFOSSILS FROM MINNESOTA
Watts, W.A.
In: Quaternary Palaeoecology, edited by M.J. Cushing and H.E. Wright. International Association for Quaternary Research (INQUA). Congress, 7th, Boulder, Col., 14 Aug - 19 Sept. 1965. *Proceedings*, v. 7. New Haven, Conn., Yale University Press, 1967: 89-87
The stratigraphic occurrence of late-glacial plant macrofossilsfrom several sites in Minnesota is described and evaluated ecologically. The chronology is based on radiocarbon dates.

5.151   1965
LATE-GLACIAL POLLEN AND PLANT MACROFOSSILS FROM SPIDER CREEK, SOUTHERN ST. LOUIS COUNTY, MINNESOTA
Baker, Richard G.
eological Society of America. Bulletin, v. 76, n. 5, May 1965: 601-610
Paleaobotanical analyses of basal sediments suggest a late-glacial tundra vegetation which was rapidly invaded by boreal forest at the beginning of the post-glacial period. Two radiocarbon dates of $22,00 \pm 600$ years from the 974 - 980cm level and $13,000 \pm 400$ years from the 887 - 890 level are older than expected, probable reflecting a certain amount of dead carbon in the sample.

5.152   1962
A LATE-GLACIAL POLLEN DIAGRAM FROM MADELIA, SOUTH-CENTRAL MINNESOTA
Jelgersma, Saskia
American Journal of Science, v. 260, n. 7, Summer 1962: 522-529
The pollen diagram is controlled by two radiocarbon dates: the transition from zone I to zone II is dated at $12,650 \pm 350$ years BP and the top of zone IV at 9300 ± 350 years BP.

5.153   1963
A LATE-GLACIAL SITE AT LOCH DROMA, ROSS AND CROMARTY
Kirk, W.; Godwin, Harry
Royal Society of Edinburgh. Transactions, v. 65, 1963: 225-249
Late-glacial deposits at Loch Droma, dated at $12,815 \pm 155$ years BP, provide minimal age for the onset of the late-glacial Interstadial.

5.154   1969
A LATE-PLEISTOCENE GLACIAL ADVANCE, BOW RIVER VALLEY, ALBERTA, CANADA
Rutter, Nathaniel W.
In: Quaternary Geology and Climate. International Association for Quaternary Research (INQUA). Congress, 7th, Boulder, Col., 14 Aug - 19 Sept. 1965. *Proceedings*, v. 16. Boulder, Col. 1968: 104-109
The minimum age for the advance may be about 9330 ± 170 years BP as dated from charcoal samples collected in a nearby area.

5.155   1965
LATE-WISCONSIN POLLEN STRATIGRAPHY AND THE GLACIAL SEQUENCE IN MINNESOTA
Cushing, Edward J.
In: International Association for Quaternary Research

(INQUA). *Congress*, 7th, Boulder, Col., 14 Aug - 19 Sept. 1965. *Abstracts of General Sessions*. Boulder, Colorado, 1965: 85
Evaluation of the glacial and pollen-stratigraphic evidence in Minnesota for the period 13,000 to 10,000 radiocarbon years ago fails to support the hypothesis that the vegetation succession was closely coordinated by climatic change with the phases of glacial activity. More than 40 radiocarbon dates are available for correlating late-Wisconsin events in Minnesota, but the precision of radiocarbon dating is insufficient to permit reliable correlation of the glacial and pollen stratigraphies

5.156   1967
LATE-WISCONSIN POLLEN STRATIGRAPHY AND THE GLACIAL SEQUENCE IN MINNESOTA
Cushing, Edward J.
In: Quaternary Palaeoecology, edited by M.J. Cushing and H.E. Wright. International Association for Quaternary Research (INQUA). Congress, 7th, Boulder, Col., 14 Aug - 19 Sept. 1965. *Proceedings*, v. 7. New Haven, Conn., Yale University Press, 1967: 59-97
More than forty radiocarbon dates are available for correlating late Wisconsin events in Minnesota, but the precision of radiocarbon dating in this time span is insufficient to permit reliable correlation of the glacial and pollen stratigraphies. A chart and a table present the radiocarbon dates and the correlation with the pollen stratigraphy.

5.157   1965
MAJOR END MORAINES IN EASTERN AND CENTRAL ARCTIC CANADA
Falconer, George; Ives, Jack D.; Löken, Olav H.
Geographical Bulletin, v. 7, n. 2, 1965: 137-153
The radiocarbon dates associated with the Cockburn Moraine system are presented in table form and analysed. The ages support the contention that the northern margin of the residual late Wisconsin ice sheet stood against the general line of an extensive moraine system between 8000 and 9000 years ago.

5.158   1951
MAZAMA AND GLACIER PEAK VOLCANIC ASH LAYERS: RELATIVE AGES
Fryxell, Roald
Science, v. 147, n. 3663, 12 March 1951: 1288-1290
Physiographic and stratigraphic evidence supports the regional correlation of two volcanic ash layers with extinct Mount Mazama at Crater Lake, Oregon and Glacier Peak in the northern Cascade Range of the Washington State. A radiocarbon age of 12,000 ± 310 years confirms the geological evidence that ash derived from the Glacier Peak eruption is substantially older than ash from the Mazama eruption of 6600 years ago.

5.159   1962
METHOD OF DEGLACIATION, AGE OF SUBMERGENCE, AND RATE OF UPLIFT WEST AND EAST OF HUDSON BAY, CANADA
Lee, H.A.
Bielutyn Peryglacjialny, n. 11, 1962: 239-245
Reviews Quaternary history of the area by correlating geological, archaeological investigations, ice recession and stratigraphic studies. The rate of land emergence has been measured by dating organic shell and bone materials related to strand lines.

5.160   1968
A MIDDLE WURM INTERSTADIAL IN SOUTHWEST WALES (DISCUSSION)
Boulton, G.S.
Geological Magazine, v. 105, n. 2, 1968: 190-191
Discusses and refutes Brian S. John conclusion about a Mid-Würm interstadial age for a sand outwash at Cil-maenllwyd (Cardiganshire) on the premise that the radiocarbon age and the pollen analysis are in contradiction. (John, *Geological Magazine*, 104, 1967: 630)

5.161   1968
A MIDDLE WURM INTERSTADIAL IN SOUTHWEST WALES; A REPLY TO G.S. BOULTON
John, Brian S.
Geological Magazine, v. 105, n. 4, 1968: 398-400
Gives some more points in support of his article (*Geological Magazine*, 104, 1967: 630), adding two more radiocarbon determinations to support his conclusion.

5.162   1964
MORAINES IN THE APPALACHIAN REGION OF QUEBEC
Gadd, N. R.
Geological Society of America. Bulletin, v. 75, n. 12, Dec. 1964: 1249-1254
The Champlain Sea is redefined. A new radiocarbon date on shells from Kinsay Falls, Quebec, gives a minimum age of 11,410 ± 150 for the so called highland front moraine and the Champlain Sea itself.

5.163   1957
MOVING PICTURE OF THE LAST ICE-AGE
Flint, Richard Foster
Natural History, v. 66, n. 4, Apr. 1957: 188-189
Account of the last invasion of the United States by glacier ice, on the basis of dating by radiocarbon. A time chart showing the 'path of the glacier's edge as timed by radiocarbon calendar' is presented.

5.164 1967
MULTIPLE DRIFT SHEETS IN SOUTHWESTERN WARD COUNTY, NORTH DAKOTA
Pettyjohn, Wayne A.
North Dakota Geological Survey. Miscellaneous Series, n. 30 (Glacial Geology of the Missouri Coteau - Midwest Friends of the Pleistocene, Field Conference, 1967, Guide Book) 1967: 123-129
Deposits at the Coteau du Missouri in Ward County have been dated by radiocarbon methods as $10,300 \pm 300$ and $10,350 \pm 300$ BP. The upper surface of the older Blue Mountain drift sheet has been oxidised and this zone can be traced in the subsurface over wide areas. Inconclusive data suggest that the Blue Mountain drift sheet is early Wisconsin and that the buried oxidised zone represents the Farmdale Stage.

5.165 1967
MURRAY SPRING, A MID-POSTGLACIAL POLLEN RECORD FROM SOUTHERN ARIZONA
Mehringer, Peter J.; Martin, Paul S.; Haynes, C. Vance
American Journal of Science, v. 265, Nov. 1967: 786-79
The pollen record from Murray Springs indicated a mid-post glacial moist interval, radiocarbon dated at 4000 to 5000 years ago.

5.166 1968
MYRTLE LAKE: A LATE AND POST-GLACIAL POLLEN DIAGRAM FROM NORTHERN MINNESOTA
Janssen, C.R.
Canadian Journal of Botany, v. 46, n. 11, Nov. 1968: 1397-1408
The various zones are dated by radiocarbon. The available dates suggest that raised bogs started to form about 3000 years ago, perhaps initiated by a worldwide change in climate.

5.167 1966
NEOGLACIAL CHRONOLOGY, NORTHEASTERN ST. ELIAS MOUNTAINS, CANADA
Denton, George H.; Stuiver, Minze
American Journal of Science, v. 264, n. 8, Oct. 1966: 577-599
In the northeastern St. Elias Mountain, Yukon, drift, morphology and stratigraphy combined with thirteen radiocarbon dates suggest that the initial widespread neoglacial advance shortly antedated 2600 to 2800 BP and that at least some major neoglacial events were essentially synchronous throughout the northern Hemisphere.

5.168 1967
NEW ^{14}C DATES FROM THE LAC ST-JEAN AREA, QUEBEC
Lassalle, Pierre; Rondot, Jehan
Canadian Journal of Earth Science, v. 4, n. 3, June1967: 563-571
Radiocarbon dates on marine shells from Quaternary deposits of the Lac St.Jean area are: Chicoutimi, 8680 ± 80 years BP; Kenogami, 8630 ± 80 years BP; St. Fulgence, 9380 ± 60 years BP; Desbiens, 9560 ± 350 years BP; and Metabetchouan, $10,060 \pm 350$ and $10,250 \pm 350$ years BP. The results of the dating have important implications on the deglaciation of the area: (1) The youngest shell dates indicate that the Champlain episode and the marine invasion of the area may have overlapped by a few hundred years. (2) The maximum age obtained on the marine shells may be a minimum age for the time of deposition of the St.Narcisse moraine: a reasonable age for the moraine would be between approximatley 10,500 and 11,000 years BP.

5.169 1961
A NEW APPROACH TO END-MORAINE CHRONOLOGY, A PRELIMINARY REPORT
Ostrem, Gunnar
Geografiska Annaler, v. 43, n. 3/4, 1961: 418-419
Radiocarbon dating of a Jotunheimen sample gave the age of 2600 ± 100 years BP for a moraine ridge hitherto supposed to date from the great 18th century glacier advance. The new age agrees well with the post-glacial climatic depression identified by Bergstrom.

5.170 1967
NEW DATA ON THE LAST SHEET GLACIATION IN SIBERIA
Troitskiy, S.L.
Akademiia Nauk, SSSR. Doklady. Earth Science Sections, v. 174, n. 1-6, Dec. 1967: 107-110
Detailed stratigraphic analysis and correlations, radiocarbon dates and other work, invalidate the old concept of the special 'Siberian' type of the latest glaciation which occurred at least 20,000 years ago. It is therefore possible to make inter-regional correlations with Europe and North America.

5.171 1967
A NEW RADIOCARBON DATE FOR WALES
Brown, J.F.; Ellis-Gruffydd, I.D.
Nature, v. 213, n. 5082, 25 March1967: 1220-1221
This communication presents the results of a radiocarbon age determination on organic material found in fluvio-glacial deposits in a kame complex at Banc - y - Warren in Cardiganshire. It supports the concept of an extensive Würm glaciation in the Irish Sea basin.

5.172 1963
NORTH ATLANTIC BIOTA AND THEIR HISTORY: A SYMPOSIUM HELD AT THE UNIVERSITY OF ICELAND, REYKJAVIK, JULY 1962, UNDER THE AUSPICES OF THE UNIVERSITY OF ICELAND AND THE MUSEUM OF NATURAL HISTORY, edited by Askell and Doris Löve
Oxford, Pergamon Press, 1963. pp. 430
A series of radiocarbon dates on a marine clay between beds of gravel and sand in southeastern Sweden, well within the area covered by the last Scandinavian ice sheet, supports the idea of an interstadial.

5.173 1956
NOTE ON ABSOLUTE CHRONOLOGY OF HUMAN EVOLUTION
Emiliani, Cesare L.
Science, v. 123, n. 3204, 25 March 1956: 924-926
Presents and explains a chart showing correlation between temperature variations of the glacial Pleistocene and continental stages, fossil hominids and culture. A time scale since Günz glacial age was established on the basis of radiocarbon dating of deep sea sediments by Rubin and Suess.

5.174 1956
A NOTE ON GLACIAL CHRONOLOGY AND THE VALIDITY OF RADIOCARBON DATES
Suess, Hans E.
Utah University. Department of Anthropology. Anthropological Papers, n. 26, Dec. 1956: 47
Indicates that some distortion of results on radiocarbon ages is possible due to change in cosmic rays activity and the size of the natural radiocarbon reservoir on the surface of the earth, but such distortion would not affect the dates by more than a few per cent.

5.175 1963
NOTES ON GLACIAL GEOLOGY, NORTHEASTERN DISTRICTS OF MACKENZIE, CANADA
Blake, Weston
Canada. Geological Survey Paper, n. 63-28, 1963. pp. 12
Radiocarbon determinations were made to establish the age of the highest beach in the area, the age of the end moraine and the rate at which the uplift of land relative to the sea has progressed. Radiocarbon determinations were made on marine shells (but for one). A table of dates is presented and the dates and position of samples are discussed.

5.176 1968
NOTES ON LATE-GLACIAL PALYNOLOGY AND GEOCHRONOLOGY AT ST. HILAIRE, QUEBEC
Terasmae, Jaan; La Salle, Pierre
Canadian Journal of Earth Science, v. 5, n. 2, 1968: 249-257
A palynological study supported by radiocarbon dates of late-glacial sediments at St. Hilaire, Quebec, indicates that the southern part of the St. Lawrence Lowland was deglaciated prior to 12,500 years BP. The late glacial episode comprises several climatic fluctuations, a probable early cool interval (Northern Boreal) more than 12,500 years BP: a relatively colder interval (tundra) about 12,500 years BP followed by another cool interval from about 12,000 years BP to about 10,000 years ago. Another relatively cold episode may have occurred about 11,000 years ago. The new studies extend the previously available palynological record in the St. Lawrence Lowland back in time by about 2,000 years and include the Champlain Sea episode.

5.177 1965
NOTES ON MORAINES AND RADIOCARBON DATES IN NORTHWEST BAFFIN ISLAND, MELVILLE PENINSULA AND NORTHEAST DISTRICT OF KEEWATIN
Craig, Bruce Gordon
Canada, Geological Survey Paper, n. 65-20, 1965. pp. 7
A linear belt of end moraines lies, both spatially and chronologically, midway in the sequence of deglaciation in the northwest quadrant of the area covered by the Wisconsin Laurentide ice-sheet. This study presents a series of new radiocarbon dates bearing on the age of these moraines, and discusses the relationship of the various segments and some aspects of their regional significance

5.178 1959
NOTES ON THE CHAMPLAIN SEA EPISODE IN THE St. LAWRENCE LOWLANDS, QUEBEC
Terasmae, Jaan
Science, v. 130, n. 3771, 7 Aug. 1959: 334-336
Palynological studies, coupled with geological investigations and radiocarbon dating, have shown that the Champlain Sea episode in the St. Lawrence Lowlands is in part contemporaneous with the Two Creeks inter-stadial of the Wisconsin Glaciation.

5.179 1957
AN OCCURRENCE OF BURIED CONIFEROUS WOOD IN THE ALTEMONT MORAINE IN NORTH DAKOTA
Moir, D.R.
North Dakota Academy of Science. Proceedings, v. 11, 1957: 69-74
Radiocarbon dating of sample of the wood, believed to be 'in situ', yield an age value of approximately 11,480 years, correlating closely with the Two Creek interstadial. The material was buried in a sandy outwash that may date the maximum advance of the Mankato ice, and suggests a climate cooler and moister than is observed at present.

Ch. 5 - Glacial Geology

5.180 1968
OLYMPIA INTERGLACIATION, PURCELL TRENCH, BRITISH COLUMBIA
Fulton, R.S.
Geological Society of America, Bulletin, v. 78, n. 8, Aug. 1968: 1075-1080
Radiocarbon dates from a conformable sequence of non-glacial deposits indicate that the Purcell Trench was not occupied by ice from at least 43,800 years BP until after 25,840 years BP. This coincides with the Olympia Interglaciation. Radiocarbon dates from outside the immediate study area give extra proof that the ice did not advance over the area after 20,000 years BP and had retreated from the Purcell Trench prior to 10,000 years BP.

5.181 1959
ON THE EXTENT OF THE LATE GLACIATION IN EASTERN ENGLAND
Suggate, R.P.; West, R.G.
Royal Society, Proceedings, series B, v. 150, n. 939, 17 Mar. 1959: 263-283
From pollen analysis of a Late-Glacial deposit in east Lincolnshire, it is inferred that the earliest sedimentation in a depression in boulder clay immediately followed melting of the ice. Radiocarbon age determinations confirm the Late-Glacial age. Dating was made of peat samples.

5.182 1964
ORIGIN OF ICE AGES: POLLEN EVIDENCE FROM ARCTIC ALASKA
Colinvaux, Paul A.
Science, v. 145, n. 3633, 14 Aug. 1964: 707-708
Pollen analysis of radiocarbon dated samples from the arctic coastal plain of Alaska shows that vegetation of 14,000 years ago reflected a climate colder than the present and that there has been a progressive warming culminating in the present cold arctic climate. The record indicates that the Arctic Ocean has been covered with ice since the Wisconsin glacial maximum.

5.183 1965
OUTLINE OF GLACIAL GEOLOGY OF ILLINOIS AND WISCONSIN
Frye, John C.; William, H.B.; Black, Robert FV
In: The Quaternary of the United States: a Review Volume for the 7th Congress of the International Association of Quaternary Research, edited by H.E. Wright and David G. Frey. Princeton, NJ, Princeton University Press, 1965: 43-61.
A revue of the state of glacial geology study in Illinois and Wisconsin. A time-space diagram shows the relation of available radiocarbon dates to the Wisconsinan glaciers.

5.184 1964
AN OUTLINE OF THE MATERIALS FOR A POST-GLACIAL BIOCLIMATIC HISTORY OF KEEWATIN, NORTH WEST TERRITORIES, CANADA
Larsen, James A.
Wisconsin University. Department of Meteorology. Technical Report, n. 15, 1964
Using radiocarbon data, a chronology of floral recolonisation of the glaciated areas is reconsidered.

5.185 1967
PALEOECOLOGY OF THE SEMINARY AND MIRROR POOLS PEAT DEPOSITS
McAndrews, John H.
In: Life, Land and Water - Conference on Environmental Studies of the Glacial Lake Agassiz Region, 1966. Proceedings. Manitoba University. Department of Anthropology. *Occasional Papers*, n. 1, 1967: 253-269
Pollen analyses show that the Lake Agassiz I-II subaerial stage occurred during the late glacial vegetational period: upland sites had boreal forests. The change to post-glacial vegetation occurred about 9000 years ago. The southern basin was not generally forested during late-glacial I-II stages nor during early post-glacial. It is possible that the area assumed a modern vegetational aspect about 7000 to 9000 years ago. Radiocarbon dates supporting this chronology were made on wood and peat.

5.186 1968
PALEOGEOGRAPHIC FEATURES AND ABSOLUTE AGE OF THE LUGA STAGE OF THE VALDAY GLACIATION ON THE RUSSIAN PLAINS
Punning, Ya.M.K.; Raukas, A.V.; Serebryannyy, I.R.
Akademiia Nauk SSSR. Doklady. Earth Science Section, v. 178, 1968: 61-63
In the ligt of radiocarbon dates it can be supposed that the Luga stage of the Valday glaciation happened in the interval from 13,200 to 12,700 years ago. The whole epoch, from the beginning of the Luga stage to the end of the Salpaussalka (Gotiglacial), spanned 3000 years. The Russian plain was finally liberated from the ice sheet at the end of the Allerød, about 11,000 years ago. A table of dates of the Gotiglacial in the Russian plain is presented.

5.187 1965
PALYNOLOGICAL STUDY OF A VERY THICK PEAT SECTION IN GREECE, AND THE WURM - GLACIAL VEGETATION IN THE MEDITERRANEAN REGION
Hammen, Th. van der; Wijnstra, T.A.; Molen, W.H. van der
Geologie en Mijnbouw, v. 44, n. 1, Jan. 1965: 37-39
A pollen diagram from Macedonia, partly dated by radiocar-

bon, suggests dense oak forests during early Holocene and steppe-like vegetation during Würm glacial, similar to conditions elsewhere in the northern Mediterranean, indicating that no proper pluvial conditions prevailed in the area during the Last Glacial.

5.188 1968
THE PATTERN AND VARIABILITY OF POSTGLACIAL UPLIFT AND ROLE OF UPLIFT IN ARCTIC CANADA
Andrews, J.T.
Journal of Geology, v. 76, n. 4, 1968: 404-425
The irregularities in post-glacial uplift are explicable in terms of distance from the former ice margin (a measure of ice thickness) and date of deglaciation. Models are developed that indicate the effect of different rates of glacial retreat on glacial uplift. A table presents information on 21 sites used in this paper, including the radiocarbon dates.

5.189 1958
PATTERNS RESULTING FROM GLACIER MOVEMENTS NORTH OF FOX BASIN, N. W.T.
Blackadar, R.G.
Arctic, v. 11, n. 3, Sept. 1958: 157-165
Radiocarbon dated archaeological material from a raised beach, 51 metres above present sea-level, suggests a rate of emergence of about 4.5 feet per century at Igloolik Island.

5.190 1968
THE PERTH READVANCE [DISCUSSION]
Jardine, William Graham
Scottish Journal of Geology, v. 4, part 2, 1968: 185-186
Discusses in terms of the radiocarbon ages presented by Sissons (*Scottish Journal of Geology*, 3, 1967: 375) and claims that the Perth readvance needs to be subdivided.

5.191 1968
THE PERTH READVANCE [DISCUSSION]
Sissons, J.B.
Scottish Journal of Geology, v. 4, part 2, 1968: 186-187
In reply to Jardine (*Scottish Journal of Geology*, 4, 1968: 185) cites two new radiocarbon dates of the latter half of post-glacial times which confirms his [Sissons] suggestion for the chronology of the Perth readvance (*Scottish Journal of Geology*, 3, 1967: 375).

5.192 1966
PHYSIOGRAPHY AND GLACIAL GEOLOGY OF BURLEIGH COUNTY, SOUTH-CENTRAL NORTH DAKOTA (Abstract)
Kume, Jack
Geological Society of America. Special Papers, n. 87, 1966: 291-292
Four advances of Pleistocene glaciation are recognised and described in the area. The Burntad Drift is dated at 10,100 ± 300 and 9990 ± 300 years BP by radiocarbon dating on mollusc shells in the drift.

5.193 1965
PHYTOGEOGRAPHY AND PALYNOLOGY OF NORTH-EASTERN UNITED STATES
Davis, Margaret B.
In: The Quaternary of the United States: a Review Volume for the 7th Congress of the International Association of Quaternary Research, edited by H.E. Wright and David G. Frey. Princeton, N. J., Princeton University Press, 1965: 377-401.
A table of radiocarbon dates from pollen sequences in Northeastern United States is presented to help examine and validate correlations of pollen sequences in the Northeast and between the Northeast and other regions.

5.194 1959
PINGO IN THE THELON VALLEY, NORTHWEST TERRITORIES: RADIOCARBON AGE AND HISTORICAL SIGNIFICANCE OF THE CONTAINED ORGANIC MATTER
Craig, Bruce Gordon
Geological Society of America, Bulletin, v. 70, 1959: 509-510
The organic material from the top of the Pingo has a radiocarbon age of 5500 ± 250 years. The age, climatological and ecological significance of this organic material indicate conditions warmer than the present and an age coinciding with that of the post-glacial thermal maximum.

5.195 1967
POLLEN ACCUMULATION RATES AT ROGERS LAKE, CONNECTICUT, DURING LATE AND POST-GLACIAL TIME
Davis, Margaret B.
Review of Palaeobotany and Palynology, v. 2, n. 1-4, 1967: 219-230
Pollen accumulation rates are estimated by dividing pollen per unit volume in 1 mL samples from a core in the lake, by the number of years represented by each sample, determined by radiocarbon.

5.196 1967
POLLEN ANALYSES OF LATE GLACIAL AND POSTGLACIAL SEDIMENTS IN IOWA
Brush, Grace S.
In: Quaternary Palaeoecology, edited by M.J. Cushing and H.E. Wright. International Association for Quaternary Research (INQUA). Congress, 7th, Boulder, Col., 14 Aug -

Ch. 5 - Glacial Geology

19 Sept. 1965. *Proceedings*, v. 7. New Haven, Conn., Yale University Press, 1967: 99-115

Pollen studies of three bogs on drift of the Des Moines lobe of north-central Iowa indicate that the climate changed from cool and moist to dry and warm during the 4500 years it took to accumulate the bog sediments. A table showing generalised sediment stratigraphy and radiocarbon dates is presented.

5.197   1960
POLLEN ANALYSIS OF THE MICHILLINDA PEATS
Sears, Paul B.; Bopp, Monika
Ohio Journal of Science, v. 60, n. 3, May 1960: 149-154

A 50 in peat layer, bracketed by two radiocarbon dates which indicate an accumulation time of about 1000 years, is examined for pollen record. It presents a classic record of the mesophytic interval (5000 to 4000 BP) and records climatic conditions accompanying the rise in the Michigan Basin just preceding the Nipissing stage of the Great Lakes.

5.198   1954
POLLEN ANALYTICAL INVESTIGATION OF A C-14 DATED ALLERØD SECTION FROM RUDS VEDBY
Krog, Harald
Dansk Geologiske Undersøgelse, v. 2, n. 80, 1954: 120-139
[Not sighted]

5.199   1952
POLLEN AND RADIOCARBON STUDIES OF ALEUTIAN SOILS
Anderson, Sv.Th.; Bank, Th. P.
Science, v. 116, n. 3004, 25 July 1952: 84-86

Studies of soil profiles have provided samples for pollen analysis and radiocarbon dating which will contribute toward a chronology of post-glacial events in the Aleutians. The investigation (Aleutian Project at Michigan University) should provide data of correlative value for geology, biology and anthropology

5.200   1967
POLLEN DIAGRAMS FROM SUB-ARCTIC CENTRAL CANADA
Nichols, Harvey
Science, v. 155, n. 3770, 31 Mar. 1967: 1665-1668

Peat from Keewatin and Manitoba contained macrofossil and palynological evidence of former latitudinal movements of the forest-tundra boundary in response to the changing location of the mean summer position of the Arctic front. Radiocarbon dating demonstrates the synchroneity of these climatic changes with those registered in northwest Europe during the past 6000 years.

5.201   1955
POLLEN PROFILES, RADIOCARBON DATING, AND GEOLOGIC CHRONOLOGY OF THE LAKE MICHIGAN BASIN
Zumbergé, James H.; Potzger, John E.
Science, v. 121, n. 3139, 25 Feb. 1955: 309-311

A layer of compacted peat perched on a 5 ft layer of sand under a 25 ft dune on the eastern shore of Lake Michigan records four major forest changes and four intermediate changes through pollen analysis. Radiocarbon age determinations giving the time placement for the forest change form the basis for an absolute chronology of the Lake Michigan basin.

5.202   1963
A POLLEN RECORD FROM ARCTIC ALASKA REACHING GLACIAL AND BEHRING LAND BRIDGE TIMES
Colinvaux, Paul A.
Nature, v. 198, n. 4880, 11 May 1963: 609-610

Discussion of five radiocarbon determinations in which the date of the deepest part of a core is younger than the other four. However, both sets of dates include the Wisconsin maximum glaciation and the land-bridge time.

5.203   1962
POLLEN SEQUENCE AT KIRSHNER MARSH, MINNESOTA
Winter, Thomas C.
Science, v. 138, n. 3639, 26 Oct. 1962: 526-528

A pollen diagram records a continuous vegetation sequence from the time of late Wisconsin ice retreat from the region. The late glacial and early post- glacial portions of the diagram are correlated with a radiocarbon dated diagram from Madelia, Minnesota. Both suggest significant climatic change at that time. The Kirshner diagram also shows advance of prairie elements in the region between 7200 and 5000 years ago.

5.204   1951
POLLEN SUCCESSION IN THE SEDIMENTS OF SINGLETARY LAKE, NORTH CAROLINA
Frey, David G.
Ecology, v. 32, n. 3, Summer 1951: 518-533

Age determinations by the radiocarbon method have been recently reported from various horizons, chiefly from Mankato maximum and post-Mankato maximum times. On the basis of this analysis, it was concluded that the Mankato maximum at Two Creek, Wisconsin, was approximately 11,000 years ago. Other dates were determined, helping to establish a chronology.

5.205 1965
THE PORT TALBOT INTERSTADIAL OF THE WISCONSIN GLACIATION
Dreimanis, Aleksis; Terasmae, Jaan
In: *International Association for Quaternary Research (INQUA). Congress,* 7th, Boulder, Col., 14 Aug - 19 Sept. 1965. *Abstracts of General Sessions.* Boulder, Colorado, 1965: 108
The second, relatively warm episode, Port Talbot II, dated by radiocarbon determinations from 35,600 to 47,690 years corresponds to the formerly defined Port Talbot interstadial.

5.206 1966
THE PORT TALBOT INTERSTADE OF THE WISONSIN GLACIATION
Dreimanis, Aleksis; Terasmae, Jaan; McKenzie, G.D.
Canadian Journal of Earth Sciences, v. 3, n. 3, 1966: 305-325
Recent test borings at the type locality of the Port Talbot Interstade, lithologic and palynologic investigations of the cores, and new radiocarbon dates, suggest that this interval was considerably longer than previously assumed, lasting from 48,000 years BP to 24,000 years BP. A table of radiocarbon dates is presented.

5.207 1965
A POSSIBLE MAIN WURM GLACIATION IN WEST PEMBROKESHIRE
John, Brian S.
Nature, v. 207, n. 4997, 7 Aug. 1965: 622-623
Two radiocarbon determinations from marine mollusc fragments in glacial outwash in Pembrokeshire have indicated that the last glaciation of West Wales from the Irish Sea probably occurred within the last 38,000 years.

5.208 1967
POST-GLACIAL CHANGES OF SEA-LEVEL IN THE WESTERN NORTH ATLANTIC OCEAN
Redfield, Alfred C.
Science, v. 157, n. 3789, 11 Aug. 1967: 687-692
Radiocarbon determinations of the age of peat indicate that in Bermuda, Southern Florida, North Carolina and Louisiana, the relative sea-level has risen at approximately the same rate, 2.5 x 10-3 foot per year (0.76 x 10-3 meter) during the past 4,000 years. It is proposed that this is the rate of eustatic change in sea-level.

5.209 1964
POST-GLACIAL CHRONOLOGY AND THE ORIGIN OF DEEP LAKE BASINS IN PRINCE EDWARD COUNTY, ONTARIO
Terasmae, Jaan; Mirymech, E.
Great Lakes Division. Publication II, University of Michigan, 1964: 161-169
[Not sighted]

5.210 1965
POST-GLACIAL CLIMATE AND ARCHAEOLOGY IN THE DESERT WEST
Baumhoff, Martin A.; Heizer, Robert F.
In: *The Quaternary of the United States: a Review Volume for the 7th Congress of the International Association of Quaternary Research,* edited by H.E. Wright and David G. Frey. Princeton, N.J., Princeton University Press, 1965: 697-707.
Various contradictory chronologies of culture, climate and rainfall based on radiocarbon dating are examined and discussed.

5.211 1955
POST-GLACIAL FORESTS IN SOUTH AND CENTRAL BRITISH COLUMBIA
Hansen, P.H.
American Journal of Science, v. 253, n. 1, 1955: 640
Discusses the post-glacial thermal maximum in the light of radiocarbon dates for the Masama eruption and the Mount Washington eruption. It may have occurred about 6000 years ago with the xerothermic interval lasting from 7500 to 3500 years ago.

5.212 1957
THE POST-GLACIAL HISTORY OF VEGETATION AND CLIMATE IN THE LABRADOR-QUEBEC REGION AS DETERMINED BY PALYNOLOGY
Grayson, John Francis
Ph.D. Dissertation. University of Michigan, 1957. pp. 278 (Abstract in: *Dissertation Abstracts,* Ann Arbor, Mich., v. 18, n. 4, 1958: 1229. L.C. number: mic. 58-921)
Radiocarbon dates are used to help construct the palynological profiles of the bogs studied.

5.213 1957
POST-GLACIAL HYPSITHERMAL INTERVAL
Deevey, Edward S.; Flint, Richard Foster
Science, v. 125, n. 3240, 1 Feb. 1957: 182-184
Restricting the previous proposed term, post-glacial thermal maximum, to a time-stratigraphic horizon or a point in time, the authors propose the adoption of Chiarugi's term Hypsithermal as a substitute for climatic optimum. The stratigraphic unit whose time was Hypsithermal is assigned to 4 pollen zones, 5 through 8 in the Danish system, Boreal through Sub-Boreal in the Blytt - Sernander terminology, dated approximately 7000 to 600 BC.

5.214 1964
POST-GLACIAL MARINE SUBMERGENCE AND EMERGENCE OF MELVILLE ISLAND, N. W.T.
Henoch, W.E.S.
Geographical Bulletin, v. 22, Nov. 1964: 105
The significance of seven radiocarbon dates of samples collected on Melville Island are discussed and used to construct a preliminary uplift curve. This curve indicates that over the last 2000 years, uplift has been negligible and that Melville Island is near isostatic equilibrium.

5.215 1960
POST-GLACIAL SUBMERGENCE OF THE GREAT BELT DATED BY POLLEN-ANALYSIS ON RADIOCARBON
Krog, Harald
In: Chronology and Climatology of the Quaternary, edited by Johs. Iversen, E. Hyyppä and S. Thorarinsson. International Geological Congress, 21st, Norden, 1960. *Reports*, part 4. Copenhagen, Berlingske Bogtrykkeri, 1960: 127-133
Radiocarbon measurements date the beginning of the submergence to ca 6600 years B.C.

5.216 1967
POST-GLACIAL UPLIFT AT TANQUARY FIORD, NORTHERN ELLESMERE ISLAND, NORTH WEST TERRITORIES
Hattersley-Smith, G.; Long, Austin
Arctic, v. 20, n. 4, 1967: 255-260
From the radiocarbon ages of samples of marine shells and peat, a post-glacial uplift curve has been constructed for the upper part of Tanquary Fiord in northern Ellesmere Island. The dates show that the head of the fiord was clear of glacial ice by at least 6500 years ago. From 6500 to 5000 years ago, isostatic uplift was at the rate of about 3.5 m per 100 years and subsequently at the rate of about 25 cm per 100 years.

5.217 1965
POSTGLACIAL EMERGENCE AT THE SOUTH END OF INUGSUIN FIORD, BAFFIN ISLAND, N. W.T.
Löken, Olav H.
Canada. Department of Mines and Technical Surveys. Geography Branch. Geological Bulletin, v. 7, n. 3-4, 1965: 243-258
A beach deposit from Inugsuin Fiord, Baffin Island, is described and its mode of formation outlined. It was possible to relate [by radiocarbon data] a number of shell samples to the sea-levels prevailing when the shells lived. An accurate emergence curve has been drawn which is similar in form to curves obtained from other areas.

5.218 1966
POSTGLACIAL ISOSTATIC MOVEMENT IN NORTHEASTERN DEVON ISLAND, CANADIAN ARCTIC ARCHIPELAGO
Müller, Fritz; Barr, W.
Arctic, v. 19, n. 3, 1966: 263-269

Raised marine feaures of the lowlands in the vicinity of Cape Sparbo were investigated. The radiocarbon dates of marine shells indicate that the area was clear of ice as early as 15,500 BP and that the most rapid isostatic uplift (approx. 6.5 m per century) took place between 9000 and 8000 years ago: the total isostatic rebound having been about 110 m. The two oldest dates (15,000 and 13,000 BP) if correct, give a rare indication of the slow onset of the isostatic uplift.

5.219 1967
A POST-WISCONSIN POLLEN SEQUENCE FROM VESTABURG BOG, MONTCALM COUNTY, MICHIGAN
Gillian, Jeanne A.; Kapp, Ronald O.; Brogue, Robert D.
Michigan Academy of Science, v. 52, part 1, 1967: 3-17
Radiocarbon determinations date the various pollen zones. The end of the spruce pollen zone is dated at $10,328 \pm 436$ years BP, the migration of Hemlock into central Michigan is dated at 7982 ± 250 years BP and the end of the Oak mixed hardwood pollen zone is dated at 3146 ± 237 years BP. These dates bracket the post-glacial fluctuation in pine which should thus be dated at approximately 5000 to 600 years BP.

5.220 1960
PRE-CLASSICAL WISCONSIN IN THE EASTERN PORTION OF THE GREAT LAKES REGION, NORTH AMERICA
Dreimanis, Aleksis
In: Chronology and Climatology of the Quaternary, edited by Johs. Iversen, E. Hyyppä and S. Thorarinsson. International Geological Congress, 21st, Norden, 1960. *Reports*, part 4. Copenhagen, Berlingske Bogtrykkeri, 1960: 108-119
It is suggested that the principal centres of glacial outflow during the Early Wisconsin were further East than during the Main or Classical Wisconsin. The dating of the various advances and retreats are supported by radiocarbon dates.

5.221 1966
PRE-VASHON PLEISTOCENE SEQUENCE IN THE CENTRAL PUGET LOWLAND, WASHINGTON (Abstract)
Easterbrook, Don J.
Geological Society of America. Special Papers, n. 87, 1966: 201

Two pre-Vashon drifts separated by interglacial sediments can be recognised. The lower earlier glaciation is dated at more than 40,000 years BP on radiocarbon dates from wood and peat samples. The younger drift was also dated at more than 40,000 years. Sand from early Vashon age is dated younger than 26,850 ± 1700, a date obtained from peat below the sand, but overlying the pre-Vashon till.

5.222 1953
THE PREHISTORIC GREAT LAKES OF NORTH AMERICA
Hough, Jack L.
American Scientist, v. 5, n. 1, March 1953: 84-110
The history of the Great Lakes as known bodies of water begins with the waning of the last major glacial stage, the Wisconsin. Radiocarbon dates help to establish a chronology of the sequence of lake stages.

5.223 1964
PRELIMINARY ACCOUNT OF THE GLACIAL HISTORY OF BATHURST ISLAND, ARCTIC ARCHIPELAGO
Blake, Weston
Canada, Geological Survey Paper, n. 64-30, 1964. pp. 8
Dates on marine shells indicate that much of the island was ice-free by 9,000 years ago, although covered by glacial ice during the classical Wisconsin. The dates supporting this conclusion are presented in table form and discussed.

5.224 1966
PRESERVATION OF VEGETATION AND PATTERNED GROUND UNDER A THIN ICE BODY IN NORTHERN BAFFIN ISLAND, N. W.T.
Falconer, George
Geographical Bulletin, v. 8, n. 2, 1966: 194-200
A thin ice body in Northern Baffin Island is undergoing rapid recession revealing undisturbed patterned ground features and vegetation. A sample of moss thus exposed has a radiocarbon age of 330 ± 75 years, and supports previous estimates of occurrence of a markedly more nival period in parts of arctic Canada two or three centuries ago.

5.225 1953
PROBABLE WISCONSIN SUBSTAGES AND LATE WISCONSIN EVENTS IN NORTHEASTERN UNITED STATES AND SOUTHEASTERN CANADA
Flint, Richard Foster
Geological Society of America, Bulletin, v. 64, n. 8, Aug. 1953: 897-919
The radiocarbon dates pertinent to the post-Cary history of the Great Lakes are shown in their approximate stratigraphic position and their validity discussed (p.917).

5.226 1965
PROBLEMS IN THE QUATERNARY PHYTO-GEOGRAPHY OF THE GREAT LAKES REGION
Cushing, Edward J.
In: The Quaternary of the United States: a Review Volume for the 7th Congress of the International Association of Quaternary Research, edited by H.E. Wright and David G. Frey. Princeton, N. J., Princeton University Press, 1965: 403-416
Generalised late- and post-Wisconsin diagrams from selected sites in the Great Lakes regions are dated by radiocarbon and a table is presented.

5.227 1965
QUATERNARY GEOLOGY OF NEW YORK
Müller, Ernest H.
In: The Quaternary of the United States: a Review Volume for the 7th Congress of the International Association of Quaternary Research, edited by H.E. Wright and David G. Frey. Princeton, N J., Princeton University Press, 1965: 99-112
Radiocarbon dates on organic material from sediments deposited after Valley Heads recession indicate that more than 12,000 years ago the ice sheet had already begun its withdrawal from the massive Valley Heads moraine. The dates listed are considered to post-date the Valley Heads maximum closely and are consistent with correlation of the Valley Heads with the Port Huron sub-stage.

5.228 1967
QUATERNARY GEOLOGY OF THE ALASKAN SHORE OF CHUKCHI SEA
McCulloch, David S.
In: The Bering Land Bridge, edited by David M. Hopkins. Stanford, Calif., Stanford University Press, 1967: 91-120
Six Pleistocene marine transgressions, two major glacial advances and two important post-glacial warm intervals are recorded. A summary in table form based on radiocarbon dates is presented.

5.229 1964
QUATERNARY GEOLOGY OF THE KENSI LOWLAND AND GLACIAL HISTORY OF THE COOK INLET REGION, ALASKA
Karlstrom, Thor N.V.
United States Geological Survey. Professional Papers, n. 443, 1964. pp. 69
A study of the Quaternary deposits of the Kensi Lowlands, interpreted within the framework of the radiocarbon dated regional drift sequence of Cook Inlet recording five glaciations and several post-Pleistocene glacial advance.

Ch. 5 - Glacial Geology

5.230 1958
QUATERNARY GEOLOGY OF THE NENANA RIVER VALLEY AND ADJACENT PARTS OF THE ALASKA RANGE
Wahrhaftig, Clyde
United States Geological Survey, Professional Papers, n. 293, part A, 1958: 1-68
A radiocarbon date of the peat deposited in the lake found behind the front Moraine of the retreating glacier indicates that the Riley Creek stage corrresponds to the continental Mankato glaciation, Naptowne on the Kauai Peninsula and the Younger Dryas in Europe.

5.231 1965
THE QUATERNARY OF NEW ENGLAND
Schafer, J.P.; Hartshorn, J.H.
In: The Quaternary of the United States: a Review Volume for the 7th Congress of the International Association of Quaternary Research, edited by H.E. Wright and David G. Frey. Princeton, N. J., Princeton University Press, 1965: 113-128
An attempt to solve the problem of the two tills in Massachusetts is made in terms of several radiocarbon dates. The age of the glacial maximum is discussed in terms of dates from various areas such as outer Cape Code, Martha's Vineyard, Totket bog (New Haven), eastern Massachusetts.

5.232 1965
THE QUATERNARY OF NORWAY
Andersen, Björn G.
In: The Quaternary, v. I, edited by Kalervo Rankama. New York, Interscience, 1965: 91-135
In this general review of the Quaternary of Norway, radiocarbon dating is used to establish the ages of various stages and sub-stages. Most of the dates were obtained from shells deposited in several moraines deposits, and all the radiocarbon dating was carried out at the Radiological Dating Laboratory of the Norwegian Institute of Technology in Trondheim, Norway.

5.233 1964
QUATERNARY SEDIMENTS AND THEIR GEOLOGICAL DATES WITH REFERENCE TO THE GEOMORPHOLOGY OF KRONPRINS OLAV KYST
Meguro, Hiroshi; Yoshida, Yoshio; Uchio, Takayasu; Kigoshi, Kunihiko; Sugawara, Ken
In: Antarctic Geology. International Symposium on Antarctic Geology, 1st, Capetown, 16-21 Sep. 1963. *Proceedings*. 1964: 73-80
As a result of radiocarbon datings on the molluscan shell fragments and foraminifera contained in the sediment, one may conclude that the retreat of the inland ice from the Ongul Islands took place at least over 2300 years BP but not earlier than 40,000 years BP and that the uplift process has lasted up to 3800 years BP or a little later. Some radiocarbon dates are not always in harmony with the topography.

5.234 1967
RADIOCARBON AGES OF SHELLS IN THE GLACIOMARINE DEPOSITS OF WESTERN ICELAND
Ashwell, Ian Y
Geographical Journal, v. 133, part 1, 1967: 48-50
Shells of sub-fossil marine mollucs, collected in the clays and sands of the lowlands of western Iceland, have been dated by radiocarbon. The chronology so established suggests that the process and chronology of the end of glaciation in western Iceland are similar to those of the Puget Lowlands in western North America.

5.235 1966
RADIOCARBON CHRONOLOGY OF LATE PLEISTOCENE DEPOSITS IN NORTHWEST WASHINGTON
Easterbrook, Don J.
Science, v. 152, n. 3723, May 1966: 765-767
Fourteen radiocarbon dates of shells and wood from late Pleistocene sediments in northwest Washington provide evidence for correlation of the Everson interstadial with the Two Creeks interval of the midcontinent and suggest possible correlations between the Sumas and Valders stadials and between the Vashon stadial and part of the Tazewell-Cary advances.

5.236 1958
A RADIOCARBON DATE FOR PENGUIN COLONIZATION OF CAPE HALLETT, ANTARTICA
Harrington, H.J.; McKellar, I.C.
New Zealand Journal of Geology and Geophysics, v. 1, n. 3, Aug. 1958: 571-576
During maxima of ice advances in the last glaciation, 10,000 to 30,000 years ago, the Ross Sea was filled by ice, so penguin colonisation of shoreline breeding sites must have occurred since that time. The age of a frozen Adelie penguin body from the base of an accumulation of penguin bodies and guano at the Cape Hallett rookery has been determined by the radiocarbon method as 1,210±70 years. The rookery was probably colonised between 400 and 700 A.D. at approximately the same time as a Northern Hemisphere warm period that stimulated a Viking expansion. There is no sign that it was temporarily abandoned at the time of a Northern Hemisphere cold period between A.D.1650 and 1750.

Ch. 5 - Glacial Geology

5.237   1954
A RADIOCARBON DATE OF PEAT FROM JAMES BAY IN QUEBEC
Potzger, John E.
Science, v. 119, n. 3104, June 1954: 908
Pollen studies of bog samples indicate that forests migrated northward during warm-dry periods of post-glacial times. A sample from the bottom level of a bog near the Rupert River gave an age of 2350 ± 200 years as marking the beginning of deposition of organic matter and doubtless of upland occupation by forests.

5.238   1968
A RADIOCARBON DATED MARINE ALGAL BED OF CHAMPLAIN SEA EPISODE NEAR OTTAWA, ONTARIO
Mott, R.J.
Canadian Journal of Earth Science, v. 5, n. 2, 1968: 319-324
A bed of marine algae (seaweed) in a sand pit in Champlain Sea sediments southwest of Ottawa, Ontario, was dated by the radiocarbon method and gave an age of 10,800 ± 150 years (GSC-570). Marine shells from above and below the algal bed gave radiocarbon ages of 10,620 ± 200 years (GSCd-587) and 10,880 ± 160 years (GSC-588) respectively. The algae and underlying shells have been excellently preserved as a result of rapid burial following deposition and remaining beneath a high water table. The similarity of the radiocarbon dates on the algae and on the shells above and below, and with previous dates on shells from other Champlain Sea deposits indicates the reliability of these shell dates.

5.239   1966
RADIOCARBON DATED POSTGLACIAL LAND UPLIFT IN NORTHERN UNGAVA, CANADA
Matthews, Barry
Nature, v. 211, n. 5054, Sept. 1966: 1164-1166
Ten radiocarbon dates of molluscs from raised terraces in Northern Ungava, Quebec, provide sufficient data for construction of a provisional uplift curve. It suggests that local ice caps and/or large valley glaciers have existed in this region during the early part of the hypsithermal interval, preventing the incursion of the sea.

5.240   1967
RADIOCARBON DATES AND THEIR SIGNIFICANCE FROM THE BRIDE MORAINE, ISLE OF MAN
King, Cuchlaine A.M.; Andrews, J.T.
Geological Journal, v. 5, part 2, 1967: 305-308
Two radiocarbon dates of 28,000 ± 1700 and 37,960 ± 1700/1408 years BP from related sites indicate that the Bride moraine would be expected to have been formed between these dates. The dates are discussed in terms of other evidence.

5.241   1966
RADIOCARBON DATES FROM ALTONIAN AND TWOCREEKIAN DEPOSITS AT SYCAMORE, ILLINOIS
Kempton, John P.
Illinois Academy of Science. Transactions, v. 59, n. 1, 1966: 39-42
Two radiocarbon dates have been determined for wood collected from an exposure in a gravel pit in the valley of East Branch Kishwaukee River at the northern edge of Sycamore III. Wood chips in the lower part of the exposure, part of the youngest drift of the Altonian Substage, were dated as older than 32,000 years BP. A log near the top of the exposure deposited during the Twocreekian Substage was dated at 12,000 ± 400 radiocarbon years BP.

5.242   1965
RADIOCARBON DATES FROM ILIAMNA LAKE, ALASKA
Detterman, Robert; Reed, Bruce L.; Rubin, Meyer
United States Geological Survey. Professional Papers, n. 525-D, 1965: D34-D36
A radiocarbon date of 8520 ± 350 years BP from a lacustrine deposit in a terrace at the west end of Iliamna lake establishes a minimum age for the second major advance of the Brooks Lake Glaciation. The advance must be considerably older than 8520 years, as other terraces are cut into the moraine as much as 49 ft above the dated terrace.

5.243   1965
RADIOCARBON DATES FROM LAKE ST. JOHN AREA, QUEBEC
Lassalle, Pierre
Science, v. 149, n. 3686, Aug. 1965: 860-862
A radiocarbon age of 8680 ± 140 years found for fossil marine shells in the Lake St. John area, Quebec, shows that marine submergence there apparently preceded the Tyrrell Sea episode in Southeastern Hudson Bay, but followed the Champlain Sea episode in the St. Lawrence lowlands.

5.244   1958
RADIOCARBON DATES FROM UPPER-EEM AND WURM-INTERSTADIAL SAMPLES
de Vries, Hessel
Eiszeitalter und Gegenwart, v. 9, 1958; 10-17
Radiocarbon dates have been obtained for Eem interglacial and Würm interglacial sections from Loopstedt (Germany) and Amersfoost (Netherlands) from North Western Europe and Austrian loess. Even the upper part of the Eemian proved to be too old to give a significant activity (age more

Ch. 5 - Glacial Geology

than 53,000 years). The results from the Würm interstadial fit well with Emiliani's paleotemperature curve.

5.245   1956
RADIOCARBON DATES OF MANKATO DRIFT IN MINNESOTA
Wright, Herbert E.; Rubin, Meyer
Science, v. 124, n. 3223, Oct. 1956: 625-626
Radiocarbon dates suggest that the surface drift at Mankato should be correlated with the Cary (Pre Two-Creek) rather than the Valders (post Two-Creek).

5.246   1957
RADIOCARBON DATES OF MANKATO DRIFT IN MINNESOTA
Leighton, Morris M.; Wright, Herbert E.
Science, v. 125, n. 3256, May 1957: 1037-1039
A reappraisal of the Mankato Drift age and of the chronology and the nomenclature of stages and substages of the Wisconsin glacial stage in the Great Lakes region on the basis of radiocarbon dating.

5.247   1967
RADIOCARBON DATES OF WISCONSIN
Black, Robert F.; Rubin, Meyer
Wisconsin Academy of Science, Arts and Letters. Transactions, v. 56, 1967-68: 99-115
Lists the radiocarbon dates older than 5000 years from Wisconsin and discusses the significance of some in terms of the interpretation of the glacial history of the State of Wisconsin.

5.248   1968
SEAWARD TERMINUS OF THE VASHON CONTINENTAL GLACIER IN THE STRAIT OF JUAN DE FUCA
Anderson, Franz E.
Marine Geology, v. 6, n. 6, 1968: 419-438
The stratigraphic units were dated by radiocarbon. The dates are presented in table form. The dates range in age from older than 30,000 to about 11,300 ± 800 years BP and encompass the period during which the Vashon continental ice reached its maximum extent both in the southern Puget Sound and in the Strait of Juan de Fuca.

5.249   1957
RADIOCARBON DATING AND POST-GLACIAL VEGETATIONAL HISTORY: SCALEBY MOSS
Godwin, Harry; Walker, Donald; Willis, E.H.
Royal Society, Proceedings, sect. B, n. 147, 1957: 352-366
Radiocarbon dating is used to make a direct test to decide whether the pollen zone boundaries and therefore the vegetational changes are synchronous or metachronous since absolute ages can be determined from organic material taken precisely at pollen-analytic zone boundaries in sites well separated from one another. The results are satisfactorily self-consistent and in conformity with the stratigraphical sequence. This indicates the suitability of the method for resolving this type of Quaternary historical problem.

5.250   1955
RADIOCARBON DATING OF PRE-MANKATO EVENTS IN EASTERN AND CENTRAL NORTH AMERICA
Flint, Richard Foster; Rubin, Meyer
Science, v. 121, n. 3149, May 1955: 649-658
A brief, critical assessment of the stratigraphic meaning of pre-Mankato samples.

5.251   1955
RADIOCARBON DATING OF THE 'COCHRANE READVANCE' IN CANADA
Karlstrom, Thor N.V.; Rubin, Meyer
Geological Society of America, Bulletin, v. 66, n. 12, Dec. 1955: 1582 (Abstract)
New data, confirmed by radiocarbon dates, give a consistent report of a glaciation and eustatic sea-level low between about 8000 and 6500 BP, which appears to correlate with the Cochrane as a post-Mankato, pre-Climatic optimum event.

5.252   1953
RADIOCARBON DATING OF THE ALLERØD PERIOD
Iversen, Johannes
Science, v. 118, n. 3053, 3 July 1953: 9-1
Radiocarbon dating indicates that the Allerød oscillation appears to be universal, at least in the Northern Hemisphere. The dating of calcareous sediments in lakes seems to be giving reliable results, within statistical error.

5.253   1959
RADIOCARBON DATING OF THE LATE GLACIAL PERIOD IN BRITAIN
Godwin, Harry; Willis, F.R.S.; Willis, E.H.
Royal Society, London, Proceedings, Sect. B, n. 150, 1959: 199-215
Radiocarbon dating by means of a proportional gas-counter containing carbon dioxide has been applied to the dating of deposits from several widely spaced sites in Great Britain. Organic samples were taken from carefully defined horizons. The datings are remarkably self-consistent and indicate that the zones as previously defined are synchronous throughout the country. A brief comparison with published radiocarbon dates from Denmark, Northwest Germany and Holland indicates that the zones are synchronous throughout the four regions. Attention is directed to five substantial

causes of error: (1) contamination by derived coal; (2) use of organic muds in which carbon has originated from photosynthesis of submerged plants in hard waters; (3) failure to observe gaps in the depositional sequence; (4) the so-called Suess effect; (5) the seepage of organic material from above into older layers.

5.254   1967
RADIOCARBON ISOCHRONES OF THE RETREAT OF THE LAURENTIDE ICE SHEET
Bryson, Reid A.; Wendland, Wayne M.
Alberta Anthropologist, v. 2, n. 1, Feb. 1968: 9-15
Using existing radiocarbon dates (published in the period 1959 - 1967) a tentative isochrone map of the position of the Laurentide ice margin at various times is constructed.

5.255   1968
RADIOCARBON ISOCHRONES ON THE DISINTEGRATION OF THE LAURENTIDE ICE SHEET
Bryson, Reid A.; Wendland, Wayne M.; Ives, Jack D.; Andrews, John T.
Arctic and Alpine Research, v. 1, n. 1, 1968: 1-14
Presents two maps on which are plotted the ice sheet perimeter at specific intervals through time utilising the existing radiocarbon and geological data.

5.256   1962
A RADIOCARBON LIMITING DATE FOR SCABLAND FLOODING
Fryxell, Roald
Northwestern Science, v. 36, n. 4, Nov. 1962: 113-119
Radiocarbon dating of organic material collected from scabland flood gravel at Wanapum Dam at 37,700 ± 900 years BP represents the first available radiocarbon dates for glacial chronology in eastern Washington. This figure represents the age of inter-stadial organic material picked up and redeposited by catastrophic discharge into the Columbia River over Frenchman Springs cataract before the furthest advance of the Okanogan lobe and thus provides a maximum limiting date for late Wisconsin scabland flooding.

5.257   1958
RADIOCARBON MEASUREMENTS OF WURM-INTERSTADIAL SAMPLES FROM JUTLAND
Tauber, Henrik; de Vries, Hessel
Eiszeitalter und Gegenwart, v. 9, July 1958: 69-71
The radiocarbon content of Würm-interstadial samples from Brorup, Jutland, has been measured in the dating laboratories in Copenhagen and Groningen. The samples were contaminated with comparatively large amounts of infiltrated younger material, but after an extraction of humic acids none of the samples showed a significant activity. This means that the interstadial at Brorup and the preceding cold period are older than 50,000 years.

5.258   1964
A RADIOCARBON-DATED POLLEN PROFILE FROM SUNBEAM PRAIRIE DOG, DARK COUNTY, OHIO
Kapp, Ronald O.; Gooding, Ansel M.
American Journal of Science, v. 262, n. 2, Feb. 1964. 259-266
The radiocarbon dates enable identification of the Two Creeks interstadial interval in the pollen diagram.

5.259   1955
RATES OF ADVANCE AND RETREAT OF THE MARGIN OF LATE-WISCONSIN ICE SHEET
Flint, Richard Foster
American Journal of Science, v. 253, n. 2, May 1955: 249-255
Rates of advance and retreat of the margin of a glacier can be calculated from the radiocarbon dates of wood debris of forests successively overridden and from other strategically situated samples. Rates calculated from certain Wisconsin glacial fluctuations in central North America are compared with those of existing glaciers and are found acceptable under the uniformitarian principles.

5.260   1965
RE-EVALUATION OF THE LENGTH OF THE PORT TALBOT INTERSTADIAL IN THE LAKE ERIE REGION, CANADA
Dreimanis, Aleksis; Vogel, John C.
In: International Conference, Radiocarbon and Tritium dating, 6th, Washington State University, Pullman, Washington, June 1-11, 1965. Proceedings. United States of America, Atomic Energy Commission, 1965. Conference No. 650652: 720-729
New radiocarbon dates, restudying of the Port Talbot Interstadial type section, and comparison with other Late Pleistocene deposits have led to some new conclusions, mainly that the Port Talbot interstadial was longer than previously assumed and may be correlated with similar long cool interstadials encountered in other areas such as the Mid-Würmian Interpleniglacial in Europe.

5.261   1966
RECENT FLUCTUATIONS OF THE SNOUT OF A GLACIER AT McBETH FIORD, BAFFIN ISLAND, N.W.T.
Harrison, D.A.
Geographical Bulletin, v. 8, n. 1, 1966: 48-58
The advance of the snout to its recent maximum was estimated from radiocarbon dates of vegetation samples

Ch. 5 - Glacial Geology

buried by the lacustrine deposits of the former ice-dammed lake.

5.262   1967
RECENT GLACIAL HISTORY OF AN ALPINE AREA IN THE COLORADO FRONT RANGE, U.S.A. I: ESTABLISHING A LICHEN-GROWTH CURVE
Benedict, James B.
Journal of Glaciology, v. 6, n. 48, Oct. 1967: 817-832
Three old surfaces were radiocarbon dated in order to extend a growth curve backward in time. The dates are discussed for validity.

5.263   1968
RECENT GLACIAL HISTORY OF AN ALPINE AREA IN THE COLORADO FRONT RANGE, U.S.A. II: DATING THE GLACIAL DEPOSITS
Benedict, James B.
Journal of Glaciology, v. 7, n. 47, Feb. 1968: 77-87
Three distinct intervals of glaciation have occurred in the last 4500 years. The earliest is dated at 2500 to 700 B.C., a later one began about A.D. 100 to 1000. The most recent is dated A.D. 1656 to 1850.

5.264   1960
THE RECESSION OF THE LAND-ICE IN SWEDEN DURING THE ALLERØD AND THE YOUNGER DRYAS AGE
Nilsson, Erik
In: Chronology and Climatology of the Quaternary, edited by Johs. Iversen, E. Hyyppä and S. Thorarinsson. International Geological Congress, 21st, Norden, 1960. *Reports,* part 4. Copenhagen, Berlingske Bogtrykkeri, 1960: 98-107
The ages obtained by varves counting is compared to the ages obtained by radiocarbon dating. These correspond quite closely, giving the following ages: limits Older Dryas Age - Allerød Age: 10040 ± 200 B.C. and Allerød Age - Younger Dryas Age: 8920 ± 160 B.C.

5.265   1964
A RECONNAISSANCE GLACIER AND GEOMORPHOLOGICAL SURVEY OF THE DUART LAKE AREA, BRUCE MOUNTAINS, BAFFIN ISLAND
Harrison, D.A.
Geographical Bulletin, v. 22, Nov. 1964: 57-71
Radiocarbon measurements help determine the age of the end moraine. Measurements were made on wind blown organic material in the recrystallised snow that forms part of the ice core of the moraine ridge. Shell dating also gives a minimum time lapse from the retreat of the outlet glacier.

5.266   1964
RESULTS OF A POLLENANALYTIC INVESTIGATION IN THE UNTERSEE NEAR LUNZ IN AUSTRIA
Burger, D.
Geologie en Mijnbouw, v. 43, n. 3, March 1966: 94-102
The vegetational history spans Late-glacial and Holocene. A date related to the period shortly before the beginning of Sub-boreal is obtained by radiocarbon at 190 cm depth, and a date relating to the Allerød oscillation is obtained at 590 cm depth.

5.267   1964
A REVIEW OF THE LATE GLACIATION IN GREAT BRITAIN
Penny, L.F.
Yorkshire Geological Society. Proceedings, v. 34: 387-411
Radiocarbon dating places the maximum extent of the ice during the last glaciation in Britain in the interval between 57,000 and 42,000 years BP, followed by a later readvance sometimes after 28,000 years BP.

5.268   1966
REVISION [OF 1964 PAPER] - SUNBEAM PRAIRIE POLLEN PROFILE
Kapp, Ronald O.; Gooding, Ansel M.
American Journal of Science, v. 264, n. 9, 1966: 743-744
Recent radiocarbon dates establish an age of 11,850 ± 100 BP for the Two Creeks forest intead of 11,400 years as previously reported. Thus, the climatic cooling which marked the end of the Two Creek interstadial occurred prior to 11,850: the Valders ice passed over the Two Creeks site at least 450 years earlier than previously assumed, and a segment of the Sunbeam Prairie bog pollen diagram which had been presumed to date from the Two Creeks interstadial is clearly from the Valders stadial.

5.269   1961
RUSSIAN SETTLEMENT AND LAND RISE IN NORDAUSTLANDET, SPITSBERGEN
Blake, Weston
Arctic, v. 14, n. 2, June 1961: 101-111
The age of a Russian hut, dated by radiocarbon at 260 ± years, confirming an age of approximately 100 years, indicate that land uplift in this area is very slight if occurring at all. The older dates of 6000 to 8000 years BP of high-level whale bone also contradict a theory of rapid uplift.

5.270   1959
SCENES IN OHIO DURING THE LAST ICE AGE
Goldthwait, Richard P.
Ohio Journal of Science, v. 59, n. 4, July 1959: 193-216

In this review article, the author notes that the numerous radiocarbon determinations which have been made are essentially correct, at last to relative order, and are internally consistent although somewhat different from the pre-conceived conventional pattern. A table of Wisconsin datings for central United States is presented.

5.271   1960
SEA-LEVEL FLUCTUATION DURING THE PAST FOUR THOUSAND YEARS
Schofield, J.C.
Nature, v. 185, n. 4716, Mar. 1960: 836
Study of constructional details of a Chenier plain along the coast of the Firth of Thames, New Zealand, together with radiocarbon dating of seven shell samples, has enabled the production of a sea-level curve. The indication is that the sea-level fell 7 ft from 2000 B.C. to about the beginning of the Christian Era and has remained relatively stable ever since.

5.272   1962
SEA-LEVELS OF THE LAST GLACIATION
Donovan, D.T.
Geological Society of America, Bulletin, v. 73, n. 10, Oct. 1962: 1297-1298
Erosion features and terrace deposits of the River Severn and Bristol Channel, England, taken together with local and other radiocarbon dates, suggests that from about 24,000 to 45,000 years ago the sea was not more than a few meters above or below its present level.

5.273   1965
SECOND AND THIRD PEARY LAND EXPEDITION, 1963 AND 1964
Knuth, Eigil
Polar Record, v. 12, n. 81, Sept. 1965: 733-738
The dominating clay-silt formation in Jørgen Brønlund Fjord are thought to have been deposited before the glacier advance and the mussels left in situ from the time of ocean cover after the glacier had retreated. Radiocarbon dating on a sample of *Mya truncata* from about 33 m above sea level gave a figure of 7290 ± 130 years BP

5.274   1961
SIGNIFICANCE OF CARBON-14 DATES FOR RANCHO LA BREA
Howard, Hildegard
Science, v. 131, n. 3402, 11 Mar. 1961: 712-714
Radiocarbon dating revealed that some 'late Pleistocene' glacial deposits are only 11,000 to 12,000 years old. Cave deposits with remain similar to those found at Rancho La Brea have been given an age of 10,000 to 12,000 years. This article provides details connected with the excavation at Rancho La Brea in order to understand the radiocarbon dates obtained from samples from three different pits.

5.275   n.d.
SIMULTANEITY OF GLACIAL AND PLUVIAL EPISODES FROM C-14 CHRONOLOGY OF THE WISCONSIN GLACIATION (Abstract)
Rubin, Meyer
United States Geological Survey, n. d.
Studies of freshwater episodes and high water stands of interior drainage lakes show that the pluvial times occur simultaneously with the maxima of glacial advances in the continental glaciers.

5.276   1963
SOME COMMENTS ON THE 'ICE-FREE REFUGIA' OF NORTHWESTERN SCANDINAVIA
Hoppe, Gunnard
In: North Atlantic Biota and their History, edited by Askell and Doris Löve. Oxford, Pergamon Press, 1963: 321-335
Use of radiocarbon dating is made to date the Würm Glaciation

5.277   1965
THE SPECTRUM OF POST-GLACIAL TIME
Schove, D.J.
In: International Association for Quaternary Research (INQUA). Congress, 7th, Boulder, Col., 14 Aug - 19 Sept. 1965. *Abstracts of General Sessions*. Boulder, Colorado, 1965: 413
The important climatic substages of post-glacial time can now be dated by clusters of radiocarbon dates, and the same substages can be identified in the standard pollen-sequences of northern Europe.

5.278   1961
STRATIGRAPHY AND AGE OF DEPOSITS IN AN EXCAVATION AT THE WRIGHT-PATTERSON AIR BASE, DAYTON, OHIO
Forsyth, Jane L.
Ohio Journal of Science, v. 61, n. 5, Sept. 1961: 315-317
A sequence of Pleistocene deposits was revealed by excavation, wood from the till was collected, identified and radiocarbon dated. It is concluded that the deposition of till in this exposure began with the last advance of the Wisconsin glacier about 21,000 years ago, with retreat starting about 17,000 years ago.

5.279   1965
STRATIGRAPHY AND CHRONOLOGY OF LATE INTERGLACIAL AND EARLY VASHON GLACIAL TIME IN THE SEATTLE AREA, WASHINGTON

Mullineaux, Donald H.; Waldron, Howard H.; Rubin, Meyer
United States Geological Survey. Bulletin, n. 1194-0, 1965
A conformable late Pleistocene sequence of interglacial and glacial sediments in the Seattle area can be divided into three units. Radiocarbon dates from the lower unit (non-glacial) indicate that the Puget glacial lake of Vashon age advanced into the area at some time after 15,000 years ago - several thousand years later than was previously believed.

5.280 1957
STRATIGRAPHY OF THE WISCONSIN GLACIAL STAGE ALONG THE NORTH-WESTERN SHORE OF LAKE ERIE
Dreimanis, Aleksis
Science, v. 126, n. 3265, 26 July 1957: 166-168
A proposal Wisconsin stratigraphy for the area north of Lake Erie is based on radiocarbon dates. These suggest that the main Wisconsin glaciation was of relatively short duration in the central and western portion of Lake Erie.

5.281 1958
STRATIGRAPHY OF WISCONSIN GLACIAL DEPOSITS OF TORONTO AREA, ONTARIO
Dreimanis, Aleksis; Terasmae, Jaan
Geological Association of Canada. Proceedings, v. 10, 1958: 119-135
Mentions one radiocarbon date from the northwest portion of Lake Erie basin of >12,600 ± 440 years BP, dating the retreat of the Ontario Erie lobe. Concludes that a long cool interval, equivalent to Port Talbot interval in the Lake Erie Region, is older than 34,000 radiocarbon years.

5.282 1967
STRATIGRAPHY, CLIMATE SUCCESSION AND RADIOCARBON OF THE LAST GLACIAL IN THE NETHERLANDS
Hammen, Th. van der; Maarleveld, Gerardus Cornelis; Vogel, John C.; Zagwijn, W.H.
Geologie en Mijnbouw, v. 46, n. 3, March 1967: 79-96
Field data, pollen analysis and some 30 radiocarbon dates have been the basis for the construction of a climatic curve for the Last Glacial in the Netherlands. The sequence in Holland is compared with the loess area of Austria. There is a close resemblance of the climatic curve with palaeotemperature curves obtained from deep-sea sediments.

5.283 1966
THE STRUCTURE OF SOME PINGOS IN THE MACKENZIE DELTA AREA, N.W.T.
Mackay, J. Rodd; Stager, John K.
Geographical Bulletin, v. 8, n. 4, 1966: 360-368
Observations and radiocarbon dates on pingos in Pleistocene sediments show that lake shoaling occurred over a period exceeding 6000 years. Pingo growth probably started less than 3000 years ago.

5.284 1967
STUDIES OF FULL-GLACIAL VEGETATION AND CLIMATE IN SOUTHEASTERN UNITED STATES
Whitehead, Donald R.
In: *Quaternary Palaeoecology*, edited by M.J. Cushing and H.E. Wright. International Association for Quaternary Research (INQUA). Congress, 7th, Boulder, Col., 14 Aug - 19 Sept. 1965. *Proceedings*, v. 7. New Haven, Conn., Yale University Press, 1967: 237-248
Data from North Carolina and Virginia suggest a vegetational change. New radiocarbon dates show the change to be a full-glacial time when rates of pollen accumulation were low. The water-level in the Bays also changed, as postglacial and pre-classical Wisconsin organic layers are below the present water level. The lakes were high during full-glacial, and the water was clear.

5.285 1964
SURFICIAL GEOLOGY OF BOOTHIA PENINSULA AND KING WILLIAM, SOMERSET AND PRINCE OF WALES ISLANDS, DISTRICT OF FRANKLIN, CANADA
Craig, Bruce Gordon
Canada, Geological Survey Paper, n. 63-44, 1964. pp. 10
Ages determined on marine shells from four localities to determine the rate and time of deglaciation of the areas.

5.286 1965
SURFICIAL GEOLOGY OF PART OF THE COCHRANE DISTRICT, ONTARIO, CANADA
Hughes, O.L.
In: *International Studies on the Quaternary*, edited by H.E. Wright and D.J. Frey. *Geological Society of America Special Papers*, n. 84 (INQUA, U.S.A. 1965): 535-565
At the end of this article, the author discusses varve correlation and radiocarbon chronology and indicates the complexity of the events involved in the Wisconsin stage. He concludes that the Cochrane readvance is of local rather than regional significance, hence does not warrant substage rank within the Wisconsin stage.

5.287 1963
SURFICIAL GEOLOGY OF VICTORIA AND STEFANSSON ISLANDS, DISTRICT OF FRANKLIN
Fyles, J.G.
Canada, Geological Survey Bulletin, n. 101, 1963. pp. 38
The dating of typical tundra vegetational remains in gravels, sands and silts beneath morainal deposits gave an age of

28,000 years BP. Radiocarbon dates suggest that the initial maximum submergence decreased in age from Northwest to Southwest, and that most of the subsequent uplift took place early in post-glacial times.

5.288   1966
THE SUSACA - INTERSTADIAL AND THE SUBDIVISION OF THE LATE-GLACIAL
Hammen, Th. van der; Vogel, John C.
Geologie en Mijnbouw, v. 45, n. 2, Feb. 1966: 33-35
During a recent pollen analytical study of a Holocene section from the Eastern Cordilliera of Columbia, an interstadial older than both the Allerød and Bølling is recognised, which lasted from about 13,700 to 13,100 years BP. A new stratigraphic table of the late-glacial, with radiocarbon dates, is presented.

5.289   1966
THE SUSACA - INTERSTADIAL AND THE SUBDIVISION OF THE LATE-GLACIAL. DISCUSSION
Dreimanis, Aleksis
Geologie en Mijnbouw, v. 45, n. 12, Dec. 1966: 445-448
Late-glacial interstadials, contemporaries with the recently introduced Susacá interstadial have been recognised for some time in the Great Lake region of North America and in Europe. It is suggested that the name Raunis Interstadial be used for the regions affected by the North European glaciation. A table of selected radiocarbon dates of the Cary-Port Huron or Cary-Mankato interstadial or its termination in central and eastern North America is presented.

5.290   1966
THE SUSACA INTERGLACIAL AND THE SUBDIVISION OF THE LATE-GLACIAL - DISCUSSION
Dreimanis, Aleksis
Geologie en Mijnbouw, v. 46, n. 12, Dec. 1967: 445-448
Geolgical evidence and radiocarbon dates from North America indicate that the Cari - Mankato Interstadial or the Cary - Port Huron, or the Bowman Interglacial, or the lake Arkona phase are correlative with the Susacá Interglacial. In Northern Europe it correlates with the Raunis Interstadial in the Eastern Baltic region and the Plyusna Interstadial in northern Russia.

5.291   1958
TAIGA-TUNDRA AND THE FULL GLACIAL PERIOD IN CHESTER COUNTY, PENNSYLVANIA
Martin, Paul S.
American Journal of Science, v. 256, n. 7, Summer 1958: 470-502
Two shallow cores from a marsh in the unglaciated piedmont of southern Pennsylvania reveal a late Pleistocene pollen sequence. Samples for radiocarbon analysis collected at 100 cm in gritty clay and 150 cm in coarse sand and clay give an age of 13,500 years, somewhat younger than the accepted age of the last Wisconsin maximum. Surface contamination may account for the discrepancy.

5.292   1957
TENTATIVE CORRELATION OF ALASKAN GLACIAL SEQUENCES
Karlstrom, Thor N.V.
Science, v. 125, n. 3237, 11 Jan. 1957: 73-74
This revision of the 1953 tentative correlation was made on the basis of new and revised radiocarbon age determinations. The new glacial chronology strongly implies astronomic climatic control with glacial events of stage rank being determined by the obliquity cycle with a periodicity of about 42,000 years.

5.293   1956
A THEORY OF THE ICE AGES
Ewing, Maurice
Science, v. 123, n. 3207, 15 June 1956: 1061-1066
Preliminary report of new ideas related to the origin of glacial climates. Radiocarbon measurements of deep-sea cores established abrupt changes in Atlantic deep-sea sediments due to temperature changes. A decrease of about 1°C per 1,000 years changed abruptly some 11,000 years ago to an increase of 10°C per 1,000 years. Then stable temperature at maximum value was reached during the interglacials of the Pleistocene.

5.294   1968
A THIN TILL IN WEST-CENTRAL SASKATCHEWAN, CANADA
Christiansen, E.A.
Canadian Journal of Earth Sciences, v. 5, n. 2, April 1968: 329-336
A thin till, 1 to 10 feet thick, named the Battleford formation, was deposited about 20,000 years ago and is separated from the underlying older till by a disconformity whose hiatus is about 15,000 years. These dates suggest that the weathering interval prior to the deposition of the Battleford formation is Farmdallian and that the Battleford formation is Woodfordian. A table of radiocarbon dates which establish this chronology is presented.

5.295   1953
TWO LATE WISCONSIN INTERSTADIAL DEPOSITS FROM ONTARIO, CANADA
Dreimanis, Aleksis
Geological Society of America, Bulletin, n. 64, v. 12, Dec. 1953: 1414

Abstract of paper presented at the November 1953 meeting of the Geological Society of America in Toronto. The radiocarbon date of 10,900 ± 400 years, if correct, suggests that the retreat of Wisconsin ice from Ontario was later than assumed.

5.296   1956
TWO LATE-GLACIAL DEPOSITS IN SOUTHERN CONNECTICUT
Leopold, Estella B.
National Academy of Science. Proceedings, v. 42, n. 11, 15 Nov. 1956: 863-867
The pollen sequence obtained from these two late-glacial deposits, one from a kettle, the Totoket Bog, the other from Durham Meadow, a formerly lacustrine basin, indicate a far more complex chronology for the early part of the New England post-glacial chronology. Radiocarbon dates (13,500 and 12,500 years) indicate that an early oscillation terminated significantly before the Two Creek interval.

5.297   1963
TWO POLLEN DIAGRAMS FROM SOUTH-EASTERN MINNESOTA: PROBLEMS IN THE REGIONAL LATE-GLACIAL AND POST-GLACIAL VEGETATIONAL HISTORY
Wright, H.V.; Winter, Thomas C.; Patten, Harvey L.
Geological Society of America. Bulletin, v. 74, 1963: 1371-1396
The radiocarbon dates and stratigraphy of the pollen records indicate a pre- Two Creek and post-Cary correlation

5.298   1968
VARIATIONS OF SOME PATAGONIAN GLACIERS SINCE THE LATE-GLACIAL
Mercer, J.H.
American Journal of Science, v. 266, n. 2, 1968: 91-109
After a late-glacial readvance, the glaciers in Argentine Patagonia receded quickly, and by 10,000 years BP at least one glacier was little larger than it is today. Three episodes of Post-glacial readvance followed, the first culminating about 4600 years BP during early Sub-boreal time, the second about 2000 years BP, during the Sub-Atlantic, and the third one during recent centuries. A table of radiocarbon dates supporting this chronology is presented.

5.299   1955
VARVES AND RADIOCARBON CHRONOLOGY APPRAISED BY POLLEN DATA
Antevs, Ernst
Journal of Geology, v. 63, n. 5, Sept. 1955: 495-499
Natural correlation with climatic age of Canadian postglacial forest types, deduced by Potzger and Courtemanche, confirms the well suggested view that the ice border oscillations at Cochrane antedated the altithermal which culminated ca 6000 years ago. The postglacial crustal rise of the James Bay County required 8000 to 10,000 years, according to Gutenberg. The Cochrane must be then correlative of the European Salpausselka stage and be some 10,000-11,000 years old. Since the ice retreat from Milwaukee to Cochrane comprised at least 7000 years, the radiocarbon age of the Two Creek forest bed at 11,400 must be much too low.

5.300   1962
VASHON GLACIATION AND LATE WISCONSIN RELATIVE SEA-LEVEL CHANGES IN THE NORTHERN PART OF THE PUGET LOWLAND, WASHINGTON
Easterbrook, Don J.
Geological Society of America, Special Paper, n. 73, Abstracts for 1962: 37
A radiocarbon date of 26,500 ± 800 years BP from wood, in Vashon, probably indicates the time of advance of the Vashon glacier. Radiocarbon dates on shells found 220 feet above sea-level mark the beginning of several hundred feet of fluvial and lacustrine sediments. The emergence is tentatively regarded as isostatic rebound. The readvance of 11,500 years BP is considered post-Vashon, possible correlating with the Valders readvance in the Great Lakes region.

5.301   1961
VEGETATION AND ITS ENVIRONMENT IN DENMARK IN THE EARLY WEICHSELIAN GLACIAL (LAST GLACIAL)
Andersen, S.T.
Denmarks Geologiske Undersøgelse, Series 2, n. 75, 1961. pp. 175
The work deals with the layers formed above Eemian Interglacial freshwater deposits in Denmark. The author supports the opinion that climatic changes may serve as chronological limits. Radiocarbon dates used to date the Amersfoot Interstadial at 64,000 years BP and the Brorup Interstadial at 59,000 years BP and other dates are discused.

5.302   1965
VOLCANIC-ASH CHRONOLOGY
Wilcox, Ray E.
In: The Quaternary of the United States: a Review Volume for the 7th Congress of the International Association of Quaternary Research, ed. by H.E. Wright and David G. Frey. Princeton, N.J., Princeton University Press, 1965: 807-816
It is apparent that many ash beds in the Quaternary deposits of the United States would be useful stratigraphic markers if properly investigated. A table of some Quaternary ash-fall deposits in the United States with radiocarbon dates is presented.

5.303   1958
WISCONSIN AGE FORESTS IN WESTERN OHIO, I. AGE AND GLACIAL EVENTS
Goldthwait, Richard P.
Ohio Journal of Science, v. 58, n. 4, May 1958: 209-219
Twenty nine dates have been determined by radiocarbon analysis for buried wood in western Ohio and immediately adjacent areas. This article is an attempt to fit known stratigraphy to dates. Since the dates are internally consistent their accuracy is not challenged. Interpretation of the stratigraphy of each collecting site and a careful study of the nature and condition of the former vegetation are presented.

5.304   1958
WISCONSIN AGE FORESTS IN WESTERN OHIO, II. VEGETATION AND BURIAL CONDITIONS
Burns, George W.
Ohio Journal of Science, v. 58, n. 4, May 1958: 220-230
Wisconsin forest remains of three distinct periods, ranging from 8500 to more than 40,000 years, are investigated from various localities in Ohio and Indiana in order to determine the nature and fate of the forest cover, and to relate the rate of periglacial tree growth to probable speed of ice advance. Ages were determined by radiocarbon dating.

5.305   1968
WISCONSIN STRATIGRAPHY AND ICE-MOVEMENT DIRECTIONS IN SOUTHEASTERN QUEBEC, CANADA (Abstract)
McDonald, Barrie Clifton
Geological Society of America. Special Papers, n. 115, 1968: 277-278
Three Wisconsin glacial phases, each antedated by a nonglacial interval are represented in the Appalachian of southeastern Quebec. The second non-glacial interval, radiocarbon dated at 54,000 and 41,500 years BP, is correlated tentatively with St. Pierre interval.

5.306   1955
WISCONSIN STRATIGRAPHY AND RADIOCARBON DATES OF SEARLES LAKE, CALIFORNIA
Flint, Richard Foster; Gale, W.A.
Geological Society of America, Bulletin, v. 66, n. 12, Dec. 1955: 1559-1560
Abstract of paper presented at the November 1955 meeting in New Orleans. Subsurface deposits of ancient Searles Lake, California, include conspicuous layers of evaporites and bodies of laminated clay and silt containing organic matter. The general character of the sequence and radiocarbon dates from critical horizons support the general correlation of glacial times with pluvial times.

5.307   1958
WISCONSIN STRATIGRAPHY OF PORT TALBOT ON THE NORTH SHORE OF LAKE ERIE, ONTARIO
Dreimanis, Aleksis
Ohio Journal of Science, v. 58, n. 2, March 1958: 65-84
Proposes a stratigraphic correlation and sequence of events. Radiocarbon dates associated with the area are presented in table form and discussed.

5.308   1965
WISCONSINAN AGE OF THE TITUSVILLE TILL (FORMERLY CALLED INNER ILLINOIAN), NORTHWESTERN PENNSYLVANIA
White, George W.; Totten, Stanley M.
Science, v. 148, n. 3667, 7 Apr. 1965: 234-235
Peat discovered below drift near Titusville, Pennsylvania, formerly called Inner Illinoian has a radiocarbon age of $31,400 \pm 2100$ years. The overlying drift, herein named Titusville Till is therefore not Illinoian but is late early Wisconsinan or late Altonian of the lake Michigan lobe classification.

5.309   1966
WISCONSINAN CHRONOLOGY AND HISTORY OF VEGETATION IN THE OGILVIE MOUNTAINS, YUKON TERRITORY, CANADA
Terasmae, Jaan; Hughes, O.L.
Palaeobotanist, v. 15, n. 1-2, 1966: 235-245
Palynology and radiocarbon dating have been used to interpret the chronology of the complex moraine sequence representing three major glacial episodes in the western Ogilvie Mountains, characterised by successive advances and retreats of valley glaciers; the youngest episode probably culminated 10,000 to 12,900 years ago. Glacial and vegetational successions in the Ogilvie Mountains can be correlated with those of north central Brooks Range, Alaska.

CHAPTER SIX
^{14}C OCEAN STUDIES

6.001   1967
^{14}C, ^{137}Cs AND ^{90}Sr PROFILES IN ANTARCTIC WATERS
Mathieu, Guy G.
In: Mixing, Diffusion and Circulation Rates in Ocean Waters. Palisades, New York, Columbia University, 1967 (Lamont Geological Observatory. Annual Progress Report, n. 8). pp. 25
Measurements of ^{14}C, ^{137}Cs and ^{90}Sr were compared with standard hydrographic information in 3 localities in the southern ocean. The results are in agreement with previous interpretation.

6.002   1958
AGE DETERMINATION OF SOUTHERN OCEAN WATERS
Brodie, J.W.; Burling, R.W.
Nature, v. 181, n. 4602, 11 Jan. 1958: 107-108
The dating of bottom water from two stations in the southern ocean suggests: (1) the value of the latitudinal component of movement of bottom water is low, (2) the total turnover time of ocean circulation is of the order of thousand of years, (3) the present oceanic circulation may be found to be complicated by factors inherited from late Pleistocene hydrological environment.

6.003   1958
THE ANTARCTIC CONVERGENCE SOUTH OF NEW ZEALAND
Garner, D.M.
New Zealand Journal of Geology and Geophysics, v. 1, n. 3, Aug. 1958: 577-596
A detailed profile of temperature and salinity across the upper layers of the Antarctic convergence is discussed in terms of water movement and mixing in this major oceanic water mass boundary. The introduction of Pacific deep water into near surface water introduces water of low radiocarbon activity from the very 'old' deep water into the surface water resulting in a very significant radiocarbon depletion of Antarctic surface water.

6.004   1967
ATMOSPHERIC - OCEANIC EXCHANGE OF CO_2 AND THE EFFECT OF INDUSTRIAL CO_2 ON OCEANIC $^{14}C/^{12}C$ RATIO
O'Brien, B.J.
New Zealand Institute of Nuclear Sciences. Report, R-47, 1967
[Not sighted]

6.005   1963
C14/C12 RATIOS IN SURFACE OCEAN WATER
Broecker, Wallace S.
In: Nuclear Geophysics. Washington DC, National Academy of Science. National Research Council, 1963: 138-148
The available data on the ratio of ^{14}C to ^{12}C in the organic carbon of the surface ocean water as a function of geographic location and time are outlined. Various implications, in particular in connection with radiocarbon dating of fossil marine organisms, are discussed.

6.006   1967
CARBON-14 DETERMINATIONS OF SEA LEVEL CHANGES IN STABLE AREAS
Shepard, Francis; Curray, Joseph R.
In: Process in Oceanography, v. 4, *The Quaternary History of the Ocean Basins.* London, Pergamon Press, 1967: 283-291
Discusses various evidence and sea-level change curves based on radiocarbon measurements. Confirms a slightly higher stand of the sea about A.D. 500.

6.007   1952
CARBON-14 MEASUREMENTS ON GEOLOGICAL SAMPLES
Kulp, J. Lawrence; Tryon, Lansing E.
Geological Society of America, Bulletin, n. 12, Dec. 1952: 1273
Abstract of paper presented at the November 1952 meeting of the Geological Society of America in Boston. Various radiocarbon measurements made during 1951 are reported. Conclusions are presented. A turn-over time of 1000 years for ocean water and the occurrence of variations in the radiocarbon concentration of the surface water are observed. This is important in the evaluation of ages derived from shell material.

6.008   1964
CARBONATE DEPOSITS AND PALEOCLIMATIC IMPLICATIONS IN THE NORTHEAST PACIFIC OCEAN
Nayudu, Y. R.
Science, v. 146, n. 3643, 23 Oct. 1964: 515-517
A narrow carbonate band consisting predominantly of

Globigerina-rich sediments is present in deep sea deposits of the Northeast Pacific Ocean, extending almost parallel to the coast of Oregon and Washington. Five radiocarbon dates in the cores from this area suggest that the greatest concentration of Globigerina-rich sediments occurred 27,000 to 12,000 years ago. This time interval corresponds roughly to the Vashon (late Wisconsin) glacial time in the Puget Lowlands. The results suggest higher carbonate sedimentation in the northeast Pacific during glacial stages.

6.009   1959
CHARACTERISTICS OF THE LATE QUATERNARY ARCTIC OCEAN
Donne, William L.; Ewing, Maurice; Menzies, R.J.
In: International Oceanographic Congress, 1959. Preprints, 1959: 19-20
Study based on water and bottom samples obtained during the International Geophysical Year (IGY) from drifting Station A. The nature, composition and radiocarbon age, etc. of the bottom material are summarised.

6.010   1965
THE COMPARISON OF RADIOCARBON AGES OF CARBONATE WITH URANIUM SERIES AGES
Thurber, David L.; Broecker, Wallace S.; Kaufman, Aaron
In International Conference on Radiocarbon and Tritium Dating, Proceedings. Pullman, Washington, Washington State University, June 1-11, 1965. United States of America, Atomic Energy Commission, 1965. Conference n. 650652: 267-382
The ^{230}Th method of dating corals is now highly developed and is capable of yielding reliable precise ages between 5000 and 250,000 years in age. Comparison of ^{230}Th/^{234}U ages with radiocarbon dates in the 5000 - 20,000 years range may allow studies of secular variations of ^{14}C/^{12}C ratio in surface ocean water.

6.011   1955-56
CORRELATION OF SIX CORES FROM THE EQUATORIAL ATLANTIC AND THE CARRIBEAN
Ericson, David B.; Goesta, Wollin
Deep-Sea Research, v. 3, 1955-56: 104-125
Curves of late Pleistocene climatic variations based on vertical distribution of planktonic Foraminifera in six cores from the Equatorial Atlantic and the Caribbean have been satisfactorily correlated. Radiocarbon determinations show that the rate of sediment accumulation took place during the latter part of Wisconsin time.

6.012   1958
A CRITICAL EVALUATION OF RADIOCARBON TECHNIQUES FOR DETERMINING MIXING RATES IN THE OCEAN AND THE ATMOSPHERE
Craig, Harmon
In: International Conference on the Peaceful Use of Atomic Energy, 2nd, Geneva, 1958. *Proceedings* (A/Conf. 15/P/1123), 1958, v. 2
From the observed distribution of carbon isotopes in the atmosphere, the biosphere and the sea, it is found that the exchange time of atmospheric carbon dioxide with the sea is about 7 years. Two models are presented of the problems of oceanic mixing which is more complicated that previously thought.

6.013   1964
THE DIATOMACEOUS SEDIMENTS OF THE GULF OF CALIFORNIA
Calvert, S.E.
Ph. D. Thesis. University of California, San Diego, 1964. pp. 265 (Abstract in: *Dissertation Abstracts*, Ann Arbor, Mich., v. 25, n. 5, 1964/5: 2927. Order n. 64-10129)
The study deals with an investigation of the distribution of the major components in the deep water sediments of the Gulf of California, the nature and origin of the laminated diatomaceous sediments and the immediate source of the silica present as biogenous opal in the sediment. Sedimentation rates determined on laminated cores by radiocarbon show that two laminae, one, opal-rich, and the other, clay-rich, are deposited in one year.

6.014   1963
EFFECT OF WIND AND CO_2 EXCHANGE ACROSS THE SEA SURFACE
Kanwisher, John
Journal of Geophysical Research, v. 68, n. 13, 1 July 1963: 3921-3927
Large quantities of carbon dioxide pass through the sea-air reservoir in short geological times. This article examines the mechanism of the movement of carbon dioxide within this reservoir.

6.015   1955
EVIDENCE OF CLIMATIC CHANGE FROM ICE ISLAND STUDIES
Crary, A. P.; Kulp, J. Lawrence; Marshall, E. W.
Science, v. 122, n. 3181, 16 Dec. 1955: 1171-1173
The Radiocarbon Dating method was used to construct a history of island build-up from the individual dirt weights obtained from the layers of a 90 feet ice core. Carbon material is found in the form of microscopic spores, fragments of woody tissues and pollen grains. The interpreta-

tion is difficult because of contamination at time of deposition, successive melts and refreezing. The evidence would indicate that we are approaching an open polar sea period.

6.016 **1968**
INTERCHANGE OF WATER BETWEEN THE MAJOR OCEANS
Broecker, Wallace S.; Yuan-Hui, Li
In: Mixing, Diffusion, and Circulation Rates in Ocean Waters. Annual Progress Report N. 9, 1967-1968. Palisades, New York, Columbia University: Lamont Geological Observatory, 1968. pp. 21
The major oceanic water types which were analysed to establish the rates at which mixing takes place are: the North-Atlantic deep water, the Pacific and Indian deep water and the warm surface water. Concentrations of radiocarbon and total dissolved carbon dioxide data and assumptions made for a mathematical model are discussed.

6.017 **1958**
THE LAMONT SEA WATER RADIOCARBON PROGRAM
Ewing, Maurice; Gerard, R.; Broecker, Wallace S.
Columbia University, New York. Lamont Geological Observatory, August 1958. pp. 24 (CU - 2 57 AT (30-1) - 1808 - Geol)
Radiocarbon measurements have been made on 40 sea water samples collected during the VEMA - 12 cruise of December 1956 - July 1957.

6.018 **1957**
THE NATURAL DISTRIBUTION OF RADIOCARBON AND THE EXCHANGE TIME OF CARBON DIOXIDE BETWEEN THE ATMOSPHERE AND THE SEA
Craig, Harmon
Tellus, v. 9, n. 1, Feb. 1957: 1-17
From the observed distribution of ^{12}C, ^{13}C, and ^{14}C in the atmosphere, the biosphere and the sea, and from the estimated production rate of ^{14}C by cosmic rays, the residence time of a carbon dioxide molecule in the atmosphere, before entering the sea is found to be between four and ten years. The average residence time of carbon dioxide in the deep sea is estimated as probably not more than about 500 years.

6.019 **1968**
NATURAL RADIOCARBON AND FISSION PRODUCTS IN ARCTIC OCEAN WATERS
Thurber, David L.; Mathieu, Guy G.; Hunkins, Kenneth
In: Mixing, Diffusion, and Circulation Rates in Ocean Waters. Annual Progress Report N. 9, 1967-1968. Palisades, New York, Columbia University: Lamont Geological Observatory, 1968. pp. 21
There are three types of Arctic Ocean waters: Arctic surface water, Atlantic water, and Arctic deep water. Radiocarbon measurements made in all areas indicate Arctic Ocean water origin. A comparison of data at time intervals before and after nuclear explosions indicate water movements. The ^{137}Cs concentrations are in concordance with predictions from radiocarbon dates.

6.020 **1960**
NATURAL RADIOCARBON IN THE ATLANTIC OCEAN
Broecker, Wallace S.; Gerard, R.; Ewing, Maurice; Heezen, Bruce C.
Journal of Geophysical Research, v. 65, n. 9, Sept. 1960: 2903-2931
In order to help understand the patterns and rates of mixing in the oceans, radiocarbon measurements have been used. The object of this research is to determine the geographic and depth variation in the Atlantic Ocean of the ratio of ^{14}C to ^{12}C. If it is assumed that factors other than the radioactive decay producing these variations could be recognised and accounted for, then the residual differences should be related to mixing rates.

6.021 **1963**
NATURAL RADIOCARBON IN THE PACIFIC AND INDIAN OCEANS
Bien, George S.; Rakestraw, N. W.; Suess, Hans E.
In: Nuclear Geophysics. Proceedings of a Conference held at Wood Holes, Mas., 7-9 June, 1962. N*uclear Sciences Series Report* No. 38, Publication 1075. National Academy of Sciences - National Research Council, Washington, D.C., 1963: 1152-160
The study of radiocarbon distribution in the oceans may be important in the solution of several problems in oceanography: (1) the rate of uptake of carbon dioxide from the atmosphere by the ocean, (2) the rate and extent of mixing from the mixed layer into the deep layer, (3) the speed of deep water movement.

6.022 **1962**
NATURAL RADIOCARBON IN THE SUB-SURFACE WATERS OF THE CARIBBEAN
Gerard, Robert D.; Broecker, Wallace S.
American Geophysical Union. Annual Meeting, 43rd, May 1962. pp. 17 [TID-15914]
The natural radiocarbon concentration of 30 subsurface water samples from the Caribbean was measured, and the results compared among different basins and with previously reported Lamont samples from the nearby Atlantic. The results indicate slower circulation and longer residence time for the deeper water of the Venezuela and Columbian basins than the Cayman and

Yucatan basins to the north.

6.023   1956
ON ASSESSING THE AGE OF DEEP OCEANIC WATER BY CARBON-14
Cooper, L.H.N.
Marine Biology Association of the United Kingdom. Journal, v. 35, 1956: 341-354
The radiocarbon method for determining the age of deep oceanic water gives ages much higher than are suggested by physical and chemical oceanographic observations. The reasons for this are explained and the significance of existing measurements of radiocarbon in deep oceanic waters is discussed.

6.024   1956
ON PALEOTEMPERATURES OF PACIFIC OCEAN BOTTOM
Emiliani, Cesare L.
Science, v. 123, n. 3194, 16 Mar. 1956: 460-461
Answer to another article which presents more detailed analysis of core needed to determinate the bottom temperature of the ocean waters at certain localities.

6.025   1966
ON THE RADIOCARBON AGE OF THE OCEAN
Miyaka, Y.; Saruhashi, K.
Papers in Meteorology and Geophysics (Tokyo), v. 17, Dec. 1966: 218-223
The radioactive decay rate of radiocarbon in the ocean is much smaller than the rate of biological uptake and regeneration of inorganic radiocarbon from organic matter. Therefore, the radiocarbon age of sea water is controlled not only by radioactive decay but also by these other factors. This is the reason for the higher age of the surface water in Polar areas. Owing to the smaller content of inorganic carbon in the surface layer, compared to the deep layer, the specific activity of radiocarbon in the deep layers tends to decrease through mixing with surface waters, which gives a higher radiocarbon age to the deep waters.

6.026   1967
ON THE TRANSFER OF RADIOCARBON IN NATURE
Nydal, Reidar
In: Radioactive Dating and Methods of Low-level Counting. Proceedings of a Symposium organised by the International Atomic Energy Agency (IAEA) in co-operation with the Joint Commission on Applied Radioactivity (ICSU and held in Monaco, 2-10 March 1967. Vienna, IAEA, 1967 (Proceedings Series): 119-128
In radiocarbon dating work, knowledge of the exchange rate of radiocarbon between various reservoirs in natures is of great interest. Artificial radiocarbon from nuclear tests performed in 1961 and 1962 has been used as a tracer to investigate the rates of mixing between the various parts of the atmosphere before the equilibrium was reached, in about four years. The amount of radiocarbon in the surface layer of the Atlantic Ocean has also been examined to make accurate measurements of the exchange rate between the atmosphere and the ocean.

6.027   1953-54
A PRELIMINARY NOTE ON THE TIME SCALE ON NORTH ATLANTIC CIRCULATION
Worthington, L.V.
Deep-Sea Research, v. 1, 1953-54: 244-251
If the layer in contact with the ocean bottom can be put aside, two types of turnover are found to be taking place in the North Atlantic. In the first, a latitudinal change, as oceanic response to climatic changes where no violence is done to the water's identity. In the other one, a vast amount of water of a single type is formed in successive years of catastrophic cold at high latitude. This does not agree with a table of Ocean's water age measurement made by Kulp in 1953.

6.028   1967
QUATERNARY SEDIMENTATION IN THE ARCTIC OCEAN
Hunkins, Kenneth; Kutschale, Henry
In: Process in Oceanography, v. 4, *The Quaternary History of the Ocean Basins*. London, Pergamon Press, 1967: 89-94
The 10 cm boundary is dated by Uranium series method at 79,000 years BP. This indicates the most recent change in pelagic deposition. Foraminifera from zones between 7 and 10 cm have been dated by radiocarbon at between 25,000 and 30,000 years BP and the 3 mm layer at 700 ± 100 years BP. This indicates that the pelagic sedimentation has continued unchanged from about 70,000 years ago to the present and implies that the ice cover has existed for that length of time.

6.029   1967
RADIOCARBON AGE OF ANTARCTIC MATERIAL
Broecker, Wallace S.
Polar Record, v. 11, n. 73, 1967: 472-473
Previous studies indicate that considerable caution must be used when dealing with radiocarbon ages on material incorporating carbon fixed by photosynthesis within the Atlantic Ocean. $^{14}C/^{12}C$ ratios in cold surface waters, above 40°N and 40°S, where the surface waters are in active communication with the waters at great depth, are considerably lower than at temperate latitudes and quite variable. This must be taken into consideration when computing the ages of fossil Antarctic materials.

6.030 1959
RADIOCARBON AGES OF OCEAN WATERS AND THE EXCHANGE TIME OF CARBON DIOXIDE FROM THE ATMOSPHERE TO THE OCEAN
(Draft only)
Fergusson, Gordon J.
In: International Oceanographique Congress, New York, 1959
Measurements of the radiocarbon specific activity of carbon dioxide from deep water samples from several locations in the South Pacific and Antarctic Ocean between latitudes 9·S and 65·S have shown that the average age of this deep ocean water is approximately 1000 years. This value can be compared to that of approximately 600 years for the North Atlantic, suppporting other data which indicates a slower turnover rate for the Pacific Ocean than for the Atlantic Ocean.

6.031 1967
RADIOCARBON AND ^{137}Cs IN THE ARCTIC OCEAN
Thurber, David L.
In: Mixing, Diffusion and Circulation Rates in Ocean Waters. Palisades, New York, Columbia University, 1967. (Lamont Geological Observatory. Annual Progress Report, n. 8). pp. 25
The circulation pattern and residence time of water in the Arctic Ocean were studied by making radiocarbon, ^{137}Cs and ^{90}Sr analyses. Results of the analyses were compared with Pacific and Atlantic ocean water dates.

6.032 1960
RADIOCARBON CONCENTRATION IN PACIFIC OCEAN WATERS
Bien, George S.; Rakestraw, N. W.; Suess, Hans E.
Tellus, v. 12, n. 4, Nov. 1960: 436-443
Results of radiocarbon determinations in surface water from the Pacific were in agreement with those reported by Rafter and Fergusson. However abnormal radiocarbon concentrations seem to exist locally for which no oceanic explanations can be given. It seems premature to draw conclusions from existing determinations as to the rate of increase of radiocarbon in surface ocean waters resulting from the uptake of radiocarbon produced in the atmosphere by atomic bomb. Radiocarbon content is shown to be decreasing from south to north. This decrease may be attributed to the radioactive decay of radiocarbon during the time of migration.

6.033 1963
RADIOCARBON DATING OF DEEP WATER OF PACIFIC AND INDIAN OCEAN
Bien, George S.; Rakestraw, N. W.; Suess, Hans E.
In: Radioactive dating. Proceedings of a Symposium organised by the International Atomic Energy Agency (IAEA), and held in Vienna, 1962. Vienna, IAEA, 1963: 159-173
Radiocarbon measurements on deep water samples of the Pacific Ocean, collected in 1958 during the International Geophysical Year (IGY), have shown that it is possible to determine true age differences of water masses by the determination of radiocarbon in the dissolved bicarbonate. The radiocarbon dating results lead to the same conclusion as those of tritium measurements from surface water: namely that the water above the thermocline takes several years to mix and that the uptake of artificial radiocarbon and tritium by the ocean over a period of one to two years affects only a thin surface layer.

6.034 1958
RADIOCARBON DATING OF THE EUSTATIC RISE IN OCEAN LEVEL
Godwin, Harry; Suggate, R.P.; Willis, E.H.
Nature, v. 181, n. 4622, 31 May 1958: 1518-1519
Radiocarbon dates confirm the conclusion reached on the basis of pollen analysis that eustatic rise was actively in process during the Boreal period and fully accomplished during the Atlantic period. Results indicate that the present level of the oceans was attained about 5500 years ago.

6.035 1967
RADIOCARBON FROM NUCLEAR WEAPON TESTS AND ITS USE AS A GEOPHYSICAL TRACER
Young, James Allen
Ph.D. Dissertation. University of Washington, Department of Chemistry, 1967. pp. 114 (Abstract in: *Dissertation Abstracts*, Ann Arbor, Mich., v. 28, n. 9B, 1968: 3758. Order No. 68-3890)
Using radiocarbon as a tracer, the rate of exchange of carbon dioxide between atmosphere and sea and the residence time of carbon dioxide molecules in the troposphere were calculated. An eddy diffusion model was developed, which was fairly successful in interpreting the distribution in time and space of excess radiocarbon in the northern hemisphere troposphere from 1963 to 1965.

6.036 1965
RADIOCARBON IN THE PACIFIC AND INDIAN OCEANS AND ITS RELATION TO DEEP WATER MOVEMENT
Bien, George S.; Rakestraw, N. W.; Suess, Hans E.
Limnology and Oceanography , 1965: R25-R37
Since 1948, radiocarbon measurements on the bicarbonate of ocean water samples from the Pacific and Indian ocean have been carried out at the Scripps Institution of Oceanography, La Jolla, California. Previous reports of the results of these investigations have been confirmed and amended by

many recent measurements. The method of extracting carbon dioxide from ocean water has been modified and perfected over the past years. The most notable results of the measurements concern the radiocarbon content of deep ocean water which can be interpreted unambiguously by considering the aging of the water during the time of movement from the Weddell Sea eastward and then northward into North Pacific Ocean.

6.037   1965
RADIOCARBON IN THE PACIFIC OCEAN AND ITS RELATION TO THE MOVEMENTS OF WATER MASSES
Bien, George S.; Rakestraw, N. W.; Suess, Hans E.
In: International Conference on Radiocarbon and Tritium Dating, Pullman, Washington, Washington State University, June 1-11, 1965. Proceedings. United States of America, Atomic Energy Commission, 1965. Conference n. 650652: 698-704
Radiocarbon measurements of both deep and surface waters show certain patterns which indicate that radiocarbon can be used as a tracer to demonstrate, at least quantitatively, the general trend of water movements, speed of total mass transport, and the extent of vertical mixing.

6.038   1958
A RADIOCARBON PROFILE IN THE TASMAN SEA
Garner, D.M.
Nature, v. 182, n. 4633, 16 Aug. 1958: 466-468
Description of the method used in collecting ocean water at various levels. The following notes are made: (1) absence of circumpolar bottom water, (2) significant increase in salinity over that found in earlier investigations, (3) measured ages (radiocarbon method) are consistent with temperature salinity diagrams.

6.039   1966
RADIOISOTOPES AND THE RATE OF MIXING ACROSS THE MAIN THERMOCLINES OF THE OCEAN
Broecker, Wallace S.
Journal of Geophysical Research, v. 71, n. 24, Dec 1966: 5827-5836
A box model of oceanic mixing is presented which permits the distribution of long-lived natural radioisotopes (^{14}C, ^{226}Ra) to be quantitatively compared with the distribution of man-made radioisotopes (^{90}Sr, ^{137}Cs). It strongly suggests that the doubts about the radiocarbon based rates of oceanic mixing are invalid.

6.040   1956
RATE OF POSTGLACIAL RISE OF SEA-LEVEL
Shepard, Francis P.; Suess, Hans E.
Science, v. 123, n. 3207, 15 June 1956: 1082-1083
A comparison of radiocarbon dates relative to depths below present sea level, established for the Dutch coast by Van Straaten (*Geologie en Mijnbouw*, 1954, 16: 247), de Vries and Barendsen (*Nature*, 1954, 174: 1138) and similar dates from the Texas and Louisiana coast indicate an approximately equal rate of rise of the sea-level at the various localities, but these measurements do not make it possible to distinguish between eustatism, tectonic movements and compaction. From the samples, there is no evidence of a higher sea stand during the climatic optimum.

6.041   1967
RATES OF SEDIMENTATION IN THE ARCTIC OCEAN
Ku, Teh-Lung; Broecker, Wallace S.
In: Progress in Oceanography, vol. 4, The Quaternary History of the Ocean Basins. New York, Pergamon Press, 1967: 95-104
The sedimentation rate in the Arctic deep sea over the last 150,000 years has undergone no significant changes. Based on both radiocarbon and uranium series isotope analyses, the deposition rate is estimated to be 0.2 cm/1000 years.

6.042   1958
THE RELATION OF DEEP SEA SEDIMENTATION RATE TO VARIATIONS IN CLIMATE
Broecker, Wallace S.; Turekian, Karl K.; Heezen, Bruce C.
American Journal of Science, v. 256, n. 7, Summer 1958: 503-517
Variations in sedimentation rates with time have been investigated in a mid-equatorial Atlantic core. Measurements of $^{14}C/^{12}C$ ratios, percent coarse fraction and percent carbonate material at frequent depth intervals permit the calculation of absolute rates of sedimentation for each of the major contributors to the sediment.

6.043   1966
REPEAT MEASUREMENTS OF TEMPERATURE, SALINITY, AND CARBON-14 DEPLETION AT AN OCEAN STATION
Houtman, Th.
New Zealand Journal of Science, v. 9, n. 2, June 1966: 457-471
Temperature and salinity were measured at six-monthly intervals to a depth of 3000 m over three years. Radiocarbon depletion was also measured on water samples from a range of depths. The results indicated the possibility of a three-year period of progressive southward movement of subtropical water in the upper layers. The radiocarbon depletion suggests predominance of vertical interchange over horizontal mixing.

6.044 1959
A SECTION OF C-14 ACTIVITIES OF SEA-WATER BETWEEN 9·S AND 66·S IN THE SOUTH-WEST PACIFIC OCEAN
Burling, R.W.; Garner, D.M.
New Zealand Journal of Geology and Geophysics, v. 2 n.4, Nov. 1959: 799-824
The radiocarbon activity of ocean water is used to study ocean water motion and circulation, sinking and upwelling.

6.045 1958
THE SIGNIFICANCE OF VARIATIONS OF LIGHT ISOTOPES ABUNDANCE IN OCEANOGRAPHIC STUDIES
Broecker, Wallace S.; Ewing, Maurice; Heezen, Bruce C.; Gerard, R.; Kulp, J. Lawrence
In: Cosmological and Geological Implications of Isotope Ratio Variations. Proceedings of an Informal Conference, Massachusetts Institute of Technology, June 13-17, 1957. Washington D.C., National Academy of Science. National Research Council. Publication n.572. Committee on Nuclear Science, Sub-committee on Nuclear Geophysics. *Nuclear Science Series Report*, n. 23,1958: 118-134
Summarises the work done at Lamont during the past three years on the applications on the radiocarbon methode to problems in oceanic circulation.

6.046 1967
SOME DEEP SEA SEDIMENTS IN THE WESTERN MEDITERRANEAN SEA
Eriksson, K. Gösta
In: Process in Oceanography, v. 4, *The Quaternary History of the Ocean Basins*. London, Pergamon Press, 1967: 267-280
Study of core 210 - Swedish Deep Sea Expedition, 1948. The chronology represented by the sediments of the core is principally based on radiocarbon dates of tests of Foraminifera and Pteropoda shell fragments. A table of radiocarbon dates is presented.

6.047 1963
STUDY OF LONG AND SHORT-TERM GEOPHYSICAL PROCESSES USING NATURAL RADIOACTIVITY
Lal, D.
In: Radioactive Dating. Proceedings of a Symposium on Radioactive Dating held in Athens, 19-23 November 1962. Vienna, International Atomic Energy Agency, 1963. (Proceedings Series): 149-157
A review of the application of the cosmic-ray produced isotopes as tracers for studying particular problems in earth sciences, such as large-scale atmospheric circulation, water circulation in the ocean and for dating the ocean sediments.

6.048 1965
TEMPERATURE DEPENDENCE OF CARBON ISOTOPE COMPOSITION IN MARINE PLANKTON AND SEDIMENTS
Sackett, William M.; Eckelman, Walter R.; Bender, Michael L.; Bé Allan, W.H.
Science, v. 148, n. 3667, 7 Apr. 1965: 235-237
Samples of marine plankton collected in high-latitude areas of the South Atlantic where surface water temperatures are near 0·C show a ^{12}C enrichment of 6 per mill relative to samples collected where temperatures are about 25·C. The organic carbon in the Drake Passage and Argentine Basin also show a ^{12}C enrichment relative to warmer areas.

6.049 1967
TRANSFER AND EXCHANGE OF THE ^{14}C BETWEEN THE ATMOSPHERE AND THE SURFACE WATER OF THE PACIFIC OCEAN
Bien, George S.; Suess, Hans E.
In: Radioactive Dating and Methods of Low-level Counting. Proceedings of a Symposium organised by the International Atomic Energy Agency in co-operation with the Joint Commission on Applied Radioactivity (ICSU) and held in Monaco, 2-10 March 1967. Vienna, IAEA, 1967. (Proceedings Series): 105-115
Determinations of radiocarbon in the bicarbonate of the surface and subsurface waters of the Pacific Ocean were carried out. These show that, due to the uptake of bomb radiocarbon, the radiocarbon content of surface water of the Pacific has been rising. During the same period, radiocarbon level in the atmosphere has increased to approximately 90% above normal.

6.050 1967
TRANSFER OF BOMB ^{14}C AND TRITIUM FROM THE ATMOSPHERE TO THE OCEAN. INTERNAL MIXING OF THE OCEAN ON THE BASIS OF TRITIUM AND ^{14}C PROFILES
Münnich, K.O.; Roether, W.
In: Radioactive Dating and Methods of Low-level Counting. Proceedings of a Symposium organised by the International Atomic Energy Agency (IAEA) in co-operation with the Joint Commission on Applied Radioactivity (ICSU) and held in Monaco, 2-10 March 1967. Vienna, IAEA, 1967. (Proceedings Series): 93-104
Presents results obtained for surface water samples and depth profiles. A maximum of both radiocarbon and tritium is shown in mid-latitudes and a drop at the equator. The data indicate a latitude-dependent exchange velocity of carbon dioxide with an average of 2.1 km/yr. which corresponds to an exchange time of 5.4 years.

CHAPTER SEVEN
14C PLEISTOCENE

7.001   1968
AGE AND CORRELATION OF PLEISTOCENE
DEPOSITS AT GARFIELD HEIGHTS
(CLEVELAND), OHIO
White, George W.
Geological Society of America. Bulletin, v. 79, n. 2, June 1968: 749-752
Radiocarbon dates and new exposures in the Mill Creek valley in Garfield Heights just southeast of Cleveland, Ohio, permit a revised correlation of the Pleistocene deposits in that area.

7.002   1968
AGE OF LATE PLEISTOCENE SHORELINE
DEPOSITS, COASTAL GEORGIA
Hoyt, John H.; Henry, Vernon J.; Weimer, R.J.
In: Means of Correlation of Quaternary Successions, edited by Roger B. Morrisson and Herbert C. Wright. International Association for Quaternary Research (INQUA). Congress, 7th, Boulder, Col., 14 Aug - 19 Sept. 1965. *Proceedings*, v. 8. Salt Lake City, University of Utah Press, 1968: 381-393
Radiocarbon dates from shoreline deposits of the Georgia coast tentatively suggest ages of approximately 48,000 to 40,000 years BP for a +4.5 m submergence, the Princess Anne shoreline, and 30,000 to 25,000 years BP for a +2 m submergence, the Silver Bluff shoreline. A small number of ages from other coastal areas of the world support the idea of two periods of submergence. These are correlated respectively with the times of glacial withdrawal known as the Port Talbot and Plum Point Farmdalian interstadials in north central North America and the Gottweigh and Paudorf intervals in Europe.

7.003   1968
AGE OF LATE-PLEISTOCENE SHORELINE
DEPOSITS, COASTAL GEORGIA
Hoyt, John H.; Henry, Vernon J.; Weimer, R.J.
In: Means of Correlation of Quaternary Successions, edited by Roger B. Morrisson and Herbert C. Wright. International Association for Quaternary Research (INQUA). Congress, 7th, Boulder, Col., 14 Aug - 19 Sept. 1965. *Proceedings*, v. 8. Salt Lake City, University of Utah Press, 1968: 505-517
Although the Ionium-Thorium method was used, radiocarbon dates were obtained to determine independently an 'absolute' chronology for the bottom sediment of Drake Passage. Results showed an apparent discrepancy by a factor of 2, radiocarbon dates being younger than the thorium dates. The radiocarbon dates could be younger because of the 6 to 14% deficiency of radiocarbon in the ocean, or may result from an erroneous assumption in the thorium dating. A table showing the two sets of dates is presented.

7.004   1967
THE AGENCY OF MAN IN ANIMAL
EXTINCTION
Hester, Jim J.
In: Pleistocene Extinctions: the Search for a Cause. International Association for Quaternary Research (INQUA). Congress, 7th, Boulder, Col., 14 Aug - 19 Sept.1965. *Proceedings*, v. 6. Boulder, Col., 1968: 169-192
Presents a table of terminal radiocarbon dates for extinct species which forms the basis of a graph comparing late Pleistocene extinctions with glacial events and population of early man. Concluded that Pleistocene man could not have caused the extinction of the North American megafauna until after natural causes had greatly reduced the population of each species. An alternative explanation is that change in ecologic niches resulting from the disappearance of the late-Wisconsin ice sheet subjected animal species to severe selection pressures.

7.005   1967
BONNEVILLE CHRONOLOGY: CORRELATION
BETWEEN THE EXPOSED STRATIGRAPHIC
RECORD AND THE SUBSURFACE
SEDIMENTARY SUCCESSION
Eardley, A.J.
Geological Society of America. Bulletin, v. 78, n. 7, Dec. 1967: 907-910
The chronology of the Pleistocene lake stages of the Bonneville basin, Utah, based on a stratigraphic study of the exposed sediment, supplemented by radiometric dates, correlates fairly well with the chronology deduced from the study of a core taken at the southeastern margin of the Great Salt Lake.

7.006   1969
THE ENVIRONMENT OF EXTINCTION OF THE
LATE PLEISTOCENE MEGAFAUNA IN THE
ARID SOUTHWESTERN UNITED STATES
Mehringer, Peter J.
In: Pleistocene Extinctions: the Search for a Cause. International Association for Quaternary Research (INQUA). Congress, 7th, Boulder, Col., 14 Aug - 19 Sept.1965. *Proceedings*, v. 6. Boulder, Col., 1968: 247-266
Concludes that because different species of the extinct late

Pleistocene megafauna occupied habitats ranging from warm semi-arid to periglacial, it seems unlikely that a single climatic cause alone is responsible for extinction. On the basis of available radiocarbon dates it is considered that the major wave of extinction was completed before 10,000 radiocarbon years ago.

7.007   1962
EVIDENCE RELATING TO THE QUATERNARY HISTORY OF THE WILDERNESS LAKES
Martin, A.R.H.
Geological Society of South Africa. Transactions and Proceedings, v. 65, part 1, Jan-June 1962: 19-42
The history of these lakes and the adjacent shorelines is tentatively sketched in terms of fluctuation, relations of land and sea levels since the Upper Pleistocene. Radiocarbon dates bracket the marine and brackish sediments between the dates 6870 ±160 and 1905 ± 60 years. Peat and gyttja were the nature of the samples dated.

7.008   1968
'FERNBANK' - A REDISCOVERED PLEISTOCENE INTERGLACIAL DEPOSIT NEAR ITHACA, NEW YORK. (Abstract)
Bloom, Arthur L.
Geological Society of America. Special Papers, n. 115, 1968: 251
The molluscan fauna, the pollen and the infinite radiocarbon date of greater than 54,000 years, and the compacted sediment, all suggest that 'Fernbank' is an interglacial deposit, possibly of Sangamon age, the only such deposit reported from New York state, exclusive of Long Island.

7.009   1967
FOUR SUPERIMPOSED LATE-PLEISTOCENE VERTEBRATE FAUNA FROM SOUTH-WEST KANSAS
Schultz, Gerald Edward
In: Pleistocene Extinctions: the Search for a Cause. International Association for Quaternary Research (INQUA). Congress, 7th, Boulder, Col., 14 Aug - 19 Sept.1965. *Proceedings*, v. 6. Boulder, Col., 1968: 321-336
Studies of a multiple collapse sink in Southern Mead County, Kansas have demonstrated the unique occurrence of four superimposed vertebrate faunas, which are assigned to Illinois, Sangamon and late Wisconsin on the basis of stratigraphic position, faunal composition and pollen evidence. The fourth fauna which lies at the top of the stratigraphic section and contains several northern mammalian species yielded a radiocarbon date of 11,000 ± 390 years BP.

7.010   1964
FRAMEWORK FOR DATING FOSSIL HOMINIDS
Oakley, Kenneth P.
London, Weidenfeld and Nicholson, 1964
The following chronological charts based on radiocarbon dates are presented: chronology of late Pleistocene and early Holocene climatic changes in Europe and the Baltic region correlated with pollen zones; chart of radiocarbon dates for middle Palaeolithic cultures and the Middle/Upper transition; stratigraphical framework and radiocarbon dating of upper Palaeolithic cultures in Africa during the last 60,000 years; dating tables of fossil hominids. The ages from the Mousterian (41,000 B.C.) to the present are based on radiocarbon determinations.

7.011   1963
GEOCHRONOLOGY OF PLUVIAL LAKE COCHISE, SOUTHERN ARIZONA II - POLLEN ANALYSIS OF A 42-METERS CORE
Martin, Paul S.
Ecology, v. 44, n. 3, Summer 1963: 436-444
A drill core from the middle of the Willcox Playa in south eastern Arizona contains Pleistocene pollen record of Lake Cochise. Most of the pollen record is utterly unlike the present natural pollen rain of the region. The very high pine pollen frequency indicate a glacial/pluvial interval in zone 2, at 2 to 1.5 m depth. Radiocarbon dates indicate a Wisconsin age for this pluvial interval

7.012   1968
GEOLOGY AND RADIOCARBON AGES OF LATE PLEISTOCENE LACUSTRINE CLAY DEPOSITS, SOUTHERN PART OF SAN JOAQUIN VALLEY, CALIFORNIA
Croft, M.G.
United States Geological Survey. Professional Papers, n. 600-B, 1968: B151-B156
Five radiocarbon dates from the southern San Joaquin Valley, California, indicate that large lakes, which existed in the valley during late Pleistocene time, may be synchronous with similar lake deposits in the Great Basin physiographic province.

7.013   1963
GEOLOGY OF THE IMURUK LAKE AREA, SEWARD PENINSULA, ALASKA
Hopkins, David Moody
United States Geological Survey. Bulletin, n. 1141-C, 1963. pp. 101
Various areas around Lake Imuruk have been radiocarbon dated. It was found that the upper member of the Kongarok gravel was probably deposited during an early Pleistocene interglacial interval. Human occupation is shown to go back to 6000 years ago.

7.014   1958
GLACIAL CHRONOLOGY OF THE NEW ZEALAND PLEISTOCENE
Gage, Maxwell; Suggate, R.D.
Geological Society of America. Bulletin, v. 69, n. 5, May 1958: 589-598
Proposes a chronology for the Pleistocene glaciation of New Zealand. So far only one radiocarbon date of 22,300 ± 350 years BP supports a comparison of the substage of Otamara with deposits of post-Sangamon, pre-'type' Wisconsin age of the Northern Hemisphere

7.015   1966
GLACIATION OF THE FRENCHMANS CAP NATIONAL PARK
Peterson, James A.
Royal Society of Tasmania. Papers and Procedures, v. 100, 1966: 117-129
The sequence of advances and retreats of glaciers in the Frenchmans Cap National Park area during the Pleistocene is examined and mapped. A radiocarbon date of 8720 ± 220 years BP gives a minimum age for the final deglaciation of the Frenchmans Cap area.

7.016   1965
ISOTOPE GEOCHEMISTRY AND THE PLEISTOCENE CLIMATIC RECORD
Broecker, Wallace S.
In: The Quaternary of the United States: a Review Volume for the 7th Congress of the International Association for Quaternary Research, edited by H.E. Wright and David G. Frey. Princeton. N.J., Princeton University Press, 1965: 737-753
The current status of isotope techniques employed for the determination of Pleistocene chronology and climate is summarised. Results derived from the application of these methods are evaluated. The conclusion is drawn that a generally reliable climatic record for the past 150,000 years has been obtained.

7.017   1966
LAKE PADUCAN OF LATE PLEISTOCENE AGE IN WESTERN KENTUCKY AND SOUTHERN ILLINOIS
Olive, Wilds W.
United States Geological Survey. Professional Papers, n. 550-D, Geological Survey Research 1966.1966: D87-D88
A Pleistocene lake in western Kentucky and southern Illinois, herein named Lake Paducah, has been dated by the radiocarbon method as late Pleistocene (Wisconsin). The date, obtained from mollusc shells in a deposit of sandy silt, is 21,080 ± 400 years BP.

7.018   1967
LATE PLEISTOCENE AND HOLOCENE FAUNAL HISTORY OF CENTRAL TEXAS
Lundelius, Ernest L.
In: Pleistocene Extinctions: the Search for a Cause. International Association for Quaternary Research (INQUA). Congress, 7th, Boulder, Col., 14 Aug - 19 Sept.1965. *Proceedings*, v. 6. Boulder, Col., 1968: 287-319.
The conclusion of this study is that the withdrawal of a number of extinct northern species in the last 10,000 years suggests that climate has become drier and / or warmer. A change to a climate with stronger seasonal contrasts, marked by hot, dry summers now characteristic of central Texas, may be sufficient to account for their disappearance. The dating of the fauna is based on radiocarbon analyses which are also discussed for validity.

7.019   1967
LATE PLEISTOCENE AND HOLOCENE SEDIMENTATION IN THE LAURENTIAN CHANNEL
Connolly, J.R.; Needham, H.D.; Heezen, Bruce C.
Journal of Geology, v. 75, n. 2, Mar. 1967: 131-147
Radiocarbon dating of sediment layers below and above a thin zone of rafted sediments enable to imply that the red tills of the Laurentian channel were deposited between 13,000 and 15,000 years ago.

7.020   1966
LATE PLEISTOCENE AND RECENT CHRONOLOGY OF PLAYA LAKES IN ARIZONA AND NEW MEXICO
Long, Austin
Unpublished Ph.D. Dissertation. University of Arizona, 1966. (Abstract in: *Dissertation Abstract*, Ann Arbor, Mich., U/AZ 27/04B/1189. Order No. 66-9782)
A stratigraphic radiocarbon study at the Willcox Playa and vicinity in Cochise County, southeastern Arizona, has revealed a sedimentary sequence reflecting the lake level chronology of ancient Lake Cochise. The climatic chronology concluded from this study is consistent with known climatic variations in the world.

7.021   1955
LATE PLEISTOCENE AND RECENT CHRONOLOGY OF SOUTH CENTRAL ALASKA
Karlstrom, Thor N.V.
Geological Society of America. Bulletin, v. 66, n. 12, Dec. 1955: 1581-1582
Abstract of paper presented at the November 1955 meeting in New Orleans. The Alaskan sequence correlates closely with the radiocarbon-dated North American chronology and European late and post-glacial chronologies as dated from Scandinavian varve sequences and from archaeologi-

cal and historical records. This argues for acceptance of properly evaluated radiocarbon dates in a near-absolute as well as relative sense.

7.022 1956
LATE PLEISTOCENE AND RECENT HISTORY OF THE CENTRAL TEXAS COAST
Shepard, Francis P.
Journal of Geology, v. 64, n. 1, Jan. 1956: 56-69
Samples from a total of 27 borings along the central Texas coast have been analysed. This work provides a basis for interpreting the history of the coastal area which, according to radiocarbon determinations, extends back almost 10,000 years. The sequence of formation stands in opposition to the hypothesis of a positive sea-level stand during the postglacial climate optimum.

7.023 1967
LATE PLEISTOCENE AND RECENT PALYNOLOGY IN THE CENTRAL SIERRA NEVADA, CALIFORNIA
Adam, David P.
In: Quaternary Palaeoecology, edited by M.J. Cushing and H.E. Wright. International Association for Quaternary Research (INQUA). Congress, 7th, Boulder, Col., 14 Aug - 19 Sept. 1965. *Proceedings*, v. 7. New Haven, Conn., Yale University Press, 1967: 273-301
The study presents a preliminary climatic history of the central Sierra Nevada for the time since the recession of the late Wisconsin glaciers. The chronology is supported by radiocarbon dates.

7.024 1959
LATE PLEISTOCENE BEDS, WELLINGTON PENINSULA
Brodie, J.W.
New Zealand Journal of Science and Technology, v. 38, n. 6, Sect. B, May 1959: 624-643
The lithology and flora of late Pleistocene beds on Wellington Peninsula and the radiocarbon ages of plant fossils from the beds indicate a cool period from 20,000 to 23,000 years ago, here named the Takapu Stadial.

7.025 1956
LATE PLEISTOCENE CLIMATE AND DEEP SEA SEDIMENTS
Ericson, David B.; Broecker, Wallace S.; Kulp, J. Lawrence; Wollin, Goesta
Science, v. 124, n. 3218, 31 Aug. 1956: 385-389
The radiocarbon dates, the climatic curve deduced from the foraminifera, and the palaeotemperature measurements, indicate that the mid-point of the major change from glacial to post-glacial conditions occurred about 11,000 years BP

and was broadly simultaneous throughout the North Atlantic. The authors discuss the most possible source of errors in the calculation of radiocarbon ages on core materials.

7.026 1968
LATE PLEISTOCENE CLIMATIC CHANGES INFERRED FROM THE STRATIGRAPHIC SEQUENCES OF JAPANESE LAKE SEDIMENTS
Horie, Shoji
In: Means of Correlation of Quaternary Successions, edited by Roger B. Morrisson and Herbert C. Wright. International Association for Quaternary Research (INQUA). Congress, 7th, Boulder, Col., 14 Aug - 19 Sept. 1965. *Proceedings*, v. 8. Salt Lake City, University of Utah Press, 1968: 311-324
Fossil evidence and radiocarbon dating point to coincidence of oligotrophic feature with evidence of cooler climates.

7.027 1949
LATE PLEISTOCENE DATES DERIVED FROM RADIOCARBON ASSAY
Flint, Richard Foster
Science, v. 109, n. 2843, 24 June 1949: 636
Current research on radiocarbon presents an opportunity to learn about actual dates of origin of carbon-bearing material less than 35,000 years old. Such dating is important for establishing the chronology of the later part of the Wisconsin age of the Pleistocene.

7.028 1966
LATE PLEISTOCENE DIATOMS FROM THE TREMPEALEAU VALLEY, WISCONSIN
Andrews, George W
United States Geological Survey. Professional Paper, n. 523A, 1966: A1-A27
The Trempealeau Valley in west-central Wisconsin contains remnants of three Pleistocene alluvial terraces ranging from early Wisconsin to late Wisconsin (Mankato) age. Near Hixton, Wis., a small deposit of diatomite occurs on the lowermost (or latest) terrace surface. Wood and organic debris associated with the diatomite have a radiocarbon age of approximately 11,000 years BP.

7.029 1958
LATE PLEISTOCENE ENVIRONMENTS AND CHRONOLOGY OF PACIFIC COASTAL ALASKA
Heusser, Calvin John
Geological Society of America. Bulletin, v. 69, n. 12, part.2, Dec. 1958: 1753-1754
The results of radiocarbon dating of basal and interbedded peats and the pollen-peat stratigraphy of some fifty sections are described. The samples came from Alexander Archipelago, Icy Cape, Prince William Sound and Katalla, Kenai peninsula and Kodiak island. Evidence of volcanism,

7.030 1961/1962
A LATE PLEISTOCENE FAUNA AND FLORA FROM UPTON WARREN, WORCESTERSHIRE
Coope, Geoffrey Russell; Shotton, Frederick William; Strachan, I.
Royal Society, London. Philosophical Transactions, Series B, Biological Science, v. 714, n. 244, 1961/1962: 379
A large fauna and a large flora are described from terrace deposits of the river Salwarpe at Upton Warren, Worcestershire. From the stratigraphy and a radiocarbon age of 42,000 years, the deposits are ascribed to the beginning of the Gottweig Interstadial immediately following the maximum of the Midland Irish Sea glaciation.

7.031 1963
LATE PLEISTOCENE FISH FROM LAKE SEDIMENTS IN SHERIDAN COUNTY, NORTH DAKOTA
Sherrod, Neil A.
North Dakota Academy of Science. Proceedings, v. 17, 1963: 32-36
The glacial geology of the area is based on radiocarbon dates. The fish was found in calcareous silty clay together with molluscs. The evidence for a late Pleistocene age is strong.

7.032 1963
LATE PLEISTOCENE FLUCTUATION OF SEA-LEVEL AND POST-GLACIAL CRUSTAL REBOUND IN COASTAL MAINE
Bloom, Arthur L.
American Journal of Science, v. 261, n. 9, Nov. 1963: 862-879
Radiocarbon dates of the marine submergence not only measure the transgression of the Presumcat Sea, but also give a maximum date for the final deglaciation of southwestern Maine.

7.033 1966
LATE PLEISTOCENE FLUCTUATIONS OF KASKAWULSH GLACIER, SOUTHWESTERN YUKON TERRITORY, CANADA
Borns, Harold W.; Goldthwait, Richard P.
American Journal of Science, v. 264, n. 8, Oct. 1966: 600-619
Radiocarbon dates on a wood sample from the outer half inch of a spruce log embedded in the end moraine drift indicate that the Kaskawulsh glacier was advancing 450 years ago and reached its terminal position by approximately A.D.1680.

7.034 1968
LATE PLEISTOCENE FOREST SUCCESSION IN NORTHERN NEW JERSEY
Harmon, Kathryn Parker
Ph.D. Dissertation. Rutgers - State University, 1968. (Abstract in: *Dissertation Abstracts*, Ann Arbor, Mich., Sec. B, v. 29, n. 6, 1968: 1942B. Order No. 68-17778)
A palaeological study using palynology to examine forest history of the Wisconsin stage surrounding a critical location just south of the terminal moraine. Two radiocarbon dates, one of $12,290 \pm 570$ BP at the B/C boundary and the other of $22,870 \pm 721$ years BP at the beginning of the Cary are rejected as being too old. This is discussed.

7.035 1958
LATE PLEISTOCENE GEOCHRONOLOGY AND THE PALEO-INDIAN PENETRATION INTO THE LOWER MICHIGAN PENINSULA
Mason, Ronald J.
University of Michigan. Museum of Anthropology. Anthropological Papers, n. 11, 1958. pp. 48
The late Pleistocene geochronology in Michigan and the basins of lakes, Michigan, Huron and Erie, based on radiocarbon dates, is presented in table form.

7.036 1964
LATE PLEISTOCENE GLACIAL CHRONOLOGY OF THE NORTH-CENTRAL BROOKS RANGE, ALASKA
Porter, Stephen C.
American Journal of Science, v. 262, n. 4, April 1964: 446-460
Radiocarbon dates indicate that the Anayaknaurak readvance occurred soon after $13,270 \pm 160$ years BP and that deglaciation following the Anivik Lake readvance began by 7241 ± 95 years BP. The Iktillik glaciation is correlated broadly with the classical Wisconsin glaciation. Moraines in tributary valleys record three post-Iktillik advances and the Fan Mountain I and II are post-hypsithermal events comparable to correlative advances noted elsewhere in the North American cordillera.

7.037 1965
LATE PLEISTOCENE GLACIAL CHRONOLOGY, NORTHEASTERN ST. ELIAS MOUNTAINS, CANADA
Denton, George H.
Ph.D. Dissertation. Yale University, 1965. pp. 100 (Abstract in: *Dissertation Abstracts*, Ann Arbor, Mich. v.26, n. 8, 1966: 4569. Order No. 65.15029)
Twenty radiocarbon dates combine to give the following late Pleistocene chronology: Shakwak glaciation (>49,000 BP), Silver non-glacial interval (>49,000 BP), Icefield

glaciation (>49,000 to ~37,700 BP), Boutellier non-glacial interval (37,700 to <30,100 BP), Kluane glaciation (30,100 to 12,500 - 7780), Slims non-glacial interval (12,500 - 9780 to 2640 BP), Neoglaciation (2640 to present).

7.038   1963
LATE PLEISTOCENE GLACIAL EVENTS AND RELATIVE SEA-LEVEL CHANGES IN THE NORTHERN PUGET LOWLANDS, WASHINGTON
Easterbrook, Don J.
Geological Society of America, Bulletin, v. 74, n. 12, Dec. 1963: 1465-1484
Radiocarbon dates and stratigraphic relationship suggest that 350 feet of emergence, 500-700 feet of submergence and emergence of 500-700 feet occurred in a period of only 1000-2000 years. These changes in relative sea-level during such a short period may have resulted from a combination of two opposed tendencies, isostatic uplift of the land due to glacial unloading and eustatic rise of sea-level, superimposed on tectonic movement.

7.039   1967
LATE PLEISTOCENE GLACIAL STRATIGRAPHY AND CHRONOLOGY, NORTHEASTERN SAINT ELIAS MOUNTAINS, YUKON TERRITORY, CANADA
Denton, Georgette; Stuiver, Minze
Geological Society of America, Bulletin, v. 78, n. 4, Apr. 1967: 485-510
Comparison of radiocarbon dated glacial events in the Yukon-Alaska (as recorded in the St Elias Mountains and Brook Range) with glacial events in Washington and British Columbia, pluvial events at Searles Lake, California, and fluctuations of the Laurentide ice sheet in the Great Lakes region, suggests, with reservations, that some major late Wisconsin climatic fluctuations in Yukon-Alaska and these other regions were broadly synchronous.

7.040   1967
LATE PLEISTOCENE HISTORY OF WOODLAND VEGETATION IN THE MOHAVE DESERT
Wells, Philip V.; Berger, Rainer
Science, v. 155, n. 3770, 31 Mar. 1967: 1640
New evidence records pluvial expansion of the pinyon-juniper zone at the close of the Wisconsin Glacial. Dating was made on plant fossils preserved in wood-rat middens and ground sloth coprolites.

7.041   1959
LATE PLEISTOCENE INVERTEBRATES OF THE NEWPORT BAY AREA, CALIFORNIA
Kanakoff, George P.; Emerson, William K.
Los Angeles County Museum Contribution to Science, n. 31, 14 Oct. 1959. pp. 47
The limited available radiocarbon evidence corroborates the conclusion that the faunal evidence precludes a post-Wisconsin age for the fauna. Fossil deposits on the lowest terraces at San Pedro and Santa Cruz, California, indicate ages greater than 30,000 years.

7.042   1960
LATE PLEISTOCENE MARINE TERRACES ON SANTA ROSA ISLAND, CALIFORNIA
Orr, Phil C.
Geological Society of America, Bulletin, v. 71, n. 7, July 1960: 1113-1120
Three wave-cut platforms with their marine and terrestrial fossil-bearing covers are described and named and are shown to be of late Pleistocene age by means of radiocarbon dating.

7.043   1965
LATE PLEISTOCENE POLLEN DIAGRAM FROM SOUTHERN CHILE
Heusser, Calvin John
In: International Association for Quaternary Research (INQUA). Congress, 7th, Boulder, Col., 14 Aug - 19 Sept. 1965. *Abstracts of General Sessions.* Boulder, Colorado, 1965: 212
Five sections collected in the Province of Llanquihue, western Patagonia resting on glaciated grounds vacated by late-Pleistocene ice as early as 16,000 BP are studied. A chronology for the sections is developed from 20 radiocarbon dates. Common to all sections is a volcanic ash horizon dated ca 9500 BP.

7.044   1967
LATE-PLEISTOCENE POLLEN STRATIGRAPHY OF WESTERN LONG ISLAND AND EASTERN STATEN ISLAND, NEW YORK
Sirkin, Leslie A.
In: Quaternary Palaeoecology, edited by M.J. Cushing and H.E. Wright. International Association for Quaternary Research (INQUA). Congress, 7th, Boulder, Col., 14 Aug - 19 Sept. 1965. *Proceedings.* v. 7. New Haven, Conn., Yale University Press, 1967:249-274
The pollen stratigraphy of bogs associated with the Harbor Hill Moraine and younger deposits on western Long Island and eastern Staten Island is described and correlated with existing work. Radiocarbon dates are used for establishing the correlation and are presented in table form.

7.045   1957
LATE PLEISTOCENE RADIOCARBON CHRONOLOGY IN IOWA
Ruhe, Robert V.; Rubin, Meyer; Scholtes, W.H.

American Journal of Science, v. 255, n. 10, Dec. 1957: 671-689
New radiocarbon dates in Iowa permit a grouping of age values and raise new problems in stratigraphy of late Pleistocene deposits.

7.046   1965
LATE PLEISTOCENE STRATIGRAPHY AND CHRONOLOGY IN SOUTHERN BRITISH COLUMBIA AND NORTH-WESTERN WASHINGTON
Armstrong, J.E.; Crandell, Dwight R.; Easterbrook, Don J.; Noble, J.B.
Geological Society of America, Bulletin, v. 76, n. 3, Mar. 1965: 321-330
Stratigraphic studies supplemented by more than 130 radiocarbon dates prompted the proposal for six geologic-climate units for the late Pleistocene sequence in southwestern British Columbia and northwestern Washington.

7.047   1965
LATE PLEISTOCENE STRATIGRAPHY, BARROW, ALASKA
Sellman, Paul V.; Brown, Jerry; Schmidt, R.A.M.
In: *International Association for Quaternary Research* (INQUA). Congress, 7th, Boulder, Col., 14 Aug - 19 Sept. 1965. *Abstracts of General Sessions*. Boulder, Colorado, 1965: 419-420
Radiocarbon dates from the re-worked upper unit at Point Barrow, Alaska, extend to 25,000 years BP and may represent a transgression that took place before the Wisconsin Maximum. The lower unit yielded several dates greater than 36,000 years.

7.048   1963
THE LATE PLEISTOCENE TERRACE DEPOSITS OF THE MEUSE
Broel, J.M.M. van den; Maarleveld, Gerardus Cornelis
Netherlands Geologishe Stichting. Meddelingen, n. s., v. 3, n. 16, 1963: 13-24
Pollen analysis provided information on the age of these terraces. Soil profiles of Allerød times are dated by the radiocarbon method at some 11,000 years BP.

7.049   1966
LATE PLEISTOCENE VEGETATION AND DEGREE OF PLUVIAL CLIMATE CHANGE IN THE CHIHUAHUAN DESERT
Wells, Philip V.
Science, v. 153, n. 3739, 26 Aug. 1966: 970
Midden deposits of Pleistocene age, accumulated by wood rats (*Neotoma*) in dry caves and rock shelters, provide a chronological record of former vegetation in the arid regions of North America.

7.050   1963
A LATE QUATERNARY CORRELATION CHART FROM NORWAY
Feyling-Hanssen, Rolf W.
Norges Geologiske Undersøkelse, Årbok for 1962, n. 273, 1963: 67-91
Presents discussion and information from pollen and other organic materials, of shoreline displacement, morainal and firn-line interpretation. Correlation is based on Blytt - Sernander sequence, the Danish pollen-zone system, de Geer's varved clay chronology and the absolute time-scale of radiocarbon dating.

7.051   1968
LATE QUATERNARY GEOLOGY OF THE SAN FELIPE AREA, BAJA CALIFORNIA, MEXICO
Walker, Theodore R.; Thompson, Robert W.
Journal of Geology, v. 76, n. 4, July 1968: 479-485
Exposures of late Pleistocene sediments near San Felipe, Baja California, Mexico provide an opportunity to observe shoreline features in ancient sediments. This is examined. The succession of events cannot at present be related with certainty to the late Quaternary timescale as the dates obtained from radiocarbon analyses of shells in this deposit are either inconsistent or beyond the limit of reliability.

7.052   1956
MICROPALEONTOLOGY AND ISOTOPIC DETERMINATIONS OF PLEISTOCENE CLIMATE
Ericson, David B.; Wollin, Goesta
Micropaleontology, v. 2, 1956: 257-270
Climatic curves derived from variations in planktonic foraminifera in three deep sea cores are compared with isotopic temperature curves drawn by Emiliani on the basis of the same cores. The radiocarbon dates in years BP according to Suess (1956) are correlated to the various curves.

7.053   1956
NEW RADIOCARBON DATES AND LATE-PLEISTOCENE STRATIGRAPHY
Flint, Richard Foster
American Journal of Science, v. 254, n. 5, May 1956: 265-287
A series of dates obtained in the Yale Chronometric Laboratory in conjunction with other dates of related samples contribute to a better understanding of related Pleistocene stratigraphy of North America. Specifically, the results fix the time of the Valders glacial maximum at around 10,700 years BP and add to the information concerning a widespread glaciation that affected North America at a time more than 30,000 years ago.

7.054   1961
NEW ZEALAND GLACIATIONS AND THE DURATION OF THE PLEISTOCENE
Gage, Maxwell
Journal of Glaciology, v. 3, n. 29, March 1961: 940-943
According to radiocarbon dates, a late Otira advance occurred in Otago somewhat earlier than 15,000 years ago, and a mid-Otira in Westland occurred about 22,300 years ago, contemporary with the Würm-Weichsel of Europe.

7.055   1968
NOTE ON RADIOMETRIC AGE DETERMINATIONS OF SAMPLES OF PEAT AND WOOD FROM TIN-BEARING QUATERNARY DEPOSITS AT SUNGEI BESI TIN MINES, KUALA LUMPUR, SELANGOR, MALAYSIA
Haile, N.S.; d'Ayob, Mohammed bin
Geological Magazine, v. 105, n. 6, 1968: 519-520
A radiocarbon age determination on wood in tin-bearing alluvium from Selangor gives a result of about 36,000 years BP. This is consistent with the view that the deposits formed during a period of rejuvenation of river systems during the Pleistocene.

7.056   1957
ON THE PLEISTOCENE SEQUENCE OF ROME, PALEOLOGIC AND ARCHEOLOGIC CORRELATIONS
Blanc, Alberto Carlo
Quaternaria, v. 4, 1957: 97-109
Defines the stratigraphic - geomorphic and palaeontologic - palaeoecologic criteria now used in connection with the relative dating of Palaeolithic sites in the Rome region, Italy. Five phases of Pleistocene glaciation are distinguished and tentatively correlated with Alpine phases.

7.057   1958
ON THE STRATIGRAPHY AND PALAEOBOTANY OF A LATE PLEISTOCENE ORGANIC DEPOSIT AT CHELFORD, CHESHIRE
Simpson, I.M.; West, R.G.
New Phytologist, v. 57, 1958: 239-250
The radiocarbon dating of the wood in the main mud bed is in agreement with the correlation of the Chelford muds suggested, indicating that the Chelford muds were deposited during the first half of the Last Glaciation. The age is 5700 years and the measurement was made with isotopically enriched carbon dioxide from the wood, by Professor de Vries at Groningen.

7.058   1960
A PALYNOLOGICAL AND GEOLOGICAL STUDY OF PLEISTOCENE DEPOSITS IN JAMES BAY LOWLANDS, ONTARIO, CANADA
Terasmae, Jaan; Hughes, O.L.
Canada, Geological Survey Bulletin, n. 62, 1960. pp. 15
The stratigraphy of Pleistocene deposits in the James Bay Lowlands is described. Layers of part silt and clay, the Missinaibi beds, situated between the glacial deposits, are given, on the basis of palynological evidence, an interstadial rank and dated by radiocarbon at between 55,000 to 64,000 years.

7.059   1968
PALYNOLOGICAL AND MINERALOGICAL ANALYSES OF A PLEISTOCENE LAKE DEPOSIT AT MORGANTOWN, WEST VIRGINIA (Abstract)
Clendening, John A.; Renton, John Jo; Parsons, Barbara M.
Geological Society of America. Special Papers, n. 115, 1968: 468
Seventy feet of alluvium from a terrace along the Monongahela river was examined. The level between 17 ft and 5 ft is assigned to the Farmdale substage, on palynological evidence. The upper portion of this interval is dated at 22,000 ± 1000 years BP by radiocarbon dating.

7.060   1965
PALYNOLOGY AND PLEISTOCENE PHYTOGEOGRAPHY OF UNGLACIATED EASTERN NORTH AMERICA
Whitehead, Donald R.
In: The Quaternary of the United States: a Review Volume for the 7th Congress of the International Association of Quaternary Research, edited by H.E. Wright and David G. Frey. Princeton. N.J., Princeton University Press, 1965: 417-432
Although various radiocarbon dates are available, reconstruction of Pleistocene vegetation and climate remains difficult.

7.061   1966
PLEISTOCENE AGE DETERMINATION FROM CALIFORNIA AND OREGON
Richards, Horace G.; Thurber, David L.
Science, v. 152, n. 3725, 20 May 1966: 1091-1092
Molluscs have been collected from Pleistocene marine deposits at Tomales Bay, California and Cape Blanco, Oregon. Dating by the radiocarbon method and $^{230}Th/^{234}K$ method suggests that the shells are at least 33,000 years old. The more probable age at the Tomales Bay locality is 50,000 years.

7.062   1959
PLEISTOCENE AND RECENT STUDIES OF WAITEMATA HARBOUR, PART 2, NORTH SHORE AND SHOAL BAY
Searle, K. J.

New Zealand Journal of Geology and Geophysics, v. 2, n. 1, Feb. 1959: 95-107

Specimens obtained from material associated with the Pupuke volcanics were dated by radiocarbon method at older than 42,000 years for wood and older than 40,000 years for peat. It is reasonable to accept that this data provide evidence for the minimum age of the lava flows and therefore that the vents were active not less than 42,000 years ago.

7.063 1968
PLEISTOCENE CLIMATE AND CHRONOLOGY IN DEEP-SEA SEDIMENTS
Ericson, David B.; Wollin, Goesta
Science, v. 162, n. 3859, 13 Dec. 1968: 1227-1234
Magnetic reversals give a time-scale of 2 million years for a complete Pleistocene with four glaciations. Dates determined by the radiocarbon, the protactinium-ionium and the protactinium methods provided an absolute time-scale from the present to about 175,000 years ago.

7.064 1953
PLEISTOCENE CLIMATIC RECORD IN A PACIFIC OCEAN CORE SAMPLE
Hough, Jack L.
Journal of Geology, v. 61, n. 3, May 1953: 252-262
A core sample from the bottom of the south-eastern end of the Pacific Ocean was dated by the W.D. Urry method. The dating for the sixth substage of the Wisconsin glaciation correlates well with the radiocarbon date for the Mankato substage.

7.065 1962
A PLEISTOCENE COLEOPTEROUS FAUNA WITH ARCTIC AFFINITIES FROM FLADBURY, WORCESTERSHIRE
Coope, Geoffrey Russell
Geological Society of London. Quarterly Journal, v. 118, 1962: 103-123
A fauna of fossil beetles from a sample of peat-like material on the base of a terrace on the river Avon at Fladbury indicates arctic conditions at the time of deposition of the peat which is dated by radiocarbon at 30,000 years BP.

7.066 1965
PLEISTOCENE DEPOSITS OF SOUTHERN ONTARIO
Dreimanis, Aleksis; Karrow, Paul F.
In: Guidebook for Field Conference G, Great Lakes - Ohio River Valley, edited by R.P. Goldthwait. International Association for Quaternary Research (INQUA), Congress, 7th., Boulder, Col., 1965: 90-100
The stratigraphy is supported by a radiocarbon dated type section of Port Talbot interstadial beds.

7.067 1965
PLEISTOCENE DEPOSITS OF THE ERIE LAKE
Goldthwait, Richard P.; Dreimanis, Aleksis; Forsyth, Jane L.; Karrow, Paul F.; White, George W.
In: The Quaternary of the United States: a Review Volume for the 7th Congress of the International Association for Quaternary Research, edited by H.E. Wright and David G. Frey. Princeton. N.J., Princeton University Press, 1965: 85-97
Various Wisconsin tills of the Erie lake and late Wisconsin drifts are dated by radiocarbon. A map of end moraines in southern Ontario showing the sites from which radiocarbon dates have been obtained is included.

7.068 1960
PLEISTOCENE DEPOSITS OF THE UPPER CLUTHA VALLEY, OTAGO, NEW ZEALAND
McKellar, I.C.
New Zealand Journal of Geology and Geophysics, v. 3, n. 3, 1960: 432-460
Five periods of moraine building and aggradation in the Wanaka - Hawea basin and Upper Clutha Valley are due to glaciations: the Hawea, Albert Town, Luggate, Lindis and Clyde advances in order of increasing age. A date of 15,100 ± 200 years on a peat section dates the youngest advance.

7.069 1960
PLEISTOCENE DEVELOPMENT OF VEGETATION AND CLIMATE IN TRISTAN DA CUNHA AND GOUGH ISLAND
Hafsten, Ulf
Bergen University. Mat-Naturv. series, n. 20, 1960. pp. 48
A peat column, 2.2 m deep, resting directly on rock, was dated by radiocarbon. The lower 0.05 m was dated at 4720 ± 130 years BP. Two dates stradling a mineral stratum at 0.80 to 1.84 m gave an average age of 2345 ± 130 years BP. This implies that the 0.35 m thick layer of peat underlying the mineral stratum took approximately as long to form as the overlying 1.80 m of peat. As the stratigraphy is very regular, this is considered impossible. It is assumed that there must be a loss of some layers.

7.070 1965
PLEISTOCENE DRAINAGE PATTERNS ON THE FLOOR OF THE CHUKCHI SEA
Creager, Joe S.; McManus, Dean A.
Marine Geology, v. 3, n. 3, 1965: 279-290
It is proposed that the lack of bathymetric continuity in a submarine valley crossing a continental shelf is the result of deltaic deposition and may be used to recognise periods of a lesser rate of sea-level rise or a period of sea-level stillstand. Radiocarbon dates are used to date these stillstands.

Ch. 7 - Pleistocenes

7.071 1957
PLEISTOCENE EMERGED MARINE PLATFORM, PORT CAMPBELL, VICTORIA
Baker, George; Gill, Edmund D.
Quaternaria, v. 4, 1957: 55-81
A late Pleistocene age is indicated for the formation of the platform, the emergence of which occurred during the last low glacial eustatic sea-level. A radiocarbon assay of one of the gastropods found abundantly, shows that it is beyond the range of radiocarbon, i.e. older than 30,000 years.

7.072 1968
PLEISTOCENE GEOLOGY AND BIOLOGY WITH ESPECIAL REFERENCE TO THE BRITISH ISLES
West, R.G.
London, Longman, 1968. pp. xiii, 377
Chapter 9 examines the various aids to chronology, this includes the radiocarbon dating method. The following tables are also presented: a time-depth graph with curves showing relative changes in sea-level, radiocarbon dated on peat (p. 148); a temperature curve for the Weichselian glacial in Denmark, including radiocarbon dates (p. 192); and a table of Weichselian radiocarbon dates (p. 285).

7.073 1964
THE PLEISTOCENE EPOCH IN DEEP-SEA SEDIMENTS
Ericson, David B.; Ewing, Maurice; Wollin, Goesta
Science, v. 146, n. 3645, 6 Nov. 1964: 723-732
Dates determined by the radiocarbon, protactinium ionium and protactionium methods provide an absolute timescale from the present, back to about 175,000 years ago. The record of the Pleistocene was the result of piecing together and correlating.

7.074 1967
PLEISTOCENE EVENTS AND CHRONOLOGY IN THE APPALACHIAN REGION OF SOUTH-EASTERN QUEBEC, CANADA
McDonald, Barrie Clifton
Ph.D. Dissertation. Yale University, 1967. pp. 215 (Abstract in: *Dissertation Abstracts*. Ann Arbor, Mich. v.28, n. 18, 1967: 234. Order No. 67-8393)
Stratigraphic evidence indicates at least three glacial phases and three non-glacial intervals antedating post-glacial times. The first glacial phase is older than 54,000 BP (St. Pierre interval). The final deglaciation of Southern Quebec is dated between 12,000 and 13,500 BP. The Highland Front Moraine was built about 12,600 BP. The marine limit of the Champlain sea was dated at $11,800 \pm 180$ BP and $11,530 \pm 160$ BP on shells at 165 m altitude.

7.075 1961
PLEISTOCENE FAUNA OF SOME MEDITERRANEAN COASTAL SITES
Higgs, E.S.
Prehistoric Society. Proceedings, n. s., v. 27, article n. 6, 1961: 144-154
Radiocarbon dating shows a relatively short time span for the late Pleistocene. The climatic periods are presented with their absolute dates based on radiocarbon measurements.

7.076 1960
PLEISTOCENE GEOLOGY OF ARCTIC CANADA
Craig, Bruce Gordon; Fyles, J.G.
Canada. Geological Survey Paper, n. 60-10, Ottawa, Queen's Printer, 1960. pp. 21
Available radiocarbon data indicates that deglaciation of northern North America took place at about the same time as the retreat of the ice sheet from northern Canada, about 7000 BP.

7.077 1961
PLEISTOCENE GEOLOGY OF ARCTIC CANADA
Craig, Bruce Gordon; Fyles, J.G.
In: Geology of the Arctic, edited by G.O. Raasch. International Symposium on Arctic Geology, 1st., Calgary, Alberta, Jan. 11-13, 1960. *Proceedings*. Toronto, University of Toronto Press, 1961: 403-420
The radiocarbon ages of post-glacial materials are presented in table form.

7.078 1955
PLEISTOCENE GEOLOGY OF EASTERN SOUTH DAKOTA
Flint, Richard Foster
United States Geological Survey. Professional Papers, n. 252, Article 128, 1955. pp. 173
The date of the Mankato maximum is inferred from radiocarbon dates at 11,000 years ago or a little earlier.

7.079 1965
PLEISTOCENE GEOLOGY OF INDIANA AND MICHIGAN
Wayne, William J.; Zumbergé, James H.
In: The Quaternary of the United States: a Review Volume for the 7th Congress of the International Association of Quaternary Research, edited by H.E. Wright and David G. Frey. Princeton. N.J., Princeton University Press, 1965: 63-84
Field studies and radiocarbon determinations provide a basis for an absolute chronology of the ice retreat in the Great Lakes region.

7.080   1966
PLEISTOCENE GEOLOGY OF THE ANAKTUVUK PASS, CENTRAL BROOKS RANGE, ALASKA
Porter, Stephen C.
Arctic Institute of North America. Technical Paper, n. 18, 1966. pp. 100
During the Itkillik glaciation, correlated loosely with late Wisconsin Glaciation of central North America, on the basis of radiocarbon dates, the Anaktuvuk pass was under ice to a minimum depth of 2000 ft.

7.081   1956
PLEISTOCENE GEOLOGY OF THE BECANCOUR MAP AREA, QUEBEC
Gadd, N.R.
Ph.D. Dissertation. University of Illinois, 1955. pp. 215 (Abstract in: *Dissertation Abstracts*. Ann Arbor, Mich., v. 16. n.3, 1956: 520. Order No. 56-812)
The age of thick deposits of peat and wood laid down towards the end of the interval which constitutes the key stratigraphic unit of the region is more than 40,000 years.

7.082   1957
PLEISTOCENE GEOLOGY OF THE DOOR PENINSULA, WISCONSIN
Thwaites, Frederick T.; Bertrand, Kenneth
Geological Society of America, Bulletin, v. 68, n. 7, July 1957: 831-880
The authors discuss the validity of radiocarbon dates obtained from samples of peat and wood from the exposed Forest Bed and present a table showing that the various dates agree only fairly well.

7.083   1965
PLEISTOCENE GEOLOGY OF THE ST. LAWRENCE LOWLANDS
Mac Clintock, Paul; Stewart, David P.
New York State Museum and Scientific Services. Bulletin, n. 394, May 1965. pp. 152
Radiocarbon dating of wood buried in sediment indicate that glacial lakes Warren (N.Y.) and Lundy (Ohio) were dammed by ice of Valders age.

7.084   1967
THE PLEISTOCENE GEOLOGY OF THE WALLKILL VALLEY
Connelly, G. Gordon; Sirkin, Leslie A.
In: Guide Book to Field Trips. New York State Geological Association, *Annual Meeting*, 39th, New Paltz, N.Y., 1967. New York City College. City University. Department of Geology, 1967: A1-A21, G1-G3
The Wallkill Valley extends from New York State into northern New Jersey. The valley's Pleistocene history began with a Wisconsin glacier that probably advanced in two lobes to the Ogdensburg - Culvers Gap moraine. As the glacier retreated, four lake stages developed. Pollen stratigraphy and radiocarbon dating indicate that the recession began prior to 15,000 years BP, suggesting correlation with the Tazewell substage of the Midwest.

7.085   1963
PLEISTOCENE GLACIAL-MARINE RELATIONS, TROIS-PISTOLES, QUEBEC (Abstract)
Lee, Hubert
Geological Society of America. Special Paper, No.73, Abstracts for 1962-1963: 195
Radiocarbon dating of shells in marine clays, underlying the delta of the St Lawrence, gives an age of 12,720 years BP and determines the time of deglaciation of the Notre Dame Mountains in that area.

7.086   1961
PLEISTOCENE GLACIATION IN KURDISTAN
Wright, Herbert E.
Eiszeitalter und Gegenwart, v. 12, 1961: 131-164
It is believed that the evidence from glaciation in the Kurdish mountains indicates that the Pleistocene climate was probably colder and wetter in the piedmonds and foothills, but that the change to a post-glacial climate much like the present was essentially complete by the time of the begining of cultivation and the establishment of permanent villages 11,000 - 9000 years ago. Radiocarbon dates of various events are given to support this theory.

7.087   1967
PLEISTOCENE HISTORY OF BERMUDA
Land, Lynton S.; Mackenzie, Fred T.; Gould, Stephen J.
Geological Society of America. Bulletin, v. 78, n. 8, 1967: 993-1006
Intergradation of aeolianites and shallow water marine biocalcarenites indicates alternations between marine and subaerial deposition in the Bermuda area during the Pleistocene. Available evidence indicates Pleistocene shore lines parallelled modern shore lines, suggesting eustatic rather than tectonic control of sea-level. Radiocarbon data indicate absolute ages from 137,000 to 37,000 years.

7.088   1960
THE PLEISTOCENE HISTORY OF THE IRISH SEA
Mitchell, G.F.
Advancement of Science, v. 17, n. 68, Nov. 1960: 313-325
Radiocarbon age of about 30,000 years dates a phase of amelioration of climate at Breda, roughly equivalent in time

Ch. 7 - Pleistocenes

to the Paudorf inter-glacial stage in Austria.

7.089 1962
PLEISTOCENE ICE VOLUMES AND SEA-LEVEL LOWERING
Donne, William L.; Farrand, William R.; Ewing, Maurice
Journal of Geology, v. 70, n. 2, 1962: 206-214
Revised estimates of late Pleistocene ice volume and resulting sea-level lowering are made. A radiocarbon date places an erosional submarine terrace at greater than 30,000 years, correlating well with the deduction for Pleistocene ice withdrawal at the maximum Illinoian stage.

7.090 1957
PLEISTOCENE MOLLUSCAN FAUNULA OF THE SYDNEY CUT, SHELBY COUNTY, OHIO
La Rocque, Aurele; Forsyth, Jane L.
Ohio Journal of Science, v. 52, n. 2, March 1957: 81-89
Radiocarbon dating of stratigraphic horizons below and above the Sydney cut (Ohio) permits the assignment of Early Wisconsin to the fossil molluscs found in a strata exposed at the bottom of the cut.

7.091 1966
PLEISTOCENE MOLLUSCS FROM CORES TAKEN FROM THE CONTINENTAL SHELF OF ARGENTINA AND CHILE
Richards, Horace G.
Quaternaria, v. 8, 1966: 253-258
Palaeontological work on cores from the continental shelf of Argentina and Chile. The first radiocarbon dates from pelecypods at 77 m is 6000 years BP. Climatic interpretation is not yet possible.

7.092 1965
PLEISTOCENE MOLLUSKS FROM CORES TAKEN FROM THE CONTINENTAL SHELF OF ARGENTINA AND CHILE
Richards, Horace G.
In: International Association for Quaternary Research (INQUA). Congress, 7th, Boulder, Col., 14 Aug - 19 Sept. 1965. *Abstracts of General Sessions*. Boulder, Colorado, 1965: 393
The radiocarbon dates of shells in water shallower than 65 fathoms vary between 11,000 and 16,000 years BP and suggest a Wisconsin age. Shells at greater depths are beyond the limit of radiocarbon dating and may indicate a shoreline of Illinois age.

7.093 1959
THE PLEISTOCENE OF FREGO - PATAGONIA; PART III, SHORELINE DISPLACEMENT
Auer, Väinö
Academia Scientiarum Fennica. Annals, s. a. III, *Geologica-Geographica*, v. 60, 1959. pp. 247
The results of peat, pollen and diatom analysis and radiocarbon dates are used for correlation and purposes of chronology.

7.094 196
PLEISTOCENE PECCARY *PLATYGONUS COMPRESSUS LECONTE* FROM SANDUSKY COUNTY, OHIO
Hoare, R.D.; Coach, J.R.; Innis, Charles; Hole, Thornton
Ohio Journal of Science, v. 64, n. 3, 1964: 207-214
The lake Warren ridge in which the peccary bones have been found is dated by radiocarbon at older than 9640 years.

7.095 1959
THE PLEISTOCENE PERIOD, ITS CLIMATE, CHRONOLOGY AND FAUNAL SUCCESSION
Zeuner, Frederick Eberard
London, Hutchinson Scientific and Technical, 1959. pp. 447
Mentions encouraging results from the method of datation by radiocarbon, agreeing closely with varve dating for the glaciation between 25,000 and 10,000 years BP for both continents, and for the post-glacial eustatic rise with a maximum intensity between 6000 and 5000 years BP (p.375).

7.096 1965
A PLEISTOCENE PHYTOGEOGRAPHICAL SKETCH OF THE PACIFIC NORTH WEST AND ALASKA
Heusser, Calvin John
In: The Quaternary of the United States: a Review Volume for the 7th Congress of the International Association for Quaternary Research, edited by H.E. Wright and David G. Frey. Princeton. N.J., Princeton University Press, 1965: 469-483
Volcanic ash assigned to the Glacier peak eruption in the Oregon Cascades is dated by radiocarbon at about 6750 BP and provides a basis for estimating the age of Zone III which is equated with a climate of maximum warmth and dryness between 4000 and 8000 BP.

7.097 1965
PLEISTOCENE POLLEN ANALYSIS AND BIOGEOGRAPHY OF THE SOUTHWEST
Martin, Paul S.; Mehringer, Peter J.
In: The Quaternary of the United States: a Review Volume for the 7th Congress of the International Association of Quaternary Research, edited by H.E. Wright and David G. Frey. Princeton. N.J., Princeton University Press, 1965: 433-451

Pollen diagrams correlated with stratigraphic units dated by radiocarbon show a full glacial record from lake beds.

7.098   1968
PLEISTOCENE RECENT STRATIGRAPHY, EVOLUTION AND DEVELOPMENT OF THE APALACHICOLA COAST, FLORIDA
Schnable, Jon F.; Goodell, H. Grant
Geological Society of America. Special Papers, n. 112, 1968. pp. 112
Pleistocene sediments deposited on an uneven Miocene surface of variable age and representing two major late Pleistocene sea-level fluctuations are examined. Radiocarbon dates and stratigraphy indicate that the upper sample probably represents a mid-Wisconsin and the lower sample a Sangamon transgression of the sea. A high stand near present sea-level between 24,000 and 40,000 years BP probably corresponds to Silver Bluff shoreline of Florida and Georgia. Sea-level was 10 to 15 feet below present between 4000 and 4500 years ago.

7.099   1963
PLEISTOCENE SEA-LEVELS, SOUTH EASTERN VIRGINIA
Oaks, Robert Q.; Coch, Nicholas K.
Science, v. 149, n. 3570, 31 May 1963: 979-983
The authors revise the stratigraphy and morphologic concepts as applied to the Atlantic coast of Virginia. Six episodes of submergence and emergence are dated from peat samples.

7.100   1963
PLEISTOCENE SEDIMENTATION AND FAUNA OF THE ARGENTINE SHELF: I, WISCONSIN SEA LEVEL AS INDICATED IN ARGENTINE CONTINENTAL SHELF SEDIMENTS
Fray, Charles; Ewing, Maurice
Academy of Natural Science, Philadelphia. Proceedings, v. 115, n. 6, 1963: 113-126
Forty-two cores from the Rio de la Plata to 50°S reveal a rather homogeneous sand layer of varying thickness underlain by one or more layers of shells. Many shells show abrasion and wear characteristic of material subject to wave action. Many of the species of molluscs and other fossils suggest a shoreline of Wisconsin age. Shells at greater depths are beyond the limit of radiocarbon and may indicate a shore line of Illinoian age. The uniformity of the depth below present sea level of the upper shell layer in cores in approximately the same depth of water suggests that there has been little if any warping of the Argentine continental shelf since late Wisconsin time.

7.101   1958
PLEISTOCENE SEQUENCE IN SOUTHERN PARTS OF THE PUGET SOUND, WASHINGTON
Crandell, Dwight R.; Mullineaux, D.R.; Waldron, Howard H.
American Journal of Science, v. 256, n. 6, June 1958: 384-397
The Pleistocene history of the Southeastern part of the Puget Sound Lowland consists of at least four episodes of glaciation separated by intervals of erosion, weathering and non-glacial sedimentation. Lowlands south of Seattle were uncovered prior to 14,000 years ago, followed by a subsequent late Wisconsin glacial advance younger than 11,300 ± 300 radiocarbon years.

7.102   1962
PLEISTOCENE SEQUENCE ON WHIDBEY ISLAND. (Abstract)
Easterbrook, Don J.
Northwestern Science, v. 36, n. 4, Nov. 1962: 128.
Paper presented at the 36th Annual meeting of the Northwest Scientific Association at Western Washington State College, Bellingham, Washington. Recent studies of the Pleistocene history of the northern part of the Puget Lowland have resulted in the recognition of the late-post-Vashon glacial events closely related to changes in sea-level. A similar succession on Whidbey Island is described. The glacio-marine drift present there was deposited 11,500 years ago.

7.103   1966
THE PLEISTOCENE SOBOBA FLORA OF SOUTHERN CALIFORNIA
Axelrod, Daniel I.
California University. Publication in Geological Sciences, v. 60, 1966. pp. 79
The radiocarbon evidence indicates that the flora is older than 38,000 years.

7.104   1968
PLEISTOCENE STRATIGRAPHY OF CAPE COD, MASSACHUSETTS (Abstract)
Oldale, Robert N.
Geological Society of America. Special Papers, n. 115, 1968: 283
Radiocarbon dates from Martha's Vineyard indicate that the ice retreat was no earlier than 15,000 years BP and fixes a minimum date for ice advance between 20,000 and 26,000 years BP.

7.105   1968
PLEISTOCENE STRATIGRAPHY OF ISLAND COUNTY: WASHINGTON

Ch. 7 - Pleistocenes

Easterbrook, Don J.
Water Supply Bulletin, Department of Water Resources, 1968. pp. 34
The oldest glacial deposits exposed in Island County belong to the Double Bluff. During the following Whidbey Interglacial floodplain sand, silt and peat were deposited. Possible correlations are mentioned. Everson glaciomarine drift was deposited on the sea floor from the melting ice. Marine shells from the deposit are dated from 11,850 to 13,010 years, setting an upper limit for the end of the Vashon.

7.106   1965
THE PLEISTOCENE STRATIGRAPHY OF SOUTHEASTERN ALBERTA, CANADA
Westgate, J.A.
In: International Association for Quaternary Research (INQUA). *Congress*, 7th, Boulder, Col., 14 Aug - 19 Sept. 1965. *Abstracts of General Sessions*. Boulder, Colorado, 1965: 500-501
The stratigraphic units are described, the Oldman drift being the youngest and dated by radiocarbon at 10,500 ± 200 years BP.

7.107   1968
PLEISTOCENE STRATIGRAPHY OF THE SASKATOON AREA, SASKATCHEWAN, CANADA
Christiansen, E.A.
Canadian Journal of Earth Sciences, v. 5, n. 5, Oct. 1968: 1167-1173
Pleistocene sediments, subdivided into two groups, the Sutherland and the Saskatoon, are examined. A hiatus of interglacial proportion between two formations, the Floral and the Battleford in the Saskatoon group, is described. Radiocarbon determinations date the hiatus as mid-Wisconsinan and the Battleford formation as Woodfordian.

7.108   1968
THE PLEISTOCENE SUCCESSION AROUND BRANDON, WARWICKSHIRE
Shotton, F.W.
Royal Society, London. Philosophical Transactions, Series B. v.254, n. 799, Dec. 1968: 387-400
Provides support for the age of between 26,000 and 10,000 years BP for a most extensive glacial phase in the area, through radiocarbon dates on organic detritus from beneath the upper till.

7.109   1955
PLEISTOCENE TEMPERATURE VARIATION IN THE MEDITERRANEAN
Emiliani, Cesare L.
Quaternaria, v. 2, 1955: 87-93

A number of stages were identified in a Mediterranean core. Stage 10 gave an age of 183,000 years by the U/Th method. Stage 2 was dated by radiocarbon at 17,200 years. From these dates an average rate of deposition of 4.3 cm per 1000 years was calculated. The temperature minimum of Stage 2 corresponds to the radiocarbon date of 17,200 years.

7.110   1955
PLEISTOCENE TEMPERATURES
Emiliani, Cesare L.
Journal of Geology, v. 63, n. 5, Sept. 1955: 538-578
Seven complete temperature cycles are shown by a Caribbean core. By extrapolation rates of sedimentation based on radiocabon data, an age of about 280,000 years is obtained for the earliest temperature minimum. Correlation with continental events suggests correspondance with the first major glaciation.

7.111   1968
PLEISTOCENE UNGULATES FROM THE BOW RIVER GRAVELS AT COCHRANE, ALBERTA
Churcher, C.S.
Canadian Journal of Earth Science, v. 5, n. 6, Dec. 1968: 1467-1468
Five ungulates are reported from gravels comprising the second major terrace above the Bow River's north bank at Cochrane, Alberta. One of these: *Equus conversidens* (extinct Mexican ass) was previously known from middle and late Pleistocene beds of the southern United States and Mexico and is here reported from the post-Wisconsin Pleistocene of Alberta and possibly Saskatchewan. Radiocarbon analysis of Bison bones from the gravels yielded two dates that averaged 11,065 BP.

7.112   1964
PLEISTOCENE VEGETATIONAL STUDIES IN THE WHITWATER BASIN, SOUTHERN INDIANA
Kapp, Ronald O.; Gooding, Ansel M.
Journal of Geology, v. 72, n. 2, May 1964: 307-326
Geologic studies in southern Indiana have disclosed an extensive Sangamon interglacial soil and Illinoian and Wisconsin interstadial silts. Radiocarbon dates of >41,000 years and field evidence establish that the interglacial humic deposits are of Sangamon age. Sangamon climatic changes, inferred from the pollen diagrams, show that the late stage of a boreal climate was followed by amelioration and eventually by the development of a thermal maximum, followed by a return to a glacial climate.

7.113   1967
THE PLEISTOCENE VERTEBRATES OF MICHIGAN
Wilson, Richard Leland

Michigan Academy of Science, Arts, and Letters. Papers, v. 52, 1967: 197-234

Four distinct Pleistocene glacial ages are recognised in North America and five glacial advances of the later Wisconsin period are recognised in Michigan. A table of radiocarbon ages of these periods: Valders, Port Huron (Mankato), Cary (Lake Border), (Tinley-Defiance), (Valparaiso) is presented. A table of locations, vertebrates, radiocarbon dates and pollen analysis is also presented.

7.114   1964
PLEISTOCENE WOOD-RAT MIDDENS AND CLIMATIC CHANGE IN MOHAVE DESERT: A RECORD OF JUNIPER WOODLANDS
Wells, Philip V.; Jorgesen, Clive
Science, v. 163, n. 3611, 13 Mar. 1964: 1171-1173
Twelve radiocarbon dates suggest that the middens were deposited between 7800 to more than 40,000 years ago. Dominance of Utah juniper and absence of pinyon pine indicate a local Pleistocene woodland climate more arid.

7.115   1968
PLEISTOCENE-RECENT BOUNDARY AND WISCONSIN GLACIAL BIOSTRATIGRAPHY IN THE NORTHERN INDIAN OCEAN
Frerichs, William E.
Science, v. 159, n. 3822, 29 Mar. 1968: 1456-1458
Includes discussion of differences in radiometric dating of deep sea cores representative of the Pleistocene-recent boundary which support the two faunal criteria presented in the article to define this boundary.

7.116   1963
POLLEN ANALYSIS OF A DEPOSIT AT RODDANO PORT, COUNTY DOWN, NORTHERN IRELAND, BEARING REINDEER ANTLER FRAGMENTS
Singh, Gurdip
Grana Palynologica, v. 4, n. 3, 1963: 466-474
Results of pollen analysis and stratigraphical evidence seem to extend the Late-Glacial occurrence of reindeer in Ireland as far back as the beginning of the Allerød period, although a radiocarbon date of a large antler fragment has given a date of $10,250 \pm 350$ years BP, which is 2000 years younger than expected. However it is not the first time that radiocarbon dates for reindeer antlers were found to be younger than expected.

7.117   1962
POLLEN FROM THE PLEISTOCENE TERRACE DEPOSITS OF WASHINGTON D.C. (Abstract)
Knox, A.S.
Pollen et Spores, v. 4, n. 2, 1962: 357-358

The Pleistocene terrace deposits on which Washington, D.C. is located, were, according to the abundant microflora contained in the sediments, deposited in freshwater during one interglacial interval. The stump of a bald cypress, standing upright in an extensive swamp is dated by radiocarbon at 38,000 years.

7.118   1962
POLLEN PROFILES OF LATE PLEISTOCENE AND RECENT SEDIMENTS AT WEBER LAKE, NORTHEASTERN MINNESOTA
Fries, Magnus
Ecology, v. 43, n. 2, 1962: 295-308
Radiocarbon of the late-glacial and post-glacial pollen stratigraphy of the Weber lake profiles are presented in table form and discussed.

7.119   1964
POSSIBLE LATE PLEISTOCENE UPLIFT, CHESAPEAKE BAY ENTRANCE
Harrison, W.; Malloy, R.J.; Rusnak, Gene A.; Terasmae, Jaan
Journal of Geology, v. 73, n. 2, March 1965: 201-229
Geologic studies, supported by radiocarbon dates, at three localities in the vicinity of Chesapeake Bay entrance provide evidence suggestive of overall crustal uplift since 15,000 years BP.

7.120   1967
PRE-OLYMPIA PLEISTOCENE STRATIGRAPHY AND CHRONOLOGY IN THE CENTRAL PUGET SOUND
Easterbrook, Don J.; Crandell, Dwight R.; Leopold, Estella B.
Geological Society of America. Bulletin, v. 78, n. 1, Jan. 1967: 13-20
The stratigraphy is described, and a tentative correlation with the Salmon Spring glaciation on the basis of radiocarbon dates on peat is discussed but not confirmed.

7.121   1967
PREHISTORIC OVERKILL
Martin, Paul S.
In: Pleistocene Extinctions: the Search for a Cause. International Association for Quaternary Research (INQUA). Congress, 7th, Boulder, Col., 14 Aug - 19 Sept.1965. *Proceedings*, v. 6. Boulder, Col., 1968: 75-120
A sudden wave of large animal extinction involving at least 200 genera, most of them lost without phyletic replacement, characterised the late Pleistocene. Although it may have occurred during time of climatic change, the event is not clearly related to climatic change.

7.122 1963
PRELIMINARY REPORT ON THE AGE AND DISTRIBUTION OF THE LATE PLEISTOCENE ICE IN NORTH CENTRAL MAINE
Borns, Harold W.
American Journal of Science, v. 261, n. 8, Oct. 1963: 738-740
A marine fossiliferous esker delta in the Bangor (Maine) area and the radiocarbon age relationship of the fossils within to fossils from Champlain Sea deposits present more evidence that an ice cap may have existed at the time that the Champlain Sea occupied the St Lawrence lowlands.

7.123 1956
THE PROBLEM OF THE COCHRANE IN LATE PLEISTOCENE CHRONOLOGY
Karlstrom, Thor N.V.
United States Geological Survey. Bulletin, n. 1021-J, 1956: 303-331
The highly controversial dating of the youngest readvance of the retreating Wisconsin ice sheet in North America is discussed along with the implications of peat and wood samples collected from the Cochrane area and dated by radiocarbon methods.

7.124 1967
THE PROBLEMS AND CONTRIBUTIONS OF METHODS OF ABSOLUTE DATING WITH THE PLEISTOCENE PERIOD
Shotton, Frederick William
Geological Society of London. Quarterly Journal, v. 122, 1967: 357-383
The basis of the radiocarbon method is discussed with special reference to sources of error in the estimations and their relative significance. Radiocarbon dates now indicate that the Würm Glaciation in the northern hemisphere had two stadia of approximately equal importance separated by two long cool interstadials. This concept is applied to an interpretation of some British deposits to which radiocarbon dates are applicable.

7.125 1965
QUATERNARY GEOLOGY OF NORTHERN GREAT PLAIN
Lemke, R.W.; Laird, W.M.; Tipton, M.J.; Lindvall, R.M.
In: The Quaternary of the United States: a Review Volume for the 7th Congress of the International Association of Quaternary Research, edited by H.E. Wright and David G. Frey. Princeton. N.J., Princeton University Press, 1965: 15-25
Six distinct advances of continental glaciers are believed to have occurred in the region during Wisconsin times. Because all drift is very similar in appearance, and because only a few radiocarbon dates are available the correlation of these advances with the Pleistocene stratigraphic sequence of the Midwestern United States is uncertain.

7.126 1965
QUATERNARY OF THE SOUTHERN GREAT PLAINS
Frye, John C.; Leonard, A. Byron
In: The Quaternary of the United States: a Review Volume for the 7th Congress of the International Association of Quaternary Research, edited by H.E. Wright and David G. Frey. Princeton. N.J., Princeton University Press, 1965: 203-216
Pleistocene deposits of the southern Great Plains have been correlated with the glacial sequence of the Mid-West by use of fossils, of distinctive buried soils, by the tracing of terraces, by use of distinctive lithologic types and to a limited extent by use of radiocarbon dating.

7.127 1956
RADIOCARBON AGE DETERMINATION OF RECENT PLEISTOCENE CONTACT IN BLOCK 126 FIELD, EUGENE ISLAND, GULF OF MEXICO
Bray, Ellis E.; Nelson, H.F.
American Association of Petroleum Geologists. Bulletin, v. 40, n. 1, Jan. 1956: 173-177
The contact between recent and Pleistocene in Eugene Island, Gulf of Mexico, is an unconformable surface between sand and clay. Sampling was made at 2 ft above and below the contact and was dated. It indicates an approximate time span of at least 15,000 years, indicating that, either deposition was slower, or part of the section is missing. Other evidence indicates erosion of the clay and unconformable contact.

7.128 1967
RADIOCARBON CHRONOLOGY IN SIBERIA
Kind, N.V.
In: The Bering Land Bridge, edited by David M. Hopkins. Stanford, Calif., Stanford University Press, 1967: 172-192
A proposed scheme for the absolute chronology of the Upper Pleistocene and Holocene is based on the results of age determinations of samples from the northern Yenisei area and on datings from the Siberian regions. A correlation table is given and a table for the absolute chronology of the Upper Pleistocene and Holocene of Siberia.

7.129 1960
RADIOCARBON DATE FOR A WOODLAND MUSC OX IN MICHIGAN
Hibbard, C.W.; Hinds, F.J.
Michigan Academy of Sciences, Arts and Letters. Papers, n. 45, 1960: 103-111

The dates of 7820 ± 450 years BP and 7070 ± 240 years BP indicate that this specimen of woodland musc ox lived in Michigan prior to the development of the Port Huron moraine

7.130   1964
RADIOCARBON DATE ON PLEISTOCENE PECCARY FIND IN SANDUSKY COUNTY, OHIO
Hoare, R.D.
Ohio Journal of Science, v. 64, n. 6, Nov. 1964: 427
A sample bone from the Pleistocene Peccary find in Sandusky County, Ohio, gave a radiocarbon age of 4290 ± 150 years. However dates obtained on plant material in this region were from 12,800 ± 250 to 8513 ± 500 years. Either the use of bone for the date gives an erroneous age, or movement of the sand in which the peccary was buried may have continued long after the retreat of the lake water.

7.131   1955
RADIOCARBON DATES AND PLEISTOCENE CHRONOLOGICAL PROBLEMS IN THE MISSISSIPPI VALLEY REGION
Horberg, Leland
Journal of Geology, v. 63, n. 3, May 1955: 278-286
Evaluates the radiocarbon dates so far available for the Mississippi Valley regions, sets up a tentative radiocarbon chronology which can be compared with various types of geological chronologies and points up uncertainties which are raised by geological lines of evidence.

7.132   1956
RADIOCARBON DATES AND PLEISTOCENE CHRONOLOGICAL PROBLEMS IN THE MISSISSIPPI VALLEY REGION: A REPLY
Leighton, Morris M.
Journal of Geology, v. 64, n. 2, Mar. 1956: 193-194
The author replies to the discussion by Ruhe (*Journal of Geology*, 64, 1956: 191) from the standpoint of stratigraphy, soil profile correlations and radiocarbon dating. A new date for wood, dated by Libby at 16,367 ± 1000 years, was made by Rubin and Suess and gave an age of 14,700 ± 400, which would support Horberg's assumption of the beginning of Cary at 4000 years ago.

7.133   1956
RADIOCARBON DATES AND PLEISTOCENE CHRONOLOGY PROBLEMS IN THE MISSISSIPPI VALLEY REGION: A DISCUSSION
Ruhe, Robert V.
Journal of Geology, v. 64, n. 2, Mar. 1956: 191-193
In his reply to Horberg (*Journal of Geology*, 63, 1955: 278) who proposes a revision of the radiocarbon chronology in Iowa developed by Ruhe and Scholtes (*Journal of Geology*, 63, 1955: 82), the author concludes that in his opinion there is no incompatibility between the radiocarbon chronologies of Iowa, Illinois, Indiana and Ohio and that radiocarbon dates offer an opportunity for a positive approach in the determination of the time factor in soil development. The negative approach of presupposing not enough time for such soil development does not seem justified in view of the foreshortening of the time concept that radiocarbon dates have already impressed on geologic thinking.

7.134   1956
RADIOCARBON DATES AND PLEISTOCENE CHRONOLOGY
Flint, Richard Foster
Geological Society of America. Bulletin, v. 67, Dec. 1956: 1814
Abstract of paper presented at the meeting of the Geological Society of America in New York, Dec. 26-30, 1956. Radiocarbon dating contributes two principal applications to Pleistocene stratigraphy: (1) the measurement of rates of pertinent geological processes, such as sedimentation rates, post-glacial rebound, rise of sea-level, changes of temperature, or rates of advance and retreat of glaciers; (2) correlation of sedimentary units in widely separated areas.

7.135   1959
RADIOCARBON DATES OF PEATS FROM NORTH-PACIFIC NORTH AMERICA
Heusser, Calvin John
Radiocarbon, v. 1, 1959: 29-34
Seventeen late-Pleistocene peat samples from North Pacific North America are dated and the pollen and peat stratigraphy used to determine the environment prevailing at, and since, the time of deposition. At a southerly Alaskan site, late glacial is dated at ca 10,800 BP and the post glacial at ca 10,000 BP and later at a northerly coastal site. A eustatic transgression is suggested during the hypsithermal interval at 5000 BP. Dates by Isotopic Inc., Westwood, New Jersey.

7.136   1951
RADIOCARBON DATING OF LATE PLEISTOCENE EVENTS
Flint, Richard Foster; Deevey, Edward S.
American Journal of Science, v. 249, n. 4, Apr. 1951: 257-300
The radiocarbon determination of age applied to a number of Upper Pleistocene samples is checked against stratigraphically dated material. The dates fall generally in the same order as the relative stratigraphic position of the samples. The results indicate that the process of deglaciation seems to have been more rapid than had been supposed. There is a checklist of samples and dates.

Ch. 7 - Pleistocenes

7.137 1956
RADIOCARBON-BASED PLEISTOCENE CORRELATIONS AND WORLDWIDE CLIMATIC CHANGE
Karlstrom, Thor N.V.
Geological Society of America, Bulletin, v. 67, n. 12, Dec. 1956: 1711, Abstract
Paper presented at the October/November 1956 meeting of the Geological Society in Minneapolis. Radiocarbon dates permit re-assessment of the concept of worldwide, contemporaneous climate history.

7.138 1956
RADIOCARBON-BASED PLEISTOCENE CORRELATIONS AND WORLDWIDE CLIMATIC CHANGE
Karlstrom, Thor N.V.
Science, v. 124, n. 3228, 9 Nov. 1956: 939
Abstract of a paper presented at the Autumn meeting of the National Academy of Science, 8-10 Nov. 1956, Washington, D.C. Radiocarbon data permit a re-assessment of the concept of world-wide, contemporaneous climatic history.

7.139 1963
RECENT RECESSION OF TROPICAL CLIFFY COASTS
Russell, R.J.
Science, v. 139, n. 3549, 4 Jan. 1963: 9-14
Elevated benches and other coastal forms give evidence of eustatic changes in sea-level. Radiocarbon ages on coral at Boundary Beach near Carnarvon, W.A., indicate that Pleistocene sea differed by less than 10 feet from present day levels, a conclusion which may indicate that Antarctic ice played little or no role in glacio-eustatic changes of sea-level.

7.140 1961
RECENT TERRACES OF TROPICAL LIMESTONE SHORES
Newell, Norman D.
Zeitschrift für Geomorphologie, Supplement 3, 1961: 87-106
Radioisotope ages of the dune rocks (including radiocarbon dates) show great variability, but they indicate that the rocks are Pleistocene in age. Tables and discussion of these ages are presented.

7.141 1956
SAN AUGUSTIN PLAINS: PLEISTOCENE CLIMATIC CHANGES
Clisby, Kathryn; Foreman, Fred; Sears, Paul B.
Science, v. 124, n. 3221, 21 Sept. 1956: 537-539
To establish a chronology of glacial episodes in the San Augustin area, sediment cores from lacustrine deposits were analysed on the basis of pollen profile. Two radiocarbon dates at the 19 ft and 28 ft level gave respective dates of $19,700 \pm 300$ and $27,000 \pm 5000$ giving the absolute dates for a series of glacial episodes.

7.142 1965
THE SIMS BAYOU LOCAL FAUNA - PLEISTOCENE OF HOUSTON, TEXAS
Slaughter, Bob H.; McClure, William L.
Texas Journal of Science, v. 17, n. 4, 1965: 404-417
On Sims Bayou, near Houston, an extensive Pleistocene deposit previously obscured by colluvium was uncovered. Fauna recovered included gastropods, pelecypods, fishes, amphibians, reptiles, and mammals. The mammals are described. A radiocarbon date of charcoal chips is in excess of 23,000 years BP. However, since the common cotton rat, *Sigmodon hispidus,* abundant in the Sims Bayou assemblage, has its oldest reported southwestern occurrence in the Moore Pit local fauna, the Houston assemblage may be placed approximately between 50,000 and 25,000 years BP.

7.143 1968
SOME BLOCKSTREAMS OF THE TOOLONG RANGE, KOSCIUSKO STATE PARK, NEW SOUTH WALES
Caine, N.; Jennings, J.N.
Royal Society of New South Wales. Journal and Proceedings, v. 101, 1968: 93-103
Fuller investigation of the blockstreams of the Toolong Range has substantiated previous interpretations placing them as part of a periglacial morphogenetic system of Upper Pleistocene age. Wood in quartz gravel in situ dated at 33,000 B.C. relates to warmer conditions just prior to periglacial phase. Other geological features imply a scarp retreat of about 33 m since 33,000 B.C.

7.144 1963
STATUS OF THE PLEISTOCENE WISCONSIN STAGE IN CENTRAL NORTH AMERICA
Flint, Richard Foster
Science, v. 139, n. 3553, 1 Feb. 1963: 402-404
A brief review of the history of the Wisconsin Stage in Pleistocene stratigraphy and of research since 1950 shows that post-Sangamon glacial drift older than the Wisconsin drift reported in the older literature is present in central North America. Known and possible stratigraphic positions are presented. Discusses the radiocarbon dates on which much of this stratigraphy is based.

7.145   1966
SURFICIAL GEOLOGY, DAWSON, LARSEN CREEK AND NASH CREEK MAP AREAS, YUKON TERRITORY
Vernon, Peter; Hughes, O.L.
Canada. Geological Survey Bulletin, n. 136, 1966. pp. 25
A Pleistocene chronology of the area is established through radiocarbon dates and permits a tentative correlation of the last glaciation of the area with that of the Alaska Range which culminates a little more than 10,000 years ago.

7.146   1964
TUNDRA RODENTS IN A LATE PLEISTOCENE FAUNA FROM THE TOFTY PLACER DISTRICT, CENTRAL ALASKA
Repenning, C.A.; Hopkins, David Moody; Rubin, Meyer
Arctic, v. 17 n.3, Sept. 1964: 177-197
Describes a rodent fauna from Alaska from late Pleistocene which, because of their great ecologic sensitivity, should have a stratigraphic significance in Alaska. Radiocarbon dating of associated woods indicate that the fossils were deposited in their present position very recently but must have been derived from sediments in the immediate area which are probably late Pleistocene.

7.147   1968
TWO LATE PLEISTOCENE RADIOCARBON DATES NEAR BUTTONWILLOW, CALIFORNIA
Manning, John C.
In: Geology and Oil Fields, West Side Southern San Joaquin Valley - AAPG, SEG, and SEPH, Pacific Section. Annual Meeting, 43rd, 1968. *Guidebook*. [Los Angeles, Calif.] American Association of Petroleum Geologists, Pacific Section, 1968: 98-99
Wood fragments in sand coming from the Kern River flood channel were dated by the radiocarbon method: the depth below the surface was 20-35 feet. The dates are 14,060 ± 450 and 13,350 ± 500 years BP and represent the youngest Pleistocene known from the area.

7.148   1968
VEGETATION AND CLIMATE IN THE UPLANDS OF SOUTH-WESTERN UGANDA DURING THE LATE PLEISTOCENE PERIOD
Morrison, M.E.S.
Journal of Ecology, v. 56, n. 2, July 1968: 363-384
Radiocarbon determinations suggest that the oldest sediments date back to about 25,000 years BP.

7.149   1961
VEGETATION, CLIMATE AND RADIOCARBON DATING IN THE LATE PLEISTOCENE OF THE NETHERLANDS; PART I, EEMIAN AND EARLY WEICHSELIAN
Zagwijn, W.H.
Netherlands Geologische Stitching. Mededelingen, n. s., n. 14, 1961: 15-45
Deals with the stratigraphy and pollen analysis of some Eemian and early Weichselian sequences in the Netherlands. The radiocarbon datings from the section dealt with are discussed, as well as correlation with other sequences in N. W. Europe. Tabulated data on the section, including the lithology of pit exposures and borings and radiocarbon ages, are included.

7.150   1966
WOOLLY RHINOCEROS FROM THE SCOTTISH PLEISTOCENE
Rolfe, W.D. Ian
Scottish Journal of Geology, v. 2, part 3, 1966: 253-258
New finds of remains of woolly rhinoceros in the Kelvin Valley, Lanarkshire, are described. A radiocarbon date on collagen gives an age of 27,550 years BP. This implies that the bone is derived from an earlier deposit or that the Kelvin Valley sands and gravels are much older than previously suggested, and are of middle/late Weichselian interstadial age.

CHAPTER EIGHT
¹⁴C ARCHAEOLOGY AFRICA

8.000   1961
ABSOLUTE AGE OF PLEISTOCENE AND HOLOCENE DEPOSITS IN THE HAUA FTEAH
McBurney, C.B.M.
Nature v. 192, n. 4803, 18 Nov. 1961: 685-686

The site seems to offer a nearly continuous culture succession from pre-Mousterian stage up to the present. The chronology suggested by the combined carbon-14 readings and stratigraphy is in general agreement with available carbon-14 dates for the corresponding cultural events at Shanidar (Kurdistan), Karakama (Afghanistan) and Isla Ilöskö (Hungary).

8.001   1962
ABSOLUTE CHRONOLOGY OF THE PALAEOLITHIC IN EASTERN LIBYA AND THE PROBLEM OF UPPER PALAEOLITHIC ORIGIN
McBurney, C.B.M.
Advancement of Science, v. 18, n. 75, Jan. 1962: 474-497
The later Pleistocene spans the range of effective radiocarbon dating.

8.002   1968
AGE OF BED 5, OLDUVAI GORGE, TANZANIA
Leakey, L.S.B.; Protsch, Reiner; Berger, Rainer
Science, v. 162, n. 3853, 1 Nov. 1968: 559-560
Various finds of Hominid remains in Olduvai Gorge, Tanzania, have focussed interest upon the age of the deposits in the sequence of six beds. After bed 1 was dated by the potassium-argon decay method, an absolute date of 10,400 years has now been obtained with the radiocarbon method from a sample of mammalian bones in bed 5.

8.003   1964
THE AGE OF THE PRETORIA WONDERBOOM
Swart, E.R.
South African Journal of Science, v. 60, n. 1, Jan. 1964: 27
The age of the Pretoria 'Wonderboom' has been estimated by radiocarbon dating and found to be at least 750 years BP and probably 1000 years BP overall. The Wonderboom is a large isolated wild fig tree (*Ficus Pretoriae Burtt-Davey*) situated just outside Pretoria.

8.004   1967
ANALYSIS AND DESCRIPTION OF ARTIFACT ASSEMBLAGES ATTRIBUTED TO THE NACHIFUKAN INDUSTRY OF ZAMBIA
Miller, Sheryl Elinor Flum
Ph.D. Dissertation. University of California at Berkeley, 1967. pp. 585
Horizons in this assemblage have been dated by the radiocarbon method. The lowest horizon which contains a macrolithic industry is dated at 23,000 years BP or older. Nachifukan I is dated from 12,000 years BP or earlier, Nachifukan II, to the sixth millenium BP and Nachifukan III a little earlier than the present millenium. The evidence indicates that the Nachifukan culture persisted almost to the present.

8.005   1958
BONE TOOLS AT THE KALKBANK MIDDLE STONE AGE SITE
Mason, Revil J.
South African Archaeological Bulletin, v. 13, n. 5, Sept. 1958: 85-88
Consistent fracturing of bones at various sites and over great periods suggests persistent sub-human and human presence. On the basis of radiocarbon determinations, a date of ca 15,000 years BP was selected as probably reliable indicating that the Pietersburg culture and the Final Sangoan culture might reasonably be expected to be of the same age.

8.006   1962
CARBON 14 CHRONOLOGY IN AFRICA SOUTH OF THE SAHARA
Clark, J. Desmond
In: Pre- and Proto-History, edited by Georges Mortelmans and Jacques Nanquin. Panafrican Congress on Prehistory and Quaternary Study, 4th, Leopoldville, 1959. *Proceedings*, Section III. Tervuren, Belgium, Musée Royal de l'Afrique Centrale, Annales, Series 8, Sciences Humaines, n. 40, 1962: 303-313
A table suggesting a correlation between the cultural succession in Sub-Saharan Africa with that of Western Europe is presented. The various radiocarbon dates on which the table is based are discussed.

8.007   1962
CARBON 14 DATES FOR THE KALOMO CULTURE
Fagan, Brian M.
South African Archaeological Bulletin, v. 17, n. 67, Sept. 1962: 196
Three dates for the Isamu Pati Mound, Kalomo / Choma Iron age project, are reported. It makes it reasonable to suppose that the Kalomo culture flourished from about A.D.

900-1300.

8.008   1965
CARBON 14 DATES FROM SOUTH WEST AFRICA
MacCalman, H.R.
South African Archaeological Bulletin, v. 20, n. 80, part IV, Dec. 1965: 215
Two radiocarbon dates are presented: (1) The Zoo Park Elephant, Windhoeck, dated 5200 ± 140 BP; this is the site of a butchery, associated with a quartz industry, and represents a single short occupation related to a single activity; (2) the Numa Entrance Shelter, Branberg, dated at 870 ± 100 BP on charcoal associated with a Wilton industry and also with copper beads.

8.009   1965
CARBON DATES FOR NIGERIA
Fagg, Bernard
Man, v. 65, article n. 8, Jan/Feb 1965: 22-23
Three dates of interest to West African archaeologists are presented with full description, all are related to the Nok culture.

8.010   1954
CARBON TEST AND SOUTH-WEST AFRICAN PAINTINGS
Breuil, Abbé, Henri
South African Archaeological Bulletin, v. 9, n. 34, June 1954: 48
Charcoal associated with cave paintings give an age of 2368 ± 200 years or between 1681 and 1281 B.C. The age of the paintings in the White Lady shelter in terms of these dates is estimated at greater than 3000 years.

8.011   1953
CARBON TEST, ZIMBABWE
Paver, F.R.
South Africa Archaeolgical Bulletin, v. 8, n. 31, Sept. 1953: 78-79
Discusses radiocarbon dates placing the age of Great Zimbabwe at between 1494 and 1264 years.

8.012   1958
THE CHIFUBWA STREAM ROCK SHELTER, SOLWEZI, NORTHERN RHODESIA
Clark, J. Desmond
South African Archaeological Bulletin, v. 13, n. 49, March 1958: 40-48
A radiocarbon determination on charcoal in the lowest part of accumulated deposit partially covering engraving on the back and side of the wall of the Chifubwa Stream rock shelter gives a date of 6310 ± 250 BP, suggesting that the earlier part of the Later Stone Age in that part of Northern Rhodesia dates to the 4th millemium B.C. (4395 ± 250 B.C.).

8.013   1962
DATING THE EMERGENCE OF MAN
Oakley, Kenneth P.
Advancement of Science, v. 18, n. 75, Jan. 1962: 415-426
Radiocarbon dating is used for the period extending back to 50,000 years. The Neanderthal from West Asia dates to 52,0000 to 35,000 at Shanidar and Mount Carmel and 57,300 ± 500 at Kalambo Falls in Kenya for an Acheulian horizon.

8.014   1966
THE DATING OF THE NAHOON FOOTPRINTS
Deacon, H.J.
South African Journal of Science, v. 62, n. 4, 1966: 111-113
Radiocarbon dating of calcareous material in the sandstone from the footprint site at Nahoon Point gave a result of 29,900 ± 410/390 years BP. This dating is in agreement with the archaeological evidence which suggests formation of this coastal 'dune rock' during the late Upper Pleistocene.

8.015   1957
A DECADE OF DISCOVERY: AFRICA
Howe, Bruce
Archaeology, v. 10, n. 4, Winter 1957: 242-243
New finds from Lybia and the Sudan are dated by radiocarbon, the first, a Neanderthal jaw at about 34,000 years BP, the second, a Neolithic culture at Shaneinab (Sudan) at 39,000 years BP.

8.016   1964
DOMBOZANGE ROCK SHELTER, MTETENGWA RIVER, BEIT BRIDGE, SOUTHERN RHODESIA. EXCAVATION RESULTS
Robinson, K.R.
Arnoldia, v. 1, n. 7, 9 June 1964: 1-13 (Series of miscellaneous publications of the National Museum of Southern Rhodesia)
A radiocarbon date on charcoal of 750 ± 100 years A.D. is the latest yet received for a late stone age site in Southern Rhodesia and fits well with the decorated rim sherds found in the horizon.

8.017   1968
THE EARLY IRON-AGE IN ZAMBIA - REGIONAL VARIANTS AND SOME TENTATIVE CONCLUSIONS
Phillipson, D.W.
Journal of African History, v. 9, n. 2, 1968: 191-211
Radiocarbon dates suggest that early iron people in Zambia

8.018 1961
AN EARLY IRON AGE SITE FOR THE SHIBI DISTRICT, SOUTHERN RHODESIA
Robinson, K.R.
South African Archaeological Bulletin, v. 16, n. 63, Sept. 1961: 75-102
Description of a site in Southern Rhodesia where Gokomere - Zimbabwe type stamped pottery have been found with a definite kraal-site occupation containing Daga structures and other finds such as glass, iron, copper, iron objects, extremely rare at other sites of this culture. Similar finds of pottery date the earlier arrival of 'Channelled Ware' people north of the Zambesi in A.D. 90 and in Southern Rhodesia between A.D. 100 and 300.

8.019 1964
EGYPT AND C14 DATING
Smith, H.S.
Antiquity, v. 38, n. 149, Mar. 1964: 32-37
The author comments on the apparent conflict between radiocarbon and 'historical' dating in 3rd millenium Egypt noted in the Egyptian dates used by Professor Libby in his comments (*Antiquity*, 1963).

8.020 1966
ENGAKURA: A REPORT ON EXCAVATION CARRIED OUT IN 1964
Sassoon, Harmo
Azania, v. 1, 1966: 79-99
Although two radiocarbon dates were obtained giving the following age: A.D. 720 ± 120 from the 30-45 cm level and A.D. 330 ± 90 for the 45-60 cm level, these dates are considered considerably earlier than expected and cannot be accepted until thoroughly confirmed by further excavation and analysis.

8.021 1961
EUROPEAN IRON WORKING IN CENTRAL AFRICA WITH SPECIAL REFERENCE TO NORTHERN RHODESIA
Fagan, Brian M.
Journal of African History, v. 2, n. 2, 1961: 199-210
Two radiocarbon dates are available from Northern Rhodesian channelled ware sites.

8.022 1963
EXCAVATIONS AT SANGA, 1957: THE PROTOHISTORIC NECROPOLIS
Nenquin, Jacques
Musee Royal de l'Afrique Centrale, Tervuren, Belgium. Annales, series IN-8e, Sciences Humaines, v. 45, 1953. pp. 272
Radiocarbon dating on bones from the excavation at Sanga, and other dates from various other necropolis in the region, place the Kisalian culture at about the 7th-9th century A.D. (p. 200-201)

8.023 1963
FURTHER HUMAN REMAINS FROM THE CENTRAL AFRICAN LATER STONE AGE
Gabel, Creighton
Man, v. 63, article n. 44, 1963: 38-42
A hot spring mound on Lochinvar Ranch in the Kafui basin produced a number of human skeletons of Northern Rhodesian Wilton cultural association. A radiocarbon date on a charcoal sample associated with one of the skeletons gave a date of 4700 ± 100 years.

8.024 1964
THE GREEFSWALD SEQUENCE: BAMBAN`DYALAN⊃ AND MAPUNGUBWE
Fagan, Brian M.
Journal of African History, v. 5, n. 3, 1964: 337-361
A tentative chronology based mainly on archaeological evidence is presented. The middle phase of the occupation at Mapungubwe is dated by radiocarbon at A.D. 1380 ± 60 and 1420 ± 60

8.025 1953
THE HAUA FTEAH FOSSIL JAW
McBurney, C.B.M.; Trevor, J.C.; Walls, L.H.
Royal Anthropological Institute of Great Britain and Ireland. Journal, v. 83, n. 1, 1953: 71-76
Chronology cannot yet be ascertained as the radiocarbon measurements of the samples collected have not yet come to hand.

8.026 1960
HUMAN ECOLOGY DURING PLEISTOCENE AND LATER TIME IN AFRICA, SOUTH OF THE SAHARA
Clark, J. Desmond
Current Anthropology, v. 1, n. 4, July 1960: 307-324
This article is an attempt to bring together the evidence of the cultural remains found in great profusion in the vast region of Africa, south of the Sahara, and to show what can legitimately be deduced from such evidence about the relationship of man and his environment throughout the various stages of prehistoric time. A number of the sites studied have been radiocarbon dated.

8.027   1965
A HUMAN FRONTAL BONE FROM THE LATE PLEISTOCENE OF THE KOM OMBO PLAIN, UPPER EGYPT
Reed, Charles A.
Man, v. 65, article n. 95, July-August 1965: 101-104
Two radiocarbon age determinations on different materials are in essential agreement. The samples, one of charcoal and one of shell of freshwater clam, were directly associated with the Sabilian artifacts and the human frontal bone. Other dates from similar material at a nearby site confirm the general validity of an age of approximately 11,000 to 11,500 years B.C. for the frontal bone.

8.028   1966
IRON AGE CHRONOLOGY IN SOUTHERN ZAMBIA
Phillipson, D.W.; Fagan, Brian M.
South African Archaeological Bulletin, v. 21, n. 83, part III, Oct. 1966: 121
A series of four radiocarbon dates from Dambwa placing the site in the eighth century raises doubts on the validity of the basal Kalundu date of A.D. 300. (Fagan, B.M., 1966. *Iron Age Cultures of Zambia I.*).

8.029   1966
THE IRON AGE OF SOUTHERN RHODESIA
Summers, Roger
Current Anthropology, v. 7, n. 4, Oct. 1966: 463-469
The Southern Rhodesia iron age falls into two clearly differentiated complexes: earlier, from A.D. 300 up to the 19th century and later from the 10th to the 19th century. Radiocarbon dates from that area are discussed.

8.030   1966
THE IRON AGE OF ZAMBIA
Fagan, Brian M.
Current Anthropology, v. 7, n. 4, Oct. 1966: 453-462
Stratigraphical evidence and radiocarbon dates from Eastern Basutoland are used to suggest a chronology for the iron age in Zambia.

8.031   1963
THE IRON-AGE SEQUENCE IN THE SOUTHERN PROVINCE OF NORTHERN RHODESIA
Fagan, Brian M.
Journal of African History, v. 4, n. 2, 1963: 157-177
A tentative iron age sequence. The chronology is based on radiocarbon dates. Includes the earliest farmers, the Kalomo culture, the Kangila people, the Ingombe Ilede site.

8.032   1968
KALAMBO FALLS PREHISTORIC SITE
Clark, J. Desmond
Cambridge University Press, 1968
In appendix G, p. 23 a summary of radiocarbon dates from excavated samples is presented in table form.

8.033   1962
THE KALAMBO FALLS PREHISTORIC SITE: AN INTERIM REPORT
Clark, J. Desmond
In: *Pre- and Proto-History*, edited by Georges Mortelmans and Jacques Nanquin. Panafrican Congress on Prehistory and Quaternary Study, 4th, Leopoldville, 1959. *Proceedings*, Section III. Tervuren, Belgium, Musée Royal de l'Afrique Centrale, Annales, Series 8, Sciences Humaines, n. 40, 1962: 195-202
A table of the Kalambo Falls correlation using radiocarbon dates is presented.

8.034   1963
THE KALOMO - CHOMA IRON AGE PROJECT (1960-1963): PRELIMINARY REPORT
Fagan, Brian M.
South African Archaeological Bulletin, v. 18, n. 69, part I, April 1963: 3-19
Three dates from Isamu Pati enable to give an estimate of ca A.D. 900 for the start of the Kalomo culture, and cover a period of some 500 years. They also reveal that there were interesting differential rates of accumulation of occupation deposit.

8.035   1965
THE LATER PLEISTOCENE CULTURES OF AFRICA
Clark, J. Desmond
Science, v. 150, n. 3698, 2 Nov. 1965: 833-847
It was during the later Pleistocene that cultures in Africa first showed significant regional specialisation. Radiocarbon dates help recognise the patterns and the spreading of traditions.

8.036   1963
LOCHINVAR MOUND: A LATER STONE AGE CAMP-SITE IN THE KAFUE BASIN
Gabel, Creighton
South African Archaeological Bulletin, v. 18, n. 70, part II, July 1963: 40-48
Description of the site and results of the excavation. Dates on two charcoal samples are in obvious conflict, giving 4700 ± 100 years at 5 ft depth and 4290 ± 150 at 8 ft depth. Contamination is probable.

8.037   1956
THE MACROLITHIC CULTURE OF FLORISBAD

Meiring, A.D.J.
Museum of Bloemfontein. Researches, v. 1, n. 9, 11 June 1956: 205-237
Radiocarbon dates support the surmise that a long period fell between the period of Lydianite tool making and dolerite tool making times.

8.038　　　　　　　　　　　　　　　　　　　1961
METHOD OF STUDYING ETHNOLOGICAL ART
Haselberg, Herta
Current Anthroplogy, v. 2, n. 4, Oct. 1961: 341
A controversial date of the horizon at the base of a rock picture helps to demonstrate a continuity of Negro art over 5000 years but is contested by E. Antevs who doubts the correlation between the archaeological deposit and the rock picture.

8.039　　　　　　　　　　　　　　　　　　　1967
THE MONK'S KOP OSSUARY
Crawford, J.R.
Journal of African History, v. 8, n. 3, 1967: 373-382
Description of the excavation of a cave used for funerary purposes in the Mtoroshanga district of Rhodesia. The earlier layer is dated by the radiocarbon method to approximately the late 13th or early 14th century A.D.

8.040　　　　　　　　　　　　　　　　　　　1962
THE NEOLITHIC CULTURE OF EAST AFRICA
Posnansky, Merrick
In:Pre- and Proto-History, edited by Georges Mortelmans and Jacques Nanquin. Panafrican Congress on Prehistory and Quaternary Study, 4th, Leopoldville, 1959. *Proceedings*, Section III. Tervuren, Belgium, Musée Royal de l'Afrique Centrale, Annales, Series 8, Sciences Humaines, n. 40, 1962: 273-281
The dates and origin of neolithic cultures in an area of the African Rift are examined. Two radiocarbon dates were produced: 3000 BP for a variant of the Stone Bowl culture at Hyrax Hill and 375 ± 175 BP at Lanet.

8.041　　　　　　　　　　　　　　　　　　　1964
A NEW FOSSIL HUMAN POPULATION FROM THE WADI HALFA AREA, SUDAN
Hewes, G.W.; Irwin, H.; Papworth, M.; Saxe, S.
Nature, v. 203, n. 4943, 25 July 1964: 341-343
Analysis of the lithic industry and radocarbon dates on burnt bones results on a tentative date for a presumably upper-Palaeolithic site west of the Nile opposite Waddi Halfa, Republic of Sudan, prior to 4000 B.C.

8.042　　　　　　　　　　　　　　　　　　　1964
NEW RADIOCARBON DATES FOR THE GWISHO SITES ON LOCHINVAR RANCH
Fagan, Brian M.
Archaeologia Zambiana, n. 1, Nov. 1964: 3
The radiocarbon dates of the Wilton sites at Gwisho Hotsprings, Lochinvar, indicate that the two sites were occupied simultaneously for a period of over a thousand years, from about 2700 B.C. to about 1700 B.C.

8.043　　　　　　　　　　　　　　　　　　　1967
NEW VIEWS ON ENGARUKA, NORTHERN TANZANIA: EXCAVATION CARRIED OUT FOR THE TANZANIAN GOVERNMENT IN 1964 AND 1966
Sassoon, Harmo
Journal of African History, v. 8, n. 2, 1967: 201-217
Radiocarbon dates suggest that the terrace site in the hillside were occupied during the first millenium A.D. and that the stone circles on the lower slopes in the valley were occupied during the 15th century A.D.

8.044　　　　　　　　　　　　　　　　　　　1968
NEW ZAMBIAN RADIOCARBON DATES
Phillipson, D.W.
Archaeologia Zambiana, n. 9, Nov. 1968: 4
Two dates from Early Iron Age levels at Gundu, near Bakota, described by Brian Fagan and T. Huffman (*Archaeologia Zambiana*, 2, 1967) are presented.

8.045　　　　　　　　　　　　　　　　　　　1968
NEW ZAMBIAN RADIOCARBON DATES
Phillipson, D.W.
Archaeologia Zambiana, n. 10, Dec. 1968: 6-8
Forty one dates from Zambian sites are presented.

8.046　　　　　　　　　　　　　　　　　　　1966
NEW ZAMBIAN RADIOCARBON DATES, FEBRUARY - JUNE 1966
Phillipson, D.W.; Fagan, Brian M.
Archaeologia Zambiana, n. 7, Aug. 1966: 6
One late stone age date at Kamusongolwa Kopje, Fasempa, and six iron age dates from Dambwa, Livingstone, Twickenham road, Lusaka, Feira and Strydom's Farm, Livingstone, are presented and discussed.

8.047　　　　　　　　　　　　　　　　　　　1962
NOK TERRACOTTAS IN WEST AFRICAN HISTORY
Fagg, B.E.B.
In:Pre- and Proto-History, edited by Georges Mortelmans and Jacques Nanquin. Panafrican Congress on Prehistory and Quaternary Study, 4th, Leopoldville, 1959. *Proceedings*, Section III. Tervuren, Belgium, Musée Royal de l'Afrique Centrale, Annales, Series 8, Sciences Humaines, n. 40, 1962: 445-450

According to radiocarbon determinations it is estimated that the date of the figurine culture can be placed in the period between about 500 B.C. and A.D. 200.

8.048   1964
NORTHERN RHODESIA 1500 YEARS AGO: RICH FINDS FROM AN IRON AGE SETTLEMENT IN THE GWEMBE VALLEY, DOWNSTREAM FROM KARIBA DAM
Fagan, Brian M.
Illustrated London News, v. 243, 20 June 1964: 988-991
The site, some thirty miles downstream of the Kariba dam presents a number of burials dated by radiocarbon between A.D. 680 and 800-900, earlier than the great period of Zimbabwe.

8.049   1959
NOTES ON THE ZIMBABWE-SOFALA PROBLEM
Mauny, Raymond
South African Archaeological Bulletin, v. 14, n. 53, March 1959: 18-20
Discusses the dating of the Zimbabwe ruins on a radiocarbon determination made on wood found in a drain in the masonery, on the basis of the too youthful appearance of the ruins and their correlation with other ruins, and also on the basis that the wood used was probably already very old at the time of construction. Also mentions contamination by radioactivity contained in the local granite.

8.050   1962
PLEISTOCENE CLIMATE AND CULTURE IN NORTH-EASTERN ANGOLA
Van Zinderen Bakker, Edward M.A.; Clark, J. Desmond
Nature, v. 196, n. 4855, 17 Nov. 1962: 639-642
Making comparison between the prehistoric cultures of Africa and Europe, it is apparent from the Angola and Kalambo Falls radiocarbon dates that the Sangoan is in large part contemporary with old Mousterian and these dates provide good evidence for a general correlation between the last pluvial in Sub-Saharan Africa and the Würm glaciation in Europe.

8.051   1959
POINTS FROM CORRESPONDENCE
Davies, J.
South African Archaeological Bulletin, v. 14, n. 53, June 1959: 56
Points out that some of the arguments used by R. Mauny (*South African Archaeol. Bull*, 1959) to prove that the dates obtained by radiocarbon dating found in a drain in the ruins of Zimbabwe is wrong as contamination by the radioactivity of the local granite would have given a date which would be too young, not too old as considered by R. Mauny.

8.052   1964
PREHISTORIC CULTURE AND PLEISTOCENE VEGETATION AT THE KALAMBO FALLS, NORTHERN RHODESIA
Clark, J. Desmond; Van Zinderen Bakker, Edward M.A.
Nature, v. 201, n. 4923, 7 Mar. 1964: 971-975
The stratigraphical and cultural sequence and the radiocarbon dates, together with pollen analysis give a picture of the culture, vegetational and climate changes in the region from 55,000 B.C. to A.D. 1590.

8.053   1963
THE PREHISTORY OF EAST AFRICA
Cole, Sonia Mary
New York, Macmillan, 1963
A review work in the search of early man in East Africa. Research works cited use radiocarbon dates and other radiometric methods to establish chronological sequences.

8.054   1954
PROVISIONAL CORRELATION OF PREHISTORIC CULTURES NORTH AND SOUTH OF THE SAHARA
Clark, J. Desmond
South African Archaeological Bulletin, v. 9, n. 33, March 1954: 3-17
Indicates the main lines along which a correlation can be effected. The correlation is presented in table form. A number of the recent dates are based on radiocarbon determinations.

8.055   1968
RADIOCARBON CHRONOLOGY OF THE IRON AGE IN SUB-SAHARAN AFRICA
Stuiver, Minze; Van der Merwe, Nicholas J.
Current Anthropology, v. 9, n. 1, Feb. 1968: 54-58
Radiocarbon dates provide the only reliable means of establishing a chronology for the transmission of Iron-age culture in Sub-Saharan Africa. The authors present the dates in table form and discuss some of these dates and their significance.

8.056   1968
RADIOCARBON DATES FOR BENIN CITY AND FURTHER DATES FOR DAIMA, N.E. NIGERIA
Connah, Graham
Historical Society of Nigeria. Journal, v. 4, n. 2, June 1968: 313-320
The dates are interesting in that they indicate that the chronological sequence suggested at the time of excavation was probably too short. They are important because they

confirm the long continuity of occupation of the site of Benin.

8.057   1967
RADIOCARBON DATES FOR DAIMA, N.E. NIGERIA
Connah, Graham
Historical Society of Nigeria. Journal, v. 3, n. 4, June 1967: 741-742
According to the radiocarbon dates presented (in table form) it seems reasonable to presume a date for the first occupation of the Daima site, somewhat before the 5th century B.C.

8.058   1967
RADIOCARBON DATES FOR KAPWIRIMBWE
Phillipson, D.W.
Archaeologia Zambiana, n. 8, 1967: 7
The fifth century A.D. is indicated for the short occupation at Kapwirimbwe

8.059   1967
RADIOCARBON DATES FROM LEOPARD'S HILL
Phillipson, D.W.
Archaeologia Zambiana, n. 8, 1967: 9
The dates, although not all consistent with stratigraphy, give further evidence for the theory of a 'long' chronology for the Central African Late Stone Age.

8.060   1966
A RADIOCARBON DATE FROM NAKAPAPULA
Phillipson, D.W.
Archaeologia Zambiana, n. 6, Feb. 1966: 4
A rockshelter at Nakapapula, near Chitambo Mission, gave an early iron age date of A.D. 770 ± 100.

8.061   1967
RADIOCARBON DATES FROM NIGERIA
Shaw, Thurston
Historical Society of Nigeria. Journal, v. 3, n. 4, June 1967: 743-751
Thirty one radiocarbon dates from Nigeria are presented in table form, assessed and discussed.

8.062   1962
RADIOCARBON DATES OF NILE SEDIMENTS
Fairbridge, Rhodes W.
Nature, v. 196, n. 4850, 13 Oct. 1962: 108-110
Radiocarbon dates help establish a tentative curve for Nile oscillations.

8.063   1968
RADIOCARBON DATING IN NIGERIA
Shaw, Thurston
Historical Society of Nigeria. Journal, v. 4, n. 3, Dec. 1968: 453-465
72 Nigerian radiocarbon dates are presented and assessed.

8.064   1964
RADIOCARBON DATING OF A LATE PALEOLITHIC CULTURE FROM EGYPT
Smith, P.E.L.
Science, v. 145, n. 3634, 21 Aug. 1964: 811
Two radiocarbon dates of about 12,000 B.C. from a new prehistoric culture from a stratified site at Kom Ombo, Upper Egypt, throw light on a deposition phase of the late Pleistocene Nile. The dates reveal that the associated blade industry is coeval with at least the later part of the Upper Paleolithic in Europe and Southwestern Asia.

8.065   1964
RADIOCARBON DATING OF IRON AGE SITES IN THE SOUTHERN TRANSVAAL
Mason, R.J.; Van der Merwe, Nicholas J.
South African Journal of Science, v. 60, n. 5, May 1964: 142
Two important Iron Age horizons in the Southern Transvaal were dated by radiocarbon. They are the first Iron Age age estimations for the Southern Transvaal and their reliability is supported by a grid of Iron Age dates in adjacent Rhodesia.

8.066   1965
RADIOCARBON DATING OF THE NOK CULTURE, NORTHERN NIGERIA
Fagg, Bernard
Nature, v. 205, n. 4967, 6 Jan. 1965: 212
Charcoal excavated from layer three from an early occupation site at Taruga, Northern Nigeria, together with another date on wood indicate that an early Nok iron age culture survived nearly five centuries.

8.067   1968
RATIFICATION AND RETROCESSION OF EARLIER SWAZILAND IRON ORE MINING RADIOCARBON DATING
Dart, Raymond A.; Beaumont, P.B.
South African Journal of Science, v. 64, n. 6, 1968: 241-246
Radiocarbon dates on carbon samples associated with stone mining tools in the basal portion of deposits at present overlying mined iron ore bedrock were dated, giving information on the antiquity of iron mining in Africa. Dates range from as far back as ca 30,000 BP.

8.068 1962
REPORT ON AN ANCIENT BURIAL GROUND, SALISBURY, SOUTHERN RHODESIA
Goodall, E.
In:Pre- and Proto-History, edited by Georges Mortelmans and Jacques Nanquin. Panafrican Congress on Prehistory and Quaternary Study, 4th, Leopoldville, 1959. *Proceedings,* Section III. Tervuren, Belgium, Musée Royal de l'Afrique Centrale, Annales, Series 8, Sciences Humaines, n. 40, 1962: 315-
Two dates obtained through radiocarbon analysis, of 670 ± 100 years BP and 800 ± 60 years BP for different burials, indicate a lengthy but reasonable time of occupation, supplying an acceptable working foundation.

8.069 1964
REPORT ON EXCAVATIONS AT POMONGWE AND TSHANGULA CAVES, MATAPOS HILLS, SOUTHERN RHODESIA
Cooke, C.K.
South African Archaeological Bulletin, v. 18, n. 71, part III, Nov. 1964: 73-151
Radiocarbon dating indicates that the earliest deposits are over 42,200 years old. A table of radiocarbon dates and cultures is presented and also a diagrammatic table of cultures, samples and dates (radiocarbon) and a table showing correlation. These dates, chronology and correlation are discussed.

8.070 1965
REVIEW OF THE PREDYNASTIC DEVELOPMENT IN THE NILE VALLEY
Arkell, A.J.; Ucko, Peter
Current Anthropology, v. 6, n. 2, Apr. 1965: 145-156
On the basis of radiocarbon dates, the development of the prehistoric Nile Valley appears almost irrelevant from the point of view of the first domestication of animals, the first cultivation of cereals, the first practice of agriculture and the first making of pottery. This article attempts to reconsider Nile Valley prehistory.

8.071 1965
SEBANZI: THE IRON AGE SEQUENCE OF LOCHINVAR AND THE TONGA
Fagan, Brian M.; Phillipson, D.W.
Royal Anthropological Institute Journal, v. 95, part 1 & 2, Jan-Dec. 1965: 253-294
Traces of Iron Age activity ranging in date from recent times to the earliest phase of the Zambian Iron Age have been discovered on Lochinvar Ranch. This is supported by two radiocarbon dates at Masili and at Lusu.

8.072 1965
SHANEINAB: AN ACCOUNT OF THE EXCAVATION OF A NEOLITHIC OCCUPATION SITE CARRIED OUT FOR THE SUDAN ANTIQUITIES SERVICE IN 1949-1950
Arkell, A.J.
London, Oxford University Press, 1953. pp. 114
A discussion of two dates, one of charcoal for the Fayum Neolithic and one of shells for the Khartoum Neolithic show a discrepancy of 800 years although the two cultures are too much connected archaeologically to accept the difference. The more ancient date on shells, at 5826 BP seems more acceptable.

8.073 1959
SOME DEVELOPMENTS IN SOUTH AFRICAN PHYSICAL ANTHROPOLOGY
Tobias, Phillip V.
In: The Skeletal Remains at Bambandyanalo, by Alexander Galloway. Johannesburgh, Witwatersrand University Press, 1959: 127-254
Although Mapungubwe and Bambandyanalo were excavated before the radiocarbon technique of age determination, samples extracted by dissection from the central core of larger fragments of charcoal were dated. The chronological sequence thus established accords with the sequence deduced from skeletal and cultural evidence. This suggests that contamination has not been a serious factor.

8.074 1961
THE SOUTHERN RHODESIAN IRON-AGE (FIRST APPROXIMATION TO THE HISTORY OF THE LAST 2000 YEARS)
Summers, Roger
Journal of African History, v. 2, n. 1, 1961: 1-13
The chronology is based on radiocarbon dates. A table recording the relevant available dates is presented.

8.075 1954
STUDY TOUR OF EARLY HOMINID SITES IN SOUTHERN AFRICA
Oakley, Kenneth P.
South African Archaeological Bulletin, v. 9, n. 35, Sept. 1954: 75-87
Samples of peat from the Florisbad site were dated at more than 41,000 years old, however later dating reported 9000 years for the next layer up and 7000 years for the next. The discrepancy could be accounted for by contamination with carbon from some other older formation.

8.076 1968
TRANSVAAL AND NATAL IRON AGE SETTLEMENTS REVEALED BY AERIAL PHOTOGRAPHY AND EXCAVATION

Mason, R.J.
African Studies, v. 27, n. 4, 1968: 167-180
Complementary to the Iron Age project, a survey is made by aerial photography of the areas in order to establish a settlement analysis. Radiocarbon dates associated with the project are presented in table form and discussed.

8.077   1954
TWO RADIOCARBON DATES FOR THE KISALIAN
Nenquin, Jacques
Antiquity, v. 35, n. 140, Dec. 1954: 322
Two dates obtained from skeletal material indicate that the 8th to 9th century A.D. may be considered as the period of the Sanga necropolis occupation.

8.078   1966
TWO RADIOCARBON DATES FROM SCOTT'S CAVE, GAMTOOS VALLEY
Deacon, H.J.
South African Archaeological Bulletin, v. 22, n. 86, part II, Sept. 1967: 51-52
Two samples of charcoal from a late stone age site in the Cape Province, South Africa, indicate a long time range of 800 years of occupation. According to archaeological observation of the deposit, the range was not expected to be more than 500 years. This suggests that the pottery phase associated with the dated charcoal may be older than generally assumed.

8.079   1954
AN UPPER PLEISTOCENE SITE AT THE KALAMBO FALLS ON THE NORTHERN RHODESIA- TANGANYAKA BORDER
Clark, J. Desmond
South African Archaeological Bulletin, v. 9, n. 34, June 1954: 51-56
Wood preserved within the actual camping floor, never before found in Africa, sent for radiocarbon dating should make it possible to obtain an absolute date for the end of the Early Stone Age.

8.080   1968
THE VALUE OF IMPORTED CERAMICS IN THE DATING AND INTERPRETATION OF THE RHODESIAN IRON AGE
Garlake, P.S.
Journal of African History, v. 9, n. 1, 1968: 13-33
The dating of evidence provided by the import of Chinese ceramics found in the important iron age centre of Zimbabwe is discussed. Radiocarbon dates give the age of period 3 and the start of period 4 at the 13th and 14th century A.D.

8.081   1966
WOODEN IMPLEMENTS FROM LATE STONE AGE SITES AT GWISHO HOT SPRINGS, LOCHINVAR, ZAMBIA
Fagan, Brian M.
Prehistoric Society. Proceedings, v. 32, n. 5, 1966: 246-261
Radiocarbon dates from Gwisho B are indicative of a sporadic occupation of the settlement for over 1000 years and place it in the middle stage of the Wilton.

8.082   1964
ZAMBIAN RADIOCARBON DATES
Phillipson, D.W.; Fagan, Brian M.
Archaeologia Zambiana, n. 1, 1964: 5-6
A list of all radiocarbon dates so far published is presented. The following sites and cultures included are as follows: (1) Kalambo Falls: early Stone Age; Sangoan, middle Stone Age; (2) Twin River Kopje: middle Stone Age; Nachifukan; Mwela Rock Shelter: late Stone Age; (3) Zambian Iron Age.

8.083   1965
ZAMBIAN RADIOCARBON DATES
Phillipson, D.W.
Archaeologia Zambiana, n. 5, Nov. 1965: 1-4
Fifty one radiocarbon dates currently available are presented together with their geographical location and their chronological distribution.

8.084   1963
ZIMBABWE, A RHODESIAN MYSTERY
Summers, Roger F.
Johannesburg, South Africa, Thomas Nelson and Sons, 1963
The first radiocarbon analysis on timber embedded in one of the walls gave an age of A.D. 591 ± 120, much earlier than had been tentatively suggested. Later dates span a period from A.D. 330 ± 150 to about A.D. 1830. The radiocarbon method of dating archaeological specimens is described in appendix 1.

CHAPTER NINE
14C ARCHAEOLOGY AMERICAS

9.001   1960
¹⁴C DATES FOR VENADO BEACH, CANAL ZONE
Lothrop, S.K.
Panama Archaeologist, n. 3, 1960: 96
[Not sighted]

9.002   1963
THE 1962 EXCAVATIONS OF THE SHERMAN PARK MOUND SITE, 39MN8, A NEWLY RADIO-CARBON DATED SITE IN SOUTH DAKOTA
Gant, E.; Hurt, W.H.
W. H. Over Museum (South Dakota State University, Vermillion). *Museum News*, v. 24, n. 1, 1963: 1-4
[Not sighted]

9.003   1968
ABORIGINAL OCCUPATION AND CHANGES IN RIVER CHANNEL ON THE CENTRAL UCAYALI, PERU
Lathrap, D.W.
American Antiquity, v. 33, n. 1, Jan, 1968: 62-79
Uses radiocarbon dates obtained for sites adjacent to the complex to suggest that the sequence represents, at the very minimum, a span of 3000, perhaps 4000 years.

9.004   1959
ABSOLUTE CHRONOLOGY IN THE CARIBBEAN AREA
Rouse, Irving; Cruxent, Jose M.; Goggin, John M.
In: *International Congress of Americanists*, 32nd, Copenhagen, 8-14 August, 1956. Proceedings. Copenhagen, Munskegaard, 1959: 508-515.
Tables of radiocarbon dates are presented and a revision of the relative chronology to conform with the radiocarbon dates is derived for Venezuala islands (offshore) and north coast Lower Orinico River. However the dates from the Lower Orinoco River do not agree with the relative chronology, hence the revised version. The relative and the revised relative chronology are presented in table form.

9.005   1953
ADDITIONAL DATA ON THE FARMINGTON COMPLEX, A STONE IMPLEMENT ASSEMBLAGE OF PROBABLY EARLY POST-GLACIAL DATE FROM CENTRAL CALIFORNIA
Treganza, A.E.; Heizer, Robert F.
University of California, Berkeley. Archaeological Survey Reports, n. 22, Archaeological Paper n. 21-26, 1953: 28-38
[Not sighted]

9.006   1956
ADDITIONAL DATES ON THE WOODRUFF OSSUARY, KANSAS
Wedel, Waldo R.; Kivett, Marvin F.
American Antiquity, v. 21, n. 4, Apr. 1956: 414-416
The Woodruff ossuary was excavated in 1946 and carbonised material collected in direct association with the archaeological and skeletal remains were submitted for dating. The results of these examinations are briefly discussed in this article together with some of their implications for Central Plains prehistory.

9.007   1967
ADDITIONAL DISCOVERIES OF FILED TEETH IN THE CAHOKIA AREA
Perino, Gregory
American Antiquity, v. 32, n. 4, Oct. 1967: 538-542
Evidence of tooth filing, North of Mexico, is found primarily in the Cahokia area in Illinois. The custom is associated with the acculturation process that took place between peoples having a Mississippian technology and those having a Late Woodland culture. Because the spread of 220 years indicated by radiocarbon dates seem an impossible figure to the author, he averages the plus and minus factors to derive a date of A.D. 1040.

9.008   1966
ADDITIONAL RADIOCARBON DATES FOR THE SAMBAQUIS OF BRAZIL
Hurt, Wesley R.
American Antiquity, v. 31, n. 3, Jan. 1966: 440-441
Three new radiocarbon dates which fit well the previously reported dates are presented.

9.009   1956
ADDITIONAL RADIOCARBON DATES, LOVELOCK CAVE, NEVADA
Cressman, L.S.
American Antiquity, v. 21, n. 3, Jan. 1956: 311-312
Radiocarbon dates published by Libby confirm earlier conclusions that the earliest date of occupation of Lovelock caves was something beyond 3100 years ago. Ages of

samples within the same level were combined as their depth were recorded for the level as a whole.

9.010   1957
AGE OF SANDIA CULTURE
Gross, Hugo
Science, v. 126, n. 3268, 16 Aug. 1957: 305-306
The author ascribes the Sandia level to early Wisconsin stadial, in agreement with Crane's first radiocarbon date (*Science*, 122, 1955: 689), on stratigraphical evidence.

9.011   1962
ALEUT-KONYAG PREHISTORY AND ECOLOGY, 1961. ARCHAEOLOGICAL INVESTIGATION ON UMNAK ISLAND, ALEUTIANS
Laughlin, William Sceva
Arctic Anthropology, v. 1, n. 1, 1962: 108-110
Describes archaeological materials excavated at the old village site of Chaluka near Nikolski. The earliest radiocarbon date, 3750 ± 180 years, is that of a charcoal sample from 60 cm depth, 60 cm above the base of the midden. Estimated occupation from about 2000 B.C. to recent.

9.012   1965
ALLIGATOR LAKE, A CERAMIC HORIZON SITE ON THE NORTWEST FLORIDA COAST
Lazarus, William C.
Florida Anthropologist, v. 18, n. 2, Dec. 1965: 83-124
A table of time relationships of culture along the Gulf Coast, based on radiocarbon dates, illustrates the pivotal nature of the Alligator Lake Site.

9.013   1957
AN ALLUVIAL SITE ON THE SAN CARLOS INDIAN RESERVATION, ARIZONA
Haury, Emil W.
American Antiquity, v. 23, n. 1, July, 1957: 2-27
Radiocarbon determinations were obtained from two sources, the University of Arizona and the University of Michigan who employ respectively the carbon black and the gas sample techniques. The dates submitted by each laboratory are, on the whole, consistent with the stratigraphy, but there is a marked discrepancy in the values. Supporting information needs to be introduced to evaluate and choose the appropriate date. This places a doubt on the validity of radiocarbon dates.

9.014   1964
ANANGULA: GEOLOGIC INTERPRETATION OF THE OLDEST ARCHEOLOGIC SITE IN THE ALEUTIANS
Black, Robert F.; Laughlin, William Sceva
Science, v. 143, n. 3612, 20 Mar. 1964: 1321-1322
Anangula preserves remains of a lamellar flake industry probably between 8000 and 12,000 years old because of a unique combination of geologic factors. The most important are the relation of land and sea in this region in the past, when the sea-level was lower than at present, and the relatively slight erosion by the sea today. The sequence of volcanic ash and soil horizon is similar to that around Nikolski when dated by the radiocarbon method.

9.015   1955
ANCIENT MAIZE AND MEXICO
MacNeish, Richard Stockton
Archaeology, v. 8, n. 2, Summer 1955: 108-115
Excavations at La Perra cave in the Canyon Diablo, yielded material which, dated by the radiocarbon method, was found to be 4445 years old, much older than previously found.

9.016   1967
ANCIENT OYSTER AND BAY SCALLOP SHELLS FROM SABLE ISLAND
Clarke, A.H.; Stanley, D.J.; Medcof, J.C.; Drinnan, R.E.
Nature, v. 215, n. 5106, 9 Sept. 1967: 1146-1148
Radiocarbon dating suggests that warm water oysters and bay scallops migrated northwards during and after the climatic thermal maximum. The bay scallops may have arrived too late to reach areas in Canada which are now favourable for them. Peat samples were used to date sea-level change.

9.017   1963
ANOTHER RADIOCARBON DATE FOR AZATLAN
Ritzenthaler, Robert
Wisconsin Archaeologist, v. 44, n. 3, 1963: 180
A refuse pit in the northeast corner of the village of Azatlan was dated at 580 ± 100 BP.

9.018   1955
ANTIQUITY OF THE SANDIA CULTURE: CARBON-14 MEASUREMENTS
Crane, H.R.
Science, v. 122, n. 3172, 14 Oct. 1955: 689-690
A mammoth tusk in association with human habitation from the Sandia Levels of Sandia cave was dated at 20,000 to 30,000 years. As this level is below the Folsom culture level, dated at 11,000 years and beyond which very little is known in American anthropology, the dating of this level is of unusual importance. However more radiocarbon measurements on other material from the same level need to be made in particular to ascertain whether the mammoth tusk is in fact contemporary with the culture level. The technique used in the measurements is described.

9.019 1965
AN ANTLER ARTIFACT FROM THE LATE PLEISTOCENE OF NORTHERN TEXAS
Slaughter, Bob H.; Hoover, B. Reed
American Antiquity, v. 30, n. 3, Jan. 1965: 351-352
A perforated pick-like antler associated with late Pleistocene fauna in Delta County was dated through associated charcoal from a burned area in the same deposit at 9550 ± 375 years BP. Interpretation of the mammals and molluscs indicate that, during the deposition of the Pleistocene alluvium, the climate was more moist and the summers cooler than at present.

9.020 1966
ARCHAEOLOGICAL AND PALAEOBOTANICAL INVESTIGATIONS IN SALTS CAVE, MAMMOTH CAVE NATIONAL PARK, KENTUCKY
Watson, Patty Jo; Yarnell, Richard A.
American Antiquity, v. 31, n. 6, 1966: 842-849
The prehistoric utilisation of Salts Cave is examined. Occupation is dated by the radiocarbon method at the last part of the last millenium B.C.

9.021 1952
THE ARCHAEOLOGICAL AND PALAEONTOLOGICAL SALVAGE PROGRAM AT THE MEDICINE CREEK RESERVOIR, FRONTIER COUNTY, NEBRASKA
Davis, E. Mott; Schultz, Bertrand C.
Science, v. 115, n. 2985, 14 March 1952: 288-289
Summary of the results of six years of scientific salvage operations by the University of Nebraska State Museum at the Medicine Creek reservoir. Judging from evidence and radiocarbon dating, the area was occupied in late-glacial times by wandering groups of bison hunters who camped repeatedly along Medicine Creek.

9.022 1949
ARCHAEOLOGICAL EXCAVATIONS AT THE HARLAN SITE, FORT GIBSON RESERVOIR, CHEROKEE COUNTY, OKLAHOMA
Bell, R.E.
Plains Archaeology Conference, Newsletter, v. 3 1949: 3-15
Excavation of five Indian mounds and two associated homes was undertaken prior to inundation by the Fort Gibson Reservoir. This paper is a progress report describing the site and excavated units. It does not give laboratory analysis, and, although radiocarbon dating is intended, no results are given.

9.023 1958
ARCHAEOLOGICAL INVESTIGATIONS OF CUICUILCO VALLEY OF MEXICO
Heizer, Robert F.; Bennyhoff, J.A.
Science, v. 127, n. 3293, 31 Jan. 1958: 232-233
The beginning of urbanism in the Central Mexican Highlands is found in a complex discovered under a lava flow in the Cuicuilco region and dated as 2040 ± 200 years on two wood charcoal samples collected from occupation deposits below the pedregal. The material is primarily pre-classic.

9.024 1957
ARCHAEOLOGICAL INVESTIGATIONS ON SOUTHAMPTON AND WALRUS ISLANDS, N.W.T.
Collins, Henry B.
Canada National Museum. Bulletin, n. 14, Annual report 1955-1956, 1957: 22-61
A date of 2060 ± 230 years was obtained on samples of charred bones from an early proto-Dorset site near the Sadlermiut site on Southampton Island.

9.025 1966
ARCHAEOLOGICAL POTENTIAL OF THE ATLANTIC CONTINENTAL SHELF
Emery, Kenneth Orriz; Edwards, R.L.
American Antiquity, v. 31, n. 5, part 1, July 1966: 733-737
The continental shelf should provide good archaeological grounds, as early man lived in the United States 11,000 years ago when most of the now submerged continental shelf was exposed. The oldest Radiocarbon dates for kitchen middens of marine refuse along the present shore appear to be younger than the oldest date for kitchen middens of non-marine content. There should be still older middens submerged out on the shelf.

9.026 1968
ARCHAEOLOGICAL SITES IN LOESS REGIONS OF THE MISSOURI DRAINAGE BASIN. PART 1: THE PROBABLE AGE OF THE ALTITHERMAL ON THE WESTERN PLAINS
Husted, Wilfred M.
In: Loess and Related Aeolian Deposits, edited by C. Bertrand Schultz and John C. Frye. International Association for Quaternary Research (INQUA), Congress, 7th, Boulder, Col., 14 Aug - 9 Sept 1965. *Proceedings*, v. 12. University of Nebraska Press, 1968: 101-106
A hiatus in the sequence of human occupation in the western plains is dated by radiocarbon at 5500 to 4500 years B.C. It corresponds to the altithermal during which much of the Western Plains were seemingly abandoned by man.

9.027 1965
ARCHAEOLOGICAL SURVEY OF THE CHIAPAS COAST, HIGHLAND AND UPPER GRIJALVA BASIN
Lowe, Garreth W.; Mason, J. Alden

In: *Handbook of Middle American Indians*, edited by Robert Wauchope, Vol. II, Part 1, *Archaeology of Southern Mesoamerica*. Austin, Texas, University of Texas Press, 1966: 195-236

The late pre-classic period is dated at 564 ± 64 B.C. to 220 ± 60 B.C. The early proto-classic period is dated at 38 ± 65 B.C. to A.D. 38 ± 45 and the late dating of the various periods is made on the basis of radiocarbon dates.

9.028   1960
AN ARCHAEOLOGICAL SURVEY OF THE MIDDLE PECOS RIVER AND THE ADJACENT LLANO ESTACADO
Jelinek, Arthur Jenkins
Ph.D. Dissertation. University of Michigan, 1960. (Abstract in: *Dissertation Abstracts* Ann Arbor, Mich. v.21, n. 8, 1960: 2074. L.C. number: mic. 60-6887)
The date of the maximal expansion of maize cultivation is situated between about 1000 and 1250 A.D.

9.029   1965
ARCHAEOLOGICAL SYNTHESIS OF OAXACA
Bernal, Ignacio
In: *Handbook of Middle American Indians*, edited by Robert Wauchope. Vol. III, *Archaeology of Southern Mesoamerica*, part 2. Austin, Texas, University of Texas Press, 1965: 788-813
Radiocarbon dates on charcoal from a hearth associated with various typical stone implements give ages of 2100 and 2000 B.C. suggesting a pre-ceramic agricultural culture.

9.030   1965
ARCHAEOLOGICAL SYNTHESIS OF THE SOUTHERN MAYA LOWLANDS
Thompson, J. Eric S.
In: *Handbook of Middle American Indians*, edited by Robert Wauchope, Vol. II, Part 1, *Archaeology of Southern Mesoamerica*. Austin, Texas, University of Texas Press, 1966: 331-359
Correlation of middle pre-classic material between Yukatan and Mamon based on radiocarbon dates.

9.031   1965
ARCHAEOLOGY AND PREHISTORY IN THE NORTHERN MAYA LOWLANDS
Wyllys, Andrew, E.
In: *Handbook of Middle American Indians*, edited by Robert Wauchope, Vol. II, Part 1, *Archaeology of Southern Mesoamerica*. Austin, Texas, University of Texas Press, 1966: 288-330
A review article which includes a number of dates to construct a chronology of the various periods.

9.032   1978
ARCHAEOLOGY AND THE EVIDENCE FOR THE PREHISTORIC DEVELOPMENT OF ESKIMO CULTURE: AN ASSESSMENT
Anderson, Douglas D.
Arctic Anthropology, v. 16, n. 1, 1979: 16-31
A chronological chart, based on radiocarbon dates, is presented on p. 17.

9.033   1960
THE ARCHAEOLOGY OF BEHRING STRAIT
Giddings, J.L.
Current Anthropology, v. 1, n. 2, Mar. 1960: 121-138
One radiocarbon date of 2258 ± 230 years ago for a solid permanently frozen wood sample and a series of radiocarbon dates by Frolich and Ralph are used to suggest a chronology of St. Laurence Island cultures. An Asian series also based on radiocarbon dates is presented. The author notes that samples which have been most continuously frozen and hence less subject to contamination in the form of root infestation and absorption of seasonal seepage in wet ground should be the most reliable for radiocarbon dating.

9.034   1964
THE ARCHAEOLOGY OF CAPE DENBIGH
Giddings, James L.
Providence, Rhode Island, Brown University Press, 1964. pp. 331
Radiocarbon dates relating to the Denbigh complex are presented in table form and discussed on p. 244-248.

9.035   1962
THE ARCHAEOLOGY OF CARCAJOU POINT, WITH AN INTERPRETATION OF THE DEVELOPMENT OF ONEOTA CULTURE IN WISCONSIN. VOL. I AND II
Hall, Robert L.
Madison, University of Wisconsin Press, 1962
Radiocarbon dating is used throughout the study to establish chronological interpretation.

9.036   1966
ARCHAEOLOGY OF LOWER CENTRAL AMERICA
Lothrop, S.K.
In: *Handbook of Middle American Indians*, Vol. IV, *Archaeological Frontiers and External Connections,* edited by Robert Wauchope. Austin, Texas, University of Texas Press, 1966: 180-208.
Radiocarbon dates show a hiatus of 19 centuries between the Monagrillo pottery, the oldest yet found in Latin America and the next dated pottery found in Panama at the Pueblo Nuevo site.

9.037   1968
ARCHAEOLOGY OF SOUTH FORK SHELTER (NV-EL-11), ELKO COUNTY, NEVADA
Heizer, Robert F.; Baumhoff, Martin A.; Clelow, C.W.
University of California, Berkeley. Department of Archaeology. Archaeological Survey Reports, n. 71, March 1968, paper n. 1: 1-58

Three radiocarbon dates from charcoal from various depths in the midden deposit at South Fork Shelter give ages of 3320 ±200 years BP at the 72 inch level, 4360 ± 300 years BP at the 120 inch level and 4310 ± 400 years BP at the 94-100 inch level indicating that the earliest occupation of the site occurred earlier than 2410 B.C. and may date back to 3000 B.C. A flexible estimate of 550 years elapsed time for the accumulation of each 12 inches of deposit is deduced.

9.038   1966
THE ARCHAEOLOGY OF THE DOMEBO SITE
Leonhardy, Frank C.
In: Domebo. A Paleo-Indian Mammoth Kill in the Prairie Plains. Museum Great Plains. Contribution, n. 1, 1966: 14-26

Discusses the chronological difference between Clovis and Plainview projectile point on the basis of radiocarbon dates and presents six radiocarbon dates for the Domebo site, with discussions. The Domebo site, Oklahoma, is assigned to the Llano complex and the radiocarbon dates are consistent with dates from other mammoth kill sites. The mammoths were killed about 11,200 years ago.

9.039   1966
ARCHAEOLOGY OF THE MUNICIPIO OF ETZATLAN, MEXICO. (AN APPLICATION OF THE DIRECT HISTORICAL APPROACH TO WEST MEXICAN ARCHAEOLOGY)
Long, S.V.
Ph.D. Dissertation. University of California, Los Angeles, 1966. pp.324 (Abstract in: *Dissertation Abstracts* Ann Arbor, Mich. v.26, sect.B, n. 1, 1966: 26. Order n.66-16811)

Radiocarbon dating of bone shell artifacts and human bones is one of the techniques used to help reconstruct the cultural history of a region from burial remains. The method is applied to a particular kind of archaeological feature: the shaft tombs from Western Mexico, Panama, Columbia, Ecuador and Peru.

9.040   1963
ARCHAEOLOGY OF THE ROSE SPRING SITE, INY-372
Lanning, Edward P.
University of California. Publications in American Archaeology and Ethnology, v. 49, n. 3, 1963: 237-308

A table presenting a tentative correlation of Western Great Basin and California sequences is based on radiocarbon dates which were determined by various other authors. This is discussed.

9.041   1961
THE ARCHAEOLOGY OF TWO SITES AT EASTGATE, CHURCHILL COUNTY, NEVADA: I, WAGON JACK SHELTER
Heizer, Robert F.; Baumhoff, Martin A.
University of California. Anthropological Records, v. 20, n. 4, 1961: 119-138

A charcoal sample from the lowest level, 206 inch, is dated, also a wood charcoal from a hearth at a depth of 72 inch is dated at 3320 ± 100 years BP indicating occupation as early as 8000 years ago. The South Fork Wagon Jack Shelter is dated at 1500 years ago by radiocarbon.

9.042   1966
ARCHAIC CULTURE ADJACENT TO THE NORTHEASTERN FRONTIERS OF MESOAMERICA
Taylor, Walter W.
In: Handbook of Middle American Indians, Vol. IV, *Archaeological Frontiers and External Connections*, edited by Robert Wauchope. Austin, Texas, University of Texas Press, 1966: 59-94

A table of cultural phases and complexes, dates and climate in northeastern Mexico is presented. The chronology is based on radiocarbon dates.

9.043   1950
THE ARCHAIC CULTURE IN THE MIDDLE SOUTH
Lewis, T.M.; Kneberg, M.
American Antiquity, v. 25, n. 2, Oct. 1959: 161-183

Radiocarbon dates seem to reinforce the argument that the Archaic Period should be extended by another 2000 years.

9.044   1966
AN ARCHAIC FRAMEWORK FOR THE HUDSON VALLEY
Funk, Robert Ellsworth
Ph. D. Dissertation. Columbia University, 1966. pp. 329 (Abstract in: *Dissertation Abstracts*, Ann Arbor, Mich. v.27, n. 8, series B, 1967. Order n. 67-00792)

Proposes a pre-ceramic sequence for eastern New York which is based almost entirely on evidence from stratified sites, aided by a small body of radiocarbon dates.

9.045   1962
ARLINGTON SPRINGS MAN
Orr, Phil C.
Science, v. 135, n. 3499, 19 Jan. 1962: 219

Bones of a man were found at a depth of 37 feet in sediments on Santa Rosa Island, California, and dated as 10,000 years BP. Two later occupational levels are dated at 7350 and 2090 years BP. No artifacts are associated with the oldest bones which are believed to be an accidental burial on the edge of a cienaga. The evidence suggests the presence of men during Wisconsin glacial period.

9.046　　　　　　　　　　　　　　　　1957
ART OF THE LOWER COLUMBIA VALLEY
Butler, B. Robert
Archaeology, v. 10, Autumn, 1957: 158-165
The chronology suggested for the onset and termination of occupation is based wherever possible on radiocarbon dates available from sites in the area. Period spans from 9000 B.C. to A.D. 1830.

9.047　　　　　　　　　　　　　　　　1965
ARTIFACT FROM DEPOSITS OF MID-WISCONSIN AGE IN ILLINOIS
Munson, Patrick J.; Frye, John C.
Science, v. 150, n. 3704, 24 Dec. 1965: 1722-1723
Discovery of an artifact of human manufacture imbedded in Roxana loess, classed as Altonian substage of the Wisconsin stage of the Pleistocene, of an age of 35,000 to 40,000 years, contributes to the determination of the age of man in the New World. The stratigraphy was dated by radiocarbon measurements on snail shells and wood.

9.048　　　　　　　　　　　　　　　　1953
ARTIFACTS WITH MAMMOTH REMAINS, NACO, ARIZONA
Haury, Emil W.; Antevs, Ernst; Lance, J.F.
American Antiquity, v. 19, n. 1, July 1953: 1-14
The Naco kill appears to be older than the Sulphur Spring Stage of the Cochise culture dated by radiocarbon at 7756 ± 370 years.

9.049　　　　　　　　　　　　　　　　1951
AN ASSESSMENT OF CERTAIN NEVADA, CALIFORNIA AND OREGON RADIOCARBON DATES
Heizer, Robert F.
American Antiquity, v. 17, part 2 (*Society for American Archaeology, Memoirs*, n. 8), 1951: 23-25
Dates from these areas are presented and discussed, in particular the dates associated with the Mount Mazama eruption which are very significant to archeological dating as Mazama pumice covers a number of archaeological cave deposits.

9.050　　　　　　　　　　　　　　　　1961
THE AVONLEA POINT
Kehoe, Thomas F.; McCorquodale, Bruce A.
Plains Anthropologist, v. 6, n. 13, Aug. 1961: 179-188
The Avonlea points dated 1500 ± 100 years BP on a sample of charcoal collected on the site at test pit 1 places this occupation approximately contemporaneous with the Besant Point. This point is sufficiently unique and temporally delimited to serve as a useful horizon marker for the early Late Prehistoric Period in the Northwestern Plains.

9.051　　　　　　　　　　　　　　　　1965
BARREN GROUND CARIBOU (*RANGIFER ARCTICUS*) FROM AN EARLY MAN SITE IN SOUTHERN MICHIGAN
Cleland, C.W.
American Antiquity, v. 30, n. 3, Jan. 1965: 350-351
The identification of a phalanx of a barren ground caribou (*Rangifer arcticus*) from the Holcombe site in Southeastern Michigan is perhaps the earliest association of man and a specific animal species in the Eastern United States. Association of this site with a beach of glacial Lake Algonquin dated by radiocarbon on wood overlain by Lake Algonquin sand places its occupation at 9200 B.C. approximately.

9.052　　　　　　　　　　　　　　　　1967
THE BERRY SITE
Canada. National Museum Bulletin, n. 206, *Anthropological series* n.72. *Contribution to Anthropology*, I, 1967: 26-53
Although the abundance of stamped collarless and low-collared rim sherd would favour the Berry site as being an early Iroquois site, it may also be a non-Iroquois site contemporary to Roebuck which is dated by the radiocarbon method as 560-100 years BP.

9.053　　　　　　　　　　　　　　　　1963
BISON AND THE PALEO INDIAN
Agogino, George A.
Anthropological Journal of Canada, 1963
The Simonson site in Iowa is the oldest site associated with *Bison Occidentalis*. The lowest level which produced projectile points, scrapers and knives in association with the bison bones is dated by radiocarbon at 6471 ± 520 years B.C.

9.054　　　　　　　　　　　　　　　　1966
***BISON OCCIDENTALIS LUCAS* FOUND AT TABER, ALBERTA, CANADA**
Trylich, C.; Bayrock, L.A.
Canadian Journal of Earth Science, v. 3, n. 7, Dec. 1966: 987-995

An articulated incomplete skeleton of a mature male *Bison occidentalis Lucas* was found in alluvial sand on the uppermost terrace of the Oldman River near Taber, Alberta. Two radiocarbon ages of 10,000 and 11,000 years and invertebrate fossil and pollen analyses show that at this time the area was free of ice and that parkland vegetation was well established in southern Alberta. The pebble chopper found with the skeleton points to human occupation at that time.

9.055　　　　　　　　　　　　　　　　1960
THE BREWSTER SITE: AN AGATE BASIN FOLSOM MULTIPLE COMPONENT SITE IN EASTERN WYOMING
Agogino, George A.; Frankforter, W.D.
Masterkey, v. 34, n. 3, Jul/Sep 1960: 102-107
Burned bones and charcoal have been collected from all cultural levels to obtain radiocarbon dates. Another site has been conservatively dated at about 7000 years old and Folsom is 10,000 years, the Agate Basin lies somewhere between these two dates.

9.056　　　　　　　　　　　　　　　　1963
A C14 DATED POLLEN PROFILE FROM WELLS MASTODON SITE NEAR RICHMOND, INDIANA
Ogden, J. Gordon; Gooding, M.A.
Manuscript, 1963
[Not sighted]

9.057　　　　　　　　　　　　　　　　1961
CADDOAN RADIOCARBON DATES
Campbell, T.N.
Texas Archaeological Society .Bulletin, v. 31, 1961: 145-151
Twenty one published radiocarbon dates from Caddoan are presented, nearly all being dates from components or sites attributed to the Gibson aspect. Seven sites in Oklahoma (Spiro, Bracket, Harlan, Hughes, Norman, Reed and McCarter), one in Louisiana (Belcher), and one in Texas (Davis), are included. A summary list is presented in table form.

9.058　　　　　　　　　　　　　　　　1965
CALENDRICS OF THE MAYA LOWLANDS
Satterthwaite, Linton
In: Handbook of Middle American Indians, edited by Robert Wauchope. Vol. III, *Archaeology of Southern Mesoamerica*, part 2. Austin, Texas, University of Texas Press, 1965: 603-631
Radiocarbon dates made at various laboratories and times, associated with Tikal are discussed in respect of the Maya Lowland calendar correlation.

9.059　　　　　　　　　　　　　　　　1959
CALIFORNIAN CULTURES AND THE CONCEPT OF AN ARCHAIC STAGE
Meighan, Clement W.
American Antiquity, v. 24, n. 3, Jan. 1959: 289-318
Recent evidence and radiocarbon dates show that the west coast of the United States was occupied as early as any other part of the country.

9.060　　　　　　　　　　　　　　　　1962
CARBON DATING OF PREHISTORIC SOOT FROM SALTS CAVES, KENTUCKY
Benington, F.; Melton, C.; Watson, Patty Jo
American Antiquity, v. 28, n. 2, Oct.,1962: 238-241
Three radiocarbon dates on samples of soot support the suggestion of a late Archaic - early Woodland placement for the prehistoric activity in Salts and Mammoth caves.

9.061　　　　　　　　　　　　　　　　1967
CARBON-14 DATES AND EARLY MAN IN THE NEW WORLD
Haynes, C. Vance
In: Pleistocene Extinctions: the Search for a Cause. International Association for Quaternary Research (INQUA). Congress, 7th, Boulder, Col., 14 Aug - 19 Sept.1965. *Proceedings*, v. 6. Boulder, Col., 1968: 267-286
Critical evaluation of radiocarbon dates of Early Man in thousand-year intervals reveals an increasing diversity of projectile point types. Radiocarbon dating of alluvial and aeolian stratigraphic sequences provide the most precise geochronological control of cultural changes and the most reliable means of correlation. The author cannot find any radiocarbon date older than 12,000 BP that can be positively related to Early Man in the New World. A table of radiocarbon dates of early man by site is presented.

9.062　　　　　　　　　　　　　　　　1965
CARBON-14 DATES AND EARLY MAN IN THE NEW WORLD
Haynes, C. Vance
In: International Conference on Radiocarbon and Tritium Dating. Pullman, Washington, Washington State University, June 1-11, 1965. *Proceedings*. United States of America, Atomic Energy Commission, 1965. Conference n. 650652: 145-164
Radiocarbon dates for early man are listed by sites and discussed. In addition to providing an absolute time-scale for geologic-climatic events radiocarbon dating has provided a means of determining cultural change with time and of comparing cultural changes during a given period in different geographic areas.

9.063   1955
CARBON-14 DATES FROM ELLSWORTH FALLS, MAINE
Byers, Douglas S.; Hadlock, Wendell S.
Science, v. 121, n. 3151, 20 May 1955: 735
A sample from middle level deposits of archaic Red Point character was dated at 3959 ± 310 years or 2005 B.C. Red Point was dated 3350 ± 400 years or 1400 B.C.

9.064   1960
CARCAJOU POINT AND THE PROBLEM OF ONEOTA ORIGINS
Hall, Robert L.
Ph.D. Dissertation. University of Wisconsin, 1960. pp. 448 (Abstract in: *Dissertation Abstract*. Ann Arbor, Mich. v.21, n. 3, 1960: 424. L.C. number: mic. 60-3204)
On the basis of absolute dating through radiocarbon analysis, external relationships and relative dating provided by artifacts, it is suggested that the prehistoric Mississippi occupation of Carcajou Point began with an Upper Mississippi culture and proceeded in the direction of a fully developed Oneota culture.

9.065   1978
CARIBOU ESKIMO ORIGINS: AN OLD PROBLEM RECONSIDERED
Burch, Ernest S.
Arctic Anthropology, v. 15, n. 1, 1978: 1-35
A chronological list of radiocarbon dates relating to the Caribou Eskimos is presented on p. 12.

9.066   1959
THE CASTLE WINDY SITE, VOLUSIA COUNTY, FLORIDA
Bullen, Ripley P.; Sleight, F.W.
Delend, Florida, The Bryant Foundation, 1959
Three dates by Lamont are used (humic acid removed).

9.067   1963
A CERAMIC SEQUENCE FOR THE PIURA AND CHIRA COAST, NORTH PERU
Lanning, Edward P.
University of California. Publications in American Archaeology and Ethnology, v. 46, n. 2, Sept. 1963: 135-284
An appendix presents and discusses radiocarbon dates for sequences from Equador which are used to establish absolute dating by analogy.

9.068   1959
CERAMICS FROM TWO PRECLASSIC PERIODS AT CHIAPA DE CORZO, CHIAPAS, MEXICO
Dixon, K.A.
New World Archaeological Foundation. Publication, n. 4, 1959. pp. 52
The radiocarbon dates for this project are presented in table form on p. 41 and discussed. It is noted that the interpretation of these dates depends on the stratigraphy and on the ceramic collection itself and span a period from 1152 - 955 B.C. to A.D. 233 - 325.

9.069   1965
CERAMICS OF OAXACA
Caso, Alfonso; Bernal, Ignacio
In: *Handbook of Middle American Indians*, edited by Robert Wauchope. Vol. III, *Archaeology of Southern Mesoamerica*, part 2. Austin, Texas, University of Texas Press, 1965: 871-895.
The pottery at the site has been divided into six periods. Period I (Upper preclassic), period II (formation period), transitional period between period IIIA and IIIB have been dated by radiocarbon at ca 650 B.C., 273 B.C. and A.D. 500 respectively. There are no radiocarbon dates for the other periods.

9.070   1958
THE CHRONOLOGICAL POSITION OF THE HOPEWELLIAN CULTURE IN THE EASTERN UNITED STATES
Griffin, James B.
University of Michigan. Museum of Anthropology. Anthropological Papers, n. 12, 1958. pp. 26
Presents a rather large series of radiocarbon dates which bears upon the origin, age and disappearance of the Hopewellian complex. This indicates that the Hopewellian culture lasted longer than previously believed by archaeologists and discounts the theories of migration.

9.071   1960
CHRONOLOGY AND CULTURE CHANGE IN THE SAN JUAN ISLANDS
Carlson, Roy L.
American Antiquity, v. 25, n. 4, Apr. 1960: 562-586
Archaeological remains from the San Juan Islands are described and discussed and some hypothesis are advanced. A few radiocarbon dates place the San Juan Phase at A.D. 1300 for the beginning of the village of Tselax at Murqueam.

9.072   1966
CHRONOLOGY OF A WEST-MEXICAN SHAFT TOMB
Long, S.V.; Taylor, R.E.
Nature, v. 212, n. 5062, 5 Nov. 1966: 651-652
The suggested age of the shaft-tomb (A.D. 250) which was based on radiocarbon measurements of three marine shell artifacts is reviewed in view of a more recent investigation into the contemporary radiocarbon assay of pre-bomb

marine shells along the West Mexican coast.

9.073   1965
CLIMATIC EPISODES AND THE DATING OF THE MISSISSIPPIAN CULTURE
Barreis, David A.; Bryson, Reid A.
Wisconsin Archaeologist, v. 46, n. 4, 1965: 203-220
Report on a new series of radiocarbon dates for prehistoric sites in Wisconsin, Iowa and Missouri (dates by Bender, Wisconsin) which are presented in table form. A timing for the cultural sequences is suggested.

9.074   1952
CLIMATIC HISTORY AND THE ANTIQUITY OF MAN IN CALIFORNIA
Antevs, Ernst
University of California, Berkeley. Department of Anthropology. Archaeological Survey Reports, n. 16, June 1952: 23-31
A number of radiocarbon dates which determine the times of the Mankato Maximum, the Provo Pluvial and the glacial maximum are discussed.

9.075   1958
COMMENTARY ON CARCAJOU CARBON-14 DATES
Hall, Robert L.
Wisconsin Archaeologist, v. 39, n. 3, 1958: 174-175
Four charcoal samples which were submitted for radiocarbon dating suggest an intermittent occupation by the same or related people for over 800 years.

9.076   1951
COMMENTS ON RADIOCARBON DATES FROM MEXICO
Terra, Helmut de
American Antiquity, v. 17, part 2 (*Society for American Archaeology, Memoirs*, n. 8), 1951: 33-36
The radiocarbon dates, which tend to substantiate the latest time concepts gained from geologic and archaeologic data, date the dawn of the Pyramid building age in the Archaic culture period, about 2500 years ago and present a new chronology for climatic changes, soils and related geologic factors known to have influenced the ancient cultures of Mexico.

9.077   1958
COMMENTS ON THE SAN JOSE RADIOCARBON DATES
Agogino, George A.; Hester, Jim J.
American Antiquity, v. 24, n. 2, 1958: 187
The finding of amaranth seeds in conjunction with charcoal which was dated at 6880±400 years ago from a hearth in the San Jose dune extends the possible cultural use of this plant for the first time into an era before the birth of Christ.

9.078   1968
CONFIGURATION OF PRE-CERAMIC DEVELOPMENT IN THE SOUTH WESTERN UNITED STATES
Irwin Williams, Cynthia
In: Contribution to South Western Prehistory. International Association for Quaternary Research (INQUA). Congress, 7th, Boulder, Col., 14 Aug - 19 Sept. 1965. *Proceedings*, v. 4. Boulder, Col., 1968
A review of the data available on prehistory in the South West, prior to the Ceramic period. Radiocarbon determinations have given dates far in excess of 20,000 BP. However contemporaneity with man-made objects is not accepted.

9.079   1964
CONTRIBUTION TO THE PREHISTORY OF VANCOUVER ISLAND
Capes, K.N.
Idaho State University Museum. Occasional Papers, v. 15, 1964. pp.119
An appendix by John G. Fyles deals with the geology of the region. Includes radiocarbon dates.

9.080   1962
CONTRIBUTIONS TO THE ARCHAEOLOGY OF THE COLUMBIA PLATEAU
Butler, B. Robert
Idaho State College Museum. Occasional Papers, v. 9, 1962. pp. 86
Sets out geological setting and climate history, and gives ages of sedimentary sections as 5400 to 7000 years BP (Libby, 1955). The present report comprises descriptions and analyses of several excavations at various stages of completion at various sites. The oldest phase of the Old Cordilleran culture, dated ca 8000 to 11,000 years BP was found to correspond to Early Dallas, on the lower Columbia. The early middle period ended 4000 year ago was coextensive with the following period of maximum warmth and dryness in the Late Middle Period and the Late Period following the latter worsening by the 9th century A.D., to historic times.

9.081   1963
THE CORN CREEK DUNES SITE, A DATED SURFACE SITE IN SOUTHERN NEVADA
Williams, Peter A.; Orlins, Robert I.
Nevada State Museum. Anthropological Papers, n. 10, 1963. pp. 66
The radiocarbon dates associated with the site are presented in table form.. One may conclude that, on the basis of

radiocarbon dates, the artifact inventory, and comparison with other contemporary sites, the Corn Creek Dune site represents an eastward extension of the Southern California Pinto Culture into the Southern Nevada culture sequence.

9.082 **1966**
CORRELATION OF ARCHAEOLOGICAL AND PALYNOLOGICAL DATA
Jelinek, Arthur Jenkins
Science, v. 152, n. 3728, 10 June 1966: 1507-1509
Analysis of a series of pollen samples and artifact materials, taken from the same excavation units in prehistoric sites in the Middle Pecos River valley of central eastern Mexico, shows meaningful correlations between some major pollen groups and categories of artifacts. The sequence was radiocarbon dated from the collagen fraction of a sample of bison bone.

9.083 **1955**
CORRELATIONS OF THE PREHISTORIC SETTLEMENTS AND DELTA DEVELOPMENT
McIntire, William C.; Morgan, James P.
Geological Society of America. Bulletin, v. 66, n. 12, Dec. 1955: 1595.
Abstract of paper presented at the November 1955 meeting of the Geological Society of America. Radiocarbon datings are being employed to verify the relative archaeological dates as well as give maximum dates for various delta masses on which the sites are located.

9.084 **1962**
COSTA RICAN ARCHAEOLOGY AND MESOAMERICA
Coe, Michael D.
Southwestern Journal of Anthropology, v. 18, n. 2, Summer 1962: 170-183
According to a chronology, partly based on radiocarbon dates, the Early Polychrome Period extends from the 4th century A.D. to about A.D. 750

9.085 **1964**
CULTURAL CHANGE AND CONTINUITY ON THE SAN DIEGO COAST
Warren, Claude Nelson
Ph.D. Dissertation. University of California, Los Angeles, 1964. pp. 277 (Abstract in: *Dissertation Abstracts* Ann Arbor, Mich., v. 25, n. 4, 1964: 270. Order No. 64-10486)
The La Jolla complex affords an opportunity to study cultural change and continuity in relative isolation. A deductive model is constructed and tested by application of the empirical data from the La Jolla complex. The reference material used includes specific radiocarbon dates among others.

9.086 **1968**
CULTURAL CHRONOLOGY OF THE GULF OF CHIRIQUÍ, PANAMA
Linares de Sapir, Olga
Smithsonian Contributions to Anthropology, v. 8, 1968. pp. 119
The early Coclé tradition is placed after A.D. 500 together with the Venado Beach (this one dated by radiocarbon). The San Lorenzo phase is estimated at A.D. 800 to A.D. 1020 in accord with a radiocarbon date of 930 ± 100 years BP (A.D. 1020). The Chiriquí phase yielded a date of 115 ± 100 years BP, obviously wrong since the site has not been occupied since the 17th century, and is estimated at A.D. 1100 to the time of conquest, A.D. 1500.

9.087 **1960**
THE CULTURAL COMPLEXES FROM THE LAGOA SANTA REGION, BRAZIL
Hurt, Wesley R.
American Anthropology, v. 62, n. 4, 1960: 565-585
A radiocarbon determination on charcoal from the cave at Lagoa Funda gave a date of 3000 ± 300 years. A human skeleton and the skull of an extinct bear were found in the bed containing charcoal 12 ft above original floor level. This may suggest the contemporaneity of the human find and extinct Pleistocene animals.

9.088 **1963**
CULTURAL DEVELOPMENT IN LOWER CENTRAL AMERICA
Bandez, Claude F.
In: Aboriginal Cultural Development in Latin America, an Interpretative Review, edited by Betty J. Meggers and Clifford Evans. *Smithsonian Miscellaneous Collection*, v. 146, n. 1, 1963: 45-54
The earliest reliable evidence of man in the entire area is dated by radiocarbon at 4850 ± 100 B.C.

9.089 **1960**
CULTURAL SEQUENCES AT THE DALLES, OREGON: AS CONTRIBUTION TO PACIFIC NORTHWEST PREHISTORY
Cressman, L.S.; Cole, David R.; Davis, Wilbur A.; Newman, Thomas M.; Sheans, Daniel J.
American Philosophical Society, Transactions, v. 5, part 10, Dec. 1960. pp.108
Radiocarbon dates serve as convenient chronological reference with the younger one marking the incidence of new cultural activity. The dates were from materials extracted from a road cut and were from 9785 ± 220 to 7875 ± 100, bracketing a depositional level.

9.090 1953
CULTURAL VARIATION WITHIN TWO WOODLAND MOUND GROUPS OF NORTHERN IOWA
Beaubien, Pierre L.
American Antiquity, v. 19, n. 1, July 1953: 56-65
Radiocarbon dates allow an extremely long period during which the group of mounds along the Mississippi could have been constructed.

9.091 1960
THE CULTURE HISTORY OF LOVELOCK CAVE, NEVADA
Grosscup, Gordon L.
California University, Berkeley. Department of Anthropology. Archaeological Survey Reports, n. 52, Oct. 1960. pp. 72
The age of the Lovelock cave deposits is discussed in terms of five radiocarbon dates, giving the following ages: early Lovelock phase from ca 2000 to 1000 B.C.; transitional phase from ca 1000 B.C. to 1 B.C. and late phase ca 1 B.C. to A.D. 900 (p.10)

9.092 1966
CULTURE PHASE DIVISIONS SUGGESTED BY TYPOLOGICAL CHANGE COORDINATED WITH STRATIGRAPHICALLY CONTROLLED RADIOCARBON DATES AT SAN DIEGO
Moriarty, James Robert
Anthropological Journal of Canada, v. 4, n. 4, 1966: 20-30
Although the chronologies are not absolute, because of the errors inherent in radiocarbon dating and human interpretation, there are enough data now for a definite date sequence of cultural traits changes from the earliest to the most recent occupants of San Diego County. The radiocarbon dates in chronological sequence are presented in table form.

9.093 1957
DANGER CAVE
Jennings, Jesse D.
American Antiquity, v. 23, n. 2, part 2, Oct. 1957: i-xiii, 1-328 (*Society for American Archaeology. Memoirs*. n. 15)
The Radiocarbon dates determined for Danger Cave are not compatible with most of the standard geologic estimates of the time. The author discusses several alternatives to resolving the problem of interpretation of both the radiocarbon and the geochronolgical dates.

9.094 1964
DATING LAKE MOHAVE ARTIFACTS AND BEACHES
Warren, Claude A.; Da Costa, John
American Antiquity, v. 30, n. 2, Oct. 1964: 206-209
Radiocarbon dates help establish that man, as known through the Lake Mohave complex, probably did not occupy the shore of Lake Mohave until after the overflow channel had been lowered by removal of soil and weathered rock and the lake had dropped to the 943 foot level, exposing the 946 foot strand for occupation.

9.095 1967
DATING OF BROKEN K PUEBLO
Martin, Paul S.
Fieldiana : Anthropology, v. 57, July 1967 (Chapter in *The Prehistory of Eastern Arizona III*, by Paul S. Martin, William A. Longacre and James N. Hill.): 139-144
Sixteen samples were sent to various radiocarbon dating laboratories. Results are discussed and shown to be disappointing. Two tables are presented: a list of radiocarbon dates plotted as ranges of time, and radiocarbon and tree ring data.

9.096 1966
DATING THE ASPECTS CULTURES
Baerreis, David A.; Bryson, Reid A.
Oklahoma Anthropological Society. Bulletin, v. 14, 1966: 105-116
Presents and discusses radiocarbon dates for the Panhandle Aspect culture in the southern Plains. The dates are presented in table form accompanied by a table of means and site averages radiocarbon dates. The dates are in general agreement with the hypothesis that the southern movement of the Panhandle people occurred as a result of the initiation of drought conditions in the north, about 1250 years ago, and lasted about 1450 years.

9.097 1968
DATING THE PREHISTORY OF OKLAHOMA
Bell, R.E.
Great Plains Journal, v. 7, n. 2, 1968: 1-11
[Not sighted]

9.098 1968
DEBERT - A PALAEO INDIAN SITE IN CENTRAL NOVA SCOTIA
MacDonald, George F.
Canada. National Museum of Anthropology Collections, n. 16, 1968. pp. 207
The geological conditions of this early man site are described. The artifacts of the site are dated with an average overall date of $10,600 \pm 47$ years BP. Cold climate animal remains suggest that the ice front was not far distant at the time.

9.099 1965
THE DEBERT ARCHAEOLOGICAL PROJECT: RADIOCARBON DATING
Stuckenrath, Robert

In: International Association for Quaternary Research (INQUA). Congress, 7th, Boulder, Col., 14 Aug - 19 Sept. 1965. *Abstracts of General Sessions*. Boulder, Colorado, 1965: 450

Charcoal in association with Palaeoindian material was dated by the radiocarbon method. Three samples gave dates equalling the northeastern Valders substage and one gave a date following in the immediate post-glacial period. Dates and position relative to other Palaeoindian dates are discussed.

9.100   1957
A DECADE OF DISCOVERY: SOUTH AMERICA
Mason, J. Alden
Archaeology, v. 10, n. 3, Autumn 1957: 234
Stone implements discovered at Ayampitin in central Argentina were dated by the radiocarbon method. The date obtained is 7070 years BP.

9.101   1957
A DECADE OF DISCOVERY: MESOAMERICA
Coe, William R.
Archaeology, v. 10, n. 3, Autumn 1957: 234
The use of radiocarbon dates enables to show that Preclassic cultures from the Guatemala valley to the valley of Mexico were well established by the middle of the second millenium B.C.

9.102   1957
A DECADE OF DISCOVERY: U.S.A.
Woodbury, Richard B.
Archaeology, v. 10, n. 3, Autumn 1957: 233
Mentions the role of radiocarbon dating in establishing chronologies for human occupation.

9.103   1963
DEER CREEK CAVE, ELKO COUNTY, NEVADA
Shutler, Mary Elizabeth; Shutler, Richard
Nevada State Museum. Anthropological Papers, n. 11, 1963. pp. 58
Seven radiocarbon dates were obtained. They seem to be generally early. It is suggested that as the cave deposit is wet and the midden contains lumps of marl, dead carbon from the lime could have been incorporated into the charcoal from the firehearth.

9.104   1955
THE DENBIGH FLINT COMPLEX IS NOT YET DATED
Giddings, J.L.
American Antiquity, v. 20, n.4, Apr. 1955: 375-376
In a discussion of the radiocarbon dates produced for the Denbigh Flint Complex, the author argues that these dates only afford no more than a stop-date forward for the complex.

9.105   1967
THE DESCRIPTIVE ARCHAEOLOGY AND CHRONOLOGY OF THE THREE SPRING BAR ARCHAEOLOGICAL SITE, WASHINGTON
Daugherty, Richard D.; Purdy, Barbara A.; Fryxell, Roald
Washington State University. Laboratory Anthropology Report, Inv. 40. Pullman, Washington State University. Laboratory of Anthropology, 1967. pp. 114
[Not sighted]

9.106   1959
DESERT SIDE-NOTCHED POINTS AS A TIME MARKER IN CALIFORNIA
Baumhoff, Martin A.; Byrne, J, S.
California University at Berkeley. Archaeological Survey Reports, n. 48, 1959: 32-65
The projectile points examined are proposed as time markers in California archaeology. A table of suggested dates for the introduction of Desert side-notched points to various localities is presented. Backing observation by radiocarbon dating, the beginning of Phase II of the Late Horizon in central California is to be dated at A.D. 1500, thus dating the presence of Desert side-notched points approximately 50 years before that.

9.107   1964
A DIRE WOLF SKELETON AND POWDER MILL CREEK CAVE, MISSOURI
Galbreath, Edwin C.
United States. Academy of Science. Transactions, v. 57, n. 4, 1964: 224-242
The skeleton of *Canis (Aenocyon) dirus* was discovered in Powder Mill Creek cave, Missouri, in 1963. Radiocarbon dates of the bones tested were $13,170 \pm 600$ year B.C.

9.108   1962
THE DISCOVERY OF EARLY HUMAN SKELETAL REMAINS NEAR LYONS FERRY, WASHINGTON
Fryxell, Roald; Daugherty, Richard D.; Baenen, James
Northwestern Science, v. 36, n. 4, Nov. 1962: 127
Abstract of paper presented at the 36th Annual meeting of the Northwest Scientific Association at Washington State College, Bellingham, Washington, 1962. The oldest human remains yet dated with certainty in the Pacific Northwest have been removed from Marmes Rockshelter near Lyons Ferry, Washington. The dating was established at between 4000 and 6000 years through stratigraphic interpretation and radiocarbon dating.

9.109 1963
THE DONALDSON SITE
Wright, J.V.; Anderson, J.E.
Canada. National Museum. Bulletin, n. 184, *Anthropological Series,* n.58, 1953. pp. 113
According to radiocarbon dates on carbonised wood, the occupation of the Donaldson site is believed to have taken place at approximately 500 B.C. One date was too young and is regarded as contaminated by recent carbon. Other radiocarbon dates at the Burley site, and Point Peninsula site are also discussed. These place the Donaldson site in the early Middle Woodland period.

9.110 1967
THE DONNELLY RIDGE SITE AND THE DEFINITION OF AN EARLY CORE AND BLADE COMPLEX IN CENTRAL ALASKA
Hadleigh-West, Frederick
American Antiquity, v. 32, n. 3, July 1967: 360-382
An early core and blade culture, the Denali complex is identified. It is compared to other complexes from Anangula Island in the Aleutians and from the Kamtchatka which have been radiocarbon dated.

9.111 1966
THE DUMAW CREEK SITE. A SEVENTEENTH CENTURY PREHISTORIC INDIAN VILLAGE AND CEMETERY IN OCEANA COUNTY, MICHIGAN
Fieldiana Anthropology, v. 56, n. 1, Dec. 1966. pp. 91
Radiocarbon dates confirm the chronological estimate based on the age of tree stumps and other temporal evidence, placing the occupation of the village site by the Dumaw Creek Indians in the period A.D. 1605 to 1620.

9.112 1956
DWARF MAMMOTHS AND MAN ON SANTA ROSA ISLAND
Orr, Phil C.
Univerity of Utah. Department of Anthropology. Anthropological Papers, n. 26, Dec. 1956: 74-81
According to radiocarbon dates it can be assumed that the dwarf mammoth was well established on the island of Santa Rosa prior to the beginning of the Wisconsinan, that man arrived at an unknown date and that the mammoth became extinct between the period of 12,000 years and 7070 years BP when no evidence of mammoth is to be found.

9.113 1968
DZIBILCHALTUN A NORTHERN MAYA METROPOLIS
Andrews, E. Wyllys
Archaeology, v. 21, n. 1, Jan. 1968: 36-47
The formative phase at Dzibilchaltun, dated by radiocarbon at 975 ± 340 B.C., is the earliest yet dated in the Maya lowlands.

9.114 1959
DZIBILCHALTUN: LOST CITY OF THE MAYA
Andrews, E.W.
National Geographic Magazine, v. 115, n. 1, Jan. 1959: 90-109
One of the earliest and possibly the latest city inhabited by the ancient Maya, at Dzibilchaltun in northern Yucatan, is dated by radiocarbon measurements on a doorway wooden beam, at A.D. 48 ± 200.

9.115 1960
THE EARLIEST POTTERY IN SOUTHEASTERN UNITED STATES, 2000-1000 B.C. AND ITS CASE AS AN INDEPENDENT INVENTION
Bullen, Ripley P.
In: Congrès International des Sciences Anthropologiques et Ethnologiques, 6th, Paris, 1960. *Proceedings* Vol. 2. Paris, Musée de l'Homme, 1963: 363-367
Based on radiocarbon dates, it is concluded that the South Eastern Fiber-tempered tradition of pottery found in Florida is older and typologically unrelated to the Woodland ceramics of North America. It seems likely to represent an independent invention of pottery making. The shape of the vessels and decorations suggest a well developed basketry industry in existence before the invention of pottery.

9.116 1967
EARLY CULTURAL REMAINS ON THE CENTRAL COAST OF PERU
Patterson, Thomas C.
In: Peruvian Archaeology: Select Readings, edited by John Howland Rowe and Dorothy Menzel. Palo Alto, California, Peek Publications, 1967: 31-39
A radiocarbon determination of $10,430 \pm 160$ years BP dates the later part or end of the period when Chivateros artifacts were made. The sample came from the Oquendo complex in the Chillón valley on the central Peruvian coast.

9.117 1965
EARLY FORMATIVE PERIOD OF COASTAL ECUADOR : THE VALDIVIA AND MACHALILLA PHASES
Meggers, Betty J.; Evans, Clifford; Estrada, Emilio
Smithsonian Contribution to Anthropology, v. 1, 1965. pp. 234
The relative and absolute dating of the Valdivia and Machobilla phases are discussed (p. 147-156). The large series of radiocarbon dates agree very well in establishing the relative duration of the two phases. The radiocarbon dates are

9.118 1961
EARLY LITHIC INDUSTRIES OF WESTERN SOUTH AMERICA
Lanning, Edward P.; Hammel, E.A.
American Antiquity, v. 27, 1961: 139-154
Five periods of lithic industries are recognised, based on the long sequence at Lauricocha in the central Peruvian highlands. On the basis of radiocarbon dates, the periods are dated as follows: Period I, ca 10,000 to 8000 B.C.; Period II, 8000 to 6000 B.C.; Period III, 6000 to 3000 B.C.; Period IV, 3000 to 1200 B.C.; Period V, 1200 B.C. to present.

9.119 1956
EARLY MAN AND FOSSIL BISON
Forbis, Richard G.
Science, v. 123, n. 3191, 24 Feb. 1956: 327-328
Radiocarbon dating was used to confirm the Folsom, Plainview, Scottsbluff succession, and a correlation of fossil bison with early types of projectile points.

9.120 1960
EARLY MAN AND THE AGE OF THE CHAMPLAIN SEA
Mason, Ronald J.
Journal of Geology, v. 68, n. 4, July 1960: 366-376
Discusses the synchronous absolute chronology for fluted points characteristic of Paleo-Indian horizons in New England, the Upper Great Lakes, the high plains and the southwestern United States. A re-interpretation based on radiocarbon is presented.

9.121 1952
EARLY MAN IN AMERICA: A STUDY IN PREHISTORY
Sellards, E.H.
Austin, Texas, University of Texas Press, 1952. pp.211
Includes a short descripion of the method of age determinations by the radiocarbon method. A number of early cultures were dated by this method, indicating that man in America had established cultures in the mountain regions and on the plains as early as about 10,000 years ago.

9.122 1964
EARLY MILLING STONE HORIZON (OAK GROVE), SANTA BARBARA, CALIFORNIA: RADIOCARBON DATES
Owen, Roger C.
American Antiquity, v. 30, n. 2, part 1, Oct. 1964: 210-213
Radiocarbon dates of 6880, 7270 and 6380 BP have been obtained for an early horizon site (Oak Grove) in the region of Santa Barbara, California. These dates permit the definite association of many cultural features which have previously been denied for the Oak Grove complex: a partial maritime economy, a core-and-flake industry, bone and shell technology, the use of asphaltum, and flexed burials. It is unlikely that the Early Horizon in Santa Barbara was sedentary to any degree.

9.123 1965
EARLY OCCUPATION IN BIGHORN CANYON, MONTANA
Husted, Wilfred M.
Plains Anthropologist, v. 10, n. 27, 1965: 7-12
Excavation at the Mangus site in Bighorn Canyon, Montana, revealed a stratum containing artifacts of the Agate Basin complex, dated by radiocarbon at 6740 B.C. and 6650 B.C. The Sorenson site contained two early cultural levels of unknown cultural affiliation, dated at 5850 B.C. and 5610 B.C. This extends the time depth of the archaeological sequence in Bighorn Canyon by 800 to 1100 years.

9.124 1960
THE EASTERN DISPERSION OF ADENA
Ritchie, William A.; Dragoo, Don W.
New York State Museum and Science Bulletin, n. 379, March 1960. pp. 80
Recent radiocarbon dates indicate that Hopewell was thriving in Illinois and Western Indiana before the end of the early Adena period, i.e. 2336 ± 250 years BP.

9.125 1959
EDITORIAL
Daniel, Glyn
Antiquity, v. 33, n. 132, Dec. 1959: 237-240
Radiocarbon dates in America are being produced for cultures and events which were thought to be undatable, except by a rough and ready guesswork calculation back from dendrochronological dates. In Europe and the Near East, however, the radiocarbon dates in many instances produce a chronology which seems at variance with archaeological dating. The editor discusses the need to evaluate all other available evidence in conjunction with the radiocabon dates in order to make an acceptable choice.

9.126 1966
ELEPHANT-HUNTING IN NORTH AMERICA
Haynes, C. Vance
Scientific American, v. 214, n. 6, June 1966: 104-112
Clovis points and mammoth remains are found together in several sites dated by radiocarbon a little earlier than 9000 B.C. By 9000 to 8000 B.C. Clovis Points are found but no mammoth remains. It is concluded that the early men who were utilising this type of point may have caused the extinction of the elephants.

9.127 1960
THE ENTRY OF MAN IN THE WEST INDIES
Rouse, Irving
Yale University. Publications in Anthropology, n. 61, 1960. pp. 26
The El Jabo complex in Western Venezuela is of considerable antiquity. El Jabo types found at a nearby site, at La Vale de Coro, were dated at $16,000 \pm 300$ BP. Three possible hypothesis concerning the entry of man in the West Indies are advanced. Several of the complexes on which these hypothesis are based have been dated by radiocarbon.

9.128 1959
ESKIMO PREHISTORY IN THE VICINITY OF POINT BARROW, ALASKA
Ford, James A.
American Museum of Natural History. Anthropological Papers, v. 47, part 1, 1959: 1-272
The few radiocarbon dates which have been thus far obtained, are not entirely consistent with the archaeological evidence. This is discussed.

9.129 1963
ESKIMOS AND ALEUTS: THEIR ORIGINS AND EVOLUTION
Laughlin, William Sceva
Science, v. 142, n. 3593, 8 Nov. 1963: 633-645
The archaeological evidence points to southern Alaska as the homeland of the Eskimo-Aleut stock. Radiocarbon dating from a cultural level underlying two layers of ash and humus places the lamellar-flake site on Anangula Island at 8245 ± 275 to 660 ± 300 years ago; the oldest known site in the Eskimo-Aleut world.

9.130 1958
ESTIMATED CORRELATIONS AND DATING OF SOUTH AND CENTRAL ANERICAN CULTURE SEQUENCES
Willey, Gordon R.
American Antiquity, v. 23, n. 4, Apr. 1958: 353-378
In this correlation, radiocarbon dating provides most of the absolute dates, however these dates have been weighted in the context of all available evidence.

9.131 1961
AN EVALUATION OF RADIOCARBON DATES FROM THE GALENA SITE, SOUTHERN TEXAS
Ring, E.R.
Texas Archaeological Society. Bulletin, n. 31, 1961: 317-325
The antiquity of the dates from the Galena shell midden presents a problem for the archaeological time concepts insofar as the Galverston Bay Focus and the Goose Creek pottery are concerned. These were established on the basis of numerous archaeological considerations but without the help of radiocarbon assays.

9.132 1959
EXCAVATIONS AT LA VENTA, TABASCO, 1955
Drucker, Philip; Heizer, Robert F.; Squier, Robert J.
Smithsonian Institution. Bureau of American Ethnology. Bulletin, n. 170, 1959. pp. 283
A table of radiocarbon dates of La Venta is presented on p. 266, with discussions. The significance of the dates in relation to Maya and other cultural patterns of Mesoamerica, to the valley of Mexico and adjacent highland regions, and to 'calendar' glyphs and other features, is also discussed.

9.133 1960
EXCAVATIONS IN THE UPPER LITTLE COLORADO DRAINAGE, EASTERN ARIZONA
Martin, Paul S.; Rinaldo, John B.
Fieldiana Anthropology, v. 51, n. 1, 1960: 1-120
Prehistoric campsites from the beaches ot Little Ortega Lake and Laguna Salada have been excavated and examined. Three radiocarbon dates enable the authors to place the date of the two pit-houses at A.D. 600-800.

9.134 1968
THE EXTENT AND CONTENT OF POVERTY POINT CULTURE
Webb, Clarence H.
American Antiquity, v. 33, n. 3, July 1968: 297-321
It is probable that Poverty Point Culture was established on the coast and the valleys between 1500 and 1000 B.C., antecedent to the Hopewell and Adena and in time contributed to Eastern Woodland origins. Charcoal and shells were dated and correlated from middens with a date from the peat directly beneath the natural levee of the Mississippi river on which the site is located.

9.135 1959
FAUNAL REMAINS FROM THE LEHNER MAMMOTH SITE
Lance, J.F.
American Antiquity, v. 25, n. 1, July 1959: 35-39
On faunal remain evidence the radiocarbon date of about 11,000 years for the Lehner Mammoth site seams logical.

9.136 1966
THE FIRE AREAS ON SANTA ROSA ISLAND, CALIFORNIA. I
Orr, Phil C.; Berger, Rainer
National Academy of Science. Proceedings, v. 56, n. 5, Nov. 1966: 1409-1416
The fire areas of the coast of Santa Rosa Island are exam-

ined. The more recent fire sites containing evidence of man in the form of broken mammoth bones and shattered skulls can be linked with the fire areas in deeper stratigraphic layers because they possess many of the same characteristics and become progressively fewer. Radiocarbon dates range from greater than 37,000 years, at a depth of 24.5 m, to 12,500 ± 250 years, at a depth of 2.45 m.

9.137   1966
THE FIRE AREAS ON SANTA ROSA ISLAND, CALIFORNIA. II
Orr, Phil C.; Berger, Rainer
National Academy of Science. Proceedings, v. 56, n. 9, Dec. 1966: 1678-1682
Discusses the colonisation of Santa Rosa by man in the light of radiocarbon dates.

9.138   1967
FISH REMAINS FROM SOUTHERN NEW ENGLAND ARCHAEOLOGICAL SITES
Waters, Joseph H.
Copeia, v. 1, 1967: 244-245
Fish skeletal materials representing 16 species were studied from seven archaeological sites, which also contained marine mollusc shells, near the coasts of Connecticut and southern Massachusetts and on islands south of Cape Cod. Strata in the sites have been dated by radiocarbon or by European artifacts, and range in age from less than 400 to more than 4100 years.

9.139   1968
FISH REMAINS FROM TWO SUBMERGED DEPOSITS IN TOMALES BAY, MARIN COUNTY, CALIFORNIA
Follett, W.I.
California Academy of Science. Occasional Papers, n. 67, 1968. pp. 8
Fish remains were taken from test borings of submerged deposits in Tomales Bay. Specimens were collected 10.7 to 12.1 feet below mean low water in sandy material mixed with charcoal and midden shell. A radiocarbon date of 340 ± 130 BP has been obtained from some of the associated material. Specimens collected 21.0 to 24.5 feet below mean low water yielded a radiocarbon date of 1700 ± 190 years on associated material. All species in the midden collection were presumably obtained by aborigines in comparatively shallow water.

9.140   1965
A FLAKE TOOL AND A WORKED ANTLER FRAGMENT FROM LATE LAKE AGASSIZ
Kenyon, W.A.; Churcher, C.S.
Canadian Journal of Earth Sciences, v. 2, n. 4, Aug. 1965: 237-246
A crude stone chopper and a worked left antler fragment were recovered from Lake Agassiz bed 11 (Rainy River District, Ontario). The antler fragment is identified as *Alces alces* and dated by radiocarbon at 5898 ± 423 B.C. This is believed to be the oldest dated report of North American *Alces* coeval with *Homo*.

9.141   1958
FLUTED POINTS AND GEOCHRONOLOGY OF THE LAKE MICHIGAN BASIN
Quimby, George I.
American Antiquity, v. 23, n. 3, Jan. 1958: 247-254
Radiocarbon dates evidence suggest an association of spruce - fir forest, mastodons and fluted points in the Lake Michigan basin area from about 10,000 to 7500 B.C.

9.142   1964
FLUTED PROJECTILE POINTS: THEIR AGE AND DISPERSION
Haynes, C. Vance
Science, v. 145, n. 3639, 25 Sept. 1964: 1408-1413
Stratigraphically controlled radiocarbon dating provides new evidence on peopling in the New World.

9.143   1964
THE FOOD GATHERING AND INCIPIENT AGRICULTURE STAGE OF PREHISTORIC MIDDLE AMERICA
MacNeish, Richard Stockton
In: *Handbook of Middle American Indians*, edited by Robert Wauchope, Vol. I, *Natural Environment and Culture*. Austin, Texas, University of Texas Press, 1966: 413-425
In terms of absolute dates for this food-gathering and incipient agriculture stage radiocarbon analysis from stratified sites in southwestern Tamaulipas has yielded a long sequence (agreeing in main with the stratigraphy) from 6600 B.C. to 1700 B.C. When dates from other areas and various other factors are taken into consideration, the period for this stage of human development in Middle America would be from 6000 ± 1000 to 1500 ± 500 B.C.

9.144   1964
THE FORMATIVE CULTURE OF THE CAROLINA PIEDMONT
Coe, Joffre Lanning
American Philosophical Society. Transactions, n.s., v. 54, part 5, 1964. pp. 130
The Doerdruck and Hardaway sites on the Pee Dee river and the Gaston site on the Roanoke river in South Carolina were excavated and examined. Six distinct cultural periods were found to have existed in the Roanoke basin, all represented in the Gaston site. A table of radiocarbon dates from the

Gaston site is presented.

9.145   1958
FURTHER DATA AND DATES FROM CERRO MANGOTE, PANAMA
McGinsey, Charles R.
American Antiquity, v. 23, n. 4, April 1958: 434
The radiocarbon date gives the first positive clue to the time of the pre-Formative to Formative transition in the Central American region.

9.146   1959
GEOLOGICAL AGE OF THE LEHNER MAMMOTH SITE
Antevs, Ernst
American Antiquity, v. 25, n. 1, July 1959: 31-39
The author concludes that on stratigraphic and climatologic grounds the radiocarbon dates determined for the Lehner site are too young by several thousand years.

9.147   1960
GEOLOGICAL SIGNIFICANCE OF A NEW RADIOCARBON DATE FOR THE LINDENMEIER SITE
Haynes, C. Vance; Agogino, George A.
Denver Museum of Natural History. Proceedings, n. 9, 1960. pp. 22
A radiocarbon date for the Folsom occupation of the Lindenmeier site is presented and discussed. It indicates that the Folsom people occupied the Two Creek erosion surface during a time of relatively dry climate, approximately 10,780 years BP, and that Folsom cultural complex was found there in contemporaneity with camel remains. This would make the Lindenmeier site the oldest Folsom occupation as yet recorded. A table presenting correlation of stratigraphy, cultures, fauna and radiocarbon dates in the southern United States is presented.

9.148   1952
GEOLOGICAL SITUATION OF THE ORLETON FARMS MASTODON
Goldthwait, Richard P.
Ohio Journal of Science, v. 52, n. 1, Jan. 1952: 5-9
Stratigraphical observations and radiocarbon dating indicate that the open pool condition resulting in the grey mud in which the mastodon was found lasted from about 15,000 to about 9000 years ago. Site in Madison County, Ohio.

9.149   1964
GEOLOGY AND ARCHAEOLOGY OF THE YARDAND FLINT STATION
Reger, R.D.; Péwé, Troy L.; Hadleigh-West, Frederick; Skarland, Ivar
Alaska University. Anthropology Papers, v. 12, n. 2, Summer 1964: 92-100
Although radiocarbon dates and a clear stratigraphy are recorded, the artifacts themselves yield little information. The artifact layer was determined by radiocarbon analysis to be 2300 ± 180 years old.

9.150   1968
THE GREAT BASIN ARCHAIC
Shutler, Richard
Eastern New Mexico University. Contributions in Anthropology, v. 1, n. 3, Dec. 1968 (*Archaic Prehistory in the Western United States*, Symposium of the Society for American Archaeologists, Santa Fe, 1968, edited by C. Irwin-Williams): 24-26
Suggests a tentative two-regional phase for a Great Basin archaic stage: the Lakeshore Ecology Phase and the Desert Phase. On the basis of radiocarbon dates, the Lakeshore Ecology Phase could have started as early as 10,000 to 12,000 years ago.

9.151   1959
HAWAIIAN ARCHAEOLOGY; FISHHOOKS
Emory, Kenneth P.; Bank, W.J.; Sinoto, Y.H.
Bernice P. Bishop Museum, Special Publication, n. 47, 1959. pp. 45
An uncontaminated date on charcoal samples from an occupation site on a sand dune at Puu Alii on the coast of Hawaii gave an age of A.D. 124 ± 60 years. This indicates an occupation of more than 800 years before the lava flow which produced the shelter at H8.

9.152   1963
THE HODGES SITE, A LATE ARCHAIC BURIAL STATION
Binford, Lewis R.
Michigan University. Museum of Anthropology. Anthropological Papers, n. 19, 1963: 124-148
The age of the site is estimated by comparison with two sites in New York which share a number of characteristics in common with the Hodges site of Saginaw County, Michigan, and which have been dated by radiocarbon at 998 ± 170 B.C. and 561 ± 250 B.C. The estimated date is approximately 650 B.C.

9.153   1963
HOPEWELL CULTURE BURIAL MOUNDS NEAR HELENA, ARKANSAS
Ford, James A.
American Museum of Natural History. Anthropological Papers, v. 50, part 1, 1963: 1-55
There is sufficient number of radiocarbon dates for Hopewell to be certain that the mound was constructed

about 2000 years ago. An evaluation of the dates is presented (p.46).

9.154   1947
HUMAN MIGRATION AND PERMANENT OCCUPATION IN THE BERING SEA AREA
Laughlin, William Sceva
In: The Bering Land Bridge, edited by David M. Hopkins. Stanford, Calif., Stanford University Press, 1967: 407-450
Radiocarbon dates on charcoal from a cultural layer establish the fact that people lived on the Unimak Island at a time when this coast was joined to Ananugula isand. There is some support for the hypothesis that the ancestors of the Aleuts and the Eskimos actually lived on the coast of the Bering Land Bridge and that the ancestral Eskimos were forced to withdraw as the water level rose while the Aleuts remained in place.

9.155   1965
HUMAN MIGRATIONS AND PERMANENT OCCUPATION IN THE BERING SEA AREA : A NORTH AMERICAN VIEW
Laughlin, William Sceva
In: International Association for Quaternary Research (INQUA). Congress, 7th, Boulder, Col., 14 Aug - 19 Sept. 1965. *Abstracts of General Sessions*. Boulder, Colorado, 1965: 283
The Bering land bridge provided two distinct kinds of migration routes into North America: coastal and interior. It also provided permanent residence. The Anangula Island unifacial core and blade industry, radiocarbon dated at 8000 BP, is most similar to pre-ceramic industries of Japan and Siberia, much less to Alaskan sites such as Lake Denbigh microblade and bi-facial complex of 5000 BP.

9.156   1968
A HUMAN SKELETON FROM SEDIMENTS OF MID-PINEDALE AGE IN SOUTHEASTERN WASHINGTON
Fryxell, Roald
American Antiquity, v. 33, n. 4, 1968: 511-514
A living site including human bones, bone midden and artifacts, at the Marmes rockshelter, southeastern Washington, is shown to be older than 11,000 years through the dating of sediments overlying the buried flood plain on which these bones occur (the dates were obtained on mussel shells). The geomorphic relationship suggests an age younger than 13,000 years.

9.157   1952
IMPLICATION OF RADIOCARBON DATINGS FOR THE ORIGIN OF THE DORSET CULTURE
Hoffman, Bernard G.
American Antiquity, v. 18, n. 1, July 1952: 15-17
Correlation of early cultures of north-eastern North America with geological events during the late Wisconsin glacial period, made possible by recent radiocarbon datings, suggests that the Dorset culture originated in north-eastern U.S.A. and southern Canada and moved northward into the Canadian Arctic with the waning of the ice sheet and the retreat of the tundra.

9.158   1954
IMPLICATIONS OF RADIOCARBON DATES FROM MIDDLE AND SOUTH AMERICA
Wauchope, Robert
Tulane University (New Orleans). Middle American Research Institute. Publications, n. 18, 1961: 19-39
Comments on dates released, (1) as they compare with archaeological stratigraphy in the Middle American zones of Monte Alban and Teotihuacan; (2) as they compare with broad archaeological sequences proposed for Middle America; (3) as they relate to the problems of correlating Mayan an European calendars; (4) as for their implication regarding developmental aspects of Meso-American and Andean prehistory. Tables and charts are presented.

9.159   1962
INTEMPERATE REFLECTIONS ON ARCTIC AND SUBARCTIC ARCHAEOLOGY
De Laguna, F.
Arctic Institute of North America. Technical Paper, n. 11, 1962: 164-169
Critical review of the one-way, south to north, cultural diffusion thesis set forth at a symposium on Prehistoric cultural relations between arctic and temperate zones of North America, held in Montreal, 1962. Argues that it is over-dependent upon radiocarbon dating. The inaccuracy of this method and the cultural evidence contradictory to radiocarbon datings are discussed and exemplified from Eskimo and Palaeoindian inventories.

9.160   1967
AN INTERPRETATION OF RADIOCARBON MEASUREMENTS ON ARCHAEOLOGICAL SAMPLES FROM PERU
Rowe, John Howland
In: Peruvian Archaeology: Selected Readings, edited by John Howland Rowe and Dorothy Menzel. Palo Alto, California, Peek Publications, 1967: 16-30
The radiocarbon dating method is the only one available for determining absolute ages in Peruvian archaeology prior to the second half of the 15th century A.D. Tables of dates and listings with discussions are presented.

9.161 1965
AN INTERPRETATION OF RADIOCARBON MEASUREMENTS ON ARCHAEOLOGICAL SAMPLES FROM PERU
Rowe, John Howland
In: International Conference on Radiocarbon and Tritium Dating. Pullman, Washington, Washington State University, June 1-11, 1965. *Proceedings.* United States of America, Atomic Energy Commission, 1965. Conference n. 650652: 187-198
The radiocarbon method is the only one available for determining absolute or chronometric ages in Peruvian archaeology prior to the second half of the 15th century of our era. There are no inscriptions, writing being unknown in ancient Peru, and no attempt has been made to set up a chronology based on tree rings. In evaluating radiocarbon measurements from this area, therefore, the only available standard is the degree of consistency observed among the measurements themselves. The existing measurements are discussed in that light.

9.162 1967
INTRODUCTION, AND AN INTERDISCIPLINARY APPROACH TO AN ARCHAEOLOGICAL PROBLEM
MacNeish, Richard Stockton
In: The Prehistory of the Tehuacan Valley, v. 1, *Environment and Subsistence*, edited by Douglas S. Byers. Published for the Robert S. Peabody Foundation, Phillips Academy, Andover, Massachusetts. Austin, Texas, University of Texas Press, 1967
Radiocarbon dating indicates that corn was cultivated before 3000 B.C. in the area south of the valley of Mexico but north of Tehuacan (Coxcatlan cave). Corn cobs were dated at 3610 ± 250 B.C. This was the impetus towards initiating full scale digs in 1961. Throughout the study, which will include four volumes, samples were carefully selected for radiocarbon analyses and checked by archaeological cross dating.

9.163 1958
AN INTRODUCTION TO THE ARCHAEOLOGY OF SOUTHEAST MANITOBA
Mac Neish, R.S.
Canada. National Museum. Bulletin, n. 157, *Anthropological Series*, n. 44, 1958. pp.184
Uses radiocarbon dates from similar sites in the Great Plains to supplement the geological evidence, and presents a tentative stratigraphic sequence of Southern Manitoba.

9.164 1964
INVESTIGATIONS IN SOUTHWEST YUKON: ARCHAEOLOGICAL EXCAVATIONS, COMPARISONS AND SPECULATIONS
MacNeish, Richard Stockton
Robert S. Peabody Foundation for Archaeology. Papers, v. 6, n. 2, 1964: 199-488
[Not sighted]

9.165 1957
THE ISLAND 35 MASTODON: ITS BEARING ON THE AGE OF ARCHAIC CULTURE IN THE EAST
Stephen, William
American Antiquity, v. 22, n. 4, April 1957: 359-372
This article contains a discussion of the radiocarbon dates obtained from wood and charcoal associated with the various tusks. Dates for the archaic culture which suggest that it preceded the Althithermal period are also discussed. The suggestion is made that the Archaic culture and the existence of at least one of the now extinct prehistoric mammals, the mastodon, overlapped in spite of the fact that no faunal remains were found in archaic sites.

9.166 1955
THE JAKETOWN SITE IN WEST-CENTRAL MISSISSIPPI
Ford, James A.; Phillips, Philip; Haag, William G.
American Museum of Natural History. Anthropological Papers, v. 45, part 1, 1955: 1-164
The radiocarbon measurements on charcoal do not agree with the estimate based on river channel sequence for the occupation of the Poverty Point site, being too young by 1000 years. Other dates on shells, although rather suspect, agree with the geological estimate.

9.167 1956
KLAMATH PREHISTORY. THE PREHISTORY AND THE CULTURE OF THE KLAMATH LAKE AREA, OREGON
Cressman, L.S.; Haag, William G.; Laughlin, William Sceva
American Philosphical Society, Transactions, v. 46, part 4, Nov. 1956: 375-514
Part 1 - the chronology of the cultural development of the Klamath area, p. 459-465. In the discussion, radiocarbon dates are used to establish the absolute chronology of the culture in the Klamath Lake area. The main three dates used are the date of the eruption of Mount Mazama, 6453 ± 165 BP, and the date of a house-pit, 430 ± 165 BP, both by radiocarbon and the date of the treaty by the Klamaths and the United States.

9.168 1956
KOLTERMAN MOUND 18 RADIOCARBON DATES
Wittry, Warren L.
Wisconsin Archaeologist, v. 32, n. 4, 1956: 133-134

The date of 1180 ± 25 years BP (776 A.D.) for Mound 18 (Otter Effigy) on the Kolterman Farm, Dodge County, Wisconsin, indicates that the Effigy Mound culture persisted for a period of time, this date applying to a relatively late manifestation.

9.169 1961
LA VICTORIA: AN EARLY SITE ON THE PACIFIC COAST OF GUATEMALA
Coe, Michael D.
Harvard University. Peabody Museum of Archaeology and Ethnology. Papers, n. 53, 1961. pp. 162
A discussion of the radiocarbon dates which have been published since 1950 and which have been confusing and intellectually inconsistent for the Formative of Middle America is presented. The absolute dating for the Victoria sequence is not very conclusive due to the lack of adequate charcoal samples and the unreliability of dates obtained from shells.

9.170 1967
LAKE-MARGIN ECOLOGIC EXPLOITATION IN THE GREAT BASIN AS DEMONSTRATED BY AN ANALYSIS OF COPROLITES FROM LOVELOCK CAVE, NEVADA
Cowan, Richard D.
University of California, Berkeley. Department of Anthropology. Archaeological Survey. Report, n. 70, April 1967, paper n. 2: 21-35
Two coprolites from Lovelock Cave, one from the exterior, one from inside were dated respectively by radiocarbon at 1210 ± 60 years BP and 145 ± 80 years BP giving a range of food use dating from perhaps A.D. 750 to the late eighteenth or early nineteenth century.

9.171 1966
LATE PLEISTOCENE RESEARCH AT DOMEBO. A SUMMARY AND INTERPRETATION
Leonhardy, Frank C.
In: Domebo. A Paleo-Indian Mammoth Kill in the Prairie Plains. Museum Great Plains. Contribution, n. 1, 1966: 51-53
Palaeoecology and palaeoclimatology of the Domebo site, Caddo County Oklahoma, during late Wisconsin times (10,000 to 11,500 years ago), have been reconstructed and suggest that not only was the late Wisconsin climate more moderate than the present climate, with somewhat more available moisture, but climatic conditions were more uniform throughout the Southern Great Plains. The principal significance of the Domebo site is that the association of mammoth and clovis points extends the distribution of the Llano Complex to the eastern margin of the southern Great Plains.

9.172 1965
LATE QUATERNARY PREHISTORY IN THE NORTH-EASTERN WOODLANDS
Griffin, James B.
In: The Quaternary of the United States: a Review Volume for the 7th Congress of the International Association of Quaternary Research, edited by H.E. Wright and David G. Frey. Princeton. N.J., Princeton University Press, 1965: 655-667
An investigation of the development of Palaeoindian and Early Archaic cultures in the Great Lakes area used radiocarbon dates to establish an estimated chronology.

9.173 1965
LATE WOODLAND CULTURES OF SOUTH-EASTERN MICHIGAN
Fitting, James E.
University of Michigan. Museum of Anthropoogy. Anthropological Papers, n. 24, 1965
Although there is only one radiocarbon date associated with the Younge tradition, there are a number of dates from northeastern North America which can be used to bracket the Younge tradition. These are presented in table form and range from A.D. 600 to A.D. 1500.

9.174 1967
THE LAUREL TRADITION AND THE MIDDLE WOODLAND PERIOD
Wright, J.V.
Canada. National Museum. Bulletin, n. 217, Anthropological Series, n. 79. 1967. pp. 175
Seven radiocarbon dates from the Northeast suggest the partial contemporaneity of Archaic, Early Woodland and Middle Woodland components.

9.175 1959
THE LEHNER MAMMOTH SITE, SOUTH EASTERN ARIZONA
Haury, Emil W.; Sayles, E.B.; Wasley, William W.
American Antiquity, v. 25, n. 1, July 1959: 1-30
The authors discuss the radiocarbon dates obtained from different laboratories, with regard to the opinion of Antevs (*American Antiquity*, 23, 1959: 434) who holds that the Lehner site pre-dates the Datil interval on account of the climate at the time and the generally accepted dates for the various other sites.

9.176 1963
THE LEVI SITE: A PALEO-INDIAN CAMPSITE IN CENTRAL TEXAS
Alexander, Herbert J.
American Antiquity, v. 28, n. 4, Apr. 1963: 510-528
Radiocarbon dates on snail and mussell shells were made.

Dates ranging from 10,000 ± 175 BP to 7350 ± 150 BP are consistent in relationship with the stratigraphy with the exception of the bottom of zone IV which is too young and was probably contaminated. The mixture of samples from all levels in zone II gives at best a mid-zone date probably slightly more recent than the median date for the zone. Several artifact types previously unknown in Palaeoindian culture were discovered at the Levi site. The uppermost zone contains artifacts assignable to Archaic and later cultures.

9.177   1978
LINDENMEIER, 1934 - 1974: CONCLUDING REPORT ON INVESTIGATIONS
Wilmsen, Edwin N.; Roberts, Frank H.H.
Smithsonian Contributions to Anthropology, n. 24, 1978. pp. 187
The Lindenmeier site is a Palaeoindian site in Northern Colorado. Four radiocarbon measurements are presented, analysed and discussed (p. 39-42).

9.178   1960
A LIST OF RADIOCARBON DATES FROM ARCHAEOLOGICAL SITES IN TEXAS
Campbell, T.N.
Texas Archaeological Society. Bulletin, v. 30, 1960: 311-320
[Not sighted]

9.179   1959
THE LITTLE HARBOR SITE CATALINA ISLAND
Meighan, Clement W.
American Antiquity, v. 24, n. 4, part 1, 1959: 383
Chronologically the radiocarbon dating of the site provides a key reference point for the placement at several southern California complexes. The date of ca 4000 years ago is the only date in this particular time range in California. The dates and other evidence are discussed.

9.180   1967
MAN AND WATER AT PLEISTOCENE LAKE MOHAVE
Davis, E.L.
American Antiquity, v. 32, n. 3, 1967: 345-353
There is high probability that Lake Mohave artifacts were coeval with moister climate associated with final strands of a Pleistocene Lake. Radiocarbon dates lie between 5000 and 8000 B.C.

9.181   1966
MESOAMERICA AND ECUADOR
Evans, Clifford; Meggers, Betty J.
In: Handbook of Middle American Indians, edited by Robert Wauchope, Vol. IV, *Archaeological Frontiers and External Connections*. Austin, Texas, University of Texas Press, 1966: 243-264
Archaeological evidence indicates that contacts must have repeatedly taken place between Mesoamerica and the west coast of South America. Radiocarbon dates are used in establishing the ages of various cultures which may have had contacts.

9.182   1966
MESOAMERICA AND THE EASTERN CARIBBEAN AREA
Rouse, Irving
In: Handbook of Middle American Indians, edited by Robert Wauchope, Vol. IV, *Archaeological Frontiers and External Connections*. Austin, Texas, University of Texas Press, 1966: 234-242
A radiocarbon date of 3770 B.C. for related people of the north coast of Venezuela suggests a time when they might have moved out into the Antilles.

9.183   1966
MESOAMERICA AND THE EASTERN UNITED STATES PREHISTORIC TIMES
Griffin, James B.
In: Handbook of Middle American Indians, edited by Robert Wauchope, Vol. IV, *Archaeological Frontiers and External Connections*. Austin, Texas, University of Texas Press, 1966: 111-131
The correlations are established on the basis of radiocarbon dates obtained from various sites both in the Eastern United States and Mesoamerica.

9.184   1961
THE MICHIGAN COLLEGE OF MINING AND TECHNOLOGY ISLE ROYAL EXCAVATIONS
Drier, Roy
In: Lake Superior Copper and the Indians; Miscellaneous Studies of Great Lakes Prehistory, edited by J.B. Griffin. University of Michigan. Museum of Anthropology. Anthropological Papers, n. 17, 1961: 1-7
A charred log at the 11 to 12 ft level was dated at 1500 B.C. A previous date at a depth of 72 in was 1047 ± 350 B.C., the first sound dates for one of the ancient copper pits in the Lake Superior area.

9.185   1965
MIXTEC WRITING AND CALENDAR
Caso, Alfonso
In: Handbook of Middle American Indians, edited by Robert Wauchope. Vol. III, *Archaeology of Southern Mesoamerica*, part 2. Austin, Texas, University of Texas Press, 1965: 948-961.

The Mixtec calendar, a regional aspect of the Mesoamerican calendar is found associated with objects that correspond to Monte Alban II and Monte Alban IIIA, the latter dated at 300-500 A.D.

9.186   1959
MODOC ROCK SHELTER: AN EARLY ARCHAIC SITE IN SOUTHERN ILLINOIS
Fowler, Melvin L.
American Antiquity, v. 24, n. 3, Jan. 1959: 257-270
Radiocarbon dates on various Archaic sites indicate that the Archaic in the eastern United States extends back to a time earlier than the Altithermal and confirms the conclusion by Roberts that there were several culture patterns already established about 10,000 years ago.

9.187   1959
MORE FINDINGS AT THE SERPENT MOUND SITE, RICE LAKE, ONTARIO
Johnston, Richard B.
Indiana Academy of Sciences. Proceedings, v. 69, 1959: 73-77
A number of burials were excavated in 1958-1959. A partial cremation from the Serpent Mound was dated at 1830 ± 200 BP or A.D. 128

9.188   1955
THE MORTLACH SITE IN THE BESANT VALLEY OF CENTRAL SASKATCHEWAN
Wettlaufer, Boyd N.
Saskatchewan. Department of Natural Resources. Anthropology Series, n. 1, 1955. pp. 84
A date on bone which seems logical for the culture should be considered as a minimal date, other dates of higher horizons were also obtained which fit well with the stratigraphic sequence established on the basis of dating by climate. The chronology goes from 1445 B.C. to A.D.1780.

9.189   1968
MUMMY CAVE - PREHISTORIC RECORD FROM ROCKY MOUNTAINS OF WYOMING
Wedel, Waldo R.; Husted, Wilfred M.; Moss, John H.
Science, v. 160, n. 3824, 12 Apr. 1968: 184-185
Radiocarbon dating of the fill of Mummy Cave in the Absaroka Mountains shows that the Soshone River must have been at its present level approximately 10,000 years ago and that it has done little downcutting since. Evidence of human occupation was found in 38 layers in 8.5 m of detrital sediment. Radiocarbon dates derived from charcoal start about 7280 B.C. for the earliest occupied layers and end at A.D. 1580.

9.190   1962
NEW AND REVISED RADIOCARBON DATES FROM BRAZIL
Hurt, W.H.
W. H. Over Museum (South Dakota State University, Vermillion). *Museum News*, v. 23, n. 11/12, 1962: 1-4
[Not sighted]

9.191   1962
NEW ENGLAND AND THE ARCTIC
Byers, Douglas S.
Arctic Institute of North America. Technical Papers, n. 11, 1962: 143-153
Evaluates main hypothesis on the Dorset origin or affinities of Eastern Palaeoindian cultures. Radiocarbon dates for selected Eskimo and Indian sites are tabulated.

9.192   1959
NEW EVIDENCE OF ANTIQUITY OF TEPEXPAN MAN AND OTHER HUMAN REMAINS FROM THE VALLEY OF MEXICO
Heizer, Robert F.
Southwest Journal of Anthropology, v. 15, 1959: 36-42
Ceramic associated with a grave in the outskirts of Mexico City indicate a Middle Preclassic date of 700 to 500 B.C. and radiocarbon dating of wood charcoal also associated with the grave gives a reading of 2525 ± 250 years BP or 568 B.C.

9.193   1955
A NEW POINT TYPE FROM HELL GAP VALLEY, EASTERN WYOMING
Agogino, George A.
American Antiquity, v. 26, 1961: 558-560
The cultural level from which this point type was excavated represents the oldest known cultural level at Hell Gap and is dated at 8890 B.C.

9.194   1963
NEW RADIOCARBON DATE FOR THE FOLSOM COMPLEX
Agogino, George A.
Current Anthropology, v. 4, n. 1, Feb. 1963: 113-114
Three dates directly assignable to the Palaeoindian Folsom Complex are presented and discussed. If these dates are chronologically significant they suggest that the Folsom culture flourished for only a short period in palaeohistory.

9.195   1960
A NEW RADIOCARBON DATE FROM SOUTH DAKOTA
Hurt, Wesley R.
W. H. Over Museum (South Dakota State University,

Vermillion). *Museum News*, v. 21, n. 1, 1960: 1-
[Not sighted]

9.196 1963
A NEW RADIOCARBON DATE ON CORN FROM THE DAVIS SITE, CHEROKEE COUNTY, TEXAS
Griffin, James B.; Yarnell, Richard A.
American Antiquity, v. 28, n. 3, Jan. 1963: 396-397
A radiocarbon date of A.D. 398 ± 175 for the Davis site was published in 1951. The material dated was charred corn cobs from a sub-mound house structure. A new radiocarbon date of A.D. 1307 ± 150 on corn cobs from the same structure would seem to be more accurate in view of certain archaeological interpretation and corn typology.

9.197 1960
NEW RADIOCARBON DATES AND THE MAYA CORRELATION PROBLEM
Satterthwaite, Linton; Ralph, Elizabeth K.
American Antiquity, v. 26, n. 1, July 1960: 165-184
Two long series of samples from Tikal Peten, Guatemala have been dated by the radiocarbon method for the purpose of limiting the range of possible correlation of the Maya calendar with the Christian calendar.

9.198 1966
NEW RADIOCARBON DATES FOR SOUTH-EASTERN FIBER-TEMPERED POTTERY
Stoltman, James B.
American Antiquity, v. 31, n. 6, Oct. 1966: 872-873
A shell midden excavated in the Savannah River Swamp is believed to represent a single occupation by Savannah River Archaic people. Two radiocarbon dates from the basal position suggest an age for the associated Stallings Plain pottery.

9.199 1960
NORTH AMERICAN BURIAL MOUNDS : THE CASE FOR INDEPENDENT INVENTION
Chard, Chester S.
In: *Congrès International des Sciences Anthropologiques et Ethnologiques*, 6th, Paris, 1960. *Proceedings* Vol. 2. Paris, Musée de l'Homme, 1963: 369-371
According to radiocarbon dates the Woodland ceramic tradition was active in the Ohio valley probably as early as 1500 B.C. and is thought to have been diffused from north eastern Asia. However the burial mounds associated with woodland ceramics are not present in the Siberian forest at early enough a date to have been ancestral to the New World complex.

9.200 1964
OCEANOGRAPHY AND MARINE LIFE ALONG THE PACIFIC COAST OF MIDDLE AMERICA
Hubbs, Carl L.; Roden, Gunnar I.
In: *Handbook of Middle American Indians*, Vol. I, *Natural Environment and Culture*, edited by Robert Wauchope. Austin, Texas, University of Texas Press, 1966: 142-186
Dating of layers in a coastal midden at Puntas Minitas, Northwestern Baja California, indicate that for thousands of years food-gathering people relied on the sea and lived along the shores.

9.201 1956
OLDEST TRACE OF EARLY MAN IN THE AMERICAS
Science, v. 124, n. 3218, 31 Aug. 1956: 396-397
Discovered near Lewisville, Denton County, Texas, by members of the Dallas Archaeological Society. Charcoal from camp fire debris, dated at 37,000 years, accompanies spear points. Palaeoindian Clovis and Llano cultural complex Sandia Cave site and Lucy, New Mexico may be as old or older. Other very early sites at Las Vegas, Nevada and Santa Rosa Island, California are dated 23,000 and 30,000 years respectively.

9.202 1963
OLMEC AND CHAVIN: REJOINDER TO LANNING
Coe, William R.
American Antiquity, v. 29, n. 1, July 1963: 101-104
A revision of the hypothesis that the New World civilisation originated in the Gulf Coast of Mexico is made, which accounts for the Chavin civilisation being the result of intrusive Olmec art and religion with an older native Peruvian tradition based on fabric construction and the worship of the condor and serpent. The author indicates the difficulty in deciding between two radiocarbon dates.

9.203 1963
OLMEC AND CHAVIN: REPLY TO MICHAEL M. COE
Lanning, Edward P.
American Antiquity, v. 29, n. 1, July 1963: 99-101
According to radiocarbon dates from combustion material of an early building stage, the large temple at Las Haldas belongs to a period well back into pre-ceramic time, contradicting the idea of Olmec derivation for Kotosh.

9.204 1967
OLMEC CIVILIZATION, VERACRUZ, MEXICO. DATING OF THE SAN LORENZO PHASE
Coe, Michael D.; Diehl, Richard A.; Stuiver, Minze
Science, v. 155, n. 3768, 17 March 1967: 1399-1401
The Olmec sculpture of the San Lorenzo phase are placed in the Early Formative period (1500 - 800 B.C.). Five or six radiocarbon dates fall within the 1200 - 900 B.C. span. This

phase therefore marks the beginning of the Olmec civilisation. Dating was done on wood charcoal from stratified hearths of the San Lorenzo phase from the river bank cut at Tenochtitlan. The dates are presented in table form.

9.205   1965
THE OLMEC STYLE AND ITS DISTRIBUTION
Coe, Michael D.
In: Handbook of Middle American Indians, edited by Robert Wauchope. Vol. III, Archaeology of Southern Mesoamerica, part 2. Austin, Texas, University of Texas Press, 1965: 739-775.
Radiocarbon dates for Complex A, La Venta, places the Olmec style during approximately 800-400 B.C. in preclassic times.

9.206   1967
ON MIGRATIONS OF HUNTERS ACROSS THE BERING LAND BRIDGE IN THE UPPER PLEISTOCENE
Muller-Beck, Hansjürgen
In: The Bering Land Bridge, edited by David M. Hopkins. Stanford, Calif., Stanford University Press, 1967: 371-408
Using the climatic history, the geographic history of the Bering land bridge, chronologies based on radiocarbon datings and palaeolithic technological development in the Old and the New Worlds during the late Pleistocene, a partial reconstruction of diffusion and migration events in the Bering region is presented.

9.207   1962
ON NEW RADIOCARBON DATES FROM THE CALIFORNIA CHANNEL ISLANDS
Orr, Phil C.
Santa Barbara Museum of Natural History. Dept. of Anthropology. Bulletin, n. 8, 1962.
[Not sighted]

9.208   1964
PALEO-INDIAN TRADITIONS. A CURRENT EVALUATION
Agogino, George A.; Rovner, Irwin
Archaeology, v. 17, n. 4, Winter 1964: 237-243
Radiocarbon dates are used to establish a tentative chronology for the various types of projectile points culture in the United States.

9.209   1962
ONION PORTAGE AND THE OTHER FLINT SITES OF THE KOBUC RIVER
Giddings, J.L.
Arctic Anthropology, v. 1, n. 1, 1962: 6-21
The Onion Portage site seems ideal for a projected test of correlation between radiocarbon dates and measures of hydration layer of obsidian. A diagram correlating river sites with sites on the coast is presented. The A.D. dates are based partly on an incomplete radiocarbon series.

9.210   1964
ORIGINS OF AGRICULTURE IN MIDDLE AMERICA
Mangelsdorf, Paul C.; MacNeish, Richard Stockton; Willey, Gordon R.
In: Handbook of Middle American Indians, edited by Robert Wauchope, Vol. I, Natural Environment and Culture. Austin. Texas, University of Texas Press, 1966: 427-445
A correlation of two regional sequences of Tamaulipas, with phase datings based largely upon radiocarbon readings is presented.

9.211   1952
THE ORLETON FARMS MASTODON
Thomas, Edward S.
Ohio Journal of Science, v. 52, n. 1, Jan. 1952: 1-5
Radiocarbon analysis of wood found in the marl immediately beneath and around the skeleton gives an age of 8420 ± 400 years ago. Site in Madison County Ohio.

9.212   1965
PACIFIC COAST ARCHAEOLOGY
Meighan, Clement W.
In: The Quaternary of the United States: a Review Volume for the 7th Congress of the International Association of Quaternary Research, edited by H.E. Wright and David G. Frey. Princeton. N.J., Princeton University Press, 1965:709-719
Investigations in Pacific Coast archaeology have established a firm chronology for early man based on radiocarbon dates extending back to 8000 years ago in scattered locations from the eastern Aleutians to southern California.

9.213   1963
PALAHNIHAN: RADIOCARBON SUPPORT FOR GLOTTO-CHRONOLOGY
Baumhoff, Martin A.; Olmsted, D.
American Anthropologist, v. 65, n. 2, Feb. 1963: 278-284
Radiocarbon dates support the hypothesis that before 2000 B.C. California was occupied by Hokan speakers, after which they were displaced by Proto-Palahnihans which relegated the Hokan group to marginal positions. Radiocarbon dates at the Lorenzen site indicate that the Palahnihan people have been in their present territory for over 33 centuries.

9.214   1960
A PALEO-INDIAN BISON-KILL IN NORTH-WESTERN IOWA

Agogino, George A.; Frankforter, W.D.
American Antiquity, v. 25, n. 3, Jan. 1960: 414-415
The lowest point of the Simonson site produced projectile points, scrapers and knives in association with *Bison Occidentalis* and was dated by radiocarbon at 6471 ± 520 BC. This and various pre-ceramic sites indicate a far heavier occupation of this region in pre-ceramic times than has been suspected in the past.

9.215   1968
PALEO-INDIAN REMAINS FROM LAGUNA DE TAGUA TAGUA, CENTRAL CHILE
Montané, Julio
Science, v. 161, n. 3846, 13 Sept. 1968: 1137-1138
Bone and stone tools associated with extinct fauna (horse and mastodon) place man's occupation in Chile at 11,380 ± 320 years BP according to a radiocarbon date on charcoal associated with the remains. This is the earliest date yet obtained for human occupation in Chile.

9.216   1966
THE PALEO-INDIAN'S GEOGRAPHY OF NOVA SCOTIA
Borns, Harold W.
In: *International Association for Quaternary Research* (INQUA), *Congress*, 7th. Boulder, Col., 1965. *Papers*. 1966: 41
Tentative radiocarbon dates from the Debert site in north Nova Scotia suggest that the palaeoindians occupied the area between 11,000 and 7000 years ago.

9.217   1965
PALEOTEMPERATURES AND CHRONOLOGY AT ARCHAEOLOGICAL CAVE SITE REVEALED BY THERMOLUMINESCENCE
Dort, Wakefield; Zeller, E.J.; Turner, M.D.; Vaz, J.E.
Science, v. 150, n. 3695, 22 Oct. 1965: 480-481
Contrasting values of remnant thermoluminescence of limestone from Jaguar Cave, east-central Idaho indicate that human occupation was limited to the duration or part of the duration of one climatic regime. Two samples of charcoal coming from near the centre of the cave, both horizontally and vertically, give dates of 10,370 ± 350 and 11,580 ± 250 years, or near the end of the Wisconsin time.

9.218   1957
PARACAS, NAZCA, AND TIAHUANACANOID CULTURAL RELATIONSHIP IN SOUTH COASTAL PERU
Strong, William Duncan
American Antiquity, v. 12, n. 4, part 2, April. 1957 (*Society for American Archaeology, Memoirs*, No. 13, Salt Lake City, 1957. pp. 48)
Recent radiocarbon dates from Ica, Nazca, and Viru Valleys are presented. The series presented is remarkably consistent.

9.219   1966
PATTERN AND PROCESS IN THE EARLY INTERMEDIATE PERIOD POTTERY OF THE CENTRAL COAST OF PERU
Patterson, Thomas C.
University of California Publications in Anthropology, v. 3 Berkeley and Los Angeles. University of California Press, 1966. pp.180
Radiocarbon dating support the stylistic correlation of early intermediate period of pottery from various sites in Peru.

9.220   1964
THE PATTERNS OF FARMING LIFE AND CIVILIZATION
Willey, Gordon R.; Eckholm, Gordon F.; Millon, Ree F.
In: *Handbook of Middle American Indians,* edited by Robert Wauchope, Vol. I, *Natural Environment and Culture*. Austin, Texas, University of Texas Press, 1966: 446-598
Absolute dates for the Preclassic period fall into two groups of radiocarbon results sharply at variance with each other. This is discussed. Classic and pre-Classic and late Classic periods are also discussed in terms of radiocarbon dates.

9.221   1966
PERSPECTIVES IN THE PREHISTORY OF KODIAK ISLAND, ALASKA
Clark, Donald W.
American Antiquity, v. 31, n. 3, Jan.1966: 358-371
Five phases in the stratigraphy of the archaeological material on Kodiak Island are described. The time range estimation is from 2500 B.C. to shortly after A.D. 1800. Radiocarbon dates were provided by R.J. Stuckenrath, Applied Science Center for Archaeology at the University of Pennsylvania.

9.222   1967
PLAINS RELATIONSHIPS OF THE FREMONT CULTURE: A HYPOTHESIS
Aikens, C. Melvin
American Antiquity, v. 32, n. 2, April 1967: 198-209
Radiocarbon dates are used to determine the origins of the Fremont Culture. It is suggested that the Fremont Culture drifted back onto the Central Plain around approximately A.D. 1400 - 1600, to become the forerunner of the Dismal River culture.

9.223   1958
PLEISTOCENE CAMPSITE NEAR LEWISVILLE, TEXAS
Crook, W.W.; Harris, R.K.

American Antiquity, v. 23, n. 3, Jan. 1958: 233-247
The absolute agreement of two different dating runs, based upon two different carbon samples, of different materials, from two different hearths in the site is considered to be significant although both are beyond the effective range of the counting apparatus, and the dates themselves are simply minimum expressions. Quantities of charcoal from these two locations have been preserved for future cross checking particularly should the radiocarbon method be extended or improved.

9.224   1964
PLEISTOCENE CHIPPED STONE TOOLS OF SANTA ROSA ISLAND, CALIFORNIA
Orr, Phil C.
Science, v. 143, n. 3603, 17 Jan. 1964: 245-244
Santa Rosa island, California, has been the site of numerous discoveries of Pleistocene man. The recent findings of a well-made chipped stone tool in situ in the mammoth beds adds further evidence. Dating of charcoal in middens in the same stratigraphy suggest an age of 12,500 to 12,620 years BP.

9.225   1956
PLEISTOCENE MAN IN FISHBONE CAVE, PERSHING COUNTY, NEVADA
Orr, Phil C.
Nevada State Museum. Dept. of Archaeology. Bulletin, n. 2, 1956: 1-20
Fishbone Cave, Pershing County, Nevada, has contributed valuable additional information to the knowledge of Pleistocene man in North America and has given the earliest radiocarbon date for human skeletal remains and perishable objects. This date together with similar dates secured from the Leonard Rock Shelter and Danger Cave, Utah, show that there is a consistency for the time when the lake waters receded sufficiently to make these caves habitable. The association of men, horses and camels in Fishbone Cave is established and stone tools from Fishbone Cave indicate that man must have been in the region from ca 11,000 to at least 6000 years BP.

9.226   1957
PLEISTOCENE, MAN AND SAN DIEGO
Carter, George F.
Baltimore, John Hopkins Press, 1957. pp. 400
A table of radiocarbon dates for Southern California is presented. The dating of soils is also based on radiocarbon determinations (p.130-136).

9.227   1965
THE POLLEN EVIDENCE FOR THE ENVIRONMENT OF EARLY MAN AND EXTINCT MAMMALS OF THE LEHNER MAMMOTH SITE: SOUTH-EASTERN ARIZONA
Mehringer, Peter J.; Haynes, C. Vance
American Antiquity, v. 31, n. 1, July 1965: 17-23
Fossil pollen is directly associated with a radiocarbon date, mammoth bones and the same stratigraphic units in which mammoth, bison, tapir and horse bones and Clovis artifacts were recovered at the Lehner site. The pollen evidence indicates that desert grassland occupied the San Pedro Valley of southeastern Arizona about 9000 BC.

9.228   1956
POSITION AND MEANING OF A RADIOCARBON SAMPLE FROM THE SHEGUIANDAH SITE, ONTARIO
Lee, T.E.
American Antiquity, v. 22, n. 1, 1956: 79
An important radiocarbon date of 7130 ± 250 years pertaining to an archaeological site in an area commonly supposed to have been under Valders (Mankato) ice is of significance as it confirms that man was present before the last ice sheet advanced over the area. This is discussed.

9.229   1966
THE POSITION OF THE DEBERT SITE WITH RESPECT TO THE EASTERN PALEO-INDIAN CONTRIBUTION OF THE DEBERT ARCHAEOLOGICAL PROJECT
Byers, Douglas S.
In: International Association for Quaternary Research (INQUA), Congress, 7th. Boulder, Col., 1965. *Papers*. 1966: 57
Discovery of a palaeoindian site near Debert, Nova Scotia, tentatively placed by preliminary radiocarbon dates at 11,000 to 9000 years ago poses a number of problems.

9.230   1965
A POSSIBLE BISON *(SUPERBISON) CRASSICORNIS* OF MID-HYPSITHERMAL AGE FROM MERCER COUNTY, NORTH DAKOTA
Brophy, John A.
North Dakota Academy of Science, Proceedings, v. 19, 1965: 214-223
The bison skulls, which were discovered in the valley of Spring Creek, Mercer County, North Dakota, were assigned on the basis of horn-core indices to the species *Bison Crassicornis*. Wood found in close association with one of the specimens was dated by radiocarbon at 5440 ± 200 years

BP which falls at about the middle of the post-glacial Hypsithermal interval, a time of relatively warm and dry climate.

9.231   1962
POST-GLACIAL MASTODON REMAINS AT TUPPERVILLE, ONTARIO
Dreimanis, Aleksis
Geological Society of America, Special Paper, n. 68, 1962: 167
Bones and teeth of mastodons were discovered in calcareous sand of a beach bar of ancient Lake St. Clair at elevation of 584 feet and 582 feet. Radiocarbon date of the gyttja found in cavities of the skull gave a date of 6230 ± 240 years BP. which can be considered as a minimum date for the mastodon and would make it one of the last mastodons that lived in the Great Lake regions; most other dated skeletons are much older.

9.232   1963
POST-GLACIAL PALYNOLOGY AND ARCHAEOLOGY IN THE NAKNAK RIVER DRAINAGE AREA, ALASKA
Heusser, Calvin John
American Antiquity, v. 29, n. 1, July 1963: 74-81
Radiocarbon dates help establish absolute dating in an attempt to construct the sequence of environments dating back to the earliest recognisable culture phase. It established that the time interval covering the material of the study is post-glacial, not pre-dating 6000 BP.

9.233   1956
POVERTY POINT, A LATE ARCHAIC SITE IN LOUISIANA
Ford, James A.; Webb, Clarence H.
American Museum of Natural History. Anthropological Papers, v. 46, part 1, 1956: 5-136
The radiocarbon dates relating to the Poverty Point site are presented in table form and figures, and discussed (p. 117). Forty two dates are presented.

9.234   1961
PRE-COLUMBIAN LITTOREA IN NOVA SCOTIA
Clarke, A.H.; Erskine, J.S.
Science, v. 134, n. 3476, 11 Aug. 1961: 393-394
Radiocarbon dating of shells of *Littorina Littorea* show that this gastropod was a native of Nova Scotia before the advent of European culture and later spread southward through human factors.

9.235   1963
A PRECERAMIC SETTLEMENT ON THE CENTRAL COAST OF PERU: ASIA UNIT 1
Engel, Frederick
American Philosophical Society. Transactions, n.s., v. 53, part 3, 1963. pp. 139
The archaeological background of the region is established with radiocarbon dates providing a chronological framework for the preceramic cultures of coastal Peru.

9.236   1965
PRECLASSIC AND CLASSIC ARCHITECTURE AT OAXACA
Acosta, Jorge R.
In: Handbook of Middle American Indians, edited by Robert Wauchope. Vol. III, *Archaeology of Southern Mesoamerica,* part 2. Austin, Texas, University of Texas Press, 1965: 814-836
One single date for the ancient city of Monte Alban first phase in the Oaxaca region gives a reading of 649 ± 170 years B.C.

9.237   1951
PREHISTORIC AMERICA: COMMENTS ON SOME C-14 DATES
Bushnell, G.H.S.
Antiquity, v. 25, n. 99, Sept. 1951: 145-149
The majority of results from radiocarbon dating are mutually consistent and are of a great help in building a coherent picture of America's past. A discussion of the cause of error which led to inconsistent dates is also included in the article.

9.238   1960
PREHISTORIC CULTURE SEQUENCES IN THE EASTERN ARCTIC AS ELUCIDATED BY STRATIFIED SITES AT IGLOOLIK
Meldgaard, Jorgen
In: Man and Culture; Selected Papers. International Congress of Anthropologic and Ethnologic Sciences, 5th, Philadelphia, 1956. Philadelphia, University of Pennsylvania Press, 1960: 588-595
Several dates on antler samples are presented. They date the pre-Dorset culture of the area.

9.239   1964
PREHISTORIC MAN IN THE NEW WORLD
Jennings, Jesse D.; Norbeck, Edward
Prehistoric Man in the New World, edited by Jesse D. Jennings and Edward Norbeck. Chicago, University of Chicago Press, 1964. pp. 633
The papers published in this volume are lengthier versions of addresses delivered at Rica University on November 9 and 10, 1962, in a symposium entitled 'Prehistoric Man in the New World'. The aim of the symposium was to present a review and appraisal of facts and theories concerning the prehistoric peoples and cultures of North and South America. Radiocarbon dating is used throughout to establish

absolute dating and chronology.

9.240   1961
PREHISTORIC MAN OF THE GREAT PLAINS
Wedel, Waldo R.
Norman, Okl., University of Oklahoma Press, 1961. pp. 355
A brief description of the method of radiocarbon dating is presented. Dates are used throughout the study, and in particular to help establish a chronology of human occupation of the Great Plains.

9.241   1966
PREHISTORIC MINERS OF SALTS CAVE, KENTUCKY
Watson, Patty Jo
Archaeology, v. 19, n. 4, Oct. 1966: 237-243
Radiocarbon dates on faecal samples and woody remains place the exploitation of Salts Cave at 1190 ± 150 BC to 290 ± 200 BC.

9.242   1966
PREHISTORIC ROCK PAINTINGS IN BAJA CALIFORNIA
Meighan, Clement W.
American Antiquity, v. 31, n. 3, Jan. 1966: 372-392
Elaborate rock paintings from several rock shelters in the central part of Baja California are assigned to the Comondu culture on the basis of artifacts found in the shelters, one of which yielded a radiocarbon date of A.D. 1435 ± 80.

9.243   1966
PREHISTORIC SPRINGS AND GEOCHRONOLOGY OF THE CLOVIS SITE, NEW MEXICO
Haynes, C. Vance; Agogino, George A.
American Antiquity, v. 31, n. 6, Oct. 1966: 812-821
From the radiocarbon dates obtained from carbonised plant remains and from their stratigraphic position, a sequence of events can be determined at the Clovis site.

9.244   1965
PREHISTORY AT PORT MOLLER, ALASKA
Workman, William
Arctic Anthropology, v. 3, n. 2, 1965: 132-153
Two radiocarbon dates from the bottom of the cultural layer in two trenches gave ages of 2680 ± 320 years BP indicating that the Hot Springs Village site was occupied by man at least as early as 2500 to 3000 years ago.

9.245   1958
PREHISTORY IN THE DISMAL LAKE AREA, N.W.T, CANADA
Harp, Elmer
Arctic, v. 11, n. 4, Dec. 1958: 219-249
A clarification of the cultural development of the Cape Dorset Eskimo and other early peoples of the Eastern Arctic, linking such groups firmly with ancestral manifestations in Alaska. Radiocarbon dating helps establish correlations.

9.246   1967
PREHISTORY OF THE PACIFIC NORTHWEST PLATEAU AS SEEN FROM THE INTERIOR OF BRITISH COLUMBIA
Sanger, David
American Antiquity, v. 32, n. 2, Apr. 1967: 186-197
Recent excavations in south-central British Columbia have revealed a 7500 year sequence which indicates cultural relationships with the subarctic and the Canadian Prairie Provinces. The chronology of the sequence was established by ten radiocarbon dates from the Lytton - Lilloget Regions, which comprise the first regional cultural chronology with any appreciable time depth from the Interior Plateau of British Columbia.

9.247   1964
PREHISTORY OF THE WEST INDIES
Rouse, Irving
Science, v. 144, n. 3618, 1 May 1964: 499-513
With the development of radiocarbon analysis it has become possible to determine the duration of the epochs and periods of Caribbean culture.

9.248   1958
PRELIMINARY ARCHAEOLOGICAL INVESTIGATIONS IN THE SIERRA DE TAMAULIPAS, MEXICO
MacNeish, Richard Stockton
American Philosophical Society. Transactions, n.s., v. 48, part 6, 1958. pp. 210
A table showing tentative assignments of relative and absolute dates (radiocarbon dates) to the cultural phases of the Sierra de Tamaulipas is presented.

9.249   1953
PRELIMINARY EXCAVATION AT THE THOMAS SITE, MARIN COUNTY
Meighan, Clement W.
University of California, Berkley. Department of Anthropology. Archaeological Survey Reports, n. 19, Jan. 1953: 1-7
Two radiocarbon dates on charcoal samples giving ages of 633 ± 200 and 911 ± 180 years BP enable the age of the mound to be deduced at roughly 1000 years with occupation until about 1800.

9.250 1951
PRELIMINARY REPORT ON THE LEONARD ROCKSHELTER SITE, PERSHING COUNTY, NEVADA
Heizer, Robert F.
American Antiquity, v. 17, n. 2, Sept. 1951: 89-98
Bat guano collected in 1930 and 1933 at the level of wooden artifacts at the open rock site which was dated by W.F. Libby and dates from grease wood atlatl foreshafts from the same layer placed the site in the category of early man (7038±350 to 866 ± 300 years).

9.251 1968
PREPARED CORE AND BLADE TRADITIONS IN THE PACIFIC NORTHWEST
Sanger, David
Arctic Anthropology, v. 5, n. 1, 1968: 92-120
Artifact collections of prepared cores and blades from about 20 sites in southern British Columbia, Washington and northern Oregon are discussed, pertinent literature of the past 20 years is reviewed, and new material is described. Radiocarbon dates, some new, on associated materials range from about 8000 B.C. to A.D 500. Two distinct, historically unrelated traditions are indicated.

9.252 1964
THE PRIMITIVE HUNTERS
Aveleyra Arroyo de Anda, Luis
In: Handbook of Middle American Indians, edited by Robert Wauchope, Vol. I, *Natural Environment and Culture.* Austin, Texas, University of Texas Press, 1966: 384-412.
A discussion of radiocarbon dates which place mammoth remains at between 16,000 and 8000 years B.C. with an inferred presence of early man in the valley of Mexico.

9.253 1965
PROBLEMS IN DATING MOHAVE ARTIFACTS
Heizer, Robert F.
Masterkey, v. 39, n. 4, Oct/Dec 1965: 125-134
Discusses the difficulty of associating radiocarbon dates on shells, tufa or other biogenic material, with recognised human artifacts in order to support the chronology of Lake Mohave artifacts.

9.254 1966
PROBLEMS IN RADIOCARBON DATING OF TEOTIHUACAN
Kovar, Anton J.
American Antiquity, v. 31, n. 3, Jan. 1966: 427-430
Radiocarbon dating is a useful tool in archaeological research but an understanding of the principles on which it is based is necessary in order to apply it intelligently and to avoid misinterpretation of the dates obtained. The author discusses various aspects such as the Suess effect, contamination, the nature of wood growth, presenting his experience in using radiocarbon dates for Teotihuacan temples.

9.255 1965
QUATERNARY HUMAN OCCUPATION OF THE PLAINS
Stephenson, Robert L.
In: The Quaternary of the United States: a Review Volume for the 7th Congress of the International Association of Quaternary Research, edited by H.E. Wright and David G. Frey. Princeton. N.J., Princeton University Press, 1965: 685-696
Three sequential culture stages are determined in the human cultural development in the Great Plains area of central North America: the Paleo-Indian stage, the Archaic stage and the Sedentary stage which includes the incipient and developed agriculturalist. The radiocarbon determinations on many of the sites studied have supported the 'guess dates' to a considerable degree.

9.256 1965
RADIOCARBON AGE OF NEVADA MUMMY
Orr, Phil C.; Berger, Rainer
Science, v. 148, n. 3676, 11 June 1965: 1466-1467
Skin tissue, bone collagen, and vegetable clothing from a well preserved Indian mummy from a dry cave in Nevada have been dated by radiocarbon. The age is about 2500 years. The ages obtained from the various samples were in close agreement.

9.257 1957
RADIOCARBON AGE OF THE DAMARISCOTTA SHELL HEAPS
Bradley, W.H.
American Antiquity, v. 22, n. 3, Jan. 1957: 296
The dates indicate in a general way that the heaps started accumulating soon after the beginning of the Christian era and may have built up rapidly, within one to a few hundred years.

9.258 1951
RADIOCARBON AGES MEASUREMENTS AND FOSSIL MAN
Terra, Helmut de
Science, v. 113, n. 2927, 2 Feb. 1951: 124-125
The date obtained from charcoal from prehistoric sites in the basin of Mexico establishes an age for early human occupation.

9.259 1965
RADIOCARBON AND GEOLOGICAL DATING OF THE LOWER FRASER CANYON ARCHAEOLOGICAL SEQUENCE

Borden, Charles B.
In: International Conference on Radiocarbon and Tritium Dating. Pullman, Washington, Washington State University, June 1-11, 1965. *Proceedings.* United States of America, Atomic Energy Commission, 1965. Conference n. 650652: 165-178
Age determinaton by means of radiocarbon analysis, combined with archaeological and geological data, have made it possible to develop for the Fraser Canyon region one of the longest continuous cultural sequences in the Western Hemisphere.

9.260   1963
RADIOCARBON ASSAYS. FIRST ANNUAL REPORT. AMERICAN BOTTOMS ARCHAEOLOGY, JULY 1, 1961 - JUNE 30, 1962
Fowler, Melvin L.
Urbana, Ill., 1963: 49-57
[Not sighted]

9.261   1964
A RADIOCARBON DATE FOR THE MACON EARTH LODGE
Wilson, Rex L.
American Antiquity, v. 30, n. 2, part 1, Oct. 1964: 202-203
A radiocarbon date of A.D. 1015 for the Macon Earth Lodge, Ocmulgee National Monument, has recently been determined. The material dated was part of a charred roof beam. This date and another from the Brown's Mount component, agree with generally accepted dates for the Macon Plateau period in central Georgia.

9.262   1961
A RADIOCARBON DATE FROM CENTRAL TEXAS
Kelly, Thomas C.
Texas Archarological Society. Bulletin, v. 31, 1961: 329-330
A sample of charcoal from the Crumley site, an Edwards Plateau Aspect site of Travis County, Texas, has been dated by Humble Oil at 3275 ± 125 years BP. This is the first radiocarbon date from a pure component of the Edwards Plateau culture.

9.263   1961
A RADIOCARBON DATE FROM GOEBEL MIDDEN, AUSTIN TEXAS
Fleming, C.B.
Texas Archarological Society. Bulletin, v. 31, 1961: 330
A sample of carbonised wood, including acorns, from a hearth in the lowest part of the midden, was dated at 4530 ± 80 years BP by Shell Development Company. The level contains arrows. This is the first radiocarbon date for the archaic of southeastern Texas.

9.264   1965
A RADIOCARBON DATE FROM THE CENTRAL COAST OF PERU
Stumer, Louis M.
American Antiquity, v. 26, n. 4, Apr. 1961: 548-550
A specimen from a bundle of rope found at Playa Grande in association with pottery of Playa Grande 1 style and the white-zoned and white-slipped varieties of the Banos de Boza style is dated A.D. 570 ± 160. This date tends to substantiate but not clearly demonstrate contemporaneity of the Mochica 3, 4, 5 sequences on the North Coast with the Playa Grande, Maranga sequence on the Central Coast and the Nazca A,B,Y development of the South Coast.

9.265   1951
A RADIOCARBON DATE ON THE DAVIS SITE IN EAST TEXAS
Krieger, Alex D.
American Antiquity, v. 17, n. 2, Sept. 1951: 144-145
The author examines radiocarbon dates available from east Texas and the conflicting results they gave. His conclusion is that radiocarbon datings present a challenge to all ideas and methods. In many cases they are confirmatory and satisfactory. When they are not, the 'accepted alignments' should be examined more critically.

9.266   1967
RADIOCARBON DATED ARCHAEOLOGICAL REMAINS ON THE NORTHERN AND CENTRAL GREAT PLAINS
Neuman, Robert W.
American Antiquity, v. 32, n. 4, Oct. 1967: 471-486
This paper consolidates and summarises the mass of radiocarbon determinations that have accumulated in various journals, newspapers and manuscripts between 1951 and 1965. The dates are derived from Palaeoindian Archaic, Early Ceramic and Late Prehistoric occupations in the Canadian Provinces of Alberta and Saskatchewan, and the Plains areas of Montana, Wyoming, Colorado, North Dakota, South Dakota, Nebraska, Kansas and Iowa. Included are 195 radiocarbon dates attesting to the presence of man on the Plains for the last 12,000 years.

9.267   1951
RADIOCARBON DATES AND EARLY MAN
Roberts, Frank H.H.
American Antiquity, v. 17, part 2 (*Society for American Archaeology, Memoirs*, n. 8), 1951: 20-22
Presents and discusses radiocarbon dates associated with early man in America. A general conclusion indicates that not only was the New World occupied 10,000 years ago but

9.268 1968
RADIOCARBON DATES AND HISTORICAL DATES; CORRELATION IN NEW WORLD ARCHAEOLOGY
Lerman, J.C.
In: *International Congress of Americanists*, 38th, Stuttgart-Munnich, 12-18 Aug. 1968. *Proceedings.* 1968
[Not sighted]

9.269 1961
RADIOCARBON DATES AND NEW WORLD CHRONOLOGY
Bushnell, Geoffrey
Antiquity, v. 35, n. 140, Dec. 1961: 286-291
Examines the contribution made by radiocarbon to the dating of the New World chronology. Ages dating back to more than 30,000 years have been obtained from the presence of early man. The radiocarbon method of dating is all the more important for the New World as other aids to dating as used in Old World archaeology are absent.

9.270 1963
RADIOCARBON DATES AND THE MAYA CORRELATION PROBLEM
Smiley, Charles H.
Nature, v. 199, n. 4892, 3 Aug. 1963: 473-474
Points out weaknesses in the interpretation of radiocarbon dates from wood from Temple I and IV and structure 10 at Tikal, Guatemala, given by L. Satterthwaite (*American Antiquity*, 21, 1956 and 26, 1960) in support of the Goodman - Thompson correlaton of the Mayan and Christian calendars.

9.271 1956
RADIOCARBON DATES AND THE MAYA CORRELATION PROBLEM
Satterthwaite, Linton
American Antiquity, v. 21, n. 4, Apr. 1956: 416-419
In spite of possible contamination at the collection stage the dates are consistent with other radiocarbon dates for comparable horizons in the northwest. Description of the samples and the horizons at which they were collected. Ages of 3959 ± 310 and 3350 ± 400 were determined.

9.272 1961
RADIOCARBON DATES FOR AZATLAN
Ritzenthaler, Robert
Wisconsin Archaeologist, v. 42, n. 3, 1961: 139
A date of 1200 ± 150 years BP (A.D. 760) is considered too early. Another date already obtained, of 320 ± 200 years BP (A.D. 1640) is considered too late. A date of A.D. 1200 would be more logical.

9.273 1967
RADIOCARBON DATES FOR EARLY 'NON RAMEY' MISSISSIPPIAN COMPLEX AT CAHOKIA
Hall, Robert L.
Society of American Archaeology. Annual Meeting, 32nd. Ann Arbor, May 1967. *List prepared for.* 1967
[Not sighted]

9.274 1961
RADIOCARBON DATES FOR SOUTHEASTERN FIBER-TEMPERED POTTERY
Bullen, Ripley P.
American Antiquity, v. 27, n. 1, July 1961: 104-106
Five new radiocarbon dates from an Archaic midden at the Palmer site on the Florida Gulf Coast confirm the previous estimate of 2000 BC for the beginning of pottery making in the South East. Correlations over hundreds of miles give extremely close results between archaeological sub-periods and radiocarbon dates.

9.275 1952
RADIOCARBON DATES FOR THE EASTERN UNITED STATES
Griffin, James B.
In: *Archaeology of the Eastern United States.* Chicago University Press, 1952: 365-370
List of dates for the Eastern United States issued by the University of Chicago, the Lamont Geological Observatory and the Michigan University Radiocarbon Laboratory, followed by discussions of the ages obtained.

9.276 1955
RADIOCARBON DATES FOR WOOD SPECIMENS FROM THE SPIRO MOUND, OKLAHOMA
Bell, Robert E.
Oklahoma Anthropological Society. Newsletter, v. 4, n. 1, 1955: 9
[Not sighted]

9.277 1965
RADIOCARBON DATES FROM A TOMB IN MEXICO
Furst, P.T.
Science, v. 147, n. 3658, 5 Feb. 1965: 612-613
The first series of radiocarbon dates on shell artifacts obtained from a deep shaft-and-chamber tomb of the type restricted, in Mesoamerica, to parts of Nayarit, Valisco and Colina in Western Mexico ranges from 2230 ± 100 to 1710 ± 80 years. Examination of the evidence indicates that, for

the present, a date equivalent to A.D. 250 should be accepted for at least one phase, possibly a late phase of the Shaft Tomb Culture and for the hollow polychrome figurines associated with the tombs.

9.278   1957
RADIOCARBON DATES FROM AN EARLY ARCHAIC DEPOSIT IN RUSSELL CAVE, ALABAMA
Miller, Carl F.
American Antiquity, v. 23, n. 1, July 1957: 84
Two dates of charcoal samples at a depth of 13 feet were obtained from different laboratories: Lamont, Michigan and U.S. Geological Survey. The two ages from Lamont and Michigan, respectively 7950±200 years and 8560±400 are not archaeologically significant and confirm placement at the beginning of early Archaic times as the associated chert projectile points suggest.

9.279   1960
RADIOCARBON DATES FROM ARCHAEOLOGICAL SITES IN OKLAHOMA
Bell, Robert E.
Oklahoma Anthropological Society. Bulletin, v. 9, 1960: 77-80
Listing of radiocarbon dates for archaeological sites located in Oklahoma. Dates determined by Michigan, Humble Oil and New Zealand Radiocarbon Dating Laboratories and also published elsewhere.

9.280   1958
RADIOCARBON DATES FROM CALIFORNIA OF ARCHAEOLOGICAL INTEREST
Heizer, Robert F.
University of California, Berkeley. Department of Anthropology. Archaeological Survey Reports, n. 44, part 1, Oct. 1958: 1-16
Examines the method of dating by radiocarbon assay and discusses the interpretation of dates. A map of sites with radiocarbon dates and a table of California dates are presented. A number of dates are discussed and assessed.

9.281   1956
RADIOCARBON DATES FROM ELLESWORTH FALLS, MAINE
Hadlock, Wendell S.; Byers, Douglas S.
American Antiquity, v. 21, n. 4, Apr. 1956: 419-420
The importance of radiocarbon evidence can scarcely be over-emphasised, but it remains true that a correct day-to-day correlation should fit reasonable interpretations of the evidence in all the other categories. Three complicating factors are discussed: contamination, date of cutting of the timber, possibility of re-usage of the timber.

9.282   1957
RADIOCARBON DATES FROM LA VENTA, TABASCO
Drucker, P.; Heizer, Robert F.; Squier, R.J.
Science, v. 126, n. 3263, 12 July 1952: 72-73
The series of radiocarbon dates are in agreement with a growing opinion that La Venta phase of Olmec culture was in existence for a long time prior to the classic Maya period.

9.283   1958
RADIOCARBON DATES FROM MANAKAWAY SITE
Suggs, Robert Carl
American Antiquity, v. 23, n. 4, part 2, Apr. 1958: 432-433
The radiocarbon dates determined by the Lamont Geological Observatory confirm that the site represents a single occupation and may be assigned to an early Bowmans Brook time level.

9.284   1958
RADIOCARBON DATES FROM NEVADA OF ARCHEOLOGICAL INTEREST
Grosscup, Gordon L.
University of California, Berkeley. Department of Anthropology. Archaeological Survey Reports, n. 44, part 1, 1958: 17-31
Twenty two archaeological samples and thirty geological samples from Nevada which have been dated by radiocarbon are presented in table form, accompanied by a map and are discussed.

9.285   1961
RADIOCARBON DATES FROM ODAISCA SITE, COLORADO
Irwin, Henry J.; Irwin, Cynthia C,
American Antiquity, v. 27, n. 1, July 1961: 114-115
Seven radiocarbon dates place the regional sequence from 3000 B.C. to A.D. 1000, validate partially the method determining complexes at the site, and date the beginning of maize agriculture in the area.

9.286   1958
RADIOCARBON DATES FROM OKLAHOMA
Bell, Robert E.
Oklahoma Anthropological Society Newsletter, v. 7, n. 3, 1958: 3-4
[Not sighted]

9.287   1963
RADIOCARBON DATES FROM OKLAHOMA AND SURROUNDING AREAS
Miller, J.
Oklahoma Anthropological Society. Bulletin, v. 11, 1963:

115-121
List of radiocarbon dates for Oklahoma sites, and surrounding areas, which have significance for the archaeology of Oklahoma.

9.288   1959
RADIOCARBON DATES FROM OKLAHOMA SITES
Bell, Robert E.
Oklahoma Anthropological Society. Newsletter, v. 8, n. 4, 1959: 2
[Not sighted]

9.289   1957
RADIOCARBON DATES FROM SANDIA CAVE; CORRECTION
Johnson, Frederick; Hibben, Frank C.
Science, v. 125, n. 3241, 8 Feb. 1957: 234-235
Note is given that the dates of two samples allegedly submitted by the late K. Bryan to Libby for dating by the radiocarbon method should be struck from the record as there is no trace of their submission.

9.290   1956
RADIOCARBON DATES FROM SANTA ROSA ISLAND. I
Orr, Phil C.
Santa Barbara Museum of Natural History. Department of Anthropology. Bulletin, v. 2, 1956: 1-10
Presents and discusses radiocarbon dates obtained from six samples collected from Indian villages on the island of Santa Rosa. The dates range from 1860 ± 340 to $29,650 \pm 2500$ years BP and are presented in table form. Three Indian cultures are postulated: Canalino, Highlander and Dune Dwellers, with a possible 'Cavauran' culture (undated) from Arlington cave.

9.291   1960
RADIOCARBON DATES FROM SANTA ROSA ISLAND. II
Orr, Phil C.
Santa Barbara Museum of Natural History. Department of Anthropology. Bulletin, v. 3, 1960. pp. 12
Another list of dates obtained from eight samples from Indian villages on Santa Rosa Island are presented and discussed. A table containing all dates so far obtained is presented with time/culture assigned to these dates. The dates range from 621 ± 1500 to 738 ± 160 years BP.

9.292   1959
RADIOCARBON DATES FROM SOUTH DAKOTA
Hurt, Wesley R.
W. H. Over Museum (South Dakota State University, Vermillion). *Museum News*, v. 20, n. 4, 1959: 1-4
[Not sighted]

9.293   1959
RADIOCARBON DATES FROM THE BULL BROOK SITE
Byers, Douglas S.
American Antiquity, v. 24, n. 4, part 1, Apr. 1959: 427-429
The dates indicate that people were living at the Bull Brook site about 7000 B.C. According to this date it appears necessary to re-examine the evidence for high-water stages in the Lake Ontario basin.

9.294   1955
RADIOCARBON DATES FROM THE HARLAN SITE, CHEROKEE COUNTY, OKLAHOMA
Bell, Robert E.
Oklahoma Anthropological Society. Newsletter, v. 5, n. 3, 1955: 6
[Not sighted]

9.295   1956
RADIOCARBON DATES FROM THE HARLAN SITE, CHEROKEE COUNTY, OKLAHOMA
Bell, Robert E.
Oklahoma Anthropological Society. Newsletter, v. 6, n. 6, 1956: 2
[Not sighted]

9.296   1957
RADIOCARBON DATES FROM THE SPIRO (CRAIG) MOUND
Bell, Robert E.
Oklahoma Anthropological Society. Newsletter, v. 6, n. 6, 1957: 3-4
[Not sighted]

9.297   1968
RADIOCARBON DATES FROM TIZAPAN, EL ALTO, JALISCO
Taylor, R.E.; Berger, Rainer
In: Excavations at Tizapan, edited by C.W. Meighan and L. Foote. Los Angeles, Latin American Center, University of California, 1968: 162-164.
The nature of the correction applied to the values obtained for four radiocarbon age determinations on charcoal derived from excavations at Tizapan. El Alto, Jalisco, Mexico, are discussed. The method used and the modification according to the Suess tree ring calibration are explained. Interpretation of the measurements suggest an occupation from ca A.D. 1000-1200 at the most.

9.298   1956
RADIOCARBON DATES IN TEOTITHUACAN
Linne, Sigvald
Ethnos, v. 21, n. 3-4, 1956: 180-193
A sample of charcoal from an incinerating pit in the ancient city of Teotihuacan, in Mexico Valley, has been dated at 236 ± 65 A.D. This does not tally so well with the date of the Tzakol epoch according to the Goodman - Martinex - Thompson correlation of Mayan and Christian calendars. Additional dating is obviously needed.

9.299   1951
RADIOCARBON DATES ON SAMPLES FROM NEW YORK STATE
Ritchie, William A.
American Antiquity, v. 17, part 2 (*Society for American Archaeology, Memoirs*, n. 8), 1951: 31-32
In terms of radiocarbon dating, New York has been the home of man for 6000 years and has participated, apparently synchronously, in most of the cultural development of the eastern United States.

9.300   1967
RADIOCARBON DATING AND ARCHAEOLOGY IN NORTH AMERICA
Johnson, Frederick
Science, v. 155, n. 3759, 13 Jan. 1967: 165-169
The history of the development of a radiocarbon chronology shows how the establishment of the times of events and the order in which they occurred has greatly improved the understanding of prehistory in North America.

9.301   1965
RADIOCARBON DATING AND THE CULTURAL SEQUENCE IN THE EASTERN UNITED STATES
Griffin, James B.
In: International Conference on Radiocarbon and Tritium Dating. Pullman, Washington, Washington State University, June 1-11, 1965. *Proceedings*. United States of America, Atomic Energy Commission, 1965. Conference n. 650652: 117-130.
In spite of the fact that there are a great many radiocarbon dates available for the Easten United States, many more are needed to have an accurate assessment of the historical development of the area. A large number of dates are from Archaic shell heaps, from middens, from mound debris and it is often difficult to tell exactly what cultural material was dated by the radiocarbon sample. Lists of the dates from Michigan, Illinois, Missouri, Iowa, Ohio, Arkansas, Georgia, Mississippi and Wisconsin are presented.

9.302   1959
RADIOCARBON DATING IN THE ARCTIC
Rainey, Froelich; Ralph, Elizabeth K.
American Antiquity, v. 24, n. 4, part 1, Apr. 1959: 365-374
A series of dates give a tentative framework for Arctic dating. The dates are listed in relation to cultural periods as defined by the Ipintak study by Larsen and Rainey. Dates by the Radiocarbon Laboratory, University of Pennsylvania, (Department of Physics and University Museum).

9.303   1953
RADIOCARBON DATING IN THE ARCTIC
Collins, Harry B.
American Antiquity, v. 18, n. 3, Jan. 1953: 197-203
Report on dates determined by Arnold and Libby. Four time-levels are represented. The interpretation, if correct, would mean that the distinct problems of Eskimo origins and early man in America must be considered in a single frame of reference.

9.304   1965
RADIOCARBON DATING IN THE SOUTHWEST UNITED STATES
Jelinek, Arthur Jenkins
In: International Conference on Radiocarbon and Tritium Dating. Pullman, Washington, Washington State University, June 1-11, 1965. *Proceedings*. United States of America, Atomic Energy Commission, 1965. Conference n. 650652: 133-144.
The extent of utilisation of radiocarbon dating in the archaeological sequence of the Southern United States is presented and discussed. Single radiocarbon dates are of little value in chronological studies, and groups of samples selected for dating must be in close association with clearly identifiable cultural materials and/or fossil remains useful in environmental reconstruction.

9.305   1951
RADIOCARBON DATING ON SAMPLES FROM THE SOUTHEAST
Webb, William S.
American Antiquity, v. 17, part 2 (*Society for American Archaeology, Memoirs*, n. 8), 1951: 30
To date archaic shell middens of the southwest of the United States, radiocarbon determinations were made on shells and antlers. The dates and the validity of the sample are discussed.

9.306   1966
RADIOCARBON DATING THE YOUNGE TRADITION
Fitting, James E.
American Antiquity, v. 31, n. 5, part 1, July 1966: 738
A series of radiocarbon dates from Younge Tradition sites in southwestern Michigan indicates that previous age esti-

mates for the four suggested phases were too conservative. Corn agriculture was practiced throughout the Younge Tradition time-range as indicated by an A.D. 700 date for the Sissung site (20MR5).

9.307   1965
RADIOCARBON DATING, BARROW, ALASKA
Brown, Jerry
In: International Association for Quaternary Research (INQUA), *Congress*, 7th, Boulder, Col., 14 Aug. - 19 Sept. 1965. *Papers*. Boulder, Col. 1965
The radiocarbon ages of some 85 samples are presented and interpreted in an attempt to clarify the complex pedologic and geomorphic history of the Barrow, Alaska area.

9.308   1961
RADIOCARBON EFFECTIVE HALF-LIFE FOR MAYA CALENDAR CORRELATIONS
Ralph, Elizabeth K.
American Antiquity, v. 27, n. 2, Oct. 1961: 229-230
A change in the value of the half-life of ^{14}C would not necessarily change the 'effective' half-life, and, on the basis of present data, no change should be made in the University of Pennsylvania dating of the Maya calendar correlation samples.

9.309   1956
RADIOCARBON, MAMMOTHS, AND MAN ON SANTA ROSA
Orr, Phil C.
Geological Society of America, Bulletin, v. 67, n. 12, Dec. 1956: 1777
Abstract of paper presented at the March 1956 meeting of the Geological Society of America in Reno. Two species of extinct mammoths occurring in marine terraces were dated at 29,650 years. Remains of fire and mammoth bone in what is believed to be a hearth were dated at 6800 years, indicating the possibility that the mammoths existed until near this time.

9.310   1964
THE RECENT AZATLAN DATE
Hurley, William M.
Wisconsin Archaeologist, v. 45, n. 3, 1964: 139-142
[Not sighted]

9.311   1956
RECENT CAVE EXPLORATIONS IN THE LOWER HUMBOLT VALLEY, NEVADA
Heizer, Robert F.
California University, Berkeley. Department of Anthropology. Archaeological Survey Reports, n. 33, March 1956, paper n. 42: 50-57
A table presents a chronology of culture levels of the Humbolt Lake region, the ages are based on radiocarbon dates.

9.312   1962
RECENT RADIOCARBON DATES FOR CENTRAL AND SOUTHERN BRAZIL
Hurt, Wesley R.
W. H. Over Museum (South Dakota State University, Vermillion). *Museum News*, v. 23, n. 11, 1962: 1-4
[Not sighted]

9.313   1953
RECENT DATES FROM TWO PALEO-INDIAN SITES ON MEDICINE CREEK, NEBRASKA
Davis, E. Mott
American Antiquity, v. 18, n. 4, Apr. 1953: 380-386
A radiocarbon date for charcoal collected, one to two feet below the top of the basal bluish clay at the Lime Creek site, Nebraska, should be regarded only as an indication of the general time interval rather than a date placed in a sequence because collection was not made in optimum conditions and only one sample was obtained.

9.314   1964
RECENT RADIOCARBON DATES FOR CENTRAL AND SOUTHERN BRAZIL
Hurt, Wesley R.
American Antiquity, v. 30, n. 1, July 1964: 25-33
Radiocarbon dates from caves and shell mounds in Central and Southern Brazil indicate occupation extending over a 10,000 year period, during which major climatic changes occurred. Correlation of these dates with geologic context of the sites indicates a close agreement with the theoretical climatic fluctuation and changes of sea-level proposed by Fairbridge.

9.315   1967
A RECONSIDERATION OF THE AGE OF LA VANTA SITE
Berger, Rainer; Graham, John A.; Heizer, Robert F.
In: Studies in Olmec Archaeology. University of California. Archaeological Research Facility. *Contribution* n. 3. 1967. pp. 24
The original samples from La Venta were rerun and four new samples were dated. The age of La Venta phase is reconsidered. A period from 1000 B.C. to 600 B.C. is advanced. This means a change of 200 years from the age of the site based upon the earlier Michigan dates.

9.316   1958
RELATIVE AND ABSOLUTE DATING OF THE SIERRA DE TAMAULIPAS SEQUENCE

MacNeish, Richard S.
American Philosophical Society. Transactions, v. 48, part 6, Dec. 1958 (Preliminary archaeological investigation in the Sierra de Tamaulipas): 193-198
A number of caves in the Sierra de Tamaulipas have been excavated. Using radiocarbon dates from materials excavated at various levels and from adjacent areas, it is possible to suggest a cultural sequence.

9.317   1956
THE RELIABILITY OF RADIOCARBON DATES FOR LATE GLACIAL AND RECENT TIMES IN CENTRAL AND EASTERN NORTH AMERICA
Griffin, James B.
Utah University. Department of Anthropology. Anthropological Papers, n. 26, Dec. 1956: 10-34
The evidence presented in this paper supports the thesis that physiographic events, the faunal sequence, the floral succession, and the prehistoric cultural stages dated by radiocarbon determinations are in essential agreement from the Eastern United States to the Rocky Mountains. It does not support the idea of a marked discrepancy between Wisconsin glacial and recent events in the East and the 'true' date of those events, and therefore does not support the contentions of Antevs and Hunt.

9.318   1961
REVIEW AND ASSESSMENT OF THE DORSET PROBLEM
Taylor, E.W.
Anthropologica, v. 1, n. 112, 1959: 24-46
A chronological summary of work bearing on the Dorset problem, one of the major problems of arctic prehistory. Chronologies based on radiocarbon dating are included in the summary.

9.319   1964
A REVIEW OF LA VENTA, TABASCO, AND ITS RELEVANCE TO THE OLMEC PROBLEM
Coe, William R.; Stuckenrath, Robert
Kroeber Anthropological Society (University of California, Berkeley). *Papers,* n. 31, Feb. 1964: 1-44
Nine charcoal samples were processed for the La Venta complex. The samples are presented and interpreted. A table summarises contrasting views.

9.320   1952
A REVIEW OF PROBLEMS IN THE ANTIQUITY OF MAN IN CALIFORNIA
Heizer, Robert F.
University of California, Berkeley. Department of Anthropology. Archaeological Survey Reports, n. 16, June 1952: 3-17
The application of the radiocarbon method for dating of early finds in California is discussed and found wanting, mainly because of the paucity of organic materials associated with the finds.

9.321   1965
A REVIEW OF RADIOCARBON DATES AND ARCHAEOLOGY ON THE NORTHERN AND CENTRAL GREAT PLAINS (Abstract)
Neuman, Robert W.
In International Conference on Radiocarbon and Tritium Dating. Pullman, Washington, Washington State University, June 1-11, 1965. *Proceedings.* United States of America, Atomic Energy Commission, 1965. Conference n. 650652: 131-132.
214 radiocarbon dates from 121 occupation attesting to man's presence in the area for the past 12,000 years are presented.

9.322   1965
REVIEW OF RADIOCARBON DATES FROM TIKAL AND THE MAYA CALENDAR CORRELATION PROBLEM
Ralph, Elizabeth K.
American Antiquity, v. 30, n. 4, Apr. 1965: 420-427
Radiocarbon dates from the inner and outer portions of a Sapote log have been determined in order to assess the growth rates of the logs from which the beams of a temple at Tikal, Guatemala were fashioned and thereby help to explain some of the differences among radiocarbon dates which were previously determined. The dates continue to support the Godman - Thompson - Martinez 'Correlation B' for temples 1 and 4.

9.323   1958
RUSSELL CAVE: NEW LIGHT ON STONE AGE LIFE
Miller, Carl F.
National Geographic Magazine, v. 113, n. 3, Mar. 1958: 426-437
The latest excavation in Russell cave, Alabama, shows that the cavern was inhabited nine thousand years ago. Radiocarbon dating was made on charcoal at 23 feet in depth. Charcoal at 14 feet dates back to 8000 years.

9.324   1966
THE ROLE OF TRANSPACIFIC CONTACTS IN THE DEVELOPMENT OF NEW WORLD PRE-COLUMBIAN CIVILIZATION
Phillips, Philip
In: Handbook of Middle American Indians, edited by Robert Wauchope, Vol. IV, *Archaeological Frontiers and External Connections.* Austin, Texas, University of Texas Press,

1966: 296-319
The author concludes that the diffusionists have not proved the unity of Far Eastern and American civilisations and cultural development. He uses radiocarbon dates to disprove the various correlations.

9.325 1967
SALVAGE ARCHAEOLOGY IN THE MISSOURI RIVER BASIN
Wedel, Waldo R.
Science, v. 156, n. 3775, 5 May 1967: 589-597
Describes the archaeological salvage which took place since 1946 in advance of reservoir construction. Various samples from various sites were radiocarbon dated to help establish a chronology of Plains history.

9.326 1967
THE SAN DIEGUITO COMPLEX: A REVIEW AND HYPOTHESIS
Warren, Claude Nelson
American Antiquity, v. 32, n. 2, Apr. 1967: 168-185
The confusing terminology surrounding the San Dieguito complex is reviewed and a critical evaluation of the content of this complex is undertaken. The radiocarbon dates associated with the earlier misconceptions are analysed and assessed.

9.327 1965
SCULPTURE AND MURAL PAINTING OF OAXACA
Caso, Alfonso
In: *Handbook of Middle American Indians*, edited by Robert Wauchope, Vol. III, *Archaeology of Southern Mesoamerica,* part 2. Austin, Texas, University of Texas Press, 1965: 849-871.
The connection of the Monte Alban culture of the Oaxaca region with the very similar culture of Monte Negro in the Mixtec region and the radiocarbon date obtained for that site confirm an antiquity of several centuries B.C.

9.328 1961
SCRIPPS ESTATE SITE, SAN DIEGO, CALIFORNIA : A LA JOLLA SITE DATED 5460- 7370 BEFORE THE PRESENT
Shumway, George; Hubbs, Carl L.; Moriarty, James Robert
New York Academy of Science. Annals, v. 93, article 3, 4 Dec. 1961: 37-131
The antiquity of the site is established by 4 critical radiocarbon dates. Dates on sound shells of the California mussels, *Mytilus Californianus,* are correlated with dates on charcoal from the same horizons.

Ch. 9 - Archaeology - Americas

9.329 1965
SCULPTURES OF THE GUATEMALA - CHIAPAS HIGHLANDS AND PACIFIC SLOPES AND ASSOCIATED HIEROGLYPHS
Miles, S.W.
In: *Handbook of Middle American Indians*, edited by Robert Wauchope, Vol. II, Part 1, *Archaeology of Southern Mesoamerica*. Austin, Texas, University of Texas Press, 1966: 237-275
Use of the date obtained from a fire pit to assign a maximum age to a jade figurine from La Venta which was defined as Olmec.

9.330 1965
SEA LEVEL CHANGES DURING THE LAST 2000 YEARS AT POINT BARROW, ALASKA
Hume, James D.
Science, v. 150, n. 3700, 26 Nov. 1965: 1165-1166
Eustatic rises of sea-level between A.D. 265 and 500 and between A.D. 1100 and 1300 caused the formation of raised beaches. After the first rise, sea-level dropped about two meters below the present level, permitting the Eskimo settlement of Birnik about A.D. 500. The second rise of the ocean flooded Birnik. At present sea-level is about 0.6 to 1.0 meters below the high water level; the ocean partially floods Birnik. This is documented by radiocarbon dates of driftwood in the beach ridge and of the Birnik culture nearby.

9.331 1965
THE SISTER'S HILL SITE: A HELL GAP SITE IN NORTH-CENTRAL WYOMING
Agogino, George A.; Galloway, Eugene
Plains Anthropologists, v. 10, n. 29, 1965: 190-195
A radiocarbon date from this site, of 9650 ± 250 B.C., agrees with late Agate Basin or early post-Agate Basin sites in the high plains

9.332 1967
SMUDGE PITS AND HIDE SMOKING: THE USE OF ANALOGY IN ARCHAEOLOGICAL REASONING
Binford, Lewis R.
American Antiquity, v. 32, n. 1, Jan. 1967: 1-12
Analogy should serve to provoke new questions about order in the archaeological record and should serve to prompt more searching investigations rather than as a means of offering interpretation. This argument is demonstrated through presentation of formal data on a class of archaeological features, namely 'smudge pits', and the documentation of their positive analogy with pits as facilities used in smoking hides. Radiocarbon dates form part of the data.

9.333   1951
SOME ADENA AND HOPEWELL RADIOCARBON DATES
Griffin, James B.
American Antiquity, v. 17, part 2 (*Society for American Archaeology, Memoirs*, n. 8), 1951: 26-29
In an effort to settle the Adena - Hopewell time relationships, relevant radiocarbon dates are examined. The results are found to be doubtfull.

9.334   1965
SOME CHARACTERISTICS OF LATE WOODLAND CULTURES OF SOUTHEASTERN MICHIGAN
Fitting, James E.
Michigan University. Museum of Anthropology. Anthropological Papers, n. 24, 1965: 165p.
Although there is only one radiocarbon date directly associated with the Younge Tradition material, there are a number of radiocarbon dates from Northeastern North America which serve as contributor dates and can be used to bracket the Younge Tradition. These are presented in graph form and discussed.

9.335   1960
SOME EARLY STONE ARTIFACT DEVELOPMENTS IN NORTH AMERICA
Sellards, E.H.
Southwestern Journal of Anthropology, v. 16, n. 2, 1960: 160-173
The five archaeological sites described, Tule Spring, Lewisville, Santa Rosa, Scripps and Texas street, show, according to radiocarbon dating, evidence of great age, from 21,000 to more than 37,000 years, that is Late Pleistocene, and contain very little stone artefacts.

9.336   1956
SOME IMPLICATIONS OF THE CARBON-14 DATES FROM A CAVE IN COAHUILA, MEXICO
Taylor, Walter W.
Texas Archaeological Society. Bulletin, v. 27, 1956: 215-234
[Not sighted]

9.337   1964
SOME INTERIM REMARKS ON THE COE-STUCKENRATH REVIEW
Heizer, Robert F.
Kroeber Anthropological Society. Papers, n. 31, Fall 1964: 45-50
Two of the nine samples discussed by Coe-Stuckenrath were redated. (Coe-Stuckenrath, *Kreuber Anthropological Sty. Papers*, 1964)

9.338   1958
SOME RADIOCARBON DATES FOR THE WISCONSIN OLD COPPER CULTURE
Ritzenthaler, Robert
Wisconsin Archaeologist, v. 39, n. 3, 1958: 173-174
Three radiocarbon dates for all three of the Old Copper cemeteries in Wisonsin are presented. Ocanto rates as the oldest at 7400 and 5600 years old; the Reigh site is 6660 ± 250 years old, and Osceola, the most recent is 3450 ± 250 years old.

9.339   1962
SOME RADIOCARBON DATES FROM CALIFORNIA'S CHANNEL ISLANDS
Meighan, Clement W.
In: U.C.L.A. Radiocarbon Conference, 28 Feb. 1962. *Papers read*. 1962
[Not sighted]

9.340   1963
SOME RECENT RADIOCARBON DATES FROM WESTERN VENEZUELA
Rouse, Irving; Cruxent, Jose M.
American Antiquity, v. 28, n. 4, Apr. 1963: 537
Two recent series of radiocarbon analyses from Muaco and Rancho Peludo in western Venezuela would seem to provide the oldest dates in the New World for stone projectile points and for pottery respectively.

9.341   1951
SOUTH AMERICAN RADIOCARBON DATES
Bird, Junius Bonton
American Antiquity, v. 17, part 2 (*Society for American Archaeology, Memoirs*, n. 8), 1951: 37-49
A number of radiocarbon dates from South American material are dicussed, evaluated and correlated with other data in order to assess the validity of the radiocarbon method of dating. A figure showing radiocarbon dates versus stratigraphic position of samples is presented as well as a table of dates for early Peruvian cultural periods based on the series of radiocarbon dates and stratigraphic data.

9.342   1955
SPECIMENS FROM SANDIA CAVE AND THEIR POSSIBLE SIGNIFICANCE
Hibben, Frank C.
Science, v. 122, n. 3172, 14 Oct. 1955: 688-689
The radiocarbon dates recently measured by Crane logically fit in with a pattern already established by other dates. It may be assumed that the human beings such as those who left their Sandia Culture were already established in the American southwest around 25,000 years BP.

9.343   1959
A SPECULATIVE FRAMEWORK OF NORTHERN NORTH AMERICAN PREHISTORY AS OF APRIL 1959
MacNeish, Richard Stockton
Anthropologica, v. 1, n. 1-2, 1959: 7-23
Reviews known dates of various prehistoric cultures in Northern North America

9.344   1965
SPLIT-TWIG FIGURINES FROM NORTHERN ARIZONA: NEW RADIOCARBON DATES
Euler, Robert C.; Olson, Alan P.
Science, v. 148, n. 3668, 16 Apr. 1965: 368-369
Recently released radiocarbon dates from split-twig figurines from Marble Canyon, Arizona, are 4095 ± 100 years ago; they substantiate previously determined dates of 3530 ± 300 and 3100 ± 110 years ago. A recently excavated site in Walnut Canyon, Arizona, extends the geographical range of the figurines. The dates of samples from this site are 3500 ± 100 and 3880 ± 90 years ago. It is hypothetised that the figurines were magico-religious artifacts related to the Pinto complex of the Desert Culture.

9.345   1958
SPLIT-TWIG FIGURINES IN THE GRAND CANYON
Schwartz, D.W.; Lange, A.L.; de Saussure, R.
American Antiquity, v. 23, n. 3, Jan. 1958: 264-274
The hypothesis that the figurines might date prior to A.D. 600 is confirmed by radiocarbon dates and places the figurines some 2,000 years earlier than anything previously found in the Grand Canyon.

9.346   1965
SPRING CREEK CAVE, WYOMING
Frison, George C.
American Antiquity, v. 31, n. 1, July 1965: 81-91
Charcoal from a hearth in Spring Creek cave dated a single component habitation in the Big Horn Basin near Ten Sleep, Wyoming at A.D. 225 ± 200. The assemblage yielded amplifies present knowledge of Late Middle period economy and technology in the Northwestern Plains.

9.347   1961
THE STANFIELD - WORLEY ROCK SHELTER
Guthe, Alfred K.
Tennessee Archaeologist, v. 17, 1961: 54-55
Dates from charcoal samples found in the Dalton Projectile Points zone are 8960 ± 400 (M-1153) and 9640 ± 450 years BP (M-1152). Evidence of Middle Mississippian, Woodland and Archaic cultures were also recovered fom higher occupation zones.

9.348   1963
THE STEUBEN VILLAGE AND MOUNDS, A MULTI-COMPONENT LATE HOPEWELL SITE IN ILLINOIS
Morse, Dan F.
Michigan University. Museum of Anthropology. Anthropological Papers, n. 21, 1963: 134
Radiocarbon dates support the relative dating of the Crew and Steuben foci but do not give a clear answer for the focus of Mound I.

9.349   1968
A STONE AGE CAMPSITE AT THE GATEWAY TO AMERICA
Anderson, D.D.
Scientific American, v. 218, n. 6, June 1968: 24-33
The Akmak tools found at Onion Portage, on the Kobuck river in northeastern Alaska, suggest an Asian relationship. A table representing the eight main trends of traditions related to the evidence of human occupation they contain correlated with a chronology supported by radiocarbon dates is presented.

9.350   1966
STRATIGRAPHY OF THE DOMEBO SITE
Albritton, Claude C.
In: Domebo. A Paleo Indian Mammoth Kill in Prairie Plains. Museum Great Plains. Contribution, n. 1, 1966: 10-13
The stratigraphy is described. The upper unit of the lower member contains mammoth bones and associated artifacts.
A tree stump rooted in the lower member yielded a radiocarbon date of 11,045 ± 647 years BP indicating a late Wisconsin age.

9.351   1960
STUART ROCKSHELTER, A STRATIFIED SITE IN SOUTHERN NEVADA
Shutler, Richard; Shutler, Mary Elizabeth; Griffith, James S.
Nevada State Museum. Anthropological Papers, n. 3, 1960. pp. 36
The cultural material recovered and radiocarbon dating on fire hearth identified within the deposit, reveal a 4000 year history of human occupation in this area of Southern Nevada.

9.352   1966
SUGGESTED REVISION FOR WEST MEXICAN ARCHAEOLOGICAL SEQUENCES
Long, S.V.; Taylor, R.E.
Science, v. 154, n. 3755, 16 Dec. 1966: 1456-1459
A review of the radiocarbon dates on published and unpub-

lished archaeological data for the West-Mexican provinces of Sinaloa, Nayarit, Jalisco and Colima has resulted in a revised tentative chronology for West Mexico.

9.353 1967
A SUMMARY, PREHISTORY AND HISTORY OF THE SIERRA PINACATA SONORA
Hayden Julian D.
American Antiquity, v. 32, n. 2, July 1967: 335-344
Phase I of the San Dieguito complex whose remains occur at almost every site in the Sierra Pinacata is dated by radiocarbon at 9350 B.C, or late pluvial time. The sites were then abandoned during the altithermal.

9.354 1956
THE T-1 SITE AT NATIVE POINT, SOUTHAMPTON ISLAND, N.W.T.
Collins, Henry B.
University of Alaska. Anthropological Papers, v. 4, n. 2, May 1956: 63-89
An early Proto-Dorset site at Native Point, Southampton Island, N.W.T., for which a radiocarbon date of 2000 ± 230 years has been obtained, affords additional evidence of a cultural continuity between the arctic flint sites and early Eskimo patterns, particularly the Dorset.

9.355 1960
TABLE ROCK PUEBLO, ARIZONA
Martin, Paul S.; Rinaldo, John B.
Fieldiana Anthropology, v. 51, n. 2, 1960: 130-294
Prehistoric Indian villages at Table Rock site were excavated near St. John, Arizona and were examined. Two dates on sections of a pinyon log were obtained. One date was determined by the Laboratory of Tree Ring Research, University of Arizona and the other by the Radiocarbon Dating Laboratory of the University of Groningen (de Vries). The excellent correspondence enabled the authors to suggest A.D. 1350 for the construction of parts of the Pueblo.

9.356 1965
TIKAL, GUATEMALA AND EMERGENT MAYA CIVILIZATION
Coe, William R.
Science, v. 147, n. 3664, 19 Mar. 1965: 1401-1419
Radiocarbon dates from various levels of the excavations at Tikal, northern Guatemala, help to establish the time period of an early complex-living at a prime Maya Indian site. The earliest found plazza and terrace construction date back to the second century B.C. The start of the classic sequence dates to 250 A.D.

9.357 1963
TIKAL: THE NORTH ACROPOLIS AND AN EARLY TOMB
Coe, William R.; McGinn, John J.
University of Pennsylvania. Expedition Bulletin, v. 5, n. 2, 1963: 24-32
Radiocarbon dating on charcoal from rubble fill gives an age which is too early by three centuries. This illustrates the need to examine very carefully the provenance of charcoal used for radiocarbon dating.

9.358 1965
TULE SPRINGS EXPEDITION
Shutler, Richard
Current Anthropology, v. 6, n. 1, Feb. 1965: 110-111
Scanty archaeological evidence corroborated by stratigraphic studies and radiocarbon dating provide proof that man first appeared in the Tule Spring area (Nevada) between 11,000 and 13,000 years ago.

9.359 1961
THE TUKTU COMPLEX AT ANAKTUVUK PASS
Campbell, John M.
University of Alaska. Anthropological Papers, v. 9, n. 2, May 1961: 61-80
Two samples, one of wood charcoal and one of charred bone were selected for the dating of this complex which is one of at least nine quite distinctive cultural components in the known Anaktuvuk archaeological sequence. The results have not yet been obtained. A guess date would be about 3000 to 4000 years old.

9.360 1964
TWENTY NEW RADIOCARBON DATES FROM MINNESOTA ARCHAEOLOGICAL SITES
Johnson, Elden
Minnesota Archaeologist, v. 26, n. 2, 1964: 34-49
A list of date of archaeological material from the state of Minnesota.

9.361 1967
TWO ANCIENT FLORIDA DUGOUT CANOES
Bullen, Ripley P.; Brooks, Harold K.
Florida Academy of Science. Quarterly Journal, v. 30, n. 2, 1967: 97-107
Wood from the Zellwood canoe, found under five feet of peat and muck in Lake Apopka, was radiocarbon dated at 1185 ± 75 years BP. Wood from the Lakeland canoe, found at the Orange Park mine under a thin layer of wind-deposited sand overlain by about four feet of fibrous peat and muck, was dated at 3040 ± 115 years; peat just above the sand layer

was dated at 2600 ± 130 years, a difference of 400 years, the sand reflecting pauses in the growth of the peat. Fluctuations of the water table in the Floridan aquifer have been superposed upon a general rise in water level correlated to a 10 ft rise in sea-level during the past 4500 years. Study of the peats in several lake basins in central Florida suggests that 3000 to 1000 years BP was the period of most rapid rise.

9.362   1963
TWO EARLY PHASES FROM THE NAKNEK DRAINAGE
Dumond, D.E.
Arctic Anthropology, v. 1, n. 2, 1963: 93-104
The dating of the sequence is based in large part on a series of 16 radiocarbon dates derived principally from fireplace remnants and is supported by stratigraphic evidence.

9.363   1965
TWO EARLY RADIOCARBON DATES FROM THE LOWER LEVELS OF WILSON BUTTE, SOUTH-CENTRAL IDAHO
Gruhn, Ruth
Tebiwa (Idaho State University Museum Notes and News), v. 8, n. 2, 1965: 57
These new dates for the Wilson Butte cave are associated with artifacts and have placed the cave within the group of earliest dated archaeological sites in North America. The dates are 13,050 and 12,550 B.C.

9.364   1959
TWO FOSSIL ELEPHANT KILL SITES IN THE AMERICAN SOUTHWEST
Haury, Emil W.
In: International Congress of Americanists, 32nd, Copenhagen, 8-14 August, 1956. Proceedings. Copenhagen, Munskegaard, 1959: 433-440
It is suggested, on the basis of geological investigation, that the age of the Laco and Lehner stations are contemporaneous and estimated at between 11,000 and 10,000 years, in the last pluvial. Radiocarbon dates on charcoal are not consistent with the geologic estimate and should be considered as suspect.

9.365   1966
TWO LATE PREHISTORIC POTTERY-BEARING SITES ON KODIAK ISLAND, ALASKA
Clark, Donald W.
Arctic Anthropology, v. 3, n. 2, 1966: 157-184
The dates associated to the site are presented and assessed and used to correct a previous set of conclusion.

9.366   1961
TWO RADIOCARBON DATES FROM THE CENTRAL BRAZOS VALLEY
Watt, F.H.
Texas Archaeological Society. Bulletin, n. 31, 1961: 327-328
Aycock shelter is a small rockshelter, north of Belton in Bell County, Texas. Charcoal collected at depths ranging from 8 to 28 in below surface were dated by Magnolia Laboratory, Dallas (FRL No. RC-24). The samples could not be accurately dated because there was too little carbon. The estimation is >10,000 years. Clark site, a midden in the Brazos River alluvium, near Waco, McLennan County, Texas, a Sanders Focus, is dated at 640 ± 150 years BP.

9.367   1961
TWO RADIOCARBON DATES FROM THE GALENA SITE OF SOUTHERN TEXAS
Ring, E.R.
Texas Archaeological Society. Bulletin, n. 31, 1961: 329
The Galena site, a shell midden, in Galena Park near Houston, Harris County, Texas, was dated by the radiocarbon dating method at 1900 ± 105 years BP and 3550 ± 115 years BP. The evaluation of these dates is presented in another article (Ring, *Texas Archaeological Bulletin*, 31, 1961: 317).

9.368   1962
USE OF ORGANIC TEMPER FOR CARBON-14 DATING IN LOWLAND SOUTH AMERICA
Evans, Clifford; Meggers, Betty J.
American Antiquity, v. 28, n. 2, Oct. 1962: 243-245
Experimental work has demonstrated the practicality of extracting sufficient organic carbon for radiocarbon dating from sherds tempered with crushed charcoal sponge spicules and siliceous bark. The sherds were from ceramic complexes on the Napo, Orinoco, and Amazon River that had been stratigraphically excavated and could be cross checked.

9.369   1963
VENEZUELAN ARCHAEOLOGY
Rouse, Irving; Cruxent, Jose M.
New Haven, Conn., Yale University Press, 1963. pp. 179
The chronology of Venezuelan archaeology is supported throughout by radiocarbon dates. An appendix lists all known dates for Venezuala, 52 of which appears to be valid and 5 which are obviously incorrect. The description of each sample is limited to the site from which it came, the complex of style with which it was associated and the period of that style in the relative chronology. A series of comments are given to explain the incorrect dates.

9.370 1967
THE VIABILITY OF PATHOGENS IN ANCIENT HUMAN COPROLITES
Tubbs, Deborah Y.; Berger, Rainer
University of California, Berkeley. Department of Anthropology. Archaeological Survey Reports, n. 70, April 1967, paper n. 5: 89-92
A one thousand year old human coprolite was investigated for viable micro-organisms common to the intestinal tract; these were found to be absent. The treatment before submitting for radiocarbon is described briefly.

9.371 1962
WEST COAST CROSS TIES WITH ALASKA
Borden, Charles B.
In: Prehistoric Cultural Relations between the Arctic and Temperal Zones of North America, edited by J.M. Campbell. Arctic Institute of North America. Technical Paper, n. 11, 1962: 9-19
Radiocarbon datings help establish a chronology of West Coast cultural development indicating that the West Coast inhabitants had cross ties with the Indians of the interior and that some cultural traits may have been passed onto the Alaskan Eskimos rather than the other way round.

9.372 1962
WYOMING MUCK TELLS OF BATTLE: ICE-AGE MAN VERSUS MAMMOTH
Irwin, Cynthia; Irwin, Henri J.; Agogino, George A.
National Geographic Magazine, v. 121, n. 6, June 1962: 828-837
The presence of a mammoth skeleton, ice-age horse and tool implements made by early man are dated at $11,280 \pm 350$ years BP by radiocarbon dating of fragements of tusk. The finds were together in a bog in Wyoming.

9.373 1965
ZAPOTEC WRITING AND CALENDAR
Caso, Alfonso
In: Handbook of Middle American Indians, edited by Robert Wauchope. Vol. III, *Archaeology of Southern Mesoamerica*, part 2. Austin, Texas, University of Texas Press, 1965: 931-947.
Writing and the calendar in the Zapotec region are the 'most ancient that have been found in Mesoamerica. The Monte Alban 'danzante' style of sculpture, contemporaneous with the Mixtec city of Yuconoo, is dated by a radiocarbon date of about 600 B.C. from the latter city.

CHAPTER TEN
14C ARCHAEOLOGY ASIA

10.001 1961
NOTE: This entry was misindexed and has been moved to Chapter 8 with the number 8.000.

10.002 1968
ANNOTATION AND CORRECTION OF THE RELATIVE CHRONOLOGY OF IRAN, 1968
Dyson, Robert H.
American Journal of Archaeology, v. 78, n. 4, Oct. 1968: 308-313
A new chronology based on radiocarbon dates is presented. New radiocarbon dates for Iran and Iraq are presented in table form.

10.003 1966
BEIDHA: 1965 CAMPAIGN
Kirkbride, Diana
Archaeology, v. 19, n. 4, Oct. 1966: 268-272
New radiocarbon dates on charcoal and carbonised pistachio nuts placed this early neolithic village at 6990 ± 160 years B.C. for level VI to 6600 ± 120 years B.C. for level II.

10.004 1966
BEIDHA: AN EARLY NEOLITHIC VILLAGE IN JORDAN
Kirkbride, Diana
Archaeology, v. 19, n. 3, June 1966: 199-207
Report on a series of excavations. This early neolithic village is placed at 6830 ± 200 years B.C. by radiocarbon dating of remains of a burnt house in level IV of the excavation.

10.005 1965
C14 DATES, BANAS CULTURE AND THE ARYANS
Agrawal, D.P.
In: International Conference on Radiocarbon and Tritium Dating, Pullman, Washington, Washington State University, June 1-11, 1965. Proceedings. United States of America, Atomic Energy Commission, 1965. Conference n. 650652: 256-263
A very consistent chronological framework of protohistoric cultures is emerging through radiocarbon dates.

10.006 1966
C14 DATES, BANAS CULTURE AND THE ARYANS
Agrawal, D.P.
Current Science, v. 35, n. 5, 5 March 1966: 114-117
An attempt is made to synthesise the archaeological data pertaining to the first half of the Dark Ages in the light of available radiocarbon dates. Radiocarbon dates for post-Harappan Chalcolithic cultures are presented in graph form.

10.007 1967
CAPE GELIDONYA: A BRONZE AGE SHIPWRECK
Bass, George F.
American Philosophical Society. Transactions, n.s. v.57, part 8, 1967. pp. 177
The shipwreck is situated on the southern coast of Turkey. It contains all the elements for bronze making. A sample of twigs from the brushwood which lined the hull was dated by radiocarbon. It indicates that the sample is not younger than 1060 B.C. and may be as old as 1250 B.C. (p.168)

10.008 1958
CARBON DATED PALEOLITHS FROM BORNEO
Harrisson, Tom
Nature, v. 181, n. 4611, 15 Mar. 1958: 792
A sample of charcoal associated with stone tools at a depth of 100 in at Niah Great Cave, West Borneo (Sarawak), gives a date of 39,600 ± 1000 years.

10.009 1960
CARBON-14 DATE FOR A NEOLITHIC SITE IN THE RUB' AL KHALI
Field, H.
Man, v. 172, article n. 214, Nov. 1960: 172
A charcoal sample associated with a camp site in Saudi Arabia gave an age of 5090 ± 200 BP indicating that this camp was occupied by neolithic hunters about 3000 B.C., contemporaneously with the civilization of Egypt, Mesopotamia, the Iranian Plateau and the Indus Valley.

10.010 1968
THE CAVE OF ALI TAPPAH AND THE EPI-PALAEOLITHIC IN N.E. IRAN.
McBurney, C.B.M.
Prehistoric Society. Proceedings, n. s., v. 34, n. 12, 1968: 385-414
Stratigraphy, absolute dating and chronology, combined with the apparent changes in the structure of the fauna provide a picture which is not inconsistent with the contemporary changes in other areas.

10.011   1958
COLLECTION OF MATERIAL DATA FROM THREE ARCHAEOLOGICAL SITES AT SHANIDAR, IRAQ
Solecki, Ralph S.
American Philosphical Society, Yearbook, 1958: 403-407
In this report to the committee on research, the author indicates that the age of Shanidar I and II near the top of layer D is about 45,000 years. This estimate is based on radiocarbon dates in charcoal.

10.012   1968
THE DATING OF ASIKLI HUJUK IN CENTRAL ANATOLIA
Todd, Ian A
American Journal of Archaeology, v. 72, n. 2, April 1968: 157-158
Five radiocarbon dates confirm the relationship of Asikli Hüyük to Çatal Hüyük and endorse the importance of the site on the basis of its obsidian industry

10.013   1958
DATING OF ZAWI CHEMI, AN EARLY VILLAGE SITE AT SHANIDAR, NORTHERN IRAQ
Solecki, Ralph S.; Rubin, Meyer
Science, v. 127, n. 3312, 20 June 1958: 1446
Radiocarbon dating of charcoal collected from a sample containing a pre-ceramic industry generally equated with early neolithic or incipient cultivation (according to Braidwood), at the open village site of Zawi Chemi Shanidar, gives the same date range as material collected in the upper stratigraphy of Shanidar cave, providing a reasonable basis for assuming contemporary or seasonal occupation at both sites.

10.014   1957
DIGGING UP JERICHO: THE RESULTS OF THE JERICHO EXCAVATIONS
Kenyon, Kathleen M.
New York, Praeger, 1957. pp. 267
Radiocarbon dates indicate that a period, when people of the plaster floor phase had long been established on this site, must be dated as long ago as the seventh millenium B.C. and, according to the evidence, the Jericho of this phase must be considered as a town. One can therefore deduce that in the sixth, and probably the seventh, millenium B.C. there was a growth of civilisation known nowhere else. Other periods of the history of Jericho are also dated on charcoal from tombs.

10.015   1968
THE DISCOVERY OF THE PALEOLITHIC IN KAMCHATKA AND THE PROBLEM OF THE INITIAL OCCUPATION OF AMERICA
Dikov, N.N.
Arctic Anthropology, v. 5, n. 1, 1968: 191-203
At Kammennyi Mys, on the southern shore of Ushki lake newly excavated levels Va, VI and VII have been dated by analogy at 14,000 to 16,000 years BP, considering that those levels are below level V which has an absolute date of 10,350 ± 350 years BP, corresponding, in the European chronological scale, to the boundary between Mesolithic and Palaeolithic. It thus is the first discovery of a Palaeolithic site in northeastern Asia. According to both the geographical location and the antiquity, this presents a new and impressive evidence for early occupation of America from Asia across its northeast territory.

10.016   1959
EARLIEST JERICHO
Kenyon, Kathleen M.
Antiquity, v. 33, n. 129, Mar. 1959: 5-9
The Neolithic times of Jericho is dated at ca 5850 ± 150 and the Mesolithic provided a date of ca 7800 B.C. Burned down beams were used for radiocarbon dating.

10.017   1968
THE EARLIEST WHEELED VEHICLE AND THE CAUCASIAN EVIDENCE
Piggott, Stuart
Prehistoric Society, Proceedings, n. s., n. 8, v. 34, 1968: 266-318
Available radiocarbon dates for the Near-Eastern evidence of wheeled vehicle are discussed and found to give a time-scale consistently too low for the historically established schemes. A chronological sequence based on radiocarbon dates is presented and discussed.

10.018   1968
EARLY BRONZE FROM NORTHEASTERN THAILAND
Solheim, William G.
Current Anthropology, v. 9, n. 1, 1968: 59-62
Nineteen charcoal samples from the site of Non Nok Tha, in northeastern Thailand, were submitted to be dated by radiocarbon; ten of the samples produced dates. Of these, three are obviously not correct for the stratigraphic layer from which the samples were collected, and another one seems very unlikely. The remaining ten dates are presented in table form. One conclusion is that bronze was being worked in northeastern Thailand nearly 1000 years before it began in Shang, China and 100 or more years before it started in the Harappa culture of the Indus valley in India.

10.019   1959
EARLY INDIA AND PAKISTAN
Wheeler, Sir Mortimer (Robert Eric Mortimer)

London, Thames and Hudson, 1959
Radiocarbon dates from various areas are collated and used to establish a chronology.

10.020 1962
EXCAVATION AT ALI KOSH, IRAN, 1961
Hole, Frank; Flannery, Kent V.
Iranica Antiqua, v. 2, 1962: 97-148
Four levels at Ali Kosh (Iran) have been dated by radiocarbon and indicate an age of about 6300 - 6500 B.C. for a zone in which a pit-house was situated. Above that, a zone of mud-brick houses with a flint assemblage different from the preceding one is dated at 5800 ± 6200 B.C.

10.021 1956
EXCAVATION IN THE QUETTA VALLEY, WEST PAKISTAN
Fairservis, Walter A.
American Museum of Natural History. Anthropological Papers, v. 45, part 2, 1956: 170-402
The time range established by radiocarbon dates on charcoal and wood collected from all levels in the excavation in the Quetta Valley conforms fairly well with previously published estimates for the Indus-Baluchistan Area.

10.022 1961
EXCAVATIONS AT HACILAR, FOURTH PRELIMINARY REPORT
Mellaart, James
Anatolian Studies, v. 11, 1961: 39-75
A series of radiocarbon dates for Hacilar form a sequence confirming the stratigraphical observations and enable the author to suggest a chronology based on these dates.

10.023 1957
EXCAVATIONS AT JERICHO, 1957
Kenyon, Kathleen M.
Palestine Exploration Quarterly, v. 89, 1957: 101-107
Report on four weeks of excavation in the summer of 1957. Two dates (with different pre-treatment) on one sample of charcoal from the floor of a house, give an age of 6770±210 B.C. which is compatible with another date of the same area. A date in the 7th. millenium B.C. may therefore be assumed for the hog-backed brick phase.

10.024 1967
EXCAVATIONS, FINDS AND SUMMARY (*IN: THE GEULA CAVES, MOUNT CARMEL*)
Wreschner, E.
Quaternaria, v. 8, 1967: 69-89
Climatic changes are dated by the radiocarbon method. It is assumed that the occupation of the cave belongs to an interstadial after the onset of the cold, Early Pluvial B, approximately 45,000 years ago.

10.025 1956
THE FIRST TOWNS?
Wheeler, Sir Mortimer (Robert Eric Mortimer)
Antiquity, v. 3, n. 119, Sept. 1956: 132-136
Radiocarbon tests carried out on snail shells and charcoal give an antiquity of 4700 B.C. for the village of Jarmo, northern Irak and at Jericho the highest level of pre-pottery neolithic give a range of 8000 to 6000 B.C.

10.026 1968
THE FUKUI MICROBLADE TECHNOLOGY AND ITS RELATIONSHIP TO NORTHEAST ASIA AND NORTH AMERICA
Hayashi, Kensaku
Arctic Anthropology, v. 5, n. 1, 1968: 128-190
Radiocarbon dates reveal a long time span of occupation with the oldest material dated at more than 32,000 years BP and the youngest at 6000 years BP.

10.027 1964
HARAPPA CULTURE: NEW EVIDENCE FOR A SHORTER CHRONOLOGY
Agrawal, D.P.
Science, v. 3609, n. 143, 28 Feb. 1964: 950-951
Radiocarbon dates suggest a total time spread of 550 years from about 2300 to 1759 B.C for the Harappa culture.

10.028 1960
HASANLU AND EARLY IRAN
Dyson, Robert H.
Archaeology, v. 13, n. 2, Summer 1960: 118-129
Presents the results of the excavation at Hasanlu and the Solduz Valley, Northwest Iran, and how they fit into the framework of history. Radiocarbon dates help establish a preliminary chronology.

10.029 1962
THE HASANLU PROJECT
Dyson, Robert H.
Science, v. 135, n. 3504, 23 Feb. 1962: 637-647
Ash, charred grains and structural timber dated by E.K. Ralph of the University of Pennsylvania help confirm the chronology obtained by typological correlation, i.e. Period V = button base phase 1151 ± 49, Period IV = grey ware 989 ± 32, Period III = triangular ware 564 ± 38.

10.030 1968
A HUNTER'S VILLAGE IN NEOLITHIC TURKEY
Perkins, Dexter; Daly, P.
Scientific American, v. 219, n. 5, 1968: 97-106
Report on the analysis of bone remains at Sukerde, an early

village site in Southwestern Turkey, trying to decide whether the inhabitants were herdsmen or hunters. The lowest level was dated by radiocarbon, on charcoal samples, at around 6500 B.C.

10.031 1962
INDIAN ARCHAEOLOGY TODAY
Sankalia, H.D.
Bombay, Asia Publishing House, 1962 (Heras Memorial Lecture, 1960) pp. 144
The radiocarbon method of dating is briefly described (p.16 to 19).

10.032 1954
THE IRAQ-JARMO PROJECT OF THE ORIENTAL INSTITUTE OF THE UNIVERSITY OF CHICAGO
Braidwood, Robert J.
Sumer, v. 10, n. 2, 1954: 120-138
The author notes the usefulness of the radiocarbon technique and indicates that a number of charcoal samples have been collected for purposes of dating.

10.033 1957
JERICHO AND ITS SETTING IN NEAR EASTERN HISTORY
Braidwood, Robert J.
Antiquity, v. 31, n. 122, June 1957: 73-81
Presents a critical appraisal of the Kenyon-Wheeler interpretation and a possible alternative. A discussion of the validity of radiocarbon dates is included.

10.034 1965
THE MEASUREMENT OF RADIOCARBON ACTIVITY AND SOME DETERMINATIONS OF AGES OF ARCHAEOLOGICAL SAMPLES
Agrawal, D.P.; Kusumgar, Sheela
Current Science, v. 34, n. 13, July 1965
Presents several radiocarbon dates for India based on archaeological samples. A new, simple, rapid and quantitative method of synthetising methane gas is briefly described.

10.035 1968
MONSOON OVER THE INDUS VALLEY DURING THE HARAPPAN PERIOD
Ramaswamy, C.
Nature, v. 217, n. 5129, 17 Feb. 1968: 628-629
Considerable reserves of groundwater in the arid regions of west Rajasthan, dated at 5000 years BP, support the conclusion that the Harappans who flourished in the Indus Valley between 2500 and 1700 B.C. lived in very much moister climatic conditions than those which exist today.

10.036 1968
NATURAL ENVIRONMENT OF EARLY FOOD PRODUCTION NORTH OF MESOPOTAMIA
Wright, Herbert E.
Science, v. 161, n. 3839, 26 July 1968: 334-339
Climatic change dated by radiocarbon at 11,000 years ago may have set the stage for primitive farming in the Zagros mountains.

10.037 1958
NEAR EASTERN PREHISTORY
Braidwood, Robert J.
Science, v. 127, n. 3312, 20 June 1958: 1419-1430
A new cluster of radiocarbon dates throws new light on Pleistocene prehistory of the Near East. It places incipient cultivation at about 10,000 BP at the onset of early postglacial times. Within 1000 years this experimental cultivation was succeeded by settled villages-farming communities.

10.038 1963
NEW DATES FOR EARLY JOMON POTTERY IN JAPAN
Oba, Toshio; Sharp, C.S.
Asian Perspective, v. 6, 1963: 75-76
The very early dates obtained for early ceramic levels at the Natsushima site near Tokyo are confirmed by two other very early dates obtained for a ceramic complex in southern Okkaido. This means that a drastic reassessment of cultural development in Asia has to be made.

10.039 1962
NEW LINKS BETWEEN WESTERN ASIA AND THE INDIA OF 4000 YEARS AGO: EXCAVATIONS IN THE HUGE DUST-HEAP OF AHAD, NEAR UDAIPUR
Sankalia, H.D.
Illustrated London News, v. 241, 1 Sept. 1962: 322-325
Radiocarbon estimates date the contact with eastern Iran at about 1500 B.C.

10.040 1966
OBSIDIAN AND EARLY CULTURAL CONTACT IN THE NEAR EAST
Renfrew, Colin; Dixon, J.E.; Cann, J.R.
Prehistoric Society. Proceedings, n. s. v.32, n. 2, 1966: 30-72
The earliest stratified finds of obsidian in the Near East are from layer C of the Shanidar Cave, as part of an Upper Palaeolithic (Baradostian) industry dated at about 30,000 years old.

10.041   1954
THE OLDEST ARCHAEOLOGICAL DATES
Solecki, Ralph S.
Sumer, v. 10, n. 2, 1954: 199-201
Two letters announcing radiocarbon dates from the excavations at Shanidar cave. This is the earliest date which has been hitherto fixed by the method.

10.042   1962
ON THE CHRONOLOGY OF EARLY VILLAGE-FARMING COMMUNITIES IN NORTHERN IRAQ
Mortensen, Peter
Sumer, v. 18, n. 1 & 2, 1962: 73-80
The chonological order indicated by the radioactive carbon determinations is proved by the stratigraphic sequence of Tell Shemshara when a pre-ceramic Jarmo-like blade tool industry was succeeded by Hassunan-Samarran levels.

10.043   1964
ON THE ORIGIN OF THE NEOLITHIC REVOLUTION
Dyson, Robert H.
Science, v. 144, n. 3619, 8 May 1964: 672-675
In this review of the book 'A History of Domesticated Animals' by Zeuner, the author discusses some radiocarbon dates of the domestication at Jarmo, Jericho and Tepe Sarab (Iran).

10.044   1978
THE PALAEOECOLOGY OF COASTAL TAMILNADU, SOUTH INDIA: CHRONOLOGY OF RAISED BEACHES
Sarma, Akkaraju
American Philosophical Society. Proceedings, v. 122, Nov. 1978: 411-426
In this study of of the palaeoecology of the costal area of Tamilnadu, South India, the dates of the terraces are based on radiocarbon dates.

10.045   1962/63
A PICTURE EMERGES: AN ASSESSMENT OF THE CARBON-14 DATINGS OF THE PROTO-HISTORIC CULTURE OF THE INDO-PAKISTAN SUB-CONTINENT
Lal, B.B.
Ancient India, v. 18/19, 1962/63: 208-221
Presents a compilation of the radiocarbon dates and their respective stratigraphic horizons for the proto-historic period of the Indo-Pakistan subcontinent. The sites considered are Kalibangan, Lothal, Mohenjo-Daro, Ahar, Navdatoli, Nevasa, Chandoli, Burzahom. The dates presented are based on 5730 ± 400 years as the half-life value of radiocarbon.

10.046   1964
PREHISTORIC ARCHAEOLOGICAL SURVEYS AND EXCAVATIONS IN AFGHANISTAN
Dupree, L.
Science, v. 146, n. 3644, 30 Oct. 1964: 638-640
Recent research backed by radiocarbon dates indicates that the foothills of northern Afghanistan may have been one of the early centres of incipient agriculture, transitional between the food gathering of the Palaeolithic and the food production of the Neolithic. In addition upper and possibly middle (Mousterian) Palaeolithic industries have been identified.

10.047   1964
PREHISTORIC FAUNA FROM SHANIDAR, IRAQ
Perkins, Dexter
Science, v. 144, n. 3625, 26 June 1964: 1565-1566
A comparative analysis of the faunal material from the Shanidar cave and the nearby Proto-Neolithic site at Zawi Chemi Shanidar indicates that sheep were domesticated there at the beginning of the ninth millenium B.C., more than 1000 years earlier than the earliest known evidence of animal husbandry. The radiocarbon age determination is $10,870 \pm 300$ years.

10.048   1963
PREHISTORY IN SHANIDAR VALLEY, NORTHERN IRAQ
Solecki, Ralph S.
Science, v. 139, n. 3551, 18 Jan. 1963: 179-193
Fresh insights into Near Eastern prehistory from Middle Palaeolithic to the Proto-Neolithic are obtained. Radiocarbon dates help to establish a framework for a chronology.

10.049   1967
THE PREHISTORY OF SOUTHWEST IRAN: A PRELIMINARY REPORT
Hole, Frank; Flannery, Kent V.
Prehistoric Society. Proceedings, n. s., v. 33, n. 2, 1967: 147-206
Radiocarbon dates are extensively used to establish a chronology for the prehistory of Southwestern Iran.

10.050   1956
THE RADIOCARBON AGE OF JERICHO
Zeuner, Frederick Eberard
Antiquity, v. 30, n. 120, Dec. 1956: 195-197
Test specimens from the Middle Bronze age, a late Chalcolithic tomb and from the complex at Khirbat Qumran were dated by the radiocarbon method. The first one gave too young an age because of the contamination and chemical alteration caused by the conditions in the tomb. The other specimen gave dates that agree with the expected ages.

Samples from the upper part of the pre-pottery neolithic complex were dated at 8200 ± 200 years BP and 7800 ± 160 BP and as they are comparatively high up in the pre-pottery sequence, it is expected that the lower layers will give considerably greater ages, thus making it evident that the Neolithic city of Jericho is the oldest city in the world and the oldest Neolithic settlement so far known.

10.051   1956
RADIOCARBON DATE FOR EARLY SOUTH ARABIA
van Beek, G.W.
American Schools of Oriental Research. Bulletin, v. 143, Oct. 1956: 6-9
A radiocarbon date on charred wooden beams which had originally served as roofing of mud-brick constructions at Hajar Bin Bumeid, Western Aden, places it within the limits 1012 to 692 B.C. giving a clear time-point established by a stratigraphically controlled excavation around which the early South Arabian chronology may be ordered.

10.052   1951
RADIOCARBON DATES AND THEIR IMPLICATIONS IN THE NEAR AND MIDDLE-EASTERN AREA, A BRIEF
Braidwood, Robert J.; Jacobsen, Thorkild; Parker, Richard A.; Weinberg, Saul S.
American Antiquity, v. 17, part 2 (*Society for American Archaeology, Memoirs*, n. 8), 1951: 52-53
The lack of radiocarbon dated material from the Near East and the Middle East is commented upon. The existing radiocarbon dates on Near and Middle-Eastern areas are presented and discussed.

10.053   1955
RADIOCARBON DATES FOR KARA KAMAR, AFGHANISTAN. UNIVERSITY OF PENNSYLVANIA II.
Coon, Carleton Stevens; Ralph, Elizabeth K.
Science, v. 122, n. 3176, 11 Nov. 1955: 921-922
Material dated: charcoal and earth from fire pits. Project: determination of cultural levels at Kara Kamar, Afghanistan. A tentative classification in 4 cultures is advanced. Age: range from older than 35,000 to 2749 ± 300. Culture 1: Mesolithic, older than 35,000 years; Culture 2: flake assemblage, white from weathering; Culture 3: Upper Palaeolithic blade tools and horse bones; Culture 4: similar assemblage to Culture 2. The Upper Palaeolithic blade culture is contemporaneous with a Middle Palaeolithic culture at Haua Fteah, Lybia.

10.054   1966
RADIOCARBON DATES IN INDIAN CHRONOLOGY
Agrawal, D.P.; Lal, D.
Uttar Pradesh, 1966
[Not sighted]

10.055   1963
RADIOCARBON DATES OF ARCHAEOLOGICAL SAMPLES
Agrawal, D.P.; Kusumgar, Sheela; Sarna, R.P.
Current Science, v. 32, 1963
Further radiocarbon dates related to archaeological sites in India are presented and discussed.

10.056   1968
RADIOCARBON DATES OF KALIBANGAN SAMPLES
Agrawal, D.P.; Kusumgar, Sheela
Current Science, v. 37, n. 4, 20 Feb. 1968: 96-99
Radiocarbon dates of samples collected at Kalibangan a well known Harappan site on the river Ghaggar are presented and discussed.

10.057   1966
RADIOCARBON DATES OF SAMPLES FROM HISTORICAL LEVELS
Agrawal, D.P.; Kusumgar, Sheela; Krishnan, M. Unni
Current Science, v. 35, n. 11, 5 June 1966: 276-278
Radiocarbon dates from the following historical sites are presented: Ahichchatra, Besnagar, Dharnikota, Nagara. The experimental procedure is briefly described.

10.058   1966
RADIOCARBON DATES OF SAMPLES FROM N.B.P. WARE AND PRE-N.B.P. WARE LEVELS
Agrawal, D.P.; Kusumgar, Sheela; Krishnan, M. Unni
Current Science, v. 35, n. 1, 5 Jan. 1966: 4-5
Samples from the following areas are presented and discussed: Ahichchatra; Atranjikhera; Rajghat and Rupar; also Chirand (Bihar), and Burzahom.

10.059   1966
RADIOCARBON DATES OF SAMPLES FROM SOUTHERN NEOLITHIC SITES
Agrawal, D.P.; Kusumgar, Sheela
Current Science, v. 35, n. 23, 5 Dec. 1966: 585-586
Radioarbon dates from the following sites are presented and discussed: Sangankallu, Hallar, Beinapalli and Narasipur. A tentative chronology of the Southern Neolithic is presented.

10.060 1965
RADIOCARBON DATES OF SOME NEOLITHIC
AND EARLY HISTORIC SAMPLES
Agrawal, D.P.; Kusumgar, Sheela
Current Science, v. 34, n. 1, 5 Jan. 1965: 42-43
Dates of samples obtained from the Neolithic site of Burzahom, Kashmir and from Hetimpur and Kaushambi, both early historic sites in Uttar Pradesh. The dates are discussed.

10.061 1967
RADIOCARBON DATES OF SOME PREHISTORIC AND PLEISTOCENE SAMPLES
Agrawal, D.P.; Kusumgar, Sheela
Current Science, v. 36, n. 21, 5 Nov. 1967: 566-568
Samples from prehistoric sites and geologic deposits are presented and discussed.

10.062 1965
RADIOCARBON DATING AND THE EXPANSION OF FARM CULTURE FROM THE NEAR EAST OVER EUROPE
Clark, J.G.
Prehistoric Society, Proceedings, n. s., v. 31, n. 4, 1965: 58-73
An examination of the many applications of radiocarbon dating to the study of the westward spread of farming from the ecologically focal region of South-West Asia to Europe and North Africa.

10.063 1966
RADIOCARBON DATING OF THE MALAYAN NEOLITHIC
Dunn, Frederick L.
Prehistoric Society. Proceedings, n. s., v. 32, 1966: 352-353
Various dates published are discussed for validity.

10.064 1967
RECENT RADIOCARBON DATES FROM RYUKYU SITES AND THEIR CHRONOLOGICAL SIGNIFICANCE
Pearson, R.
In: Archaeology of the Eleventh Pacific Science Congress, edited by W.G. Solheim, University of Hawaii. Social Science Research Institute, 1967 (*Asian and Pacific Archaeology series*, n. 1): 19-23
Description and assessment of dates, mostly on shell samples, from the Ryukyu Islands. A tentative chronology is discussed.

10.065 1966
RECENT TRENDS IN THE PRE- AND PROTO-HISTORIC ARCHAEOLOGY OF SOUTH ASIA
Dales, George F.
American Philosophical Society. Proceedings, v. 110, n. 2, 1966: 130-139
Points out the importance of radiocarbon dating for determining the chronological framework of South Asian pre- and protohistory, in particular for the Harappan civilisation, in providing the most dramatic confirmations and modifications of the conjectured dates made on archaeological evidence alone.

10.066 1968
A REVIEW OF THE CHRONOLOGY OF AFGHANISTAN, BALUCHISTAN AND THE INDUS VALLEY.
Dales, George F.
American Journal of Archaeology, v. 74, n. 4, Oct. 1968: 305-307
List of new radiocarbon dates available for Kashmir neolithic, South Indian neolithic and chalcolithic, and Indus Valley civilisation.

10.067 1957
SEVEN CAVES: ARCHAEOLOGICAL EXPLORATION IN THE MIDDLE EAST
Coon, Carleton Stevens
London, Jonathan Cape, 1957. pp. 323
The author used radiocarbon dating to correlate the stratigraphies of various caves excavated in the Middle East. Eight groups belonging to different cultures stratigraphically delimited are dated from $11,860 \pm 840$ BP to 1220 ± 230 BP or from Mesolithic seal hunters to Islamic period. These dates suggest that blade usage had already begun in the eastern part of the northern Eurasiatic zone early in the ice age and that the Upper Palaeolithic way of life in that area may be as old as in France.

10.068 1957
SHANIDAR CAVE
Solecki, Ralph S.
Scientific American, v. 197, n. 5, Nov. 1957: 58-64
This rocky shelter in Iraq has been inhabited by man for 100,000 years. Firebeds in layers B and C give dates of 12,000 years (Middle stone age) and 29,000 to 36,00 respectively. Layer D, 45,000 years, contains neanderthal remains.

10.069 1964
SHANIDAR CAVE, A LATE PLEISTOCENE SITE
Solecki, Ralph S.
In: International Association of Quaternary Research (INQUA), Congress, 6th, Warsaw, 1961. Report. Lodz, 1964: 413-422
The investigations of Shanidar cave indicate that man occu-

pied this cave from the middle Palaeolithic to recent times, a period of 50,000 years. Thirteen radiocarbon dates support this evidence.

10.070   1961
SHANIDAR CAVE, A LATE PLEISTOCENE SITE IN NORTHERN IRAQ
Solecki, Ralph S.
In: International Association of Quaternary Research (INQUA), *Congress*, 6th, Warsaw, 1961. *Abstracts*. 1961: 144
On the basis of radiocarbon dates, coupled with the data on the fossil pollens, there is a firm basis for climatic changes in this part of the Near-East to about the beginning of the last glaciation.

10.071   1955
SHANIDAR CAVE, A PALAEOLITHIC SITE IN NORTHERN IRAQ, AND ITS RELATIONSHIP TO THE STONE AGE SEQUENCE OF IRAQ.
Solecki, Ralph S.
Sumer, v. 11, n. 1, 1955:14-38
Two samples of carbon dated by Dr. Suess gave dates of 29,500 ± 1500 years at a depth of 10 feet and older than 34,000 years at a depth of 15 feet.

10.072   1955
SHANIDAR CAVE, A PALEOLITHIC SITE IN NORTHERN IRAQ
Solecki, Ralph S.
Smithsonian Institution. Annual Report for the year ended June 30, 1954. 1955: 389-425
Description of the site and its environment. A radiocarbon date on charcoal samples dates the bottom of layer B at 12,000 ± 400 years BP; a date on a hearth at a depth of 10 ft in layer C gives 27,500 ± 1500 and on a hearth at 15 ft is greater than 34,000 years.

10.073   1955
THE SHANIDAR CHILD
Solecki, Ralph S.
Archaeology, v. 8, n. 3, Autumn 1955: 169-175
Several layers of Shanidar cave were dated by the radiocarbon method. The following ages were obtained: lower part of layer B, 12,000 years BP; layer C, Baradost horizon, between older than 34,000 years BP and 29,500±1500 years BP.

10.074   1960
SOLDUZ VALLEY, IRAN: PISDELI TEPE
Dyson, Robert H.; Young, T.C.
Antiquity, v. 34, n. 133, Mar. 1960: 19-28
The authors support the evidence for chronological placement of Pisdeli Ware by radiocarbon dates.

10.075   1966
A SUGGESTED CHRONOLOGY FOR AFGHANISTAN, BALUCHISTAN AND THE INDUS VALLEY
Dales, George F.
In: Chronology in Old World Archaeology. Chicago, University of Chicago Press, 1966: 257-284
[Not sighted]

10.076   1967
SURVEY OF EXCAVATIONS IN IRAN DURING 1965-1966: GHAR-I KHAR AND GANJ-I-DAREH TEPE
Smith, P.E.L.
Iran, v. 5, 1967: 138-139
The typology of the flint artifacts is estimated to belong to the early Neolithic tradition, confirmed by a radiocarbon date on charcoal of 8450 ± 150 B.C. A higher level gives a date of 6960 ± 170 B.C. making this level contemporary with basal Jarmo horizon in Iraqi Kurdistan.

10.077   1968
SURVEY OF EXCAVATIONS IN IRAN DURING 1967: GANJ DAREH TEPE
Smith, P.E.L.
Iran, v. 6, 1968: 158-160
A radiocarbon date of 6960 ± 170 B.C. suggests that some of the deposits may represent the elusive VIIIth millenium.

10.078   1962
TAKING THE PREHISTORY OF HASANLU AREA BACK ANOTHER FIVE THOUSAND YEARS: SIXTH AND FIFTH MILLENIUM SETTLEMENTS IN SOLDUZ VALLEY, PERSIA
Young, T.C.
Illustrated London News, v. 124, n. 6431, 3 Nov. 1962: 707-709
The two areas Hajji Firuz and Tepe Sarab in the Kermanshah plain show typological parallel, borne out by the radiocarbon date on level II at Hajji Firuz of 5226 ± 86 years B.C. calculated on the radiocarbon half-life of 5800 years. At Dalma Tepe a mid-fifth millennium date is confirmed by radiocarbon. This poses a question of cultural relationships.

10.079   1961
THREE ADULT NEANDERTHAL SKELETONS FROM SHANIDAR CAVE, NORTHERN IRAQ
Solecki, Ralph S.
Sumer, v. 17, n. 122, 1961: 71-96
According to radiocarbon dating, the base of the Baradostian (layer C) appears to be close to 35,000 years old. There seems to be a cultural hiatus between this culture and the top

of the Mousterian, dated at 50,000 ± 4000. The top of layer D at 5.1 m below 0 datum was dated 46,000 ± 1500 years BP. This indicates that the Shanidar I Neandethal is about 46,000 years old.

10.080   1958
THE TOMB OF A KING OF PHRYGIA DISCOVERED INTACT, 2700 YEARS OLD TREASURES FROM THE HEART OF THE GREAT TUMULUS OF GORDION
Young, R.S.
Illustrated London News, v. 232, 17 May 1958:828-831
Radiocarbon tests on samples from the tomb walls have indicated an age of about 2700 years, suggesting that the timber was cut in the second half of the eighth century B.C.

10.081   1957
TWO NEANDERTHAL SKELETONS FROM SHANIDAR CAVE
Solecki, Ralph S.
Sumer, v. 13, n. 122, 1957: 59-60
The skeletons found in layer D of the excavation at Shanidar cave must be older than the date assigned by radiocarbon dating to the bottom of layer C, i.e. older than 34,000 years old and could be assumed to be 45,000 to 50,000 years old.

10.082   1961
ZAWI CHEMI SHANIDAR, A POST-PLEISTOCENE VILLAGE SITE IN NORTHERN IRAQ
Solecki, Rose Lilian
In: *International Association of Quaternary Research* (INQUA), *Congress*, 6th, Warsaw, 1961. *Abstracts*. 1961: 144-145
On the basis of the archaeological, botanical, zoological and radiocarbon evidence, Zawi Chemi represents one of the earliest village sites known. It is just post-Pleistocene in date.

10.083   1964
ZAWI CHEMI SHANIDAR, A POST-PLEISTOCENE VILLAGE SITE IN NORTHERN IRAQ
Solecki, Rose Lilian
In: *International Association of Quaternary Research* (INQUA), *Congress*, 6th, Warsaw, 1961. *Reports*. Lodz, 1964: 405-412
The lowest layer of the Zawi Chemi Shanidar site represents a very old village dating back to 10,870 ± 300 BP (radiocarbon date). The occupation is placed in the proto-Neolithic period.

CHAPTER ELEVEN
^{14}C ARCHAEOLOGY EUROPE

11.001   1967
THE ^{14}C CHRONOLOGY OF THE CENTRAL AND SE EUROPEAN NEOLITHIC
Quitta, Hans
Antiquity, v. 41, n. 164, Dec. 1967: 263-270
Since the application of radiocarbon dating to European Neolithic cultures, there has been much discussion about the relative validity of 'long' or 'short' chronologies. The author deals here with the critical areas of central and South-East Europe, sets out the long chronology obtained by radiocarbon dates and discusses the methodological problems involved.

11.002   1968
ABSOLUTE CHRONOLOGY OF THE NEOLITHIC AND ANEOLITHIC PERIODS IN CENTRAL AND SOUTH-EASTERN EUROPE
Neustupny, Evzen
Slovensky Archeologia, v. 16, 1968: 19-60
Discussion of absolute chronological systems which have been proposed. A long discussion on the radiocarbon method of age determination is included. Lists of radiocarbon dates for Central and South Eastern Europe are presented in graph and table form, followed by a table of corrections of selected radiocarbon dates and a chronology of four key areas based on comparison of uncorrected radiocarbon dates.

11.003   1965
ADVANCES IN DATING THE PALEOLITHIC
(Abstract)
Vogel, John C.
In: International Conference on Radiocarbon and Tritium Dating, Pullman, Washington, Washington State University, June 1-11, 1965. Proceedings. United States of America, Atomic Energy Commission, 1965. Conference n. 650652: 224
Viewed in the light of the most recent evidence on climatic variations during the last glaciation, the chronology of the Paleolithic needs considerable revision. Due to the lack of good samples radiocarbon can as yet not give any detailed information about the sequence of events during the Mousterian, but a quite consistent pattern is emerging for the Upper Palaeolithic. The date of the transition between the two complexes can be set at ca 38,000 years BP.

11.004   1960
ANATOLIA AND THE BALKANS
Mellaart, James
Antiquity, v. 34, n. 136, Dec. 1960: 270-278
The author describes his misgivings about the conventional dating of Balkan cultures and their correlation with Anatolia and proposes an alternative which reconciles recent radiocarbon dates from the Balkans with new Anatolian evidence.

11.005   1965
THE ANTIKYTHERA SHIPWRECK RECONSIDERED
Weinberg, G.D.; Grave, V.R.; Edwards, G.R.; Robinson, H.S.; Throckmorton, Peter; Ralph, Elizabeth K.
American Philosophical Society. Transaction, v. 55, part 3, June 1965. pp. 48
A radiocarbon date obtained from a wooden fragment of the Antikythera wreck gives a date of 155 ± 42 B.C. calculated with the 5568 half life and 220 ± 43 B.C. calculated with the 5730 half life. This latter is considered to be closer to the true value. This is earlier than the estimated date of the sinking as it represents the date of the cutting of the tree, and if large logs were used and the sample dated comes from the centre of a log, it would be earlier than the cutting of the tree by an amount equal to the age of the tree.

11.006   1963
ARCHAEOLOGICAL DISCOVERIES IN THE RAISED BOGS OF THE SOMERSET LEVELS, ENGLAND
Dewar, H.S.L.; Godwin, Harry
Prehistoric Society. Proceedings, n. s., v. 29, 1963: 17-49
Four time sequences are utilised in these studies of the Somerset raised bogs, bog stratigraphy, pollen analytic zoning, archaeological typology and radiocarbon dating, the latter giving a good approximation to the absolute time scale.

11.007   1959
ARCHAEOLOGY OF THE FARTHEST NORTH
Knuth, Eigil
In: International Congress of Americanists, 32nd, Copenhagen, 8-14 August, 1956. Proceedings. Copenhagen, Munskegaard, 1959: 561-573.
Two cultures on the Pearyland Peninsula in the northeast corner of Greenland, at a place called Princess Ingeborg Halvø were dated by radiocarbon determinations on driftwood at 3840 ± 170 for the older and 2830 ± 130 years for the younger indicating an interval of 1000 years between two immigration periods.

11.008   1967
A BARROW-CEMETERY OF THE SECOND MILLENIUM B.C. IN WILTSHIRE, ENGLAND
Christie, Patricia M.
Prehistoric Society. Proceedings, n. s., v. 33, n. 12, 1967: 336-366
On the evidence of the pottery and radiocarbon dates for two phases, the site was used as a cemetery from ca 2000 B.C. for at least 1000 years and probably longer.

11.009   1955
BEAKER TYPES AND THEIR DISTRIBUTION IN THE NETHERLANDS
Van der Waals, J.D.; Glasbergen, W.
Palaeohistoria, v. 4, 1955: 5-46
Because beakers represent the 'international' element par excellence in Dutch prehistory, a detailed examination of pottery and graves is presented. Radiocarbon determinations suggest a date about 22,000 B.C. for the early specimens, 1700 to 1500 B.C. for the true Bell beakers series.

11.010   1960
BRITTANY
Giot, Pierre Roland; L'Helgouach, J.; Briard, J.
London, Thames and Hudson, 1960. pp. 272 (*Ancient Peoples and Places*, v. 13)
The successive cultures of stone builders in Brittany are dated from a neolithic culture starting before 3500 B.C., date of button axes at Curnic, 3140 ± 60 B.C. confirmed at the corbelled passage graves of Ile Carn, Ploudalmezeau, Finistère to a middle-bronze age dated on the wooden flooring of the central tomb of the Kervingar Barrow at Plouerzel, 1350 ± 50 B.C. finally to early iron age from small barrows on the plateau of the Morbihan.

11.011   1961
C-14 DATES FOR SITES IN THE MEDITERRANEAN AREAS
Kohler, Ellen L.; Ralph, Elizabeth K.
American Journal of Archaeology, v. 65, 1961: 357-367
Lists of radiocarbon dates of special interest to prehistorians and historians of the Mediterranean area. The following sites are dated: from Anatolia, Hacilar, Beycesultan, Gordion; from Greece, Eutresis, Lerna, Pylos, Chios.

11.012   1961
C-14 DATES FOR THE MALTESE EARLY NEOLITHIC
Evans, John D.
Antiquity, v. 35, n. 138, June 1961: 143-144
Charcoal from beneath the floor of megalithic temples at Ta Hagrat Mgarr, Malta, give higher date than expected (2700 ± 150 B.C.). The author indicates that there is need for redating the earlier phase of Period I in Malta.

11.013   1965
C-14 DATES FROM MESOLITHIC SITES IN CENTRAL EUROPE
Bandi, H.G.
In: International Conference on Radiocarbon and Tritium Dating, Pullman, Washington, Washington State University, June 1-11, 1965. *Proceedings*. United States of America, Atomic Energy Commission, 1965. Conference n. 650652: 225-231.
Discussion of the dating of some Mesolithic sites in Central Europe

11.014   1960
C-14 DATING AND THE NEOLITHIC IN IRELAND
Watts, W.A.
Antiquity, v. 34, n. 134, June 1960: 111-116
The radiocarbon dates produced in the Radiocarbon Dating Laboratory of the University of Dublin, with special reference to the Neolithic in Ireland are presented. A full new Neolithic chronology based on radiocarbon dating is still not achieved, but the new results and studies today bring this achievement nearer.

11.015   1960
C-14 NEOTHOLICS FROM FRANCE
Coursaget, Jean; Giot, Pierre Roland; Le Run, J.
Antiquity, v. 34, n. 134, June 1960: 147-148
These dates contribute to the absolute, and ever relative chronology of the Neolithic culture of Western Europe, inducing a 'long' chronology instead of a 'short' one as was in favour in many archaeological circles.

11.016   1963
CARBON, MALTA AND THE MEDITERRANEAN
Trump, David
Antiquity, v. 34, n. 148, Dec. 1968: 302-303
A series of five radiocarbon dates lead to a reappraisal of the chronology of Maltese prehistory and the Mediterranean prehistory.

11.017   1965
CHRONOLOGIES OF OLD WORLD ARCHAEOLOGY
Ehrich, Robert W., ed.
Chicago, University of Chicago Press, 1965. pp. 557
A nw volume to replace 'Relative Chronologies in Old World Archaeology' in which radiocarbon is the biggest single factor for a review of ancient chronologies.

Ch. 11 - Archaeology - Europe

11.018 1962
CHRONOLOGY OF THE NEOLITHIC IN EUROPE
(Editorial)
Daniel, Glyn
Antiquity, v. 36, n. 141, Mar. 1962: 81-82
Comments on radiocarbon dates applied to the chronology of the Neolithic in Europe.

11.019 1966
THE CLIMATE, ENVIRONMENT AND INDUSTRIES OF STONE AGE GREECE: PART II
Higgs, E.S.; Vita-Finzi, Claudio
Prehistoric Society, Proceedings, n. s, v. 32, n. 1, 1966: 1-29
Radiocarbon dates for an excavation of Asprochaliko are discussed as they do not correlate with the stratigraphy. It is concluded that they do not date either the deposition of the layers of the industries.

11.020 1967
THE CLIMATE, ENVIRONMENT AND INDUSRIES OF STONE AGE GREECE: PART III
Higgs, E.S.; Vita-Finzi, Claudio; Harris, D.R.; Fagg, A.E.
Prehistoric Society. Proceedings, n. s, v. 33, n. 1, 1967: 1-29
A high lake level defined by hearths within the Katrista cave is dated at between 20,800 ± 810 to 20,200 ± 480 BP contemporary with the Bradenburg stage in Northern Europe. Cave deposits contained a continuous sequence of near horizontal hearth complexes containing advanced palaeolithic industries. One date, one meter below datum, gave an age of 13,400 ± 210 years BP.

11.021 1967
COLONIALISM AND MEGALITHISMUS
Renfrew, Colin
Antiquity, v. 41, n. 164, Dec. 1967: 276-288
On the basis of radiocarbon dates, the author proposes a multilinear evolutionary hypothesis in European prehistoric archaeology.

11.022 1965
A COMMENT ON THE C-14 DATA OF EASTERN MIDDLE EUROPEAN PALEOLITHIC, WITH SUGGESTIONS FOR FURTHER STANDARDIZATION OF RADIOCARBON AGE DETERMINATIONS
Vertes, L.
In: International Conference on Radiocarbon and Tritium Dating, Pullman, Washington, Washington State University, June 1-11, 1965. *Proceedings*. United States of America, Atomic Energy Commission, 1965. Conference n. 650652: 210-223
Contradictory results of radiocarbon analysis from adjacent geographical areas for the period between 30,000 and 45,000 radiocarbon years are assessed. Various suggestions to increase the scientific reliability and potential usefulness of radiocarbon age determinations are made.

11.023 1965
THE COPPER AGE IN SOUTHEASTERN EUROPE
Waterbolk, H.T.
In: International Conference on Radiocarbon and Tritium Dating. Pullman, Washington, Washington State University, June 1-11, 1965. *Proceedings*. United States of America, Atomic Energy Commission, 1965. Conference n. 650652: 243-244
The supposed priority of copper working in the Iberian Peninsula over that in Southeastern Europe cannot be maintained in view of the radiocarbon dates.

11.024 1966
THE CRUCK-BUILT BARN OF MIDDLE LITTLETON, WORCESTERSHIRE
Horn, Walter; Charles, F.W.B.; Berger, P.
Society of Architectural Historians. Journal, v. 25, n. 4, Dec. 1966: 221-239
Radiocarbon dating refutes the assumption that the roof of the barn had been rebuilt. An appendix presents and discusses the dates used.

11.025 1968
CRUCKS IN THE WEST BERKSHIRE AND OXFORD REGION
Fletcher, John
Oxoniensia, v. 33, 1968: 71-88
This is a list and record of specimens of Cruck buildings for the West Berkshire and Oxford region. A table of radiocarbon dates for the North Berkshire Crucks is presented with discussions.

11.026 1968
THE DATE OF THE 'RED LADY' OF PAVILAN
Oakley, Kenneth P.
Antiquity, v. 42, n. 168, Dec. 1968: 306-307
A skeleton dicovered in 1823 in Pavilan cave on the coast of Gower Peninsula, Glamorganshire, is dated from bone powder drilled out from the bones. 16,500 B.C. makes it the oldest human skeletal material to have been dated by radiocarbon on internal carbon.

11.027 1966
DATING OF THE WILSFORD SHAFT
Ashbee, Paul
Antiquity, v. 40, n. 159, Sept. 1966: 227-228

The age of the Wilsford shaft is related to the ages of shafts in sites from the South of England through radiocarbon dates.

11.028 1960
THE EARLIEST SETTLEMENTS OF EUTRESIS; SUPPLEMENTARY EXCAVATIONS, 1958
Caskey, John L.; Caskey, Elizabeth G.
Hesperia, Journal of Classical Studies at Athens, v. 30, 1960: 126-167
Two radiocarbon dates on carbonised wood from the western area and the central area gave ages around 2500 B.C. Archaeological comparison with other sites would estimate dates earlier in the third millemium B.C.

11.029 1967
AN EARLY NEOLITHIC SETTLEMENT ON COYGAN ROCK, CARMATHENSHIRE
Wainwright, G.J.
Antiquity, v. 41, n. 161, Mar. 1967: 66
A radiocarbon date obtained from charred walnut shells found in a pit of early Neolithic occupation confirms the conception of a community of culture in the Irish Sea area around the beginning of the 3rd millenium B.C.

11.030 1965
AN EARLY NEOLITHIC VILLAGE IN GREECE
Rodden, Robert J.
Scientific American, v. 212, n. 4, Apr. 1965: 82-92
Excavation at Nea Nikodemia in Northern Greece has uncovered a community of 8000 years ago. Its remains suggest that agriculture and animal husbandry came to Europe earlier than had been supposed. Samples of organic material yielded a radiocarbon date of 6220 ± 150 B.C. This is the earliest date as yet assigned to neolithic material from Europe.

11.031 1964
A EUROPEAN LINK WITH ÇATAL HUYUK: THE 7TH MILLENIUM SETTLEMENT OF NEA NIKOMEDEIA IN MACEDONIA. PART I: SITE AND POTTERY
Rodden, Robert J.; Rodden, J.M.
Illustrated London News, n. 244, 11 Apr. 1964: 564-567
A date of 6220 ± 150 B.C. for an assemblage including pottery may change the present understanding of the origin and spread of this art.

11.032 1965
THE EXCAVATION OF A NEOLITHIC ROUND BARROW AT PITNACREE, PORTSHIRE, SCOTLAND
Coles, John P.; Simpson, D.D.A.
Prehistoric Society. Proceedings, n. s, v. 31, article n. 3, 1965: 23-57
A radiocarbon date of the old land surface gives an age of 2860 ± 90 B.C. and is in accord with the series of absolute dates indicating Neolithic settlement of Britain and Ireland.

11.033 1959
EXCAVATION OF ELATEIA, 1959
Weinberg, Saul S.
Hesperia, Journal of Classical Studies at Athens, v. 29, 1959: 158-209
Radiocarbon dates indicate that pottery production was well under way in the Near East by 6000 B.C. but give no evidence of its appearance in Greece before 5500 B.C. Other dated phases are also discussed.

11.034 1954
EXCAVATIONS AT STAR CARR: AN EARLY MESOLITHIC SITE AT SEAMER NEAR SCARBOROUGH, YORKSHIRE
Clark, John Graham Douglas
Cambridge, Cambridge University Press, 1954. pp. 200
The age determined by radiocarbon measurements on a sample of birchwood falls well within the range of dates previously accepted for the pre-Boreal period on the basis of the de Geer's varve chronology.

11.035 1962
EXCAVATIONS AT THE MAGLEMOSIAN SITES AT THATCHAM, BERKSHIRE, ENGLAND
Wymer, John
Prehistoric Society. Proceedings, v. 28, n. 13, 1962: 329
The evidence of radiocarbon analyses of charcoal from the ancient surface and of wood from the algal marl strongly suggests that the main occupation falls within the earlier part of the pre-boreal to boreal.

11.036 1951
EXCAVATIONS AT THE MESOLITHIC SITE OF STAR CARR, YORKSHIRE, 1949-1950
Clark, Grahame
Archaeology, v. 4, n. 2, Summer 1951: 61-70
The average of 2 radiocarbon tests of birchwood samples from the archaeological level of Soames gives a date of 9488 ± 350 years BP, confirming the date obtained using the de Geer geochronological system.

11.037 1962
EXCAVATIONS OF THE EARLY NEOLITHIC SITES OF NEA NIKOMEDIA, GREEK MACEDONIA (1961 SEASON)
Rodden, Robert J.
Prehistoric Society. Proceedings, v. 28, n. 11, 1962: 267-

An early radiocarbn date of the first settlement phase of Nea Nikodemia suggests approximate contemporaneity with the time of primary village - farming community in Western Asia.

11.038 1962
A FRESH SERIES OF RADIOCARBON DATES FROM FRANCE
Coursaget, Jean; Giot, Pierre Roland; Le Run, J.
Antiquity, v. 36, n. 141, Mar. 1962: 139-141
Dates from Neolithic levels are presented and discussed. Some dates which appear 3500 or 4000 years too old could be explained by the fact that the builders of a tomb used fossilised wood from peat bogs. The authors conclude that there is no difficulty in obtaining satisfactory dates from satisfactory material.

11.039 1960
FURTHER EXCAVATIONS AT A MESOLITHIC SITE AT OAKHANGER, SELBORNE, HANTS
Rankine, W.F.; Dimbleby, G.W.
Prehistoric Society. Proceedings, n. s., v. 26, article n. 12, 1960: 246-262
Radiocarbon dating of phase II level indicates no or little break in occupation and points to the Atlantic period as the time of occupation.

11.040 1967
FURTHER RADIOCARBON DATES FROM STONEHENGE
Atkinson, R.J.C.
Antiquity, v. 41, n. 161, Mar. 1967: 63-64
Three new dates from antler specimen help conclude that the most probable date for the Stonehenge II/IIIa transition lies in the 17th century B.C. and places the IIIb/IIIc transition to the mid-12th century B.C.

11.041 1967
GEOMORPHOLOGY AND STRATIGRAPHY OF THE PALEOLITHIC SITE OF BUDINO
Butzer, Karl W.
Eiszeitalter und Gegenwart, v. 18, 1967: 82-103
Radiocarbon dates and stratigraphic analysis give a middle Würmian age for the site in Budiño, in the middle Louro Valley of Galicia, Spain.

11.042 1966
THE HEARTHS OF THE UPPER PERIGORDIAN AND AURIGNACIAN HORIZONS AT THE ABRI PATAUD, LES EYZIES (DORDOGNE) AND POSSIBLE SIGNIFICANCE
Movius, Hallam L.
American Anthropologist, v. 68, n. 2, part 2, Apr. 1966: 296-325
The author discusses radiocarbon dates for several horizons and finds that most are either too young or too old to fit in with the cultural occupation at these levels.

11.043 1968
HOMOLKA AN ENEOLITHIC SITE IN BOHEMIA
Erich, Robert W.; Pleslová-Stiková, Emilie
American School of Prehistoric Research. Peabody Museum. Harvard University. Bulletin, n. 24, 1968
The site of Homolka is in the Klodono County in central Bohemia. Three radiocarbon dates are presented. One of these dates, which is estimated valid, places the site occupation at approximately 2300 B.C. The other two dates are considered much too late and the samples were most probably contaminated.

11.044 1965
HOW LONG DID UNMIXED FOOD COLLECTION SURVIVE IN NORTHERN EUROPE? - A CASE STUDY OF ARCHAEOLOGICAL OBSERVATION, RADIOCARBON DETERMINATION AND ARCHAEOLOGICAL INFERENCE
Moberg, Carl-Axel
In: International Conference on Radiocarbon and Tritium Dating, Pullman, Washington, Washington State University, June 1-11, 1965. *Proceedings*. United States of America, Atomic Energy Commission, 1965. Conference n. 650652: 245-255
The latest survival of unmixed food-collecting in Scandinavia is discussed for the period 5050 ± 45 years ago and for north Norway, 750 ± 45 years ago.

11.045 1965
THE KITCHEN MIDDEN SITE AT WESTWARD HO!, DEVON, ENGLAND : ECOLOGY, AGE AND RELATION TO CHANGES IN LAND AND SEA LEVEL
Churchill, D.M.
Prehistoric Society. Proceedings, n. s., v. 31, article n. 5, 1965: 74-84
Radiocarbon dating from the topmost layer of the kitchen midden gives an age of 6585 ± 130 BP, other evidence indicates that there has been no measurable tectonic displacement over the past 6500 years.

11.046 1954
LAKE-STRATIGRAPHY, POLLEN-ANALYSIS AND VEGETATIONAL HISTORY
Walker, Donald; Godwin, Harry
In: Excavation at Star Carr; an Early Neolithic Site at Seamer near Scarborough, Yorkshire, by Clark, J.G.D.,

Cambridge University Press, 1954: 25-69
A radiocarbon analysis of birch wood from the occupation platform gave an age of 9488 ± 350 years BP which confirmed the result achieved by pollen analysis, placing Star Carr occupation close to the transition between Zone IV and V and within the chronology of North-West Europe.

11.047   1951
THE LASCAUX CAVES
Movius, Hallam L.
American Antiquity, v. 17, part 2 (*Society for American Archaeology, Memoirs*, n. 8), 1951: 50-51
A charcoal bearing horizon at Lascaux was dated by radiocarbon at 15,516 ± 900 years BP or possibly from Magdalenian times, but the age relationship with the wall paintings and the engravings cannot be solved on the basis of the information available.

11.048   1951
LASCAUX CHARCOAL
Baghoorn, Elso S.; Movius, Hallam L.
Science, v. 114, n. 2961, 28 Sept. 1951: 333
Discussion of the identification of charcoal fragments from the Lascaux caves which were used in radiocarbon determinations.

11.049   1957
LATE-GLACIAL HUMAN CULTURES IN THE NETHERLANDS
Hijszeler, C.C.W.J.
Geologie en Mijnbouw, v. 19, 1957: 288-299
Ground traces of a tent found at Usselerveen (Netherlands) was excavated and a layer, correlated to the Ussels stratigraphic layer, was dated and found to be younger than the top of the Allerød in the valley profile. Nine dates showing correlations with periods in various areas of western Europe are presented in table form on p. 296-297.

11.050   1964
LEBOUS
Arnal, J.; Martin-Gravel, H.; Sangmeister, E.
Antiquity, v. 38, n. 151, Sept. 1964: 191-200
A chalcolithic castle in the south of France is one of the earliest monuments to have an architectural style so exceptional that it is difficult to give it an exact date. It contains a large amount of Fontbouisse ware in the primary archaeological layer (this culture is dated by radiocarbon between 2600 and 2300 B.C.)

11.051   1968
LERNA IN THE EARLY BRONZE AGE
Caskey, J.L.
American Journal of Archaeology, v. 72, n. 4, Oct. 1968: 313-316
A late Neolithic date of 3021 ± 58 B.C. is obtained by radiocarbon dating for period II of the site of Lerna, 10 km south of Argos.

11.052   1962
LONG BARROWS, CHRONOLOGY AND CAUSEWAYED CAMPS
Case, Humphrey
Antiquity, v. 36, n. 148, Sept. 1962: 212-216
The British Neolithic chronology is discussed in the context of radiocarbon dates.

11.053   1961
MORE ON UPPER PALAEOLITHIC ARCHAEOLOGY: COMMENT
Valoch, Karel
Current Anthropology, v. 2, n. 5, Dec. 1961: 448
The author comments on the paper by Movius (*Current Anthropology*, 1960: 355-391) and supplements it by a report on recent investigations of the loess of the Brno region and the cave profiles in the Moravian Karst radiocarbon dated by de Vries.

11.054   1961
MORE ON UPPER PALAEOLITHIC ARCHAEOLOGY: COMMENTS
Kozlewski, Janusz K.
Current Anthropology, v. 2, n. 5, Dec. 1961: 436-437
The author comments on the paper by Movius (*Current Anthropology*, 1960: 355) and offers a few additional observations from the perspective of East-Central and Eastern Europe.

11.055   1961
MORE ON UPPER PALAEOLITHIC ARCHAEOLOGY: COMMENTS
Kukla, Jiri; Klima, Bohuslav
Current Anthropology, v. 2, n. 5, Dec. 1961: 437-439
The authors comment on the paper by Movius (*Current Anthropology*, 1960: 355) incorporating new data in the form of results of recent radiocarbon analyses of samples from an open section at Dolni Vestonic (Unter Wistenitz) brickyard.

11.056   1961
MORE ON UPPER PALAEOLITHIC ARCHAEOLOGY: COMMENTS
Muller-Beck, Hansjürgen
Current Anthropology, v. 2, n. 5, Dec. 1961: 439-444
The author comments on the paper by Movius (*Current Anthropology*, 1960: 355) and presents a table of the main stages of Upper Pleistocene and Western Central Europe,

including radiocarbon dates and a culture sequence correlation.

11.057   1961
MORE ON UPPER PALAEOLITHIC ARCHAEOLOGY: COMMENTS
Brandtner, F.
Current Anthropology, v. 2, n. 5, Dec. 1961: 427-428
The author comments on the paper by Movius (*Current Anthropology*, 1960: 355) and concludes that a discussion of radiocarbon dates without a clear stratigraphic framework is not useful and only causes confusion. He contests a number of the dates presented on various grounds.

11.058   1961
MORE ON UPPER PALAEOLITHIC ARCHEOLOGY: COMMENTS
Gross, Hugo
Current Anthropology, v. 2, n. 5, Dec. 1961: 428-434
The author comments on the paper by Movius (*Current Anthropology*, 1960: 355) and reassesses various points by discussing the validity of both correlations and radiocarbon dates.

11.059   1961
MORE ON UPPER-PALAEOLITHIC ARCHAEOLOGY: A REPLY
Movius, Hallam L.
Current Anthropology, v. 2, n. 5, Dec. 1961: 448-454
The author replies to various comments on his paper by Brandtner, Gross and Kozlewski (*Current Anthropology*, 1961: 427, 436) and comments on the archaeologic sequence while refraining from commenting on the geological aspect until the experts in the field have thoroughly assessed the field evidence. He also comments on the new radiocarbon dates.

11.060   1967
MORTUARY HOUSES AND FUNERAL RITES IN DENMARK
Kjaerum, Poul
Antiquity, v. 41, n. 163, Sept. 1967: 190-196
Two Neolithic mortuary houses at Tustrup and Ferslev, in Denmark were excavated. The author discusses their dating and their relationship to Scandinavian Passage Graves in general. A radiocarbon date was obtained from the bark of wall posts and shows the contemporaneity of passage graves in Scandinavia, in the continental Funnel-Beaker area and in the Western Megalith area.

11.061   1967
NEAR EASTERN, MEDITERRANEAN AND EUROPEAN CHRONOLOGY
Thomas, Homer L.
In: Studies in Mediterranean Archaeology, vol. 17, 1967. pp. 175 and charts
Presents the historical, archaeological, pollen analytical and geohronological evidence. Explores the implications of radiocarbon dates for the archaeological chronology of Europe. There is a bibliography of radiocarbon dates, p. 158-160.

11.062   1963
NEOLITHIC BOWS FROM SOMERSET, ENGLAND, AND THE PREHISTORY OF ARCHERY IN NORTH-WEST EUROPE
Clark, John Graham Douglas
Prehistoric Society. Proceedings, n. s., v. 29, article n. 3, 1963: 50-98
The bows from Ashcott and Meare, Somerset, found near the base of a dark humidified peat where other Neolithic implements were found give ages of 2665 ± 120 and 2690 ± 120 B.C. which is consistent with a Neolithic age.

11.063   1963
NEOLITHIC CHARCOAL FROM HEMBURY
Fox, Aileen
Antiquity, v. 36, n. 147, Sept. 1963: 228-229
Three samples of charcoal from a Neolithic settlement at Hembury hillfort, stored in the British Museum since 1935 were dated. The results are highly satisfactory as the dates are consistent and corroborate with each other within the $\pm$ 150 years margin. They establish on a firm basis that Neolithic settlement in Southwestern England existed in the early part of the 4th millenium.

11.064   1968
A NEOLITHIC GOD-DOLLY FROM SOMERSET, ENGLAND
Coles, John
Antiquity, v. 42, n. 168, Dec. 1968: 275-277
A hermaphroditic figure carved from a solid piece of ashwood, discovered under a Neolithic trackway, the Bell B track at Westhay, was dated at 2890 ± 100 B.C. from timber of the overlying track.

11.065   1962
THE NEOLITHIC OF CAMBRIDGESHIRE FENS
Clark, John Graham Douglas; Godwin, Harry
Antiquity, v. 36, n. 141, Mar. 1962: 10-23
A stratified sequence of deposits containing artifacts and fossil pollen at Peacock's Farm, Shippea Hill, Cambridgeshire, is investigated. It provided new samples for pollen analysis and radiocarbon dating. The series of radicabon dates were broadly self-consistent but also agree with the dating for pollen zones, archaeology and stratigraphy.

11.066 1963
THE NEOLITHIC PERIOD IN GREECE
Weinberg, Saul S.
Current Anthropology, v. 4, n. 4, Oct. 1963: 377
The author presents some radiocarbon dates for the Neolithic period in Greece which have come to hand since the publication of the 'index of radiocarbon dates' by R. Jelinek (*Current Anthropology*, 1962: 451-475).

11.067 1956
NEOLITHIC PERIOD IN SWITZERLAND AND DENMARK
Troels-Smith, J.
Science, v. 124, n. 3227, 2 Nov. 1956: 876-879
Radiocarbon dating of two dwelling places in Switzerland and Denmark embedded in Quaternary sediments (making dating by pollen analysis possible) indicate that the oldest agriculture economies in Switzerland and Denmark started almost simultaneously about 2740 ± 90 and 2620 ± 80 B.C. respectively.

11.068 1965
A NEOLITHIC SITE ON IPING COMMON, SUSSEX, ENGLAND
Keef, P.A.M.; Wimer, J.J.; Dimbleby, G.W.
Prehistoric Society. Proceedings, n. s., v. 31, article n. 6, 1965: 85-92
The flint industry from Iping Common is of the pure Maglemosian tradition of the Mesolithic period and can be equated to one of the earlier phase of Oakhanger, dated by radiocarbon at 4340 ± 120 B.C.

11.069 1961
NEW RADIOCARBON DATES FROM FRANCE
Coursaget, Jean; Giot, Pierre Roland; Le Run, J.
Antiquity, v. 35, n. 138, June 1961: 147-148
The dates presented mainly relate to the later prehistory of France.

11.070 1959
THE NURAGHI OF SARDINIA
Liliu, G.
Antiquity, v. 33, n. 129, Mar. 1959: 32-38
The current knowledge of the Sardinian Nuragic civilisation is summarised. A wooden bath used in the walls of the ground floor chamber in the central tower of Su Naraxi, Barumini gives a date of 1270 ± 200 B.C. which places this tower between 1470 and 1070 B.C. The author considers the second date as more probable.

11.071 1961
ON A FIND OF A PREBOREAL DOMESTIC DOG (*CANIS FAMILIARIS L.*) FROM STAR CARR, YORKSHIRE, WITH REMARKS ON OTHER MESOLITHIC DOGS
Degerbal, Magnus
Prehistoric Society. Proceedings, n. s, v. 27, article n. 3, 1961
The site is dated by pollen analysis as dating back to the end of the Preboreal, or by radiocarbon method at 9488 ± 350 BP thus being older by about 100 years than the oldest Maglemosian site.

11.072 1965
PALAEOLITHIC RADIOCARBON DATES FROM SOUTH-WESTERN EUROPE AND THE MEDITERRANEAN BASIN
Smith, Philip E.L.
In: International Conference on Radiocarbon and Tritium Dating, Pullman, Washington, Washington State University, June 1-11, 1965. *Proceedings*. United States of America, Atomic Energy Commission, 1965. Conference n. 650652: 199-209
The chronological frameworks for the Palaeolithic industries in Southwestern Europe and the Mediterranean basin as created on the basis of radiocarbon analysis are presented and discussed.

11.073 1965
THE PALEOLITHIC SITE OF MOLDOVA V ON THE MIDDLE DNIESTER (USSR)
Ivanova, I.K.; Chernysh, A.P.
Quaternaria, v. 7, 1965: 197-218
Radiocarbon dating is used to date various levels, the oldest one proves to be more than 48,000 years BP.

11.074 1959
PALEO-ESKIMO CULTURE IN DISKO BUGT, WEST GREENLAND
Larsen Helge; Meldgaard, Jorgen
Meddeleser om Grønland, v. 161, n. 2, 1958. pp. 75
Radiocarbon measurements from peat layers containing implement types of the Sarqaq culture gave a date of 790 ± 100 years B.C. and from charcoal the date was 810 ± 100 years B.C. indicating that the Sarqaq culture was in existence at Disko Bugt around the period 7-900 B.C. The Sarqaq culture is dated 3250 ± 110 years BP on charcoal from fireplace and the Dorset culture at 2016 ± 250 years by comparison with the Denbigh Flint Complex.

11.075   1958
A POSSIBLE SOURCE FOR THE VOGELHERD AURIGNACIAN
Müller-Beck, Hans Jürgen
Arctic Anthropology, v. 5, n. 1, 1968: 48-61
Radiocarbon dates place layer 6 at Nietoperzowa cave (Cracow, Poland) at 38,160 ± 1250 years BP and also indicate that the Pardero ended before about 25,000 years BP.

11.076   1964
PREHISTORIC DISC-WHEELS IN THE NETHERLANDS; WITH A NOTE ON THE EARLY IRON AGE DISC WHEELS FROM EZINGE
Van der Waals, J.D.
Palaeohistoria, v 10, 1964: 103-146
A radiocarbon chronology of disc wheels and neolithic culture is presented in this study of disc wheels in the Netherlands.

11.077   1968
PREHISTORIC ROADS AND TRACKS IN SOMERSET, ENGLAND
Coles, John M.; Hibbert, F. Alan
Prehistoric Society. Proceedings, n. s., v. 34, n. 6, 1968: 238
Tracks at Honegore and the Abbot's Way are dated by radiocarbon as being of the same age although they occupy different stratigraphic positions. The results are discussed. The age of 2800 B.C. makes them the oldest known wooden roadways in Europe.

11.078   1960
PREHISTORIC WOODEN TRACKWAYS OF THE SOMERSET LEVELS: THEIR CONSTRUCTION, AGE AND RELATION TO CLIMATIC CHANGE
Godwin, Harry
Prehistoric Society. Proceedings, n. s., v. 26, 1960: 1-36
This paper represents the results of applying pollen analysis, radiocarbon dating and peat-stratigraphical studies to a specific archaeological problem in southwestern Europe.

11.079   1967
PRIMITIVE HUNTER AND PLEISTOCENE EXTINCTION IN THE SOVIET UNION
Vereschagin, N.K.
In: Pleistocene Extinctions: the Search for a Cause. International Association for Quaternary Research (INQUA). Congress, 7th, Boulder, Col., 14 Aug - 19 Sept.1965. *Proceedings*, v. 6. Boulder, Col., 1968: 365-398
The extent of hunting by early man in the USSR is shown in the remains found in sites of the Upper Paleolithic which are dated by radiocarbon at 9000 to 14,000 years old. However the main reason for the absolute extinction of animals of the mammoth complex and for the reduction of range in some species is the change in climate and terrain, especially the change in the region of winter weather. The destructive effect of man supplemented and intensified the influence of climatic factors.

11.080   1967
THE QUATERNARY OF THE BRITISH ISLES
West, R.G.
In: The Quaternary, v. II, edited by Kalervo Rankama. New York, Interscience, 1967: 2-87
A review study which brings together a large number of surveys made of the Quaternary in the British Isles. It notes that the radiocarbon dating method has been extensively used in problems of stratigraphy, vegetational history, changes in the land-to-sea levels and archaeology. Also presents a table of Weichsel radiocarbon dates.

11.081   1968
A RADIOCARBON DATE FOR A WEST SCOTTISH NEOLITHIC SETTLEMENT
Scott, J.G.
Antiquity, v. 42, n. 168, Dec. 1968: 296-297
Charcoal associated with sherds of Rothesay type vessels confirm the predominantly late Neolithic character of the ware, retrieved from a disused pebble quarry at Townhead, Rothesay, Isle of Bute.

11.082   1960
RADIOCARBON DATES AND UPPER PALAEOLITHIC ARCHAEOLOGY IN CENTRAL AND WESTERN EUROPE
Movius, Hallam L.
Current Anthropology, v. 1, n. 5-6, Sept.-Nov. 1960: 355-391
A review article which collates radiocarbon dates from Central and Western Europe in order to establish a chronology for the last 60,000 years. Part I of this paper summarises the chronological results with reference to the Fourth Glacial (Würm/Weichsel) stage; part II attempts to interpret the classic Upper Palaeolithic sequence of South-Western France in terms of this chronology and in the light of the available radiocarbon dates.

11.083   1967
RADIOCARBON DATES FOR THE WILLERBY WOLD LONG BARROW
Manby, T.G.
Antiquity, v. 41, n. 164, Dec. 1967: 306-307
The Willerby dates obtained on charcoal samples are in accord with the pollen evidence which places the construction of this long barrow in the cool, moist conditions appropriate to zone VIIa of the Atlantic Period, around 3000 B.C.

It indicates that the cremation aspect of Yorkshire long barrows had already developed by the opening of the third millenium B.C.

11.084   1967
RADIOCARBON DATES FROM CORFU, GREECE
Sordinas, Augustus
Antiquity, v. 41, n. 161, Mar. 1967: 64
Two dates obtained from ashes for two Early Neolithic levels at the site of Sidari on the north-west tip of Corfu are reported.

11.085   1959
THE RADIOCARBON DATES FROM DURRINGTON WALLS
Piggott, Stuart
Antiquity, v. 33, n. 132, Dec. 1959: 289-290
The author gives the reason why a date obtained from charcoal beneath the chalk rubble mound of the Henge of Durrington Walls is not satisfactory.

11.086   1960
RADIOCARBON DATES FROM WINDMILL HILL
Smith, I.
Antiquity, v. 34, n. 135, Sept. 1960: 212-213
Charcoal samples were collected from three separate layers. Each date refers to the content of a layer which has accumulated over a considerable period of time. These dates form a series which accord not only with the facts of stratigraphy but also from evidence for elsewhere and place the Neolithic Windmill Hill camp at 2950 ± 150 B.C. to 1540 ± 150 B.C.

11.087   1967
RADIOCARBON DATES ON OCCUPATION SITES OF PLEISTOCENE AGE IN THE USSR
Klein, R.G.
Arctic Anthropology, v. 4, n. 2, 1967: 224-226
List of 41 assayed samples compiled from recent Soviet publications.

11.088   1965
RADIOCARBON DATING AND THE SPREAD OF AGRICULTURE IN THE OLD WORLD
Clark, Grahame
In: International Conference on Radiocarbon and Tritium Dating, Pullman, Washington, Washington State University, June 1-11, 1965. *Proceedings*. United States of America, Atomic Energy Commission, 1965. Conference n. 650652: 232-242
By providing a chronological guide radiocarbon dating has opened the way to a new interpretation of the basic archaeological material. In particular it has placed the early spread of farming at least 1000 years earlier than was believed previously.

11.089   1965
RADIOCARBON DATING AND THE SPREAD OF FARMING ECONOMY
Clark, John Graham Douglas
Antiquity, v. 39, n. 153, Mar. 1965: 45-48
Illustrates how far the radiocarbon determinations available as of October 1964 are adequate to throw light on the expansion of farming. A consistent, coherent pattern is revealed defining the primary zone in Iran and Greece, the early expansion and the drive north.

11.090   1959
RADIOCARBON DATING OF PREHISTORIC WOODEN TRACKWAYS
Godwin, Harry; Willis, H.H.
Nature, v. 184, n. 4685, 15 Aug. 1959: 490-491
All trackways dated so far belong to the same archaeological period, namely the Bronze Age to early Iron Age transition. This helps to place a period of climatic deterioration approximately 600 B.C.

11.091   1961
RADIOCARBON DATING OF THE FALLAHOGY LANDNAM PHASE
Smith, A.G.; Willis, E.H.
Ulster Journal of Archeology, v. 24/25, 1961/1962: 16-24
The radiocarbon dates covering the clearance and farming stage of the Landnam phase are discussed. They are broadly in accordance with the interpretation of the pollen analytic data.

11.092   1964
THE RADIOCARBON DATING OF THE FUSSELL'S LODGE LONG BARROW
Ashbee, Paul
Antiquity, v. 38, n. 150, June 1964: 139-140
The dates obtained from carbonised pieces of oak found in the enclosure fit in with previous dates for the early phase of the British Neolithic.

11.093   1959
THE RADIOCARBON DATING OF THE NUTBANE LONG BARROW
Mallett Vatcher, Faith de
Antiquity, v. 33, n. 132, Dec. 1959: 289
The date, obtained from charcoal resulting from the burning of oak posts, of 2721 ± 150 B.C., indicates an earlier position for the Windmill cultures.

11.094 1959
RADIOCARBON DATING OF THE PILTDOWN SKULL AND JAW
de Vries, Hessel; Oakley, Kenneth P.
Nature, v. 184, n. 4682, 25 July 1959: 224-226
Radiocarbon dating has confirmed that the Piltdown skull (human) is post Pleistocene, probably less than 8000 years old; and that the Piltdown mandible (Orang-Utang) is younger rather than older.

11.095 1952
RADIOCARBON DATING RESULTS FROM THE OLD WORLD
McBurney, C.B.M.
Antiquity, v. 26, n. 101, Mar. 1952: 35-40
Old world results so far available vary quite considerably in their agreement with historically and geologically deduced dates. One practical issue of importance is a closer estimate of the relative accuracy of different kinds of organic materials, and the exact contribution of such sources of contamination as are likely to be met in the field.

11.096 1967
RADIOCARBON DATING, NEOLITHIC AND DARK AGES SETTLEMENTS ON HIGH PEAK, SIDMOUTH, DEVON
Pollard, Sheila H.H.
Devon Archaeological Society. Proceedings, v. 25, 1967: 41
[Not sighted]

11.097 1963
RECENT FINDS OF DANISH PREHISTORIC PLOUGHING IMPLEMENTS
Steensberg, A.
In: *Congrès International des Sciences Anthropologiques et Ethnologiques*, 6th, Paris, 1960. Vol.2. Paris, Musée de l'Homme, 1963: 363-367
A sample of wood from the plough-beam of a prehistoric plough of the Ard type, found in a Danish bog was dated by radiocarbon at 350 ± 100 B.C. that is the Early Celtic Iron age.

11.098 1954
RECENT WORK AT STONEHENGE
Piggott, Stuart
Antiquity, v. 28, n. 112, Dec. 1954: 221-224
Charcoal from an Aubrey hole gave a radiocarbon dating which indicated a probable range of dates between 2123 and 1573 B.C.

11.099 1968
SALIAGOS: A NEOLITHIC SITE IN THE CYCLADES
Renfrew, Colin; Evans, John D.
Archaeology, v. 21, n. 4, Oct. 1968: 262-271
The dating suggested by pottery for Saliagos is confirmed by radiocarbon dating at around 4000 B.C. Bracelet fragments made from *Spondylus* shells were used for radiocarbon analysis.

11.100 1958
THE SEMERMIUT EXCAVATIONS 1955
Mathiassen, Therkel
Meddeleser om Grønland, v. 161, n. 3, 1958. pp. 52
Investigation at Sermermiut proved the existence of two different and independent Palaeoeskimo cultures, the Sarqaq and the Dorset cultures, before the Neo-Eskimo bearers of the Thule culture occupied Disko Bugt. The three bearing horizons were separated by sterile layers, this indicates that the regions was unoccupied in the periods between the Sarqaq and the Dorset and after the Dorset had disappeard. The Sarqaq culture is dated at 3250 ± 110 years BP by radiocarbon measurements on charcoal from fireplaces and the Dorset culture is dated at 2016 ± 250 years BP by comparison with the Denbigh Flint complex.

11.101 1958
THE SKULDELEV SHIPS
Olsen, O.; Crumlin Perdersen, O.
Acta Archaeologica, v. 29, 1958: 161-175
Sheep's wool recovered from Ship II's caulking was dated at 910 ± 100 AD and thin oak branches at 960 ± 100 A.D. This agrees well with the typological features and is sufficient evidence for an approximate dating of the Skuldelev ships to between the end of the 9th and the middle of the 11th century A.D., the peak and the end of the Viking period.

11.102 1961
SKORBA, MALTA AND THE MEDITERRANEAN
Trump, David
Antiquity, v. 35, n. 140, Dec. 1961: 300-303
A discussion of the interpretation of the sequence of cultures in Malta in the context of the absolute chronology of the Neolithic period and the role which Malta played in the prehistory of the Central and West Mediterranean. A radiocarbon date of 2690 ± 150 B.C. ascribed to the Zebbug phase suggests an initial date fo the human occupation of Malta within the fourth millenium.

11.103   1955
SOME RADIOCARBON DATES FROM THE RAISED BOG NEAR EMMEN (NETHERLAND)
Van Zeist, Willem
Palaeohistoria, v. 4, 1955: 113-118
A peat monolith collected in the raised bog east of the town of Emmen was examined and radiocarbon determinations were made by Prof. Hessel de Vries. The dates are presented and analysed. A diagram of pollen profiles correlated with radiocarbon dates is presented.

11.104   1950
STAR CARR (editorial notes)
Crawford, O.G.S.
Antiquity, v. 24, n. 95, Sept. 1950: 113
Report of the first absolute date from the radiocarbon method, obtained from wood found in the Mesolithic settlement at Star Carr, excavated by Graham Clark, which agrees widely with dates already accepted for the Boreal period.

11.105   1962
THE STRATIGRAPHY OF THE MESOLITHIC SITES III AND IV AT THATCHAM, BERKSHIRE, ENGLAND
Churchill, D.M.
Prehistoric Society. Proceedings, n. s., v. 28, n. 14, 1962: 362-370
Occupation at the site is dated at between 7890 ± 160 B.C. to 7540 ± 160 B.C. at the start of the boreal period.

11.106   1967
THE TARTARIA TABLETS
Hood, M.S.F.
Antiquity, v. 41, n. 162, June 1967: 99-113
The discovery of three tablets in a Neolithic context at Tartaria in Romania has brought into sharper focus the discrepancy between dates based upon archaeological correlations and those based upon radiocarbon for the Neolithic of Southeast Europe.

11.107   1968
THE TARTARIA TABLETS
Hood, M.S.F.
Scientific American, v. 218, n. 5, May 1968: 30-37
The author discusses the validity of the radiocarbon dates for Neolithic Europe north of the Mediterranean in trying to resolve the difficult problem of the Tartaria tablets. These tablets, although bearing signs greatly resembling those of Summerian tablets, were discovered in the level of the Vinca culture which is Neolithic and therefore probably more than 1000 years older than the oldest Summerian tablets.

11.108   1968
THE TARTARIA TABLETS: A CHRONOLOGICAL ISSUE
Neustupny, Evzen
Antiquity, v. 42, n. 165, Mar. 1968: 32-35
The radiocarbon dates obtained for the beginning of the Aegean early bronze age are discussed. When correction for the influence of the Earth magnetic field is applied, the dates agree with archaeological correlations. The dates for the layer of the Tartaria tell in which clay tablets were found do not fit this chronology. It is shown that these layers were mixed.

11.109   1964
TWO RADIOCARBON DATES FROM A CLYDE-SOLWAY CHAMBERED CAIRN
McKie, Euan W.
Antiquity, v. 38, n. 149, Mar. 1964: 52-54
Both samples were collected some 7 in above the sterile underlying soil and were therefore laid down some time after the construction of the tomb. The dates suggest a span of as much as a thousand years for the use of the tomb.

CHAPTER TWELVE
14C ARCHAEOLOGY OCEANIA

12.001 1965
ABORIGINAL CARBON DATES FROM TASMANIA
Reber, Grote
Mankind, v. 6, n. 6, Nov. 1965: 264-268
Charcoal samples from the bottom layer of a number of kitchen middens in Tasmania were dated by radiocarbon. The results suggest that a minimum antiquity of between eight and nine thousand years is indicated for man in Tasmamania.

12.002 1955
ABORIGINAL MIDDEN SITES IN WESTERN VICTORIA DATED BY RADIOCARBON ANALYSIS
Gill, Edmund D.
Mankind, v. 5, n. 2, Sept. 1955: 51-55
An account of the significance of the sites from which two Australian samples which were submitted for dating to W.F. Libby. One is from Koroit beach north-west of Warrnambool, Victoria, the other one is from a kitchen midden near Goose Lagoon in Western Victoria.

12.003 1967
ADDITIONAL RADIOCARBON DATES FOR WESTERN POLYNESIA
Davidson, J.M.; Green, R.C.; Buist, A.G.; Peters, K.M.
Polynesian Society. Journal, v. 76, n. 2, June 1967: 223-230
Fourteen radiocarbon dates from archaeological sites in western Polynesia are presented and discussed.

12.004 1962
ADDITIONAL RADIOCARBON DATES FROM HAWAII AND THE SOCIETY ISLANDS
Emory, Kenneth P.
Polynesian Society. Journal, v. 71, n. 1, March 1962: 105-106
The dates are from the Puu Alii sand dune site on the island of Hawaii and range from 0 to 1835 years ago. The reason for this inconsistency is unknown. Other dates from Waiakukini cave shelter on Hawaii and Nualolo bluff shelter on Kauai and five dates from the Society Islands are also presented. These latter are quite late and suggest that the sites belong to the late period.

12.005 1962
ADVANCING FRONTIERS IN AUSTRALIAN ARCHAEOLOGY
Mulvaney, D.J.
Oceania, v. 33, n. 2, Dec. 1962: 135-138
Preliminary notes on the results of two excavations at two sites in South Central Queensland on Mount Moffatt cattle station are presented. At Kenniff cave, the oldest date for Australian cultural material is reported: 10,950 ± 170 B.C.

12.006 1955
THE AGE OF THE KEILOR MAN, AUSTRALIA
Gill, Edmund D.
Anthropos, v. 50, 1955: 417
It is clear that the Keilor Man lived during a recession of the last glaciation. A radiocarbon date of 8500 ± 25 years confirms the geological results and is also consistent with other radiocarbon datings in the same area.

12.007 1965
ARCHAEOLOGICAL AND GEOMORPHOLOGICAL INVESTIGATIONS ON MOUNT MOFFAT STATION, QUEENSLAND, AUSTRALIA
Mulvaney, D.J.
Prehistoric Society. Proceedings, v. 31, n. 8, 1965: 147-212
The chronology is based on 14 radiocarbon dates which are presented in table form.

12.008 1964
ARCHEOLOGICAL EXCAVATION OF ROCK SHELTER NO. 6, FROMM'S LANDING, SOUTH AUSTRALIA
Mulvaney, D.J.; Lawton, E.H.; Twidale, C.R.
Royal Society of Victoria. Proceedings, v. 77, part 2, 1964: 479-516
Radiocarbon dates of 1000 ± 91 B.C. and 1220 ± 94 B.C. establish the antiquity of a Murray River flood which is the highest on record.

12.009 1960
ARCHAEOLOGICAL EXCAVATIONS AT FROMM'S LANDING ON THE LOWER MURRAY RIVER
Mulvaney, D.J.
Royal Society of Victoria. Proceedings, v. 72, part 2, 1960: 53-85
Refers and correlates to various radiocarbon dates from other areas

12.010 1956
ARCHAEOLOGICAL EXCAVATIONS IN NEW CALEDONIA
Gifford, E.W.; Shutler, Dick
University of California. Anthropological Records, v. 18, n. 1, 1956: 1-148
A list and table of radiocarbon dates are presented and discussed. From this, rates of accumulation are deducted. The author notes that with the increase in number of radiocarbon dates from various areas of Oceania, a chronology of the region should emerge on a more stable basis.

12.011 1962
ARCHAEOLOGICAL EXCAVATIONS ON THE AIRE RIVER, OTWAY PENINSULA, VICTORIA
Mulvaney, D.J.
Royal Society of Victoria. Proceedings, v. 75, part 1, 1962: 1-15
Two small rock shelters were excavated. Both sites were stratified to a depth of several feet and radiocarbon dating upon charcoal from a depth of 6 feet gave an age estimation of 370 ± 45 years.

12.012 1962
ARCHAEOLOGICAL FIELD SURVEY WORK IN NORTHERN NEW SOUTH WALES
McBryde, Isabel
Oceania, v. 33, n. 1, Sept. 1962: 12-17
Although two occupation phases were dated for Seelands, (1920 ± 120 B.C. and A.D. 1040 ± 80) there is no definite evidence suggesting to which should the art of the site be assigned. A dated level at the Clarence Valley containing large flaked chopping and scraping tools dated 1920 ± 120 B.C., below a level containing microlith, dated at A.D. 1040 ± 80, suggest that for the Clarence Valley, the microlithic industry is comparatively recent.

12.013 1968
ARCHAEOLOGICAL INVESTIGATIONS IN THE GRAMAN DISTRICT
McBryde, Isabel
Archaeology and Physical Anthropology in Oceania, v. 3, n. 2, July 1968: 77-93
The radiocarbon date for site I indicates a period of occupation from the fourth millenium B.C. to the first century B.C. A discrepancy of over a thousand years for the same level in another trench only 7 feet away is unexplained but if it can be confirmed, this would make the assemblage in this trench the oldest backed-blade industry for New South Wales and probably Australia as a whole.

12.014 1962
ARCHAEOLOGY OF EASTER ISLAND. REPORT ON THE NORWEGIAN ARCHAEOLOGICAL EXPEDITION TO EASTER ISLAND AND THE EAST PACIFIC. VOL 1
Heyerdahl, Thor; Ferdon, Edwin N.
Monograph of the School of American Research and the Museum of New Mexico, Santa Fe. n. 24, London, Allen and Unwin, 1962
[Not sighted]

12.015 1961
THE ARCHAEOLOGY OF NUKU HIVA, MARQUESAS ISLANDS, FRENCH POLYNESIA
Suggs, Robert Carl
American Museum of Natural History. Anthropological Papers, v. 49, part 1, 1961: 1-205
The time scale is based on radiocarbon dates which are presented in table form. These dates are also discussed and compared with New Zealand dates from Moa Hunter sites.

12.016 1964
ARCHAEOLOGY ON EASTER ISLAND : A REVIEW ARTICLE
Duff, Roger
Polynesian Society. Journal, 73, n. 1, March 1964: 78-83
A three phase sequence involving at least 1100 years for the Ahu platforms on Easter Island, is based on radiocarbon dates.

12.017 1961
AUSTRALIAN RADIOCARBON DATES
Mulvaney, D.J.
Antiquity, v. 35, n. 137, Mar. 1961: 37-39
Discusses the significance of some of the first Australian radiocarbon dates. The paucity of material finds will make reliance on radiocarbon data very important.

12.018 1967
CHRONOLOGY OF HAWAIIAN FISH-HOOKS
Sinoto, Yosihiko H.
Polynesian Society. Journal, v. 71, n. 2, June 1962: 162-166
The importance of fish-hooks in reconstructing the past history of an area of Polynesia is outlined. In conjunction with radiocarbon dates it is possible to establish a true time-space perspective analysis.

12.019 1967
COASTAL DUNES AND FIELD ARCHAEOLOGY IN S.E. AUSTRALIA
Coutts, J.F
Archaeology and Physical Anthropology in Oceania, v. 2, n. 1, April 1967: 28

Pleistocene dune sequence at Wilson Promontory contains elements of so-called Bondaian culture dated by radiocarbon at 6550 ± 100 BP to 3060 ± 100 BP.

12.020   1960
THE CONFIRMATION OF AN EASTER LEGEND THROUGH ARCHAEOLOGY
Smith, Carlyle S.
In: Congrès International des Sciences Anthropologiques et Ethnologiques 6th, Paris, 1960. *Proceedings* Vol.2. Paris, Musée de l'Homme, 1963: 369
The legend of the traditional battle between the Hanau Momoku and the Hanau Eepe on Easter Island is confirmed by archaeological investigation. The date of ca 1680 based upon genealogical studies is confirmed by radiocarbon.

12.021   1957
CULTURE SUCCESSION IN SOUTHERN AUSTRALIA FROM LATE PLEISTOCENE TO THE PRESENT
Tindale, Norman B.
South Australian Museum Records, v. 13, n. 1, April 1957: 1-49
Summarises the work in Adelaide since 1928 to establish the archaeological culture succession. Gives newly available radiocarbon dates for the Kartan culture of Late Pleistocene, tracing changes through Tartangan, Pirrian and Mudukian cultures. It was noted that in Australia, freshwater shells are providing data very closely similar to modern wood standard.

12.022   1968
A DATED CULTURE SEQUENCE FOR THE SOUTH SYDNEY REGION OF NEW SOUTH WALES
Megaw, J. Vincent S.
Current Anthropology, v. 9, n. 4, Oct. 1968: 325-329
An aboriginal site at Curracurrang has been excavated by the author and radiocarbon dated. The dates are presented in table form and discussed. The span of time represented is from 5500 B.C. to 1750 A.D.

12.023   1957
A DATED TARTANGAN IMPLEMENT SITE FROM CAPE MARTIN, SOUTH-EAST OF SOUTH AUSTRALIA
Tindale, Norman B.
Royal Society of South Australia. Transactions, v. 80, May 1957: 109-123
Records the findings, at Cape Martin, South Australia, of an aboriginal camp-site of the Tartangan Culture which has been dated, by a radiocarbon test at the Dominion Physical Laboratory, Lower Hutt, N.Z., as having been occupied in 8700 ± 120 years BP.

12.024   1959
DATING AUSTRALIAN PREHISTORY
Mulvaney, D.J.
Nature, v. 184, n. 4691, 26 Sept. 1959: 918
The Devon rock shelter in the Lower Murray valley of South Australia excavated to a depth of 18 ft in 1928 was dated on mussel shells by radiocarbon at 4250 ± 180 years. It contains Pirris (unifaced points), known from surface collections all over the continent but not since European settlements. Two stratigraphic levels excavated in the same area (Fromm's Landing), level 4, 6 ft from the surface, were dated respectively at 3240 ± 80 years and 4850 ± 100 on mussel shells. They also contained Pirris and Microliths.

12.025   1955
DATING OF NEW ZEALAND'S PREHISTORY
Golson, J.
Polynesian Society. Journal, v. 64, n. 1, March 1955: 113-136
Notes the major contribution that radiocarbon dating will provide to establish New Zealand Prehistoric chronology.

12.026   1968
DATING OF RECENT LOW SEA-LEVEL AND MAORI ROCK CARVING, ONGARI POINT
Schofield, J.C.
Earth Science Journal (Waikato Geological Society), v. 2, n. 2, 1968: 167-174
A sea-level of at least 1.5 ft below the present one, and contemporaneous rock carvings have a radiocarbon date of 180 ± 50 years BP

12.027   1957
A DECADE OF DISCOVERY: PACIFIC
Spoehr, Alexander
Archaeology, v. 10, n. 3, Autumn 1957: 236-237
According to radiocarbon dating it appears that the Pacific Islands were settled much earlier than previously thought. The following early dates are presented: New Caledonia, 847 B.C.; Marianas, 1527 B.C.; Fiji, 46 B.C.; Hawaii, A.D.1004.

12.028   1968
EARLY POTTERY OBJECTS FROM FIJI
Birks, Lawrence; Birks, Helen
Polynesian Society. Journal, v. 77, n. 3, Sept. 1968: 296-299
The earliest occupation zone, yielding pottery fragments, in the coastal sand dune at Sigatoka in the South West of Viti Levu, Fiji, was dated by radiocarbon determination on a sample of charcoal 51 ± 90 B.C.

12.029 1967
EARLY STONE AXES IN ARNHEM LAND
White, Carmel
Antiquity, v. 41, n. 162, June 1967: 149-152
The radiocarbon dates suggest that man was in Australia twenty thousand years ago, during the maximum Würm glaciation, and that he then was making edge-ground axes, for so long regarded as a diagnostic feature of the Neolithic.

12.030 1965
EXCAVATIONS IN THE ROYAL NATIONAL PARK, NEW SOUTH WALES: A FIRST SERIES OF RADIOCARBON DATES FROM THE SYDNEY DISTRICT
Megaw, J. Vince
Oceania, v. 35, n. 3, March 1965: 202-207
A series of radiocarbon dates on undisturbed wood and charcoal samples in a rock shelter at Curracurrang is presented.

12.031 1958
FOREST HISTORY AND NEW ZEALAND PREHISTORY
McKelvey, P.J.
New Zealand Science Review, v. 16, n. 3-4, 1958: 28-32
Radiocarbon dating has helped in interpreting forest development and in recognition of stages in forest succession. The various factors which influenced the development and fluctuations of forests in New Zealand are examined.

12.032 1967
FOSSIL MAN IN AUSTRALIA
Macintosh, N.W.G.
Australian Journal of Science, v. 30, n. 3, 3 Sept. 1967: 86-98
An overview of the state of knowledge about fossil man in Australia. Radiocarbon dating helps put back the arrival of man in Australia as early as 30,000 years.

12.033 1968
THE GEOGRAPHICAL BACKGROUND TO THE ARRIVAL OF MAN IN AUSTRALIA AND TASMANIA
Jones, Rhys
Archaeology and Physical Anthropology in Oceania, v. 3, n. 3, Oct. 1968: 186-215
In this review of early man in Australia and Tasmania and of the physical and climatic background to his arrival on the continent the author uses radiocarbon dates to establish the antiquity of the presence of man in Australia, antedating the colonisation of the Americas by some 10,000 years.

12.034 1959
GEOLOGICAL AND ARCHAEOLOGICAL INTERPRETATION OF A SECTION IN RANGITOTO ASH ON MOTUTAPU ISLAND, AUCKLAND
Brothers, R.N.; Golson, J.
New Zealand Journal of Geology and Geophysics, v. 2, n. 3, 1959: 569-577
A foredune section at Pig Bay, Motutapu, has been excavated to expose non-volcanic beach sand below a single Rangitoto ash layer which, in turn, is overlain by a series of wind-blown and water-lain sediments containing reworked ash. The date of the eruption was determined by dating of carbon samples at A.D. 188 ± 50 years. The sediment above the Rangitoto ash include a number of well defined human occupation levels some of which have yielded adzes of Moa-hunter type.

12.035 1957
GEOLOGICAL TECHNIQUES IN DATING NEW ZEALAND PREHISTORY
Kear, D.
New Zealand Science Review, v. 15, n. 9-10, 1957: 76-79
A shell bed on the Hauraki Plains, about 10 ft above sea-level, has been radiocarbon dated at 2270 ± 70 years BP. This retreat of the sea, which can be recognised by several independent methods, is a suitable base for New Zealand's archaeological time scale. The use of the radiocarbon method of dating is illustrated.

12.036 1967
GREEN GULLY BURIAL
Bowler, J.M.; Mulvaney, D.J.; Casey, D.A.; Darragh, T.A.
Nature, v. 213, n. 5072, 14 Jan. 1967: 152-154
Human bones and implements in the Keilor Terrace of the Marybyrnong river, near Melbourne, provided new evidence of the antiquity of man in Australia. Radiocarbon dates of two samples of charcoal collected near the bones, 3 ft 6 below and 1 ft above the bodies, gave dates of 8155 ± 130 BP. These dates are discussed with regard to the stratigraphic sequence.

12.037 1967
HORTICULTURE IN NEW GUINEA HIGHLANDS: C14 DATING
Lampert, Ronald J.
Antiquity, v. 41, n. 164, Dec. 1967: 307-309
Radiocarbon dates recently published for New Guinea indicate that horticulture has been practised in the highlands for a considerable time. The author comments on the relation of this knowledge to other recent research in this area.

12.038   1966
THE LAST DECADE IN NEW ZEALAND ARCHAEOLOGY, PART I AND II
Golson, J.; Gathercole, P.W.
New Zealand Archaeological Association Newsletter, v. 9, n. 1, March 1966: 4-19
Reviews the progress of archaeology in New Zealand since 1950. On the evidence of radiocarbon dates, Moa hunter settlements were well established along the Eastern Seaboard from Auckland to Bluff, by about A.D. 1200-1350.(see *Antiquity*, 35, 1962)

12.039   1958
LES MARQUISES : PREHISTORY OF POLYNESIA
Shapiro, H.L.
Natural History, v. 67, n. 5, May 1958: 262-271
A small series of radiocarbon dates for the Hawaiian Islands, Easter Island, Rapa and the Marquesas provide a basis for a chronology of Polynesian prehistory. More dates are needed from Tahiti, Samoa and Tonga.

12.040   1957
MARIANAS PREHISTORY: ARCHAEOLOGICAL SURVEY AND EXCAVATIONS ON SAIPAN, TINIAN AND ROTA
Spoehr, Alexander
Fieldiena: Anthtropology, v. 48, 24 June 1957:
A radiocarbon date from a site on Tinian gives a result of A.D. 845 ± 145, the first evidence found of the duration of the Latte building period. Another date from Chalan Piao on Saipan, if assumed reliable, indicates that the Marianas were settled by 1527 ± 200 B.C.

12.041   1961
MOOREAN ARCHAEOLOGY: A PRELIMINARY REPORT
Green Roger
Man, v. 61, article n. 200, 1961: 169-173
The first dates obtained from samples of charcoal from the base of cultural deposits located in Papeto'ai village suggest an occupation of only 800 years, which do not satisfy those who, on theoretical ground, predict a much longer period of settlement in the Society Islands.

12.042   1961
NEOLITHIC DIFFUSION RATES
Edmonson, Munro S.
Current Anthropology, v. 2, n. 2, Apr. 1961: 71-102
Presents a model for the diffusion rate of culture traits, using radiocarbon dates and archaeological evidence as a basis. The radiocarbon dates are described and assessed for errors.

12.043   1965
NEW ABORIGINAL CARBON DATES FROM TASMANIA
Reber, Grote
Mankind, v. 6, n. 6, June 1967: 435-437
The new dates present ages from 1000 to 3000 years, other sites indicate a much longer occupation. Some dates are definitely erroneous due to equipment fault.

12.044   1966
NEW DATES FOR PYRAMID VALLEY MOAS
Gregg, D.R.
New Zealand Archaeological Association Newsletter, v. 9, n. 4, Dec. 1966: 155-159
Recently determined radiocarbon dates have enabled definite ages to be assigned to some of the Pyramid valley moas. This has led to an interpretation of the history of the Pyramid Valley deposit.

12.045   1961
NEW RADIOCARBON DATES FOR AUSTRALIA
McBryde, Isabel
Antiquity, v. 35, n. 140, Dec. 1961: 312-313
Two dates from a rock shelter overlooking the river Clarence, near Grafton on the north coast of New South Wales, provide the earliest evidence for human occupation and date the uniface pebble and microlithic industry in the region.

12.046   1967
A NOTE ON CARBON DATES FOR HORTICULTURE IN THE NEW GUINEA HIGHLANDS
Golson, J.; Lampert, Ronald J.; Wheeler, J.M.; Ambrose, W.R.
Polynesian Society Journal, v. 76, n. 3, Sep. 1967: 369-371
Two radiocarbon dates relating to an early part of the stratigraphy in a drain cutting on the Manton property, near Mount Hagen, give dates for horticulture well before the period at which the sweet potato is assumed, on present evidence, to have arrived in New Guinea. Whatever the crop grown at the Manton site 2000 years before, this cultivation has important implications for recent theories of the evolution of Highland agricultural systems.

12.047   1968
ON THE RELIABILITY OF CHARCOAL FOR RADIOCARBON DATING NEW ZEALAND ARCHAEOLOGICAL SITES
Trotter, Michael M.
New Zealand Archaeological Association Newsletter, v. 11, n. 2, June 1968: 86-88
It is the opinion of the author that charcoal is not reliable for dating archeological remains as dates obtained from charcoal relate not to the time of human occupation of the site but

12.048 1967
ORIGINS OF THE MELANESIANS
Shutler, Mary Elizabeth; Shutler, Richard
Archaeology and Physical Anthropology, v. 2, n. 2, July 1967: 91-99
A review of the hypotheses concerning the origin of the Melanesians. Several radiocarbon dates support the presence of Melanesian culture in the islands of the New Hebrides as early as 900 B.C.

12.049 1962
AN OUTLINE OF EASTER ISLAND ARCHAEOLOGY
Smith, Carlyle S.
Asian Perspectives, v. 6, n. 1-2, Summer-Winter 1962: 239-243
The time spans for three periods are established by radiocarbon dates: Early period from before 400 to about 1100 A.D.: Middle period from ca 1100 to 1680 A.D. Late period from ca 1680 A.D. to 1868 A.D. (1868 dates the arrival of the Missionaries).

12.050 1961
PEOPLING IN THE PACIFIC IN THE LIGHT OF RADIOCARBON DATING
Shutler, Richard
Asian Perspective, v. 5, n. 2, Winter 1961: 201-216
An attempt to evaluate radiocarbon dates from the Pacific and their significance in constructing a geological framework of the migration into this wide expanse of ocean (Review of a paper presented at the 10th, Pacific Science Congress, Honolulu, August 1961).

12.051 1966
PIG BONES FROM TWO ARCHAEOLOGICAL SITES IN THE NEW GUINEA HIGHLANDS
Bulmer, Susan
Polynesian Society Journal, v. 75, n. 4, Dec. 1966: 504-505
Radiocarbon dating suggests that the shelters were inhabited at periods from approximately 8400 ± 140 B.C. to some considerable time after 2890 ± 140 B.C.

12.052 1964
THE PLEISTOCENE COLONIZATION OF AUSTRALIA
Mulvaney, D.J.
Antiquity, v. 38, n. 151, Sept. 1964: 263-267
Summarises the positive evidence for the peopling of the continent of Australia, in the light of radiocarbon dates.

to the time at which the plant died.

12.053 1966
A POSSIBLE TREND IN RADIOCARBON DATES IN AUSTRALIA
McCarthy, Fred D.
Australian Institute of Aboriginal Studies Newsletter, v. 2, n. 3, Jan. 1966: 27
A hundred or so radiocarbon dates available from prehistoric sites in Australia provide evidence about the possible migration routes or the direction of spread of the Aborigines over the continent.

12.054 1966
THE PREHISTORY OF THE AUSTRALIAN ABORIGINE
Mulvaney, D.J.
Scientific American, v. 214, n. 3, Mar. 1966: 84-93
Radiocarbon dating helps establish the fact that man first reached Australia at least 16,000 years ago.

12.055 1968
PRELIMINARY REPORT OF EXCAVATION ON WATON ISLAND
Specht, Jim
Polynesian Society. Journal, v. 77, n. 2, June 1968: 117-134
A tentative reconstruction of the site history, based on radiocarbon dates, subsoil contouring and observed stratigraphy is offered.

12.056 1966
PROVENANCE AND AGE OF THE KEILOR CRANIUM: OLDEST KNOWN HUMAN SKELETAL REMAINS IN AUSTRALIA
Gill, Edmund D.
Current Anthropology, v. 7, n. 5, Dec. 1966: 581-584
The stratigraphic provenance of the Keilor cranium is established and radiocarbon dates of materials in a similar stratigraphical position are presented, indicating an age of $18,000 \pm 500$ years BP

12.057 1966
RADIOCARBON DATES (AUSTRALIA)
Australian Institute of Aboriginal Studies. Newsletter, v. 2, n. 4, Oct. 1966: 50
Corrections to radiocarbon dates in the list which appeared in the Institute Newsletter, v. 2, n. 2, 1966.

12.058 1961
A RADIOCARBON DATE FROM MT. WELLINGTON
Golson, J.
New Zealand Archaeological Association Newsletter, v. 4, n. 2, March 1961: 51
The date of 1430 ± 40 A.D. falls well within the period of

Archaic occupation on the nearby island of Motutapu and close to the date of Archaic site at Mercury Bay.

12.059   1965
RADIOCARBON DATES FOR ARCHAEOLOGICAL SITES IN THE CLARENCE VALLEY, NORTHERN NEW SOUTH WALES
McBryde, Isabel
Oceania, v. 35, n. 4, June 1965: 260-266
The radiocarbon dates obtained for samples from various sites in the Clarence Valley of New South Wales are discussed.

12.060   1955
RADIOCARBON DATES FOR AUSTRALIAN ARCHAEOLOGICAL AND GEOLOGICAL SAMPLES
Gill, Edmund D.
Australian Journal of Science, v. 18, n. 2, 21 Oct. 1955: 49-52
A chart presents in table form the results of radiocarbon assays of samples of late Quaternary age from Australia. It endeavours to interpret radiocarbon dating for Australia.

12.061   1966
RADIOCARBON DATES FOR NORTHERN NEW SOUTH WALES
McBryde, Isabel
Antiquity, v. 40, n. 160, Dec. 1966: 285-292
Further news on radiocarbon dating in Northern New South Wales. The significance of the dates for the regional sequence and for the cultural succession for Eastern Australia as a whole is discussed.

12.062   1967
RADIOCARBON DATES FOR SIGATOKA, FIJI
Birks, Lawrence; Birks, Helen
New Zealand Archaeological Association Newletter, v. 10, n. 2, June 1967: 69
Five new dates for Sigatoka, Fiji, are presented. They appear to confirm the view previously advanced that the early deposit in the rock shelter site on Yanuka Island would antedate the Ohine site near the mouth of the Sigatoka River.

12.063   1967
RADIOCARBON DATES FOR SVAI'I, WESTERN SAMOA
Buist, A.G.
New Zealand Archaeological Association Newsletter, v. 10, n. 2, June 1967: 70
Three dates are presented. The first two dates confirm the conclusion reached on field evidence that village B at Ologogo was occupied in the contact period and that village C is prehistoric.

12.064   1965
RADIOCARBON DATES FOR WESTERN SAMOA
Green, R.C.; Davidson, J.K.
Polynesian Society. Journal, v. 74, n. 1, March 1965: 63-69
Seven charcoal samples from archaeological investigations in Western Samoa were dated by radiocarbon. The dates are presented in table form and discussed.

12.065   1967
RADIOCARBON DATES FROM CURRACURRANG CAVE, NEW SOUTH WALES
Megaw, J. Vincent S.
Australian Institute of Aboriginal Studies. Newsletter, v. 2, n. 5, Apr. 1967: 26-30
Five samples of charcoal were submitted to the radiocarbon dating laboratory at the Gakushuin University, Tokyo. The dates form a consistent pattern with ages ranging from 2230 ± 80 to 1430 ± 90 years BP, and one modern, which fits with the other dates already obtained. A table of 16 dates for the Curracurrang cave is presented and the emerging chronology discussed.

12.066   1968
RADIOCARBON DATES FROM GUAM, MARIANA ISLANDS
Reinman, Fred M.
Polynesian Society. Journal, v. 77, n. 1, March 1968: 80-82
As a result of an archaeological survey and series of test excavations on the Island of Guam in 1965-1966, twenty four samples were dated by radiocarbon. The samples came from Inarajan Village, Talofofo River, Nomna Bay, Fouha Bay and Pulantat. The dates are presented in table form. The material dated is mostly charcoal, and some pottery.

12.067   1964
RADIOCARBON DATES FROM NEW GUINEA
Bulmer, Susan
Polynesian Society Journal, v. 73, n. 3, Sept. 1964: 327-328
A series of dates for prehistoric material from New Guinea is presented. The earliest date $10,350 \pm 140$ BP is further support for the thesis of a late Pleistocene age for human settlement of the then connected Australia - New Guinea landmass.

12.068   1967
RADIOCARBON DATES FROM NORTH OTAGO
Trotter, Michael M.
New Zealand Archaeological Association Newsletter, v. 10, n. 4, Dec. 1967: 137-142

The dates presented are assessed; site descriptions are included. The author warns of the danger of accepting radiocarbon ages uncritically.

12.069 [1962 ?]
RADIOCARBON DATES FROM RAP ITI, RAIVAVAE AND THE MARQUESAS
Smith, C.S.
Norwegian Archaeological Expedition to Easter Island and the East Pacific. Report, n. 2, [1962 ?]
[Not sighted]

12.070 1964
RADIOCARBON DATES OF INTEREST TO AUSTRALIAN ARCHAEOLOGISTS
Tindale, Norman B.
Australian Journal of Science, v. 27, n. 1, July 1964: 24
Dates from the Noola rock shelter, Lake Menindee, Cape Northumberland, and Shellharbour are discussed.

12.071 1967
RADIOCARBON DATING AND MAN IN SOUTH-EAST ASIA, AUSTRALIA AND THE PACIFIC
Shutler, Richard
In: *Archaeology of the Eleventh Pacific Science Congress*, edited by W.G. Solheim. University of Hawaii. Social Science Research Institute, 1967 (Asian and Pacific Archaeology series, n. 1): 79-87
Discusses dated cultural sequences in these areas.

12.072 1965
RADIOCARBON DATING AND PALAEO-ECOLOGY OF THE AITAPE FOSSIL HUMAIN REMAINS
Hossfeld, Paul S.
Royal Society of Victoria. Proceedings, v. 78, part 2, 1965: 161-165
Fossils associated with human cranial remains from the Aitape region, New Guinea, show that the enclosing mudstone was deposited in a coastal mangrove swamp. Radiocarbon dating of organic materials in the fossiliferous mudstone gave an age of about 5000 years BP.

12.073 1965
RADIOCARBON DATING IN THE PACIFIC
Shutler, Richard
In: *International Conference on Radiocarbon and Tritium Dating*, Pullman, Washington, Washington State University, June 1-11, 1965. *Proceedings*. United States of America, Atomic Energy Commission, 1965. Conference n. 650652: 264-276
A number of cultural sequences have already been established for the Pacific Island areas and some of these securely dated by the radiocarbon process. The earliest evidence for man in the Pacific areas, based on radiocarbon dating comes from the Philippines, Sarawak and Australia.

12.074 1966
RECENT AUSTRALIAN RADIOCARBON DATES
Australian Institute of Aboriginal Studies. Newsletter, v. 2, n. 3, Jan. 1966: 20-26
A list of archaeological dates for various areas of Australia.

12.075 1967
RECENT AUSTRALIAN RADIOCARBON DATES
Polach, Henry A.
Australian Institute of Aboriginal Studies Newsletter, v. 2, n. 6, Oct. 1967: 20-35
A list of radiocarbon dates for Australian prehistory.

12.076 1961
REPORT ON AUSTRALIA AND MELANESIA
McCarthy, Fred D.
Asian Perspective, v. 5, n. 2, 1961: 141-155
An examination of datable material and radiocarbon dates is presented with a list of the existing radiocarbon dates so far obtained from sites of undoubted human occupation.

12.077 1966
REPORT ON EXCAVATION IN THE SOUTH SYDNEY DISTRICT, 1964-1965
Megaw, J. Vincent S.
Australian Institute of Aboriginal Studies. Newsletter, v. 2, n. 3, Jan. 1966: 4-15
Radiocarbon dates indicate the general termini ante and post quem for the backed blade industry of 2360 ± 90 BP and 840 ± 70 BP. Comparable figures from New England (I. McBryde) are 300-500 BP.

12.078 1964
SEA-LEVEL CHANGES IN THE PAST 6000 YEARS: POSSIBLE ARCHAEOLOGICAL SIGNIFICANCE
Shepard, Francis P.
Science, n. 143, v. 3606, 7 Feb. 1964: 574-576
Evidence from many stable areas indicates that sea-level has risen slowly during the past 6000 years, with a total change of about 6 meters. Since the same period is also important in the history of man, the rise in sea-level explains the widespread submergence of building sites and other human relics along coasts where ancient man lived. A correlation of elevation and ages obtained from radiocarbon dates from various areas of America, the Pacific Islands and Australia document this sea-level rise.

12.079   1955
SIX FIJIAN RADIOCARBON DATES
Gifford, E.W.
Polynesian Society Journal, v. 64, n. 2, June 1955: 240
The dates were made on charcoal and are from two sites on Viti Levu in Lautauka and Ra provinces.

12.080   1961
THE STONE AGE OF AUSTRALIA
Mulvaney, D.J.
Prehistoric Society. Proceedings, n. s, v. 27, article n. 4, 1961: 56-107
The radiocarbon dates associated with various sites in Australia are presented in table form and discussed.

CHAPTER THIRTEEN
14C CONFERENCES

13.001 1960
THE 1959 CARBON-14 SYMPOSIUM AT GRONINGEN
Waterbolk, H.T.
Antiquity, v. 34, n. 133, March 1960: 14-18
The results of the Radiocarbon Conference at Groningen, 15-19 September 1959, are summarised and some of the problems of discrepancy are examined.

13.002 1965
CARBON-14 AND TRITIUM DATING
Olson, Edwin A.; Chatters, Roy M.
Science, v. 150, n. 3702, 10 Dec. 1965: 1488-1492
A review of the conference which took place in Pullman, Washington, from 7-11 June 1965.

13.003 1959
CARBON-14 DATING CONFERENCE AT GRONINGEN, SEPTEMBER 14-19, 1959
Godwin, Harry
Nature, v. 184, n. 4696, Oct. 1959: 1365-1366
Survey of methods in use at various laboratories: scintillation counting (Saclay-paraldehyde), Methanol (Trinity College, Dublin). Fluctuation in radiocarbon content of atmosphere in the past 1200 years, glaciation dating, co-ordination of work between laboratories. Acceptance of oxalic acid standard of U.S. Bureau of Standards.

13.004 1954
CARBON-14 SYMPOSIUM IN COPENHAGEN, SEPTEMBER 1-4, 1954
Godwin, Harry
Nature, v. 174, n. 4436, 6 Nov. 1954: 868
Various methods of counting are compared: solid carbon and various gas countings (carbon dioxide, acetylene, methane). Sources of error and standardisation of publishing results are also discussed.

13.005 [1962 ?]
THE FIFTH INTERNATIONAL CONFERENCE ON RADIOCARBON DATING
Rubin, Meyer
Geotimes, n. d. (1962 ?): 32
A report on the Conference held at the University of Cambridge, England, July 23-28, 1962.

13.006 1965
INTERNATIONAL CONFERENCE ON RADIOCARBON AND TRITIUM DATING, PROCEEDINGS
Compiled by Roy M. Chatters and Edwin A. Olson.
International Conference on Radiocarbon and Tritium Dating, Pullman, Washington, Washington State University, June 1-11, 1965. *Proceedings*. United States of America, Atomic Energy Commission, 1965. Conference n. 650652. pp. 784
Individual papers on a wide range of topics have been indexed individually.

13.007 1968
LOESS AND RELATED EARLIER DEPOSITS OF THE WORLD
Schultz, Bertrand C.; Frye, John C.
Loess and Related Earlier Deposits of the World, edited by C. Bertrand Schultz and John C. Frye. International Association for Quaternary Research (INQUA). Congress, 7th, Boulder, Col., 14 Aug - 19 Sept 1965. *Proceedings*, v. 12. Lincoln, Neb., University of Nebraska Press, 1968: pp. 367
Papers and reports, several of which contain mention of Radiocarbon dating. Individual papers are indexed separately.

13.008 1954
LOW LEVEL COUNTING KEY TO ADVANCES IN RADIOCARBON DATING
Kulp, J. Lawrence
Nucleonics, v. 12, n. 12, Dec. 1954: 19-21
Correlation of the many recent advances in radiocarbon dating was achieved at the conference held in Andover, Mass., in October 1954. The basic assumptions and problems in radiocarbon dating were summarised, counting techniques were described, and interpretations of dating results were presented.

13.009 1967
RADIOACTIVE DATING AND METHODS OF LOW LEVEL COUNTING
Radioactive Dating and Methods of Low-level Counting. Proceedings of a Symposium organised by the International Atomic Energy Agency (IAEA) in co-operation with the Joint Commission on Applied Radioactivity (ICSU) and held in Monaco, 2-10 March 1967. Vienna, IAEA, 1967. (Proceedings Series). pp. 744
Fifty seven papers on radiocarbon dating and methods of low level counting are presented. Separate abstracts are presented for each relevant paper.

13.010   1967
RADIOACTIVE DATING AND METHODS OF LOW-LEVEL COUNTING
Walton, Alan
Antiquity, v. 41, n. 164, Dec. 1967: 317-318
Report on a Symposium on Radioactive Dating and Methods of Low Level Counting held in Monaco, 2-10 March 1967. Some of the problems discussed are: dating of bone samples, application of bomb-produced radiocarbon to geochemical and geophysical processes.

13.011   1958
RADIOCARBON DATES
Science, v. 127, n. 3307, 16 May 1958: 1166
Notice of formation of a Committee for Distribution of Radiocarbon Dates at the Conference on Radiocarbon dating, Andover, Massachusetts, October 1956. Peabody Foundation, Box 7, Andover. In charge of the committee was Frederick Johnson, R.J. Peabody Foundation.

13.012   1957
RADIOCARBON DATING
Johnson, Frederick; Arnold, James R.; Flint, Richard Foster
Science, v. 125, n. 3241, 8 Feb. 1957: 240-242
Report of the International Conference on Radiocarbon Dating held at Andover, Massachusetts, 1-4 October, 1956. The problems of mixing times and contemporary assay were discussed, together with variations in accuracy of radiocarbon dates. Various applications were discussed: regional stratigraphy and chronology, climatic variations, sea-level changes, development of human culture, time scale. Technical developments and methods were also discussed.

13.013   1959
RADIOCARBON DATING
Nature, v. 184, n. 4700, 28 Nov. 1959: 1686-1687
Announcement of the publication of Radiocarbon, a yearly supplement of the American Journal of Science, entirely devoted to the publication of date lists and allied items.

13.014   1955
RADIOCARBON DATING CONFERENCE IN CAMBRIDGE
Levi, Hilde
Nature, v. 176, n. 4485, 15 Oct. 1955: 727-728
The conference was almost entirely devoted to the discussion of new methods, e.g. gas counting of carbon dioxide, acetylene and methane, and liquid scintillation counting in which the sample to be counted is incorporated into a scintillator. Discussion of error and error computation and standardisation of the form of publication of date lists were also discussed.

13.015   1962
RADIOCARBON DATING. FIFTH INTERNATIONAL CONFERENCE AT CAMBRIDGE, JULY 23-28, 1962
Godwin, Harry
Nature, v. 195, n. 4845, 8 Sept. 1962: 943-945
The following topics were discussed: redetermination of half-life of radiocarbon agreement on 5730 ± 40; measure of material of known age in order to determine initial activity as constant; comparison with tree ring dating; confirmation of past deviations; decision not to adopt new half-life value and to make statement of age in years BP.

13.016   1960
RADIOCARBON SUPPLEMENT: AMERICAN JOURNAL OF SCIENCE
Putman, J.L.
International Journal of Applied Radiation and Isotopes, v. 7, n. 7, 1960: 336
Announces the publication of the Radiocarbon Supplement to the American Journal of Science as a forum for the publication of radiocarbon measurements.

CHAPTER FOURTEEN
14C DATE LISTS

14.001   1965
14C DATING. I
Hsu, Yuin-Chi; Huang, Chia-Yi; Hsy, Yi-Chuan
Chinese Journal of Physics (Taiwan), v. 3, Oct. 1965: 120-122
This date list covers many of the datings made by using a 1000 mL carbon dioxide-filled proportional counter with a multiple anode councidence ring counter.

14.002   1963
AGE DETERMINATIONS BY THE RADIOCARBON METHOD. IV
Vinogradov, A.P.; Devirts, A.L.; Dobkina, E.I.; Markova, N.G.V.
Geochemistry, 1963, n. 9: 823-840
Method: liquid scintillation counting on benzene. Projects: Upper Pleistocene chronology for European Russia and dating of recent fossil soils. Materials dated: peat and wood.

14.003   1968
ALL-UNION GEOLOGICAL INSTITUTE RADIOCARBON DATES I
Arslanov, Kh. A.; Gromova, L.I.; Rudneyev. Yu. A.
Radiocarbon, v. 10, n. 2, 1968: 448-450
Method: Liquid scintillation counting on benzene.

14.004   1967
ANU RADIOCARBON DATE LIST I
Polach, Henry A.; Stipp, Jerry J.; Golson, J.; Lovering, John F.
Radiocarbon, v. 9, 1967: 15-27
Proportional gas counting on methane and also liquid scintillation counting on benzene. Radiocarbon half-life: 5568 years. Results expressed as years BP and calendar years. Short description of laboratory and methods included.

14.005   1968
ANU RADIOCARBON DATE LIST II
Polach, Henry A.; Golson, J.; Lovering, John F.; Stipp, J. J.
Radiocarbon, v. 10, n. 2, 1968: 179-199
Measurements carried out between January and August 1967. Half life: 5568 years. Liquid scintillation counting on benzene.

14.006   1962
ARIZONA RADIOCARBON DATES III
Damon, Paul E.; Long, Austin
Radiocarbon, v. 4, 1962: 239-249
Proportional gas counting on carbon dioxide.

14.007   1963
ARIZONA RADIOCARBON DATES IV
Damon, Paul E.; Long, Austin; Sigalove, Joel J.
Radiocarbon, v. 5, 1963: 283-301
Reporting period: Dec. 1961 - Oct. 1962. Proportional gas counting on carbon dioxide. Radiocarbon half-life: 5570 years. The sample descriptions are presented in a classified sequence.

14.008   1964
ARIZONA RADIOCARBON DATES V
Damon, Paul E.; Haynes, C. Vance; Long, Austin
Radiocarbon, v. 6, 1964: 91-107
Reporting period: Oct. 1962 - Nov. 1963. Sample descriptions are presented in a classified sequence. Description of pretreatment technique included.

14.009   1966
ARIZONA RADIOCARBON DATES VI
Haynes, C. Vance; Damon, Paul E.; Grey, Donald C.
Radiocarbon, v. 8, 1966: 1-21
Reporting period: 1 November 1963 - 15 November 1965. Radiocarbon half-life: 5568 years. Results expressed as years BP and calendar years.

14.010   1967
ARIZONA RADIOCARBON DATES VII
Haynes, C. Vance; Grey, Donald C.; Damon, Paul E.; Bennett, Richmond
Radiocarbon, v. 9, 1967: 1-14
Reporting period: 15 November 1965 to 15 June 1966. Proportional gas counting on carbon dioxide. Radiocarbon half-life: 5568. Results expressed as years BP and calendar years.

14.011   1964
BERLIN RADIOCARBON MEASUREMENTS I
Kohl, G.; Quitta, Hans
Radiocarbon, v. 6, 1964: 308-317
Reporting period: since 1961. Proportional gas counting on acetylene. Radiocarbon half-life: 5568 years. Results expressed in years BP and calendar years.

14.012   1966
BERLIN RADIOCARBON MEASUREMENTS II
Kohl, G.; Quitta, Hans
Radiocarbon, v. 8, 1966: 27-45

Reporting period: since 1963. Radiocarbon half-life: 5570 years. Results expressed as years BP and calendar years.

14.013   1959
BERN RADIOCARBON DATES I
Oeschger, Hans; Schwartz, U.; Gfeller, Chr.
Radiocarbon, v. 1, 1959: 133-143
Reporting period: up to Summer 1958. Proportional counting on acetylene.

14.014   1961
BERN RADIOCARBON DATES II
Gfeller, Chr.; Oeschger, Hans; Schwartz, U.
Radiocarbon, v. 3, 1961: 15-25
Reporting period: Spring 1959 to Summer 1960. Proportional gas counting on acetylene. Radiocarbon half life: 5568 years.

14.015   1963
BERN RADIOCARBON DATES III
Gfeller, Chr.; Oeschger, Hans
Radiocarbon, v. 5, 1963: 305-311
Reporting period: Summer 1960 - Summer 1962. Proportional gas counting on methane. Radiocarbon half-life: 5568 years.

14.016   1965
BERN RADIOCARBON DATES IV
Oeschger, Hans; Riesen, T.
Radiocarbon, v. 7, 1965: 1-9
Reporting period: since Summer 1962. Proportional gas counting on methane. Results expressed in years BP and calendar years.

14.017   1966
BERN RADIOCARBON DATES V
Oeschger, Hans; Riesen, T.
Radiocarbon, v. 8, 1966: 22-26
Reporting period: 1965. Proportional gas counting on carbon dioxide. Results expressed as years BP and calendar years.

14.018   1967
BERN RADIOCARBON DATES VI
Oeschger, Hans; Riesen, T.
Radiocarbon, v. 9, 1967: 28-34
Reporting period: from Summer 1965.

14.019   1967
BIRMINGHAM UNIVERSITY RADIOCARBON DATES I
Shotton, Frederick William; Blundell, D.J.; Williams, R.E.G.
Radiocarbon, v. 9, 1967: 35-37
Reporting period: since October 1966. Proportional gas counting on methane. Radiocarbon half-life: 5570 years. Results expressed in years BP and calendar years.

14.020   1968
BONN RADIOCARBON MEASUREMENTS I
Scharpenseel, H.W.; Pietig, F.; Tamers, Murray A.
Radiocarbon, v. 10, n. 1, 1968: 8-28
Period covered: From 1966 to date. Radiocarbon half-life: 5588 years. Method: liquid scintillation counting on benzene. Materials: mainly soils. Projects: Some archaeological dates and modern samples of plant material to monitor radiocarbon from nuclear weapon contamination.

14.021   1959
BRITISH MUSEUM NATURAL RADIOCARBON MEASUREMENTS I
Barker, Harold; Mackey, C.J.
Radiocarbon, v. 1, 1959: 81-86
Proportional gas counting on acetylene. Short discussion of technique and apparatus used, background and standard. Radiocarbon half-life: 5568 ± 30 years.

14.022   1960
BRITISH MUSEUM RADIOCARBON MEASUREMENTS I
Barker, Harold; Mackey, John
British Museum Quarterly, v. 22, n. 3/4, Apr. 1960: 94-101
Method: proportional gas counting on acetylene. A description of the method and of the application of the standard deviation is included.

14.023   1960
BRITISH MUSEUM NATURAL RADIOCARBON MEASUREMENTS II
Barker, Harold; Mackey, C.J.
Radiocarbon, v. 2, 1960: 26-30
Proportional gas counting on acetylene. Radiocarbon half-life: 5568 ± 30 years.

14.024   1961
BRITISH MUSEUM RADIOCARBON MEASUREMENTS II
Barker, Harold; Mackey, John
British Museum Quarterly, v. 23, n. 4, June 1961: 118-123
Period reported: to May 1960

14.025   1961
BRITISH MUSEUM NATURAL RADIOCARBON MEASUREMENTS III
Barker, Harold; Mackey, John
Radiocarbon, v. 3, 1961: 39-45

Radiocarbon half-life: 5568 ± 30 years. Proportional gas counting on acetylene.

14.026   1963
BRITISH MUSEUM RADIOCARBON MEASUREMENTS III
Barker, Harold; Mackey, John
British Museum Quarterly, v. 27, 1963: 48-58
Period reported: June 1960 to Aug. 1961. Radiocarbon half life: 5568 years. Results expressed in years BP (1950) and Christian Calendar.

14.027   1963
BRITISH MUSEUM NATURAL RADIOCARBON MEASUREMENTS IV
Barker, Harold; Mackey, John
Radiocarbon, v. 5, 1963: 104-108
Reporting period: to the end of 1961. Radiocarbon half-life: 5568 years. Results reported both in years BP and in terms of Christian calendar.

14.028   1968
BRITISH MUSEUM NATURAL RADIOCARBON MEASUREMENTS V
Barker, Harold; Mackey, John
Radiocarbon, v. 10, n. 1, 1968: 1-7
Period covered: Sept. 1962 to August 1964. Radiocarbon half-life: 5570 years. Method: measurements on acetylene gas with proportional counter. Projects: archaeological samples.

14.029   1959
CAMBRIDGE UNIVERSITY NATURAL RADIOCARBON MEASUREMENTS I
Godwin, Harry; Willis, E.H.
Radiocarbon, v. 1, 1959: 63-75
Reporting period, 1958. Proportional counting of carbon dioxide. Projects: British glacial and late glacial periods.

14.030   1960
CAMBRIDGE UNIVERSITY NATURAL RADIOCARBON MEASUREMENTS II
Godwin, Harry; Willis, E.H.
Radiocarbon, v. 2, 1960: 62-72
Reporting period 1959. Proportional gas counting on carbon dioxide.

14.031   1961
CAMBRIDGE UNIVERSITY NATURAL RADIOCARBON MEASUREMENTS III
Godwin, Harry; Willis, E.H.
Radiocarbon, v. 3, 1961: 60-76
Reporting period: 1960

14.032   1961
CAMBRIDGE UNIVERSITY NATURAL RADIOCARBON MEASUREMENTS IV. NUCLEAR-WEAPON TESTING AND THE ATMOSPHERIC RADIOCARBON CONCENTRATION
Godwin, Harry; Willis, E.H.
Radiocarbon, v. 3, 1961: 77-80
Measurements of annual rings of *Populus nigra*, 1953 - 1959. Measurements of tropospheric radiocarbon activity in 1960.

14.033   1962
CAMBRIDGE UNIVERSITY NATURAL RADIOCARBON MEASUREMENTS V
Godwin, Harry; Willis, E.H.
Radiocarbon, v. 4, 1962: 57-70
Reporting period: 1961. Proportional gas counting on carbon dioxide.

14.034   1964
CAMBRIDGE UNIVERSITY NATURAL RADIOCARBON MEASUREMENTS VI
Godwin, Harry; Willis, E.H.
Radiocarbon, v. 6, 1964: 116-137
Proportional gas counting on carbon dioxide.

14.035   1965
CAMBRIDGE UNIVERSITY NATURAL RADIOCARBON MEASUREMENTS VII
Godwin, Harry; Willis, E.H.; Switsur, V.R.
Radiocarbon, v. 7, 1965: 205-212
Reporting period: 1966 - 1964. Proportional gas counting on carbon dioxide.

14.036   1966
CAMBRIDGE UNIVERSITY NATURAL RADIOCARBON MEASUREMENTS VIII
Godwin, Harry; Switsur, V.R.
Radiocarbon, v. 8, 1966: 390-400
Reporting period: 1964 - 1965. Proportional gas counting on carbon dioxide.

14.037   1968
CANADIAN ARCHAEOLOGICAL RADIOCARBON DATES
Wilmeth, Rosco
National Museum of Canada. Bulletin, v. 232, Anthropological Series n.87, 1968: 68-127
Published and unpublished Canadian radiocarbon dates are compiled in a new list and classified according to the geographical location of their respective sites.

14.038   1955
CARBON-14 AGE DETERMINATIONS AT THE UNIVERSITY OF SASKATCHEWAN

McCallum, K.G.
Royal Society of Canada. Transactions, Third series, section 4, v. 49, 1955: 31-39
The original Libby technique has been used to determine ages of materials at the University of Saskatchewan. Some of the limitations and difficulties which were encountered are discussed. A list of some radiocarbon dates on a number of the materials which were investigated is appended.

14.039   1959
CARBON-14 DATING IN PISA
Ferrara, G.; Reinharz, M.; Tongiorgi, E.
Radiocarbon, v. 1, 1959: 103-110
Reporting period: 1958. Proportional counting on acetylene. Description of the techniques and apparatus used are given. Ages not corrected for Suess effect.

14.040   1961
CARBON-14 DATING IN PISA - II
Ferrara, G.; Fornaca-Rinaldi, G.; Tongiorgi, E.
Radiocarbon, v. 3, 1961: 99-104
Proportional gas counting on acetylene.

14.041   1952
CHICAGO RADIOCARBON DATES III
Libby, Willard F.
Science, v. 116, n. 3024, 12 Dec. 1952: 673-681
Period covered: September 1952 - September 1952. Radiocarbon half-life: 5568. Counting time: 48 hours.

14.042   1954
CHICAGO RADIOCARBON DATES IV
Libby, Willard F.
Science, v. 119, n. 3083, 27 June 1954: 119
Period covered: September 1952 to September 1953. Radiocarbon half-life: 5568 years.

14.043   1954
CHICAGO RADIOCARBON DATES V
Libby, Willard F.
Science, v. 120, n. 3123, 5 Nov. 1954: 733-742
Period covered: September 1953 to September 1954. Radiocarbon half-life: 5568 years.

14.044   1965
CONTRIBUCIONES DEL LABORATORIO DE GEOCRONOMETRIA - PT. 3. COMPENDO DE EDADES DE RAIOCARBONE DE MUESTRAS MEXICANAS DE 1962 A 1964.
Valencia, Josefina; Fries, Carl
Mexico Universita. Nacional Automana Instituto Geologia. Boletin, n. 73, 1965: 135-191
Since the publication of Boletin 64 of this series in 1962, 127 additional age samples have been collected from Baja California to Yucatan, including some dredged from the sea bottom off the coasts of Mexico. These are tabulated here together with brief descriptions of the samples and discussions of significance of the results. All dates will be published in Radiocarbon.

14.045   1953
COPENHAGEN NATURAL RADIOCARBON MEASUREMENTS I
Anderson, Ernest C.; Levi, Hilde; Tauber, Henrik
Science, v. 118, n. 3053, 3 July 1953: 6-9
Method: double screen wall counter. Radiocarbon half-life: 5568. Errors: standard deviation of counting data. Material dated: mainly gyttja. Samples: late glacial period.

14.046   1956
COPENHAGEN NATURAL RADIOCARBON MEASUREMENTS II
Tauber, Henrik
Science, v. 124, n. 3227, 2 Nov. 1956: 879-881
Period Covered: February 1953 to May 1955. Method: solid carbon as per Libby's. Projects: neolithic period in Switzerland and Denmark.

14.047   1960
COPENHAGEN NATURAL RADIOCARBON MEASUREMENTS III, CORRECTIONS TO RADIOCARBON DATES MADE WITH THE SOLID CARBON TECHNIQUE
Tauber, Henrik
Radiocarbon, v. 2, 1960: 5-111
A short discussion on the Suess effect in Denmark is included.

14.048   1960
COPENHAGEN NATURAL RADIOCARBON MEASUREMENTS IV
Tauber, Henrik
Radiocarbon, v. 2, 1960: 12-25
Proportional gas counting of carbon dioxide. Radiocarbon half-life: 5568 ± 30 years. Results expressed in years BP.

14.049   1962
COPENHAGEN RADIOCARBON DATES V
Tauber, Henrik
Radiocarbon, v. 4, 1962: 27-34
Reporting period: 1956-1961. Radiocarbon half-life: 5568 ± 30 years. Results expressed in years BP.

14.050   1964
COPENHAGEN RADIOCARBON DATES VI
Tauber, Henrik
Radiocarbon, v. 6, 1964: 215-225

Reporting period: up to November 1963. Radiocarbon half-life 5570 ± 30 years. Results expressed in years BP and calendar years.

14.051 1966
COPENHAGEN RADIOCARBON DATES VII
Tauber, Henrik
Radiocarbon, v. 8, 1966 213-234
Reporting period: up to November 1965. Radiocarbon half-life: 5570. Results expressed as years BP and calendar years.

14.052 1967
COPENHAGEN RADIOCARBON ME ASUREMENTS VIII GEOGRAPHIC VARIATIONS IN ATMOSPHERIC C14 ACTIVITY
Tauber, Henrik
Radiocarbon, v. 9, 1967: 246-256
Reporting period 1956 to 1966. Materials dated: cereal and grass (northern hemisphere). Project: measurements of bomb-produced radiocarbon.

14.053 1968
COPENHAGEN RADIOCARBON DATES IX
Tauber, Henrik
Radiocarbon, v. 10, n. 2, 1968: 295-327
A selected number of measurements made up to October 1967. Radiocarbon half-life: 5570 years.

14.054 1961
DUBLIN RADIOCARBON DATES I
McAuley, I.R.; Watts, W.A.
Radiocarbon, v. 3, 1961: 26-38
Reporting period: early 1958 to early 1960. Liquid scintillation counting on methyl alcohol.

14.055 1962
FIRST RADIOCARBON DATES FROM THE USSR
Chard, Chester S.
Arctic Anthropology, v. 1, n. 1, 1962: 84-86
Dating between September 1959 and April 1960 at the Radiocarbon Laboratory of the Institute of Archaeology in Leningrad. Method: Liquid scintillation counting.

14.056 1966
FLORIDA STATE UNIVERSITY RADIOCARBON DATES I
Stipp, Jerry J.; Knauer, G.A.; Goodell, H. Grant
Radiocarbon, v. 8, 1966: 46-53
Reporting period: since February 1963. Liquid scintillation counting on benzene. Radiocarbon half-life: 5570 years. Results expressed as years BP and calendar years. A short description of procedures is included.

14.057 1967
FLORIDA STATE UNIVERSITY RADIOCARBON DATES II
Knauer, G.A.; Martin, J.R.; Goodell, H. Grant; Phelps, D.S.
Radiocarbon, v. 9, 1967: 38-42
Reporting period: from September 1965. Proportional gas counting on acetylene.

14.058 1962
GAKUSHUIN NATURAL RADIOCARBON MEASUREMENTS I
Kigoshi, Kunihiko; Tomikura, Yoshio; Endo, Kunihiko
Radiocarbon, v. 4, 1962: 84-94
Reporting period 1959-1961. Proportional gas counting on acetylene.

14.059 1963
GAKUSHUIN NATURAL RADIOCARBON MEASUREMENTS II
Kigoshi, Kunihiko; Endo, Kunihiko
Radiocarbon, v. 5, 1963: 109-171
Reporting period: November 1961 to October 1962. Radiocarbon half-life: 5570±30 years. Proportional gas counting on acetylene.

14.060 1964
GAKUSHUIN NATURAL RADIOCARBON MEASUREMENTS III
Kigoshi, Kunihiko; Lin, Der-Hwang; Endo, Kunihiko
Radiocarbon, v. 6, 1964: 197-207
Reporting period: Nov. 1962 - Oct. 1963. Gas proportional counting on acetylene. Radiocarbon half-life: 5570 ± 30 years. Results expressed in years BP.

14.061 1965
GAKUSHUIN NATURAL RADIOCARBON MEASUREMENTS IV
Kigoshi, Kunihiko; Kobayashi, Hiromi
Radiocarbon, v. 7, 1965: 10-23
Reporting period: November 1963 - October 1964. Radiocarbon half-life: 5570±30 years. Results expressed in years BP and calendar years.

14.062 1966
GAKUSHUIN NATURAL RADIOCARBON MEASUREMENTS V
Kigoshi, Kunihiko; Kobayashi, Hiromi
Radiocarbon, v. 8, 1966: 54-73
Reporting period: September 1964 to October 1965. Radiocarbon half-life: 5570±30 years. Results expressed as years BP and calendar years.

14.063 1967
GAKUSHUIN NATURAL RADIOCARBON MEASUREMENTS VI

Kigoshi, Kunihiko
Radiocarbon, v. 9, 1967: 43-62
Radiocarbon half-life: 5570 ± 30. Results expressed as years BP and calendar years.

14.064 1965
GEOCHRON LABORATORIES INC. RADIOCARBON MEASUREMENTS
Krueger, Harold; Weeks, C. Francis
Radiocarbon, v. 7, 1965: 47-53
Reporting period: 1964. Short description of equipment and procedure is included.

14.065 1966
GEOCHRON LABORATORIES INC. RADIOCARBON MEASUREMENTS II
Krueger, Harold W.; Weeks, C. Francis
Radiocarbon, v. 8, 1966: 142-160
Reporting period: 1964-1965. Results expressed as years BP and calendar years.

14.066 1968
GEOLOGICAL INSTITUTE RADIOCARBON DATES I - III
Cherdyntsev, V.V.; Alekseyev, V.A.; Kind, N.V.; Forova, V.S.; Sulerzhitskiy, L.D.
Radiocarbon, v. 10, n. 2, 1968: 419-445
Method: Liquid scintillation counting on benzene. A brief description of the chemical treatment is given.

14.067 1962
GEOLOGICAL SURVEY OF CANADA RADIOCARBON DATES I
Dyck, Willy; Fyles, J.G.
Radiocarbon, v. 4, 1962: 13-26
Reporting date, January to November 1962. Radiocarbon half-life: 5568 ± 30 years. Proportional gas counting on carbon dioxide. A short description of apparatus and procedure is included.

14.068 1963
GEOLOGICAL SURVEY OF CANADA RADIOCARBON DATES II
Dyck, Willy; Fyles, J.G.
Radiocarbon, v. 5, 1963: 39-55
Reporting period: December 1961 to November 1962. Results expressed in years BP.

14.069 1963
GEOLOGICAL SURVEY OF CANADA RADIOCARBON DATES 1 & 2
Dyck, Willy; Fyles, J.G.
Canada. Geological Survey of Canada. Paper, n. 63-21. pp. 31
Reports the first age determinations in the radiocarbon laboratory of the Geological Survey of Canada, completed from January to November 1961. Apparatus and procedures are briefly described. Half-life of radiocarbon used is 5568 ± 30. Method: carbon dioxide proportional counting.

14.070 1964
GEOLOGICAL SURVEY OF CANADA RADIOCARBON DATES III
Dyck, Willy; Fyles, J.G.
Radiocarbon, v. 6, 1964: 167-181
Proportional gas counting on carbon dioxide. Radiocarbon half-life: 5570 ± 30 years. A new pretreatment on shell samples was introduced, which removes the outer 20%, because measurements on samples which had only had the outer 10% removed gave consistently older ages on the innermost fraction.

14.071 1965
GEOLOGICAL SURVEY OF CANADA RADIOCARBON DATES IV
Dyck, Willy; Fyles, J.G.; Blake, Weston
Radiocarbon, v. 7, 1965: 24-45
Radiocarbon half-life: 5568 ± 30 years. Proportional gas counting on carbon dioxide. Results expressed in years BP and calendar years. Description of a new 5 L counter and laboratory procedures are included.

14.072 1966
GEOLOGICAL SURVEY OF CANADA RADIOCARBON DATES V
Dyck, Willy; Lowdon, J.A.; Fyles, J.G.; Blake, Weston
Radiocarbon, v. 8, 1966: 96-127
Reporting period: 1965. Radiocarbon half-life: 5568 ± 30 years. Results expressed as years BP and calendar years. A number of measurements were made to test the validity of radiocarbon dates from bones.

14.073 1967
GEOLOGICAL SURVEY OF CANADA RADIOCARBON DATES VI
Lowdon, J.A.; Fyles, J.G.; Blake, Weston
Radiocarbon, v. 9, 1967: 156-197
Reporting period: October 1965 to September 1966. Radiocarbon half-life: 5568 ± 30 years. Results expressed as years BP and calendar years.

14.074 1967
GEOLOGICAL SURVEY OF CANADA RADIOCARBON DATES VI
Lowdon, J.A.; Fyles, J.G.; Blake, Weston
Geological Survey of Canada. Department of Energy,

Mines and Resources. Papers, n. 67-2, part B, 1967. pp. 42 Reports ages determined in the Radiocarbon Dating Laboratory of the Geological Survey of Canada between December 1965 and November 1966. (Reprinted from *Radiocarbon*, v. 9, 1967: 156-197)

14.075   1968
GEOLOGICAL SURVEY OF CANADA RADIOCARBON DATES VII
Lowdon, J.A.; Blake, Weston
Geological Survey of Canada. Department of Energy, Mines and Resources. Papers, n. 68-2, part B, 1963
Report ages determined in the Radiocarbon Dating Laboratory of the Geological Survey between December 1966 and November 1967. (Reprinted from *Radiocarbon,* v. 10, 1968: 207-245)

14.076   1968
GEOLOGICAL SURVEY OF CANADA RADIOCARBON DATES VII
Lowdon, J.A.; Blake, Weston
Radiocarbon, v. 10, n. 2, 1968: 207-248
Period covered: 1967. Half-life: 5568 ± 30 years. Lists of average background and standard counting, and tests for radiocarbon contamination in shell samples are presented.

14.077   1962
GEOLOGICAL SURVEY OF FINLAND RADIOCARBON MEASUREMENTS I
Hyyppa, E.; Hoffren, Vaino; Isola, A.
Radiocarbon, v. 4, 1962: 81-83
Proportional gas counting on carbon dioxide. Radiocarbon half-life: 5570 years.

14.078   1963
GEOLOGICAL SURVEY OF FINLAND RADIOCARBON MEASUREMENTS II
Hyyppa, E.; Isola, A.; Hoffren, Vaino
Radiocarbon, v. 5, 1963: 302-304
Method: proportional gas counting on carbon dioxide. Results expressed as years B.C.

14.079   1964
GEOLOGICAL SURVEY OF FINLAND RADIOCARBON MEASUREMENTS III
Hyyppa, E.; Toivonen, A.V.P.; Isola, A.
Radiocarbon, v. 6, 1964 110-111
Proportional gas counting on carbon dioxide.

14.080   1966
GIF NATURAL RADIOCARBON MEASUREMENTS II
Delibrias, Georgette; Guillier, M.T.; Labeyrie, Jean
Radiocarbon, v. 8, 1966: 74-95
Reporting period: 1963-October 1965. Radiocarbon half-life: 5570 years. Results expressed as years BP and calendar years. The laboratories of Saclay and Gif-sur-Yvette have joined together under the name of Gif.

14.081   1966
GIF-SUR-YVETTE NATURAL RADIOCARBON MEASUREMENTS I
Coursaget, Jean; Le Run, J.
Radiocarbon, v. 8, 1966: 128-141
Reporting period: 1958 to March 1963. Radiocarbon half-life: 5570 years. Proportional gas counting on carbon dioxide. Results expressed as years BP and calendar years.

14.082   1958
GRONINGEN RADIOCARBON DATES II
de Vries, Hessel; Barendsen, G.W. ; Waterbolk, H.T.
Science, v. 127, n. 3290, 17 Jan. 1958: 129-137
Period covered: August 1954 to March 1956. Method: Proportional couting on carbon dioxide. Radiocarbon half-life: 5570 years. Projects: geological problems, geological samples from northern Europe, archaeological samples from Europe, Africa and America.

14.083   1958
GRONINGEN RADIOCARBON DATES III
de Vries, Hessel; Waterbolk, H.T.
Science, v. 128, n. 3338, 19 Dec. 1958: 1550-155
Period covered: March 1956 - August 1957. Method: proportional gas counting of carbon dioxide. Ages reported in years BP.

14.084   1963
GRONINGEN RADIOCARBON DATES IV
Vogel, John C.; Waterbolk, H.T.
Radiocarbon, v. 5, 1963: 163-202
A random selection of dates measured since 1958. Radiocarbon half-life: 5570 years. Results expressed in years BP.

14.085   1964
GRONINGEN RADIOCARBON DATES V
Vogel, John C.; Waterbolk, H.T.
Radiocarbon, v. 6, 1964: 349-369
Reporting period: 1964.

14.086   1967
GRONINGEN RADIOCARBON DATES VI
Vogel, John C.; Zagwijn, W.H.
Radiocarbon, v. 9, 1967: 63-106
Dates related to the chronology of the last glaciation. Results expressed as years BP and calendar years.

14.087 1967
GRONINGEN RADIOCARBON DATES VII
Vogel, John C.; Waterbolk, H.T.
Radiocarbon, v. 9, 1967: 107-155
Proportional gas counting on carbon dioxide. All dates are archaeological dates. Descriptions of laboratory and equipment are included.

14.088 1962
HANNOVER RADIOCARBON MEASUREMENTS I
Wendt, Immo; Schneekloth, Heinrich; Budde, Enno
Radiocarbon, v. 4, 1962: 100-108
Reporting period: to the end of 1960. Proportional gas counting on acetylene.

14.089 1962
HANNOVER RADIOCARBON MEASUREMENTS II
Geyh, Mebus A.; Schneekloth, Heinrich; Wendt, Immo
Radiocarbon, v. 4, 1962: 137-143
Reporting period: 1961. Proportional gas counting on acetylene.

14.090 1964
HANNOVER RADIOCARBON MEASUREMENTS III
Geyh, Mebus A.; Schneekloth, Heinrich
Radiocarbon, v. 6, 1964: 251-268
Reporting period: January 1962 - March 1963. Proportional gas counting on acetylene. Results expressed in years BP and calendar years.

14.091 1967
HANNOVER RADIOCARBON MEASUREMENTS IV
Geyh, Mebus A.
Radiocarbon, v. 9, 1967: 198-217
Reporting period: March 1963 to July 1966. Proportional gas counting on acetylene and ethane. Radiocarbon half-life: 5568 years. Results expressed as years BP and calendar years.

14.092 1967
HANNOVER RADIOCARBON MEASUREMENTS V
Geyh, Mebus A.
Radiocarbon, v. 9, 1967: 218-236
Reporting period: March 1963 - July 1966.

14.093 1957
HEIDELBERG NATURAL RADIOCARBON MEASUREMENTS I
Münnich, K.O.
Science, v. 126, n. 3266, 2 Aug. 1957: 194-199
Period: March 1954 to 1956. Method: Proportional gas counting. Standard: 19th century wood. Ages reported in years.

14.094 1957
HUMBLE OIL COMPANY RADIOCARBON DATES I
Brannon, H.R.; Daughtry, A.C.; Perry, D.; Simons, L.H.; Whitaker, W.W.; Williams, Milton
Science, v. 125, n. 3239, 25 Jan. 1957: 147-150
Method: proportional counting of carbon dioxide. Material: carbonized wood and plant material, bones, antlers, charcoal. Age reported in years. Radiocarbon half-life: 5568 ± 30. Projects: archaeological problems, absolute chronology of sedimentary events.

14.095 1957
HUMBLE OIL COMPANY RADIOCARBON DATES II
Brannon, H.R.; Simons, L.H.; Perry, D.; Daughtry, A.C.; McFarland, E.
Science, v. 125, n. 3254, 10 May 1957: 919-923
Method: proportional counting of carbon dioxide. Radiocarbon half-life: 5568 ± 30 years. Ages reported in years. Projects: geological problems, late Quaternary development of Southern Louisiana, sea-level changes. Material dated: wood, peat, shells.

14.096 1968
INSTITUT ROYAL DU PATRIMOINE ARTISTIQUE RADIOCARBON DATES I
Schreurs, Anne Nicole
Radiocarbon, v. 10, n. 1, 1968: 29-35
Dating was started in 1966. Method: Gas counting of carbon dioxide. A description of the apparatus and method is given. Uses an Oeschger counter.

14.097 1965
INSTITUTO VENEZOLANO DE INVESTIGACIONES CIENTIFICAS NATURAL RADIOCARBON MEASUREMENTS I
Tamers, Murray A.
Radiocarbon, v,7, 1965: 54-65
Reporting period: 1964. Liquid scintillation counting on benzene. Radiocarbon half-life: 5568 years. Results expressed in years BP and calendar years.

14.098 1966
INSTITUTO VENEZOLANO DE INVESTIGACIONES CIENTIFICAS NATURAL RADIOCARBON MEASUREMENTS II
Tamers, Murray A.

Radiocarbon, v. 8, 1966: 204-212
Reporting period: December 1964 to November 1965. Liquid scintillation counting on benzene. Radiocarbon half-life: 5568 years. Results expressed as years BP and calendar years.

14.099   1967
INSTITUTO VENEZOLANO DE INVESTIGACIONES CIENTIFICAS NATURAL RADIOCARBON MEASUREMENTS III
Tamers, Murray A.
Radiocarbon, v. 9, 1967: 237-24
Reporting period: 1966. Liquid scintillation dating on benzene. Radiocarbon half-life: 5568 years. Results reported as years BP and calendar years.

14.100   1968
ISOTOPES RADIOCARBON MEASUREMENTS
Buckley, J.D.; Trautman, Milton A.; Willis, E.H.
Radiocarbon, v. 10, n. 2, 1968: 246-294
Gas proportional counting on carbon dioxide.

14.101   1961
ISOTOPES, INC. RADIOCARBON MEASUREMENTS I
Walton, Alan; Trautman, Milton A.; Friend, James P.
Radiocarbon, v. 3, 1961: 47-59
Reporting period 1957 - 1960. Proportional gas counting on carbon dioxide. Radiocarbon half-life: 5568 ± 30 years. Brief description of pretreatment and counting techniques, instrumentation, and age calculation method.

14.102   1962
ISOTOPES, INC. RADIOCARBON MEASUREMENTS II
Trautman, Milton A.; Walton, Alan
Radiocarbon, v. 4, 1962: 35-42
Reporting period: 1961. Proportional gas counting on carbon dioxide.

14.103   1968
ISOTOPES, INC. RADIOCARBON MEASUREMENTS III
Trautman, Milton A.
Radiocarbon, v. 5, 1963: 62-79
Reporting period 1961 and 1962.

14.104   1964
ISOTOPES, INC. RADIOCARBON MEASUREMENTS IV
Trautman, Milton A.
Radiocarbon, v. 6, 1964: 269-279
Reporting period: 1962 - 1963. Proportional gas counting on carbon dioxide. Results expressed in years BP and calendar years.

14.105   1966
ISOTOPES, INC. RADIOCARBON MEASUREMENTS V
Trautman, Milton A.; Willis, Eric H.
Radiocarbon, v. 8, 1966: 161-203
Reporting period: 1963 - 1965. Proportional gas counting on carbon dioxide. Results reported as years BP and calendar years.

14.106   1968
KHLOPIN INSTITUTE RADIOCARBON DATES I
Arslanov, Kh. A.
Radiocarbon, v. 10, n. 2, 1968: 446-447
Seven dates presented.

14.107   1966
KOLN RADIOCARBON MEASUREMENTS I
Schwabedissen, H.; Freundlich, J.
Radiocarbon, v. 8, 1966: 239-247
Reporting period: middle 1963 to May 1964. Proportional gas counting on carbon dioxide. Radiocarbon half-life: 5570 years. Results expressed as years BP and calendar years. Short description of sample preparation is included.

14.108   1960
LA JOLLA NATURAL RADIOCARBON MEASUREMENTS I
Hubbs, Carl L.; Bien, George S.; Suess, Hans E.
Radiocarbon, v. 2, 1960: 197-223
Reporting period: August 1957 to 1959. Proportional gas counting on acetylene. A classified subject index of the bearings of each date on various events and processes is included.

14.109   1962
LA JOLLA NATURAL RADIOCARBON MEASUREMENTS II
Hubbs, Carl L.; Bien, George S.; Suess, Hans E.
Radiocarbon, v. 4, 1962: 204-238
Reporting period: mid-1957 to 1959. A subject index relating to the dates is included.

14.110   1963
LA JOLLA NATURAL RADIOCARBON MEASUREMENTS III
Hubbs, Carl L.; Bien, George S.; Suess, Hans E.
Radiocarbon, v. 5, 1963: 254-272
Reporting period: 1962. Three different counters were used and results compared.

14.111 1965
LA JOLLA NATURAL RADIOCARBON MEASUREMENTS IV
Hubbs, Carl L.; Bien, George S.; Suess, Hans E.
Radiocarbon, v. 7, 1965: 66-117
Reporting period: 1963-1964. Radiocarbon half-life: 5570 years. Results expressed in years BP and calendar years. The measurements are reported in a structured sequence under 7 headings. A number of the measurements reported were made to test the reliability and precision of dates.

14.112 1967
LA JOLLA NATURAL RADIOCARBON MEASUREMENTS V
Hubbs, Carl L.; Bien, George S.
Radiocarbon, v. 9, 1967: 261-294
Reporting period: December 1964 - November 1966. Proportional gas counting on acetylene. Results reported as years BP and calendar years. Short description of methods included.

14.113 1951
LAMONT NATURAL RADIOCARBON MEASUREMENTS I
Kulp, J. Lawrence; Feely, Herbert W.; Tryon, Lansing E.
Science, v. 114, n. 2970, 30 November 1951: 565-568
Radiocarbon measurements on natural carbon-bearing materials of geological and archaeological interest. Radiocarbon half-life 5568. Table 1 gives data on known samples used for calibration.

14.114 1950
LAMONT NATURAL RADIOCARBON MEASUREMENTS II
Broecker, Wallace S.; Kulp, J. Lawrence; Tucek, C.S.
Science, v. 124, n. 3213, 27 July 1956: 154-165
Period covered: 1953. Method: black carbon and carbon dioxide gas counting. Projects: geological, archaeological and deep sea cores. Ages given in years BP.

14.115 1952
LAMONT NATURAL RADIOCARBON MEASUREMENTS III
Kulp, J. Lawrence; Tryon, Lansing E.; Eckelman, Walter R.; Snell, W.A.
Science, v. 116, n. 3016, 17 Oct. 1952: 404-414
Period covered November 1951 - August 1952. An advance in technique permits dating back to 30,000 years, an increase of 5000 years. A discussion of the factors to be considered to construct a satisfactory hypothesis for the anomaly shown by shell samples is included (shells give dates which are too old).

14.116 1957
LAMONT NATURAL RADIOCARBON MEASUREMENTS IV
Broecker, Wallace S.; Kulp, J. Lawrence
Science, v. 126, n. 3287, 27 Dec. 1957: 1324-1334
Period covered; October 1955 - March 1957. Method: carbon dioxide gas proportional counting. Material dated: peat, wood, limnic peat, organic peat, marine shells, oolite sand, charcoal, land shells, ash, tree rings, lithoid tufa.

14.117 1958
LAMONT NATURAL RADIOCARBON MEASUREMENTS V
Olson, Edwin A.; Broecker, Wallace S.
Radiocarbon, v. 1, 1959, 1-28
Reporting period: February 1957 - July 1958. Main categories of research: North American glacial geology, pluvial lake level, relative sea-level changes. Discussions on the following topics are presented: age calculations, new methods of chemical pretreatment designed to detect and overcome possible sample contamination, the effect of material type on the reliability of a radiocarbon date.

14.118 1959
LAMONT RADIOCARBON MEASUREMENTS VI
Broecker, Wallace S.; Olson, Edwin A.
Radiocarbon, v. 1, 1959: 111-132
Dating of samples of known age. The measurements were made in order to gain an understanding of the distribution of radiocarbon within the dynamic carbon reservoir both today and at times in the past. A discussion of natural radiocarbon variations and of the best way to express it follows together with a discussion of the sources of variation in radiocarbon.

14.119 1961
LAMONT NATURAL RADIOCARBON MEASUREMENTS VII
Olson, Edwin A.; Broecker, Wallace S.
Radiocarbon, v. 3, 1961: 141-175
Reporting period: July 1958 - November 1960. Proportional gas counting on carbon dioxide.

14.120 1961
LAMONT RADIOCARBON MEASUREMENTS VIII
Broecker, Wallace S.; Olson, Edwin A.
Radiocarbon, v. 3, 1961: 176-204
A list containing only samples of known age. The measurements were made in order to gain an understanding of the distribution of radiocarbon within the dynamic carbon reservoir.

14.121   1962
LOUVAIN NATURAL RADIOCARBON MEASUREMENTS I
Dossin, J.M.; Deumer, J.M.; Capron, P.C.
Radiocarbon, v. 4, 1962: 95-99
Reporting period: May to December 1961. Proportional gas counting on carbon dioxide. Radiocarbon half-life: 5760 years. Results expressed in years BP.

14.122   1964
LOUVAIN NATURAL RADIOCARBON MEASUREMENTS II
Deumer, J.M.; Gilot, E.; Capron, P.C.
Radiocarbon, v. 6, 1964: 160-166
Reporting period: July 1962 - Oct. 1963. Proportional gas counting on methane. Radiocarbon half-life: 5570 years. Result expressed in years BP and calendar years.

14.123   1965
LOUVAIN NATURAL RADIOCARBON MEASUREMENTS III
Gilot, E.; Ancion, N.; Capron, P.C.
Radiocarbon, v. 7, 1965: 118-122
Reporting period: 1964. Radiocarbon half-life: 5570 years. Results expressed in years BP and calendar years.

14.124   1967
LOUVAIN NATURAL RADIOCARBON V
Gilot, E.
Radiocarbon, v. 9, 1967: 295-300
Reporting period: 1966. Radiocarbon half-life: 5570 years. Results expressed as years BP and calendar years.

14.125   1968
LOUVAIN NATURAL RADIOCARBON MEASUREMENTS VI
Gilot, E.
Radiocarbon, v. 10, n. 1, 1968: 55-60
Period covered: 1966-1967. Method: Gas proportional counting on methane. Radiocarbon half-life: 5570 years. Projects: palynological studies. Materials dated: peat.

14.126   1966
LOUVAIN RADIOCARBON MEASUREMENTS IV
Gilot, E.; Ancion, N.; Capron, P.C.
Radiocarbon, v. 8, 1966: 248-255
Proportional gas counting on methane. Radiocarbon half-life: 5570 years. Results expressed as years BP and calendar years.

14.127   1965
MIAMI NATURAL RADIOCARBON CORRECTIONS I-III
Östlund, H. Göte; Bowman, Albert L.; Rusnak, Gene A.
Radiocarbon, v. 7, 1965: 153-155
A systematic error of calculation has been included in all radiocarbon age figures released from this laboatory. A table of corrected ages is presented.

14.128   1962
MIAMI NATURAL RADIOCARBON MEASUREMENTS I
Östlund, H. Göte ; Bowman, Albert L.; Rusnak, Gene A.
Radiocarbon, v. 4, 1962: 51-56
Brief description of the laboratory and technique is included.

14.129   1963
MIAMI RADIOCARBON MEASUREMENTS II
Rusnak, Gene A.; Bowman, Albert L.; Östlund, H. Göte
Radiocarbon, v. 5, 1963: 23-33
Proportional gas counting on carbon dioxide. Radiocarbon half-life: 5568 ± 30 years. Results expressed in years BP.

14.130   1964
MIAMI NATURAL RADIOCARBON MEASUREMENTS III
Rusnak, Gene A.; Bowman, Albert L.; Östlund, H. Göte
Radiocarbon, v. 4, 1964: 208-214
Projects: deep sea cores and coastal deposits.

14.131   1964
MONACO RADIOCARBON MEASUREMENTS I
Thommeret, J.; Rapaire, J.L.
Radiocarbon, v. 6, 1964: 194-196
Reporting period: from 1962. Proportional gas counting on carbon dioxide. Radiocarbon half-life: 5570 ± 30 years. Results expressed in years BP

14.132   1966
MONACO RADIOCARBON MEASUREMENTS II
Thommeret, J.; Thommeret, Y.
Radiocarbon, v. 8, 1966: 286-291
Radiocarbon half-life: 5568 ± 30 years. Results expressed as years BP and calendar years.

14.133   1968
NANCY NATURAL RADIOCARBON MEASUREMENTS I
Coppens, R.; Durand, G.L.A.; Guillet, B.
Radiocarbon, v. 10, n. 1, 1968: 119-123
First dating from 1964. Method: Gas proportional counting on carbon dioxide. Half-life: 5560 ± 30. A short description of the apparatus and method of counting is given.

Ch. 14 - Date Lists

14.134   1963
NATIONAL PHYSICAL LABORATORY RADIOCARBON MEASUREMENTS I
Callow, W.J.; Baker, M.J.; Pritchard, Daphne H.
Radiocarbon, v. 5, 1963: 34-38
Reporting period: from end of 1961. Proportional gas counting on carbon dioxide. Radiocarbon half-life: 5568 years.

14.135   1964
NATIONAL PHYSICAL LABORATORY RADIOCARBON MEASUREMENTS II
Callow, W.J.; Baker, M.J.; Pritchard, Daphne H.
Radiocarbon, v. 6, 1964: 25-30
Reporting period: Nov. 1962 - Nov. 1963. Radiocarbon half-life: 5568 years. Results expressed in years BP and calendar years.

14.136   1965
NATIONAL PHYSICAL LABORATORY RADIOCARBON MEASUREMENTS III
Callow, W.J.; Baker, M.J.; Hassall, Geraldine I.
Radiocarbon, v. 7, 1965: 156-161
Reporting period: to the end of November 1964. Includes the expressions used when calculating results.

14.137   1965
NATIONAL PHYSICAL LABORATORY RADIOCARBON MEASUREMENTS III (Abstract)
Callow, W.J.; Baker, M.J.; Hassall, Geraldine I.
In: International Conference on Radiocarbon and Tritium Dating. Pullman, Washington, Washington State University, June 1-11, 1965. *Proceedings.* United States of America, Atomic Energy Commission, 1965. Conference n. 650652: 393-395.
Age given relative to A.D. 1950. Half-life of radiocarbon: 5568. Correction made for fractionation, NBS oxalic acid is used as a contemporary reference standard.

14.138   1966
NATIONAL PHYSICAL LABORATORY RADIOCARBON MEASUREMENTS IV
Callow, W.J.; Barker, M.J.; Hassall, Geraldine I.
Radiocarbon, v. 8, 1966: 340-347
Reporting period: to the end of November 1965. Radiocarbon half-life: 5568 years. Results expressed as years BP and calendar years.

14.139   1968
NATIONAL PHYSICAL LABORATORY RADIOCARBON MEASUREMENTS V
Callow, W.J.; Hassall, Geraldine I.
Radiocarbon, v. 10, n. 1, 1968: 115-118
The method of calculating and techniques of measurement are unchanged.

14.140   1953
NEW ZEALAND AGE MEASUREMENTS I
Fergusson, Gordon J.; Rafter, T.A.
New Zealand Journal of Science and Technology, v. 35, n. 1, Sect. B, July 1953: 127-128
Living and dead carbon runs interspersed with samples of unknown age. Radiocarbon half-life: 5568 +- 30 years. Ages in years. Material dated: charcoal, wood and peat.

14.141   1955
NEW ZEALAND C-14 AGE MEASUREMENTS II
Fergusson, Gordon J.; Rafter, T.A.
New Zealand Journal of Science and Technology, v. 36, n. 4, Sect. B, Jan. 1955: 371-374
Use of carbon dioxide filled proportional gas counter. Age up to 35,000 years and more. Radiocarbon half-life 5568 ± 30. 18 samples. Material dated: charcoal, peat, fossil, wood, ash. Projects: geological samples.

14.142   1955
NEW ZEALAND C-14 AGE MEASUREMENTS III
Fergusson, Gordon J.; Rafter, T.A.
New Zealand Journal of Science and Technology, v. 38, n. 7, Sect. B, July 1957: 732-749
Period covered; May 1955 - May 1956. Proportional counting of carbon dioxide. Error takes into account uncertainty in counting for background, modern sample and unknown sample and uncertainty in the current value of ^{14}C half-life. Radiocarbon half-life of 5568 ± 30 years. Material dated: wood, peat, charcoal, shells, seal bone. Projects: geological and archeological.

14.143   1959
NEW ZEALAND ^{14}C AGE MEASUREMENTS IV
Fergusson, Gordon J.; Rafter, T.A.
New Zealand Journal of Geology and Geophysics, v. 2, n. 1, February 1959: 208-241
Period covered: 1956 to July 1958. Method: proportional counting of carbon dioxide. Standard: modern wood. Correction of 100 years for samples of surface ocean water to account for the Suess effect.

14.144   1963
NEW ZEALAND NATURAL RADIOCARBON MEASUREMENTS I-V
Grant-Taylor, T.C.; Rafter, T.A.
Radiocarbon, v. 5, 1963: 118-162
Comprises dates which were published previously (1953, 1955, 1957, 1959 and 1962).

14.145 1962
NEW ZEALAND RADIOCARBON AGE MEASUREMENTS V
Grant-Taylor, T.C.; Rafter, T.A.
New Zealand Journal of Geology and Geophysics, v. 5, n. 2, May 1962: 331-359
Tabulated list of 117 samples dated by radiocarbon method between July 1958 and 1959. Method: proportional counting of carbon dioxide gas. Projects: Holocene studies, Pleistocene studies, archaeology, volcanology. Dates based on half-life of 5760 ± 50 years

14.146 1968
OAK RIDGE ASSOCIATED UNIVERSITIES RADIOCARBON DATES II
Noakes, John Edward; Kim, Stephen M.; Fisher, F.
Radiocarbon, v. 10, n. 2, 1968: 346-349
Method: Liquid scintillation counting with benzene. Half-life: 5570 years.

14.147 1967
OAK RIDGE INSTITUTE OF NUCLEAR STUDIES RADIOCARBON DATES I
Noakes, John Edward; Kim, Stephen M.; Akers, L.K.
Radiocarbon, v. 9, 1967: 309-315
Liquid scintillation counting on benzene. Radiocarbon half-life: 5570 years. Results expressed as years BP and calendar years.

14.148 1964
OHIO WESLEYAN UNIVERSITY NATURAL RADIOCARBON MEASUREMENTS I
Ogden, J. Gordon; Hay, Ruth J.
Radiocarbon, v. 6, 1964: 340-348
Proportional gas counting on methane. Radiocarbon half-life: 5568 years. Results expressed in years BP and calendar years. A short description of the laboratory and operational procedures is included.

14.149 1965
OHIO WESLEYAN UNIVERSITY NATURAL RADIOCARBON MEASUREMENTS II
Ogden, J. Gordon; Hay, Ruth J.
Radiocarbon, v. 7, 1965: 166-173
Reporting period: since November 1963. Proportional gas counting on methane. Radiocarbon half-life: 5568 years. Results expressed in years BP and calendar years.

14.150 1967
OHIO WESLEYAN UNIVERSITY NATURAL RADIOCARBON MEASUREMENTS III
Ogden, J. Gordon; Hay, Ruth J.

Radiocarbon, v. 9, 1967: 316-332
Reporting period to November 1966. Proportional gas counting on methane. Radiocarbon half-life: 5568 years. Results expressed as years BP and calendar years.

14.151 1965
PACKARD INSTRUMENT COMPANY RADIOCARBON DATES I
Kowalski, Sandra J.
Radiocarbon, v. 7, 1965: 200-204
Reporting period: since Autumn 1963. Liquid scintillation counting on benzene. Short description of preparation and counting method included.

14.152 1966
PACKARD INSTRUMENT COMPANY RADIOCARBON DATES II
Kowalski, Sandra J.; Schrodt, Ariel G.
Radiocarbon, v. 8, 1966: 386-389
Liquid scintillation counting on benzene. Radiocarbon half-life: 5570 ± 30 years. Results expressed as years BP and calendar years. Short description of chemical procedure included.

14.153 1950
RADIOCARBON DATES (SEPTEMBER 1, 1950)
[Chicago]
Arnold, James R.; Libby, Willard F.
University of Chicago. Institute for Nuclear Studies, 1950. pp. 15
[Not sighted]

14.154 1951
RADIOCARBON DATES [Chicago]
Arnold, James R.; Libby, Willard F.
Science, v. 113, n. 2327, 2 Feb. 1951: 111-120
Dates based on half-life of 5568 ± 30. Period covered: June 1950 to February 1951. Counting time limited to 48 hours. Material dated: wood, charcoal, wheat and barley grain, peat, mud. Archaeological and geological projects.

14.155 1951
RADIOCARBON DATES II [Chicago]
Libby, Willard F.
Science, v. 114, n. 2960, 21 Sept. 1951: 291-296
Period covered: February 1951 to September 1951. Half-life of radiocarbon 5568 ± 30 years. Counting times maximum 48 hours. Dating of Mesopotamian, West Asian and American archaeology and glacial geology. Sample types: wood, charcoal, peat, mud, organic matter, wheat and barley grain, linen wrapping of Dead Sea scrolls, bone.

14.156   1968
RADIOCARBON DATES FROM SOVIET LABORATORIES, 1 JANUARY 1962 - 1 JANUARY 1966. COMMISSION FOR THE STUDY OF THE QUATERNARY PERIOD. AKADEMIA NAUK SSSR.
Radiocarbon, v. 10, n. 2, 1968: 417-418
323 dates radiocarbon dates determined between 1962 and 1965 are presented by various laboratories (see under their individual names). Radiocarbon half-life: 5568 years.

14.157   1964
RADIOCARBON DATES FROM THE LABORATORY OF THE GEOLOGICAL INSTITUTE, ACADEMY OF SCIENCES, U.S.S.R.
Cherdyntsev, V.V.; Alekseyev, V.A.; Kind, N.V.; Forova, V.S.; Sulerzhitskiy, L.D.
Geochemistry, 1964, n. 1: 268-274
Method: liquid scintillation counting on benzene. A brief description of the chemical treatment is given. Materials dated: wood and soils. Projects: archaeological and geological dating.

14.158   1966
RADIOCARBON DATES IN THE VERNADSKY INSTITUTE, I-IV
Vinogradov, A.P.; Devirts, A.L.; Dobkina, E.I.; Markova, N.G.
Radiocarbon, v. 8, 1966: 292-323
Reporting period 1956-1964. Proportional gas counting on carbon dioxide and ethane. Radiocarbon half-life 5568±30 years. Results expressed as years BP and calendar years.

14.159   1967
RADIOCARBON DATES OBTAINED THROUGH GEOGRAPHICAL BRANCH FIELD OBSERVATION
Andrews, J.T.; Drapier, Lyn
Canada. Department of Energy, Mines and Resources. Geographical Branch. Geology Bulletin, v. 9, n. 2, 1967: 115-168
Radiocarbon dates are given for samples obtained, mainly in the Northwest Territories, by Geographical Branch research parties during the period 1950 to 1966.

14.160   1967
RADIOCARBON DATES OF THE GEOGRAPHICAL BRANCH, 1950-1966, PART I
Andrews, John T.
Geographical Bulletin, v. 9, n. 2, 1967: 115-162
Radiocarbon dates for samples obtained, in the Northern Territories, by Geographical Branch Research Parties. The dates were provided by various laboratories, but were not all made public through the usual channels.

14.161   1955
RADIOCARBON DATES [University of London]
Zeuner, Frederick Eberard
University of London. Institute of Archaeology. Annual Report, 11th, 1955: 43-50
Date list.

14.162   1962
RADIOCARBON DATING AT THE UNIVERSITY OF WASHINGTON I
Dorn, Thomas Felder; Fairhall, A.W.; Schell, William Raymond; Takashima, Y.
Radiocarbon, v. 4, 1962: 1-12
Part I, description of apparatus and method of proportional gas counting on methane. Radiocarbon half-life: 5568 years. Part II dating of tree rings to investigate radiocarbon activity variations with time. Part IV: a note concerning the mechanism of oxidation of radiogenic radiocarbon in the atmosphere.

14.163   1963
RADIOCARBON DATING AT THE UNIVERSITY OF WASHINGTON II
Fairhall, A.W.; Schell, William Raymond
Radiocarbon, v. 5, 1963: 80-81
Decription of new counter for dating on methane.

14.164   1966
RADIOCARBON DATING AT THE UNIVERSITY OF WASHINGTON III
Fairhall, A.W.; Schell, William Raymond; Young, James Allen
Radiocarbon, v. 8, 1966: 498-506
Reporting period: from 1962. Proportional gas counting on methane. Radiocarbon half-life: 5568 years. Results expressed as years BP and calendar years. Contamination by tritium from the 1961-1962 bomb tests is noted.

14.165   1965
RADIOCARBON DATING IN THE SOVIET UNION
Butomo, S.V.
Radiocarbon, v. 7, 1965: 223-228
Reporting period: 1959-1961. Ethyl-benzol technique.

14.166   1963
RADIOCARBON DATING : RESULTS
Kusumgar, Sheela; Lal, D.; Sarna, R.P.
Indian Academy of Science. Proceedings, Section A, v. 58, n. 3, Sept. 1963: 141-152
The first series of radiocarbon dates of samples of archaeological and geological interest, measured by the Radiocar-

bon Laboratory of the Tata Institute of Fundamental Research are presented. Method: Acetylene gas counting. Half-life of radiocarbon: 5780 ± 40 years and 5568 ± 30 years, hence two dates for each sample. Dates refer to years BP (1950).

14.167   1954
RADIOCARBON MEASUREMENTS OF THE U.S. GEOLOGICAL SURVEY
Suess, Hans E.; Rubin, Meyer
Geological Society of America, Bulletin, v. 65, n. 12, Dec. 1954: 1311
Abstract of paper presented at the November 1954 meeting of the Geological Society of America. The main objects of the dates were the dating of pre-Mankato material and the establishing of an absolute time scale for glacial events in the later part of the Wisconsin age.

14.168   1961
REPORT: 12 OCTOBER 1960 - OCTOBER 1961
Godwin, Harry
Cambridge. University of Cambridge. Sub-department of Quaternary Research, 1961. pp. 6
Includes a report on the radiocarbon laboratory and the various dates produced in the context of the research to which they apply.

14.169   1962
REPORT: 13 OCTOBER 1961 - OCTOBER 1962
Godwin, Harry
Cambridge. University of Cambridge. Sub-department of Quaternary Research, 1962. pp. 6
The laboratory organised an informal International Conference on Radiocarbon Dating, held during 13-28 July 1962 in Cambridge. Dates determined by the radiocarbon laboratory are reported and appraised.

14.170   1963
REPORT: 14 OCTOBER 1962 - OCTOBER 1963
Godwin, Harry
Cambridge. University of Cambridge. Sub-department of Quaternary Research, 1963. pp. 7
Radiocarbon dates for late glacial (Late Wechselian) site, Mesolithic and Neolithic in Britain are reported.

14.171   1967
RHODESIA RADIOCARBON MEASUREMENTS III
Sheppard, J.G.; Swart, E.R.
Radiocarbon, v. 9, 1967: 382-386
Results expressed as years BP and calendar years.

14.172   1966
RHODESIAN RADIOCARBON MEASUREMENTS II
Sheppard, J.G.; Swart, E.R.
Radiocarbon, v. 8, 1966: 423-429
Reporting period: from December 1963. Proportional gas counting on acetylene.

14.173   1964
RIKEN NATURAL RADIOCARBON MEASUREMENTS I
Yamasaki, Fumio; Hamada, Tatsuji; Fujiyama, Chikako
Radiocarbon, v. 6, 1964: 112-115
Radiocarbon half-life: 5568 years.

14.174   1966
RIKEN NATURAL RADIOCARBON MEASUREMENTS II
Yamasaki, Fumio; Hamada, Tatsuji; Fujiyama, Chikako
Radiocarbon, v. 8, 1966: 324-229
Proportional gas counting on carbon dioxide. Radiocarbon half-life: 5568 years. Results expressed as years BP and calendar years.

14.175   1967
RIKEN NATURAL RADIOCARBON MEASUREMENTS III
Yamasaki, Fumio; Hamada, Tatsuji; Fujiyama, Chikako
Radiocarbon, v. 9, 1967: 301-308
Reporting period: 1966. Proportional gas counting on carbon dioxide. Radiocarbon half-life: 5568 years. Results expressed as years BP and calendar years.

14.176   1968
RIKEN NATURAL RADIOCARBON MEASUREMENTS IV
Yamasaki, Fumio; Hamada, Tatsuji; Fujiyama, Chikako
Radiocarbon, v. 10, n. 2, 1968: 333-345
Method: dates obtained by counting carbon dioxide in a 2.7 L. stainless steel counter. Radiocarbon half-life: 5568 years.

14.177   1964
SACLAY NATURAL RADIOCARBON MEASUREMENTS I
Delibrias, Georgette; Guillier, M.T.; Labeyrie, Jean
Radiocarbon, v. 6, 1964: 233-250
Reporting period: 1956 - 1962. Proportional gas counting on carbon dioxide. Results expressed in years BP and calendar years.

14.178   1965
SACLAY NATURAL RADIOCARBON MEASUREMENTS II

Ch. 14 - Date Lists

Delibrias, Georgette; Guillier, M.T.; Labeyrie, Jean
Radiocarbon, v. 7, 1965: 236-244
Reporting period: 1963. Radiocarbon half-life: 5570 years. Results expressed in years BP and calendar years.

14.179   1964
SHARP LABORATORIES MEASUREMENTS I
Ellis, John G.; Sharp, Rodman A.
Radiocarbon, v. 6, 1964: 108-109
Samples dated mainly for the purpose of intercomparison with other established laboratories. Description of apparatus and method included. Proportional gas counting on methane.

14.180   1964
SMITHSONIAN INSTITUTION RADIOCARBON MEASUREMENTS I
Sigalove, Joel J.; Long, Austin
Radiocarbon, v. 6, 1964: 182-188
Reporting period: from Spring 1963. Proportional gas counting on methane. Short description of laboratory techniques is included.

14.181   1965
SMITHSONIAN INSTITUTION RADIOCARBON MEASUREMENTS II
Long, Austin
Radiocarbon, v. 7, 1965: 245-256
Reporting period: Nov. 1963 - Nov. 1964. Results expressed in years BP and calendar years.

14.182   1966
SMITHSONIAN INSTITUTION RADIOCARBON MEASUREMENTS III
Long, Austin; Mielke, James E.
Radiocarbon, v. 8, 1966: 413-422
Reporting period: 1965

14.183   1967
SMITHSONIAN INSTITUTION RADIOCARBON MEASUREMENTS IV
Long, Austin; Mielke, James E.
Radiocarbon, v. 9, 1967: 368-381
Reporting period: 1966. Proportional gas counting on methane. Results expressed as years BP and calendar years.

14.184   1960
SOCONY MOBIL RADIOCARBON DATES I
Bray, Ellis E.; Burke, W.J.
Radiocarbon, v. 2, 1960: 97-111
Proportional gas counting on methane. Short description of preparation is given. Radiocarbon half-life: 5568 years.

14.185   1961
SOME HIGHLIGHTS FROM THE NATURAL RADIOCARBON DATINGS OF THE LA JOLLA LABORATORY
Hubbs, Carl L.
Science, v. 134, n. 3488, 3 Nov. 1961: 1430
Survey of the contributions made by the La Jolla radiocarbon dating supervised by Hans E. Suess.

14.186   1964
SOUTHERN RHODESIAN RADIOCARBON MEASUREMENTS I
Robins, P.A.; Swart, E.R.
Radiocarbon, v. 6, 1964: 31-36
Reporting period, from Oct. 1962. Proportional gas counting on acetylene.

14.187   1967
SOVIET ARCHAEOLOGICAL RADIOCARBON DATES III
Chard, Chester S.; Powers, Roger
Arctic Anthropology, v. 5, n. 1, 1967: 224-233
The listing contains all radiocarbon dates on archaeological samples determined by Soviet laboratories between January 1962 and January 1966, excluding those on Pleistocene occupation sites published by Klein (1967). For dates prior to 1962, see Chard and Workman 1965.

14.188   1965
SOVIET RADIOCARBON DATES II
Chard, Chester S.; Workman, William B.
Arctic Anthropology, v. 3, n. 1, 1965: 146-150
Includes dates up to January 1968. Some evaluations are included.

14.189   1957
STOCKHOLM NATURAL RADIOCARBON MEASUREMENTS
Östlund, H. Göte
Science, v. 126, n. 3272, 13 Sept. 1957: 493-497
Period covered: 1955, 1956 - May 1957. Method: black carbon with double screenwall counter. Standard: oak tree ring 1850. Ages reported as years BP. Material dated: wood, charcoal, peat, plant sediments, gyttja, resin, mammoth remains.

14.190   1959
STOCKHOLM NATURAL RADIOCARBON MEASUREMENTS II
Östlund, H. Göte
Radiocarbon, v. 1, 1959: 35-44
Proportional counting of carbon dioxide. Radiocarbon half-life: 5568.

14.191   1960
STOCKHOLM NATURAL RADIOCARBON MEASUREMENTS III
Östlund, H. Göte; Engstrand, Lars G.
Radiocarbon, v. 2, 1960: 186-196
Radiocarbon half-life: 5568 ± 30 years.

14.192   1962
STOCKHOLM NATURAL RADIOCARBON MEASUREMENTS IV
Engstrand, Lars G.; Östlund, H. Göte
Radiocarbon, v. 4, 1962: 115-136
Proportional gas counting on carbon dioxide. Radiocarbon half-life: 5568 ± 30 years. Results expressed in years BP.

14.193   1963
STOCKHOLM NATURAL RADIOCARBON MEASUREMENTS V
Östlund, H. Göte; Engstrand, Lars G.
Radiocarbon, v. 5, 1963: 203-227
Reporting period: Jan.- Nov. 1962. Proportional gas counting on carbon dioxide.

14.194   1965
STOCKHOLM NATURAL RADIOCARBON MEASUREMENTS VI
Engstrand, Lars G.
Radiocarbon, v. 7, 1965: 257-290

14.195   1967
STOCKHOLM NATURAL RADIOCARBON MEASUREMENTS VII
Engstrand, Lars G.
Radiocarbon, v. 9, 1967: 387-438
Results expressed as years BP and calendar years.

14.196   1966
TARTU RADIOCARBON DATES I
Liiva, A.; Ilves, E.; Punning, J.B.
Radiocarbon, v. 8, 1966: 430-441
Reporting period: from 1959. Liquid scintillation counting on methanol or benzene. Radiocarbon half-life: 5568 ± 30 years. Results expressed as years BP and calendar year. A short description of the procedure is included.

14.197   1968
TARTU RADIOCARBON DATES I (REVISIONS)
Liiva, A.; Ilves, E.; Punning, J.M.
Radiocarbon, v. 10, n. 2, 1968: 465
Revisions to dates published in 1966.

14.198   1968
TARTU RADIOCARBON DATES II
Punning, J.M.; Ilves, E.; Liiva, A.
Radiocarbon, v. 10, n. 1, 1968: 124-130
Period covered: 1956 to 1961. Method: liquid scintillation counting on benzene. Radiocarbon half-life: 5568 ± 30 years.

14.199   1968
TARTU RADIOCARBON DATES III
Punning, J.M.; Liiva, A.; Ilves, E.
Radiocarbon, v. 10, n. 2, 1968: 379-383
Equipment used is the same as described in Tartu I and II.

14.200   1963
TATA INSTITUTE RADIOCARBON DATE LIST I
Kusumgar, Sheela; Lal, D.; Sarna, R.P.
Radiocarbon, v. 5, 1963: 273-282
Reporting period: from August 1961. Proportional gas counting on acetylene. Radiocarbon half-life: 5568 years.

14.201   1964
TATA INSTITUTE RADIOCARBON DATE LIST II
Agrawal, D.P.; Kusumgar, Sheela; Lal, D.; Sarna, R.P.
Radiocarbon, v. 6, 1964: 226-232
Proportional gas counting an acetylene. Radiocarbon half-life: 5568 years. Results expressed in years BP and in calendar years.

14.202   1965
TATA INSTITUTE RADIOCARBON DATE LIST III
Agrawal, D.P.; Kusumgar, Sheela; Lal, D.
Radiocarbon, v. 7, 1965: 291-295
Proportional gas counting on acetylene. Radiocarbon half-life: 5568 years. Results expressed in years BP and calendar years.

14.203   1966
TATA INSTITUTE RADIOCARBON DATE LIST IV
Agrawal, D.P.; Kusumgar, Sheela
Radiocarbon, v. 8, 1966: 442-452
Reporting period: from September 1964. Proportional gas counting on methane. Radiocarbon half-life: 5568 years. Results expressed as years BP and calendar year. Short description of procedure included.

14.204   1968
TATA INSTITUTE RADIOCARBON DATE LIST V
Agrawal, D.P.; Kusumgar, Sheela
Radiocarbon, v. 101, n. , 1968: 131-143
Method: Gas counting on methane. Radiocarbon half-life: 5568 years.

Ch. 14 - Date Lists

14.205 1968
TBILISI RADIOCARBON DATES I
Burchladze, A.A.
Radiocarbon, v. 10, n. 2, 1968: 466-467

14.206 1964
TEXAS A. & M. UNIVERSITY RADIOCARBON DATES I
Noakes, John Edward; Stipp, Jerry J.; Hood, Donald W.
Radiocarbon, v. 6, 1964: 189-193
Liquid scintillation counting on benzene. Radiocarbon half-life: 5568 years. Results expressed in years BP and calendar years.

14.207 1963
TEXAS BIO-NUCLEAR RADIOCARBON MEASUREMENTS I
Chandler, John B; Kinningham, Russell; Masey, Don S.
Radiocarbon, v. 5, 1963: 56-61
Reporting period: from December 1961. Liquid scintillation counting on benzene. Radiocarbon half-life: 5568 years. Results expressed in years BP.

14.208 1959
TRONDHEIM NATURAL RADIOCARBON MEASUREMENTS I.
Nydal, Reidar
Radiocarbon, v. 1, 1959: 76-80
Reporting period: July 1957 to September 1958. Proportional counting of carbon dioxide. Radiocarbon half-life: 5568 years. A short description of the technique and apparatus used is included.

14.209 1960
TRONDHEIM NATURAL RADIOCARBON MEASUREMENTS II
Nydal, Reidar
Radiocarbon, v. 2, 1960: 82-96
Reporting period: September 1958 - December 1959. Proportional gas counting on carbon dioxide. Radiocarbon half-life: 5570 years. Description of counters, pretreatment and calculations are included.

14.210 1962
TRONDHEIM NATURAL RADIOCARBON MEASUREMENTS III
Nydal, Reidar
Radiocarbon, v. 4, 1962: 160-181
Reporting period: January 1960 to December 1961. Proportional gas counting on carbon dioxide .Radiocarbon half-life: 5570 years. Results expressed in years BP.

14.211 1964
TRONDHEIM NATURAL RADIOCARBON MEASUREMENTS IV
Nydal, Reidar; Lovseth, Knut; Skullerud, Kari E.; Holm, Marianne
Radiocarbon, v. 6, 1964: 280-290
Reporting period: 1962 - 1963. Proportional gas counting on carbon dioxide. Radiocarbon half-life: 5568 years. Results expressed in years BP and calendar years.

14.212 1954
U.S. GEOLOGICAL SURVEY RADIOCARBON DATES I
Suess, Hans E.
Science, v. 120, n. 3117, 24 Sept. 1954: 467-473
Period covered: October 1953 - April 1964. Method: acetylene gas counting. Radiocarbon half-life: 5568 ± 30 years. Projects: pre-Mankato sub-stage of the last glaciation. Material dated: wood, peat and shells.

14.213 1955
U.S. GEOLOGICAL SURVEY RADIOCARBON DATES V
Suess, Hans E.
Science, v. 121, n. 3145, 8 Apr. 1955: 481-488
Period covered: May to October 1954. Projects: the establishment of a pre-Mankato absolute time scale. The dating of foraminifera shells contained in deep sea sediments for which the environmental temperature of growth had been determined by the $^{16}O/^{18}O$ ratio leads to the most direct evidence of temperature variation with time.

14.214 1955
U.S. GEOLOGICAL SURVEY RADIOCARBON DATES III
Rubin, Meyer; Suess, Hans E.
Science, v. 123, n. 3194, 16 Mar. 1965: 442-448
Period covered: February 1955 - June 1955. Material dated: deep sea cores, carbonates, wood, charcoal, oolite, shells, peat, organic silt, fern mould, limestone, bones, antlers.

14.215 1958
U.S. GEOLOGICAL SURVEY RADIOCARBON DATES IV
Rubin, Meyer; Alexander, Corrinne
Science, v. 127, n. 3313, 27 June 1958: 1476-1487
Period covered: July 1955 - November 1956. Method: acetylene gas counting. Ages reported in years.

14.216 1960
U.S. GEOLOGICAL SURVEY RADIOCARBON DATES V
Rubin, Meyer; Alexander, Corrinne

Radiocarbon, v. 2, 1960: 129-185
Reporting period up to 1959. Proportional counting on acetylene.

14.217   1961
U.S. GEOLOGICAL SURVEY RADIOCARBON DATES VI
Rubin, Meyer; Berthold, Sarah M.
Radiocarbon, v. 3, 1961: 86-98
Period reported: 1960. Radiocarbon half-life 5568 ± 30 years. Proportional gas counting on acetylene.

14.218   1964
U.S. GEOLOGICAL SURVEY RADIOCARBON DATES VII
Ives, Patricia C.; Levin, Betsy; Robinson, Richard D.; Rubin, Meyer
Radiocarbon, v. 6, 1964: 37-76
Reporting period: 1961 - 1963. Proportional gas counting on acetylene. Radiocarbon half-life: 5568 ± 30 years.

14.219   1965
U.S. GEOLOGICAL SURVEY RADIOCARBON DATES VIII
Levin, Betsy; Yves, Patricia C.; Oman, Charles L.; Rubin, Meyer
Radiocarbon, v. 7, 1965: 372-398
Reporting period: 1963 and 1964. Radiocarbon half-life 5568 ± 30 years. Results expressed in years BP and calendar years.

14.220   1967
U.S. GEOLOGICAL SURVEY RADIOCARBON DATES IX
Ives, Patricia C.; Levin, Betsy; Oman, Charles L.; Rubin, Meyer
Radiocarbon, v. 9, 1967: 505-529
Reporting period: 1965 and 1966. Proportional gas counting on acetylene. Radiocarbon half-life: 5568 ± 30 years. Results expressed as years BP and calendar years.

14.221   1965
U.S.D.A. SEDIMENTATION LABORATORY RADIOCARBON DATES I
McDowell, L.L.; Ryan, M.E.
Radiocarbon, v. 7, 1965: 174-178
Liquid scintillation counting on benzene. Radiocarbon half-life: 5570 years. Results expressed in years BP and calendar years. Short descriptions of the benzene synthesis and of the equipment used are included.

14.222   1962
UCLA RADIOCARBON DATES I
Fergusson, Gordon J.; Libby, Willard F.
Radiocarbon, v. 4, 1962: 109-114
Reporting period, from August 1961. Radiocarbon half-life: 5568 ± 30 years. Proportional gas counting on carbon dioxide.

14.223   1963
UCLA RADIOCARBON DATES II
Fergusson, Gordon J.; Libby, Willard F.
Radiocarbon, v. 5, 1963: 1-22
Radiocarbon half-life: 5568 years.

14.224   1964
UCLA RADIOCARBON DATES III
Fergusson, Gordon J.; Libby, Willard F.
Radiocarbon, v. 6, 1964 318-339
Reporting period: 1963. Proportional gas counting on carbon dioxide. Radiocarbon half-life: 5568 years. Results expressed in years BP and calendar years.

14.225   1965
UCLA RADIOCARBON DATES IV
Berger, Rainer; Fergusson, Gordon J.; Libby, Willard F.
Radiocarbon, v. 7, 1965: 336-371
Reporting period: 1964. Proportional gas counting on carbon dioxide. Radiocarbon half-life: 5568 years. Results expressed in years BP and calendar years.

14.226   1966
UCLA RADIOCARBON DATES V
Berger, Rainer; Libby, Willard F.
Radiocarbon, v. 8, 1966: 467-497
Reporting period: 1965. Proportional gas counting on carbon dioxide. Radiocarbon half-life: 5568 years. The samples are reported in a classified list.

14.227   1967
UCLA RADIOCARBON DATES VI
Berger, Rainer; Libby, Willard F.
Radiocarbon, v. 9, 1967: 477-504
Reporting period: 1966. Proportional gas counting on carbon dioxide. Radiocarbon half-life: 5568. Results expressed as years BP and calendar years.

14.228   1968
UCLA RADIOCARBON DATES VII
Berger, Rainer; Libby, Willard F.
Radiocarbon, v. 10, n. 1, 1968: 149-160
Period covered: first half of 1967. Method: gas proportional counting on carbon dioxide. Radiocarbon half-life: 5568 years.

14.229   1968
UCLA RADIOCARBON DATES VIII
Berger, Rainer; Libby, Willard F.
Radiocarbon, v. 10, n. 2, 1968: 402-416
Period covered: second half of 1967. Method: gas proportional counting on carbon dioxide. Radiocarbon half-life: 5568 years.

14.230   1958
UNIVERSITY OF ARIZONA RADIOCARBON DATES
Wise, E.N.; Shutler, Dick
Science, v. 127, n. 3289, 10 Jan. 1958: 72-78
Period covered: 1955. Method: black carbon (Libby's method modified by Ballario). Ages reported in radiocarbon years BP. Material dated: charcoal, charred wood, wood

14.231   1959
UNIVERSITY OF ARIZONA RADIOCARBON DATES II
Shutler, Richard; Damon, Paul E.
Radiocarbon, v. 1, 1959: 59-62
Reporting period: May 1957 to May 1958. Solid carbon method. Contains a short description of the techniques used.

14.232   1968
UNIVERSITY OF BIRMINGHAM RADIOCARBON DATES II
Shotton, Frederick William; Blundell, D.J.; Williams, R.E.
Radiocarbon, v. 10, n. 2, 1968: 200-206
Half-life: 5570 ± 30 years. Method: uses a 6 L counter enclosed in a double ring of Geiger tubes.

14.233   1966
UNIVERSITY OF KIEL RADIOCARBON MEASUREMENTS I
Willkomm, H.; Erlenkeuser, H
Radiocarbon, v. 8, 1966: 235-238
Reporting period: first half of 1965. Proportional gas counting on carbon dioxide. Radiocarbon half-life: 5570 years. Results expressed as years BP and calendar years.

14.234   1967
UNIVERSITY OF KIEL RADIOCARBON MEASUREMENTS II
Willkomm, H.; Erlenkenser, H.
Radiocarbon, v. 9, 1967: 257-260
Proportional gas counting on carbon dioxide. Radiocarbon half-life 5568 ± 30 years. Results reported as years BP and calendar years.

14.235   1968
UNIVERSITY OF KIEL RADIOCARBON III
Wilkomm, H.; Erlenkeuser, H.
Radiocarbon, v. 10, n. 2, 1968: 328-332
Method: use of a 4.5 L carbon dioxide counter. Radiocarbon half-life: 5570 years.

14.236   1968
UNIVERSITY OF LUND RADIOCARBON DATES I
Håkansson, Soren
Radiocarbon, v. 10, n. 1, 1968: 36-54
Dating started in 1966. Method: proportional counting on carbon dioxide. Radiocarbon half-life: 5568 years. A short description of the apparatus and procedure is given. Projects: mostly archaeological. Materials dated: wood, peat and charcoal.

14.237   1955
UNIVERSITY OF MICHIGAN RADIOCARBON DATES I
Crane, H.R.
Science, v. 124, n. 3224, 12 Oct. 1955: 664-672
Period covered: 1950-1954. Method: carbon black (Libby's method) to 1952, then carbon dioxide and Geiger Counter. Ages expressed in years.

14.238   1958
UNIVERSITY OF MICHIGAN RADIOCARBON DATES II
Crane, H.R.; Griffin, James B.
Science, v. 127, n. 3206, 9 May 1958: 1098-1105
Method: carbon dioxide counting by Geiger Counter. Material dated: charcoal, wood, tusk, animal bones, textiles, cordage, bird skin, vegetable material.

14.239   1958
UNIVERSITY OF MICHIGAN RADIOCARBON DATES III
Griffin, James B.; Crane, H.R.
Science, v. 128, n. 3332, 7 Nov. 1958: 1117-1123
A description of the technique is given. Use of CO_2 filled Geiger Counters, CS_2 quenched. Age given in years. Samples: lake bottom muck, wood, charcoal, bones, human faeces, shells (freshwater and marine) corn cobs, human bones.

14.240   1959
UNIVERSITY OF MICHIGAN RADIOCARBON DATES IV
Crane, H.R.; Griffin, James B.
Radiocarbon, v. 1, 1959: 173-198
For description of method, see list III.

14.241   1960
UNIVERSITY OF MICHIGAN RADIOCARBON DATES V
Crane, H.R.; Griffin, James B.
Radiocarbon, v. 2, 1960: 31-48
Same method as for list II and IV

14.242   1961
UNIVERSITY OF MICHIGAN RADIOCARBON DATES VI
Crane, H.R.; Griffin, James B.
Radiocarbon, v. 3, 1961: 105-129
Reporting period: 1960. For description of the method, see Michigan list III.

14.243   1962
UNIVERSITY OF MICHIGAN RADIOCARBON DATES VII
Crane, H.R.; Griffin, James B.
Radiocarbon, v. 4, 1962: 183-203
Reporting period: from December 1960.

14.244   1963
UNIVERSITY OF MICHIGAN RADIOCARBON DATES VIII
Crane, H.R.; Griffin, James B.
Radiocarbon, v. 5, 1963: 228-253
Reporting period: from Dec. 1961. Radiocarbon half-life: 5568 years. Results expressed in years BP.

14.245   1964
UNIVERSITY OF MICHIGAN RADIOCARBON DATES IX
Crane, H.R.; Griffin, James B.
Radiocarbon, v. 6, 1964: 1-24
Reporting period: from Dec. 1962. Radiocarbon half-life: 5570 years. Results expressed as years BP and calendar years.

14.246   1965
UNIVERSITY OF MICHIGAN RADIOCARBON DATES X
Crane, H.R.; Griffin, James B.
Radiocarbon, v. 7, 1965: 123-152
Reporting period: since 1963. Proportional gas counting on carbon dioxide. Radiocarbon half-life: 5570 years. Results expressed in years BP and calendar years.

14.247   1966
UNIVERSITY OF MICHIGAN RADIOCARBON DATES XI
Crane, H.R.; Griffin, James B.
Radiocarbon, v. 8, 1966: 256-285
Reporting period: since December 1964. Radiocarbon half-life: 5570 years. Results expressed as years BP and calendar years.

14.248   1968
UNIVERSITY OF MICHIGAN RADIOCARBON DATES XII
Crane, H.R.; Griffin, James B.
Radiocarbon, v. 10, n. 1, 1968: 61-114
Period covered: from 1965. Method: gas counting of carbon dioxide with Geiger counter. Radiocarbon half-life: 5570 years.

14.249   1965
UNIVERSITY OF NEW SOUTH WALES RADIOCARBON DATES I
Green, J.H.; Harris, Josephine; Neuhaus, John William George; Sewell, D.K.B.; Watson, Maureen
Radiocarbon, v. 7, 1965: 162-165
Proportional gas counting on carbon dioxide. Radiocarbon half-life: 5568 years. Results expressed in years BP and calendar years.

14.250   1955
UNIVERSITY OF PENNSYLVANIA RADIOCARBON DATES I
Ralph, Elizabeth K.
Science, v. 121, n. 3136, 4 Feb. 1955: 149-151
Equipment essentially the same as developed by Anderson, Arnold and Libby. Error includes the standard error and the ± 0.1 uncertainty for the time-scale. Counting time average 48 hours. Samples mostly charcoal.
[Note: University of Pennsylvania Radiocarbon Dates II is in Chapter 10 under the number 10.053]

14.251   1959
UNIVERSITY OF PENNSYLVANIA RADIOCARBON DATES III
Ralph, Elizabeth K.
Radiocarbon, v. 1, 1959: 45-58
Reporting period: 1958. Proportional counting of carbon dioxide. Reported age: years BP. Radiocarbon half-life used: 5568. Archaeological dates.

14.252   1961
UNIVERSITY OF PENNSYLVANIA RADIOCARBON DATES IV
Ralph, Elizabeth K.; Ackerman, Robert E.
Radiocarbon, v. 3, 1961: 4-14
Radiocarbon dates already previously reported by Rainey and Ralph (*American Antiquity*, 24, 1959: 365-374). BP dates calculated from A.D. 1957 for solid carbon and A.D. 1958 for carbon dioxide dates.

Ch. 14 - Date Lists

14.253   1962
UNIVERSITY OF PENNSYLVANIA RADIOCARBON DATES V
Ralph, Elizabeth K.; Stuckenrath, Robert
Radiocarbon, v. 4, 1962: 144-159
Reporting period: 1959 and 1960. Radiocarbon half life: 5568 ± 30 years. Results expressed in years BP.

14.254   1963
UNIVERSITY OF PENNSYLVANIA RADIOCARBON DATES VI
Stuckenrath, Robert
Radiocarbon, v. 5, 1963: 82-103
Reporting period: 1961 and 1962. Radiocarbon half-life: 5568 ± 30 years. Results expressed in years BP.

14.255   1965
UNIVERSITY OF PENNSYLVANIA RADIOCARBON DATES VII
Ralph, Elizabeth K.; Michael, Henry N.; Gruninger, John
Radiocarbon, v. 7, 1965: 179-186
A program of parallel dating by dendrochronology and radiocarbon dating of sequoia and bristlecone pines is presented.

14.256   1965
UNIVERSITY OF PENNSYLVANIA RADIOCARBON DATES VIII
Stuckenrath, Robert; Ralph, Elizabeth K.
Radiocarbon, v. 7, 1965: 187-199
Reporting period: 1963-1964. Radiocarbon half-life: 5568 years. Results expressed in years BP and calendar years.

14.257   1966
UNIVERSITY OF PENNSYLVANIA RADIOCARBON DATES IX
Stuckenrath, Robert; Coe, William R.; Ralph, Elizabeth K.
Radiocarbon, v. 8, 1966: 348-385
Reporting period up to November 1965. Radiocarbon half-life: 5568 years. Results expressed as years BP and calendar years.

14.258   1967
UNIVERSITY OF PENNSYLVANIA RADIOCARBON DATES X
Stuckenrath, Robert
Radiocarbon, v. 9, 1967: 333-345
Reporting period: to November 1966. Radiocarbon half-life: 5568. Results expressed as years BP and calendar years.

14.259   1964
UNIVERSITY OF ROME CARBON-14 DATES II
Alessio, M.; Bella, F.
Radiocarbon, v. 6, 1964: 77-90
Reporting period: from 1958. Proportional gas counting on carbon dioxide. Radiocarbon half-life: 5568 ± 30 years. Results expressed in years BP and calendar years. A short description of apparatus, counter characteristics and carbon dioxide preparation are included.

14.260   1965
UNIVERSITY OF ROME CARBON-14 DATES III
Alessio, M.; Bella, F.
Radiocarbon, v. 7, 1965: 213-222
Reporting period: December 1963 - September 1964. Proportional gas counting on carbon dioxide. Radiocarbon half-life: 5568 ± 30 years. Results expressed in years BP and calendar years.

14.261   1966
UNIVERSITY OF ROME CARBON-14 DATES IV
Alessio, M.; Bella, F.
Radiocarbon, v. 8, 1966: 401-412
Reporting period: December 1964 - October 1965. Radiocarbon half-life: 5568 ± 30 years. Results expressed as years BP and calendar years.

14.262   1967
UNIVERSITY OF ROME CARBON-14 DATES V
Alessio, M.; Bella, F.; Bachechi, F.; Cortesi, Cesare
Radiocarbon, v. 9, 1967: 346-367
Reporting period: December 1965 - October 1966. Proportional gas counting on carbon dioxide. Radiocarbon half-life: 5568 ± 30 years. Results expressed as years BP and calendar years.

14.263   1968
UNIVERSITY OF ROME CARBON-14 DATES VI
Alessio, M.; Bella, F.; Cortesi, Cesare; Graziadei, B.
Radiocarbon, v. 10, n. 2, 1968: 350-364
Period covered: Dec. 1966 to Nov. 1967. Half-life: 5568 ± 30 years.

14.264   1960
UNIVERSITY OF SASKATCHEWAN DATES II
McCallum, K.J.; Dyck, Willy
Radiocarbon, v. 2, 1960: 73-81
Reporting period 1956-1959. Proportional gas counting on acetylene.

14.265   1962
UNIVERSITY OF SASKATCHEWAN RADIOCARBON DATES III
McCallum, K.J.; Wittenberg, J.
Radiocarbon, v. 4, 1962: 71-80

Material dated: charcoal, wood, marl, shells, organic silt, gyttja.

14.266   1965
UNIVERSITY OF SASKATCHEWAN RADIOCARBON DATES IV
McCallum, K.J.; Wittenberg, J.
Radiocarbon, v. 7, 1965: 229-235
Material dated: Charcoal, wood, marl, shells, organic silt, gyttja.

14.267   1968
UNIVERSITY OF SASKATCHEWAN RADIOCARBON DATES V
McCallum, K.J.; Wittenberg, J.
Radiocarbon, v. 10, n. 2, 1968: 365-378
The samples presented are mostly geologic.

14.268   1968
UNIVERSITY OF TEXAS AT AUSTIN RADIOCARBON DATES VI
Valastro, S.; Davis, E. Mott; Rightmire, Craig T.
Radiocarbon, v. 10, n. 2, 1968: 384-401
Reports dating projects made in the year ending Nov. 1967. Method: liquid scintillation counting with benzene. Radiocarbon half-life: 5568 years.

14.269   1962
UNIVERSITY OF TEXAS RADIOCARBON DATES I
Stipp, Jerry J.; Davis, E. Mott; Noakes, John E; Hoover, Tom E.
Radiocarbon, v. 4, 1962: 43-50
Liquid scintillation counting on benzene. Radiocarbon half-life: 5568 years. Ages expressed in years BP.

14.270   1964
UNIVERSITY OF TEXAS RADIOCARBON DATES II
Tamers, Murray A.; Pearson, Frederick Joseph; Davis, E. Mott
Radiocarbon, v. 6, 1964: 138-159
Reporting period: Feb.- Nov. 1963. Liquid scintillation counting on benzene solutions with toluene. Radiocarbon half-life: 5568 years. Description of apparatus. Three possible sources of error were made the subject of special studies: radon, isotope effect and quenching.

14.271   1965
UNIVERSITY OF TEXAS RADIOCARBON DATES III
Pearson, Frederick Joseph; Davis, E. Mott; Tamers, Murray A.; Johnstone, Robert W.
Radiocarbon, v. 7, 1965: 296-314
Liquid scintillation counting on benzene. Radiocarbon half-life: 5568 years. Results expressed in years BP and calendar years.

14.272   1966
UNIVERSITY OF TEXAS RADIOCARBON DATES IV
Pearson, Frederick Joseph; Davis, E. Mott; Tamers, Murray A.
Radiocarbon, v. 8, 1966: 453-466
Liquid scintillation counting of benzene. Radiocarbon half-life: 5568 years. Results expressed as years BP and calendar years. A number of bone samples were dated to determine the suitability of this material for radiocarbon dating.

14.273   1967
UNIVERSITY OF TEXAS RADIOCARBON DATES V
Valastro, S.; Pearson, Frederick Joseph; Davis, E. Mott
Radiocarbon, v. 9, 1967: 439-453
Method: Liquid scintillation counting on benzene. Radiocarbon half-life: 5560 years. Results expressed as years BP and calendar years.

14.274   1968
UNIVERSITY OF TOKYO RADIOCARBON MEASUREMENTS I
Sato, Jun; Sato, Tomoko; Suzuki, Hisashi
Radiocarbon, v. 10, n. 1, 1968: 144-148
Period covered: Sept. 1966 to July 1967. Method: Gas proportional counting on carbon dioxide. Half-life: 5570 ± 30 years. A brief description of the apparatus and method is given.

14.275   1965
UNIVERSITY OF WISCONSIN RADIOCARBON DATES I
Bender, Margaret M.; Bryson, Reid A.; Baerreis, David A.
Radiocarbon, v. 7, 1965: 399-407
Reporting period: from 1963. Proportional gas counting on methane. Radiocarbon half-life: 5568 years. Results expressed in years BP and calendar years.

14.276   1966
UNIVERSITY OF WISCONSIN RADIOCARBON DATES II
Bender, Margaret M.; Bryson, Reid A.; Baerreis, David A.
Radiocarbon, v. 8, 1966: 522-533
Reporting period: since November 1964. Proportional gas counting on methane. Radiocarbon half-life: 5568 years. Results expressed as years BP and calendar years.

14.277 1967
UNIVERSITY OF WISCONSIN RADIOCARBON DATES III
Bender, Margaret M.; Bryson, Reid A.; Baerreis, David A.
Radiocarbon, v. 9, 1967: 530-544
Radiocarbon half-life: 5568 years. Results expressed as years BP and calendar years.

14.278 1968
UNIVERSITY OF WISCONSIN RADIOCARBON DATES IV
Bender, Margaret M.; Bryson, Reid A.; Baerreis, David A.
Radiocarbon, v. 10, n. 1, 1968: 161-168
Period covered: since Nov. 1966. Half-life: 5568 years. See Wisconsin II for description of procedure.

14.279 1968
UNIVERSITY OF WISCONSIN DATES V
Bender, Margaret M.; Bryson, Reid A.
Radiocarbon, v. 10, n. 2, 1968: 473-478
Dates obtained since August 1967. Procedures described in Wisconsin II. Half-life: 5568 years.

14.280 1959
UPPSALA NATURAL RADIOCARBON MEASUREMENTS I
Olsson, Ingrid U.
Radiocarbon, v. 1, 1959: 87-102
Reporting period: 1957-1958. Proportional gas counting on carbon dioxide. Short description of pretreatment is given. Radiocarbon half-life: 5570 years. Results expressed in years BP.

14.281 1960
UPPSALA NATURAL RADIOCARBON II
Olsson, Ingrid U.
Radiocarbon, v. 2, 1960: 112-128
Reporting period: 1959. Proportional gas counting on carbon dioxide. Radiocarbon half-life: 5570 years. Results expressed in years BP. A short description of pretreatment method is included.

14.282 1961
UPPSALA NATURAL RADIOCARBON MEASUREMENTS III
Olsson, Ingrid U.; Cazeneuve, Horacio; Gustavsson, John; Karlén, Ingvar
Radiocarbon, v. 3, 1961: 97 81-85
Reporting period: 1960. Radiocarbon half-life: 5570 years. Results reported in years BP.

14.283 1964
UPPSALA NATURAL RADIOCARBON MEASUREMENTS IV
Olsson, Ingrid U.; Kilicci, Serap
Radiocarbon, v. 6, 1964: 291-307
Proportional gas counting on carbon dioxide. Radiocarbon half life: 5570 years. Results expressed in years BP and calendar years.

14.284 1965
UPPSALA NATURAL RADIOCARBON MEASUREMENTS V
Olsson, Ingrid U.; Piyamy, Piya
Radiocarbon, v. 7. 2965: 315-330
Proportional counting on carbon dioxide. Radiocarbon half-life: 5570. Results expressed in years BP and calendar years.

14.285 1965
UPPSALA NATURAL RADIOCARBON MEASUREMENTS VI
Olsson, Ingrid U.; Karlén, Ingvar
Radiocarbon, v. 7, 1965: 331-335
Reporting period: since 1962. Samples measured to determine the increase of the $^{14}C/^{12}C$ ratio due to explosion of nuclear devices.

14.286 1967
UPPSALA NATURAL RADIOCARBON MEASUREMENTS VII
Olsson, Ingrid U.; Stenberg, Allan; Göksu, Yeter
Radiocarbon, v. 9, 1967: 454-470
Proportional gas counting on carbon dioxide. Radiocarbon half-life: 5570 years. Results expressed as years BP and calendar years.

14.287 1967
UPPSALA RADIOCARBON MEASUREMENTS VIII
Stenberg, Allan; Olsson, Ingrid U.
Radiocarbon, v. 9, 1967: 471-476
Samples measured since Autumn 1964 to determine the increase of the $^{14}C/^{12}C$ ratio due to explosion of nuclear devices.

14.288 1968
VERNADSKY INSTITUTE RADIOCARBON DATES IV - V
Vinogradov, A.P.; Devirts, A.L.; Dobkina, E.I.; Markova, N.G.V.
Radiocarbon, v. 10, n. 2, 1968: 451-464
Method: liquid scintillation counting on benzene. Projects: Upper Pleistocene chronology for European Russia and dating of recent fossil soils. Materials dated: peat and wood.

14.289 1966
VICTORIA NATURAL RADIOCARBON I
Bermingham, Anne
Radiocarbon, v. 8, 1966: 507-521
Reporting period: from the end of 1963. Proportional gas counting on carbon dioxide. Radiocarbon half-life: 5568 years. Results expressed in years BP and calendar years.

14.290 1968
WASHINGTON STATE UNIVERSITY NATURAL RADIOCARBON MEASUREMENTS I
Chatters, Roy M.
Radiocarbon, v. 10, n. 2, 1968: 479-498
The laboratory started operation in November 1962, using a Sharp Laboratory Inc., CDL-14 system based upon the methane method of Fairhall, Shell and Takashima (1961). Radiocarbon half-life: 5568 years.

14.291 1953
YALE NATURAL RADIOCARBON MEASUREMENTS I: PYRAMID VALLEY, NEW ZEALAND, AND ITS PROBLEMS
Blau, J.H.; Deevey, Edward S.; Gross, Marsha S.
Science, v. 118, n. 3056, 24 July 1953: 118
A description of the establishment of the laboratory is followed by a discussion of the possible sources of errors, particularly in the dating of carbonaceous material. The contamination produced by the atomic bomb test at Eniwetok is mentioned. The dating of a Moa deposit at Pyramid Valley, New Zealand, is described in full as the methodologic interest of the series consists in the possibility of comparison of radiocarbon analyses based on carbonates and organic carbon at the same stratigraphic horizons.

14.292 1955
YALE NATURAL RADIOCARBON MEASUREMENTS II
Preston, R.S.; Person, Elaine; Deevey, Edward S.
Science, v. 122, n. 3177, 18 Nov. 1955
Method: acetylene counting as per Suess. Projects: North American geology, Alaskan Little Ice Age, Caribbean archaeology. Material dated: wood, peat, gyttja, shells, charcoal. There is a discussion of the procedures in calculating standards error.

14.293 1957
YALE NATURAL RADIOCARBON MEASUREMENTS III
Barendsen, G.W.; Deevey, Edward S.; Gralenski, L.J.
Science, v. 126, n. 3279, 1 Nov. 1957: 908-919
Period covered: July 1955 - March 1957. Method: carbon dioxide gas proportional counting, duplicate 24 hours run on each counter. Ages reported as years BP. Material dated: wood, charcoal, peat, antler, gyttja, marl, bone.

14.294 1959
YALE NATURAL RADIOCARBON MEASUREMENTS IV
Deevey, Edward S.; Gralenski, L.J.; Hoffren, Vaino
Radiocarbon, v. 1, 1959: 144-172
Reporting period: up to 1958.

14.295 1960
YALE NATURAL RADIOCARBON MEASUREMENTS V
Stuiver, Minze; Deevey, Edward S.; Gralenski, L.J.
Radiocarbon, v. 2, 1960: 49-61
Some repetition of Yale IV. Mainly gas counting on carbon dioxide.

14.296 1961
YALE NATURAL RADIOCARBON MEASUREMENTS VI
Stuiver, Minze; Deevey, Edward S.
Radiocarbon, v. 3, 1961: 126-140
Includes a series of ^{14}C assays of lake waters and other lacustrine materials normalised for ^{13}C content according to the formulation by Broecker and Olson (Lamont VIII).

14.297 1962
YALE NATURAL RADIOCARBON MEASUREMENTS VII
Stuiver, Minze; Deevey, Edward S.
Radiocarbon, v. 4, 1962: 250-262
Radiocarbon half-life: 5570 ± 30 years.

14.298 1963
YALE NATURAL RADIOCARBON MEASUREMENTS VIII
Stuiver, Minze; Deevey, Edward S.; Rouse, Irving
Radiocarbon, v. 5, 1963: 312-341
Radiocarbon half-life: 5570 ± 30 years

AUTHOR INDEX

Ackerman, Robert E.: **14**.252
Acosta, Jorge R.: **9**.236
Adam, David P.: **7**.023
Aegerter, S.K.: **2**.306
Agogino, George A.: **9**.053, 055, 077, 147, 193, 194, 208, 214, 243, 331, 372
Agrawal, D.P.: **3**.181, 252; **10**.005, 006, 027, 034, 054, 055, 056, 057, 058, 059, 060, 061; **14**.201, 202, 203, 204
Ahmad, N.: **5**.105
Aikens, C. Melvin: **9**.222
Aitken, M.J.: **2**.196, 197; **3**.217
Akers, L.K.: **3**.048, 055; **14**.147
Akers, Lawrence: **3**.295
Albritton, Claude C.: **9**.350
Alder, B.: **3**.133, 134, 274
Alder, Bernhard: **3**.275, 309
Aldous, K.J.: **3**.110, 205
Alekseyev, V.A.: **14**.066, 157
Alessio, M.: **2**.115; **14**.259, 260, 261, 262, 263
Alessio, Marisa: **3**.054
Alexander, Corrinne: **14**.215, 216
Alexander, G. W.: **2**.022
Alexander, Herbert J.: **9**.176
Alhonen, Pentti: **4**.315
Alimen, Marie-Henriette: **4**.298, 303
Allegri, Lucia: **3**.054
Allen, J.R.L.: **4**.400
Allison, Ira S.: **4**.291; **5**.046
Alvarez, Julian: **3**.152
Ambrose, W.R.: **3**.003; **12**.046
An.: **1**.008, 013, 014, 017, 031; **2**.001, 203, 213, 253; **3**.007, 045, 099, 128, 204, 258; **4**.284, 285; **9**.052, 201; **12**.057, 074; **13**.009, 011, 013
Ancion, N.: **4**.128; **14**.123, 126
Andersen, Björn G.: **5**.084, 232
Andersen, S.T.: **5**.301
Anderson, D.D.: **9**.349
Anderson, Douglas D.: **9**.032
Anderson, Ernest C.: **2**.007, 122, 179, 207, 252, 316; **3**.154, 166, 176, 325; **14**.045
Anderson, Franz E.: **5**.248
Anderson, G.M.: **2**.049, 249; **3**.276
Anderson, J.E.: **9**.109
Anderson, Sv.Th.: **5**.034, 199
Anderson, T.W.: **5**.073

Andrews, E. Wyllys: **9**.113
Andrews, E.W.: **9**.114
Andrews, George W: **7**.028
Andrews, J.T.: **1**.027; **4**.426; **5**.188, 240; **14**.159
Andrews, John T.: **4**.247; **5**.142, 255; **14**.160
Antevs, Ernst: **4**.144, 454; **5**.068, 069, 071, 299; **9**.048, 074, 146
Apelgot, S.: **3**.332
Arkell, A.J.: **8**.070, 072
Armstrong, J.E.: **7**.046
Armstrong, W.D.: **3**.080, 100, 105, 292
Arnal, J.: **11**.050
Arnold, James R.: **2**.007, 010, 033, 058, 122; **3**.082, 154, 176, 313; **4**.115; **13**.012; **14**.153, 154
Arrhenius, Gustaf: **3**.178; **4**.014, 118, 218
Arslanov, Kh. A.: **3**.263, 314; **4**.140; **14**.003, 106
Artemiev, V.V.: **3**.161, 162
Ashbee, Paul: **11**.027, 092
Ashwell, Ian Y: **5**.234
Atkinson, R.J.C.: **11**.040
Atlury, C.R.: **2**.199
Audric, B.N.: **3**.021, 177, 179, 347
Auer, Väinö: **7**.093
Aveleyra Arroyo de Anda, Luis: **9**.252
Axelrod, Daniel I.: **7**.103
d'Ayob, Mohammed bin: **7**.055

Baadsgaard, H.: **2**.165
Bachechi, F.: **14**.262
Back, William: **2**.141; **4**.066, 360, 361, 387
Baden, Howard P.: **3**.132
Baenen, James: **9**.108
Baerreis, David A.: **9**.096; **14**.275, 276, 277, 278
Baghoorn, Elso S.: **11**.048
Baillie, L.A.: **3**.078, 079
Bainbridge, Arnold D.: **2**.162
Baker, George: **7**.071
Baker, M.J.: **2**.060; **14**.134, 135, 136, 137
Baker, Richard G.: **5**.151
Ballario, C.: **3**.009
Ballentine, Robert: **2**.150
Bandez, Claude F.: **9**.088
Bandi, H.G.: **11**.013
Bandy, Orvill C.: **4**.405
Bank, Th. P.: **5**.199
Bank, W.J.: **9**.151
Banks, Maxwell R.: **4**.041, 056
Bannister, Bryant: **2**.111
Barendsen, G.W.: **3**.183, 202, 261, 283; **14**.293
Barendsen, G.W.: **14**.082
Baret, C.: **3**.209
Barghoorn, E.S.: **4**.136, 192

Barker, Harold: **2**.242; **3**.272, 284, 285; **14**.021, 022, 023, 024, 025, 026, 027, 028
Barker, M.J.: **14**.138
Barklay, F.R.: **2**.284
Barnes, Peter W.: **4**.163
Barr, W.: **5**.218
Barreis, David A.: **9**.073
Bartlett, Alexandra S.: **4**.136
Bartlett, H.A.: **5**.105
Bartlett, H.H.: **2**.233; **3**.249
Baskerville, Charles A.: **4**.145
Bass, George F.: **10**.007
Baumgart, I.L.: **4**.409
Baumhoff, Martin A.: **5**.210; **9**.037, 041, 106, 213
Baxter, M.S.: **2**.060; **3**.033
Bayrock, L.A.: **4**.168; **9**.054
Bé Allan, W.H.: **6**.048
Beaubien, Pierre L.: **9**.090
Beaumont, P.B.: **8**.067
Beetham, N.: **4**.264
Behrens, E. William: **4**.380
Bell, J.: **3**.097
Bell, R.E.: **9**.022, 097
Bell, Robert E.: **9**.276, 279, 286, 288, 294, 295, 296
Bell, S.V.: **4**.412
Bella, F.: **2**.115; **14**.259, 260, 261, 262, 263
Bender, Margaret M.: **2**.167; **14**.275, 276, 277, 278, 279
Bender, Michael L.: **6**.048
Bender, V.R.: **4**.204
Benedict, James B.: **4**.331; **5**.262, 263
Beneventano, M.: **3**.009
Beng, Hans-Jürgen: **4**.084
Benington, F.: **9**.060
Bennett, Richard: **4**.061
Bennett, Richmond: **14**.010
Bennyhoff, J.A.: **9**.023
Benson, G.T.: **4**.017
Bent, A.M.: **4**.258
Berg, Thomas E.: **4**.458
Berger, P.: **11**.024
Berger, Rainer: **1**.018; **2**.022, 231, 232, 271; **3**.096, 199, 271, 277; **4**.086; **7**.040; **8**.002; **9**.136, 137, 256, 297, 315, 370; **14**.225, 226, 227, 228, 229
Berglund, B.E.: **4**.194, 195, 196, 197
Bermingham, Anne: **14**.289
Bernal, Ignacio: **9**.029, 069
Berner, R.A.: **4**.366
Bernstein, Richard B.: **3**.317
Bernstein, William: **2**.150; **3**.319
Berry, Elmer G.: **2**.193
Berry, L.: **4**.412
Berthold, Sarah M.: **14**.217

Berti, A.A.: **5**.073
Bertrand, Kenneth: **7**.082
Bien, George S.: **2**.061, 158; **4**.366; **6**.021, 032, 033, 036, 037, 049; **14**.108, 109, 110, 111, 112
Bigeleisen, Jacob: **2**.127
Binford, Lewis R.: **9**.152, 332
Bird, Junius Bonton: **2**.257; **9**.341
Birks, Helen: **12**.028, 062
Birks, Lawrence: **12**.028, 062
Biscaye, P.E.: **2**.293
Bishop, W.W.: **5**.131, 148
Bjerkenes, Clara: **3**.319
Black, L.D.: **3**.201
Black, Robert F.: **4**.455, 458; **5**.049, 090, 110, 183, 247; **9**.014
Blackadar, R.G.: **5**.189
Blackburn, G.: **4**.336
Blackburn, Kathleen B.: **5**.045
Blake, Weston: **4**.155, 232, 289, 290, 349, 350, 371; **5**.060, 175, 223, 269; **14**.071, 072, 073, 074, 075, 076
Blanc, Alberto Carlo: **5**.066; **7**.056
Blanchard, F.R.: **3**.156
Blau, J.H.: **14**.291
Blifford, I.H.: **2**.014, 024
Bliss, Wesley L.: **3**.248
Bloom, Arthur L.: **4**.182, 283, 433; **7**.008, 032
Blundell, D.J.: **14**.019, 232
Bohnanberger, Otto: **4**.019
Boissonneau, A.N.: **5**.094, 095
Bolin, Bert: **2**.078, 189; **4**.156
Bonatti, Enrico: **4**.118
Bonatti, Henrico: **4**.218
Bonis, Samuel: **4**.019
Bopp, Monika: **5**.197
Borch, C.C. von der: **4**.221
Borden, Charles B.: **9**.259, 371
Bordes, Francois: **3**.208
Borns, Harold W.: **4**.015; **7**.033, 122; **9**.216
Boughey, A.S.: **4**.096
Boulton, G.S.: **5**.136, 160
Bowen, R.N.C.: **2**.261
Bowler, J.M.: **4**.294, 311; **12**.036
Bowman, Albert L.: **14**.127, 128, 129, 130
Bowsker, A.L.: **5**.107
Brabant, J.M.: **2**.120
Bradley, C.W.: **4**.059
Bradley, W.H.: **4**.451; **9**.257
Braidwood, Robert J.: **2**.009, 184; **10**.032, 033, 037, 052
Brandtner, F.: **11**.057
Brandtner, Friedrich: **4**.072
Brannon, H.R.: **2**.092, 093, 251; **3**.238; **14**.094, 095
Bray, Ellis E.: **4**.165, 166, 341; **7**.127; **14**.184

Author Index

Bray, J. Roger: **2.**012, 023, 298; **5.**100
Breuil, Abbé, Henri: **8.**010
Briard, J.: **11.**010
Briggs, Lyman J.: **3.**337
Bright, R.C.: **4.**269
Britt, Claude: **2.**205
Broda, E.: **3.**116, 341
Brodie, J.W.: **6.**002; **7.**024
Broecker, Wallace S.: **2.**017, 035, 046, 086, 117, 137, 248, 254, 257, 274; **3.**039, 247, 286, 348; **4.**048, 055, 123, 314, 323, 324, 334, 357, 364, 365, 367; **5.**003, 097; **6.**005, 010, 016, 017, 020, 022, 029, 039, 041, 042, 045; **7.**016, 025; **14.**114, 116, 117, 118, 119, 120
Broel, J.M.M. van den: **7.**048
Brogue, Robert D.: **5.**219
Brooks, Harold K.: **9.**361
Brophy, John A.: **4.**410; **5.**064; **9.**230
Brothers, R.N.: **12.**034
Brotzen, F.: **5.**121
Brown, F.: **3.**184
Brown, G.F.: **4.**092
Brown, J.: **4.**087
Brown, J.F.: **5.**171
Brown, Jerry: **4.**342, 359; **7.**047; **9.**307
Brown, Sarborn C.: **3.**035
Brown, W.C.: **3.**297
Brownell, Gordon L.: **3.**052, 053
Bruggemann: **4.**407
Brush, Grace S.: **5.**196
Bryson, Reid A.: **4.**321; **5.**033, 254, 255; **9.**073, 096; **14.**275, 276, 277, 278, 279
Bucha, V.: **2.**077
Buchanan, Donald L.: **2.**053; **3.**189
Buckley, J.D.: **14.**100
Buckley, Jane T.: **5.**028
Budde, Enno: **14.**088
Buist, A.G.: **3.**246; **12.**003, 063
Bullen, Ripley P.: **9.**066, 115, 274, 361
Bulmer, Susan: **12.**051, 067
Burch, Ernest S.: **9.**065
Burchladze, A.A.: **14.**205
Burger, D.: **5.**266
Burke, W.H.: **3.**032, 187
Burke, W.J.: **14.**184
Burling, R.W.: **6.**002, 044
Burns, George W.: **5.**304
Burton, Virginia L.: **2.**050
Bushnell, G.H.S.: **9.**237
Bushnell, Geoffrey: **9.**269
Butler, B. Robert: **9.**046, 080
Butler, Patrick: **4.**245
Butomo, S.V.: **3.**161, 162; **14.**165

Buttlar, H., von: **3.**167
Butzer, Karl W.: **11.**041
Buyske, D.A.: **3.**085
Byers, Douglas S.: **4.**117; **9.**063, 191, 229, 281, 293
Byers, F.M.: **4.**171
Byrne, J, S.: **9.**106

$\mathbb{C}$aine, N.: **4.**060; **7.**143
Caini, Vasco: **3.**243
Callendar, G.S.: **2.**187
Callow, W.J.: **2.**060; **14.**134, 135, 136, 137, 138, 139
Calvert, S.E.: **4.**008, 238; **6.**013
Calvin, Melvin: **3.**142
Campbell, C.A.: **2.**132; **4.**035, 456
Campbell, D.A.: **4.**173
Campbell, John M.: **3.**259; **9.**359
Campbell, T.N.: **9.**057, 178
Cann, J.R.: **10.**040
Capes, K.N.: **9.**079
Capron, P.C.: **3.**221; **4.**128; **14.**121, 122, 123, 126
Cardini, L.: **4.**244
Carlson, Paul R.: **4.**216
Carlson, Roy L.: **9.**071
Carnan, R.D.: **3.**329
Carr, D.R.: **3.**075
Carson, Charles E.: **4.**344
Carter, George F.: **9.**226
Case, Humphrey: **11.**052
Casey, D.A.: **12.**036
Caskey, Elizabeth G.: **11.**028
Caskey, J.L.: **11.**051
Caskey, John L.: **11.**028
Caso, Alfonso: **9.**069, 185, 327, 373
Castle, L.: **3.**023
Caswell, R.S.: **2.**120
Cazeneuve, Horacio: **14.**282
Chagnon, Jean-Yves: **4.**032
Chandler, John B: **14.**207
Chapman, V.J.: **4.**209
Chard, Chester S.: **9.**199; **14.**055, 187, 188
Charles, F.W.B.: **11.**024
Chatters, Roy M.: **3.**203; **4.**464; **13.**002, 006; **14.**290
Cheatum, E.P.: **4.**231
Cherdyntsev, V.V.: **14.**066, 157
Chernysh, A.P.: **11.**073
Chesterman, C.W.: **4.**024
Childe, V. Gordon: **2.**087
Christiansen, E.A.: **4.**078; **5.**294; **7.**107
Christie, Patricia M.: **11.**008
Christman, David R.: **3.**115
Chudy, M.: **3.**291
Church, R.E.: **4.**026

Churcher, C.S.: **4**.206; **7**.111; **9**.140
Churchill, D.M.: **4**.108, 186; **11**.045, 105
Clark, Donald W.: **9**.221, 365
Clark, Grahame: **11**.036, 088
Clark, J. Desmond: **8**.006, 012, 026, 032, 033, 035, 050, 052, 054, 079
Clark, J.G.: **10**.062
Clark, John Graham Douglas: **11**.034, 062, 065, 089
Clarke, A.H.: **9**.016, 234
Clayton, G.D.: **3**.082
Cleland, C.W.: **9**.051
Clelow, C.W.: **9**.037
Clendening, John A.: **7**.059
Clisby, Kathryn: **7**.141
Coach, J.R.: **7**.094
Coachman, L.K.: **2**.136, 223
Coch, Nicholas K.: **7**.099
Coe, Joffre Lanning: **9**.144
Coe, Michael D.: **9**.084, 169, 204, 205
Coe, William R.: **9**.101, 202, 319, 356, 357; **14**.257
Coetzee, J.A.: **4**.261
Cole, David R.: **9**.089
Cole, G.A.: **4**.058
Cole, Sonia Mary: **3**.190; **8**.053
Coleman, James M.: **4**.203, 384
Coles, John: **11**.064
Coles, John M.: **11**.077
Coles, John P.: **11**.032
Colinvaux, Paul A.: **4**.312; **5**.017, 182, 202
Collier, Donald: **2**.040, 181, 246
Collier, J.: **3**.126
Collins, Gary: **4**.081
Collins, Harry B.: **9**.303
Collins, Henry B.: **9**.024, 354
Commission for the Study of the Quaternary Period. Akademia Nauk SSSR.: **14**.156
Connah, Graham: **8**.056, 057
Connelly, G. Gordon: **7**.084
Connolly, J.R.: **7**.019
Cook, S.F.: **2**.204; **3**.016, 196
Cook, Sherburne F.: **3**.321
Cooke, C.K.: **8**.069
Cooke, H.B.S.: **5**.059
Coon, Carleton Stevens: **10**.053, 067
Coope, Geoffrey Russell: **7**.030, 065
Cooper, G.R.: **5**.116
Cooper, L.H.N.: **6**.023
Coppens, R.: **14**.133
Corliss, J.B.: **4**.076
Cornwall, I.W.: **3**.066
Cortesi, Cesare: **3**.009; **14**.262, 263
Costin, Alec B.: **4**.229, 408
Coulter, H.W.: **4**.337

Coursaget, Jean: **11**.015, 038, 069; **14**.081
Coutts, J.F: **12**.019
Cowan, Clyde: **2**.199
Cowan, Richard D.: **9**.170
Craig, Bruce Gordon: **4**.439; **5**.177, 194, 285; **7**.076, 077
Craig, Harmon: **2**.048, 055, 123, 138, 175; **3**.171; **6**.012, 018
Crandell, D.P.: **5**.096
Crandell, Dwight R.: **4**.293, 437; **7**.046, 101, 120
Crane, H.R.: **2**.065; **3**.020, 050, 051, 071; **9**.018; **14**.237, 238, 239, 240, 241, 242, 243, 244, 245, 246, 247, 248
Crary, A. P.: **6**.015
Crathorn, A.R.: **3**.064, 114, 343
Crawford, J.R.: **8**.039
Crawford, O.G.S.: **3**.094; **11**.104
Creager, Joe S.: **4**.148, 158, 459; **7**.070
Cressman, L.S.: **3**.350; **9**.009, 089, 167
Crevecoeur, E.H.: **3**.221
Croft, M.G.: **7**.012
Crook, W.W.: **9**.223
Crosby, James W.: **3**.203; **4**.464
Crowe, C.: **2**.056
Crumlin Perdersen, O.: **11**.101
Cruxent, Jose M.: **9**.004, 340, 369
Cumming, G.L.: **2**.165
Curran, S.C.: **4**.105
Curray, Joseph R.: **4**.051, 160, 182, 189, 191; **6**.006
Currie, L.A.: **3**.228
Curtiss, L.F.: **2**.151
Cushing, Edward J.: **2**.266; **5**.155, 156, 226

Da Costa, John: **9**.094
Daddario, Joseph J.: **4**.434
Dahlman. R.C.: **2**.039
Dales, George F.: **10**.065, 066, 075
Dalrymple. D: **3**.038
Daly, P.: **10**.030
Damon, Paul E.: **2**.133, 214, 225, 288, 300; **4**.058, 065; **14**.006, 007, 008, 009, 010, 231
Daniel, Glyn: **3**.090, 091; **9**.125; **11**.018
Daniels, R.B.: **4**.031
Dansgaard, Willi: **2**.083, 223
Danyluk,S.S.: **3**.245
Darragh, T.A.: **12**.036
Dart, Raymond A.: **8**.067
Daugherty, Richard D.: **9**.105, 108
Daughtry, A.C.: **2**.093, 251; **14**.094, 095
Daughtry, Perry D.: **2**.092
Davidson, J.K.: **12**.064
Davidson, J.M.: **12**.003
Davidson, Jack D.: **3**.220

Author Index

Davidson, Jackson B.: **3**.149
Davidson, Jackson D.: **3**.340
Davies, J.: **8**.051
Davis, E. Mott: **3**.137; **9**.021, 313; **14**.268, 269, 270, 271, 272, 273
Davis, E.L.: **9**.180
Davis, Margaret B.: **4**.252, 445; **5**.147, 193, 195
Davis, Wilbur A.: **9**.089
Dawson, W.A.: **4**.459
de Geer, Ebba Hult: **4**.080, 099; **5**.070
de Jong, Jan D.: **4**.306
De Laguna, F.: **9**.159
de Marco, A.: **3**.009
de Saussure, R.: **9**.345
de Vries, A.E.: **2**.157, 282; **3**.282
de Vries, Hessel: **2**.032, 057, 096, 098, 299; **3**.111, 173, 183, 236, 241, 261, 282, 300, 302; **4**.343; **5**.034, 065, 066, 244, 257; **11**.094; **14**.082, 083
de Vries, Hessel : **2**.223; **3**.202
Deacon, H.J.: **3**.257; **8**.014, 078
Deacon, J.: **1**.001
Decker, R.W.: **4**.019
Deevey, Edward S.: **1**.029; **2**.036, 178, 237, 296; **3**.095, 109, 216, 232; **4**.109, 252, 254, 370; **5**.213; **7**.136; **14**.291, 292, 293, 294, 295, 296, 297, 298
Deffeyes, Kenneth S.: **4**.004
Degens, E.T.: **2**.054
Degerbøl, Magnus: **11**.071
Delaney, C.F.G.: **3**.279
Delibrias, Georgette: **3**.069, 070, 209; **14**.080, 177, 178
Delorme, L.D.: **5**.075
Denton, George H.: **4**.015; **5**.029, 167; **7**.037
Denton, Georgette: **7**.039
Detterman, R.L: **5**.107
Detterman, Robert: **5**.242
Deumer, J.M.: **14**.121, 122
Deuser, W.G.: **2**.054
Devirts, A.L.: **2**.089, 113; **4**.006, 070; **5**.044, 145; **14**.002, 158, 288
Dewar, H.S.L.: **11**.006
Diehl, Richard A.: **9**.204
Diethorn, Ward: **3**.188
Dietrich, Jacob E.: **3**.148
Dietz, W.T.: **4**.176
Diez, M.: **3**.081
Dikov, N.N.: **10**.015
Dimbleby, G.W.: **3**.219; **11**.039, 068
Dinsdale, B.: **3**.096
Dixon, J.E.: **10**.040
Dixon, K.A.: **9**.068
Dobbs, Horace E.: **3**.242

Dobkina, E.I.: **2**.089, 113; **5**.044; **14**.002, 158, 288
Dodd, John: **4**.081
Donaldson, J.W.: **4**.209
Donne, William L.: **6**.009; **7**.089
Donner, J.J.: **5**.076
Donovan, D.T.: **5**.272
Dorn, Thomas Felder: **3**.280; **14**.162
Dort, Wakefield: **9**.217
Dossin, J.M.: **14**.121
Douglas, David L.: **2**.168; **3**.185
Douglas, L.A.: **4**.063, 453
Doyle, Michael V.: **4**.219
Dragoo, Don W.: **9**.124
Drake, David E.: **4**.163
Drapier, Lyn: **14**.159
Dreimanis, Aleksis: **4**.129, 179, 214, 462; **5**.016, 065, 127, 205, 206, 220, 260, 280, 281, 289, 290, 295, 307; **7**.066, 067; **9**.231
Drever, R.W.P.: **3**.022
Drier, Roy: **9**.184
Drinnan, R.E.: **9**.016
Drobinski, J.C.: **3**.307, 308
Droste, John B.: **5**.005
Drozhzhin, V.M.: **3**.161, 162
Drucker, P.: **9**.282
Drucker, Philip: **9**.132
Ducker, Alfred: **5**.021
Duff, Roger: **12**.016
Dugas, Doris J.: **2**.110
Dumond, D.E.: **9**.362
Duncan, John R.: **4**.216
Dunn, Frederick L.: **10**.063
Dunning, John R.: **2**.144
Dunshee, Bryant: **3**.100
Dupree, L.: **10**.046
Durand, G.L.A.: **14**.133
Dury, G.H.: **1**.002, 003, 004
Dutro, J.T.: **5**.107
Dyck, Willy: **2**.262, 278, 279; **3**.119; **14**.067, 068, 069, 070, 071, 072, 264
Dyson, Robert H.: **10**.002, 028, 029, 043, 074

Eardley, A.J.: **4**.375; **7**.005
Easterbrook, Don J.: **5**.080, 221, 235, 300; **7**.038, 046, 102, 105, 120
Eckelman, Walter R.: **2**.219; **6**.048; **14**.115
Eckholm, Gordon F.: **9**.220
Eden W.J.: **4**.046
Edmonson, Munro S.: **12**.042
Edwards, B.: **3**.306
Edwards, G.: **2**.053
Edwards, G.R.: **11**.005

Edwards, R.L.: **9**.025
Ehhalt, D.: **2**.289; **4**.457
Ehrich, Robert W., ed.: **11**.017
Eichler, R.: **2**.049
Eidinoff, Maxwell Leigh: **3**.182
Einarsson, T.: **4**.053
Elliott, M.J.: **2**.284
Ellis, John G.: **3**.333; **14**.179
Ellis-Gruffydd, I.D.: **5**.171
Elsasser, Walter: **2**.099, 100
Elson, John A.: **5**.077, 124
Elwood, B.: **4**.327
Emerson, William K.: **7**.041
Emery, Kenneth Orriz: **4**.009, 033, 136, 341, 396, 397, 398, 431; **5**.059; **9**.025
Emiliani, Cesare L.: **4**.244, 442; **5**.173; **6**.024; **7**.109, 110
Emory, Kenneth P.: **9**.151; **12**.004
Endo, Kunihiko: **2**.312; **14**.058, 059, 060
Engel, Frederick: **9**.235
Engelkemeir, Antoinette G.: **2**.154; **3**.098
Engstrand, Lars G.: **3**.197; **14**.191, 192, 193, 194, 195
Erich, Robert W.: **11**.043
Erickson. N.: **4**.452
Ericson, David B.: **4**.101, 401; **6**.011; **7**.025, 052, 063, 073
Eriksson, E.: **2**.078, 200
Eriksson, K. Gösta: **2**.270; **3**.324; **4**.104, 399; **6**.046
Erlenkenser, H.: **14**.234
Erlenkeuser, H.: **14**.233, 235
Erskine, J.S.: **9**.234
Estas, A.H.: **5**.022
Estrada, Emilio: **9**.117
Euler, Robert C.: **9**.344
Evans, Clifford: **9**.117, 181, 368
Evans, Ernest D.: **4**.165, 166
Evans, John D.: **11**.012, 099
Evans, R.D.: **2**.150
Ewart, Anthony: **4**.441
Ewing, G.C.: **4**.076
Ewing, Maurice: **4**.123, 401; **5**.293; **6**.009, 017, 020, 045; **7**.073, 089, 100

F aas, Richard W.: **4**.243
Fagan, Brian M.: **1**.022, 023, 024, 025, 026; **8**.007, 021, 024, 028, 030, 031, 034, 042, 046, 048, 071, 081, 082
Fagg, A.E.: **11**.020
Fagg, B.E.B.: **8**.047
Fagg, Bernard: **8**.009, 066
Fairbridge, Rhodes W.: **2**.079; **4**.098, 226, 393, 466; **5**.036; **8**.062
Fairhall, A.W.: **2**.013, 172, 255; **3**.010; **4**.452; **14**.162, 163, 164

Fairservis, Walter A.: **10**.021
Falconer, George: **5**.142, 157, 224
Faris, Barbara: **3**.139
Farnham, R.S.: **4**.199, 248
Farrand, William R.: **4**.278, 314, 335; **5**.123; **7**.089
Faul, Henry: **2**.186
Fauth, H.: **2**.173
Faye, John C.: **5**.049
Feely, Herbert W.: **2**.293; **3**.335; **14**.113
Feffay, Henry: **3**.152
Feigelson, Philip: **3**.220
Feltz, H.R.: **3**.225
Fenton, T.E.: **4**.176
Ferdon, Edwin N.: **12**.014
Ferguson, C. Wesley: **2**.037, 114
Fergusson, Gordon J.: **2**.008, 028, 029, 030, 031, 198, 228, 229, 267; **3**.005, 029, 062, 127, 278, 342, 345, 346; **4**.378; **6**.030; **14**.140, 141, 142, 143, 222, 223, 224, 225
Fernald, Arthur T.: **4**.345, 382; **5**.101
Ferrara, G.: **14**.039, 040
Ferrians, Oscar J.: **5**.063, 109, 139
Feth, J.H.: **4**.358
Feyling-Hanssen, Rolf W.: **4**.133, 403; **7**.050
Field, H.: **10**.009
Fisher, F.: **14**.146
Fisk, H.H.: **4**.143, 184
Fitting, James E.: **9**.173, 306, 334
FitzPatrick, E.A.: **5**.120
Flamm, E.J.: **2**.208
Flannery, Kent V.: **10**.020, 049
Fleming, C.A.: **4**.029, 340
Fleming, C.B.: **9**.263
Fletcher, John: **3**.264; **11**.025
Flint, Richard Foster: **1**.029; **2**.105, 119, 246; **3**.095; **4**.072, 170, 188, 420; **5**.081, 092, 163, 213, 225, 250, 259, 306; **7**.027, 053, 078, 134, 136, 144; **13**.012
Florini, J.: **3**.085
Focken, Charles M.: **3**.019, 266, 268
Folinsbee, R.E.: **2**.165
Follett, W.I.: **9**.139
Folliari, M.: **5**.066
Fonselius, Stig: **3**.195
Forbis, Richard G.: **9**.119
Ford, James A.: **9**.128, 153, 166, 233
Foreman, Fred: **7**.141
Fornaca-Rinaldi, G.: **14**.040
Forova, V.S.: **2**.256; **14**.066, 157
Forsyth, Jane L.: **4**.022; **5**.009, 047, 278; **7**.067, 090
Foster, H.L.: **5**.056
Fowler, Melvin L.: **9**.186, 260
Fox, Aileen: **11**.063

Author Index

Frankforter, W.D.: **9**.055, 214
Fratelli, P.: **2**.115
Fray, Charles: **7**.100
Frazier, David E.: **4**.379
Fredskild, Bent: **4**.282
Freedman, Arthur J.: **3**.166
Frerichs, William E.: **7**.115
Freundlich, J.: **3**.290; **14**.107
Frey, David G.: **4**.241; **5**.204
Friedman, Gerald M.: **4**.233
Friedman, Irving: **2**.162, 313; **4**.360
Friedman, Lewis: **2**.150
Friend, James P.: **14**.101
Fries, Carl: **14**.044
Fries, Magnus: **4**.237, 429; **5**.140; **7**.118
Frison, George C.: **9**.346
Fromm, E.: **4**.007
Frye, John C.: **4**.220; **5**.031, 035, 040, 054, 183; **7**.126; **9**.047; **13**.007
Fryxell, Roald: **5**.038, 158, 256; **9**.105, 108, 156
Fujiyama, Chikako: **14**.173, 174, 175, 176
Fulton, R.S.: **5**.180
Funk, Robert Ellsworth: **9**.044
Funt, B.L.: **3**.201, 245, 310
Furst, P.T.: **9**.277
Fyles, J.G.: **4**.155, 440, 465; **5**.287; **7**.076, 077; **14**.067, 068, 069, 070, 071, 072, 073, 074

Gabel, Creighton: **8**.023, 036
Gabrielsen, Gunnar: **4**.212
Gadd, N.R.: **5**.162; **7**.081
Gage, Maxwell: **4**.040; **7**.014, 054
Gajda, R.T.: **5**.123
Galbreath, Edwin C.: **9**.107
Gale, W.A.: **4**.420; **5**.306
Galloway, Eugene: **9**.331
Galloway, R.W.: **4**.095
Gant, E.: **9**.002
Garlake, P.S.: **8**.080
Garner, D.M.: **6**.003, 038, 044
Garrison, Gail C.: **4**.268
Garrison, Louis E.: **4**.397
Garutt, V.E.: **4**.106
Gates, David M.: **2**.106
Gathercole, P.W.: **12**.038
Gefeller, G.: **3**.327
Geiss, J.: **3**.327
Gejvall, N.G.: **3**.197
Genunche, Ana: **1**.005, 006
Genutene, I.K.: **4**.354
Geoffrey, J.D.: **2**.165
Gerard, R.: **6**.017, 020, 045

Gerard, Robert D.: **6**.022
Gerasimov, I.P.: **5**.052
Geyh, Mebus A.: **2**.173, 272; **3**.106, 235; **14**.089, 090, 091, 092
Gfeller, Chr.: **14**.013, 014, 015
Gibbs, J.A.: **3**.218
Giddings, J.L.: **4**.141; **9**.033, 104, 209
Giddings, James L.: **9**.034
Gifford, E.W.: **12**.010, 079
Gile, L.H.: **4**.013
Giletti, Bruno J.: **3**.289
Gill, Edmund D.: **2**.240; **4**.003, 038, 039, 040, 041, 042, 056, 068, 120, 300, 309, 310, 313, 339, 347, 348, 389, 404, 447; **7**.071; **12**.002, 006, 056, 060
Gillian, Jeanne A.: **5**.219
Gilot, E.: **4**.128; **14**.122, 123, 124, 125, 126
Ginsburg, R.N.: **4**.318
Giot, Pierre Roland: **11**.010, 015, 038, 069
Glasbergen, W.: **11**.009
Glass, H.D.: **4**.220; **5**.040
Godwin, Harry: **2**.082, 152; **3**.260; **4**.301; **5**.133, 153, 249, 253; **6**.034; **11**.006, 046, 065, 078, 090; **13**.003, 004, 015; **14**.029, 030, 031, 032, 033, 034, 035, 036, 168, 169, 170
Goel, Parmatma S.: **2**.101; **3**.339; **4**.089
Goesta, Wollin: **6**.011
Goggin, John M.: **9**.004
Göksu, Yeter: **3**.112; **14**.286
Goldberg, Leo: **2**.004
Goldin, A.S.: **3**.307, 308
Goldsmith, P.: **2**.284; **3**.184
Goldsmith, Victor: **4**.111
Goldthwait, Richard P.: **4**.124; **5**.048, 270, 303; **7**.033, 067; **9**.148
Golik, Abraham: **4**.159
Golson, J.: **3**.056; **12**.025, 034, 038, 046, 058; **14**.004, 005
Gonzales, E.: **4**.188
Goodall, E.: **8**.068
Goodell, H. Grant: **4**.240; **7**.098; **14**.056, 057
Gooding, Ansel M.: **4**.328, 329; **5**.111, 258, 268; **7**.112
Gooding, M.A.: **9**.056
Gordu, S.: **3**.085
Gorsline, Donn S.: **4**.163
Gould, H.R.: **4**.018
Gould, Stephen J.: **7**.087
Graham, John A.: **9**.315
Gralenski, L.J.: **14**.293, 294, 295
Grant-Taylor, T.C.: **14**.144, 145
Grave, V.R.: **11**.005
Gravenor, C.P.: **4**.327
Gray, James: **2**.042, 043
Grayson, John Francis: **5**.212

Graziadei, B.: **14**.263
Green Roger: **12**.041
Green, D.H.: **5**.105
Green, H.F.: **3**.184
Green, H.H.: **2**.239
Green, J.H.: **3**.097; **14**.249
Green, R.: **4**.152
Green, R.C.: **12**.003, 064
Gregg, D.R.: **12**.044
Gregory, J.N.: **2**.315
Grey, Donald C.: **2**.133, 288, 300; **14**.009, 010
Grey, James: **2**.286
Griffin, James B.: **3**.049; **9**.070, 172, 183, 196, 275, 301, 317, 333; **14**.238, 239, 240, 241, 242, 243, 244, 245, 246, 247, 248
Griffith, James S.: **9**.351
Grivenko, V.A.: **4**.178
Gromova, L.I.: **3**.263; **4**.140; **14**.003
Gross, Hugo: **9**.010; **11**.058
Gross, M.G.: **4**.459
Gross, Marsha S.: **2**.178; **14**.291
Grosscup, Gordon L.: **1**.007; **9**.091, 284
Grosse, Aristid V.: **2**.097, 144, 179, 252, 295
Grosswald, M.G.: **5**.044
Grubich, Donald N.: **4**.248
Gruhn, Ruth: **9**.363
Gruninger, John: **14**.255
Grunlund, Jean H.: **2**.146
Gugluer, S.M.: **4**.459
Guilday, John E.: **4**.074
Guillet, B.: **14**.133
Guillier, M.T.: **14**.080, 177, 178
Gulbransen, E.A.: **2**.308
Gustavsson, John: **14**.282
Guthe, Alfred K.: **9**.347

Haag, William G.: **9**.166, 167
Hackett, J.E.: **5**.011, 129
Hadleigh-West, Frederick: **9**.110, 149
Hadlock, Wendell S.: **9**.063, 281
Hafsten, Ulf: **5**.058; **7**.069
Hageman, Frerich T.: **2**.042, 043, 286
Haigh, C.P.: **3**.150
Haile, N.S.: **7**.055
Hails, John R.: **4**.406
Håkansson, Soren: **14**.236
Hall, George Frederick: **4**.154, 176
Hall, N.T.: **3**.326
Hall, Robert L.: **2**.291; **9**.035, 064, 075, 273
Halstead, Carl: **5**.043
Halstead, E.C.: **4**.438
Hamada, Tatsuji: **14**.173, 174, 175, 176

Hamermesh, B.: **2**.129
Hamill, W.H.: **2**.154
Hamilton, E.I.: **3**.014
Hamilton, Thomas M.: **4**.202
Hammel, E.A.: **9**.118
Hammen, Th. van der: **4**.023; **5**.039, 053, 187, 282, 288
Hammen, Th. van der : **4**.188, 422
Handshaw, Bruce B.: **2**.313
Hansbury, Elizabeth: **2**.095
Hansbury, V.N.: **2**.094
Hansen, Henry P.: **4**.071, 090
Hansen, P.H.: **5**.211
Hansen, Sigurd: **4**.302
Hanshaw, Bruce B.: **2**.141; **3**.225; **4**.066, 360, 361, 387
Harbaugh, John W.: **3**.330
Harding, J.R.: **3**.076
Hardy, P.H.: **2**.068
Harford, L.B.: **4**.311
Haring, A.: **3**.282
Harmon, Kathryn Parker: **7**.034
Harp, Elmer: **9**.245
Harp, G.D.: **2**.013, 172
Harring, A.: **2**.157, 282
Harrington, H.J.: **5**.236
Harris, D.R.: **11**.020
Harris, G.M.: **2**.202
Harris, Josephine: **14**.249
Harris, R.K.: **9**.223
Harrison, D.A.: **5**.261, 265
Harrison, W.: **4**.376, 392; **7**.119
Harrisson, Tom: **10**.008
Hart, S.R.: **2**.218
Hartman, Richard T.: **4**.135
Hartshorn, J.H.: **5**.072, 231
Hasagewa, Hiroishi: **2**.161, 280, 310
Haselberg, Herta: **8**.038
Haselton, George M.: **5**.088
Hassall, Geraldine I.: **14**.136, 137, 138, 139
Hattersley-Smith, G.: **5**.216
Haury, Emil W.: **9**.013, 048, 175, 364
Hawkins, R.C.: **2**.145; **3**.211
Hawley, J.W.: **4**.013
Hay, Richard L.: **4**.373
Hay, Ruth J.: **14**.148, 149, 150
Hayashi, Kensaku: **10**.026
Hayden Julian D.: **9**.353
Hayes, F. Newton: **2**.094, 095; **3**.130, 145, 153, 154, 158, 159, 160, 163, 222
Haynes, C. Vance: **3**.028, 287, 288; **4**.058, 139, 299; **5**.165; **9**.061, 062, 126, 142, 147, 227, 243; **14**.008, 009, 010
Healy, James: **4**.417
Heezen, Bruce C.: **4**.401; **6**.020, 042, 045; **7**.019

Heidelberger, Charles: **3.**142
Heinrich, F.: **3.**167
Heintz, Anatol: **4.**106, 225
Heinzelin, J. de: **4.**177
Heizer, Robert F.: **2.**166, 204; **3.**016, 073, 321; **5.**210; **9.**005, 023, 037, 041, 049, 132, 192, 250, 253, 280, 282, 311, 315, 320, 337
Helley, Edward J.: **4.**100
Hemmingsen, E.: **2.**136, 223
Hemon, A.F.: **3.**180
Henoch, W.E.S.: **5.**214
Henry, Vernon J.: **4.**169; **7.**002, 003
Herman, Y.: **4.**125, 295
Herrington, H.B.: **5.**132
Hester, Jim J.: **1.**016; **7.**004; **9.**077
Heusser, Calvin John: **4.**010, 265, 356; **7.**029, 043, 096, 135; **9.**232
Hewes, G.W.: **8.**041
Heyerdahl, Thor: **12.**014
Heymann, D.: **2.**090
Hibbard, C.W.: **7.**129
Hibben, Frank C.: **9.**289, 342
Hibbert, F. Alan: **11.**077
Hiebert, R.D.: **3.**108, 163
Higashamura, T.: **2.**131
Higgs, E.S.: **7.**075; **11.**019, 020
Hijszeler, C.C.W.J.: **11.**049
Hinds, F.J.: **7.**129
Hoare, R.D.: **7.**094, 130
Hodge, C.A.H.: **4.**094
Hoffman, Bernard G.: **9.**157
Hoffmeister, J.E.: **4.**428
Hoffren, Vaino: **14.**077, 078, 294
Holdetahl, Olaf: **4.**138, 210, 388
Hole, Frank: **3.**073; **10.**020, 049
Hole, Thornton: **7.**094
Holm, Marianne: **14.**211
Holmes, G.W.: **5.**056
Holmes, Mark L.: **4.**158
Honda, Masatake: **4.**115
Hontermans, F.G.: **3.**327
Hood, Donald W.: **3.**025, 026, 027, 060; **14.**206
Hood, M.S.F.: **11.**106, 107
Hoover, B. Reed: **9.**019
Hoover, Tom E.: **14.**269
Hopkins, David Moody: **4.**075, 141; **5.**025, 056, 061; **7.**013, 146
Hoppe, Gunnard: **4.**429; **5.**276
Hopson, Clifford A.: **4.**204
Horberg, Leland: **4.**279; **7.**131
Horie, Shoji: **4.**448; **7.**026
Horn, Walter: **11.**024
Horney, Amos G.: **3.**271

Horrocks, D.L.: **3.**088
Hossfeld, Paul S.: **12.**072
Hough, Jack L.: **5.**078, 222; **7.**064
Houtermans, Jan: **2.**126, 190
Houtman, Th.: **6.**043
Howard, Hildegard: **5.**274
Howe, Bruce: **8.**015
Hoyt, John H.: **4.**169, 406; **7.**002, 003
Hsu, Yuin-Chi: **14.**001
Hsy, Yi-Chuan: **14.**001
Huang, Chia-Yi: **14.**001
Hubbs, Carl L.: **3.**322; **9.**200, 328; **14.**108, 109, 110, 111, 112, 185
Hubee, N. King: **4.**175
Huber, B.: **2.**114; **3.**167
Hughes, E.E.: **2.**147, 149; **3.**123
Hughes, O.L.: **5.**098, 286, 309; **7.**058, 145
Hulston, John R.: **3.**170; **4.**286
Hume, James D.: **9.**330
Hunkins, Kenneth: **6.**019, 028
Hunt, Charles B.: **2.**245, 283
Hunter, R.F.: **2.**145; **3.**211
Hurley, Patrick M.: **2.**156
Hurley, William M.: **9.**310
Hurt, W.H.: **9.**002, 190
Hurt, Wesley R.: **9.**008, 087, 195, 292, 312, 314
Hussey, K.M.: **4.**337
Husted, Wilfred M.: **9.**026, 123, 189
Hutchinson, G.E.: **2.**178, 185
Hyne, Norman J.: **4.**240
Hyyppa, E.: **14.**077, 078, 079

Ilves, E.: **14.**196, 197, 198, 199
Immamura, Gakuro: **2.**269
Ingerson, Earl: **4.**119
Inghram, Mark G.: **2.**142, 143, 154
Inman, D.L.: **4.**076
Innis, Charles: **7.**094
Inoue, Aoi: **2.**268
Irving, William N.: **4.**321; **5.**033
Irwin Williams, Cynthia: **9.**078
Irwin, Cynthia: **9.**372
Irwin, Cynthia C,: **9.**285
Irwin, H.: **8.**041
Irwin, Henri J.: **9.**372
Irwin, Henry J.: **9.**285
Isbell, A.F.: **3.**027, 060
Isbell, A.T.: **3.**025
Isbell, W.S.: **3.**026
Isola, A.: **14.**077, 078, 079
Ivanova, I.K.: **11.**073
Iversen, Johannes: **4.**390; **5.**252

Ives, Jack D.: **5.**142, 157, 255
Ives, John David: **5.**085
Ives, Patricia C.: **14.**218, 220
Iyengar, T.S.: **3.**102

Jacobsen, T.: **2.**009
Jacobsen, Thorkild: **10.**052
Jansen, Hans S.: **2.**067, 073, 084, 112; **3.**012, 169, 172, 233
Janssen, C.R.: **5.**166
Jardine, William Graham: **5.**190
Jeffay, Henry: **3.**063
Jelgersma, Saskia: **4.**162, 276; **5.**152
Jelinek, Arthur Jenkins: **1.**015; **9.**028, 082, 304
Jelley, J.V.: **2.**284
Jenkinson, David S.: **3.**312
Jenks, G.M.: **2.**047
Jennings, J.N.: **4.**040; **7.**143
Jennings, Jesse D.: **3.**250; 093, 239
Jensen, E.H.: **2.**016
John, Brian S.: **5.**067, 161, 207
Johnson, Elden: **9.**360
Johnson, Frederick: **1.**010; **2.**153, 246, 250; **3.**077, 092, 093, 129, 267; **9.**289, 300; **13.**012
Johnson, R.H.: **4.**427
Johnston, Richard B.: **9.**187
Johnston, William H.: **3.**029
Johnston, William M.: **2.**295
Johnstone, Robert W.: **14.**271
Jones, Rhys: **12.**033
Jones, W.M.: **2.**116
Jope, E.M.: **2.**103
Jordan, Pierre: **3.**318
Jorgesen, Clive: **7.**114
Judson, Sheldon: **4.**222
Junge, C.: **2.**183

Kalab, B.: **3.**341
Kallberg, Per: **3.**002
Kamen, Martin D.: **2.**124, 125, 212
Kanakoff, George P.: **7.**041
Kandel, R.J.: **3.**222
Kanwisher, John: **2.**195; **6.**014
Kaplan, I.R.: **2.**015; **4.**239
Kapp, Ronald O.: **4.**045, 250; **5.**219, 258, 268; 112
Karlén, Ingvar: **2.**258; **3.**002, 083, 122; **14.**282, 285
Karlstrom, Thor N.V.: **5.**037, 093, 229, 251, 292; **7.**021, 123, 137, 138
Karrow, Paul F.: **7.**066, 067
Katsui, Y.: **2.**107
Kauffman, Erle G.: **5.**018
Kaufman, Aaron: **2.**086; **4.**324; **6.**010
Kaye, C.A.: **4.**192; **5.**057

Kear, D.: **12.**035
Keef, P.A.M.: **11.**068
Keeling, Charles D.: **2.**088
Kehoe, Thomas F.: **9.**050
Keith, M.L.: **2.**049, 249; **3.**276
Kelley, R.: **3.**085
Kelly, Thomas C.: **9.**262
Kempton, John P.: **5.**011, 129, 241
Kendall, R.L.: **4.**114
Kennedy, William R.: **3.**148
Kenyon, Kathleen M.: **10.**014, 016, 023
Kenyon, W.A.: **9.**140
Kerney, M.P.: **5.**149
Kerr, Vernon N.: **2.**094, 095; **3.**145, 158, 160
Khotinskiy, N.A.: **4.**006; **5.**145
Kibilda, Z.A.: **4.**354
Kigoshi, Kunihiko: **2.**161, 265, 277, 280, 294, 310, 311, 312; **5.**233; **14.**058, 059, 060, 061, 062, 063
Kilicci, Serap: **3.**002; **14.**283
Kim, Stephen M.: **3.**047, 048, 055, 295; **14.**146, 147
Kind, N.V.: **4.**325; **5.**002; **7.**128; **14.**066, 157
King, Cuchlaine A.M.: **5.**028, 240
Kingsley, David H.: **4.**117
Kinningham, Russell: **14.**207
Kirk, W.: **5.**133, 153
Kirkbride, Diana: **10.**003, 004
Kirschenbaum, A.D.: **2.**179, 252
Kitchener, J.A.: **3.**306
Kivett, Marvin F.: **9.**006
Kjaerum, Poul: **11.**060
Kjartansson, Gudmundur: **4.**053
Kjellberg, G.: **3.**178; **4.**014
Klassen, R.W.: **4.**418; **5.**075
Klein, R.G.: **11.**087
Klima, Bohuslav: **11.**055
Klovning, Ivar: **5.**058
Knauer, G.A.: **14.**056, 057
Kneberg, M.: **9.**043
Kneller, William A.: **4.**045
Knox, A.S.: **7.**117
Knox, F.B.: **2.**198
Knuth, Eigil: **5.**273; **11.**007
Kobayashi, Hiromi: **14.**061, 062
Kocharov, G.E.: **2.**020
Kohl, G.: **14.**011, 012
Kohler, Ellen L.: **11.**011
Kohman, Truman P.: **1.**020, 021; **2.**101; **3.**339; **4.**089
Kokta, L.: **3.**237
Kondo, Y.: **2.**107
Königsson, Lars König: **4.**034
Konstantinov, B.P.: **2.**020
Korff, S.A.: **2.**129
Korpela, K.: **5.**119

Author Index

Koteff, Carl: **5**.072
Kouts, H.J.: **2**.211
Kovar, Anton J.: **9**.254
Kowalski, Sandra J.: **14**.151, 152
Kozlewski, Janusz K.: **11**.054
Krasnov, I.I.: **4**.140
Kraybill, H.L.: **2**.178
Krieger, Alex D.: **2**.247; **9**.265
Kriewaldt, M.: **4**.402
Krinsley, Daniel Bernard: **5**.108
Krishnan, M. Unni: **10**.057, 058
Kritchevsky, D.: **3**.244
Krog, Harald: **4**.200; **5**.198, 215
Krueger, Harold: **14**.064
Krueger, Harold W.: **3**.226; **14**.065
Krueger, Paul: **3**.328
Ku, Teh-Lung: **6**.041
Kucera, C.L.: **2**.039
Kukla, Jiri: **11**.055
Kulm, L.D.: **4**.216
Kulp, J. Lawrence: **2**.044, 071, 091, 139, 176; **3**.039, 074, 075, 107, 286, 335; **6**.007, 015, 045; **7**.025; **13**.008; **14**.113, 114, 115, 116
Kume, Jack: **5**.192
Kupsh, W.O.: **4**.369
Kuroya, H.: **2**.310
Kusumgar, Sheela: **3**.181, 252, 281; **10**.034, 055, 056, 057, 058, 059, 060, 061; **14**.166, 200, 201, 202, 203, 204
Kutschale, Henry: **6**.028

L'Helgouach, J.: **11**.010
La Marche, Valmore C.: **4**.100; **5**.010
La Rocque, Aurele: **7**.090
La Salle, Pierre: **4**.032; **5**.176
Labeyrie, Jean: **3**.069, 070; **14**.080, 177, 178
Laborel, Jacques: **4**.381
Ladenburg, Rudolf: **2**.003
Lagomaximo, R.J.: **2**.293
Laird, W.M.: **7**.125
Lal, B.B.: **10**.045
Lal, D.: **2**.026, 080, 220; **3**.181, 281, 320; **6**.047; **10**.054; **14**.166, 200, 201, 202
Lamb, H.H.: **4**.449
Lambert, R. St. J.: **3**.289
Lampert, Ronald J.: **12**.037, 046
Lance, J.F.: **9**.048, 135
Land, Lynton S.: **7**.087
Lanfway, C.C.: **3**.134
Lange, A.L.: **9**.345
Langford-Smith, Trevor: **1**.002; **4**.391
Langway, Chester C.: **3**.133, 274, 275, 309

Lanning, Edward P.: **9**.040, 067, 118, 203
Larsen Helge: **11**.074
Larsen, James A.: **4**.321; **5**.033, 184
Lasca, N.P.: **4**.273
Lassalle, Pierre: **5**.168, 243
Lathrap, D.W.: **9**.003
Latter, Albert L.: **2**.072, 292
Laughlin, William Sceva: **9**.011, 014, 129, 154, 155, 167
Lawton, E.H.: **12**.008
Lazarus, William C.: **9**.012
Le Run, J.: **11**.015, 038, 069; **14**.081
Leakey, L.S.B.: **8**.002
Lee, Abel: **3**.251
Lee, H.A.: **5**.130, 159
Lee, Hubert: **7**.085
Lee, T.E.: **9**.228
Leger, Concèle: **3**.065, 209
Leighton, Morris M.: **5**.032, 064, 113, 115, 246; **7**.132
Lemke, R.W.: **7**.125
Leonard, A. Byron: **7**.126
Leonhardy, Frank C.: **9**.038, 171
Leopold, Estella B.: **4**.075, 183, 272; **5**.296; **7**.120
Lerman, J.C.: **2**.128; **9**.268
Leroi-Gourhan, A.: **4**.242
Levi, Hilde: **1**.011, 012; **3**.325; **13**.014; **14**.045
Levin, Betsy: **14**.218, 219, 220
Lewis, C.F.M.: **5**.073
Lewis, T.M.: **9**.043
Libby, Willard F.: **2**.005, 007, 010, 022, 025, 097, 140, 154, 155, 170, 179, 199, 216, 226, 227, 231, 236, 241, 243, 252, 271, 285, 295, 316; **3**.004, 098, 144, 176, 178, 192, 253, 255, 256, 271, 331; **4**.014, 086; **14**.041, 042, 043, 153, 154, 155, 222, 223, 224, 225, 226, 227, 228, 229
Lichti-Federovitch, Sigrid: **4**.161
Liestøl, Olav: **5**.055
Liiva, A.: **14**.196, 197, 198, 199
Likins, Robert C.: **2**.193
Liliu, G.: **11**.070
Lin, Der-Hwang: **14**.060
Linares de Sapir, Olga: **9**.086
Lindenbaum, Arthur: **3**.292
Lindvall, R.M.: **7**.125
Lingenfelter, R.E.: **2**.208, 209
Linne, Sigvald: **2**.234; **9**.298
List, R.J.: **2**.290
Livingstone, B.G.R.: **5**.144
Livingstone, D.A.: **4**.256; **5**.004, 022, 144
Llano, G.A.: **4**.223
Lockhart, Helen S.: **3**.052, 053
Lodge, James P.: **2**.061
Löken, Olav H.: **5**.015, 157, 217

Long, Austin: **2**.133, 300; **3**.336; **4**.027, 028; **5**.216; **7**.020; **14**.006, 007, 008, 180, 181, 182, 183
Long, J.V.P.: **2**.019; **3**.021, 177, 179, 347
Long, S.V.: **9**.039, 072, 352
Loosemore, W.R.: **3**.114
Loosli, H.H.: **2**.306; **3**.274, 275
Lothrop, S.K.: **9**.001, 036
Löve, Axel: **5**.172
Löve, Doris: **5**.172
Love, J.D.: **4**.130
Lovering, John F.: **14**.004, 005
Lovseth, Knut: **2**.121; **14**.211
Lowdon, J.A.: **14**.072, 073, 074, 075, 076
Lowe, Garreth W.: **9**.027
Lundelius, Ernest L.: **7**.018
Lundqvist, Gösta: **4**.142; **5**.117
Lundqvist, Jan: **4**.305
Luyanas, V.Yu.: **4**.354
Lyon, C.J.: **4**.376

Maarleveld, Gerardus Cornelis: **4**.422; **5**.039, 282; **7**.048
Mac Clintock, Paul: **7**.083
Mac Neish, R.S.: **9**.163
MacCalman, H.R.: **8**.008
MacDonald, George F.: **9**.098
Machta, Lester: **2**.042, 043, 192, 286, 290
Macintosh, N.W.G.: **12**.032
Mackay, Colin: **2**.188
Mackay, J. Rodd: **4**.266; **5**.283
Mackenzie, Fred T.: **7**.087
Mackey, C.J.: **14**.021, 023
Mackey, John: **14**.022, 024, 025, 026, 027, 028
MacNeil, F.S.: **4**.075
MacNeish, Richard S.: **9**.316
MacNeish, Richard Stockton: **9**.015, 143, 162, 164, 210, 248, 343
Maddock, A.G.: **2**.021
Mageda, T.: **4**.244
Magnusson, E.: **5**.118
Mallett Vatcher, Faith de: **11**.093
Malloy, R.J.: **7**.119
Manby, T.G.: **11**.083
Mangelsdorf, Paul C.: **9**.210
Mann, W.B.: **2**.145, 147, 149, 260; **3**.117, 123, 211
Manning, John C.: **7**.147
Manov, G.G.: **2**.151
Marin, Paul S.: **5**.128
Markova, N.G.: **2**.113; **4**.006, 070; **5**.145; **14**.158
Markova, N.G.V.: **14**.002, 288
Marlow, W.F.: **2**.147, 260
Marshall, E. W.: **6**.015
Marthinussen, M.: **4**.050, 077

Martin, A.R.H.: **7**.007
Martin, E.L.: **4**.004, 318
Martin, J.R.: **14**.057
Martin, Paul S.: **4**.181; **5**.165, 291; **7**.011, 097, 121; **9**.095, 133, 355
Martin-Gravel, H.: **11**.050
Martishchenko, L.V.: **2**.113
Masey, Don S.: **14**.207
Mason, J. Alden: **9**.027, 100
Mason, R.J.: **8**.065, 076
Mason, Revil J.: **8**.005
Mason, Ronald J.: **7**.035; **9**.120
Mathews, W.H.: **4**.044
Mathiassen, Therkel: **11**.100
Mathieu, Guy G.: **6**.001, 019
Matson, F.R.: **3**.046, 228
Matthews, Barry: **4**.190; **5**.013, 082, 091, 135, 239
Mauny, Raymond: **8**.049
May, Irving: **3**.140
Maycock, P.F.: **5**.013
Mc Andrews, J.H.: **4**.199
McAndrews, John F.: **4**.253
McAndrews, John H.: **5**.185
McAuley, I.R.: **3**.279; **14**.054
McBryde, Isabel: **12**.012, 013, 045, 059, 061
McBurney, C.B.M.: **4**.244; **8**.000, 001, 025; **10**.010; **11**.095
McCabe, William J.: **3**.170; **4**.286
McCallum, G. John: **3**.061, 062
McCallum, John: **3**.104
McCallum, K.G.: **14**.038
McCallum, K.J.: **2**.132; **4**.035, 173; **14**.264, 265, 266, 267
McCarthy, Fred D.: **12**.053, 076
McClure, William L.: **7**.142
McCorquodale, Bruce A.: **9**.050
McCready, C.C.: **3**.089
McCulloch, David: **5**.061
McCulloch, David S.: **4**.423; **5**.018, 228
McDaniel, E.W.: **3**.020
McDonald, Barrie Clifton: **5**.050, 305; **7**.074
McDowell, L.L.: **3**.143, 262; **14**.221
McFarland, E.: **4**.184, 346, 355; **14**.095
McFarlane, A.S.: **3**.030
McGinn, John J.: **9**.357
McGinsey, Charles R.: **9**.145
McGregor, Ronald L.: **4**.052
McIntire, William C.: **9**.083
McKay, C.: **2**.259
McKellar, I.C.: **5**.236; **7**.068
McKelvey, P.J.: **12**.031
McKenzie, G.D.: **5**.206
McKie, Euan W.: **11**.109
McManus, Dean A.: **4**.148, 158; **7**.070

Author Index

McMurray, W.R.: **2.**217
McNutt, Charles H.: **1.**009
Mead, J.F.: **2.**022, 271
Medcof, J.C.: **9.**016
Medlock, R.W.: **2.**260
Megaw, J. Vince: **12.**030
Megaw, J. Vincent S.: **12.**022, 065, 077
Meggers, Betty J.: **9.**117, 181, 368
Meguro, Hiroshi: **5.**233
Mehringer, Peter J.: **4.**198, 262; **5.**165; **7.**006, 097; **9.**227
Meighan, Clement W.: **3.**301; **9.**059, 179, 212, 242, 249, 339
Meinschein, W.G.: **3.**032, 187
Meiring, A.D.J.: **8.**037
Meldgaard, Jorgen: **9.**238; **11.**074
Melhuish, W.H.: **3.**174, 175; **4.**286
Mellaart, James: **10.**022; **11.**004
Melton, C.: **9.**060
Menzies, R.J.: **6.**009
Mercer, J.H.: **5.**298
Merrilees, D.: **4.**208
Merrill, R.S.: **3.**234
Merrill, S. Arthur: **4.**033
Michael, Henry N.: **2.**206; **14.**255
Mielke, James E.: **14.**182, 183
Miles, S.W.: **9.**329
Miller, B.B.: **4.**335
Miller, Carl F.: **9.**278, 323
Miller, J.: **9.**287
Miller, John P.: **4.**030
Miller, Robert: **4.**293
Miller, Robert D.: **5.**138
Miller, Sheryl Elinor Flum: **8.**004
Miller, Warren W.: **2.**150; **3.**035, 124, 228
Milliman, John D.: **4.**396
Millon, Ree F.: **9.**220
Mirymech, E.: **5.**209
Mitani, S.: **2.**310
Mitchell, G.F.: **4.**271, 416; **7.**088
Miwa, T.: **2.**310
Miyaka, Y.: **6.**025
Moar, N.T.: **4.**085, 257
Moberg, Carl-Axel: **11.**044
Mohapatra, G.C.: **4.**047
Mohler, F.L.: **3.**298
Moir, D.R.: **5.**179
Molen, W.H. van der: **5.**187
Moljk, A.: **3.**022
Montané, Julio: **9.**215
Mook, W.G.: **2.**128
Mooney, H.A.: **5.**010
Moore, David G.: **5.**026
Moore, Ruth: **3.**168

Moreno y Moreno, Augusto: **3.**239
Morgan, James P.: **9.**083
Moriarty, James Robert: **9.**092, 328
Morrison, A.: **4.**180; **5.**051
Morrison, M.E.S.: **7.**148
Morrison, Robert B.: **4.**217, 322; **5.**125
Morrisson, M.E.S.: **4.**430
Morse, Dan F.: **9.**348
Mortensen, Peter: **10.**042
Moscicki, Wldzimierz: **3.**011, 212, 213
Moss, John H.: **9.**189
Mott, R.J.: **4.**419; **5.**075, 238
Movius, Hallam L.: **2.**235; **3.**006; **11.**042, 047, 048, 059, 082
Muir, Janis A.: **3.**214
Mullenders, W.: **4.**128
Müller, Ernest H.: **5.**083, 227
Müller, Fritz: **4.**330; **5.**012, 062, 218
Müller-Beck, Hans Jürgen: **11.**075
Muller-Beck, Hansjürgen: **9.**206; **11.**056
Mullineaux, D.R.: **7.**101
Mullineaux, Donald H.: **4.**293; **5.**279
Multer, H.G.: **4.**428
Mulvaney, D.J.: **12.**005, 007, 008, 009, 011, 017, 024, 036, 052, 054, 080
Munk, W.: **2.**126
Münnich, K.O.: **2.**057, 085, 263, 304, 305; **4.**054, 093, 174; **6.**050, 051; **14.**093
Munson, Patrick J.: **9.**047
Murray, K.: **3.**030
Murray, Raymond C.: **4.**222

Naff, George E.: **5.**038
Naidu, A.S.: **4.**333
Nakai, Naboyuki: **2.**296
Nakao, Akira: **3.**189
Nakaparksin, S.: **3.**038
Nakea, A.: **2.**053
Nasmith, H.: **4.**044
Nayudu, Y. R.: **6.**008
Needham, H.D.: **7.**019
Nelson, C. Hans: **4.**216
Nelson, H.F.: **7.**127
Nelson, J.G.: **4.**207
Nenquin, Jacques: **8.**022, 077
Neuhaus, John William George: **3.**097, 103; **14.**249
Neuman, Robert W.: **9.**266, 321
Neustadt, M.I.: **4.**005, 205
Neustupny, Evzen: **2.**077; **11.**002, 108
Neville, O.K.: **3.**043
Newell, Norman D.: **4.**182; **7.**140
Newman, Thomas M.: **9.**089

Newman, W.A.: **4**.160, 182
Newman, Walter S.: **4**.164, 393
Ney, E.T.: **2**.099
Nichols, D.R.: **4**.249
Nichols, Harvey: **4**.274; **5**.200
Nichols, Robert L.: **4**.132; **5**.079
Nickoloff, Nick: **3**.334
Nier, Alfred O.: **2**.150, 308; **3**.296
Niering, W.A.: **4**.264
Nilsson, Erik: **5**.264
Nissenbaum, A.: **2**.015; **4**.239
Noakes, John E: **14**.269
Noakes, John E.: **3**.025
Noakes, John Edward: **3**.026, 027, 047, 048, 055, 060, 194, 295; **14**.146, 147, 206
Noble, J.B.: **7**.046
Nobofusa, Saito: **1**.020, 021
Nohara, N.: **2**.131
Norbeck, Edward: **9**.239
Norris, L.D.: **2**.142, 143
Novskiy, V.A.: **4**.140
Nutt, David C.: **2**.223, 264
Nydal, Reidar: **2**.121, 134, 159, 301; **3**.240, 265, 338; **6**.026; **14**.208, 209, 210, 211
Nygaard, K.J.: **3**.147
Nystrom, R.F.: **3**.297

O'Brien, B.J.: **6**.004
O'Sullivan, J.B.: **4**.337
Oakley, Kenneth P.: **2**.104; **3**.072, 273; **7**.010; **8**.013, 075; **11**.026, 094
Oaks, Robert Q.: **7**.099
Oba, Toshio: **10**.038
Oda, M.: **2**.161
Oeschger, Hans: **2**.306; **3**.133, 134, 165, 274, 275, 309, 327; **14**.013, 014, 015, 016, 017, 018
Ogden, J. Gordon: **4**.134, 320, 328, 329, 363, 414; **9**.056; **14**.148, 149, 150
Olausson, Eric: **4**.424
Oldale, Robert N.: **5**.072; **7**.104
Olive, Wilds W.: **7**.017
Oliviero, Vincent T.: **3**.340
Olmsted, D.: **9**.213
Olsen, O.: **11**.101
Olson, Alan P.: **9**.344
Olson, Edwin A.: **2**.035, 102, 254, 257, 274; **3**.229, 230, 247, 334, 348; **4**.048, 055, 365; **13**.002, 006; **14**.117, 118, 119, 120
Olsson, Ingrid U.: **2**.045, 174, 258, 270, 287, 314; **3**.002, 031, 058, 059, 083, 112, 122, 198, 200, 236, 243, 324; **4**.133, 193, 289, 290, 371; **14**.280, 281, 282, 283, 284, 285, 286, 287

Oltz, Donald F.: **4**.250
Oman, Charles L.: **14**.219, 220
Orlins, Robert I.: **9**.081
Orr, Phil C.: **4**.055, 323; **7**.042; **9**.045, 112, 136, 137, 207, 224, 225, 256, 290, 291, 309
Osborne, A.R.: **2**.284
Osburn, J.O.: **3**.146
Oshio, T.: **2**.310
Östlund, Göte H.: **3**.037, 195; **4**.001
Östlund, H. Göte: **4**.353; **14**.127, 129, 130, 189, 190, 191, 192, 193
Östlund, H. Göte : **2**.057; **14**.128
Ostrem, Gunnar: **4**.288; **5**.169
Oswalt, Wendell, H.: **1**.019
Ott, Donald G.: **3**.145, 158, 160
Owen, Roger C.: **9**.122

Page, Neil R.: **5**.014
Palmer, C.E.: **3**.331
Pande, G.S.: **3**.101
Pandow, Mary: **2**.188, 259
Pannekock, A.J.: **4**.276
Pantin, H.M.: **4**.374
Papworth, M.: **8**.041
Parham, A.G.: **3**.184
Parker, Richard A.: **2**.009; **10**.052
Parkinson, G.B.: **3**.117
Parry, W.T.: **4**.110
Parsons, Barbara M.: **7**.059
Partridge, Jeannette: **4**.002
Patten, Harvey L.: **5**.297
Patterson, R.L.: **2**.014, 024
Patterson, Thomas C.: **9**.116, 219
Patty, F.A.: **3**.082
Paul, Catherine M.: **3**.115
Paul, E.A.: **2**.132; **4**.035, 173
Paver, F.R.: **8**.011
Pearson, Frederick Joseph: **2**.297; **3**.120, 141, 344; **4**.057, 119; **14**.270, 271, 272, 273
Pearson, R.: **10**.064
Pecora, William T.: **3**.001
Peisach, M.: **3**.305
Pennack, Robert W.: **4**.113
Pennington, W.: **4**.255
Penny, L.F.: **5**.267
Perino, Gregory: **9**.007
Perkins, Dexter: **10**.030, 047
Perrin, R.: **4**.094
Perry, D.: **2**.251; **14**.094, 095
Person, Elaine: **14**.292
Peters, K.M.: **12**.003
Peterson, James A.: **7**.015

Author Index

Peterson, M.N.A.: **4**.366
Peterson, N.V.: **4**.228
Pettyjohn, Wayne A.: **5**.164
Pevear, D.R.: **4**.236
Péwé, T.L.: **4**.026, 223
Péwé, Troy L.: **4**.116; **5**.006; **9**.149
Pflaker, George: **4**.287
Phelps, D.S.: **14**.057
Phillips, Philip: **9**.166, 324
Phillipson, D.W.: **8**.017, 028, 044, 045, 046, 058, 059, 060, 071, 082, 083
Pichat, L.: **3**.209
Pietig, F.: **14**.020
Piggott, Stuart: **10**.017; **11**.085, 098
Pikley, Orrin H.: **4**.236
Piyamy, Piya: **14**.284
Plafker, George: **4**.460
Plass, Gilbert G.N.: **2**.052
Plazin, John: **3**.084
Pleslová-Stiková, Emilie: **11**.043
Plesset, Milton S.: **2**.072, 110, 292
Ploey, J. de: **4**.308
Polach, Henry A.: **2**.180; **3**.056, 131, 254; **12**.075; **14**.004, 005
Polevaya, N.I.: **3**.263
Pollard, Sheila H.H.: **11**.096
Poole, E. Grey: **2**.059; **5**.008, 137
Poole, Lynn: **2**.059
Porter, James Warren: **3**.017
Porter, Stephen C.: **5**.029, 104; **7**.036, 080
Posnansky, Merrick: **8**.040
Potzger, John E.: **3**.216; **5**.141, 201, 237
Povinec, P.: **3**.291
Powers, H.A.: **4**.461
Powers, Roger: **14**.187
Prest, V.K.: **4**.419
Preston, R.S.: **14**.292
Pringle, R.W.: **3**.157, 201, 245, 310
Pritchard, Daphne H.: **14**.134, 135
Prosser, N.J.D.: **3**.083
Protsch, Reiner: **8**.002
Pullar, W.A.: **4**.417
Punning, J.B.: **14**.196
Punning, J.M.: **14**.197, 198, 199
Punning, Ya.M.K.: **5**.186
Purdom, C.E.: **2**.034
Purdy, Barbara A.: **9**.105
Purdy, Edward: **4**.367
Putman, J.L.: **13**.016

Quennerstedt, N.: **4**.429
Quimby, George I.: **9**.141

Quitta, Hans: **11**.001; **14**.011, 012

Rafter, T.A.: **2**.002, 028, 029, 030, 031, 074, 075, 076, 224; **3**.034, 044, 207, 223, 231, 294; **14**.140, 141, 142, 143, 144, 145
Ragland, James B.: **3**.057
Rainey, Froelich: **2**.182, 246; **3**.015, 018; **9**.302
Rakestraw, N.W.: **6**.021, 032, 033, 036, 037
Ralph, Elizabeth K.: **2**.041, 069, 206; **3**.018; **9**.197, 302, 308, 322; **10**.053; **11**.005, 011; **14**.250, 251, 252, 253, 255, 256, 257
Rama: **2**.080
Ramaswamy, C.: **10**.035
Ramaty, R.: **2**.160
Ramsden, D.: **2**.148; **3**.125, 349
Rankama, Kalervo: **4**.112
Rankine, W.F.: **11**.039
Rapaire, J.L.: **14**.131
Rapkin, E.: **3**.155, 218
Raukas, A.V.: **5**.186
Reber, Grote: **12**.001, 043
Redfield, Alfred C.: **4**.021, 280; **5**.208
Reed, Bruce L.: **5**.242
Reed, Charles A.: **8**.027
Reed, E.C.: **4**.455
Reeves, C.C.: **4**.110, 251, 421
Reger, R.D.: **9**.149
Reid, Allan F.: **2**.144, 179, 252
Reid, James C.: **3**.142
Reinharz, M.: **3**.013; **14**.039
Reinman, Fred M.: **12**.066
Renaud, André: **3**.274, 275, 309
Renfrew, Colin: **10**.040; **11**.021, 099
Rennie, D.A.: **2**.132; **4**.035
Rennie, R.J.: **4**.173
Renton, John Jo: **7**.059
Repenning, C.A.: **7**.146
Revelle, Roger: **2**.051, 135
Reynolds, S.A.: **2**.168
Richards, Horace G.: **7**.061, 091, 092
Richardson, J.L.: **4**.067
Richmond, Gerald M.: **5**.038, 106
Riesen, T.: **14**.016, 017, 018
Rigg, G.B.: **4**.018
Rightmire, Craig T.: **4**.368; **14**.268
Rinaldo, John B.: **9**.133, 355
Rinehart, C. Dean: **4**.175
Ring, E.R.: **9**.131, 367
Ritchie, J.C.: **4**.161
Ritchie, William A.: **9**.124, 299
Ritzenthaler, Robert: **9**.017, 272, 338
Rivard, N.: **4**.223

Rivera, J.: **2.**068
Roberts, Frank H.H.: **2.**064, 070; **9.**177, 267
Roberts, W.A.: **3.**224
Robie, Richard A.: **4.**279
Robins, P.A.: **3.**121; **14.**186
Robinson, H.S.: **11.**005
Robinson, K.R.: **8.**016, 018
Robinson, Richard D.: **14.**218
Rocco, Gregory G.: **2.**117
Rodden, J.M.: **11.**031
Rodden, Robert J.: **11.**030, 031, 037
Rodegker, Waldtraut: **3.**215
Roden, Gunnar I.: **9.**200
Roether, W.: **2.**085, 289; **4.**093; **6.**050, 051
Rogers, Betty S.: **3.**130, 153, 158
Rogers, W.H.: **3.**222
Rohrer, Willis L.: **4.**147
Rolfe, W.D. Ian: **7.**150
Romanova, E.N.: **3.**161, 162
Rondot, Jehan: **5.**168
Roosma, Aino: **4.**073
Ropp, Gus A.: **3.**149
Rosen, A.A.: **2.**118; **3.**191
Ross, Alexander: **2.**271
Ross, H.H.: **3.**024, 151
Ross, J.F.: **2.**022, 271
Rouse, G.E.: **4.**044
Rouse, Irving: **1.**029; **3.**095; **9.**004, 127, 182, 247, 340, 369; **14.**298
Rovner, Irwin: **9.**208
Rowe, John Howland: **1.**030; **9.**160, 161
Royse, Chester F.: **4.**215
Ruben, Samuel: **2.**212
Rubin, Meyer: **2.**118, 141, 193, 222, 313; **3.**001, 191; **4.**012, 021, 031, 033, 066, 092, 097, 136, 204, 221, 272, 293, 358, 360, 361, 387, 423, 431, 460; **5.**005, 027, 049, 140, 242, 245, 247, 250, 251, 275, 279; **7.**045, 146; **10.**013; **13.**005; **14.**167, 214, 215, 216, 217, 218, 219, 220
Rudenko, S.I.: **3.**161, 162
Rudnev, Yu.P.: **4.**140
Rudneyev. Yu. A.: **14.**003
Ruhe, Robert V.: **4.**025, 167, 176, 307, 332; **5.**102, 114, 122; **7.**045, 133
Rusnak, Gene A.: **4.**164, 392; **7.**119; **14.**127, 128, 129, 130
Russell, R.J.: **7.**139
Rutter, Nathaniel W.: **5.**154
Ruxton, B.P.: **4.**088, 377
Ryan, M.E.: **3.**143, 262; **14.**221

Sackett, W.M.: **3.**038
Sackett, William M.: **6.**048
Sadarangani, S.H.: **3.**102
Saito, K.: **2.**161
Salmi, Martti: **4.**362
Samos, George: **3.**323
Sanders, Phyllis: **3.**130
Sando, C.H.S.: **5.**116
Sanger, David: **9.**246, 251
Sangmeister, E.: **11.**050
Sankalia, H.D.: **10.**031, 039
Sarma, Akkaraju: **10.**044
Sarna, R.P.: **3.**252; **10.**055; **14.**166, 200, 201
Saro, S.: **3.**291
Saruhashi, K.: **6.**025
Sassoon, Harmo: **8.**020, 043
Sato, Jun: **14.**274
Sato, Tomoko: **14.**274
Satoh, Hiroguki: **4.**292
Satterthwaite, Linton: **9.**058, 197, 271
Saucier, Roger T.: **4.**224
Saunders, G.E.: **5.**103
Saxe, S.: **8.**041
Sayles, E.B.: **9.**175
Schaeffer, O.A.: **2.**090
Schafer, J.P.: **5.**231
Scharpenseel, H.W.: **14.**020
Schell, William R.: **3.**138
Schell, William Raymond: **2.**013, 171, 172; **3.**010; **14.**162, 163, 164
Schlander, P.F.: **2.**136
Schmidt, R.A.M.: **7.**047
Schmoll, H.R.: **5.**063
Schnable, J.N.: **4.**126
Schnable, Jon F.: **7.**098
Schneekloth, Heinrich: **14.**088, 089, 090
Schnitker, Detmar: **4.**236
Schofield, J.C.: **4.**394; **5.**271; **12.**026
Scholander, Per Fredrik: **2.**223
Scholl, David W.: **4.**383, 385, 386
Scholtes, W.H.: **4.**332; **5.**114; **7.**045
Schove, D.J.: **5.**277
Schreurs, Anne Nicole: **14.**096
Schrodt, Ariel G.: **14.**152
Schubert, Jack: **3.**080, 292
Schuch, R.L.: **3.**163
Schulert, Arthur: **2.**035
Schultz, Bertrand C.: **4.**415; **9.**021; **13.**007
Schultz, Gerald Edward: **5.**074; **7.**009
Schultz, Hyman: **3.**227, 228
Schwabedissen, H.: **14.**107
Schwartz, D.W.: **9.**345
Schwartz, U.: **14.**013, 014
Schwebel, A.: **2.**120

Author Index

Scott, Glenn R.: **5**.138
Scott, J.G.: **11**.081
Searle, E.J.: **4**.463
Searle, K. J.: **7**.062
Sears, Paul B.: **4**.073, 267; **5**.126, 197; **7**.141
Seitz, H.: **2**.293
Selander, Robert K.: **4**.043
Seliga, M.: **3**.291
Seliger, H.H.: **2**.260
Sellards, E.H.: **9**.121, 335
Sellman, P.V.: **4**.087
Sellman, Paul V.: **4**.151, 342; **7**.047
Sellstedt, H.: **3**.197
Serebryannyy, I.R.: **4**.070; **5**.186
Sewell, D.K.B.: **14**.249
Shapiro, H.L.: **12**.039
Shapiro, I.: **3**.135
Shapiro, I.L.: **3**.244
Sharma, V.K.: **3**.281
Sharp, C.S.: **10**.038
Sharp, Rodman A.: **3**.333; **14**.179
Sharpe, Jack: **3**.206
Shaw, Thurston: **8**.061, 063
Sheans, Daniel J.: **9**.089
Shepard, Francis: **6**.006
Shepard, Francis P.: **4**.051; 160, 182, 395; **5**.026; **6**.040; **7**.022; **12**.078
Sheppard, Herbert: **3**.215
Sheppard, J.G.: **14**.171, 172
Sherrod, Neil A.: **7**.031
Shibata, S.: **2**.161, 310
Shidei, T.: **2**.131
Shima, Makoto: **4**.235
Shleien, Bernard: **3**.307, 308
Shotton, F.W.: **7**.108
Shotton, Frederick William: **5**.007; **7**.030, 124; **14**.019, 232
Shubert, Jack: **3**.105
Shuliya, K.S.: **4**.354
Shumway, George: **9**.328
Shutler, Dick: **12**.010; **14**.230
Shutler, Mary Elizabeth: **9**.103, 351; **12**.048
Shutler, Richard: **9**.103, 150, 351, 358; **12**.048, 050, 071, 073; **14**.231
Sigalove, Joel J.: **14**.007, 180
Sigmond, R.S.: **3**.265
Sim, Victor Wallace: **4**.230
Simons, L.H.: **14**.094, 095
Simonson, G.H.: **4**.031
Simpson, D.D.A.: **11**.032
Simpson, I.M.: **7**.057
Simpson, J.A.: **2**.210
Sinex, F. Marrott: **3**.139

Singer, Leon: **3**.100
Singh, Gurdip: **4**.275; **7**.116
Sinoto, Y.H.: **9**.151
Sinoto, Yosihiko H.: **12**.018
Sirkin, Leslie A.: **7**.044, 084
Sissons, J.B.: **5**.099, 191
Skarland, Ivar: **9**.149
Skeels, Margaret Anne: **4**.213
Skinner, B.J.: **4**.012
Skinner, Brian J.: **4**.221
Skinner, H.C.W.: **4**.012
Skullerud, Kari E.: **14**.211
Slats, W.: **2**.282
Slaughter, Bob H.: **4**.231; **7**.142; **9**.019
Sleight, F.W.: **9**.066
Sliepcevic, A.: **3**.036
Smiley, Charles H.: **9**.270
Smith, A.G.: **4**.275; **11**.091
Smith, C.S.: **3**.269; **12**.069
Smith, Carlyle S.: **12**.020, 049
Smith, G.I.: **4**.187
Smith, George I.: **4**.435, 446
Smith, H.S.: **8**.019
Smith, I.: **11**.086
Smith, P.E.L.: **8**.064; **10**.076, 077
Smith, Paul V.: **4**.234
Smith, Philip E.L.: **11**.072
Smith, W.G.: **4**.384
Smith, William G.: **4**.203
Snell, W.A.: **14**.115
Sobering, S.: **3**.201, 310
Solecki, Ralph S.: **4**.242; **10**.011, 013, 041, 048, 068, 069, 070, 071, 072, 073, 079, 081
Solecki, Rose Lilian: **10**.082, 083
Solheim, William G.: **10**.018
Soman, S.D.: **3**.102
Sordinas, Augustus: **11**.084
Spaulding, A.C.: **3**.316
Specht, Jim: **12**.055
Spiridonova, Ye.A.: **4**.140
Spoehr, Alexander: **12**.027, 040
Squier, R.J.: **9**.282
Squier, Robert J.: **9**.132
Squires, Donald F.: **4**.064
Srdoc, D.: **3**.036
Srodon, Andrzej: **4**.371
Stager, John K.: **5**.283
Stahl, W.: **2**.173
Stalker, S. MacS.: **4**.150
Stander, L.O.: **2**.217
Stanley, D.J.: **9**.016
Starik, I.E.: **3**.161, 162, 314
Stearns, Charles E.: **4**.132

Stearns, Harold T.: **4**.121
Steece, F.V.: **5**.112
Steele, Robert: **3**.084, 319
Steensberg, A.: **11**.097
Stenberg, Allan: **2**.258, 314; **3**.112; **14**.286, 287
Stephen, William: **9**.165
Stephens, N.: **4**.201, 430
Stephenson, Robert L.: **9**.255
Stevens, Nelson P.: **4**.165, 166
Stevens, W.H.: **2**.145
Stewart, David P.: **7**.083
Stewart, Gordon: **2**.313
Stipp, J.J.: **3**.025
Stipp, Jerry J.: **2**.180; **3**.027, 047, 060, 126, 131; **14**.004, 005, 056, 206, 269
Stoiber, R.F.: **4**.019
Stoltman, James B.: **9**.198
Story, James A.: **4**.352
Strachan, I.: **7**.030
Stranks, D.R.: **2**.202; **3**.311
Strong, William Duncan: **9**.218
Stuckenrath, Robert: **2**.069; **3**.040, 210; **9**.099, 319; **14**.253, 254, 256, 257, 258
Stuiver, Minze: **2**.062, 191, 194, 296, 302, 303; **3**.067, 236, 269; **4**.015, 109, 229, 372, 383, 385, 386, 433, 434; **5**.023, 167; **7**.039; **8**.055; **9**.204; **14**.295, 296, 297, 298
Stumer, Louis M.: **9**.264
Suess, Hans E.: **2**.038, 051, 061, 081, 114, 126, 135, 158, 162, 177, 191, 220, 221, 230, 275, 276, 281; **3**.164, 193; **4**.326; **5**.001, 174; **6**.021, 032, 033, 036, 037, 040, 049; **14**.108, 109, 110, 111, 167, 212, 213, 214
Sugawara, Ken: **5**.233
Suggate, R.D.: **7**.014
Suggate, R.P.: **4**.149, 185, 246, 277, 444; **5**.181; **6**.034
Suggs, Robert Carl: **9**.283; **12**.015
Sugihara, Thomas T.: **3**.144
Sulerzhitskiy, L.D.: **2**.256; **14**.066, 157
Sullivan, Geraldine R.: **2**.050
Summers, Roger: **8**.029, 074
Summers, Roger F.: **8**.084
Suzuki, Hisashi: **14**.274
Swain, J.L.: **3**.199
Swanston, Douglas N.: **4**.146
Swart, E.R.: **2**.011; **3**.087, 121; **8**.003; **14**.171, 172, 186
Sweeton, F.H.: **2**.047
Swift, Donald J.P.: **4**.117; **5**.059
Switsur, V.R.: **14**.035, 036

Taggart, M.S.: **3**.238
Takashima, Y.: **3**.010; **14**.162

Tamers, Murray A.: **2**.297; **3**.042, 065, 081, 126, 141, 303, 304, 315; **4**.157, 316, 317; **14**.020, 097, 098, 099, 270, 271, 272
Tamm, C.O.: **4**.353
Tauber, Henrik: **2**.163, 164, 201, 304; **3**.086, 293; **4**.107; **5**.257; **14**.045, 046, 047, 048, 049, 050, 051, 052, 053
Taylor, Dwight W.: **2**.222; **4**.130, 423
Taylor, E.W.: **9**.318
Taylor, R.E.: **2**.231, 232; **3**.096, 199, 277; **9**.072, 297, 352
Taylor, R.S.: **5**.089
Taylor, Walter W.: **9**.042, 336
Te Punga, Martin T.: **4**.122, 351
Tedrow, John Charles Fremont: **4**.036, 063, 407, 411, 453
Telegadas, K.: **2**.290
Terasmae, Jaan: **4**.082, 083, 266, 281, 426; **5**.098, 176, 178, 205, 206, 209, 281, 309; **7**.058, 119
Terra, Helmut de: **9**.076, 258
Thatcher, Leyland: **4**.092
Thielen, C.: **4**.317
Thilo, L.: **4**.093
Thom, Bruce G.: **4**.229
Thomas, Edward S.: **9**.211
Thomas, G.S.: **3**.048
Thomas, Homer L.: **11**.061
Thommeret, J.: **14**.131, 132
Thommeret, Y.: **14**.132
Thompson, B.N.: **4**.020
Thompson, J. Eric S.: **9**.030
Thompson, Robert W.: **7**.051
Thorarinsson, Sigurdur: **4**.053, 450
Throckmorton, Peter: **11**.005
Thurber, David: **4**.367
Thurber, David L.: **6**.010, 019, 031; **7**.061
Thurmond, John T.: **4**.297
Thwaites, Frederick T.: **5**.024; **7**.082
Tindale, Norman B.: **4**.041; **12**.021, 023, 070
Ting, William S.: **4**.086
Tipton, M.J.: **7**.125
Tobias, Phillip V.: **8**.073
Todd, Ian A: **10**.012
Toivonen, A.V.P.: **14**.079
Tolbert, Bert M.: **3**.142
Tomikura, Yoshio: **2**.294, 311; **14**.058
Tongiorgi, E.: **4**.244; **14**.039, 040
Totten, Stanley M.: **5**.308
Tracey, J.I.: **4**.182
Trautman, Milton A.: **14**.100, 101, 102, 103, 104, 105
Treganza, A.E.: **9**.005
Trevor, J.C.: **8**.025
Troels-Smith, J.: **11**.067
Troitskiy, S.L.: **5**.170
Trotter, Michael M.: **12**.047, 068

Author Index

Truesdail, R.W.: **4**.131
Trump, David: **11**.016, 102
Trylich, C.: **9**.054
Tryon, Lansing E.: **3**.107, 335; **6**.007; **14**.113, 115
Tsukada, Matsuo: **4**.254
Tubbs, Deborah Y.: **9**.370
Tucek, C.S.: **3**.247; **14**.114
Turchinetz, W.: **3**.157, 245, 310
Turekian, Karl K.: **2**.273; **6**.042
Turkevich, Anthony: **2**.286
Turnbull, S.H.: **3**.083
Turner, Judith: **3**.008
Turner, M.D.: **9**.217
Turner, Ruth D.: **4**.227
Twidale, C.R.: **4**.270; **12**.008
Tykva, R.: **3**.237

Uchio, Takayasu: **5**.233
Ucko, Peter: **8**.070
Upson, Joseph E.: **4**.272
Ustinov, V.I.: **4**.178

Valastro, S.: **14**.268, 273
Valencia, Josefina: **14**.044
Valoch, Karel: **11**.053
Van Andel, Tj.H.: **4**.381
van Beek, G.W.: **10**.051
Van der Merwe, Nicholas J.: **3**.041, 067, 068, 186; **8**.055, 065
Van der Waals, J.D.: **11**.009, 076
Van Slyke, Donald D.: **3**.084
Van Straaten, L.M.J.U.: **4**.338
Van Zeist, Willem: **4**.413; **11**.103
Van Zinderen Bakker, Edward M.A.: **4**.091, 260, 263; **5**.020, 143; **8**.050, 052
Vander Stricht, A.: **3**.221
Vanderhaeghe, G.: **3**.013
Vasari, Y.: **4**.425
Vaugham, David Evan: **3**.136
Vaz, J.E.: **9**.217
Vaze, P.K.: **3**.102
Veeh, H.H.: **4**.182
Vereschagin, N.K.: **11**.079
Vernon, Peter: **7**.145
Vertes, L.: **11**.022
Viereck, Leslie A.: **5**.019
Vinogradov, A.P.: **2**.089, 113; **4**.049, 178; **5**.134, 145; **14**.002, 158, 288
Vita-Finzi, Claudio: **4**.062; **11**.019, 020
Vogel, John C.: **2**.063, 128, 263, 289, 305; **4**.054, 172, 174, 422, 457; **5**.053, 260, 282, 288; **11**.003; **14**.084, 085, 086, 087

Volchok, Herbert L.: **2**.091
Vucetich, C.G.: **4**.417

Wada, Masami: **2**.268
Wagner, Frances J.E.: **4**.011
Wagner, Marie R.: **3**.156
Wagner, William Philip: **5**.041
Wahrhaftig, Clyde: **5**.230
Wainwright, G.J.: **11**.029
Wait, Robert L.: **4**.432
Waldron, Howard H.: **5**.279; **7**.101
Walker, Donald: **5**.249; **11**.046
Walker, Philip C.: **4**.135
Walker, Philip H.: **4**.443
Walker, Theodore R.: **7**.051
Walls, L.H.: **8**.025
Walton, Alan: **2**.046, 060, 137; **3**.033; **4**.364; **13**.010; **14**.101, 102
Walton, G.F.: **4**.407, 411
Wanke, H.: **4**.326
Warren, Claude A.: **9**.094
Warren, Claude Nelson: **9**.085, 326
Washburn, A.L.: **4**.153, 372
Wasley, William W.: **9**.175
Watanabe, Kazue: **4**.137
Waterbolk, H.T.: **11**.023; **13**.001; **14**.082, 083, 084, 085, 087
Waters, Aaron, C.: **4**.204
Waters, Joseph H.: **9**.138
Watson, Maureen: **14**.249
Watson, Patty Jo: **9**.020, 060, 241
Watt, D.E.: **2**.148; **3**.125
Watt, F.H.: **9**.366
Watts, R.J.: **3**.108
Watts, W.A.: **4**.269; **5**.150; **11**.014; **14**.054
Wauchope, Robert: **9**.158
Wayne, William J.: **7**.079
Weaver, Kenneth, F.: **3**.337
Webb, Clarence H.: **9**.134, 233
Webb, William S.: **9**.305
Webber, P.J.: **4**.426
Weber, Florence Robinson: **4**.116
Webster, Ruth M.: **4**.081
Wedel, Waldo R.: **9**.006, 189, 240, 325
Weeks, C. Francis: **14**.064, 065
Wehr, Larry: **4**.081
Weidick, Auker: **4**.296
Weimer, R.J.: **7**.002, 003
Weinberg, G.D.: **11**.005
Weinberg, S.S.: **2**.009
Weinberg, Saul S.: **10**.052; **11**.033, 066

Weinberger, Arthur J.: **3**.149
Weinhouse, Sydney: **2**.144, 179, 252
Weis, Paul E.: **5**.038
Weiss, H.G.: **3**.135
Welinder, Stig: **4**.436
Wells, Philip V.: **7**.040, 049, 114
Wendland, Wayne M.: **5**.254, 255
Wendorf, Fred: **4**.030
Wendt, Immo: **2**.173, 272; **14**.088, 089
Wenner, Carl Gösta: **4**.079
Wessels, Vincent E.: **4**.352
West, R.G.: **5**.181; **7**.057, 072; **11**.080
Westgate, J.A.: **4**.462; **7**.106
Wetherill, George W.: **2**.215
Wettlaufer, Boyd N.: **9**.188
Wheeler, J.M.: **12**.046
Wheeler, Richard P.: **1**.009
Wheeler, Sir Mortimer (Robert Eric Mortimer): **10**.019, 025
Whitaker, W.W.: **2**.092, 093, 251; **14**.094
White, Carmel: **12**.029
White, D.E.: **4**.057
White, George W.: **5**.005, 308; **7**.001, 067
Whitehead, Donald R.: **4**.219; **5**.284; **7**.060
Whitehouse, F.P.: **4**.041
Whiteman, A.J.: **4**.412
Whitmore, Frank C.: **5**.059
Wickman, Frans E.: **2**.307, 309
Wigley, L.: **4**.431
Wigley, R.L.: **4**.136
Wijnstra, T.A.: **5**.187
Wilcox, Ray E.: **4**.461; **5**.302
Wilding, L.P.: **3**.270
Wilkes, Owen: **3**.299
Wilkomm, H.: **14**.235
Willey, Gordon R.: **9**.130, 210, 220
William, H.B.: **5**.183
William, H.R.: **4**.220
Williams, D.L.: **2**.094; **3**.153, 222
Williams, John R.: **5**.139
Williams, M.A.J.: **4**.016
Williams, Milton: **2**.092, 093, 251; **3**.238; **14**.094
Williams, Peter A.: **9**.081
Williams, R.E.: **14**.232
Williams, R.E.G.: **14**.019
Willis, E.H.: **2**.021, 027, 238, 244, 304; **4**.094, 211; **5**.249, 253; **6**.034; **11**.091; **14**.029, 030, 031, 032, 033, 034, 035, 100
Willis, Eric H.: **14**.105
Willis, F.R.S.: **5**.253
Willis, H.H.: **11**.090
Willis, Harry H.: **2**.130
Willkomm, H.: **14**.233, 234

Willman, H.B.: **5**.031, 035, 040, 049, 054
Wilmeth, Rosco: **14**.037
Wilmsen, Edwin N.: **9**.177
Wilson, Alex T.: **2**.066
Wilson, H.W.: **2**.148
Wilson, Rex L.: **9**.261
Wilson, Richard Leland: **7**.113
Wimbush, D.J.: **4**.229
Wimer, J.J.: **11**.068
Winkler, Ebhard M.: **4**.319
Winkler, J.R.: **2**.099
Winter, Thomas C.: **5**.203, 297
Wise, E.N.: **14**.230
Wiseman, John D.A.: **4**.069, 103
Wittenberg, J.: **14**.265, 266, 267
Wittry, Warren L.: **9**.168
Woldstedt, Paul: **4**.304
Wolf, Karl H.: **4**.001
Wolfe, John A.: **4**.352
Wolfgang, Richard L.: **2**.188, 259, 295; **3**.144
Wollin, Goesta: **7**.025, 052, 063, 073
Wood, L.: **2**.140
Woodburry, Nathalie F.S.: **1**.028
Woodbury, Richard B.: **9**.102
Woodcock, Alfred H.: **4**.102
Workman, William: **9**.244
Workman, William B.: **14**.188
Worsley, P.: **5**.136
Worthington, L.V.: **6**.027
Wreschner, E.: **10**.024
Wright, E.: **3**.023
Wright, H. E.: **4**.199
Wright, H.V.: **5**.297
Wright, Herbert E.: **3**.118; **4**.037, 258; **5**.030, 102, 140, 146, 245, 246; **7**.086; **10**.036
Wright, J.V.: **9**.109, 174
Wyatt, E.I.: **2**.168
Wyllys, Andrew, E.: **9**.031
Wymer, John: **11**.035

Yaffe, L.: **2**.146
Yamada. O.: **2**.131
Yamakoshi, K.: **2**.161, 310
Yamasaki, Fumio: **14**.173, 174, 175, 176
Yanko, W.H.: **3**.297
Yankwich, Peter E.: **3**.142
Yarnell, Richard A.: **9**.020, 196
Yoshida, Yoshio: **5**.233
Young, I.J.: **4**.452
Young, James Allen: **2**.255; **6**.035; **14**.164
Young, R.S.: **10**.080
Young, T.C.: **10**.074, 078

Author Index

Yuan, L.C.L.: **2**.211
Yuan-Hui, Li: **6**.016
Yves, Patricia C.: **14**.219

Zagwijn, W.M.: **4**.259, 422; **5**.034, 282; **7**. 149; **14**.086
Zarrina, Ye. P.: **4**.140
Zastawny, Andrezey: **3**.011
Zavel, F.S.: **3**.113
Zbarski, S.H.: **3**.100
Zeller, E.J.: **9**.217
Zeuner, Frederick: **2**.006, 109; **7**. 095; **10**.050; **14**.161
Zharkov, A.P.: **3**.314
Zoltai, S.C.: b2.018; **5**.086, 132
Zonneveld, J.I.S.: **4**.127
Zumbergé, James: **5**042, 141, 201; **7**.079

SUBJECT INDEX

Abbot's Way site: **11**.077
Aboriginal art: **12**.012
Abri Pataud, Dordogne: **11**.042
Absaroka Mountains, Wyoming: **9**.189
Absolute age: **2**.107, 113; **3**.002; **4**.067, 106; **5**.186; **8**.000; **9**.161
Absolute chronology: **4**.006, 007, 018, 099, 289, 323, 324; **5**.001, 002, 042, 201; **7**.073, 079, 128; **9**.004, 062, 120, 167; **11**.002, 006, 102
Absolute dates: **3**.256
Absolute dating: **2**.105, 261; **3**.001, 018, 190, 273; **4**.119, 138, 217, 301; **5**.028; **7**.050; **9**.064, 067, 130, 232, 316; **10**.010; **11**.032, 104
Abyssal zone: **6**.018
Acceptability: **3**.090; **11**.042
Acculturation: **9**.007
Accuracy: **2**.140, 184, 214; **3**.003, 004, 041, 139, 253, 266, 278, 325; **4**.456; **9**.196; **11**.095; **13**.012
Acetone: **3**.287
Acetylene: **2**.230; **3**.027, 039, 047, 048, 055, 087, 114, 135, 136, 143, 177, 193, 194, 239, 252, 285, 343, 347; **13**.004, 014
Acheulian: **8**.013
Acid treatment: **3**.271
Activated carbon: **3**.336
Activity (volcanic): **4**.204
Adan Weiss Peak, Wyoming: **4**.147
Adena culture: **9**.124, 134, 333
Adobe: **10**.020
Adzes: **12**.034
Aeolian deposits: **4**.117, 118, 167, 415, 428; **7**.087; **13**.007
Aerial photography in archeaology: **8**.076
Agate Basin: **9**.331
Agate Basin complex: **9**.055, 123
Age: **4**.318; **6**.025; **7**.103
Age corrections: **2**.084
Age limit: **3**.266
Age reporting: **3**.251
Agricultural revolution: **10**.037
Agriculture: **8**.070; **9**.029, 210, 255, 285, 306; **10**.046; **11**.030, 037, 067, 088; **12**.046
Ahab site: **10**.039
Ahichchatra site, Uttar Pradesh: **10**.057, 058

Ahklun Mountains, Alaska: **5**.104
Air: **3**.307, 308
Air circulation: **2**.192
Air composition: **2**.004
Air pollution: **2**.061; **4**.131
Air reservoir: **6**.014
Air sampling: **2**.001; **4**.131
Aire River, Victoria: **12**.011
Aitape site, New Guinea: **12**.072
Aitonava bog, Finland: **4**.362
Alaska earthquake: **4**.287
Algae: **4**.396, 451; **5**.238
Ali Kosh site, Iran: **10**.020
Ali Tappeh cave, Iran: **10**.010
Allerød: **4**.138, 205, 430; **5**.045, 068, 069, 070, 198, 252, 264, 266
Alligator Lake site, Florida: **9**.012
Alluvial plains: **4**.224, 311
Alluvium: **4**.016, 030, 031, 062, 139, 147, 198, 202, 231; **7**.059
Alpha particles: **3**.290
Alpine areas: **5**.262
Altemont moraine, North Dakota: **5**.179
Altitude: **2**.228, 229; **3**.331
Amaranth seeds: **9**.077
Amberat: **4**.086
Amersfoot: **5**.301
Amplifiers: **3**.097
Anaktuvuk pass, Alaska: **7**.080
Analogy: **9**.332
Anangula Island, Alaska: **9**.155
Anangula site, Alaska: **9**.014
Anatolian Mountains: **4**.084
Anderson Bay, Canada: **5**.060
Andrea Bog, Minnesota: **2**.266
Animal bones: **5**.045; **7**.150; **9**.136
Animal husbandry: **10**.047
Animal remains: **4**.045, 223
Anions: **3**.203
Anomalies: **3**.286; **6**.032
Anomalous age: **2**.178, 248, 249, 274, 297; **4**.152; **5**.141; **6**.023; **7**.116; **9**.357
Antarctic Ocean: **6**.001, 029
Antarctic convergence: **6**.003
Anthropogenic factor: **3**.008
Anticoincidence circuits: **2**.176; **3**.103, 107, 115, 147, 237, 240, 338, 343
Anticoincidence counters: **3**.183
Antikythera shipwreck: **11**.005
Antilles: **9**.182
Antiquities: **3**.259
Antlers: **7**.116; **9**.019, 140, 238; **11**.040
Apalachicola coast, Florida: **7**.098

Apatite: **3**.288
Appalachian: **5**.050, 162, 305
Apparent age: **2**.231
Applications: **2**.017, 064, 065, 071, 102, 139, 156, 176, 234, 236, 237, 242, 244, 261, 315; **3**.012, 014, 074, 086, 129, 192, 219, 232, 254, 255, 260, 273, 284, 302, 337; **4**.094, 360; **9**.320; **13**.010
Applied Science Center for Archaeology: **3**.015
Aquatic shells: **4**.177
Aquatic vegetation: **4**.058, 081
Aquifers: **2**.085, 272; **4**.066, 119, 157, 172, 299, 316, 361, 387; **9**.361
Archaeological dating: **2**.018, 019, 059, 104, 108, 166, 168, 182, 197; **3**.016, 094, 137, 197, 253; **9**.125, 360; **10**.082; **11**.028, 106
Archaeological features: **9**.332
Archaeological material: **9**.011
Archaeological methodology: **3**.321, 326
Archaeological samples: **2**.009, 106, 113, 233; **3**.094, 103, 161, 162, 252, 253, 289; **9**.160, 284; **10**.034
Archaeological sites: **3**.196, 234; **8**.004, 020, 032, 066; **9**.011, 022, 026, 040, 044, 049, 075, 138, 164, 168, 325, 336, 360; **10**.057, 061; **11**.087; **12**.022, 023, 040, 053; 055, 059, 064
Archaeological specimen: **8**.084
Archaeological survey: **12**.066
Archaeology: **1**.007; **2**.064, 108, 197, 235, 273; **3**.018, 069, 076, 086, 129, 217, 219, 248, 321, 350; **4**.242; **9**.027, 268; **11**.061, 080; **12**.007, 010, 014, 035, 049, 057, 074, 075
Archaic period: **4**.139; **9**.042, 043, 044, 059, 060, 063, 076, 150, 152, 165, 172, 174, 176, 186, 198, 233, 255, 266, 274, 278, 301, 305, 347; **12**.058
Architecture: **9**.236
Arctic: **9**.245, 318
Arctic Alaska: **4**.453
Arctic Archipelago: **5**.218
Arctic Canada: **7**.077
Arctic Ocean: **5**.182; **6**.009, 019, 028, 031
Arctic regions: **3**.259; **4**.036, 063, 243, 312, 344, 452; **7**.076; **9**.159, 302, 303
Argon: **3**.296
Arid climates: **4**.129; **9**.230
Arid environment: **4**.086, 223
Arid lands: **2**.289; **4**.061, 316; **7**.049
Art: **9**.046
Artesian water: **4**.058
Artesian well: **4**.457
Artifacts: **4**.139; **8**.027; **9**.019, 047, 081, 082, 149, 176, 253, 350; **11**.065; **12**.018
Artificial radioactivity: **2**.026, 034, 043, 072, 080, 094, 121, 162, 213, 221, 253, 254, 293; **3**.164; **4**.364; **6**.039
Aryans: **10**.005, 006
Ash: **10**.029; **11**.084
Asikli Hüyük, Anatolia: **10**.012
Asphalt: **9**.122
Asprochaliko cave: **11**.019
Assay: **2**.007, 093
Assemblages (archaeology): **10**.020
Assessment: **3**.294
Assiniboine River, Canada: **4**.418
Assumptions: **2**.010
Astrophysics: **2**.020
Atlantic (period): **4**.244, 259, 416; **6**.034; **11**.039, 083
Atlantic Coast: **4**.432
Atlantic Ocean: **4**.396; **6**.016, 020
Atmosphere: **1**.031; **2**.012, 013, 026, 027, 030, 031, 032, 043, 051, 068, 074, 094, 121, 122, 129, 133, 134, 158, 161, 162, 164, 171, 183, 192, 194, 199, 213, 220, 221, 241, 263, 265, 290, 292, 293, 310; **4**.115, 131; **6**.012, 018, 026, 035, 049, 051
Atmospheric carbon: **2**.179, 276; **3**.082
Atmospheric carbon dioxide: **2**.014, 024, 035, 048, 051, 052, 060, 078, 088, 123, 126, 133, 135, 172, 187, 188, 200, 248, 265, 267, 278; **3**.127, 291, 323, 342; **4**.156
Atmospheric chemistry: **2**.021
Atmospheric circulation: **2**.028, 029, 255; **3**.342; **4**.156, 378; **6**.047
Atmospheric mixing: **2**.267
Atmospheric nitrogen: **2**.025
Atmospheric radiocarbon: **2**.001, 011, 023, 028, 029, 042, 063, 073, 076, 089, 190, 225, 251, 270, 271, 275, 277, 280, 281, 294, 298, 300, 302, 303, 304, 311, 312
Atmospheric reservoir: **2**.057, 063, 072
Atomic Weapons Research Establishment, Great Britain: **3**.122
Atranjikhara site, India: **10**.058
Aubray hole: **11**.098
Auckland Isthmus, New Zealand: **4**.209
Aurignacian: **11**.042
Australia: **12**.052, 060
Australian Aborigines: **3**.019; **12**.002, 022, 043, 053, 054
Australian samples: **1**.003
Australites: **4**.300, 313
Automatic control: **3**.005
Automatic counters: **3**.020
Avian species: **4**.043
Avoca River, New Zealand: **4**.149

Avonlea point: **9**.050
Axel Heiberg Island, Canada: **4**.285, 330; **5**.062
Axes: **12**.029
Ayampitin site, Argentina: **9**.100
Aycock shelter, Texas: **9**.366
Azatlan site, Mexico: **9**.272
Azatlan, Mexico: **9**.017

B ackground radiations: **2**.096, 110; **3**.021, 022, 061, 062, 107, 111, 183, 221, 325
Badentarbet bog, Scotland: **4**.449
Balkan cultures: **11**.004
Baltic sea: **4**.007, 070
Banas culture, Rajasthan: **10**.005, 006
Banff National Park, Alberta: **4**.462
Baobab: **2**.011
Baradostian assemblage: **10**.073, 079
Baradostian industry: **10**.040
Barium carbonate: **2**.143, 168; **3**.080, 105, 148, 305, 323
Bark: **4**.459
Barns: **11**.024
Barnstable Marsh, Massachusetts: **4**.245
Barrier islands: **4**.169
Barriers: **4**.259
Barrows: **11**.008, 010
Basalt: **4**.132
Basketry: **9**.115
Batch mode: **3**.334
Bathurst Island, North West Territories: **4**.155
Battleford formation: **5**.294
Bavendale, New Zealand: **4**.085
Bay scallops: **9**.016
Beach ridges: **4**.171
Beaches: **4**.340, 344; **5**.217
Bear: **9**.087
Beavers: **4**.268; **5**.057
Becancour map area: **7**.081
Beidha site, Jordan: **10**.003, 004
Beinapalli site, India: **10**.059
Belemnite standard: **3**.170
Bell B Track, Westhay: **11**.064
Bell Beakers series: **11**.009
Belt cave, Iran: **10**.067
Benin City, Nigeria: **8**.056
Benzene: **3**.021, 026, 027, 048, 055, 060, 065, 131, 141, 194, 295, 304, 314, 315
Benzene synthesis: **3**.025, 042, 047, 126, 135, 143, 262
Bering Land Bridge: **5**.025, 202; **9**.155, 206
Bering Sea: **4**.148
Berry site, Canada: **9**.052
Beryllium isotopes: **2**.033
Besnagar site, India: **10**.057

Besont point: **9**.050
Beta counting: **2**.154
Beta particles: **3**.144
Beta rays: **2**.120
Beycesultan, Turkey: **11**.011
Bicarbonate: **2**.049, 054, 137, 141, 272; **3**.276; **4**.316, 457; **6**.036
Big Horn basin: **9**.346
Biogenic origin: **3**.270; **4**.112
Biogenous fines: **4**.163
Biogenous material: **6**.013
Biogeography: **7**.097
Biological communities: **3**.276
Biological samples: **3**.085, 215
Biology: **3**.207
Biosphere: **1**.031; **2**.030, 031, 094, 133, 185, 263, 278, 279, 316; **4**.378; **6**.012, 018
Biostratigraphy: **7**.115
Biota: **5**.018
Biotope: **2**.307
Birnik culture: **9**.330
Bisitun cave, Iran: **10**.067
Bisons: **7**.111; **9**.021, 053, 054, 082, 214, 227, 230
Black Sea: **4**.178
Blade culture: **10**.042, 053, 067
Blade industry: **8**.064
Blockstreams: **7**.143
Blood: **3**.081
Bogoslov island, Alaska: **4**.171
Bogs: **4**.021, 056, 113, 180, 209, 282, 356, 362, 371, 449; **5**.051, 074, 134, 196, 212, 237; **9**.372; **11**.006, 065, 078, 097
Bølling: **4**.314
Bomb effect: **2**.022, 028, 029, 030, 031, 032, 034, 035, 043, 060, 066, 072, 074, 076, 085, 094, 121, 158, 159, 162, 164, 196, 201, 216, 228, 229, 231, 255, 271, 286, 290, 293, 301; **3**.164, 184, 342; **4**.093, 174, 364; **6**.032, 049, 050, 051; **9**.072; **13**.010
Bondaian culture: **12**.019
Bone artifacts: **9**.039
Bone dating: **2**.204, 262, 297; **3**.140, 197, 271, 272; **11**.026; **13**.010
Bone implements: **8**.005; **9**.215
Bone middens: **9**.156
Bone organic matter: **3**.140
Bone samples: **3**.140
Bone shell technology: **9**.122
Bones: **3**.016, 028, 072, 139, 226; **4**.074, 329, 349, 350; **7**.094, 130; **8**.002, 041; **9**.048, 053, 055, 082, 227, 231; **12**.051
Boracic acid: **3**.111
Boreal: **4**.212, 413; **5**.176, 213; **6**.034; **11**.035, 104, 105

Subject Index

Boreal climate: **7**.112
Boreal fauna: **4**.074
Boreal forests: **4**.052
Boreholes: **3**.133
Botany Bay Island, South Carolina: **4**.406
Boulder Clay: **5**.007, 008
Boundary Beach, Western Australia: **7**.139
Bow River Valley, Alberta: **5**.154
Bow River, Alberta: **4**.150; **7**.111
Bows: **11**.062
Brackish water: **4**.108
Brazos Valley, Texas: **9**.366
Breccia: **4**.020
Bride moraine, Isle of Man: **5**.240
Bridge River ash: **4**.044
Bridge River, British Columbia: **4**.462
Brigantine City Barrier, New Jersey: **4**.434
Bristlecone pine: **2**.037, 038, 114, 184
British Museum. Research Laboratory. London: **14**.026
Broad River Valley, Tasmania: **4**.060
Broken K Pueblo, Arizona: **9**.095
Bronze: **10**.018
Bronze age: **3**.183; **10**.007, 050; **11**.010, 090, 108
Brooks Lake Glaciation: **5**.242
Brooks Range, Alaska: **4**.256; **5**.107; **7**.036
Brorup: **5**.301
Bubble chambers: **3**.029
Budiño site, Galicia: **11**.041
Bull Brooke site, Ontario: **9**.293
Burial mounds: **9**.153, 187, 199
Burial remains: **9**.039
Burial sites: **8**.068; **9**.152
Burials: **8**.048, 068; **9**.192
Buried peat: **4**.046
Buried samples: **3**.229; **5**.261
Buried soils: **4**.022, 046, 199; **5**.061, 120; **7**.126
Burley site, Canada: **9**.109; **14**.037
Burzahom site, Kashmir: **10**.058, 060
Byam Martin Island, North West Territories: **4**.155

C_{14}/C_{12} ratio: **4**.365; **6**.010, 042
Caddoan culture: **9**.057, 276, 296
Cahokia site, Illinois: **9**.007, 273
Cainozoic: **4**.056
Calcareous concretions: **4**.374
Calcareous sands: **4**.001, 333
Calcareous sediments: **2**.178; **5**.252; **8**.014
Calcite: **4**.055, 428
Calcium carbonate: **3**.288; **4**.064, 146, 307
Calcium oxide: **3**.238
Calculations: **3**.172, 233; **13**.014
Caldera: **4**.448

Calendars: **9**.185
Calibration: **2**.191; **3**.318
Calibration curve: **2**.038
Caliche: **2**.233; **3**.249; **4**.368
Camp fire: **9**.201
Camp sites: **8**.036; **9**.223, 349; **10**.009
Canadian Arctic: **9**.238
Canadian shield: **5**.124
Canal Zone, Panama: **9**.001
Cape Blanco, Oregon: **7**.061
Cape Campbell, New Zealand: **4**.374
Cape Cod, Massachusetts: **4**.021
Cape Gelidonya, Turkey: **10**.007
Cape Northumberland, South Australia: **12**.070
Capillary tubes: **3**.013
Carbon: **2**.049, 050; **3**.084, 170, 176, 189, 296; **4**.307
Carbon black: **9**.013
Carbon content: **3**.196
Carbon cycle: **1**.008; **2**.203, 253, 254, 277, 307; **4**.109
Carbon dioxide: **2**.004, 012, 013, 014, 031, 052, 054, 112, 113, 116, 135, 136, 151, 154, 177, 183, 189, 195, 198, 230, 251, 255, 256, 259, 286, 301; **3**.011, 031, 032, 034, 035, 037, 044, 050, 052, 054, 060, 088, 097, 101, 102, 103, 117, 166, 174, 175, 180, 182, 202, 211, 212, 213, 225, 238, 240, 241, 247, 261, 278, 283, 288, 297, 307, 323, 338, 341, 345, 346; **4**.058, 065, 286, 316, 378; **6**.004, 018, 030, 035, 036; **13**.004, 014; **14**.001
Carbon dioxide cycle: **2**.309; **3**.247, 286; **6**.014
Carbon disulphide: **2**.151; **3**.117, 182, 211, 212, 213
Carbon inventory: **2**.276
Carbon isotopes: **2**.013, 015, 053, 054, 055, 083, 138, 282, 307, 308; **3**.142, 292, 317; **4**.109, 457; **5**.023; **6**.048
Carbon isotopes distribution: **6**.012
Carbon monoxide: **2**.259
Carbon reservoir: **2**.275, 276; **3**.276
Carbonaceous sediments: **2**.055; **3**.268
Carbonate: **2**.002, 046, 086, 271; **3**.140, 226, 344; **4**.066, 286
Carbonate rocks: **2**.138, 272; **4**.335
Carbonate sediments: **4**.012, 368; **6**.008
Carbonates: **2**.045, 055; **3**.120; **4**.163; **6**.010
Carcajou Point site, Wisconsin: **9**.035, 064, 075
Caribou: **9**.051
Carolina Piedmont, South Carolina: **9**.144
Carrizo Sand, Texas: **4**.057
Carson Desert, Nevada: **5**.125
Cary: **5**.030, 090, 102, 124
Castle Windy site, Florida: **9**.066
Catalysts: **3**.048, 055, 135

Subject Index

Cathode tubes: **3**.179
Causewayed camps: **11**.052, 086
cave art: **8**.010; **11**.047
caves: **4**.055, 073; **9**.049, 256, 311, 316; **10**.024
Cayman Basin: **6**.022
Cellulose: **2**.066; **3**.227, 228; **4**.314
Cemeteries: **9**.338; **11**.008
Cenozoic: **4**.075; **5**.025, 125
Central Plains: **9**.006
Ceramic culture: **9**.012
Ceramic period: **10**.038
Ceramics: **3**.199, 277; **4**.139; **8**.080; **9**.067, 068, 069, 115, 192, 199, 368
Cereals: **8**.070
Ceremonial centres: **9**.319
Chagvan Bay, Alaska: **5**.104
Chalk: **4**.014
Chaluka site, Aleutians: **9**.011
Chambered cairns: **11**.109
Champlain Sea: **5**.162
Champlain Sea episode: **5**.050, 176, 238, 243; **7**.074, 122; **9**.120
Channel Islands, California: **9**.339
Channelled ware: **8**.018
Charcoal: **3**.046, 087, 183, 196, 287; **4**.017, 019, 030, 062, 074, 095, 226, 292, 450; **5**.033, 154; **8**.008, 010, 012, 016, 023, 027, 036, 072, 073, 078; **9**.019, 023, 029, 037, 041, 055, 077, 087, 099, 134, 151, 154, 166, 169, 189, 192, 201, 217, 224, 249, 258, 262, 298, 313, 328, 346, 357, 359, 362, 366; **10**.003, 008, 009, 011, 013, 021, 023, 025, 030, 032, 072; **11**.028, 047, 048, 063, 074, 081, 083, 085, 086, 092, 093, 098, 100; **12**.001, 011, 030, 036, 047, 079
Charred bone: **9**.359
Charred grains: **10**.029
Charred timber: **10**.016, 051
Charred wood: **3**.183, 227, 228; **4**.020, 175, 228, 377, 461; **9**.024, 184, 203; **10**.003, 004
Chavin civilisation: **9**.202
Chavin culture: **9**.203
Chelford muds: **7**.057
Chemical composition: **3**.196
Chemical dating: **3**.016
Chemical processes: **2**.095, 127, 189; **3**.026, 048, 055, 060, 126, 131, 139, 152, 176, 193, 196, 263, 279, 295, 297, 310, 317, 320, 323, 334, 336
Chemistry: **3**.207
Chemung River valley, United States: **4**.207
Cheniers: **4**.394; **5**.271
Cherangani Hills, Kenya: **4**.091; **5**.143

Chernozem: **4**.035
Chert: **9**.278
Chiapa de Corzo site, Mexico: **9**.068
Chifubwa Stream rock shelter, Zambia: **8**.012
Chios, Greece: **11**.011
Chipped stone tools: **9**.224
Chira coast, Peru: **9**.067
Chirand site, Bihar: **10**.058
Chiriquí phase, Panama: **9**.086
Chivateros tradition: **9**.116
Choctawhatchee Bay, Florida: **4**.111, 240
Choppers: **9**.054
Christchurch Formation: **4**.185
Christchurch Metropolitan area, New Zealand: **4**.277
Christian calendar: **9**.197, 270, 298
Chronology: **2**.017, 037, 064, 103, 235; **3**.006, 049, 129, 137, 173, 186, 248, 299, 321, 337; **4**.034, 071, 090, 101, 128, 138, 139, 144, 217, 224, 253, 279, 306, 308, 312, 315, 322, 325, 418, 437, 443; **5**.003, 020, 029, 034, 054, 060, 065, 066, 141, 150, 169, 199, 209, 210, 222, 235, 237, 279, 296, 302, 309; **6**.046; **7**.005, 010, 016, 021, 023, 029, 035, 043, 049, 063, 093, 095, 123, 131, 132, 133, 134, 145; **8**.000, 001, 006, 019, 024, 028, 055, 056, 057, 073, 074; **9**.031, 035, 042, 070, 071, 084, 089, 092, 100, 102, 125, 150, 161, 172, 179, 188, 208, 212, 217, 240, 247, 248, 269, 270, 304, 318, 325, 352, 369, 371; **10**.002, 010, 017, 019, 022, 027, 028, 029, 042, 048, 049, 051, 054, 064, 065, 075; **11**.001, 003, 014, 015, 016, 017, 018, 052, 053, 054, 055, 056, 057, 058, 059, 061, 072, 082, 088; **12**.010, 018, 020, 025, 035, 039; **13**.012
Chukchi Sea: **4**.148; **7**.070
Chusca Mountains, New Mexico: **4**.258
Cil-maenllwyd, Wales: **5**.067, 160
Civilisation: **9**.202
Clarence River, New South Wales: **12**.045
Clarence Valley, New South Wales: **12**.059
Clark site, Texas: **9**.366
Classic period: **9**.220, 236, 282
Classification: **5**.031
Clay: **4**.046, 199, 373; **6**.013; **7**.012
Clay minerals: **4**.220
Clays: **4**.016, 137, 143, 284, 319
Clear Lake, Oregon: **4**.017
Climate: **4**.065, 187, 200, 263, 267, 274, 330, 333, 394, 411, 413; **5**.020, 061, 074, 128, 210, 212, 284; **7**.025, 052, 060, 063, 069, 086, 095, 148, 149; **8**.050; **9**.019, 171, 180; **10**.035;

Subject Index
11.019
Climatic classification: **5**.282; **7**.052
Climatic cycles: **2**.017, 023, 052, 079, 081, 105, 172, 194, 225, 276, 306; **4**.067, 069, 072, 073, 074, 075, 086, 096, 103, 120, 123, 124, 125, 134, 183, 188, 214, 226, 242, 260, 270, 282, 295, 301, 320, 323, 324, 325, 334, 357, 391, 422, 447, 454, 466; **5**.001, 002, 003, 004, 033, 034, 036, 062, 075, 147, 155, 173, 196, 197, 200, 203, 277, 292, 301; **6**.011, 015, 042; **7**.016, 018, 020, 023, 026, 039, 046, 049, 088, 112, 114, 121, 137, 138, 141; **8**.052; **9**.073, 074, 080, 314; **10**.024, 036, 070; **11**.003, 079; **13**.012
Climatic dating: **9**.188
Climatic environment: **5**.018, 116; **7**.006
Climatic periods: **7**.075
Closed systems: **2**.013; **3**.339
Clovis points: **9**.038, 126, 171, 201, 227
Clovis site: **9**.243
Coahuila, Mexico: **9**.336
Coast of emergence: **4**.021
Coast of submergence: **7**.032
Coastal deposits: **5**.217
Coastal dunes: **4**.076
Coastal plains: **4**.075
Coastal shelves: **7**.071
Coasts: **4**.070, 276, 336
Coasts of emergence: **4**.059, 273; **5**.217
Coasts of submergence: **4**.164, 376, 380, 383, 384, 385, 386, 393, 432, 433, 434; **5**.215, 243; **7**.002; **12**.078
Cochise culture: **9**.048
Cochrane: **5**.071, 251; **7**.123
Cockburn moraine, Canada: **5**.157
Coclé tradition, Panama: **9**.086
Coincidence circuits: **3**.148; **14**.001
Collagen: **2**.271; **3**.028, 197, 226, 288; **9**.082, 256
Collection: **3**.132
Columbia Plateau, United States: **5**.038
Columbian Basin: **6**.022
Combustion: **2**.078, 126, 187; **3**.030, 057, 085, 100, 132, 215, 224, 338, 340
Committee for the Distribution of Radiocarbon Dates: **3**.258; **13**.011
Comondu culture: **9**.242
Compaction: **4**.235; **6**.040
Comparison: **3**.154
Computer language: **3**.110
Computer processing: **3**.058, 059, 146, 156
Computer software: **3**.059, 110, 146, 156
Conferences: **13**.001, 002, 006, 012

Conformity: **11**.063
Consistency: **9**.161; **11**.063
Contact (geology): **4**.392
Contaminants: **3**.270, 287, 300
Contamination: **2**.109, 213, 233, 270, 274; **3**.036, 112, 139, 140, 191, 198, 200, 203, 229, 230, 248, 249, 250, 325, 336, 348; **4**.001, 335, 363, 387; **5**.068, 253, 257, 291; **6**.015; **8**.036, 049, 051, 073, 075; **9**.033, 103, 254, 271, 281; **10**.050; **11**.095
Continental shelves: **4**.033, 136, 158, 163, 184, 189, 236, 240, 333, 374, 396, 397; **5**.059, 084; **7**.070, 091, 092, 100; **9**.025
Continental slopes: **4**.215
Conversion: **2**.152; **3**.027
Conversion factor: **2**.153
Cook inlet, Alaska: **5**.229
Copper River Basin, Alaska: **5**.063, 109
Copper age: **10**.006, 050, 066; **11**.023, 043, 050
Copper mines: **9**.184
Coprolites: **4**.086; **9**.170, 370
Cordilleran culture: **9**.080
Core and blade culture: **9**.110, 155, 251
Core drilling: **4**.087; **5**.293
Cores: **7**.064
Corn: **2**.167, 291; **9**.162, 306
Corn Creek Dunes site, Nevada: **9**.081
Correction: **2**.005, 099, 100, 313; **3**.004, 098, 264, 299, 344; **4**.316, 363, 456; **9**.196, 289, 297, 315
Correction factors: **3**.033
Correlation: **2**.045, 060, 062, 103, 105, 128, 298, 302; **4**.079, 087, 088, 130, 202, 263, 271, 297, 370, 401, 447; **5**.014, 040, 069, 076, 107, 128, 170, 193, 290; **7**.001, 050, 056, 073, 093, 120, 125, 132, 137, 138; **8**.038, 054; **9**.040, 058, 082, 130, 175, 183, 197, 268, 271, 274, 281, 308; **11**.004, 019, 056, 058, 095; **12**.009, 078
Cosmic rays: **2**.003, 010, 025, 033, 073, 081, 084, 090, 091, 097, 098, 100, 101, 109, 129, 133, 155, 179, 196, 207, 209, 211, 221, 227, 241, 252, 275, 276, 281, 295; **3**.061, 062, 138; **4**.089, 115; **5**.174; **6**.018, 047
Cosmogenic origin: **3**.339
Cosmology: **1**.020, 021
Coteau du Missouri, North Dakota: **5**.164
Counters: **2**.096; **3**.111
Counting efficiency: **3**.242
Counting equipment: **3**.014, 044
Counting methods: **2**.058, 065, 130; **3**.043, 154, 165, 212, 213, 335; **13**.004
Counting time: **3**.221

Cowichan Bay, British Columbia: **4**.438
Cowichan Valley, Vancouver Island: **5**.043
Coxcatlan cave, Tehuacan valley, Mexico: **9**.162
Cracking process (chemical): **3**.038, 194
Crater Lake, Oregon: **4**.461, 462
Crater lakes: **4**.237; **5**.017
Craters: **2**.101
Cremation: **9**.187; **11**.083
Crooked River: **4**.232
Cruck buildings: **11**.025
Crumley site, Texas: **9**.262
Crustal movements: **4**.376, 448
Cultivation: **7**.086; **8**.070; **9**.028, 162; **10**.037
Cultural adaption: **10**.044
Cultural changes: **9**.061, 062, 071, 085, 092, 255
Cultural chronology: **9**.246
Cultural complex: **9**.042, 194
Cultural contact: **9**.181, 324; **10**.039, 040; **12**.063
Cultural deposits: **9**.154; **12**.041
Cultural evolution: **9**.088, 167; **10**.038
Cultural expansion: **8**.035; **9**.159; **10**.062; **11**.088, 089
Cultural history: **9**.039
Cultural material: **9**.304
Cultural periods: **9**.302, 317
Cultural remains: **8**.026; **9**.087, 301; **12**.005
Cultural sequence: **8**.052; **9**.081, 089, 119, 243, 247, 259, 301, 316, 359; **10**.029; **11**.056, 102; **12**.016, 061, 073
Cultural traits: **9**.092, 371; **12**.042
Culture succession: **12**.021
Cultures: **3**.301; **5**.210; **8**.054; **9**.032, 096, 130, 238, 267, 299, 311; **11**.007, 049; **12**.071
Curracurrang cave, New South Wales: **12**.022, 030, 065

Daima, Nigeria: **8**.056, 057
Dallas Bay, Greenland: **5**.079
Dalma Tepe, Iran: **10**.078
Dalton Projectile Points: **9**.347
Damariscotta Shell Heap, Maine: **9**.257
Dambwa site, Zambia: **8**.028
Danger cave, Utah: **9**.093
Dark Ages: **11**.096
Dark current noise: **3**.163
Data processing: **3**.096, 110, 233
Date Lists: **14**.044, 075
Date lists: **1**.009, 015; **3**.256, 267; **13**.013, 016
Datil interval: **9**.175
Dating: **2**.157, 173; **3**.006, 066, 076, 118, 168, 234, 321; **5**.047; **9**.096
Dating methods: **3**.017, 018, 049, 073, 268, 293, 330; **5**.081; **9**.013
Davis site, Texas: **9**.196, 265

de Geer's chronology: **4**.079, 080, 099, 138: **5**.070; **7**.050; **11**.036
de Vries effect: **3**.103
Dead carbon: **2**.224, 256; **4**.335
Debert Archaeological Project: **9**.099
Debert site, Nova Scotia: **9**.098, 216, 229
Decay constant: **2**.215
Deception Bay, Quebec: **5**.013, 082
Decontamination: **3**.112
Deductive models: **9**.085
Deep Spring Lake, California: **4**.366
Deep sea: **2**.175
Deep sea cores: **2**.091; **3**.293, 324; **4**.101; **7**.052, 115
Deep sea sediments: **2**.226; **3**.075; **4**.003, 008, 069, 103, 104, 125, 341, 399, 424, 442; **5**.293; **6**.008, 011, 013, 041, 042, 046, 047; **7**.003, 063, 064, 073
Deep water: **6**.021, 023, 029, 030, 033
Deer Creek cave, Nevada: **9**.103
Dehydration: **3**.317
Deltaic deposits: **4**.143, 184, 379
Deltas: **4**.355; **9**.083
Denali complex, Alaska: **9**.110
Denbigh Flint complex, Alaska: **9**.034, 104; **11**.074, 100
Dendrochronology: **1**.001; **2**.037, 038, 073, 084, 111, 114, 133, 171, 182, 184, 226, 281, 300; **3**.185, 299; **4**.100; **5**.010; **9**.034, 095, 297; **13**.015
Denekamp: **5**.053
Denudation: **4**.375
Deposition: **4**.084, 207, 337, 341, 365; **5**.287
Depositional environment: **4**.143, 222, 239
Desert culture: **9**.326, 344
Deserts: **4**.013, 092; **5**.210
Devon Island, Canadian Arctic Archipelago: **5**.218
Devon rock shelter, Lower Murray Valley, South Australia: **12**.024
Dharnikota site, India: **10**.057
Diachronism: **5**.249
Diagenesis: **5**.026
Diastrophism: **4**.348; **6**.040
Diatomaceous sediments: **4**.238; **6**.013
Diatomaceous silica: **4**.008; **7**.028
Diatoms: **4**.081, 238, 362
Diborane: **3**.143
Diffusion (archaeology): **9**.159; **12**.042
Diffusion models: **6**.035
Diluents: **3**.245, 283, 344
Dilution: **2**.251
Diphenyl oxazol: **3**.130, 163, 283, 347
Dire wolf: **9**.107
Discrepancy: **2**.150, 172, 206, 283; **3**.004, 210; **8**.019, 075; **9**.322; **13**.001

Subject Index

Disintegration rate: **2**.120, 211
Dismal Lake, North West Territories, Canada: **9**.245
Dismal River culture: **9**.222
Distribution: **2**.217
Dolni Vestonic brickyard, Czechoslovakia: **11**.055
Dolomite: **4**.004, 012, 110, 221, 233, 366, 421
Dombozange rock shelter, Southern Rhodesia: **8**.016
Domebo site, Oklahoma: **9**.038, 171, 350
Domestication: **8**.070; **10**.043, 047; **11**.030, 071
Donaldson site, Canada: **9**.109; **14**.037
Donnelly Ridge site, Alaska: **9**.110
Dorset culture, North America: **9**.024, 157, 191, 245, 318, 354; **11**.074, 100
Drainage basins: **4**.375
Drake Passage: **7**.003
Draved forest, Denmark: **4**.390
Drift (geology): **4**.176; **5**.005, 042, 112, 122, 221; **7**.033
Drift (instrumentation): **3**.149
Driftwood: **4**.026, 077, 153, 289, 349, 350, 372, 391; **5**.044; **11**.007
Drill cores: **4**.103, 118, 125, 183, 341, 399, 424, 442; **5**.017, 291; **7**.091, 092
Duck Mountain, Manitoba: **5**.075
Dugout canoes: **9**.361
Dumaw Creek site, Michigan: **9**.111
Duncan map area, British Columbia: **4**.438
Dunes: **9**.151; **12**.019
Durham Meadow, Connecticut: **5**.296
Durrington Walls, Wiltshire: **11**.085
Dutch coast: **4**.051
Dzibilchaltun site, Mexico: **9**.113

Early Polychrome period: **9**.084
Early historic: **10**.060
Early towns: **10**.012, 014
Earth movements: **4**.108, 460; **11**.045
Earth-flow: **4**.046
Earthquakes: **4**.287, 460
East Africa: **8**.053
East Pacific: **12**.069
Easter Island: **12**.049
Eastern Caribbean area: **9**.182
Eastern Hemisphere: **9**.269
Eastern United States: **9**.275
Echuca, Victoria: **4**.311
Ecology: **4**.113; **5**.194
Edwards Plateau culture, Texas: **9**.262
Eemian: **5**.244, 301; **7**.149
Efficiency: **2**.131; **3**.021, 078, 079, 108, 147, 211, 218, 236, 347
Effigy Mound culture: **9**.168
El Jabo complex, Venezuela: **9**.127

Elateia site, Greece: **11**.033
Electromagnetic interference: **2**.180
Electronic equipment: **3**.097, 169, 338
Elephant tusk: **4**.047
Elk: **5**.045
Engabreen, Norway: **5**.055
Engakura, Africa: **8**.020
Engineering geology: **4**.116, 145
Ennadai Lake, North West Territories: **4**.274, 321
Enrichment: **6**.048
Environment: **2**.213; **9**.227, 232; **11**.019
Environmental changes: **4**.114, 134
Environmental impact: **2**.155
Environmental reconstruction: **9**.304
Epilimnion: **4**.109
Eqe Bay, North West Territories: **5**.028
Equatorial zone: **6**.042
Equilibrium: **6**.026
Erosion: **4**.062, 176, 443; **7**.143; **9**.014
Errors: **2**.005, 069, 082, 083, 086, 119, 137, 237, 256, 291, 297, 304; **3**.028, 077, 104, 175, 198, 200, 203, 221, 248, 250, 258, 269, 286, 304, 325, 336, 350; **4**.316, 399, 456; **5**.068, 253; **7**.124; **9**.103, 237, 254, 281; **12**.042; **13**.004, 014
Eskers: **7**.122
Eskimo culture: **9**.318
Eskimo-Aleuts (Alaska people): **9**.011, 129, 154
Eskimos (North American people): **9**.032, 065, 128, 154, 159, 191, 245, 303, 330, 354, 371; **11**.100
Estimating: **3**.316
Estoril Beach, Mozambique: **4**.096
Estuaries: **4**.033, 243
Estuarine deposits: **4**.108, 185, 275
Ethane: **2**.113; **3**.167
Ethanol: **3**.147
Ethanolamine: **3**.152
Ethyl alcohol: **3**.347
Ethylbenzene: **3**.314
European calendar: **9**.158
Eustasy: **4**.040, 041, 098, 120, 232, 404; **5**.104; **6**.034, 040; **7**.087
Eustatic curve: **4**.346; **5**.036
Eustatic movements: **4**.039, 186, 247, 280, 348, 376, 380, 393; **5**.208; **7**.038, 135, 139
Eustatism: **4**.121
Eutresis site, Greece: **11**.011, 028
Evaluation: **3**.049; **9**.013
Evaporation: **2**.289
Evaporite deposits: **4**.435; **5**.023
Evaporite minerals: **4**.233
Everson: **5**.235
Evidence: **5**.113, 114, 115; **9**.125

Evolution: **3**.168
Excavations (archaeology): **9**.022, 080, 164; **10**.023; **12**.010, 077
Exchange mechanism: **2**.032, 063; **6**.014
Exchange rate: **2**.135, 177, 183, 189, 230, 292, 308; **6**.026
Exchange reactions: **3**.323
Exchange time: **2**.026, 031, 134, 267; **6**.012, 051
Extinct fauna: **9**.087, 215; **12**.044

F actor of merit: **3**.106
Faeces: **9**.241
Fall-out: **2**.117, 213, 216, 217, 290, 296; **3**.331
Fallahogy Landnam phase: **11**.091
Far Eastern civilisations: **9**.324
Farmdale: **5**.064, 164; **7**.059
Farmdallian: **5**.294
Farming: **8**.031; **9**.220; **10**.037, 042, 062; **11**.088, 089, 091
Farmington complex, California: **9**.005
Fault scarps: **4**.149
Fauna: **3**.185; **4**.284, 374; **7**.009, 030, 075, 146; **9**.019
Faunal extinction: **1**.016; **4**.129, 181, 208, 214; **7**.004, 006, 018, 121; **9**.126; **11**.079
Faunal specimens: **5**.006; **7**.142; **9**.135
Faunal succession: **4**.415; **7**.095; **9**.317
Fenlands, Great Britain: **4**.211
Fernbank, New York (state): **7**.008
Fernslev site, Denmark: **11**.060
Fiber tempered pottery: **9**.198, 274
Fiber tempered tradition: **9**.115
Field methods: **3**.040, 225; **4**.087
Field work: **3**.061, 133, 134, 286, 309; **4**.285; **5**.042
Figtree: **8**.003
Figurines: **8**.047; **11**.064
Fiji: **12**.039
Finnmark, Norway: **4**.077
Fire sites: **9**.136
Firn: **7**.050
Firth of Thames, New Zealand: **4**.394
Fish: **7**.031; **9**.138
Fishbone cave, Nevada: **9**.225
Flake assemblage: **10**.053
Flake tools: **9**.140; **12**.012
Flandrian: **4**.463
Flexed inhumation: **9**.122
Flint: **9**.209, 354; **10**.020; **11**.068
Floating tree ring chronology: **2**.114
Flood plains: **4**.116, 139, 207, 231, 345
Floods: **4**.100; **12**.008
Flora: **4**.085, 250, 301; **7**.030, 103
Floral evolution: **5**.134
Floral succession: **5**.184; **9**.317

Florida coast: **4**.051, 383, 384, 385
Florida keys: **4**.428
Floridan aquifer: **4**.066; **9**.361
Florisbad site, South Africa: **8**.037, 075
Flotation: **3**.270
Flow charting: **3**.058, 059
Flow rate: **3**.120; **4**.057, 172
Fluctuation: **4**.189
Fluted points: **9**.120
Fluvial deposits: **4**.185, 294, 354, 440
Fluvial environment: **2**.248, 249; **3**.191, 276
Fluvial sediments: **5**.300
Fluvioglacial deposits: **5**.171
Flåmsdalen valley, Norway: **5**.058
Folsom complex: **9**.055
Folsom culture: **9**.018, 119, 194
Fontbouisse ware: **11**.050
Food chain: **3**.276
Food gathering: **9**.143, 170, 200; **10**.046; **11**.044
Foraminifera: **3**.112, 200, 324; **4**.075, 190, 399, 405; **5**.233; **6**.028, 046; **7**.025
Forest Bed: **7**.082
Forest succession: **4**.037, 129, 135, 281, 320, 321, 328; **5**.010, 017, 033, 201, 219; **7**.034, 040, 114; **9**.141; **12**.031
Forests: **4**.369; **5**.303, 304
Formic acid: **3**.317
Fort Lawrence, Nova Scotia: **4**.376
Fortran: **3**.058, 059, 110
Fosheim Peninsula, North West Territories: **4**.230
Fossil assemblage: **7**.142
Fossil beaver: **5**.061
Fossil beetles: **7**.065
Fossil bison: **9**.119
Fossil bones: **3**.028, 288; **4**.047; **5**.075; **8**.025
Fossil carbon: **2**.173; **3**.082
Fossil dunes: **4**.064
Fossil elephants: **9**.364
Fossil fuel: **2**.056, 061, 078, 188, 195, 224, 251
Fossil hominids: **3**.273; **5**.173; **7**.010; **8**.002
Fossil lakes: **4**.183, 322, 323, 324, 358, 420; **5**.023; **7**.020
Fossil man: **12**.032
Fossil molluscs: **5**.132
Fossil plants: **5**.151
Fossil pollen: **10**.070; **11**.065
Fossil shells: **4**.244; **5**.243
Fossil trees: **4**.232
Fossil vertebrates: **4**.415
Fossil water: **4**.299
Fossil wood: **4**.226, 351; **11**.038
Fossilisation: **2**.204
Fossils: **3**.268; **4**.150; **7**.024, 126; **9**.304
Fouha Bay, Guam: **12**.066

Subject Index

Foxe Basin, North West Territories: **5**.015, 028, 142, 189
Fractionation: **2**.053, 054, 112, 132, 207; **3**.336
Fraser Canyon, Canada: **9**.259
Frego, Patagonia: **7**.093
Fremont culture: **9**.222
Frenchmans Cap National Park, Tasmania: **7**.015
Freshwater: **2**.046, 137; **4**.368
Freshwater environment: **2**.049
Freshwater shells: **8**.027
Frobisher Bay, Canada: **4**.190
Fromm's Landing site, South Australia: **12**.008, 009
Frozen soils: **4**.087
Fuel combustion: **2**.051
Fukui rockshelter, Japan: **10**.026
Fulvic acid: **4**.035
Funeral customs: **11**.060
Funnel beaker pottery: **11**.060
Furnace: **3**.087
Fussell's Lodge, Great Britain: **11**.092

Galena site, Texas: **9**.131, 367
Galverston Bay Focus, Texas: **9**.131
Gamtoos Valley, South Africa: **8**.078
Ganima Wadi, Tripolitania: **4**.062
Ganj Dareh Tepe, Iran: **10**.076, 077
Gardner Spring site, New Mexico: **4**.013
Gas counters: **3**.167, 180, 211, 325, 343
Gas counting: **2**.113, 212; **3**.020, 034, 043, 052, 098, 099, 102, 114, 136, 188, 212, 213, 252, 333, 341; **9**.013; **13**.004, 014
Gas proportional counters: **2**.260; **3**.011, 106, 138, 166, 213, 214, 313, 318, 345
Gas proportional counting: **2**.230; **3**.019, 031, 032, 036, 039, 044, 064, 097, 115, 187, 189, 193. 237, 239, 318, 329, 338; **14**.001, 022
Gas purification: **3**.241
Gas sampler: **3**.134
Gas sampling: **3**.133, 134
Gases: **2**.264; **3**.170
Gastropodes: **7**.071; **9**.234
Geiger counters: **2**.116, 142, 144, 146, 150, 151, 154, 260; **3**.005, 009, 022, 035, 050, 071, 098, 114, 116, 117, 124, 144, 174, 175, 182, 212, 213, 243, 327, 343
Gelatin: **3**.139
Gels: **3**.089
Geochemistry: **2**.219, 273; **4**.299, 360
Geochronology: **1**.002, 003, 004, 020, 021; **2**.006, 139, 218, 261; **3**.001, 118, 253, 330; **4**.105, 139, 148, 262, 299, 359; **5**.176; **9**.093, 141, 166; **11**.061
Geographic areas: **9**.062

Geological factors: **2**.140
Geological processes: **4**.207; **13**.010
Geological samples: **2**.009, 044, 113, 119, 245; **3**.253, 289, 294; **4**.097, 142; **5**.199; **9**.284
Geological Survey of Canada. Radiocarbon dating laboratory: **3**.119
Geological terraces: **4**.016, 030, 062, 097, 121, 150, 297, 331, 411, 443, 444, 460; **5**.138, 242; **7**.028, 048, 059, 065, 117, 126, 140
Geological time: **3**.330
Geology: **1**.020, 021; **2**.219; **4**.079; **7**.078
Geomagnetic field: **4**.152
Geomagnetism: **2**.077, 084, 099, 100, 160, 268, 276, 277, 280
Geomorphological cycle: **4**.096
Geomorphology: **4**.153, 154, 229, 266, 311, 412, 437; **5**.082, 110; **9**.307; **12**.007
Geophysics: **2**.219
Geula caves, Mount Carmel: **10**.024
Ghar-i Khar cave, Iran: **10**.076
Gherangani, Kenya: **4**.263
Gibson aspect: **9**.057
Gillis Lake, Nova Scotia: **5**.144
Glacial advance: **4**.224, 410; **5**.011, 044, 071, 093, 099, 105, 129, 154, 220, 228, 298; **7**.015, 033, 054, 068, 104
Glacial chronology: **4**.080, 099, 295, 430; **5**.008, 027, 032, 035, 039, 040, 041, 044, 046, 053, 072, 077, 081, 082, 083, 084, 087, 089, 090, 092, 094, 095, 099, 101, 103, 104, 106, 107, 108, 111, 113, 114, 115, 124, 126, 127, 128, 129, 130, 143, 146, 156, 160, 161, 167, 168, 170, 174, 185, 186, 190, 191, 192, 204, 213, 221, 232, 234, 241, 242, 245, 246, 247, 248, 253, 256, 262, 263, 268, 275, 280, 285, 286, 288, 289, 290, 291, 292, 295, 305, 309; **6**.015; **7**.014, 027, 036, 037, 097
Glacial climates: **5**.293
Glacial cycles: **2**.023, 079; **5**.029, 037, 229, 259, 273, 275, 278; **7**.039, 074
Glacial deposits: **4**.296, 354, 410; **5**.103, 119, 164, 167, 207, 234, 245, 246, 274, 279, 281; **7**.008, 102, 105, 125
Glacial features: **4**.411; **5**.084, 085, 086, 087
Glacial geology: **2**.244;
Glacial ice: **2**.136; **5**.223
Glacial intervals: **2**.194; **7**.124
Glacial lakes: **4**.187, 410, 435; **5**.038, 050, 077, 087, 093, 094, 095, 125, 132, 185, 222, 306; **7**.017, 083
Glacial lobes: **5**.031, 035, 072, 102
Glacial maximum: **5**.231; **9**.074

Glacial periods: **4**.136, 168; **5**.003, 306; **7**.025
Glacial retreat: **4**.011, 138, 180, 190, 246, 247, 273, 289, 438, 439, 454; **5**.004, 005, 011, 013, 022, 028, 050, 051, 052, 060, 068, 069, 070, 071, 076, 079, 085, 086, 091, 093, 097, 098, 099, 129, 159, 168, 177, 184, 188, 220, 233, 254, 255, 264, 265, 273, 285; **7**.015, 032, 076, 079, 084, 085, 089, 104, 105, 136
Glacial stades: **4**.050
Glacial stages: **5**.039, 059
Glacial till: **4**.146, 146; **5**.011, 129, 231, 294, 308; **7**.108
Glacial unloading: **7**.038
Glaciation theory: **5**.003, 036
Glaciations: **2**.105, 171; **4**.033; **5**.004, 012, 014, 019, 020, 037, 104, 112, 122, 207, 229; **7**.056, 063, 086, 101, 110, 120, 125, 126
Glacier Bay, Alaska: **4**.124; **5**.048
Glacier ice: **2**.264; **3**.133, 134; **5**.043
Glacier Peak eruption: **4**.018; **5**.158; **7**.096
Glacier Peak, Oregon: **4**.461
Glaciers: **5**.006, 047, 055, 259, 261, 298
Glass Mountain, California: **4**.024
Glass vials: **3**.218
Globigerina ooze: **6**.008
Glotto-chronology: **9**.213
Goebel Midden, Texas: **9**.263
Goodman - Martinex - Thompson correlation: **9**.298
Goose Creek site, Texas: **9**.131
Goose Lagoon, Victoria: **12**.002
Gordion Tumulus, Turkey: **10**.080
Gordion, Turkey: **11**.011
Graman district: **12**.013
Granite: **8**.049
Grantsburg lobe: **5**.140
Graphite: **3**.339
Grasses: **2**.167; **5**.075
Grassland: **4**.267, 369
Gravel: **4**.025, 095, 168; **7**.013, 111, 143
Gravity separation: **3**.270
Great Bahama Banks: **4**.318, 367
Great Basin: **4**.323, 324; **9**.150, 170
Great Lakes: **5**.030, 073, 078, 197, 222, 225; **7**.066
Great Lakes region: **4**.037, 281; **5**.147, 220; **9**.172, 231
Great Plains: **4**.415; **9**.026, 038, 050, 121, 163, 171, 222, 240, 255, 266, 321, 325, 350
Great Salt Lake: **7**.005
Groundwater: **2**.085, 141, 198, 248, 272, 313; **3**.120, 344; **4**.054, 057, 061, 066, 092, 093, 157, 172, 316, 317, 360, 457, 464; **10**.035
Groundwater movement: **4**.119, 157, 361
Groundwater recharge: **4**.061, 066, 157, 172
Growth (plants): **3**.264

Growth rate: **4**.055, 248, 365; **5**.262
Guano: **9**.250
Guatemala valley: **9**.101
Gubik formation: **4**.337; **5**.110
Gudbrandsdalen Valley, Norway: **4**.225
Guerrero Negro, Mexico: **4**.076
Gulf of Maine: **4**.431
Gulf of Mexico: **4**.165, 234
Gundu site, Zambia: **8**.044
Gwisho site, Zambia: **8**.042, 081
Gyttja: **7**.007; **9**.231
Günz glaciation: **5**.173

Hacilar site, Turkey: **10**.022; **11**.011
Hackberry lake: **4**.267
Hajji Firuz, Iran: **10**.078
Half-life: **2**.047, 115, 116, 120, 142, 143, 144, 145, 146, 147, 148, 149, 150, 151, 152, 153, 154, 168, 171, 172, 174, 215, 243, 287; **3**.083, 095, 122, 123, 198; **9**.308; **13**.015
Hallar site, India: **10**.059
Hamilton River, Labrador: **4**.180
Hanau Eepe (Easter Island people): **12**.020
Hanau Momoku (Easter Island people): **12**.020
Handbook: **3**.056
Harappa culture: **10**.027
Harappan civilisation: **10**.065
Harappan period: **10**.035
Harappan sites, Punjab: **10**.056
Hard water: **2**.178; **4**.363; **5**.253
Harlan site, Oklahoma: **9**.022, 294, 295
Harper River, New Zealand: **4**.149
Hartstown bog, Pennsylvania: **4**.135
Hasanlu, Iran: **10**.078
Haua Fteah cave, Lybia: **8**.025; **8**.000
Hawaiian islands: **12**.039
Health hazard: **2**.034, 216
Hearths: **9**.029, 041, 077, 103, 346, 351; **11**.020, 042
Heartwood: **2**.066; **3**.264
Hekla volcanoe, Iceland: **4**.450
Hell Gap Valley, Wyoming: **9**.193
Hell Gap, Wyoming: **9**.331
Hembury Hillfort, Devon: **11**.063
Hemisphere: **2**.026
Henge: **11**.040, 085
Herdsmen: **10**.030
Heron Island, Queensland: **4**.001
Hetimpur site, Uttar Pradesh: **10**.060
Hieroglyphs: **9**.329
High cave, Tangier: **10**.067
High Peak, Devon: **11**.096
Hillforts: **11**.063

Subject Index

Historic period: **10**.057
Historical dating: **8**.019
Hodges site, Michigan: **9**.152
Hokan people: **9**.213
Holcombe site: **9**.051
Holocene: **2**.245; **4**.005, 006, 013, 029, 049, 068, 127, 147, 158, 159, 160, 161, 162, 163, 164, 165, 182, 202, 205, 313, 315, 352, 357, 382, 391, 393, 396, 400, 404, 432; **5**.187, 229, 266; **7**.010, 019, 098, 127, 128; **8**.000; **10**.069, 083; **11**.094
Hominids: **3**.001; **8**.075
Homolka site, Bohemia: **11**.043
Honegore site, Somerset: **11**.077
Hopewell culture: **9**.070, 124, 134, 153, 333, 348
Horizontal mixing: **4**.364; **6**.043
Horne Lake, British Columbia: **4**.440
Horticulture: **12**.037, 046
Hosterman's pit, Pennsylvania: **4**.074
Hot Springs Village site, Alaska: **9**.244
Hot Springs, Wyoming: **4**.147
Hotu cave, Iran: **10**.067
Hudson Bay: **4**.011, 232; **5**.015, 142, 159
Hudson Straits: **5**.091
Hudson Valley, New York: **9**.044
Huefarno Park, Colorado: **5**.018
Human bones: **8**.023, 027; **9**.039, 045, 087, 156; **12**.036
Human habitation: **9**.018
Human migrations: **9**.155, 206
Human occupation: **4**.073; **7**.013; **8**.068; **9**.003, 020, 026, 046, 054, 083, 094, 100, 102, 123, 139, 155, 156, 189, 215, 217, 240, 244, 255, 258, 314, 321, 347, 351, 362, 363; **10**.015, 024, 026, 044, 069; **11**.029, 102; **12**.041, 067, 071, 076
Human population: **2**.159; **8**.041
Human remains: **2**.104; **9**.192; **12**.072
Human settlements: **8**.048; **10**.050; **11**.096; **12**.027
Humans: **2**.022, 035
Humbolt Lake, Nevada: **9**.311
Humic acid: **3**.229, 287, 348; **4**.035; **5**.257
Humus: **2**.132, 249; **4**.035, 094, 173, 313
Hunter-gatherers: **4**.139
Hunters: **9**.021, 206; **10**.009, 030
Hunting: **9**.252
Huron basin: **5**.098
Hyamine: **3**.101
Hydration: **2**.054, 107, 189
Hydrocarbons: **4**.166, 222, 234
Hydrochloric acid: **3**.288
Hydrogen: **2**.162
Hydrogenation: **3**.010, 320
Hydroisochronic maps: **2**.141

Hydrological cycle: **4**.109
Hydrology: **4**.360
Hydrosphere: **2**.220, 221
Hypolimnion: **4**.109
Hypsithermal interval: **4**.067, 071, 171, 181, 244, 249, 256, 268, 344, 347, 356, 466; **5**.003, 010, 014, 062, 082, 091, 139, 194, 211, 213, 239, 299; **6**.040; **7**.022, 036, 096; **9**.016, 026, 165, 186, 230, 353

Ice: **2**.223; **3**.274, 275, 309; **7**.089
Ice Age theory: **2**.081
Ice advance: **5**.102
Ice ages: **5**.182, 293; **9**.372
Ice cores: **3**.134; **4**.288; **5**.265; **6**.015
Ice movement (glaciation): **4**.431; **5**.062, 089, 090, 094, 095, 100, 179, 180, 224, 227, 236, 295, 304
Ice sheet: **4**.438; **5**.015, 038, 255
Icebergs: **2**.223
Icefield ranges, Yukon Territory: **4**.124
Ihumatao volcanoe, New Zealand: **4**.463
Ikpik Bay, North West Territories: **5**.028
Iliamna Lake, Alaska: **5**.242
Illinoian: **5**.101, 111, 112; **7**.100
Impurities: **3**.036, 241
Imuruk Lake, Alaska: **7**.013
Inarajan Village, Guam: **12**.066
Incipient agriculture: **9**.143; **10**.013
Indian Ocean: **4**.333; **6**.016
Indians of North America: **9**.111, 133, 184, 256, 355, 371
Indo-Pakistan subcontinent: **10**.045
Indus civilisation: **10**.027, 056
Indus-Baluchistan region: **10**.021
Industrial fuels: **2**.051, 126, 187; **6**.004
Industrial revolution: **2**.230
Industrial wastes: **2**.118; **3**.191
Ingombe Ilede site, Rhodesia: **8**.031
Inorganic carbon: **2**.193
Insects: **5**.116
Institute of Physics, University of Uppsala, Sweden: **3**.122
Instrumentation: **1**.031; **2**.242, 243; **3**.010, 029, 030, 032, 050, 065, 085, 097, 098, 099, 102, 103, 106, 107, 108, 117, 119, 124, 133, 153, 154, 182, 187, 202, 205, 206, 215, 236, 239, 240, 243, 261, 274, 280, 281, 304, 307, 308, 313, 319, 327, 333, 338, 342, 345
Intensity: **2**.003, 090
Interglacial stages: **4**.072, 426; **5**.003, 117, 118, 119; **7**.013, 112

Interglaciations: **4**.139
International Conference on Radiocarbon Dating, Andover, Massachusetts, 1-4 October, 1956: **13**.012
International Radiocarbon Conference, Fifth, Cambridge, July 23-28, 1962: **13**.015
Interpretation: **2**.111; **3**.040, 077, 086; **5**.113, 114, 115; **9**.093, 160, 161, 196, 254, 280
Interstadials: **4**.189; **5**.121, 205, 256, 260, 289; **7**.058
Inugsuin Fjord, Baffin Island: **5**.217
Inventory: **6**.051
Invertebrate: **7**.041
Ion exchange: **3**.203
Ionisation chambers: **2**.120, 168; **3**.053, 185
Ionium deficiency dating: **2**.226
Irish Sea Glaciation: **7**.030
Irish Sea glaciers: **5**.007
Iron: **3**.041, 067, 068, 186
Iron age: **1**.024; **3**.186; **8**.007, 017, 018, 021, 028, 029, 030, 031, 034, 044, 046, 048, 055, 060, 065, 066, 071, 074, 076, 080; **11**.010, 076, 090, 097
Iron alloys: **3**.067
Iron ores: **8**.067
Iroquois: **9**.052
Isamu Pati Mound, Zambia: **8**.007, 034
Isla Ilosko site, Hungary: **8**.001
Island 35: **9**.165
Isostasy: **4**.190, 247, 273, 372; **5**.050, 091, 097, 130, 214, 218, 300; **7**.038
Isostatic curve: **5**.135
Isotope effect: **3**.038, 141
Isotopes: **2**.004; **6**.047
Isotopic abundance: **2**.309; **3**.296, 298
Isotopic composition: **2**.048, 049, 248, 274, 289; **4**.178, 368
Isotopic enrichment: **2**.055, 155, 157, 179, 202, 309; **3**.233, 282, 317
Isotopic fractionation: **2**.196, 291; **3**.247, 304; **4**.109
Isotopic ratio: **2**.002, 045, 046, 051, 112, 137, 138, 167, 171, 172, 174, 185, 189, 202, 224, 307, 313; **3**.170, 198, 247, 264, 344; **4**.057, 177, 286, 457; **5**.023; **6**.004, 005, 020, 045
Isotopic substitution: **2**.127
Itasca, Minnesota: **4**.253
Iyatayet Valley, Alaska: **4**.141

Jaguar cave, Idaho: **9**.217
Jaketown site, Mississippi: **9**.166
Jarmo culture: **10**.042
Jarmo site, Iraq: **10**.025, 032
Jericho site, Jordan: **10**.025

Jericho, Jordan: **10**.023
Jørgen Brønlund Fjord, Greenland: **5**.273
Jotunheimen National Park, Norway: **5**.169

Kalambo Falls site, Zambia/Tanzania: **8**.013, 032, 033, 079
Kalibangan site, Punjab: **10**.056
Kalkbank site, South Africa: **8**.005
Kalomo culture, Zambia: **1**.024; **8**.007, 031, 034
Kames: **5**.171
Kammennyi Mys site, Kamchatka: **10**.015
Kangila (African people): **8**.031
Kara Kamar cave, Afghanistan: **8**.000; **10**.067
Karst: **11**.053
Kartan culture, South Australia: **12**.021
Kaskawulsh glacier, Yukon Territory: **7**.033
Katrista cave, Greece: **11**.020
Kauai Peninsula, Alaska: **5**.230
Kaushambi site, Uttar Pradesh: **10**.060
Keewatin District, North West Territories: **5**.033, 089, 184
Keilor terrace, Victoria: **12**.006, 036, 056
Kelvin Valley, Scotland: **7**.150
Keniff cave, Queensland: **12**.005
Kensi Lowland, Alaska: **5**.229
Kern River, California: **7**.147
Kill sites: **8**.008; **9**.038, 171, 214, 350, 364
Killik River, Alaska: **4**.411
King glacier, Tasmania: **5**.105
King of Phrygia: **10**.080
Kisalian culture, Zaire: **8**.022, 077
Kishwaukee River, Illinois: **5**.241
Kitchen middens: **9**.025, 037; **11**.045; **12**.001, 002
Klamath Lake, California: **9**.167
Knives: **9**.053, 214
Known age: **2**.007, 009, 010, 038, 069, 194, 257, 297; **3**.068, 092, 185; **13**.015
Kodiak Island, Alaska: **9**.221
Kolterman Mounds, Wisconsin: **9**.168
Koroit beach, Victoria: **12**.002
Kotosh site, Peru: **9**.203
Kotzebue Sound, Alaska: **4**.423
Kronprins Olav Kyst, Antarctica: **5**.233
Kurdish mountains: **7**.086
Kuusamo District, Finland: **4**.425

La Brea Tar Pit, California: **3**.185
La Jolla complex: **9**.085
La Jolla Radiocarbon Laboratory: **3**.322
La Perra cave, Mexico: **9**.015
La Vale·de Coro, Venezuela: **9**.127
La Venta complex, Mexico: **9**.337
La Venta site, Mexico: **9**.132, 205, 315, 319, 329

Subject Index

La Victoria site, Guatemala: **9.**169
Laboratory description: **3.**011, 020, 051, 231, 266, 322; **14.**038
Laboratory equipment: **3.**005, 104
Laboratory manuals: **3.**142
Laboratory methods: **1.**031; **2.**041, 144, 157, 180, 196, 236, 238, 262, 315; **3.**009, 010, 027, 030, 031, 032, 033, 034, 037, 041, 042, 044, 047, 048, 050, 051, 054, 055, 057, 060, 067, 071, 074, 075, 080, 085, 092, 098, 099, 100, 102, 103, 104, 107, 112, 117, 121, 122, 125, 131, 132, 135, 136, 151, 153, 154, 157, 163, 169, 181, 182, 186, 187, 201, 202, 205, 209, 214, 215, 222, 223, 226, 227, 228, 238, 239, 240, 246, 252, 254, 255, 256, 257, 261, 262, 263, 270, 271, 272, 274, 279, 281, 283, 285, 287, 292, 295, 299, 300, 304, 307, 309, 310, 313, 315, 317, 320, 322, 329, 332, 334, 335, 336, 338, 344, 345, 348; **4.**235, 290; **10.**031, 034, 057
Lac St.Jean, Quebec: **5.**168
Laco site, Arizona: **9.**364
Lacustrine deposits: **4.**005, 056, 102, 130, 188, 193, 199, 267, 284, 354, 363, 375; **5.**134, 252, 261, 300; **7.**012, 026, 059, 117
Lacustrine environments: **2.**248; **4.**109, 344; **7.**026
Lacustrine sediments: **2.**086; **5.**073
Lacustrine terraces: **4.**448
Lagoe Funda cave, Brazil: **9.**087
Lagoons: **4.**033, 164, 283
Laguna de Tagua-tagua, Chile: **9.**215
Laguna Salada, Arizona: **9.**133
Laguna San Raphael, Southern Chile: **5.**083
Lake Agassiz: **4.**410; **5.**077, 087, 124, 185; **9.**140
Lake Algonquin: **5.**127
Lake Arkona, Ohio: **4.**179
Lake Bonneville: **4.**322, 323, 324, 358; **5.**093; **7.**005
Lake Cochise, Arizona: **7.**011, 020
Lake District, Great Britain: **4.**255
Lake dwellings: **2.**114
Lake Erie: **5.**073, 307; **7.**035
Lake Erie Lobe: **7.**067
Lake Eyre, South Australia: **4.**270
Lake Geneva, Wisconsin: **5.**090
Lake George, New South Wales: **4.**095
Lake Grassmere, United States: **5.**127
Lake Huron: **5.**016; **7.**035
Lake Huron glacial lobe: **5.**093
Lake Ishango, Zaire: **4.**177
Lake Lahontan: **4.**073, 322, 323, 324; **5.**093, 125
Lake Lundy: **5.**126, 127
Lake Menindee, New South Wales: **12.**070

Lake Michigan: **5.**031, 102, 141; **7.**035
Lake Michigan basin: **9.**141
Lake Michigan lobe: **5.**030
Lake Minong, Ontario: **5.**087
Lake Mohave, California: **9.**180, 253
Lake Naivasha, Kenya: **4.**067
Lake Ontario basin: **9.**293
Lake Paducah, Kentucky/Illinois: **7.**017
Lake Sarkkilanjarvi, Finland: **4.**315
Lake Searles: **4.**183, 446; **5.**023
Lake Shippewa, Indiana: **4.**319
Lake St Clair: **5.**127
Lake Superior: **9.**184
Lake Superior region, Ontario: **5.**086
Lake Tapps, Washington: **4.**437
Lake Tuborg, North West Territories: **4.**027, 028
Lake Victoria: **4.**114
Lake Warren: **4.**179, 214; **5.**127
Lake Weber, Minnesota: **7.**118
Lake Whittlesey, Ohio: **4.**179
Lake-level: **4.**095, 323, 324, 358
Lakes: **4.**028, 254, 319, 451
Laminae: **4.**238, 459; **6.**013
Lamont Geochronogical Observatory Radiocarbon Laboratory: **3.**335
Lamont Geochronological Observatory Radiocarbon Laboratory: **9.**283
Lamont Sea Water Radiocarbon Program: **6.**017
Lampasas River, Texas: **4.**231
Land ice: **4.**346, 465; **5.**157, 177
Landslides: **4.**009, 032, 375, 427
Lapaneva bog, Finland: **4.**362
Larsen Creek, Canada: **7.**145
Lascaux caves, France: **11.**047, 048
Last Glaciation: **2.**272; **4.**029, 042, 422; **5.**116, 170, 187, 282
Late-glacial period: **4.**023, 194, 195, 200, 201, 250, 252, 400, 403; **5.**018, 072, 085, 131, 132, 133, 134, 144, 145, 147, 148, 149, 150, 151, 152, 153, 176, 181, 185, 195, 196, 203, 266, 267, 288, 297; **7.**116, 118; **9.**021, 317; **11.**049
Latitude: **2.**164, 228, 229, 301; **3.**331
Laurel tradition: **9.**174
Laurentian channel: **7.**019
Laurentide ice sheet: **5.**254; **7.**039
Lauricocha site, Peru: **9.**118
Lava flows: **4.**017; **7.**062
Leaching: **3.**287
Lebous fortified village, France: **11.**050
Lehner Mammoth site, Arizona: **9.**135, 146, 175, 227, 364
Leonard rock shelter, Nevada: **9.**250

Leopard's Hill site, Zambia: **8**.059
Lerna, Greece: **11**.011, 051
Levees: **4**.143
Levi site, Texas: **9**.176
Lewisville site, Pennsylvania: **9**.335
Lichens: **5**.262
Lignin: **3**.287; **4**.314
Lime: **2**.222; **3**.269
Lime Creek site, Nebraska: **9**.313
Limestone: **2**.248; **3**.344; **4**.316, 381, 428; **7**.140
Limestone caves: **4**.048, 074; **11**.053
Limnology: **2**.296
Lindenmeier site, Colorado: **9**.147, 177
Linsley Pond, New England: **4**.109
Liquid scintillation: **3**.007, 245
Liquid scintillation counters: **3**.042, 063, 101, 130, 147, 148, 149, 150, 161, 162, 209, 314
Liquid scintillation counting: **1**.017; **2**.094, 095, 131; **3**.027, 030, 047, 048, 055, 057, 060, 065, 078, 079, 081, 088, 126, 141, 143, 145, 146, 147, 151, 152, 153, 154, 155, 156, 157, 159, 160, 163, 178, 194, 201, 218, 220, 222, 224, 242, 279, 283, 303, 304, 306, 310, 312, 314, 315, 319; **13**.003, 014
Liquid scintillators: **3**.158, 159, 160
Liquids: **3**.013
Literature reviews: **1**.020, 021; **3**.155, 302; **9**.061
Lithic industry: **8**.004, 008, 037, 041; **9**.014, 118, 122, 129, 209, 224, 354; **10**.012, 026, 042; **11**.019, 020, 068; **12**.013, 045, 077
Lithium: **3**.285
Lithium aluminium hydride: **3**.297
Lithium carbide: **3**.087
Lithology: **7**.126
Little Harbor site, Catalina Island: **9**.179
Little Ice Age: **4**.015, 356, 408; **5**.048
Little Ortega Lake, Arizona: **9**.133
Littorina Littorea: **9**.234
Llano complex: **9**.038, 171, 201
Llanquihue Province, Patagonia: **7**.043
Loam: **4**.331
Loch Droma, Scotland: **5**.133, 153
Lochinvar Mound, Northern Rhodesia: **8**.036
Lochinvar Ranch, Northern Rhodesia: **8**.023
Loess: **4**.130, 154, 167, 218, 220, 343; **5**.040, 122, 244; **13**.007
Long barrows: **11**.052, 083, 092, 093
Long Island, New York: **7**.044
Long lived isotopes: **6**.039
Long lived radiocarbon: **2**.212
Lorenzen site, California: **9**.213
Lovelock cave, Nevada: **9**.009, 091, 170
Low activity: **2**.168; **3**.005, 013, 150, 176

Low background radiations: **3**.349
Low-level counting: **2**.155; **3**.020, 022, 027, 029, 064, 106, 144, 165, 166, 167, 178, 185, 215, 236, 243, 327, 349; **13**.008, 009, 010
Lower Columbia Valley: **9**.046
Lower Mississippi Valley: **4**.224
Lucy: **9**.201
Lynn Lake, Manitoba: **4**.274

Machobilla phase, Ecuador: **9**.117
Mackenzie District, Canada: **5**.175
Macon Earth Lodge, Georgia: **9**.261
Macrofossils: **5**.150, 200
Madelia, Minnesota: **5**.152
Magdalenian: **11**.047
Maglemosian: **11**.035, 068
Maglemosian site: **11**.071
Magnetic field: **2**.077
Magnetic moment: **2**.310
Magnetism: **2**.227
Maize: **9**.015, 028, 196, 285
Malaya: **10**.063
Mammals: **5**.074; **7**.111; **8**.002
Mammoth bones: **9**.350
Mammoth cave, Kentucky: **9**.060
Mammoth Lake, California: **4**.175
Mammoth Mountain, California: **4**.175
Mammoths: **4**.106, 139, 206, 213, 225; **5**.059; **9**.018, 038, 048, 112, 126, 171, 227, 252, 309, 372; **11**.079
Man: **3**.168
Manakaway site, Connecticut: **9**.283
Mandible: **11**.094
Mangere volcanoe, New Zealand: **4**.463
Mangus site, Montana: **9**.123
Mankato: **5**.068, 069, 070, 124, 140, 179, 204, 230; **7**.064, 078; **9**.074
Marengo moraine, Wisconsin: **5**.090
Marine clay: **5**.172; **7**.085
Marine deposits: **5**.234; **7**.061, 087
Marine environment: **2**.049, 231, 232
Marine fossils: **4**.190; **7**.042, 122
Marine geology: **4**.240
Marine molluscs: **4**.146, 244; **5**.135, 142, 207
Marine sediments: **2**.270; **4**.108, 143, 149, 185, 210, 240, 297, 372, 398, 423, 440; **5**.025, 084, 238; **6**.028
Marine shells: **2**.092, 231, 232; **3**.183; **4**.050, 155, 230, 347, 386, 406, 436, 465; **5**.015, 072, 168, 175, 216, 217, 223, 243; **9**.072, 328
Marine terraces: **4**.059; **5**.108, 135, 239; **7**.042
Maritime economy: **9**.122
Marl: **2**.233; **3**.249; **4**.056, 199, 284, 335; **5**.005; **9**.103,

Subject Index

Marmes rock shelter, Washington: **9**.108, 156
Marquesa islands: **12**.039
Marshes: **4**.283; **5**.291
Marsupialia: **4**.002, 208
Mass spectrometers: **2**.142, 143, 154
Mass spectrometry: **2**.132, 149, 167; **3**.123, 170, 298
Mastodons: **4**.023, 129, 213, 214, 250, 328, 329; **5**.059; **9**.141, 148, 165, 211, 231
Matanuska glacier, Alaska: **5**.139
Mathematical methods: **3**.172
Maximum age: **2**.297; **3**.221; **9**.083
Mayan calendar: **9**.058, 158, 197, 270, 298, 308, 322
Mayas: **9**.030, 031, 113, 114, 132, 270, 271, 282, 356
Maybeso Valley, Prince of Wales Island: **4**.146
Measurements: **3**.142
Megalithic monuments: **11**.012
Megaliths: **11**.021, 060
Melanesian culture: **12**.048
Meltwater: **2**.223
Memory effect: **3**.167
Mesoamerica: **9**.027, 030, 031, 132, 181
Mesolithic: **10**.016, 053, 067; **11**.013, 034, 036, 039, 068, 104, 105
Meson pulse: **3**.329
Mesophytic interval: **5**.197
Meteorites: **2**.101; **3**.339; **4**.089, 115, 326
Meteorology: **2**.030, 192
Methane: **2**.162, 179; **3**.010, 032, 103, 131, 181, 187, 188, 280, 320; **10**.034; **13**.004, 014
Methane cycle: **2**.185
Methane synthesis: **3**.038, 334, 336
Methanol: **3**.279, 297; **13**.003
Method description: **1**.015; **3**.056; **5**.024; **13**.012
Methodological problems: **11**.001
Methyl alcohol: **3**.279
Methyl borate: **3**.245
Meuse River: **7**.048
Michigan basin: **5**.098, 197
Micro-palaeontology: **4**.401
Microliths: **4**.313; **12**.012, 024, 045
Middens: **9**.103, 134, 274, 301
Middle Ages: **2**.288
Middle Mississippian: **9**.347
Middle Palaeolithic: **10**.048, 053
Migrations: **4**.076; **9**.070; **12**.052, 053, 054
Mineral alterations: **4**.373
Minimum age: **3**.226; **4**.311
Mining: **8**.067
Miocene: **4**.392
Missinaibi beds: **7**.058
Missinaibi River, Ontario: **4**.082
Mississippi deltaic plain: **4**.355
211; **11**.035

Mississippi River: **9**.090
Mississippi River delta: **4**.379
Mississippi technology: **9**.007
Mississippi Valley: **7**.131
Mississippian culture: **9**.073
Mississippian tradition: **9**.276, 296
Mixing: **2**.080, 140
Mixing rate: **2**.030, 122, 175, 251; **3**.331; **6**.030
Mixtec (Mesoamerican people): **9**.185, 327, 373
Moa bones: **2**.076
Moa hunters: **2**.076; **12**.015, 034, 038, 044
Moaning cave, California: **4**.055
Models: **2**.013, 292; **5**.097; **6**.039
Modern age: **4**.222, 234
Modern carbon: **2**.014, 061
Modoc rock shelter, Illinois: **3**.046; **9**.186
Mohave Desert, California: **4**.198
Moldova V site, USSR: **11**.073
Mollusc shells: **2**.249; **4**.200, 226, 231; **5**.233, 234; **7**.017, 090; **9**.138
Molluscs: **3**.112; **4**.075, 130; **5**.239; **7**.031, 091, 092
Mologa - Shiksha Lake, USSR: **4**.139
Monagrillo pottery: **9**.036
Monasterian transgression: **4**.209
Monk's Kop ossuary, Rhodesia: **8**.039
Mono Lake, California: **4**.175
Monte Alban culture: **9**.327
Monte Alban site, Oaxaca: **9**.185, 236, 373
Monte Alban, Mexico: **9**.158
Monte Negro culture: **9**.327
Montezuma Well, Arizona: **4**.058
Monuments: **11**.050
Moorean archaeology: **12**.041
Moraine systems: **5**.157
Moraines: **4**.161, 288, 465; **5**.006, 009, 019, 055, 086, 090, 117, 137, 142, 162, 169, 177, 232, 240, 242, 265, 287, 309; **7**.033, 034, 050, 067, 068
Morphology: **4**.330; **5**.167
Mortar: **3**.069, 070, 269
Mortlach Site, Saskatchewan: **9**.188
Mortuary houses: **11**.060
Mosquera, Columbia: **4**.023
Mounds: **8**.036; **9**.022, 090, 249; **11**.085
Mount Carmel, Israel: **8**.013
Mount Field National Park, Tasmania: **4**.060
Mount Kosciusko, New South Wales: **4**.229, 408
Mount Lamington, Papua: **4**.088
Mount Lofty Range (South Australia): **4**.270
Mount Mazama eruption: **4**.291; **5**.158, 211; **9**.049, 167
Mount Mazama, Oregon: **4**.132, 461, 462
Mount Rainier, Washington: **4**.204, 293
Mount St. Helens, Washington: **4**.462

Mount Washington eruption: **5**.211
Mountain lakes: **4**.113
Mousterian: **7**.010; **8**.050; **10**.046, 079; **11**.003
Mowbray Swamp, Tasmania: **4**.056
Mud: **4**.159, 166, 209, 352; **5**.023, 045, 253; **7**.057
Mud flow: **4**.373
Mudukian culture, South Australia: **12**.021
Muir Inlet, Alaska: **5**.088
Mummies: **9**.256
Mummy cave, Wyoming: **9**.189
Murray River, Australia: **12**.008
Musc ox: **7**.129
Muscotah marsh, Delaware: **4**.052
Museum of Applied Sciences of Victoria. Radiocarbon dating laboratory: **3**.266
Mussel shells: **9**.176
Mussels: **5**.273; **12**.024

Naco kill, Arizona: **9**.048
Nagara site, India: **10**.057
Nahoon site, South Africa: **8**.014
Nakapapula rock shelter, Zambia: **8**.060
Narasipur site, India: **10**.059
Nash Creek, Canada: **7**.145
Native Point site, Southampton Island: **9**.354
Natsushima site, Japan: **10**.038
Nea Nikodemia site, Greece: **11**.030, 031, 037
Neanderthal man: **8**.013, 015; **10**.068, 079, 081
Necropolis: **8**.022, 077
Negro art: **8**.038
Nenana River Valley, Alaska: **4**.345; **5**.230
Neolithic: **1**.024; **4**.298; **8**.015, 040, 072, 078; **10**.003, 004, 009, 013, 016, 025, 030, 043, 046, 047, 048, 050, 059, 060, 063, 066, 068, 076; **11**.001, 002, 010, 012, 014, 015, 018, 029, 030, 032, 037, 038, 046, 051, 052, 060, 062, 063, 064, 065, 066, 067, 068, 076, 077, 081, 084, 086, 092, 096, 099, 106, 107; **12**.029, 042
Neutron flux: **2**.207, 209
Neutrons: **2**.025, 096, 098, 129, 180, 192, 211, 252, 268; **3**.062
New England: **5**.147
New England coast: **9**.138
New Guinea: **12**.037
New Guinea Highlands: **12**.046, 051
New Haven Harbor, Connecticut: **4**.272
New Zealand prehistory: **12**.031
Newberry volcano, Oregon: **4**.132, 228
Newport Bay, California: **7**.041
Niah Great cave, West Borneo: **10**.008
Nietoperzowa cave, Poland: **11**.075
Niger delta: **4**.400

Nile River: **4**.016, 226
Nile valley: **8**.062, 064, 070
Nipissing: **5**.016
Nitration: **3**.287
Nitric acid: **3**.339
Nitrogen: **2**.033, 192; **3**.072, 296
Niwot Ridge, Colorado: **4**.331
Nok culture, Nigeria: **1**.024; **8**.009, 047, 066
Nomenclature: **5**.032
Nomna Bay, Guam: **12**.066
Non Nok Tha site, Thailand: **10**.018
Non-glacial periods: **4**.082; **5**.061, 096
Noola rock shelter, New South Wales: **12**.070
Nordaustlandet, Spitzbergen: **4**.289, 349, 350; **5**.269
North America: **2**.303; **9**.343
North Atlantic Ocean: **4**.280; **7**.025
North Loup River, Nebraska: **5**.138
North Pacific Ocean: **6**.036
North Truro Quadrangle, Massachusetts: **5**.072
North-American Cordillera: **7**.036
Northeastern Pacific: **4**.216
Northern Black Polished Ware: **10**.058
Northern Florida Coast: **4**.126
Northern Hemisphere: **2**.067, 183, 201, 267; **4**.174; **5**.167
Northwest Alaska: **5**.061
Northwestern Europe: **4**.123
Nualolo bluff shelter, Kauai: **12**.004
Nuclear explosions: **2**.027, 110, 121, 192, 203, 213, 217, 253, 254, 265, 290, 293; **6**.019
Nuclear radiation shields: **3**.111, 147, 333
Nuclear weapons: **2**.031; **6**.035
Nucleons: **3**.062
Nuclides: **2**.296; **4**.115
Nuku Hiva, Marquesas Islands: **12**.015
Nullarbor Plain, Western Australia: **4**.002
Numa Entrance shelter, Namibia: **8**.008
Nuragic civilisation: **11**.070
Nutbane site, Wiltshire: **11**.093

Oak Grove site, California: **9**.122
Oakhanger site, Hampshire: **11**.039
Obsidian: **4**.024; **10**.012, 040
Obsidian hydration dating: **2**.107; **9**.209
Ocala limestone, Georgia: **4**.387
Ocanto site, Wisconsin: **9**.338
Occupation site: **9**.151
Ocean atmosphere interaction: **2**.028, 029, 123, 134, 135, 177, 189, 255, 265, 299; **6**.004, 012, 050
Ocean bottom: **2**.079; **4**.148, 398; **6**.002, 024
Ocean circulation: **2**.073, 122; **4**.163; **6**.001, 002, 007, 019, 022, 027, 031, 036, 037, 044, 047
Ocean water: **2**.031, 032, 075, 122, 135, 171, 195, 263; **3**.075, 195, 294; **6**.005, 007, 009, 010,

Subject Index

016, 019, 020, 023, 025, 036
Ocean water circulation: **6**.030
Oceanic carbon dioxide: **2**.078, 200
Oceanic mixing: **2**.122, 133, 251, 299; **6**.001, 003, 012, 016, 020, 021, 030, 031, 037, 039, 050
Oceanic reservoir: **2**.072; **6**.014
Oceanography: **2**.017; **6**.045; **9**.200
Oceans: **2**.013, 051, 251; **6**.018, 051
Ocmulgee National Monument, Georgia: **9**.261
ODaiska site, Colorado: **9**.285
Odiorne Point, New Hampshire: **4**.376
Ogilvie Mountains, Yukon Territory: **5**.309
Ohakea Terrace, New Zealand: **4**.351
Ohine site, Fiji: **12**.062
Ohio River valley: **7**.066; **9**.199
Oil shales: **4**.451
Old Copper culture: **9**.338
Older Dryas: **4**.430; **5**.084
Oldest Dryas: **5**.143
Oldman River, Alberta: **9**.054
Olmec civilisation: **9**.132, 202, 204
Olmec culture: **9**.203, 205, 282, 319, 329, 337
Ologogo site, Western Samoa: **12**.063
Olympia interglaciation: **5**.096, 180
Oneota culture: **9**.035, 064
Ongari Point, New Zealand: **12**.026
Onion Portage site, Alaska: **9**.209
Onion Portage, Alaska: **9**.349
Oolites: **4**.236, 367, 396
Ooliths: **4**.318
Oolitic sand: **4**.318
Ooze: **4**.014, 401; **5**.117
Opal: **3**.270; **6**.013
Oquendo complex, Peru: **9**.116
Organic carbon: **2**.138, 307
Organic matter: **2**.015, 118, 263, 316; **3**.028, 046, 057, 081, 160, 191, 196, 219, 292, 312, 348; **4**.001, 035, 081, 087, 116, 165, 166, 188, 229, 237, 238, 239, 293, 353, 369, 386, 398, 407, 411, 419, 438; **5**.012, 022, 117, 159, 171, 194, 227, 237, 253, 256, 265, 287, 306; **7**.028, 057, 108; **9**.253, 256, 320; **11**.095; **12**.072
Organic ooze: **4**.451
Organic sediments: **2**.266
Organic soils: **5**.129
Organic solvents: **3**.244
Organic temper: **9**.368
Orleton Farms, Madison County, Ohio: **9**.148, 211
Osceola site, Wisconsin: **9**.338
Oslofjord District, Norway: **4**.210
Ossuaries: **8**.039; **9**.006
Oxalic acid: **3**.002
Oxalic acid standard: **2**.084; **3**.103, 171; **13**.003
Oxidation: **3**.100, 224, 226
Oxygen: **2**.033, 049; **3**.201, 296
Oxygen isotopes: **2**.083, 223
Oyster shells: **4**.033, 352
Oysters: **9**.016

Pacific coast: **9**.212
Pacific Islands: **12**.073
Pacific North-West: **4**.018
Pacific Ocean: **6**.016, 037; **12**.050
Paintings: **9**.327
Palaeoatmosphere: **2**.136
Palaeobotany: **2**.079; **4**.426
Palaeoclimates: **2**.136; **4**.251, 408; **5**.023
Palaeoclimatology: **4**.242; **5**.010, 201
Palaeoecology: **4**.243; **5**.185; **10**.044; **12**.072
Palaeoeskimos: **11**.074, 100
Palaeogeography: **5**.186
Palaeoindian site: **4**.117
Palaeoindians: **9**.022, 053, 098, 099, 120, 123, 126, 159, 172, 176, 177, 191, 194, 201, 208, 214, 215, 216, 229, 255, 266, 290, 313, 331, 350
Palaeolithic: **4**.242; **7**.010, 056; **9**.206; **10**.015, 046, 069, 071, 072; **11**.003, 019, 020, 022, 041, 042, 072, 073
Palaeoliths: **10**.008
Palaeomagnetism: **2**.226, 227, 280
Palaeontology: **5**.007
Palaeopedology: **4**.307
Palaeosols: **4**.139, 167, 199, 202
Palaeotemperatures: **4**.244, 424, 442; **6**.024; **7**.025, 110; **9**.217
Palaeovegetation: **4**.258
Palaeozoic: **4**.335
Palahnihan people: **9**.213
Palmer site, Florida: **9**.274
Paludal environment: **4**.283
Palynological dating: **5**.200; **7**.058, 059
Palynology: **2**.036; **4**.082, 083, 128, 285, 419, 426; **5**.073, 176, 178, 193; **7**.023, 034, 060
Paracas culture: **9**.218
Paraffin wax: **3**.111
Paraldehyde: **3**.209, 319
Park counties, Wyoming: **4**.147
Parksville, British Columbia: **4**.440
Particulate matter: **2**.061; **3**.082
Passage graves: **11**.010, 060
Patagonia: **5**.298
Paudorf: **5**.053; **7**.088
Pavilan cave, Wales: **11**.026
Peacock's Farm, Cambridgeshire: **11**.065

Subject Index

Peat: **2.**233; **3.**216, 249; **4.**005, 010, 018, 021, 034, 044, 049, 050, 051, 053, 056, 071, 077, 085, 136, 145, 164, 180, 192, 203, 209, 219, 235, 245, 248, 268, 276, 280, 283, 292, 312, 342, 344, 352, 367, 371, 386, 392, 396, 414, 416, 429, 431, 440, 450; **5.**007, 008, 011, 045, 066, 089, 096, 107, 119, 129, 140, 181, 197, 200, 208, 216, 221, 237, 308; **7.**007, 029, 055, 062, 065, 069, 072, 081, 082, 099, 120, 123, 135; **8.**075; **9.**016, 134, 361; **11.**062, 078
Peat bogs: **4.**205, 248, 416; **11.**038
Peat deposits: **5.**185
Pebble tools: **12.**045
Pecatorica drift, Illinois: **5.**064
Peccary: **7.**094, 130
Pecos River, Texas: **9.**028
Pedogenesis: **4.**022, 036, 336; **7.**133
Pedology: **3.**219; **4.**013, 359, 456; **9.**307
Pelecypods: **4.**190; **7.**091
Penguins: **5.**236
Periglacial zone: **4.**117, 408; **5.**021, 304
Perigordian: **11.**042
Permafrost: **4.**106, 151, 249, 453; **5.**061
Permeability maps: **2.**141
Perth Readvance: **5.**190, 191
Petroleum: **4.**398
Photomultiplier tubes: **3.**065, 242
Photosynthesis: **2.**159, 259; **6.**029
Physical processes: **2.**189
Physics: **2.**197; **3.**326
Physiography: **4.**339
Phytogeography: **5.**193, 226; **7.**060, 096
Pickerel Lake, South Dakota: **4.**269
Piltdown mandible: **11.**094
Piltdown skull: **11.**094
Pingos: **5.**012, 056, 194, 283
Pinto complex, California: **9.**344
Pinto culture, California: **9.**081
Pintwater cave: **4.**086
Pirri points: **12.**024
Pirrian culture, South Australia: **12.**021
Pisdeli Tepe, Iran: **10.**074
Pit-dwellings: **10.**020
Pits: **11.**029
Piura coast, Peru: **9.**067
Plainview culture: **9.**119
Plainview points: **9.**038
Plankton: **4.**125
Plano culture: **9.**331
Plant remains: **3.**287; **4.**085, 250, 340; **7.**024
Plants: **5.**150
Plaster: **3.**269

Plateau (electronics): **3.**106
Playa: **4.**198
Plazza: **9.**356
Pleistocene: **1.**016; **2.**103, 105, 245; **3.**185; **4.**005, 013, 025, 029, 064, 080, 083, 087, 090, 102, 130, 137, 147, 227, 236, 239, 258, 270, 296, 300, 313, 326, 332, 391, 397, 401, 405, 428, 464; **5.**004, 025, 037, 039, 042, 043, 049, 066, 074, 075, 081, 104, 125, 154, 173, 192, 221, 228, 235, 236, 260, 272, 274, 278, 279, 282, 283, 291, 293; **6.**011; **8.**000, 014, 027, 035, 050, 079; **9.**019, 047, 087, 171, 180, 206, 223, 225, 226; **10.**037, 061, 069, 082; **11.**079, 087; **12.**019, 067
Pliocene: **4.**137, 341, 392
Plough: **11.**097
Pluvial lakes: **4.**073, 110, 251, 322, 358, 420, 421; **5.**046; **7.**011
Pluvials: **4.**251, 316, 366, 420; **5.**039, 275, 306; **7.**039, 040, 049; **8.**050; **9.**074, 353, 364
Podzol: **4.**035, 094, 313, 321, 353; **5.**033
Point Barrow, Alaska: **7.**047; **9.**307
Point Peninsula site, Canada: **9.**109
Pollen: **4.**075; **5.**051, 151
Pollen analysis: **2.**105; **3.**008, 216; **4.**010, 034, 049, 052, 073, 079, 081, 084, 086, 107, 129, 134, 135, 161, 181, 183, 194, 195, 196, 197, 198, 242, 250, 252, 253, 254, 255, 256, 257, 258, 259, 260, 261, 262, 263, 267, 268, 272, 275, 288, 292, 312, 315, 354, 362, 363, 370, 414, 416, 429, 430; **5.**017, 058, 075, 134, 144, 145, 147, 155, 165, 166, 181, 182, 185, 187, 193, 195, 196, 197, 198, 199, 201, 202, 203, 204, 212, 219, 249, 266, 277, 282, 291, 296, 297, 299; **6.**034; **7.**029, 043, 044, 048, 050, 093, 097, 112, 116, 135; **8.**052; **9.**082, 227; **11.**006, 046, 061, 067, 078, 083, 091
Pollen diagrams: **2.**266; **4.**071, 107, 183, 264, 265, 266, 271, 445; **5.**022, 128, 152, 200, 258, 268; **9.**056; **11.**103
Pollen profiles: **4.**022
Pollen sequence: **4.**328
Pollen stratigraphy: **5.**057, 156; **7.**118
Pollen zones: **4.**083, 256; **7.**010; **11.**065
Pollution: **2.**118; **3.**191
Polyethylene vials: **3.**218
Polynesian Prehistory: **12.**039
Pomongwe cave, Southern Rhodesia: **8.**069
Pontine Marshes, Italy: **5.**066
Population migrations: **12.**050
Port Huron: **5.**030, 102, 227

Subject Index

Port Huron moraine: **7**.129
Port Talbot: **4**.206, 219; **5**.065, 205, 206, 260; **7**.066
Post-classic period: **9**.310
Post-glacial period: **2**.103; **3**.006; **4**.018, 037, 040, 041, 044, 045, 060, 071, 081, 083, 084, 124, 133, 149, 155, 185, 196, 197, 200, 201, 216, 225, 249, 265, 268, 272, 274, 275, 277, 278, 279, 280, 281, 282, 283, 319, 340, 347, 390, 410, 413, 427, 433, 444, 449, 454, 462; **5**.002, 014, 026, 033, 057, 058, 134, 144, 145, 165, 166, 169, 188, 191, 195, 196, 199, 203, 208, 209, 210, 212, 214, 215, 216, 217, 218, 219, 228, 237, 277, 287, 296, 297, 298; **7**.025, 074, 077, 086, 095, 118; **9**.005, 099, 230, 232; **10**.037
Post-palaeolithic: **10**.010
Post-pluvial period: **4**.218; **5**.128
Potassium: **2**.097, 215; **3**.296
Potassium argon dating: **2**.165, 214; **8**.002
Potomac River, United States: **4**.097
Pottery: **8**.018, 070, 078; **9**.036, 069, 115, 131, 198, 219, 264, 340, 365; **11**.008, 009, 031, 033, 099; **12**.028
Poverty Point culture: **9**.134
Poverty Point site, Louisiana: **9**.166, 233
Powder Mill Creek cave, Missouri: **9**.107
Prairie: **4**.052
Pre-Columbian civilisations: **9**.324
Pre-Dorset culture: **9**.238
Pre-Wisconsinan: **4**.419
Pre-boreal: **4**.413; **11**.034, 035, 071
Pre-ceramic period: **9**.078, 214
Pre-ceramic period, Peru: **9**.235
Pre-classic period: **9**.023, 030, 068, 101, 113, 117, 144, 145, 169, 202, 204, 205, 220, 236
Pre-treatment: **2**.045, 270; **3**.136, 227, 228
Precambrian: **4**.112
Precipitates: **3**.080
Precipitation (meteorology): **4**.320
Precision: **3**.039, 113, 221, 318; **11**.005
Prehistoric fauna: **10**.047
Prehistoric man: **2**.104, 105; **3**.046, 168; **7**.004, 121; **8**.013, 026, 032, 053, 054; **9**.025, 045, 047, 051, 059, 060, 061, 062, 074, 082, 088, 094, 098, 108, 119, 120, 121, 127, 133, 136, 137, 140, 142, 143, 159, 180, 183, 201, 212, 224, 225, 226, 228, 238, 239, 240, 244, 250, 252, 258, 267, 269, 299, 303, 304, 309, 320, 321, 323, 326, 342, 343, 355, 358, 365, 370, 372; **10**.061, 068; **11**.079; **12**.001, 006, 013, 033, 036, 052, 053, 054, 078

Prehistoric tools: **9**.372
Prehistory: **2**.087, 244; **3**.006, 118, 190; **8**.050, 070; **9**.006, 079, 097, 184, 237, 300, 325; **10**.037, 049, 065; **11**.012, 016, 021, 069, 097; **12**.025, 063
Presumcat sea, Maine: **7**.032
Pretoria Wonderboom: **8**.003
Primitive farming: **10**.036
Primitive implements: **8**.081; **11**.062; **12**.023
Primitive roads: **11**.077
Prince Edward County, Ontario: **5**.209
Prince Patrick Island, Canada: **4**.407
Princess Ingeborg Halvø site, Greenland: **11**.007
Proglacial lakes: **5**.063, 097, 109
Projectile points: **9**.038, 050, 053, 106, 119, 126, 141, 142, 193, 208, 214, 278, 340
Proportional counters: **3**.022, 034, 054, 083, 188, 202, 235, 236, 241, 261, 278, 280, 285, 290, 339, 346, 349
Proportional counting: **3**.240, 307
Protactinium ionium dating: **7**.063, 073
Proto-Neolithic: **10**.083
Proto-history: **3**.006; **8**.022; **10**.005, 045, 065
Protons: **2**.210
Pteropoda: **6**.046
Publications: **13**.013
Puget Sound: **5**.248
Puget glacial lake: **5**.279
Pulantat, Guam: **12**.066
Pulbeena Swamp, Tasmania: **4**.056
Pulse amplifiers: **3**.097
Pulse analysis: **3**.097, 329
Pulse height: **3**.078, 079, 130, 145
Pulse height analysers: **3**.166
Pulse height analysis: **2**.131; **3**.158
Pumice: **4**.019, 024, 071, 175, 204, 228, 291, 292, 349, 350, 417, 441; **9**.049
Punched cards: **1**.019, 028; **3**.096, 258
Pupuke volcanics: **7**.062
Purcell Trench: **5**.180
Purification: **3**.037, 291, 300, 338
Puu Alii, Hawaii: **12**.004
Pylos, Greece: **11**.011
Pyramids: **9**.076
Pyroclasts: **4**.293

Qu'Appelle River, Canada: **4**.418
Quantitative analysis: **2**.006, 212
Quartz: **8**.008
Quartz tubes: **3**.106
Quaternary: **2**.045, 214, 242; **3**.260; **4**.003, 007, 029, 038, 043, 053, 069, 075, 079, 085, 091, 096, 098, 101, 103, 104, 137, 139, 141, 151,

170, 183, 184, 186, 187, 188, 189, 190, 191, 193, 194, 195, 196, 197, 198, 208, 210, 217, 224, 233, 262, 271, 284, 294, 295, 296, 297, 298, 299, 301, 302, 303, 304, 305, 306, 307, 308, 309, 310, 311, 312, 325, 334, 346, 348, 374, 377, 389, 391, 411, 415, 417, 418, 423, 425, 426, 435, 437, 441, 447, 455; **5.**002, 020, 037, 038, 110, 121, 122, 135, 168, 226, 227, 229, 231, 232, 249, 302; **6.**009, 028; **7.**067; **9.**172, 255; **11.**067, 080; **12.**060

Quenching: **3.**023, 024, 151, 160, 175, 179, 180, 201, 244, 304

Quenching agents: **3.**023, 242

Quenching circuits: **3.**116, 117

Radiation counters: **3.**206

Radioactive Dating and Methods of Low Level Counting Symposium held in Monaco, 2-10 March 1967: **13.**010

Radioactive decay: **2.**085, 097, 099, 100, 110, 215; **3.**165, 316, 330; **6.**020, 032

Radioactive isotopes: **2.**218; **3.**165, 207

Radioactive tracers: **2.**022, 170; **3.**045, 082, 100, 342; **4.**156; **6.**026, 035, 037, 047

Radioactivity: **1.**020, 021; **2.**050, 220, 221; **3.**084

Radiocarbon: **1.**008; **2.**025, 026, 116, 124, 147, 148, 150, 154, 160, 181, 188, 208, 209, 211, 215, 252; **3.**081, 108, 302; **4.**089

Radiocarbon activity: **2.**228, 229, 314

Radiocarbon assay: **2.**007, 010, 082, 092, 095, 119; **3.**089, 238, 341

Radiocarbon chronology: **2.**206; **5.**052; **9.**300

Radiocarbon concentration: **2.**024, 030, 032, 035, 043, 051, 056, 057, 060, 062, 073, 074, 089, 126, 137, 171, 172, 179, 190, 198, 217, 222, 228, 229, 232, 254, 262, 271, 277, 278, 279, 280, 287, 299, 303, 304, 306, 310, 311, 312, 316; **3.**184, 195, 247; **4.**174, 364, 452; **6.**003, 007

Radiocarbon Conference, Andover, Mass., October 1954: **13.**008

Radiocarbon content: **2.**027, 031, 109, 133, 158, 161, 199, 227, 230, 263, 284, 286, 305; **3.**189; **4.**058, 368; **6.**022

Radiocarbon cycle: **2.**259

Radiocarbon dates: **1.**001, 002, 003, 004, 005, 006, 007, 015, 016, 018, 022, 023, 024, 025, 026, 027, 028, 029, 030; **2.**036, 043, 075, 076, 108, 234, 235, 240, 269; **3.**125, 164, 263, 350; **4.**038, 051, 068, 071, 072, 077, 101, 104, 125, 129, 133, 134, 139, 142, 152, 167, 189, 192, 247, 251, 253, 261, 262, 266, 279, 293, 294, 302, 303, 305, 306, 309, 310, 325, 332, 335, 338, 346, 359, 370, 382, 383, 388, 389, 396, 406, 435, 455; **5.**002, 020, 029, 034, 035, 037, 039, 044, 048, 054, 060, 063, 078, 080, 084, 093, 099, 123, 124, 128, 130, 135, 141, 147, 150, 155, 156, 163, 175, 193, 228, 247, 270, 277, 282, 288, 301, 303, 307; **6.**006, 020, 041; **7.**004, 006, 010, 018, 035, 037, 043, 044, 052, 067, 070, 072, 075, 086, 118, 135, 140, 144, 145; **8.**006, 029, 030, 031, 032, 033, 042, 045, 053, 054, 056, 057, 061, 063, 065, 074, 082, 083; **9.**001, 004, 032, 040, 044, 049, 057, 058, 061, 062, 065, 067, 070, 081, 091, 092, 095, 101, 115, 117, 124, 132, 137, 153, 158, 169, 178, 181, 190, 191, 195, 207, 209, 218, 222, 226, 251, 260, 266, 267, 273, 275, 276, 279, 280, 284, 286, 287, 288, 290, 291, 294, 295, 296, 301, 306, 307, 310, 311, 312, 319, 321, 322, 332, 333, 334, 339, 341, 343, 360, 369, 371; **10.**006, 017, 019, 027, 034, 045, 052, 054, 055, 063; **11.**002, 011, 014, 015, 016, 018, 021, 022, 023, 038, 052, 056, 061, 066, 069, 072, 080, 082, 087, 089, 095, 103; **12.**004, 009, 010, 015, 016, 017, 039, 042, 049, 053, 057, 059, 060, 061, 064, 065, 066, 068, 069, 070, 074, 075, 076, 080; **14.**037, 159, 160

Radiocarbon dating: **1.**013, 014; **2.**016, 037, 041, 058, 070, 071, 086, 102, 125, 130, 139, 156, 165, 168, 174, 176, 182, 184, 205, 234, 236, 237, 238, 239, 240, 241, 244, 246, 247, 250, 285, 288; **3.**012, 014, 035, 066, 071, 074, 109, 113, 153, 172, 187, 192, 198, 204, 208, 210, 217, 232, 235, 254, 255, 256, 265, 281, 284, 289, 303; **7.**072; **9.**280; **12.**025; **13.**002, 009

Radiocarbon Dating Conference, Cambridge, 1955: **13.**014

Radiocarbon Dating Conference, University of Cambridge, England, 1962: **13.**005

Radiocarbon dating laboratories: **1.**009; **3.**019, 204, 328

Radiocarbon Dating Laboratory. University of Rome: **3.**054

Radiocarbon distribution: **2.**013, 020, 039, 072, 123, 163, 207, 243, 253, 254; **4.**378; **6.**021

Radiocarbon exchange: **2.**013

Radiocarbon half-life: **2.**115

Radiocarbon production: **2.**140, 253, 268, 277

Radiocarbon reservoir: **2.**190, 227; **5.**174; **6.**026

Subject Index

Radiocarbon variations: **2.**032, 062, 075, 194, 201, 258, 275, 279, 305, 306; **3.**173; **5.**100; **13.**003
Radiocarbon years: **2.**206
Radioisotopes: **2.**163
Radiometric dating: **2.**006, 008, 033, 165, 182, 186, 218, 219; **3.**289, 330; **4.**090, 093, 105, 170; **8.**053
Radiometric maps: **5.**255
Radionuclides: **1.**020, 021; **2.**213; **3.**138
Radium: **2.**097
Radon: **3.**183, 241, 290, 300, 304, 336
Rain: **2.**294
Rainfall: **4.**092; **5.**210
Raised beaches: **4.**011, 146, 179, 190, 278, 289, 290, 349, 350, 444, 466; **5.**082, 135, 189, 239; **9.**330; **10.**044
Raised bogs: **3.**008; **4.**052, 271, 413; **5.**166; **11.**006, 103
Rajghat site, Punjab: **10.**058
Range (extremes): **2.**214; **3.**278
Rangifer arcticus: **9.**051
Rangitoto eruption, New Zealand: **12.**034
Reaction time: **3.**334
Reactors (chemistry): **3.**334
Rebound: **5.**218
Red Lady of Pavilan: **11.**026
Red Lake Bog, Minnesota: **4.**248
Red Maple Swamps, Connecticut: **4.**264
Reducing conditions: **4.**111, 239
Reefs: **4.**412
Regression (geology): **4.**096, 211, 240, 276, 277
Reigh site, Wisconsin: **9.**338
Relative chronology: **9.**004
Relative dating: **3.**190; **7.**056; **9.**064, 083, 316, 348
Reliability: **2.**016, 087, 174, 269; **3.**037, 113, 136, 198; **4.**081, 094; **5.**238; **11.**022; **12.**047
Religious artifacts: **9.**344
Rensselaer Bay, Greenland: **5.**079
Reporting: **2.**130, 238; **3.**044, 045, 056, 077, 091, 104, 110, 294; **13.**004, 013, 015, 016
Reproducibility: **3.**037, 101, 175, 304
Reservoir exchange: **2.**275
Reservoirs: **2.**020, 024, 030, 056, 134, 208, 292; **3.**308
Residence time: **2.**123, 175, 190; **3.**331; **4.**457; **6.**018, 022, 035
Reviews: **2.**040; **4.**191; **5.**270; **9.**343; **12.**016; **13.**001, 002
Rhenium isotope: **2.**215
Rhodesia: **8.**065
Rhyolite: **4.**020
Ridgetown Island, Ontario: **4.**179
Riding Mountain, Manitoba: **5.**075
Rift Valley, Africa: **8.**040
Rio Grande Valley: **4.**025
River Annan, Scotland: **4.**257

River Avon, Worcestershire: **7.**065
River features: **9.**166
River terraces: **4.**432; **7.**111; **9.**054
Riverina Plain, New South Wales: **4.**294, 391
Roanoke Basin, South Carolina: **9.**144
Rock engravings: **8.**012; **12.**026
Rock paintings: **8.**038; **9.**242
Rock shelters: **9.**108, 242, 347; **12.**008, 011, 030, 045, 062
Rocks: **2.**050
Rocky Mountains: **5.**038, 106
Rocky Mountains Piedmont: **4.**279
Roebuck site, Canada: **9.**052
Roger Lake, Connecticut: **4.**252; **5.**195
Romerike District, Norway: **4.**210
Roots: **2.**039; **9.**033
Rope: **9.**264
Rose Spring site, California: **9.**040
Rota Island, Marianas: **12.**040
Rothesay Ware: **11.**081
Round barrows: **11.**032
Royal National Park, New South Wales: **12.**030
Rubble: **11.**085
Rubidium isotope: **2.**215
Rubidium strontium dating: **2.**165
Ruds Vedby, Denmark: **5.**198
Rupar site, Punjab: **10.**058
Russell cave, Alabama: **9.**278, 323
Russian plain: **5.**186
Ruthenium catalyst: **3.**010, 334, 336
Ryukyu sites, Japan: **10.**064

Saanich Inlet, British Columbia: **4.**459
Sadlermiut site, Southampton Island: **9.**024
Saguenay River, Canada: **4.**032
Sahab Tepe, Iran: **10.**078
Saint Lawrence Lowlands, Canada: **4.**083; **5.**176; **7.**122
Saint Lawrence Valley, Canada: **4.**046
Saint Lawrence delta, Canada: **7.**085
Saipan Island, Marianas: **12.**040
Saliagos site: **11.**099
Salinity: **3.**195; **6.**003, 038, 043
Salt Flat, West Texas: **4.**233
Salt marshes: **4.**164, 431
Salt water: **4.**012, 387
Salts cave, Kentucky: **9.**020, 060, 241
Salvage archaeology: **9.**021, 325
Samoa: **12.**039
Sample changers: **3.**005, 149
Sample collection: **2.**224; **3.**040, 046, 056, 127, 267, 291, 308, 309, 342; **6.**038
Sample description: **3.**040, 267, 301

Sample handling: **3**.112, 309
Sample pre-treatment: **3**.119, 210; **4**.193
Sample preparation: **2**.243; **3**.043, 057, 089, 119, 128, 136, 200, 224, 225, 263, 291, 295, 309, 318, 324, 335; **4**.087
Sample storage: **3**.040, 200, 218
Sample storing: **3**.112
Sample submission: **3**.056, 128
Samplers: **3**.308
Samples: **3**.093
Sampling: **2**.095; **3**.210, 257, 274
Sampling procedures: **2**.243
San Augustin Plains: **7**.141
San Dieguito complex, California: **9**.326, 353
San Joaquin Valley, California: **7**.012, 147
San Lorenzo phase, Mexico: **9**.204
San Lorenzo phase, Panama: **9**.086
San Pedro valley, Arizona: **4**.284
Sand: **4**.056, 143; **7**.100
Sanders Focus: **9**.366
Sandia cave, New Mexico: **9**.018, 201, 289, 342
Sandia culture: **9**.010, 342
Sandstone: **4**.239; **8**.014
Sanga necropolis, Zaire: **8**.077
Sanga site, Zaire: **8**.022
Sangakallu site, India: **10**.059
Sangamon: **4**.218, 426; **5**.101, 111; **7**.008, 098
Sangoan: **8**.050
Santa Rosa Island, California: **9**.112, 290, 291, 335
Sapelo Island, Georgia: **4**.169
Sapwood: **2**.066
Sarqaq culture, North America: **11**.074, 100
Saskatoon Low, Saskatchewan: **4**.078
Scablands: **5**.256
Scandinavian ice sheet: **5**.172
Scarps: **7**.143
Schatzman, Fred: **3**.007
Schweizer's reagent: **3**.227, 228
Scientific research: **3**.015
Scintillators: **2**.095; **3**.021, 063, 088, 145, 161, 162, 163, 167, 201, 244, 279, 310, 311, 314, 315; **13**.014
Scintillometers: **3**.021, 106, 311, 313
Scottish Readvance: **5**.103
Scottsbluff culture: **9**.119
Scrapers: **9**.053, 214
Screen wall counters: **2**.176, 243; **3**.009, 107, 213, 223, 313, 325, 349
Scripps Estate site, San Diego, California: **9**.328, 335
Sculptures: **9**.327, 329
Sea water: **2**.117; **4**.027, 028; **6**.017, 026, 038, 043, 048, 049
Sea-level: **2**.079; **4**.009, 021, 033, 039, 040, 041, 050, 051, 068, 075, 096, 098, 108, 120, 121, 122, 126, 127, 136, 160, 162, 169, 182, 184, 185, 186, 189, 191, 192, 203, 230, 236, 240, 241, 245, 247, 272, 275, 276, 277, 280, 334, 338, 340, 346, 347, 352, 355, 376, 380, 381, 383, 384, 385, 386, 392, 393, 394, 395, 396, 397, 402, 433, 434, 463, 466; **5**.025, 026, 028, 035, 059, 189, 208, 251, 271, 272, 273, 300; **6**.006, 034, 040; **7**.002, 007, 022, 038, 070, 071, 087, 089, 095, 098, 099, 102, 139; **9**.014, 016, 314, 330, 361; **10**.044; **11**.045, 080; **12**.026, 035, 078; **13**.012
Seals: **4**.223; **5**.006
Searles Lake, California: **4**.187, 420, 435; **5**.306
Seasonal variations: **2**.027; **4**.174, 365
Seasons: **3**.331
Secular variations: **2**.190, 276, 277, 279; **6**.010
Sediment transport: **4**.216
Sedimentary environment: **4**.111, 243
Sedimentary processes: **4**.216
Sedimentary rate: **4**.008, 030, 031, 037, 067, 069, 114, 123, 134, 148, 158, 163, 235, 237, 238, 251, 357, 363, 398, 405, 421; **6**.011, 028, 041, 042; **7**.109, 110; **8**.062
Sedimentation: **2**.194; **4**.004, 014, 163, 169, 218, 221, 400; **5**.022, 026, 037, 283; **7**.100, 127; **13**.007
Sediments: **3**.293; **4**.081, 084, 087, 118, 139, 148, 165, 178, 226, 234, 237, 252, 260, 328, 340, 352, 366, 367, 369, 401, 405, 423, 435, 446, 447, 455, 459, 462; **5**.018, 037, 118, 196, 227, 235, 279; **7**.001, 007, 051, 066, 142; **11**.067
Seeds: **9**.077
Semermiut site, Greenland: **11**.100
Sensitivity: **3**.013
Separation: **3**.046
Sequoia: **3**.280
Serpent Mound site, Ontario: **9**.187
Seward Peninsula, Alaska: **7**.013
Shaft Tomb culture: **9**.277
Shaft and chamber tombs: **9**.277
Shaft tombs: **9**.039, 072; **11**.027
Shaneinab site, Sudan: **8**.015
Shanidar cave, Iraq: **4**.242; **8**. 000, 013; **10**.011, 013, 040, 041, 047, 068, 069, 070, 071, 072, 073, 079, 081, 082
Shanidar valley, Iraq: **10**.048
Shawnigan map area, British Columbia: **4**.438
Sheep: **10**.047
Sheguiandah site, Ontario: **9**.228
Shell River, Canada: **4**.418

Subject Index

Shell artifacts: **9.**039, 072, 277
Shell middens: **9.**008, 131, 139, 198, 200, 257, 305, 314
Shellharbour, New South Wales: **12.**070
Shells: **2.**222, 262, 270; **3.**075, 271; **4.**064, 111, 153, 159, 184, 227, 241, 245, 246, 289, 349, 350, 355, 372, 380, 396, 402, 404, 412, 439; **5.**013, 028, 074, 084, 136, 138, 162, 232, 235, 238, 265, 285; **6.**007; **7.**051, 061, 085, 100; **8.**072; **9.**016, 047, 134, 166, 169, 234, 253; **10.**025, 064; **11.**099; **12.**021, 024, 035
Shelters: **12.**051
Sherds: **9.**368; **11.**081
Sherman Park Mound site, Dakota: **9.**002
Sheyenne delta, North Dakota: **4.**410
Shield (geology): **4.**112, 381
Shielding: **3.**237, 240, 349
Shore platforms: **7.**042
Shoreline: **4.**003, 038, 121, 133, 200, 201, 309, 310, 348, 389, 403; **5.**097; **7.**050, 051
Siberia: **9.**199
Sidari site: **11.**084
Sierra Nevada del Cocuy, Columbia: **4.**188
Sierra Nevada, California: **7.**023
Sierra de Taumalipas, Mexico: **9.**248
Sigatoka site, Fiji: **12.**028
Significance: **2.**005; **12.**017
Silica: **6.**013
Silica alumina catalyst: **3.**135, 143
Silica gel: **3.**135
Silts: **2.**266; **4.**049, 151; **5.**118, 129
Silty clay: **7.**031
Simonson site, Iowa: **9.**053, 214
Sims Bayou, Texas: **7.**142
Singletary Lake, North Carolina: **5.**204
Sinkhole: **5.**074
Sirsa Valley, India: **4.**047
Sissung site, Michigan: **9.**306
Site history: **12.**055
Skede Mose, Sweden: **4.**034
Skeletal remains: **3.**072; **8.**077; **9.**108; **12.**056
Skeletons: **9.**054, 107; **10.**079, 081
Skudelev ships: **11.**101
Skulls: **4.**404; **9.**230, 231; **11.**094; **12.**056, 072
Skutz Falls, British Columbia: **4.**438
Slate: **4.**112
Small volume detectors: **3.**159
Smoky Lake, Alberta: **4.**327
Snail shells: **2.**193; **9.**176
Snails: **9.**047; **10.**025
Sodium bicarbonate: **6.**033, 049
Sodium carbonate: **3.**148
Sodium hydroxide: **3.**195

Soft water: **4.**363
Software: **3.**058
Soil dating: **2.**132; **3.**219; **4.**036, 063, 307, 321, 343; **5.**021; **9.**226
Soil horizons: **4.**353
Soil profiles: **5.**056, 199; **7.**048, 132
Soils: **2.**039; **3.**061, 270; **4.**013, 035, 154, 173, 294, 307, 321, 373, 407, 453; **13.**007
Solar activity: **2.**020, 081, 210, 225, 298, 302, 306; **5.**100
Solar cycles: **2.**020, 209, 210; **5.**100
Solar flares: **2.**096, 098; **3.**111
Solar radiations: **2.**208
Solid carbon counting: **3.**144, 213, 223; **13.**004
Solid state counting: **2.**144; **3.**071
Solifluction: **4.**408
Solutes: **3.**130, 145, 177
Solvents: **3.**042, 130, 245, 304, 310, 315
Somerset Levels: **11.**006, 078
Somino Lake area, USSR: **5.**134
Soot: **9.**060
South Fork Shelter, Nevada: **9.**037
South-West Kansas: **7.**009
Southeastern United States: **5.**284
Southern Rhodesia: **8.**074
Southern Transvaal, South Africa: **8.**065
Southern hemisphere: **2.**067; 183, 267
Spear points: **9.**201
Spear thrower: **9.**250
Speciation: **4.**043
Specific activity: **2.**122
Specimens: **2.**245
Spider Creek, Minnesota: **5.**151
Spiro (Craig) Mound, Oklahoma: **9.**296
Spiro Mound, Oklahoma: **9.**276
Spit: **4.**026
Split-twig figurines: **9.**344, 345
Spores: **6.**015
Spring Creek Valley, North Dakota: **9.**230
Spring Creek cave, Wyoming: **9.**346
Squibnoket Cliff, Martha's Vineyard: **4.**414
St. Lawrence Lowlands, Canada: **4.**082
Stable isotopes: **2.**138, 307; **4.**057, 368
Stalling Plains pottery: **9.**198
Standard deviation: **2.**069; **3.**210; **14.**022
Standard error: **2.**114
Standard year of reference: **3.**095
Standardisation: **3.**221; **11.**022
Standards: **2.**131; **3.**033, 298
Standards (radiocarbon): **2.**002, 057, 083, 238, 260; **3.**002, 105, 171, 222
Stanfield - Worley rock shelter, Tennessee: **9.**347
Star Carr site, Yorkshire: **11.**034, 036, 046, 071, 104
Staten Island, New York: **7.**044

Statigraphic units: **4**.447
Statistical analysis: **3**.005, 316
Statistical errors: **3**.037; **4**.302
Stenben village, Illinois: **9**.348
Stockholm Natural 14C Station: **4**.142
Stone age: **8**.005, 012, 016, 023, 036, 046, 059, 078, 079, 081; **9**.005, 323, 349; **10**.071; **12**.080
Stone artifacts: **9**.335
Stone circles: **8**.043
Stone implements: **2**.107; **8**.067; **9**.005, 029, 054, 100, 215
Stone tools: **10**.008; **12**.029
Stonehenge, Wiltshire: **11**.040, 098
Strait of Juan de Fuca: **5**.248
Stratified sequence: **9**.044; **11**.065
Stratigraphic correlation: **4**.104, 217, 442; **5**.077, 307; **9**.108, 146; **12**.056
Stratigraphic evidence: **4**.126; **5**.302; **7**.116; **9**.010, 362; **10**.022
Stratigraphic units: **4**.139, 435; **5**.213; **7**.046, 097, 106
Stratigraphy: **2**.105, 245; **3**.350; **4**.010, 088, 108, 275, 283, 306, 308, 315, 322, 332, 414, 416, 419, 421, 422, 435, 437; **5**.007, 012, 032, 057, 080, 121, 141, 155, 167, 250, 279, 280, 281, 303, 305, 306; **7**.044, 045, 046, 047, 053, 069, 074, 098, 099, 105, 106, 107, 120, 125, 127, 132, 134, 135, 136, 144, 146; **9**.013, 047, 142, 148, 149, 163, 188, 341, 350; **10**.010; **11**.006, 019, 046, 053, 054, 055, 056, 057, 058, 059, 065, 078, 080, 105; **12**.055; **13**.012
Stratosphere: **2**.024, 074, 163, 192, 258, 265, 284, 286, 292, 293, 301; **3**.184, 308, 331, 342
Strontium isotopes: **2**.026, 080, 201
Stuart rock shelter, Nevada: **9**.351
Stylistic correlation: **9**.219
Su Nuraxi site, Sardinia: **11**.070
Sub-Atlantic: **4**.413
Sub-Boreal: **4**.244, 259, 413; **5**.213, 266
Sub-Saharan Africa: **8**.006, 055
Submerged deposits: **9**.139
Subsidence: **4**.287
Substages: **5**.049
Subsurface water: **6**.022
Suess effect: **2**.056, 057, 060, 155, 196; **5**.253; **9**.254
Sugei Besi tin mines, Malaysia: **7**.055
Sukerde site, Turkey: **10**.030
Sulphur: **4**.178
Sulphur deposits: **2**.015
Sulphuric acid: **3**.225, 317
Sumas: **5**.235
Summer lake, Oregon: **4**.291
Sunbeam Prairie Bog, Ohio: **5**.258, 268

Sunnybrook till, Ontario: **4**.206
Sunspots: **2**.062, 128, 281, 288, 298, 302
Superquiet sun: **2**.194
Surface water: **2**.031, 122, 295; **3**.195; **6**.005, 010, 026, 029, 032, 049
Surficial deposits: **5**.094, 095
Surficial geology: **4**.308, 345, 438; **5**.286
Swan River District, Western Australia: **4**.186
Synchronism: **5**.249, 253; **9**.120

Ta Hagrat Mgarr temples, Malta: **11**.012
Table Rock Pueblo site, Arizona: **9**.355
Takapu Stadial: **7**.024
Talofofo River Guam: **12**.066
Tanana River valley, Alaska: **4**.382; **5**.101
Tanner Basin, California: **4**.163
Tapirs: **9**.227
Tarns: **4**.255
Tartangan culture, South Australia: **12**.021, 023
Tartaria Tell, Rumania: **11**.106, 107, 108
Tartaria tablets: **11**.106, 107, 108
Tata Institute of Fundamental Research. Radiocarbon Laboratory, Bombay: **3**.281
Taupo area, New Zealand: **4**.441
Tazewell: **5**.030, 102; **7**.084
Techniques: **2**.102, 242; **13**.003
Tectonic movements: **4**.122, 276, 311; **7**.038
Tectonics: **4**.404
Tectonism: **4**.246, 287, 444
Teeth: **4**.206; **5**.059; **9**.231
Tell Hasanlu, Iran: **10**.028, 029
Tell Shemshara, Iraq: **10**.042
Tell el-Sultan, Jordan: **10**.014, 016, 023, 033, 050
Temperate fauna: **4**.074
Temperate regions: **2**.201; **9**.159
Temperature: **2**.023, 195; **3**.175; **6**.003, 024, 043; **7**.109
Temperature variations: **2**.225, 278, 279; **4**.091, 263, 301, 320, 442; **5**.293; **7**.052
Temples: **9**.203; **11**.012
Teotihuacan, Mexico: **9**.158
Tepexpan man: **9**.192
Tephra: **2**.214; **4**.450
Termini ante quem: **12**.077
Termini post quem: **12**.077
Terrace (architecture): **9**.356
Terracotta: **8**.047
Terrestrial fossils: **7**.042
Terrestrial sediments: **4**.297
Terrigenous fines: **4**.163
Tertiary: **4**.137
Tesuque Valley, New Mexico: **4**.030
Texas coast: **5**.026

Subject Index

Texas street site: **9**.335
Thames Valley, Great Britain: **5**.116
Theory: **2**.102
Thermal areas: **2**.198; **3**.170; **4**.286
Thermal diffusion: **2**.157
Thermal diffusion separation: **2**.282
Thermoclines: **6**.039
Thermoluminescence: **2**.182, 197; **9**.217
Thermoluminescence dating: **3**.199
Thomas site, California: **9**.249
Thompson Creek Watershed, Iowa: **4**.031
Thorium: **2**.215
Thorium isotopes: **6**.010
Three Spring Bar site, Washington: **9**.105
Thule Springs site, Nevada: **4**.262
Thule culture: **11**.100
Tidal flats: **4**.259
Tides: **4**.376
Tiger Hills, Manitoba: **4**.161
Tihuanacanoid culture: **9**.218
Tikal site, Guatemala: **9**.058, 197, 270, 271, 322, 356, 357
Till: **4**.419; **5**.122, 136, 278
Tillite: **7**.019
Timber: **3**.264; **9**.281, 322; **10**.029, 080
Time correlation (geology): **5**.081
Time of residence: **4**.353
Time scale: **5**.001; **13**.012
Time stratigraphic units: **5**.049
Tinian Island, Marianas: **12**.040
Titanium: **4**.014
Tizapan site, Mexico: **9**.297
Toluene: **3**.063, 130, 152, 158, 163, 283, 310, 347
Tomales Bay, California: **7**.061; **9**.139
Tombs: **10**.014; **11**.109
Tonga: **12**.039
Toolong Range, New South Wales: **7**.143
Toronto area, Ontario: **5**.281
Totoket Bog, Connecticut: **5**.296
Towada caldera, Japan: **4**.292
Towns: **10**.025, 050
Trace analysis: **4**.242
Trace elements: **2**.273
Tracers: **2**.028, 029, 124, 198, 254
Tradition: **8**.035
Transgression: **4**.034, 050, 159, 164, 211, 212, 275, 276, 277, 342, 394, 396, 423; **5**.135, 228; **7**.032, 047
Transport: **2**.163
Transport processes: **2**.121
Traps: **3**.225
Travertine: **4**.365
Tree ring calibration: **9**.297

Tree rings: **2**.048, 066, 067, 128, 161, 194, 294, 303, 304, 310, 311; **3**.264, 280
Trees: **2**.112
Tregaron Moss, Cardiganshire: **3**.008
Trempelean Valley, Wisconsin: **7**.028
Trinity River, Texas: **4**.297
Tritium: **2**.025, 026, 080, 162, 215, 284, 286, 294, 295; **3**.081, 099, 108, 115, 175, 184, 192, 320, 336; **4**.092, 093, 288; **6**.051
Tritium dating: **13**.002
Tritons: **2**.210
Tropical regions: **4**.188, 451
Troposphere: **2**.074, 080, 094, 159, 163, 164, 228, 229, 258, 265, 292, 301; **3**.308, 342; **4**.174, 364, 378, 452; **6**.035
True age: **2**.098, 191, 206, 257; **3**.229, 344; **6**.033
Tshangula cave, Southern Rhodesia: **8**.069
Tucson Basin, Arizona: **4**.061
Tufa: **4**.299, 358, 446; **9**.253
Tuktu complex, Alaska: **9**.359
Tule Spring area, Nevada: **4**.299; **9**.358
Tule Spring site, Nevada: **9**.335
Tumulus: **10**.080
Tundra: **4**.036, 063, 407, 453; **5**.033, 176, 287
Tunguska meteor: **2**.089, 128, 199
Tusks: **4**.206, 225; **9**.018
Tustrup site, Denmark: **11**.060
Tuto tunnel, Greenland: **3**.274
Two Creek: **4**.199; **5**.127, 140, 179, 204, 235, 258, 268
Two Creek Forest Beds: **4**.314; **5**.024, 068, 069, 070
Typological correlation: **10**.029
Typology: **9**.092, 196; **11**.006
Tyrrell Sea: **4**.011; **5**.130
Tyrrell Sea episode: **5**.243

U.S. Bureau of Standards: **13**.003
Ucáyali complex, Peru: **9**.003
Ula River, Lithuanian SSR: **4**.354
Unconformities: **4**.151; **5**.294; **7**.127
Underwater archaeology: **10**.007
United States National Bureau of Standards: **2**.260; **3**.122, 298
University of Dublin. Radiocarbon Dating Laboratory: **11**.014
Unmak island, Alaska: **4**.171
Untersee, Austria: **5**.266
Uplift: **4**.011, 246, 247, 278, 287, 290, 296, 371, 380, 403; **5**.016, 044, 091, 135, 159, 175, 188, 214, 216, 218, 233, 239, 269, 287; **7**.038, 119
Upper Boulder Clay: **5**.136
Upper Mississippi Valley: **4**.455; **5**.122

Upper Mississippi culture: **9**.064
Upper Palaeolithic: **8**.001, 041, 064; **10**.040, 053, 067; **11**.003, 053, 054, 055, 056, 057, 058, 059, 079, 082
Upper Pleistocene: **2**.015; **7**.136, 143; **8**.001, 064; **11**.056; **12**.021
Upwelling: **2**.231, 232; **6**.044
Ural, USSR: **4**.006
Uranium: **2**.215
Uranium isotope dating: **6**.010, 041; **7**.003
Uranium thorium - lead dating: **2**.165
Uranium thorium dating: **2**.086; **4**.423
Urbanism: **9**.023

Vacuum apparatus: **3**.134
Vacuum systems: **3**.036, 175, 205, 214
Valday: **5**.186
Valders: **4**.445; **5**.068, 069, 070, 124, 235, 268; **7**.053, 083; **9**.099
Valdivia phase ,Ecuador: **9**.117
Validation: **2**.184
Validity: **2**.069, 076, 109, 119, 155, 193, 245, 283, 297, 313; **3**.003, 041, 091, 093, 190, 208, 248, 249, 256, 258, 259, 284, 293, 348, 350; **4**.001, 026, 144, 192, 458; **5**.068, 069, 070, 113, 114, 115, 174, 225; **6**.029; **7**.018, 034; **9**.013, 093, 104, 125, 146, 169, 196, 223, 265, 313, 341, 369; **10**.033, 063; **11**.004, 005, 019, 022, 058, 107; **12**.068
Valley Head moraine: **5**.227
Valley of Mexico: **9**.101
Vancouver Island: **9**.079
Variations: **2**.073, 088, 091, 110, 310, 312
Varve dating: **4**.079, 144, 302; **5**.045, 286, 299; **7**.095
Varves: **4**.238, 339, 459; **5**.105, 118
Vashon: **5**.248
Vashon glacier: **5**.300; **7**.105
Vegetation: **4**.262, 274; **5**.056, 144, 187, 212, 224, 303, 304; **7**.049, 060, 069, 148, 149
Vegetational sequence: **3**.008; **4**.006, 052, 113, 128, 153, 188, 194, 195, 196, 197, 253, 265, 312, 390, 425, 445; **5**.021, 155, 185, 203, 203, 249, 284, 297, 309; **8**.052; **11**.046, 080
Vegetational zones:198, 265; **5**.144
Vehicles: **10**.017
Velocity: **4**.061
Venado Beach, Panama: **9**.001, 086
Venezuela Basin: **6**.022
Vertebrates: **7**.009, 113
Vertical mixing: **2**.299; **6**.043
Vertical movement: **4**.397

Vestaburg bog, Michigan: **5**.219
Vials: **3**.065
Victoria Island, North West Territories: **4**.155
Vikings: **11**.101
Villages: **7**.086; **10**.003, 004, 013, 037, 042; **11**.037
Vinca culture: **11**.107
Vogelherd cave, Germany: **11**.075
Volcanic ash: **4**.015, 019, 044, 071, 088, 102, 204, 215, 216, 279.,336, 345, 373, 377, 409, 417, 461, 462; **5**.158, 302; **7**.043, 096; **9**.014; **12**.034
Volcanic eruptions: **4**.019, 175, 177, 417, 441, 450, 461, 463
Volcanic glass: **4**.216
Volcanism: **2**.256; **4**.020, 088, 237, 409
Volcanoes: **4**.171

Wachapreague, Virginia: **4**.164
Wagon Jack Shelter, Nevada: **9**.041
Waiakukini cave, Hawaii: **12**.004
Wall effect: **3**.022
Wallkill Valley, United States: **7**.084
Wasaga Beach, Ontario: **4**.335
Water: **2**.173, 222, 289; **3**.170, 225; **6**.024; **9**.180
Water atmosphere interface: **2**.054
Water flow: **4**.057
Water masses: **6**.033
Water movement: **6**.002, 003, 021, 032, 043, 044
Water sampling: **3**.203
Water-level: **4**.067
Waton island: **12**.055
Wattle and daub: **3**.277
Wave erosion: **4**.009
Weathering: **4**.377
Weddell Sea: **6**.036
Weichselian: **5**.021, 034, 118, 136, 137, 301; **7**.072, 149, 150; **11**.080, 082
Wells Mastodon site, Indiana: **4**.329; **9**.056
Western Hemisphere: **9**.142, 202, 239, 259, 268, 269
Western Polynesia: **12**.003
Wet chemistry: **3**.189
Wet combustion: **3**.292
Wet sieving: **3**.046
Whalebones: **4**.289
Wheels: **10**.017; **11**.076
Whidbey Island: **5**.080
Whidbey interglacial: **7**.105
White Lady shelter: **8**.010
White Pass cave, Arabia: **10**.067
Whixall Moss, Shropshire: **3**.008
Wilderness Lakes: **7**.007
Willcox Playa, Arizona: **7**.011, 020

Subject Index

Willerby Wold site: **11**.083
Wilsford Shaft: **11**.027
Wilson Butte, Idaho: **9**.363
Wilton culture (South Africa): **8**.023
Windmill Hill camp, England: **11**.086
Windmill Hill culture, England: **11**.093
Winter Harbour Moraine, North West Territories: **4**.465
Wisconsin Laurentide ice sheet: **4**.439
Wisconsin Till: **7**.067
Wisconsinan: **2**.091, 225, 266; **4**.010, 095, 121, 150, 184, 198, 206, 218, 219, 224, 231, 240, 241, 284, 314, 320, 324, 339, 346, 357, 418, 420, 423; **5**.001, 005, 009, 011, 012, 015, 017, 030, 031, 032, 035, 049, 052, 054, 062, 063, 074, 081, 093, 101, 102, 105, 109, 111, 113, 114, 115, 121, 122, 123, 129, 138, 139, 141, 142, 155, 156, 157, 163, 164, 177, 178, 182, 183, 202, 204, 205, 206, 220, 222, 223, 225, 226, 245, 246, 247, 250, 251, 256, 259, 270, 275, 280, 281, 284, 286, 295, 299, 300, 303, 304, 305, 306, 307, 308, 309; **6**.008, 011; **7**.011, 027, 028, 034, 036, 039, 040, 047, 057, 064, 074, 084, 090, 092, 098, 100, 101, 107, 112, 113, 115, 123, 125, 144, 145; **9**.010, 045, 047, 112, 157, 171, 217, 317
Wood: **2**.007, 062, 063, 073, 093, 281, 300, 311; **3**.227, 228; **4**.017, 024, 050, 122, 147, 149, 184, 268, 314, 329, 355, 356, 362, 373, 376, 404, 432; **5**.009, 016, 061, 066, 067, 096, 105, 140, 179, 185, 221, 235, 241; **7**.033, 055, 057, 062, 082, 083, 117, 123, 143, 147; **8**.049, 079; **9**.033, 047, 051, 241, 322, 350, 361; **10**.021; **11**.034, 046, 070, 097, 101, 104; **12**.030
Wood standard: **2**.069; **12**.021
Wooden trackways: **11**.064, 077, 078, 090
Woodfordian: **4**.220; **5**.294; **7**.107
Woodland culture: **9**.007, 060, 090, 109, 115, 134, 172, 173, 174, 199, 347
Woodruff ossuary, Kansas: **9**.006
Wool: **11**.101
Woolly rhinoceros: **4**.106; **7**.150
World's Fairway complex, New York: **4**.145
Wostenholme Fjord, Greenland: **5**.108
Wright Valley, Antarctica: **4**.227
Würmian: **4**.339; **5**.007, 020, 066, 067, 160, 161, 171, 173, 207, 244, 257, 260, 276; **7**.124; **8**.050; **11**.003, 041, 082; **12**.029

Yanuka Island site, Fiji: **12**.062
Yardand Flint Station, Alaska: **9**.149
Yarra delta, Victoria: **4**.120
Yenissei River, Siberia: **4**.325
Young Dryas: **5**.230
Younge Tradition: **9**.173, 306, 334
Younger Dryas: **4**.138, 430; **5**.084, 264
Yucatan Basin: **6**.022
Yuconoo site, Mexico: **9**.373
Yukon-Koyukuk lowlands, Alaska: **4**.116

Zambian archaeology: **8**.045, 046, 058, 059, 060, 083
Zapotec (Mesoamerican people): **9**.373
Zawi Chemi Shanidar site, Iraq: **4**.242; **10**.013, 083
Zebbug phase: **11**.102
Zero year age BP: **2**.152
Zimbabwe site: **8**.011, 049, 051
Zion National Park, Utah: **4**.375
Zoo Park Elephant site, Namibia: **8**.008

LOCATION INDEX

Abisko (Sweden): **2.**314
Academy of Sciences of the USSR: **14.**002, 158, 288
Academy of Sciences of the USSR. Geological Institute. Radiocarbon laboratory: **14.**157
Adelphi (Ohio): **5.**009
Afghanistan: **10.**046, 053, 066, 067, 075
Africa: .1.022, 023, 024, 025, 026; **4.**016, 114, 226, 261, 308; **5.**004, 020; **7.**010; **8.**006, 014, 032, 035, 037, 038, 041, 048, 050, 052, 053, 054, 055, 062, 064, 066, 067, 070, 077, 084
Africa (Sub-Saharan): **1.**001; **8.**026, 029
Ahar (Rajasthan): **10.**045
Aitape (Papua New Guinea): **4.**404
Aitkin (Minnesota): **4.**199
Alabama: **9.**278, 323
Alaska: **4.**010, 015, 026, 063, 075, 087, 116, 124, 141, 146, 151, 171, 249, 256, 265, 287, 312, 337, 342, 344, 345, 356, 359, 382, 411, 423, 453, 460; **5.**017, 037, 048, 061, 063, 088, 101, 104, 107, 109, 110, 139, 182, 199, 229, 230, 242, 292; **7.**013, 021, 029, 036, 039, 047, 080, 096, 135, 146; **9.**011, 014, 034, 104, 110, 128, 129, 149, 154, 155, 209, 212, 221, 232, 244, 245, 301, 307, 318, 330, 349, 359, 362, 365, 371
Alberta: **4.**044, 150, 279, 327, 462; **5.**041, 154; **7.**106, 111; **9.**054, 266
Aleutian Islands (Alaska): **4.**171; **5.**199; **9.**011, 014, 110, 212
All-Union Geological Institute (USSR). Laboratory for Quaternary Geology: **14.**003
Amazon River: **9.**368
Americas: **4.**072; **9.**018, 062, 237, 260, 266; **12.**078
Amersfoot (Netherland): **5.**244
Amesbury (Wiltshire): **11.**040, 085, 098
Anaktuvuk Pass (Alaska): **9.**359
Anangula Island (Alaska): **9.**129, 154
Anatolia: **11.**004
Angola: **1.**024; **8.**050
Ann Arbor (Michigan): **14.**237, 238, 239, 240, 241, 242, 243, 244, 245, 246, 247, 248
Antarctic Ocean: **2.**075, 088; **6.**002, 030
Antarctica: **4.**223, 227, 261; **5.**006, 233, 236
Appalachians: **7.**074
Arabia: **4.**092; **10.**067

Arctic Alaska: **5.**202
Arctic Archipelago (North West Territories): **5.**223
Arctic Canada: **5.**224
Arctic Ocean: **3.**195; **6.**041
Argentine: **5.**298; **7.**091, 092, 093, 100; **9.**100
Argentine Basin: **6.**048
Arizona: **2.**303; **4.**058, 061, 284; **5.**165; **7.**011, 020; **9.**013, 048, 094, 095, 133, 175, 227, 344, 345, 355, 364; **14.**006, 007, 008, 009, 010, 230, 231
Arkansas: **9.**153, 301
Ashcott (Somerset): **11.**062
Asia: **10.**046, 053; **11.**089
Atescosa County (Texas): **4.**057
Atlantic Coast: **4.**164; **7.**099
Atlantic Ocean: **3.**195; **4.**069, 431; **5.**293; **6.**011, 026, 027, 030, 042
Auckland (New Zealand): **4.**209, 463; **12.**038
Austin (Texas): **9.**263; **14.**268, 269, 270, 271, 272, 273
Australia: **1.**002, 003, 004; **2.**240; **3.**019; **4.**001, 003, 012, 038, 039, 041, 042, 056, 060, 068, 095, 120, 208, 229, 270, 294, 300, 309, 310, 311, 313, 336, 339, 347, 348, 389, 391, 408, 443, 466; **5.**105; **7.**071, 139, 143; **12.**001, 002, 005, 006, 007, 008, 009, 011, 012, 013, 017, 019, 022, 023, 024, 029, 030, 032, 033, 036, 043, 045, 052, 053, 054, 056, 057, 059, 061, 065, 070, 071, 073, 074, 075, 076, 077, 078, 080; **14.**004, 005, 249, 289
Australian National University. Radiocarbon Dating Research Laboratory: **14.**004, 005
Austria: **4.**343; **5.**244, 266
Avebury (Wiltshire): **11.**086, 093
Azatlan (Mexico): **9.**310
Azores: **4.**237

Baffin Island (Canada): **5.**060
Baffin Island (North West Territories): **4.**190, 426; **5.**028, 142, 177, 217, 261, 265
Baja California (Mexico): **4.**076; **7.**051; **9.**200, 242
Bakota (Zambia): **8.**044
Balkans: **11.**004
Baltic: **5.**290
Baltic region: **7.**010
Baluchistan: **10.**066, 075
Bambandyalano (South Africa): **8.**024, 073
Banc-y-Warren (Wales): **5.**171
Barnstable (Massachusetts): **4.**021
Barrow (Alaska): **4.**087, 337, 342, 359
Barumini (Sardinia): **11.**070
Basutoland: **8.**030

Location Index

Bathurst Island (North West Territories): **5.**223
Beeri (Israel): **4.**239
Beira (Mozambique): **4.**096
Beitbridge (Zimbabwe): **8.**016
Belgium: **4.**128; **14.**096, 121, 122, 123, 124, 125, 126
Bell County (Texas): **9.**366
Belton (Texas): **9.**366
Beltrami County (Minnesota): **4.**248
Bend (Oregon): **4.**132
Bering Straits: **9.**033
Berkshire: **11.**035, 105
Berlin (F.R.G.): **14.**011, 012
Bermuda: **4.**280; **5.**208; **7.**087
Bern (Switzerland): **14.**013, 014, 015, 016, 017, 018
Besant Valley (Saskatchewan): **9.**188
Bighorn Canyon (Montana): **9.**123
Bihar: **10.**058
Birmingham (England): **14.**019, 232
Bjørnøya (Sweden): **4.**193
Blekinge (Sweden): **4.**194, 195, 196, 197
Bluff (New Zealand): **12.**038
Bolivia: **1.**030
Bombay (India): **14.**166, 200, 201, 202, 203, 204
Bonn: **14.**020
Boothia Peninsula (North West Territories): **5.**285
Borneo: **10.**008
Brandon (England): **7.**108
Brazil: **4.**381; **9.**008, 087, 190, 312, 314
Bristol Channel (England): **5.**272
British Columbia: **4.**044, 438, 440, 459, 462; **5.**043, 180, 211, 248; **7.**046; **9.**079, 246, 251
British Museum. Research Laboratory: **14.**021, 022, 023, 024, 025, 026, 027, 028
British West Indies: **4.**373
Brittany: **11.**010
Brno (Moravia): **11.**053
Brooks Range (Alaska): **7.**080
Brorup (Denmark): **5.**257
Bruce Mountains (North West Territories): **5.**265
Brunswick (Georgia): **4.**387, 432
Brussels: **14.**096
Bucharest: **1.**005, 006
Burleigh County (North Dakota): **5.**192
Burzahom (Kashmir): **10.**045
Buttonwillow (California): **7.**147

Caddo County (Oklahoma): **9.**038, 171
Cahokia (Illinois): **9.**007
California: **3.**185; **4.**008, 024, 055, 059, 100, 131, 139, 163, 175, 183, 187, 198, 341, 366, 398, 420, 435, 446; **5.**010, 274, 306; **7.**012, 023, 040, 041, 042, 061, 103, 114, 147; **9.**005, 040, 045, 049, 059, 074, 085, 092, 106, 112, 120, 122, 136, 137, 139, 179, 201, 207, 212, 213, 224, 226, 249, 253, 280, 290, 291, 309, 320, 326, 328, 335, 339; **14.**108, 109, 110, 111, 112, 185, 222, 223, 224, 225, 226, 227, 228, 229
California Channel Islands: **9.**207
Cambridge (England): **14.**029, 030, 031, 032, 033, 034, 035, 036, 168, 169, 170
Cambridge (Massachusetts): **14.**064, 065
Cambridge University. Sub-Department of Quaternary Research: **14.**029, 030, 031, 032, 033, 034, 035, 036, 168, 169, 170
Cambridgeshire: **11.**065
Canberra (Australia): **14.**004, 005
Canyon Diablo (Mexico): **9.**015
Cape Breton Island (Nova Scotia): **4.**419
Cape Breton Plateau (Nova Scotia): **5.**022
Cape Cod (Massachusetts): **4.**245; **5.**231; **7.**104; **9.**138
Cape Denbigh (Alaska): **4.**141
Cape Hallett (Antarctica): **5.**236
Cape Martin (South Australia): **12.**023
Cape Province (South Africa): **8.**078
Caracas (Venezuela): **14.**097, 098, 099
Cardiganshire (Wales): **3.**008; **5.**067, 160, 171
Caribbeans: **6.**022; **7.**110; **9.**004, 247
Carmathenshire: **11.**029
Carnarvon (Western Australia): **7.**139
Cascade Range (Washington): **5.**158
Catalina Island (California): **9.**179
Catron County (Mexico): **7.**141
Central Africa: **8.**021, 022
Central America: **4.**019; **9.**036, 039, 042, 084, 088, 117, 130, 145, 158, 169
Central Europe: **2.**272; **11.**013, 022, 053, 054, 055, 056, 057, 058, 059
Centre County (Pennsylvania): **4.**074
Centre National de la Recherche Scientific. Centre des Faibles Radioactivités: **14.**080, 081, 177, 178
Centre Scientific de Monaco. Laboratoire de Radioactivité Appliquée: **14.**131, 132
Centre de Physique Corpusculaire, Héverlé Louvain: **14.**121, 122, 123, 124, 125, 126
Cerro Mangote (Panama): **9.**145
Chalottenlund Slot (Denmark): **3.**045
Cherokee County (Oklahoma): **9.**294, 295
Cherokee County (Texas): **9.**196
Chesapeake Bay (Virginia): **7.**119
Cheshire: **5.**136, 137; **7.**057
Chester County (Pennsylvania): **5.**291
Chiapas (Mexico): **9.**068
Chiapas Highlands (Guatemala): **9.**329

Chiapas coast (Mexico): **9**.027
Chicago: **13**.155
Chicago (Illinois): **14**.041, 042, 043, 153, 154
Chile: **5**.083; **7**.043, 091, 092; **9**.215
Chillicott (Ohio): **5**.009
Chillón, Peru: **9**.116
Chitambo Mission (Zambia): **8**.060
Christchurch (New Zealand): **4**.085, 277
Chukchi Sea: **4**.158
Churchill County (Nevada): **9**.041
Clarence Valley (New South Wales): **3**.019; **12**.012
Clark County (Nevada): **4**.299
Cleveland (Ohio): **7**.001
Clonsast (Ireland): **4**.416
Cochise County (Arizona): **7**.020
Cochrane (Alberta): **4**.150; **7**.111
Cochrane District (Ontario): **5**.286
Colima Province (Mexico): **9**.277, 352
Collector (New South Wales): **4**.095
College Station (Texas): **14**.206
Colorado: **2**.099; **4**.113, 331; **5**.018, 262, 263; **9**.147, 177, 266, 285; **14**.207
Colorado Springs (Colorado): **14**.207
Columbia: **4**.023, 188; **5**.288; **9**.039
Columbia plateau: **9**.080
Columbia River: **5**.256
Columbia University. Lamont-Doherty Geological Observatory: **14**.113, 114, 115, 116, 117, 118, 119, 120
Columbus (Ohio): **4**.268
Congo Republic: **4**.308
Connecticut: **4**.252, 264, 272, 283, 433; **5**.195, 296; **9**.138, 283; **14**.291, 292, 293, 294, 295, 296, 297, 298
Connecticut Natural Area: **4**.264
Cook Inlet (Alaska): **5**.037
Coorong (South Australia): **4**.012, 221
Copenhagen (Denmark): **14**.045, 046, 047, 048, 049, 050, 051, 052, 053
Copper River Basin (Alaska): **4**.249
Corfu (Greece): **11**.084
Corry (Pennsylvania): **5**.005
Costa Rica: **9**.084
County Down (Ireland): **4**.275; **7**.116
County Durham (England): **5**.045
County Mead (Kansas): **7**.009
Coygan Rock (Carmathenshire): **11**.029
Crater Lake, (Oregon): **5**.158
Cuicuilco (Mexico): **9**.023
Culbertson County (Texas): **4**.233
Cyclades: **11**.099
Cyrenaica (Lybia): **8**.000
Czechoslovakia: **3**.291; **11**.053, 055

Dakota: **9**.002, 292
Dallas (Texas): **14**.184
Damariscotta (Maine): **9**.257
Dark County (Ohio): **5**.258, 268
Darlington (England): **5**.045
Dawson (Yukon Territory): **7**.145
Debert (Nova Scotia): **4**.117
Delaware (Ohio): **14**.148, 149, 150
Delaware River: **4**.052
Delta County (Texas): **9**.019
Denmark: **2**.201; **3**.045; **4**.107, 200, 302, 390; **5**.034, 257, 301; **11**.060, 067, 097; **14**.045, 046, 047, 048, 049, 050, 051, 052, 053
Denton County (Texas): **9**.201
Derbyshire: **4**.427
Devon: **11**.045, 063, 096
Disko Bugt (Greenland): **11**.074, 100
Dismal Lake (North West Territories): **9**.245
District of Keewatin (North West Territories): **4**.274
Dniester River (USSR): **11**.073
Dodge County (Wisconsin): **9**.168
Door Peninsula (Wisconsin): **7**.082
Dordogne (France): **11**.042, 047, 048
Downers Grove (Illinois): **14**.151, 152
Drake Passage: **6**.048
Duart Lake (North West Territories): **5**.265
Dublin (Ireland): **14**.054
Dumfriesshire (Scotland): **4**.257; **5**.131, 148
Dutch Coast: **6**.040
Dzibilchaltun (Mexico): **9**.113, 114

East Africa: **8**.017, 040
Easter Island: **12**.014, 016, 020, 049
Eastern Europe: **11**.022, 053, 054, 056, 059
Eastern Hemisphere: **11**.017
Echuca (Victoria): **4**.311
Ecuador: **9**.039, 117, 181
Egypt: **4**.457; **8**.019, 027, 062, 064, 070; **10**.009
El Alto (Jalisco): **9**.297
El Salvador: **4**.254
Elko County (Nevada): **9**.037, 103
Ellesmere Island (North West Territories): **4**.027, 028, 230
Ellesworth Falls (Maine): **9**.063, 281
Emmen (The Netherlands): **11**.103
Engaruka (Tanzania): **8**.043
England: **3**.008; **5**.045, 149, 181, 272; **7**.057; **11**.006, 008, 024, 027, 034, 035, 036, 039, 040, 045, 046, 062, 063, 064, 065, 068, 071, 077, 078, 083, 085, 086, 092, 093, 094, 098, 104, 105; **14**.019, 021, 022, 023, 024, 025, 026, 027, 028, 029, 030, 031, 032,

Location Index

033, 034, 035, 036, 134, 135, 136, 137, 138, 139, 168, 169, 170, 232
Equatorial Africa: **4**.263
Estonia: **14**.196, 197, 198, 199
Eugene Island (Gulf of Mexico): **7**.127
Europe: **3**.173; **4**.072, 080, 090, 099, 123, 430, 454; **5**.052, 057, 076, 146, 200, 244; **7**.021; **10**.062; **11**.001, 002, 013, 015, 018, 021, 022, 023, 030, 031, 037, 038, 042, 050, 060, 061, 066, 067, 069, 070, 072, 082, 084, 088, 089, 095, 099, 106, 107, 108
Ezinge (The Netherlands): 11.076

F airbanks (Alaska): **4**.151
Fargo (North Dakota): **4**.410
Fayum (Republic of Soudan): **8**.072
Federal Republic of Germany: **5**.021; **14**.011, 012, 020, 088, 089, 090, 091, 092, 093, 107, 233, 234, 235
Fiji: **12**.027, 028, 062, 079
Finland: **4**.315, 362, 425; **5**.117, 119; **14**.077, 078, 079
Firth of Thames (New Zealand): **5**.271
Fladbury (Worcestershire): **7**.065
Florida: **4**.004, 066, 111, 240, 280, 361, 386, 428, 451; **7**.098; **9**.012, 066, 115, 274, 361; **14**.056, 057, 127, 128, 129, 130
Florida Bay: **4**.004
Florida State University. Radiocarbon Dating Laboratory: **14**.056, 057
Florisbad (South Africa): **4**.260
Fort Gibson Reservoir (Oklahoma): **9**.022
Foxton (New Zealand): **4**.122
France: **4**.303; **11**.010, 015, 038, 042, 047, 048, 050, 069, 082; **14**.080, 081, 133, 177, 178
French Polynesia: **12**.015
Frontier County (Nebraska): **9**.021

G akushuin University, Tokyo: **14**.058, 059, 060, 061, 062, 063
Galicia: **11**.041
Gallejaure (Sweden): **5**.118
Garfield Heights (Ohio): **7**.001
Garlock fault (California): **4**.446
Geological Survey of Canada: **14**.067, 068, 069, 070, 071, 072, 073, 074, 075, 076
Geological Survey of Finland: **14**.077, 078, 079
Georgia: **4**.169, 236, 387, 432; **7**.002, 003; **9**.261, 301
Georgian SSR: **14**.205
Germany: **4**.304; **11**.075
Gif-sur-Yvette (France): **14**.080, 081, 177, 178
Gisborne (New Zealand): **4**.417
Glamorganshire: **11**.026

Glossop (Great Britain): **4**.427
Gordium (Turkey): **10**.080
Gormanston (Tasmania): **4**.339
Gota River (Sweden): **5**.121
Gothenburg (Norway): **4**.212
Gough Island: **7**.069
Goulburn valley (Victoria): **4**.294
Gower Peninsula (Glamorganshire): **11**.026
Grafton (New South Wales): **12**.045
Graman (New South Wales): **12**.013
Grand Canyon (Arizona): **9**.345
Grand Fall (Labrador): **4**.180
Great Barrier Reef (Queensland): **4**.001
Great Britain: **2**.082, 242, 244, 284; **3**.184, 260; **4**.108, 211, 255, 257, 301, 427, 429; **5**.007, 008, 067, 099, 116, 120, 131, 133, 136, 137, 148, 149, 153, 160, 161, 240, 253, 267; **7**.030, 065, 072, 108, 124, 150; **11**.025, 032, 052, 080, 090, 096; **14**.161
Great Lakes (North America): **7**.079; **9**.120
Great Lakes region (North America): **5**.226, 246
Great Plains (United States): **7**.126
Greece: **5**.187; **11**.011, 019, 020, 028, 030, 031, 033, 037, 051, 066, 084, 089, 099
Greefswald (South Africa): **8**.024
Greenland: **2**.223; **3**.274; **4**.153, 273, 282, 296, 372; **5**.079, 108, 273; **9**.318; **11**.007, 074, 100
Greenwich (Connecticut): **9**.283
Grijalda basin (Mexico): **9**.027
Groningen (Netherlands): **14**.082, 083, 084, 085, 086, 087
Guam (Marianas): **12**.066
Guatemala: **4**.019, 254; **9**.058, 101, 169, 197, 270, 271, 322, 329, 356, 357
Gulf of California: **4**.008; **6**.013
Gulf of Mexico: **4**.166; **7**.127
Gulf of Panama: **4**.159
Gwembe valley (Zimbabwe): **8**.048

H ajar Bin Bumeid (South Arabia): **10**.045
Hampshire: **11**.039
Hannover (F.R.G.): **14**.088, 089, 090, 091, 092
Hasanlu (Iran): **10**.028, 029
Hauraki Plains (New Zealand): **12**.035
Hawaii: **4**.102; **9**.151; **12**.004, 018, 027
Hawera, (New Zealand): **4**.029
Heidelberg (F.R.G.): **13**.093
Helena (Arkansas): **9**.153
Herault (France): **11**.050
Herbert (Saskatchewan): **4**.369
Héverlé Louvain (Belgium): **14**.121, 122, 123, 124, 125, 126
Himachal Pradesh (India): **4**.047

Hixton (Wisconsin): **7.**028
Hokkaido: **4.**448
Holstenborg (Greenland): **4.**296
Horsnund (Spitzbergen): **4.**371
Houston (Texas): **7.**142; **14.**094, 095
Hudson Bay (Canada): **4.**321; **5.**130
Hudspeth County (Texas): **4.**233
Humble-Esso Production Research Company: **14.**094, 095

Iberian Peninsula: **11.**023
Ica valley (Peru): **9.**218
Iceland: **4.**053, 450; **5.**234
Idaho: **9.**217, 363
Igloolik Island (North West Territories): **5.**189; **9.**238
Illinois: **3.**046; **4.**220; **5.**011, 042, 054, 064, 129, 183, 241; **7.**017, 133; **9.**007, 047, 124, 186, 273, 301, 348; **14.**041, 042, 043, 151, 152, 153, 154
India: **4.**047, 333; **10.**005, 006, 019, 034, 035, 039, 044, 045, 054, 055, 056, 057, 058, 059, 060, 061; **14.**166, 200, 201, 202, 203, 204
Indian Ocean: **6.**021, 033; **7.**115
Indiana: **4.**319, 329; **5.**042, 111, 304; **7.**079, 112, 133; **9.**056, 124
Indus Valley: **10.**009, 066, 075
Inglefield Land (Greenland): **5.**079
Institut National Polytechnique de Nancy. Centre de Recherches Radiogéologiques: **14.**133
Institut Royal du Patrimoine Artistique: **14.**096
Institute for Ancient and Modern History (Berlin): **14.**011, 012
Institute of Nuclear Sciences (New Zealand): **14.**140, 141, 142, 143, 144, 145
Institute of Physical and Chemical Research (Riken): **14.**173, 174, 175, 176
Institute of Zoology and Botany. Academy of Sciences of the Estonian S.S.R.: **14.**196, 197, 198, 199
Ioannina (Greece): **11.**019
Iowa: **4.**031, 154, 176, 332; **5.**102, 114, 122, 196; **7.**045, 133; **9.**053, 073, 090, 214, 301
Iping Common (Sussex): **11.**068
Ipintak (Alaska): **9.**301
Iran: **10.**002, 010, 020, 028, 029, 036, 043, 049, 067, 074, 076, 077, 078; **11.**089
Iranian Plateau: **10.**009
Iraq: **4.**242; **8.**013; **10.**011, 013, 025, 032, 040, 041, 042, 043, 047, 048, 068, 069, 070, 071, 072, 073, 079, 081, 082, 083
Ireland: **4.**201, 271, 275, 416; **7.**116; **11.**014, 032, 091; **14.**054

Irish Sea: **5.**207; **7.**088
Island County (Washington): **7.**105
Isle of Bute (Scotland): **11.**081
Isle of Man (Great Britain): **5.**240
Isortok River (North West Territory): **4.**426
Isotopic Inc.: **14.**101, 102, 103, 104, 105
Israel: **4.**239; **8.**013; **10.**024
Italy: **5.**066; **7.**056; **11.**070; **14.**039, 040, 259, 260, 261, 262, 263

Jackson (Wyoming): **4.**130
Jalisco Province (Mexico): **9.**297, 352
James Bay (Ontario): **7.**058
James Bay (Quebec): **5.**237
James Bay County (North West Territories): **5.**299
Japan: **2.**161, 280, 311, 312; **4.**137, 292, 448; **7.**026; **10.**026, 038, 064; **14.**058, 059, 060, 061, 062, 063, 173, 174, 175, 176, 274
Jarmo (Irak): **10.**043
Jericho (Jordan): **10.**014, 016, 023, 033, 043, 050
Jordan: **10.**003, 004, 014, 016, 023, 025, 033, 043, 050
Jutland (Denmark): **4.**390; **5.**257
Jylland (Denmark): **4.**200

Kafue River (Zambia): **8.**036
Kalambo Falls (Zimbabwe/Tanzania): **8.**050, 052
Kalibangan (Punjab): **10.**045
Kamchatka (USSR): **9.**110; **10.**015
Kansas: **5.**074; **7.**009; **9.**006
Kapwirimbwe (Zambia): **8.**058
Kara Kamar (Afghanistan): **10.**053
Kariba Dam (Zimbabwe): **8.**048
Kashmir: **10.**045, 058, 060, 066
Keewatin District (North West Territories): **5.**142, 177, 200
Kensington (New South Wales): **14.**249
Kentucky: **7.**017; **9.**020, 060, 241
Kenya: **4.**091, 263; **5.**143; **8.**013
Khartoum (Republic of Soudan): **8.**072
Kiel (F. R. G.): **14.**233, 234, 235
Kihniö (Finland): **4.**362
King Valley (Tasmania): **5.**105
King William Island (North West Territories): **5.**285
Kinsay Falls (Quebec): **5.**162
Kirkuk (Iraq): **10.**032
Kirshner Marsh (Minnesota): **5.**203
Klodono County (Bohemia): **11.**043
Kobuc River (Alaska): **9.**209
Kodiak Island (Alaska): **9.**365
Kolterman Farm (Wisconsin): **9.**168
Kom Ombo (Egypt): **8.**064
Kom Ombo Plain (Egypt): **8.**027

Location Index

Kosaka (Japan): **4.**292
Kosciusko State Park (New South Wales): **7.**143
Krueger Enterprises, Inc. Geochron Laboratories Division: **14.**064, 065
Kurdistan: **7.**086;
Kuwana District (Japan): **4.**137
Kvam (Norway): **4.**225
Kyushu (Japan): **10.**026
Köln (F.R.G.): **14.**107

La Jolla (California): **9.**085; **14.**108, 109, 110, 111, 112, 185, 224
La Venta (Mexico): **9.**282
Laboratory of Nuclear Geology. University of Pisa (Italy): **14.**040
Labrador: **4.**180, 232; **5.**051, 085, 212
Lagoa Santa (Brazil): **9.**087
Lake Algonquin (Canada): **5.**097
Lake Erie (Canada): **5.**260, 280
Lake Kutchura (Hokkaido): **4.**448
Lake Melville District (Labrador): **4.**232
Lake Michigan: **5.**024
Lake Michigan basin: **5.**201
Lake Mohave (Arizona): **9.**094
Lake Mohave (Nevada): **9.**094, 180
Lake Searles (California): **7.**039
Lake St. John (Quebec): **5.**243
Lanarkshire (Scotland): **7.**150
Lapland: **5.**117
Las Haldas (Peru): **9.**203
Las Vegas (Nevada): **9.**201
Las Vegas Valley (Nevada): **4.**262
Lautauka Province (Fiji): **12.**079
Leningrad: **14.**003, 055, 106
Les Eyzies (Dordogne): **11.**042
Lethbridge (Alberta): **5.**041
Lewes (Sussex): **11.**094
Lewisville (Texas): **9.**201, 223
Libya: **4.**062
Lincolnshire: **5.**181
Lithuanian S.S.R.: **4.**354
Llano Estacado (Texas): **9.**028
Llcyn Peninsula (Wales): **5.**103
Lochinvar (Zambia): **8.**071, 081
Lochinvar Ranch (Zambia): **8.**042
Lockerbie (Scotland): **5.**131, 148
Logan County (Ohio): **4.**134, 328
London: **14.**021, 022, 023, 024, 025, 026, 027, 028, 161
Long Beach (North Carolina): **4.**219
Loopstedt (Germany): **5.**244
Los Angeles (California): **3.**185; **4.**131
Lothal (Maharashtra): **10.**045

Louisiana: **4.**143, 184, 203, 280, 334, 346; **5.**035, 208; **9.**057, 233
Louisiana Coast: **6.**040
Lower Hutt (New Zealand): **14.**140, 141, 142, 143, 144, 145
Lund (Sweden): **14.**236
Lunz (Austria): **5.**266
Lusu (Zambia): **8.**071
Lybia: **8.**000, 001, 015, 025;
Lynn County (Texas): **4.**110
Lyons Ferry (Washington): **9.**108
Lytton-Lilloget Region (British Columbia): **9.**246

MacLennan County (Texas): **9.**366
Macedonia: **5.**187; **11.**030, 031, 037
Mackenzie Delta (North West Territories): **4.**266; **5.**012, 283
Madelia (Minnesota): **5.**203
Madison (Wisconsin): **14.**275, 276, 277, 278, 279
Madison County (Ohio): **9.**148, 211
Maharashtra: **10.**045
Maine: **7.**032, 122; **9.**063, 257, 281
Malaysia: **7.**055
Malmö (Sweden): **4.**436
Malta: **11.**012, 016, 102
Mammoth Cave National Park (Kentucky): **9.**020
Manitoba: **4.**161, 274; **5.**075, 124, 200; **9.**163
Mapungubwe (South Africa): **8.**024, 073
Marble Canyon (Arizona): **9.**344
Marianas: **12.**027, 040, 066
Maribyrnong River (Victoria): **12.**006, 036, 056
Marin County (California): **9.**139, 249
Marquesas Islands: **12.**015, 069
Martha's Vineyard (Massachusetts): **4.**414; **5.**231; **7.**104
Masili (Zambia): **8.**071
Massachusetts: **4.**021, 192, 245, 414; **5.**072, 231; **7.**104; **9.**138; **14.**064, 065
Matapos Hills (Rhodesia): **8.**069
McKenzie District (North West Territories): **4.**439
McMurdo Sound (Antarctica): **4.**223, 227
Mead County (Kansas): **5.**074
Meare (Somerset): **11.**062
Medicine Creek (Nebraska): **9.**021, 313
Mediterranean: **7.**109; **11.**061
Mediterranean Basin: **11.**016, 072, 102
Mediterranean Sea: **2.**270; **4.**104, 399, 424; **6.**046
Mediterranean coast: **7.**075
Melanesia: **12.**076
Melbourne (Victoria): **14.**289
Melville Island (North West Territories): **4.**465; **5.**177, 214
Melville Peninsula (North West Territories): **5.**142

Mercer County (North Dakota): **9**.230
Mesa Verde: **4**.218
Mesoamerica: **9**.069, 101, 182, 183, 185, 236, 277, 327, 373
Mesopotamia: **10**.009
Mesters Vig District (Greenland): **4**.153, 273, 372
Mexico: **4**.076, 380; **7**.051, 141; **9**.015, 017, 023, 027, 029, 030, 039, 068, 069, 072, 076, 082, 101, 113, 114, 132, 143, 158, 162, 185, 192, 200, 204, 205, 210, 236, 242, 248, 252, 254, 258, 272, 277, 282, 297, 298, 310, 315, 316, 319, 327, 329, 336, 337, 352, 373; **14**.044
Mexico Universita. Instituto Geologia. Laboratorio de Cronometria: **14**.044
Miami (Florida): **14**.127, 128, 129, 130
Michigan: **4**.213, 250; **5**.030, 042, 197, 219; **7**.035, 079, 113, 129; **9**.051, 111, 152, 173, 184, 301, 306; **14**.237, 238, 239, 240, 241, 242, 243, 244, 245, 246, 247, 248
Michigan City (Indiana): **4**.319
Micronesia: **4**.160, 182
Middle America: **9**.220
Middle East: **10**.052
Middle Littleton (Worcestershire): **11**.024
Middlesex (England): **14**.134, 135, 136, 137, 138, 139
Middleton Island (Alaska): **4**.287
Midlands (Great Britain): **5**.008
Midwest (United States): **7**.125, 126
Mie Prefecture (Japan): **4**.137
Mill Creek Valley (Ohio): **7**.001
Minnesota: **2**.266; **4**.199, 248, 253; **5**.030, 102, 140, 150, 151, 152, 155, 156, 166, 203, 245, 246, 297; **7**.118; **9**.360
Mississippi: **9**.166, 301; **14**.221
Mississippi Valley: **7**.132, 133
Missouri: **9**.073, 107, 301
Missouri River basin: **9**.325
Mobil Oil Corporation: **14**.184
Mohave Desert (California): **4**.183; **7**.040, 114
Monaco: **14**.131, 132
Montana: **4**.279; **9**.123, 266
Montcalm County (Michigan): **5**.219
Montezuma National Monument (Arizona): **4**.058
Montignac (Dordogne): **11**.047, 048
Moravia: **11**.053
Morayshire (Scotland): **5**.120
Morgantown (West Virginia): **7**.059
Moscow: **14**.002, 156, 157, 158, 288
Moscow (Washington): **4**.464
Motutapu Island (New Zealand): **12**.034, 058
Mound County (Texas): **4**.110
Mount Carmel (Israel): **10**.024
Mount Hagen (New Guinea New Guinea): **12**.046
Mount Moffatt cattle Station (Queensland): **12**.005, 007
Mount Wellington (New Zealand): **12**.058
Mozambique: **4**.096
Mtoroshanga District (Rhodesia): **8**.039
Muaco (Venezuela): **9**.340
Murray Springs (Arizona): **5**.165

Naco (Arizona): **9**.048
Nagasaki Prefecture, Kyushu (Japan): **10**.026
Naknek River (Alaska): **9**.232
Naknek (Alaska): **9**.362
Namibia: **8**.008
Nancy (France): **14**.133
Napo River: **9**.368
Natal (South Africa): **8**.076
National Museum. Department of Natural Science (Copenhagen): **14**.045, 046, 047, 048, 049, 050, 051, 052, 053
National Physical Laboratory. Division of Radiation Science (Teddington, England): **14**.134, 135, 136, 137, 138, 139
Natural Sciences Laboratory of the Royal University (Netherlands): **14**.082, 083, 084, 085, 086, 087
Navdatoli (Maharashtra): **10**.045
Nayarit Province (Mexico): **9**.277, 352
Nazca valley (Peru): **9**.218
Near East: **10**.017, 037, 052, 062; **11**.061
Nebraska: **4**.202; **5**.138; **9**.021, 266, 313
Netherlands: **14**.082, 083, 084, 085, 086, 087
Nevada: **1**.007; **4**.073, 198, 262, 299; **5**.010; **9**.009, 041, 049, 081, 091, 094, 103, 170, 180, 201, 225, 250, 256, 284, 311, 351, 358
New Caledonia: **12**.010, 027
New England: **4**.109, 393; **9**.120, 191
New Guinea: **12**.067
New Hampshire: **4**.376
New Haven (Connecticut): **4**.272; **14**.291, 292, 293, 294, 295, 296, 297, 298
New Hebrides (now Vanuatu): **12**.048
New Jersey: **4**.393, 434; **7**.034, 084; **14**.100, 101, 102, 103, 104, 105
New London (Connecticut): **4**.264
New Mexico: **4**.013, 025, 030, 258; **9**.010, 119, 194, 201, 243, 289, 342
New Orleans (Louisiana): **4**.143
New South Wales: **3**.019; **4**.095, 391, 408, 443; **7**.143; **12**.012, 013, 022, 030, 045, 059, 061, 065, 070, 077; **14**.249
New York: **4**.145
New York (state): **4**.207; **5**.227; **7**.008, 044, 083, 084;

Location Index

9.044, 299; **14**.113, 114, 115, 116, 117, 118, 119, 120
New Zealand: **2**.198; **3**.170, 231, 246; **4**.003, 020, 029, 038, 041, 085, 122, 149, 185, 209, 246, 277, 340, 351, 374, 389, 394, 409, 417, 441, 444, 447, 463; **5**.271; **7**.014, 024, 054, 062, 068; **12**.015, 025, 026, 031, 034, 035, 038, 044, 047, 058, 068; **14**.140, 141, 142, 143, 144, 145
Niedersachsisches Landesamt für Bodenforschung: **14**.088, 089, 090, 091, 092
Nigeria: **1**.024; **8**.009, 056, 057, 061, 063, 066
Nome (Alaska): **4**.075, 265
Nordaustlandet Island (Spitsbergen): **5**.269
North Africa: **10**.062
North America: **1**.016; **3**.173; **4**.080, 090, 099, 129, 320, 322, 323, 324, 370, 454; **5**.029, 057, 059, 071, 076, 078, 093, 128, 215, 220, 222, 250, 259, 290; **7**.019, 049, 053, 060, 076, 123, 144; **9**.032, 080, 121, 126, 135, 141, 142, 146, 222, 239, 300, 303, 334, 343
North Atlantic: **5**.208
North Bay (Ontario): **5**.098
North Branch (Minnesota): **5**.140
North Carolina: **4**.219, 280; **5**.204, 208, 284
North Dakota: **4**.410; **5**.164, 179, 192; **7**.031; **9**.230, 266
North Otago (New Zealand): **12**.068
North West Territories: **1**.027; **4**.011, 027, 028, 155, 190, 230, 266, 274, 330, 407, 426, 439, 465; **5**.012, 028, 033, 089, 142, 175, 177, 184, 189, 214, 217, 223, 261, 265, 283, 285, 287, 299; **9**.024, 245, 354; **14**.159, 160
Northeast Pacific Ocean: **6**.008
Northeastern United States: **5**.193
Northern Hemisphere: **2**.030, 074, 088, 126; **5**.252; **6**.035
Northern Rhodesia: **8**.021, 023, 031
Northern United States: **5**.225
Norway: **3**.265; **4**.050, 077, 138, 210, 212, 225, 289, 388; **5**.014, 055, 058, 084, 169, 232, 269; **7**.050; **11**.044; **14**.208, 209, 210, 211
Norwegian Institute of Technology. Radiological Dating Laboratory: **14**.208, 209
Notre Dame Mountains (Quebec): **7**.085
Nova Scotia: **4**.117, 376, 419; **5**.022, 144; **9**.098, 216, 229, 234

Oak Ridge Associated Universities. Radiocarbon Dating Laboratory (Tennessee): **14**.146, 147
Oak Ridge (Tennessee): **14**.146, 147
Oaxaca (Mexico): **9**.029, 069, 185, 236, 327, 373
Ober Fellabrun (Austria): **4**.343
Oceana County (Michigan): **9**.111

Ohio: **4**.022, 134, 179, 268, 328; **5**.009, 042, 047, 258, 268, 270, 278, 303, 304; **7**.001, 083, 090, 094, 130, 133; **9**.148, 211, 301; **14**.148, 149, 150
Ohio Valley (United States): **9**.199
Ohio Wesleyan University. Department of Botany and Bacteriology: **14**.148, 149, 150
Okkaido: **10**.038
Oklahoma: **9**.022, 038, 057, 097, 171, 276, 279, 286, 287, 288, 294, 295, 296
Olduvai Gorge (Tanzania): **8**.002
Ongul Islands (Antarctica): **5**.233
Ontario: **4**.046, 082, 083, 179, 206, 214, 335; **5**.016, 065, 086, 087, 094, 095, 098, 124, 127, 132, 209, 238, 281, 286, 295, 307; **7**.058, 066, 067; **9**.140, 187, 228, 231, 293
Oregon: **4**.017, 132, 216, 228, 291, 461; **5**.046, 096, 158; **7**.061, 096; **9**.049, 089, 167, 251
Oregon coast: **6**.008
Orinoco River: **9**.004, 368
Otago (New Zealand): **7**.054, 068
Otaniemi (Finland): **14**.077, 078, 079
Ottawa: **4**.046; **5**.238; **14**.067, 068, 069, 070, 071, 072, 073, 074, 075, 076
Otway Peninsula (Victoria): **12**.011
Oxford (Mississippi): **14**.221

Pacific Coast: **2**.231, 232; **4**.121; **9**.169
Pacific Islands: **12**.078
Pacific Ocean: **1**.018; **4**.014, 118; **6**.021, 024, 030, 032, 033, 044, 049
Pacific region: **12**.071
Packard Instrument Co, Inc. Low level Counting Laboratory: **14**.151, 152
Paekakeriki (New Zealand): **4**.340
Pakistan: **10**.019, 021, 027
Palisades (New York): **14**.113, 114, 115, 116, 117, 118, 119, 120
Panama: **9**.001, 036, 039, 086, 145
Papeto'ai (Society Islands): **12**.041
Papua New Guinea: **4**.038, 088, 404; **12**.037, 046, 051, 072
Patagonia: **7**.043, 093
Patagonia (Argentine): **5**.298
Peary Land (Greenland): **5**.273
Pearyland Peninsula (Greenland): **11**.007
Pecos River Valley (Mexico): **9**.082
Pembrokeshire (Wales): **5**.207
Pennsylvania: **4**.074, 135, 207; **5**.005, 291, 308; **9**.335; **14**.250, 251, 252, 253, 254, 255, 256, 257, 258
Pershing County (Nevada): **9**.225, 250

Peru: **1**.030; **9**.003, 039, 067, 116, 118, 160, 161, 202, 203, 218, 219, 235, 264, 341
Peten (Guatemala): **9**.058, 197, 270, 271, 322, 356, 357
Philadelphia (Pennsylvania): **14**.250, 251, 252, 253, 254, 255, 256, 257, 258
Philippines: **12**.073
Pig Bay (New Zealand): **12**.034
Pisa (Italy): **14**.039, 040
Pitnacree (Portshire): **11**.032
Plant remains: **4**.045
Playa Grande (Peru): **9**.264
Point Barrow (Alaska): **4**.026, 063; **9**.128, 330
Point Peron (Western Australia): **4**.466
Poland: **11**.075
Polynesia: **12**.039, 055
Porsi (Lapland): **5**.117
Port Campbell (Victoria): **4**.300, 313; **7**.071
Port Talbot (Ontario): **5**.307
Portshire (Scotland): **11**.032
Pretoria (South Africa): **8**.003
Pribilof Islands (Alaska): **5**.017
Prince Olav Coast (Antartica): **5**.233
Prince of Wales Island (Alaska): **4**.146
Prince of Wales Island (North West Territories): **5**.285
Pueblo Nuevo (Panama): **9**.036
Puget Lowlands (Washington): **7**.102
Puget Sound (Washington): **4**.437; **5**.300; **7**.038, 101, 120
Pullman (Washington): **4**.464; **13**.162, 290
Punjab: **10**.027, 035, 045, 056, 058
Puntas Minitas (Baja California): **9**.200
Pyramid Valley (New Zealand): **12**.044

Quebec: **5**.212
Quebec (province): **4**.032, 082, 083; **5**.013, 050, 051, 082, 085, 091, 135, 162, 168, 176, 178, 243, 305; **7**.074, 081, 085
Queensland: **4**.001; **12**.005, 007
Queenstown (Tasmania): **4**.339
Quetta Valley (Pakistan): **10**.021

Ra Province (Fiji): **12**.079
Radioactive Dating Laboratory, Stockholm: **14**.189, 190, 191, 192, 193, 194, 195
Radioisotopes and Radiations Laboratory. Washington State University: **14**.162, 290
Rainy River District (Ontario): **9**.140
Rajasthan: **10**.006, 039, 045
Rancho La Brea (California): **5**.274
Rancho Peludo (Venezuela): **9**.340
Rangitikei Valley (New Zealand): **4**.351
Rapahoe (New Zealand): **4**.444
Recife (Brazil): **4**.381

Red Sea: **4**.125
Republic of Sudan: **4**.016; **8**.041, 072
Rhodesia: **1**.024; **8**.017, 029, 039, 080; **14**.171, 172
Rice Lake (Ontario): **9**.187
Richmond (Indiana): **9**.056
Riken (Tokyo): **14**.173, 174, 175, 176
River Severn (England): **5**.272
Rochester (Indiana): **4**.329
Romania: **11**.106, 107, 108
Rome: **7**.056; **14**.259, 260, 261, 262, 263
Ross County (Ohio): **5**.009
Ross and Cromarty shire (Scotland): **5**.133, 153
Rothesay (Scotland): **11**.081
Rotoiti (New Zealand): **4**.020
Rotorua (New Zealand): **4**.417
Rovaniemi (Finland): **5**.119
Rub' al Khali (Saudi Arabia): **10**.009
Rumania: **1**.005, 006
Russia: **5**.290
Ruwenzori Range (Uganda): **5**.004
Ryukyu Islands (Japan): **10**.064

Sable Island (Canada): **9**.016
Saginaw County (Michigan): **9**.152
Sahara: **1**.024; **4**.298
Saint Paul Island (Alaska): **5**.017
Salisbury (Rhodesia): **8**.068; **14**.171, 172, 186
San Carlos Indian Reservation (Arizona): **9**.013
San Diego (California): **9**.226, 328; **14**.108, 109, 110, 111, 112, 185, 224
San Diego County (California): **9**.092
San Felipe (Baja California): **7**.051
San Jose: **9**.077
San Juan Islands: **9**.071
San Pedro Valley (Arizona): **9**.175, 227
Sandusky County (Ohio): **7**.094, 130
Santa Barbara (California): **9**.122
Santa Cruz (California): **4**.059
Santa Rosa Island (California): **7**.042; **9**.045, 112, 136, 137, 201, 224, 290, 291, 309
Sarawak: **10**.008; **12**.073
Sardinia: **11**.070
Sarnia (Ontario): **5**.016
Saskatchewan: **4**.078, 168, 369; **5**.294; **7**.107; **9**.188, 266; **14**.038, 264, 265, 266, 267
Saskatchewan Research Council: **14**.038, 264, 265, 266
Saskatoon (Saskatchewan): **4**.078; **14**.038, 264, 265, 266, 267
Saudi Arabia: **10**.009
Savannah River: **9**.198
Scandinavia: **5**.276; **7**.021; **11**.044, 101
Schleswig-Holstein (F.R.G.): **5**.021

Location Index

Science Museum of Victoria. Radiocarbon Dating Laboratory: **14**.289
Scotland: **4**.201, 257, 429, 449; **5**.099, 120, 131, 133, 148, 153; **7**.150; **11**.032, 081, 109
Scripps Institution of Oceanography: **14**.108, 109, 110, 111, 112, 185, 224
Seamer (Yorkshire): **11**.034, 036, 046, 071
Seattle (Washington): **5**.279; **13**.163, 164
Selangor (Malaysia): **7**.055
Selborne (Hampshire): **11**.039
Sharp Laboratories: **14**.179
Shelby County (Ohio): **7**.090
Sheridan County (North Dakota): **7**.031
Shetland Islands (Scotland): **4**.429
Shibi District (Zimbabwe): **8**.018
Shippea Hill (Cambridgeshire): **11**.065
Shropshire: **3**.008; **5**.136, 137
Siberia: **4**.106, 325; **5**.002, 170; **7**.128; **10**.015
Sidmouth (Devon): **11**.096
Sierra Pinacata Sonora: **9**.353
Sierra de Taumalipas (Mexico): **9**.316
Sigatoka River (Fiji): **12**.062
Silver Lake (Ohio): **4**.134, 328
Sinaloa Province (Mexico): **9**.352
Siskiyou County (California): **4**.024
Skeldal (Greenland): **4**.273
Smithsonian Institution. C14 Laboratory: **14**.180, 181, 182, 183
Society Islands: **12**.004, 041
Sognefjord (Norway): **5**.058
Solduz Valley (Iran): **10**.028, 029, 074, 078
Solwezi (Zambia): **8**.012
Somerset: **11**.006, 062, 064, 077, 078
Somerset Island (North West Territories): **5**.285
South Africa: **2**.289; **4**.260; **8**.003, 005, 024, 065, 073, 075, 076, 078
South America: **4**.080, 127, 188; **9**.100, 118, 130, 158, 161, 203, 239, 341, 368
South Arabia: **10**.045
South Asia: **10**.065
South Australia: **4**.012, 221, 270, 336; **12**.008, 009, 019, 021, 023, 024
South Carolina: **4**.406; **9**.144
South Dakota: **4**.269; **5**.112; **7**.078; **9**.195, 266
South Pacific: **2**.075
Southampton Island (North West Territories): **9**.024, 354
Southeast Asia: **10**.062; **12**.071
Southeastern Europe: **11**.023
Southern Canada: **5**.225
Southern Europe: **11**.072
Southern Florida: **5**.208
Southern Hemisphere: **2**.030, 074
Southern Rhodesia: **14**.186

Southwest Africa: **8**.010
Spain: **11**.041
Spitsbergen: **4**.133
Spitzbergen: **4**.289, 290, 349, 350, 371, 403; **5**.044, 269
St. Elias Mountain (Canada): **5**.167; **7**.037
St. Hilaire (Quebec): **5**.176
St. John (Arizona): **9**.355
St. Laurence lowlands (Canada): **5**.178, 243
St. Louis County (Minnesota): **5**.151
St. Mathieu de Treviers (Herault): **11**.050
St. Vincent: **4**.373
Stefansson Island (North West Territories): **5**.287
Stockholm: **13**.189, 190, 191, 192, 193, 194, 195
Sudan: **4**.412; **8**.015
Suriname: **4**.127
Sussex: **11**.068, 094
Svai'i (Western Samoa): **12**.063
Svartisen (Norway): **5**.055
Swaziland: **8**.067
Sweden: **2**.314; **3**.031; **4**.034, 142, 193, 194, 195, 196, 197, 305, 436; **5**.118, 121, 172, 264; **14**.189, 190, 191, 192, 193, 194, 195, 236, 280, 281, 282, 283, 284, 285, 286, 287
Switzerland: **2**.114; **11**.067; **14**.013, 014, 015, 016, 017, 018
Sycamore (Illinois): **5**.241
Sydney (New South Wales): **12**.077
Sydney (Ohio): **4**.022

Tabasco (Mexico): **9**.205, 282, 315, 319, 329, 337
Tabasco Province (Mexico): **9**.132
Taber (Alberta): **9**.054
Tallahassee (Florida): **14**.056, 057
Tama County (Iowa): **4**.154
Tamilnadu (India): **10**.044
Tangier: **10**.067
Tanzania: **8**.002, 033, 043, 079
Tartu (Estonia): **14**.196, 197, 198, 199
Taruga (Nigeria): **8**.066
Tasman Sea: **6**.038
Tasmania: **4**.056, 060, 339; **5**.105; **7**.015; **12**.001, 033, 043
Tata Institute of Fundamental Research (Bombay): **14**.166, 200, 201, 202, 203, 204
Taumalipas (Mexico): **9**.143, 210
Taupo (New Zealand): **4**.409, 417, 441
Tbilisi (USSR): **13**.205
Tbilisi University. Radiocarbon Laboratory: **14**.205
Teddington (Middlesex): **14**.134, 135, 136, 137, 138, 139
Teindland (Scotland): **5**.120
Teledyne Isotopes (New Jersey): **14**.100
Ten Sleep (Wyoming): **9**.346

Tennessee: **9**.347; **14**.146, 147
Teotihuacan (Mexico): **9**.254
Tepe Sarab (Iran): **10**.043
Terry County (Texas): **4**.110
Texas: **4**.057, 110, 231, 233, 251, 297, 397, 421; **7**.018, 142; **9**.019, 028, 057, 131, 176, 178, 196, 201, 223, 262, 263, 265, 366, 367; **14**.094, 095, 184, 206, 268, 269, 270, 271, 272, 273
Texas A & M University. Department of Oceanography: **14**.206
Texas Coast: **6**.040; **7**.022
Texas-Bio-Nuclear (Kaman Nuclear): **14**.207
Thailand: **10**.018
Thatcham (Berkshire): **11**.035, 105
The Netherlands: **4**.162, 172, 276, 306, 338, 413, 422, 457; **5**.034, 053, 282; **7**.149; **11**.009, 049, 076, 103
The Norwegian Institute of Technology. Radiological Dating Laboratory: **14**.210, 211
The Tonga (Zambia): **8**.071
Thelon Valley (North West Territory): **5**.194
Thule (Greenland): **3**.274; **5**.108
Titusville (Pennsylvania): **5**.308
Tofty Placer District, Alaska: **7**.146
Tokyo: **2**.312; **10**.038; **14**.058, 059, 060, 061, 062, 063, 173, 174, 175, 176, 274
Toronto (Ontario): **5**.281
Toten (Norway): **4**.225
Totket Bog (New Haven): **5**.231
Transvaal (South Africa): **8**.065, 076
Transylvania: **11**.106, 107, 108
Travis County (Texas): **9**.262
Trinity College. Department of Botany: **14**.054
Tristan da Cunha Island: **7**.069
Trnava (Czechoslovakia): **3**.291
Trois-Pistoles (Quebec): **7**.085
Troms (Norway): **5**.084
Trondheim (Norway): **3**.265; **14**.208, 209, 210, 211
Tucson (Arizona): **14**.006, 007, 008, 009, 010, 230, 231
Tupperville (Ontario): **9**.231
Turkey: **4**.084; **10**.007, 022, 030, 080; **11**.011

U daipur (Rajahstan): **10**.039
Uganda: **5**.004; **7**.148
Umiat (Alaska): **4**.256, 265
Umnak Island (Aleutians): **9**.011
Ungava (Quebec): **5**.013, 091, 135, 239
Ungava Peninsula (Quebec): **5**.051, 082, 085
Unimak Island (Alaska): **9**.154
United States Department of Agriculture. Sedimentation Laboratory: **14**.221

United States Geological Survey: **14**.167, 212, 213, 214, 215, 216, 217, 218, 219, 220
Universitat Bonn. Institute fürBodenkunde: **14**.020
University College of Rhodesia and Nyassaland. Department of Chemistry: **14**.186
University of Arizona. Laboratory of Isotope Geochemistry: **14**.006, 007, 008, 009, 010, 230, 231
University of Bern. Institute of Physics: **13**.013, 014, 015, 016, 017, 018
University of Birmingham. Department of Geology: **14**.019, 232
University of California, Los Angeles. Institute of Geophysics: **14**.222, 223, 225,226, 227, 228, 229
University of California, San Diego: **14**.108, 109, 110, 111, 112, 185, 224
University of Chicago: **14**.041, 042, 043, 154
University of Chicago. Institute for Nuclear Studies: **14**.153
University of Heidelberg. C-14 Laboratory: **14**.093
University of Kiel. Institute for Pure and Applied Physics: **14**.233, 234, 235
University of Köln. Institut of Ancient and Modern History. C14 Laboratory: **14**.107
University of London. Institute of Archaeology: **14**.161
University of Miami. Rosensteil School of Marine and Atmospheric Science: **14**.127, 128, 129, 130
University of Michigan: **14**.237, 238, 239, 240, 241, 242, 243, 244, 245, 246, 247, 248
University of New South Wales. Department of Nuclear and Radiation Chemistry: **14**.249
University of Pennsylvania. Dept. of Physics: **14**.250, 251, 252, 253, 254, 255, 256, 257, 258
University of Pisa. Laboratory of Nuclear Geology.: **14**.039
University of Rhodesia (Zimbabwe). Gulbenkian Radiocarbon Dating Laboratory: **14**.171, 172
University of Rome. Radiocarbon Dating Laboratory: **14**.259, 260, 261, 262, 263
University of Saskatchewan. Saskatchewan Research Council: **14**.267
University of Texas at Austin. Radiocarbon Laboratory: **14**.268, 269, 270, 271, 272, 273
University of Tokyo. Carbon Dating Laboratory: **14**.274
University of Uppsala. Institut of Physics: **14**.280, 281, 282, 283, 284, 285, 286, 287
University of Washington. Department of Chemistry: **14**.163, 164
University of Wisconsin. Radiocarbon Laboratory of the Center for Climatic Research: **13**.275,

Location Index

276, 277, 278, 279
Uppsala (Sweden): **3**.031; **14**.280, 281, 282, 283, 284, 285, 286, 287
Upton Warren (Worcestershire): **7**.030
USSR: **4**.005, 006, 049, 140, 178, 325; **5**.002, 134, 145, 186; **9**.110; **11**.073, 079, 087; **14**.002, 003, 106, 156, 157, 158, 165, 187, 188, 196, 197, 198, 199, 288
Utah: **4**.375; **7**.005; **9**.093
Uttar Pradesh: **10**.058, 060

V.G. Khlopin Institute. Radiology Laboratory: **14**.106
Valisco Province (Mexico): **9**.277
Vancouver Island (British Columbia): **4**.440, 459; **5**.043
Velsen (The Netherlands): **4**.338
Venezuela: **4**.157, 317; **9**.004, 127, 182, 340, 369; **14**.097, 098, 099
Venezuelan Institute of Scientific Investigation. Ecology Department: **14**.097, 098, 099
Vera Cruz (Mexico): **9**.204
Vernadski Institute of Geochemistry (Moskow): **14**.002, 158, 288
Victoria: **2**.240; **4**.042, 120, 294, 300, 311, 313, 347; **7**.071; **12**.002, 006, 011, 036, 056; **14**.289
Victoria Island (North West Territories): **5**.287
Victoria Land (Antarctica): **4**.458; **5**.006
Virginia: **4**.164; **5**.284; **7**.099, 119
Viru valley (Peru): **9**.218
Viti Levu (Fiji): **12**.028, 079

Waco (Texas): **9**.366
Waddi Halfa (Republic of Sudan): **8**.041
Wairakei (New Zealand): **2**.198
Waitemata Harbour (New Zealand): **7**.062
Wales: **3**.008; **5**.067, 103, 160, 161, 171, 207; **11**.026, 029
Walnut Canyon (Arizona): **9**.344
Walrus Island (North West Territories): **9**.024
Wanganui, (New Zealand): **4**.029
Ward County (North Dakota): **5**.164
Warrnambool (Victoria): **12**.002
Warwickshire (England): **7**.108
Washington (state): **4**.018, 204, 215, 216, 293, 437, 452, 462, 464; **5**.096, 158, 235, 248, 256, 279, 300; **7**.038, 046, 101, 102, 105, 120; **9**.105, 108, 156, 251; **14**.162, 163, 164, 290
Washington coast: **6**.008
Washington DC: **7**.117; **14**.167, 180, 181, 182, 183, 212, 213, 214, 215, 216, 217, 218, 219, 220
Weier (Denmark): **4**.107
Wellington Peninsula (New Zealand): **7**.024

West Africa: **8**.009, 047
West Indies: **9**.127, 247
West Virginia: **7**.059
Western Australia: **4**.002, 186, 402, 466; **7**.139
Western Europe: **11**.015, 053, 054, 055, 056, 057, 058, 059
Western Samoa: **12**.063, 064
Westhay (Somerset): **11**.064
Westward Ho! (Devon): **11**.045
Westwood (New Jersey): **13**.100, 101, 102, 103, 104, 105
Whidbey Island (United States): **7**.102
Wilson Promontory (South Australia): **12**.019
Wiltshire (England): **11**.008, 040, 085, 086, 093, 098
Windhoek (Namibia): **8**.008
Wisconsin: **4**.222, 314; **5**.042, 064, 090, 183, 247; **7**.028, 082; **9**.035, 073, 075, 168, 301, 338; **14**.275, 276, 277, 278, 279
Worcestershire (England): **7**.030, 065; **11**.024
Wright-Patterson Air Base (Ohio): **5**.278
Wyoming: **4**.130, 147; **9**.055, 189, 193, 266, 331, 346, 372

Yaku Island (Japan): **2**.161, 280
Yale University. Radiocarbon Laboratory: **14**.291, 292, 293, 294, 295, 296, 297, 298
Yavapai County (Arizona): **4**.058
York County (Ontario): **4**.206
Yorkshire (England): **11**.034, 036, 046, 071, 083, 104
Yukatan (Mexico): **9**.030, 113, 114
Yukon (Alaska): **4**.015; **5**.167
Yukon Territory (Canada): **4**.124; **5**.309; **7**.033, 039, 145; **9**.164

Zagros Mountains (Iran): **10**.036
Zagros Mountains (Iraq): **4**.242; **8**.013; **10**.011, 013, 032, 040, 041, 047, 068, 069, 070, 071, 072, 073, 079, 081, 082, 083
Zambia: **1**.024; **8**.004, 007, 012, 017, 028, 030, 033, 034, 036, 042, 044, 045, 058, 059, 060, 071, 079, 081, 082, 083
Zambia / Tanzania border (Africa): **8**.032